T0134742

Advances in Intelligent Systems and Computing

Volume 821

Series editor

Janusz Kacprzyk, Polish Academy of Sciences, Warsaw, Poland
e-mail: kacprzyk@ibspan.waw.pl

The series "Advances in Intelligent Systems and Computing" contains publications on theory, applications, and design methods of Intelligent Systems and Intelligent Computing. Virtually all disciplines such as engineering, natural sciences, computer and information science, ICT, economics, business, e-commerce, environment, healthcare, life science are covered. The list of topics spans all the areas of modern intelligent systems and computing such as: computational intelligence, soft computing including neural networks, fuzzy systems, evolutionary computing and the fusion of these paradigms, social intelligence, ambient intelligence, computational neuroscience, artificial life, virtual worlds and society, cognitive science and systems, Perception and Vision, DNA and immune based systems, self-organizing and adaptive systems, e-Learning and teaching, human-centered and human-centric computing, recommender systems, intelligent control, robotics and mechatronics including human-machine teaming, knowledge-based paradigms, learning paradigms, machine ethics, intelligent data analysis, knowledge management, intelligent agents, intelligent decision making and support, intelligent network security, trust management, interactive entertainment, Web intelligence and multimedia.

The publications within "Advances in Intelligent Systems and Computing" are primarily proceedings of important conferences, symposia and congresses. They cover significant recent developments in the field, both of a foundational and applicable character. An important characteristic feature of the series is the short publication time and world-wide distribution. This permits a rapid and broad dissemination of research results.

More information about this series at http://www.springer.com/series/11156

Sebastiano Bagnara · Riccardo Tartaglia
Sara Albolino · Thomas Alexander
Yushi Fujita
Editors

Proceedings of the 20th Congress of the International Ergonomics Association (IEA 2018)

Volume IV: Organizational Design and Management (ODAM), Professional Affairs, Forensic

Editors
Sebastiano Bagnara
University of the Republic of San Marino
San Marino, San Marino

Riccardo Tartaglia
Centre for Clinical Risk Management
and Patient Safety, Tuscany Region
Florence, Italy

Sara Albolino
Centre for Clinical Risk Management
and Patient Safety, Tuscany Region
Florence, Italy

Thomas Alexander
Fraunhofer FKIE
Bonn, Nordrhein-Westfalen
Germany

Yushi Fujita
International Ergonomics Association
Tokyo, Japan

ISSN 2194-5357 ISSN 2194-5365 (electronic)
Advances in Intelligent Systems and Computing
ISBN 978-3-319-96079-1 ISBN 978-3-319-96080-7 (eBook)
https://doi.org/10.1007/978-3-319-96080-7

Library of Congress Control Number: 2018950646

This Springer imprint is published by the registered company Springer Nature Switzerland AG
The registered company address is: Gewerbestrasse 11, 6330 Cham, Switzerland

Preface

The Triennial Congress of the International Ergonomics Association is where and when a large community of scientists and practitioners interested in the fields of ergonomics/human factors meet to exchange research results and good practices, discuss them, raise questions about the state and the future of the community, and about the context where the community lives: the planet. The ergonomics/human factors community is concerned not only about its own conditions and perspectives, but also with those of people at large and the place we all live, as Neville Moray (Tatcher et al. 2018) taught us in a memorable address at the IEA Congress in Toronto more than twenty years, in 1994.

The Proceedings of an IEA Congress describes, then, the actual state of the art of the field of ergonomics/human factors and its context every three years.

In Florence, where the XX IEA Congress is taking place, there have been more than sixteen hundred (1643) abstract proposals from eighty countries from all the five continents. The accepted proposal has been about one thousand (1010), roughly, half from Europe and half from the other continents, being Asia the most numerous, followed by South America, North America, Oceania, and Africa. This Proceedings is indeed a very detailed and complete state of the art of human factors/ergonomics research and practice in about every place in the world.

All the accepted contributions are collected in the Congress Proceedings, distributed in ten volumes along with the themes in which ergonomics/human factors field is traditionally articulated and IEA Technical Committees are named:

I. Healthcare Ergonomics (ISBN 978-3-319-96097-5).
II. Safety and Health and Slips, Trips and Falls (ISBN 978-3-319-96088-3).
III. Musculoskeletal Disorders (ISBN 978-3-319-96082-1).
IV. Organizational Design and Management (ODAM), Professional Affairs, Forensic (ISBN 978-3-319-96079-1).
V. Human Simulation and Virtual Environments, Work with Computing Systems (WWCS), Process control (ISBN 978-3-319-96076-0).

VI. Transport Ergonomics and Human Factors (TEHF), Aerospace Human Factors and Ergonomics (ISBN 978-3-319-96073-9).
VII. Ergonomics in Design, Design for All, Activity Theories for Work Analysis and Design, Affective Design (ISBN 978-3-319-96070-8).
VIII. Ergonomics and Human Factors in Manufacturing, Agriculture, Building and Construction, Sustainable Development and Mining (ISBN 978-3-319-96067-8).
IX. Aging, Gender and Work, Anthropometry, Ergonomics for Children and Educational Environments (ISBN 978-3-319-96064-7).
X. Auditory and Vocal Ergonomics, Visual Ergonomics, Psychophysiology in Ergonomics, Ergonomics in Advanced Imaging (ISBN 978-3-319-96058-6).

Altogether, the contributions make apparent the diversities in culture and in the socioeconomic conditions the authors belong to. The notion of well-being, which the reference value for ergonomics/human factors is not monolithic, instead varies along with the cultural and societal differences each contributor share. Diversity is a necessary condition for a fruitful discussion and exchange of experiences, not to say for creativity, which is the "theme" of the congress.

In an era of profound transformation, called either digital (Zisman & Kenney, 2018) or the second machine age (Bnynjolfsson & McAfee, 2014), when the very notions of work, fatigue, and well-being are changing in depth, ergonomics/human factors need to be creative in order to meet the new, ever-encountered challenges. Not every contribution in the ten volumes of the Proceedings explicitly faces the problem: the need for creativity to be able to confront the new challenges. However, even the more traditional, classical papers are influenced by the new conditions.

The reader of whichever volume enters an atmosphere where there are not many well-established certainties, but instead an abundance of doubts and open questions: again, the conditions for creativity and innovative solutions.

We hope that, notwithstanding the titles of the volumes that mimic the IEA Technical Committees, some of them created about half a century ago, the XX Triennial IEA Congress Proceedings may bring readers into an atmosphere where doubts are more common than certainties, challenge to answer ever-heard questions is continuously present, and creative solutions can be often encountered.

Acknowledgment

A heartfelt thanks to Elena Beleffi, in charge of the organization committee. Her technical and scientific contribution to the organization of the conference was crucial to its success.

References

Brynjolfsson E., A, McAfee A. (2014) The second machine age. New York: Norton.
Tatcher A., Waterson P., Todd A., and Moray N. (2018) State of science: Ergonomics and global issues. Ergonomics, 61 (2), 197–213.
Zisman J., Kenney M. (2018) The next phase in digital revolution: Intelligent tools, platforms, growth, employment. Communications of ACM, 61 (2), 54–63.

Sebastiano Bagnara
Chair of the Scientific Committee, XX IEA Triennial World Congress
Riccardo Tartaglia
Chair XX IEA Triennial World Congress
Sara Albolino
Co-chair XX IEA Triennial World Congress

Organization

Organizing Committee

Riccardo Tartaglia (Chair IEA 2018)	Tuscany Region
Sara Albolino (Co-chair IEA 2018)	Tuscany Region
Giulio Arcangeli	University of Florence
Elena Beleffi	Tuscany Region
Tommaso Bellandi	Tuscany Region
Michele Bellani	Humanfactor[x]
Giuliano Benelli	University of Siena
Lina Bonapace	Macadamian Technologies, Canada
Sergio Bovenga	FNOMCeO
Antonio Chialastri	Alitalia
Vasco Giannotti	Fondazione Sicurezza in Sanità
Nicola Mucci	University of Florence
Enrico Occhipinti	University of Milan
Simone Pozzi	Deep Blue
Stavros Prineas	ErrorMed
Francesco Ranzani	Tuscany Region
Alessandra Rinaldi	University of Florence
Isabella Steffan	Design for all
Fabio Strambi	Etui Advisor for Ergonomics
Michela Tanzini	Tuscany Region
Giulio Toccafondi	Tuscany Region
Antonella Toffetti	CRF, Italy
Francesca Tosi	University of Florence
Andrea Vannucci	Agenzia Regionale di Sanità Toscana
Francesco Venneri	Azienda Sanitaria Centro Firenze

Scientific Committee

Sebastiano Bagnara (President of IEA2018 Scientific Committee)	University of San Marino, San Marino
Thomas Alexander (IEA STPC Chair)	Fraunhofer-FKIE, Germany
Walter Amado	Asociación de Ergonomía Argentina (ADEA), Argentina
Massimo Bergamasco	Scuola Superiore Sant'Anna di Pisa, Italy
Nancy Black	Association of Canadian Ergonomics (ACE), Canada
Guy André Boy	Human Systems Integration Working Group (INCOSE), France
Emilio Cadavid Guzmán	Sociedad Colombiana de Ergonomia (SCE), Colombia
Pascale Carayon	University of Wisconsin-Madison, USA
Daniela Colombini	EPM, Italy
Giovanni Costa	Clinica del Lavoro "L. Devoto," University of Milan, Italy
Teresa Cotrim	Associação Portuguesa de Ergonomia (APERGO), University of Lisbon, Portugal
Marco Depolo	University of Bologna, Italy
Takeshi Ebara	Japan Ergonomics Society (JES)/Nagoya City University Graduate School of Medical Sciences, Japan
Pierre Falzon	CNAM, France
Daniel Gopher	Israel Institute of Technology, Israel
Paulina Hernandez	ULAERGO, Chile/Sud America
Sue Hignett	Loughborough University, Design School, UK
Erik Hollnagel	University of Southern Denmark and Chief Consultant at the Centre for Quality Improvement, Denmark
Sergio Iavicoli	INAIL, Italy
Chiu-Siang Joe Lin	Ergonomics Society of Taiwan (EST), Taiwan
Waldemar Karwowski	University of Central Florida, USA
Peter Lachman	CEO ISQUA, UK
Javier Llaneza Álvarez	Asociación Española de Ergonomia (AEE), Spain
Francisco Octavio Lopez Millán	Sociedad de Ergonomistas de México, Mexico

Contents

Organizational Design and Management (ODAM)

Professional Affairs

Forensic

Organizational Design and Management (ODAM)

The Making of Robust Schedules: Strategies to Dealing with Uncertainties

Laetitia Flamard[1,2(✉)], Adelaide Nascimento[1], Pierre Falzon[1], and Ghislaine Tirilly[2]

[1] CRTD, Cnam, Equipe d'ergonomie, 41 rue Gay Lussac, 75005 Paris, France
laetitia.flamard@wanadoo.fr

[2] SNCF, Agence d'Accompagnement des Managers, SNCF, 2 place aux Etoiles, 93633 La Plaine St Denis, France

Abstract. The abstract should summarize the contents of the paper in short terms, i.e. 150–250 words.

Keywords: Scheduling · Robustness · Management of constraints

1 Introduction

Scheduling means assigning resources to tasks so as to meet objectives in dynamic unstable environments [1]. It is regarded as a key function in production management [2, 3]. Companies use it for leverage to adapt to business competition and to the requirements of their ever more demanding customers [2].

Empirical research has characterised scheduling activities as a continuous management of constraints [4, 5]. These constraints arise at both strategic levels, which define objectives, and operational levels, which reflect the current state of production [6]. Iterative reschedulings are needed in order to cope with the dynamic context of production [4].

In particular, we are interested in staff scheduling, i.e. the assignment of human resources to tasks during given periods of work [7, 8]. Staff scheduling is particularly important when the need for human resources varies over a working day in order to satisfy the demand for production of goods or services [7, 9]. Its output is the work schedules of the employees and determines part of their working conditions. Many studies underline the importance of taking into account social criteria such as the characteristics and preferences of operators [7, 10]. However, this work remains most often underpinned by a managerial vision of work according to which resources must be made profitable [11].

Given the complexity of scheduling situations, this function is often distributed among several schedulers according to the time horizon and the service they are in charge of [5, 12].

The multiplicity of actors involved in scheduling, together with complex production objectives, can in some cases lead constraints to conflict [4, 13]. To deal with such situations, schedulers endeavour to make sure their scheduling is robust. Drawing from Billault, Moukrim and Sanlaville [14], we will use the term robustness for describing the

S. Bagnara et al. (Eds.): IEA 2018, AISC 821, pp. 3–12, 2019.
https://doi.org/10.1007/978-3-319-96080-7_1

complete process by which scheduling is constructed in the presence of uncertainties. Robustness is understood as the ability of a system to resist to uncertainties [15].

The term can take on several different meanings according to how the scheduling activity is modelled [16]. Taking a static approach, robustness equates to stability, i.e. scheduling is considered robust if it stays constant despite changes in the production environment [16]. In this approach, rescheduling is considered as degrading robustness. Conversely, in a dynamic approach, rescheduling is necessary to pursue the design of production plans in acceptable conditions. Here the term robustness is often associated with flexibility, i.e. schedulers have some freedom to adapt production plans, while maintaining a level of performance near or equivalent to that of the initial scheduling [17]. In this case the solutions adopted by the schedulers are robust if they successfully deal with uncertainty and perturbations while ensuring set objectives are met [15, 16].

Nevertheless, the taking into account of a single criterion of robustness seems to be reductive in view of the dynamic environment in which the scheduling process is designed. Currently, we know of no work suggesting that in a context of dynamic scheduling, the criteria of robustness change as the scheduling advances. Thus, concerns relating to robustness are often limited to production issues and the analysis of scheduling activities does not highlight the specificities relating to human resources management.

A research was conducted in the French public rail transport service and more particularly in the scheduling unit responsible for workers allocation in sales units. The aim of this communication presents schedulers' strategies to ensure robustness of staff scheduling in order to avoid vacancies in order to propose the conditions to support schedulers' activity whose objective is to design robust schedules.

2 Method

2.1 Context

This research was conducted, in the French public rail transport service (Groupe Public Ferroviaire, GPF), France's sole passenger rail transport operator until 2019. Specifically, it took place in a production sales unit responsible for service provision to customers at 17 different high-street counters staffed by a pool of 130 commercial agents.

A request to address this problem was made by the manager of the in order to analyse the organization of planning, i.e. the staff scheduling, with regard to the recurrent occurrence of "unstaffed positions", i.e. sales positions that were not open to customers, because no commercial agent could be assigned to them. This was adversely affecting both the performance of the unit (loss of turnover, customer dissatisfaction) and the health of the agents and their managers (e.g. high absenteeism).

2.2 The Scheduling Process

The scheduling process concerns the assignment of commercial agents to work shifts at sales points to ensure planned service production. Inside this unit, three levels of management could be considered as schedulers insofar as they assigned resources to positions at different times and with different scope:

- Strategic level. The three senior managers made decisions based on yearly trends in human and physical resources and needs;
- Tactical level. The two controllers oversaw the organisation of the resources to meet the production objectives set for each sales counter;
- Operational level. The operational managers were in charge of managing the assignment of resources over smaller geographical areas on the actual day the service was offered.

Figure 1 shows the scheduling support by the different scheduler levels.

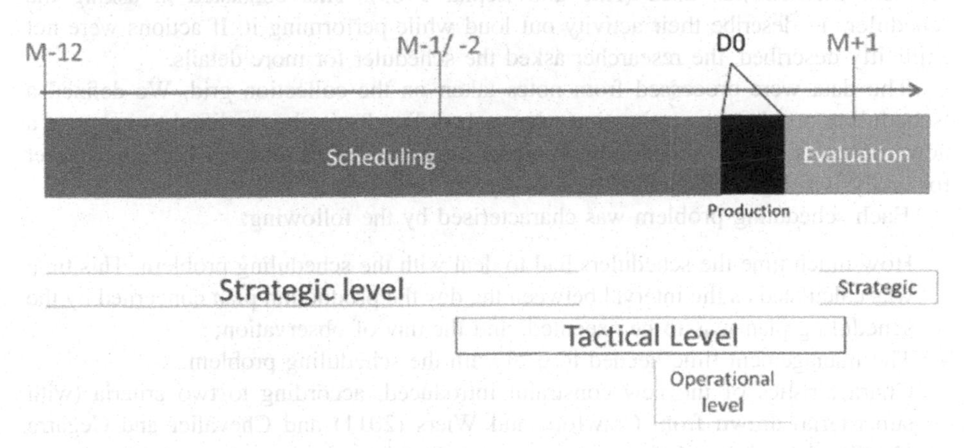

Fig. 1. Course of scheduling activity by management level involved

In this article, we are interested in the schedulers at the tactical level, who are in charge of reducing the number of unstaffed positions on a daily basis by rescheduling production plans, in particular to respond to the specific context of labour relations. This management level comprised two schedulers:

- *Agent I* had worked as a scheduler for 10 years, and in this unit for 5 years. He had previously been a commercial agent.
- *Agent P* had been employed for 1 month at the time the observations were made. He had no other prior experience as a scheduler but had worked in the company as a human resources records manager.

2.3 Data Collection and Analysis

The method we used was ergonomic on-site analysis, with observation of the activity deployed by the schedulers at the tactical level.

First, open observation sessions totalling 8 h were performed between July and September 2015 on two schedulers in order to understand their role and their positioning with respect to the actors at other levels who were involved in scheduling. These observations also served to build a systematic observation grid for which we drew on the framework proposed by MacCarthy et al. (2001). The production environment was taken into account by recording the characteristics of the human and physical resources. The distribution of tasks among schedulers and how they saw their work were also integrated into our analysis.

Second, six periods of systematic on-site observation were performed between October and December 2015. These enabled us to cover all the variability of the scheduling process, and to take into account time constraints in the analysis. The verbal protocol method was used (Hoc and Leplat 1983). This consisted in asking the schedulers to describe their activity out loud while performing it. If actions were not explicitly described, the researcher asked the scheduler for more details.

The data were processed from notes taken on the collection grid. We defined a 'scheduling problem' as a situation when scheduling has to be modified to integrate a new constraint. All the scheduling problems were transcribed into an Excel spreadsheet for analysis using a set method devised by the researcher.

Each scheduling problem was characterised by the following:

- How much time the schedulers had to deal with the scheduling problem. This time was calculated as the interval between the day the production plan concerned by the scheduling plan was to be executed, and the day of observation;
- The management time needed to deal with the scheduling problem;
- Characteristics of the new constraint introduced, according to two criteria (with subcriteria) drawn from Crawford and Wiers (2011) and Chevalier and Cegarra (2008): the origin of the new constraint (i.e. how it came to enter the design planning process) and the nature of the new constraint which refers to the dimensions of production;
- The results of the action taken to deal with the scheduling problem.

3 Results

In the first part of the results, we will identify the diversity of constraints faced by tactical-level schedulers and then analyse the strategies they put in place to deal with them.

3.1 A Permanent Management of Constraints

In all, 150 scheduling problems were identified (99 for Agent I and 51 for Agent P). The Table 1 presents the characteristics of the constraints that generated the 150 scheduling problems, according to their origin and nature.

Table 1. Characteristics of constraints generating scheduling problems by nature and origin.

Origin	Nature		
	Human	Operational	Total
Prescribed	38	29	**67**
Unforeseen	38	1	**39**
Constructed	5	8	**13**
Deduced	12	19	**31**
Total	**93**	**57**	**150**

Concerning the nature of the constraint, two thirds of the introduced constraints were related to human dimensions and one third to operational dimensions. Constraints of different natures had to be met throughout the design process. By contrast, on the day the production plans were executed, only constraints of a human nature were integrated. Figure 2 presents the proportion of constraints introduced in the scheduling according to their nature and time of occurrence.

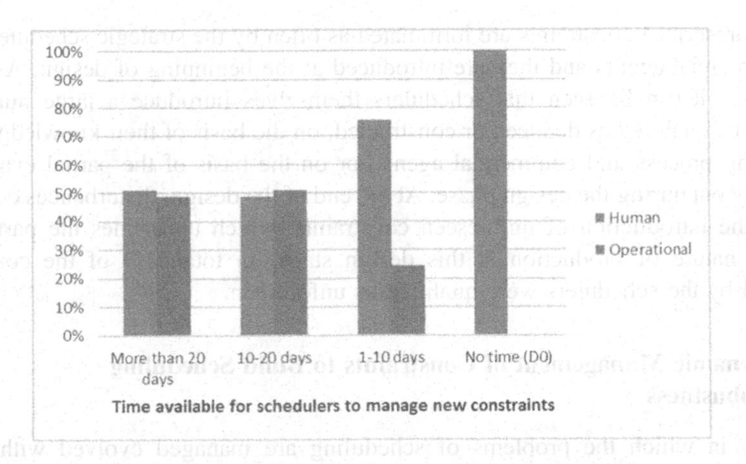

Fig. 2. Nature of the constraints introduced according to the advancement of scheduling

Throughout the design process, schedulers have to articulate the different dimensions of production.

Concerning the origin of the constraint, the new constraints, i.e. those causing the scheduling problem, were introduced in different ways during the design stage. They broke down as follows: 13 constructed, 31 deduced, 39 unforeseen, and 67 prescribed.

Figure 3 shows the proportion of constraints managed by schedulers according to their origin and time of occurrence.

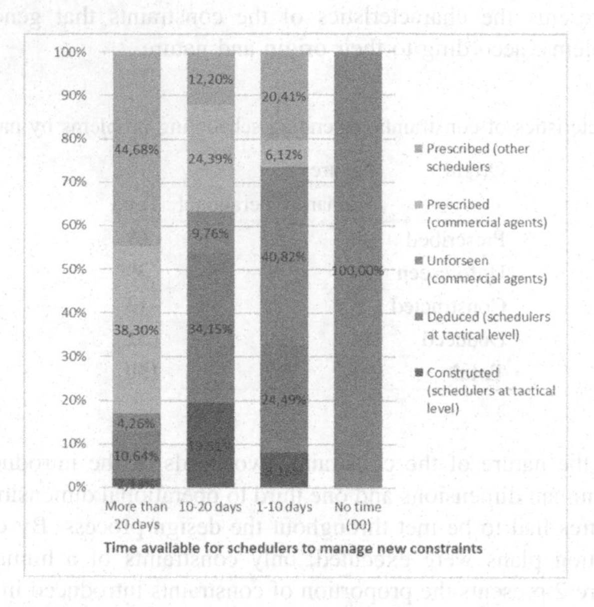

Fig. 3. Origin of the constraints introduced according to the advancement of scheduling

The prescribed constraints are formulated as often by the strategic schedulers as by the commercial agents and they are introduced at the beginning of design. As design progresses, it can be seen that schedulers themselves introduce a large number of constraints, qualified as deduced or constructed, on the basis of their knowledge of the scheduling process and commercial agents, or on the basis of the partial evaluations they carry out during the design phase. At the end of the design, disturbances occur and lead to the introduction of unforeseen constraints, which underlines the particularly unstable nature of production at this design stage. In total, 1/3 of the constraints managed by the schedulers were qualified as unforeseen.

3.2 Dynamic Management of Constraints to Build Scheduling Robustness

The way in which the problems of scheduling are managed evolved with design progress and the type of constraints that schedulers must manage. The implementation of these different strategies leads to the dynamic construction of robustness which takes different forms and concerns several temporalities.

At the Start of the Design Process by Tactical Schedulers

In meeting the prescribed constraints of an operational nature that were formulated, the schedulers took into account the human dimension by endeavouring to allow for the life-work balance of the commercial agents. In addition, the schedulers accepted some adaptations requested by commercial agents, i.e. constraints of a human nature. These decisions might oblige schedulers to arbitrate preventively between these constraints and those of an operational nature. The following example shows how Agent P opted to assign a mobile agent who had no position on Monday afternoon after a rest period: "*There, as I've got the choice, I'm going to let him start on Monday at 1:30 pm. That way he'll have a week-end of two and a half days*" (Agent P). This strategy enables the schedulers to reduce the uncertainty in the availability of the commercial agents. In the present context of labour unrest, if the commercial agents were satisfied with their assignments, there would be less risk of them being absent from work. Schedulers seek to build so-called "immediate robustness" through stable decisions that they will not reschedule.

Despite such adjustments, schedulers did not meet all the human constraints that were formulated. They also implement strategies of less compromise, i.e. they arbitrated preventively in favour of constraints of an operational nature. Five constraints of a human nature were not met, and the schedulers postponed the treatment of two others. In the following example, the scheduler chooses to postpone the management of this constraint when he will have more information on the progress of the design: *"In theory I can grant him his two weekends. For now I have some slack, but you never know, if it gets complicated I might need it. I'll put that aside, we'll see how it goes"* (Agent P). Indeed, if he makes the decision immediately he may not be able to manage future constraints. With this type of strategy, schedulers seek to maintain flexibility and maintain robustness until the implementation of schedules, in other words to build a "delayed robustness".

Using Forward Planning to Maintain Scheduling Robustness

As the design of production advanced, the schedulers also looked ahead, i.e. they foresaw that changes would occur in the data they had to design the scheduling. Schedulers perform a diagnostic-prognostic activity. This forward planning resulted in the schedulers formulating 26 deduced constraints and 12 constructed constraints between 1 and 20 days before execution of the production plans. Integrating the resulting constraints compelled the schedulers to partially reschedule the production plans. To deal with these scheduling problems, they had to create extra leeway. In the following example, the scheduler rescheduled to anticipate the possible unavailability of one of the commercial agents: *"So why am I bringing in Myriam as an extra? I know very well that Catherine [a commercial agent] is going to want a day off for union work, so when I can, I put her on as an extra and look for someone else."* (Agent I).

When the decisions arising from their forward planning were 'efficient', it was when they had the necessary leeway to make them. The management of this type of constraint thus resulted in the new constraint being met without generating any unstaffed positions. Following the anticipation made by the scheduler in the previous example, he avoided the occurrence of an unstaffed position: *"Catherine's just got time off for union work. Since I planned for it, all I have to do now is validate what I planned to allow for her change."* (Agent I).

Overall, the management of 15 out of the 39 unforeseen constraints generated no unstaffed positions because the schedulers had anticipated the evolution of the production environment. This type of strategy allows schedulers to stabilize their decision and maintain robustness for the end of the design, i.e. build a "deferred robustness".

At the same time, schedulers implement mutually beneficial strategies, i.e. schedulers make the choice to satisfy additional requests from sales agents (e.g. a change of schedule or the agent for a day off) in exchange for additional opportunities, which they used to manage situations in which unstaffed positions might be generated. In the example below, after the scheduler had agreed to modify a commercial agent's assignment to meet a family commitment, he wanted to modify another of his assignments to prevent a position becoming unstaffed: *"As the other day I agreed to change him so he could start later, now I'm going to call him to see if he'll agree to do this shift. That way I can be sure to open the counter on Friday."* (Agent I). At the same time, this type of decision allows schedulers to gain flexibility in designing future schedules, which we have identified as building "sustainable robustness".

Making Considered Sacrifices to Limit the Degradation of Scheduling Robustness

A few days before the implementation of production plans, the schedulers managed unforeseen constraints that had not been anticipated. However, given the narrow leeway the schedulers had, and the large number of unforeseen events that could arise, they could not always avoid unstaffed positions. To limit the ensuing adverse effects, they had to consider which commercial and economic activity they wanted to maintain, and which they were ready to sacrifice. This meant arbitrating between constraints of an operational nature.

To make this choice, several criteria were taken into account by the controllers, such as hours and 'sales potential' estimated as a function of turnover and expected customer flows. Each of these criteria was weighed according to the situation to be managed. In the following example, the scheduler took into account both the 'sales potential' of the counter and the shift times of the unstaffed positions: *"So I've unstaffed a position, that is I've taken off a commercial agent who was an extra at the station – on a Saturday it's like unstaffed a position – so as to let the agent have the weekend and put him on a shift Monday. I use him on the opening shift at the station. I had to cover him. But they'll be one agent short on the closing shift though, because I had to staff the position at B5."* (Agent I).

Although some of the decisions made by the schedulers generated unstaffed positions, arbitration let them choose which position would be unstaffed, i.e. where. The aim was both to limit the adverse impact on the health of the commercial agents and ensure that positions were staffed at specific counters.

At the end of the design process, schedulers may not have enough leeway to arbitrate responsively. In all, 17 scheduling problems led to this type of situation. They were characterised by the advent of unforeseen constraints of a human nature some days before or on the day when production plans were to be executed. Thus about half of all the unforeseen constraints could not be managed by the schedulers.

4 Discussion- Conclusion

Results confirm that schedulers have to manage a multitude of dynamic constraints [4, 5] that correspond to the needs of the company, the operators and the schedulers.

Schedulers adapt their strategies according to the diversity of the constraints they have to manage, in particular time-related ones, and to the production's context. The strategies employed by the schedulers make it possible to integrate the human and social dimension into scheduling choices. Schedulers take operators' preferences into account, negotiate with them, and foresee what may happen. They attempt to anticipate disturbances that may arise in this uncertain context. The implementation of these strategies leads the schedulers to carry out numerous rescheduling, which confirms previous findings [4, 5].

Our results allow us to provide elements to define robustness in the field of personnel scheduling. Unlike other works [14–17], our results show that the robustness is multidimensional. The analysis of the strategies implemented by schedulers shows that the robust contract of scheduling takes different forms, depending on each other. Schedulers seek both to build stable solutions that they won't go back on, while making some flexible decisions to be able to re-order until the day the schedules are implemented.

In addition, robustness concerns relate to several temporalities. Schedulers seek both to build:

- An "immediate robustness" linked to progress at time t of scheduling
- A "deferred robustness", i.e. the schedulers are already building up resources so that scheduling remains robust until they are implemented.
- A "durable robustness", i.e. the schedulers build up potential resources for the design of future schedules.

Despite these strategies, some vacancies remain due to a failure to maintain robustness at the end of the scheduling's process. These results lead us to focus on the conditions, especially collective conditions of the robustness' construction to dealing with uncertainties. From shared knowledge of the production management process, and discussion of the various underlying rationales, the unit's actors would be better able to adjust their organisational practice to reduce dysfunction due to unstaffed positions.

Moreover, given the number of constraints formulated by the executors of the plans, especially at the end of their design, their involvement in the scheduling process should not be overlooked. Indeed, both schedules have consequences on the way operators perceive the quality of their working conditions, and the constraints formulated by the operators have effects on the quality of the schedules designed.

In order to reduce uncovered positions, the organization should therefore also recognize operators as the designer of their own schedules. Meeting schedulers and operators would contribute to "organizational work" [18] aimed at re-engineering scheduling rules to satisfy all stakeholders.

References

1. Baker KR (1974) Introduction to sequencing and scheduling. Wiley, Hoboken
2. MacCarthy BL, Wilson JR, Crawford S (2001) Human performance in industrial scheduling: a framework for understanding. Hum Fact Ergon Manuf 11(4):299–320
3. Sanderson PM (1989) The human planning and scheduling role in advanced manufacturing systems: an emerging human factor domain. Hum Fact 31(6):635–666
4. Cegarra J (2008) A cognitive typology of scheduling situations: a contribution to laboratory and field studies. Theor Issues Ergon Sci 9:201–222
5. Guerin C, Hoc JM, Mebarki N (2012) The nature of expertise in industrial scheduling- strategic and tactical process, constraint and object management. Int J Ind Ergon 42(5):457–468
6. Hoc JM, Mebarki N, Cegarra J (2004) L'assistance à l'opérateur humain pour l'ordonnancement dans les ateliers manufacturiers. Le Travail Humain 67(2):181–208
7. Ernst AT, Jiang H, Krishnamoorthy M, Sier D (2004) Staff scheduling and rostering: a review of applications, methods and models. Eur J Oper Res 153:3–27
8. Meisels A, Kaplansky E (2004) Iterative restart technique for solving timetabling problems. Eur J Oper Res 153:41–50
9. Lezaun M, Pérez G, Sainz de la Maza E (2010) Staff rostering for the station personnel of the railway company. J Oper Res Soc 61:1104–1111
10. Schaerf A, Meisels A (1999) Solving employee timetabling problems by generalized local search. In: Congress of the Italian association for artificial intelligence. Springer, Berlin, pp 380–389
11. Van den Bergh J, Beliën J, De Bruecker P, Demeulemeester E, De Boeck L (2013) Personnel scheduling: a literature review. Eur J Oper Res 226:367–385
12. De Snoo C, Van Wezel W (2014) Coordination and task interdependence during schedule adaptation. Hum Fact Ergon Manuf Serv Ind 24(2):139–151
13. Alfares H (2004) Survey, categorization, and comparison of recent tour scheduling literature. Ann Oper Res 127:145–175
14. Billault JC, Moukrim A, Sanlaville E (2008) Flexibility and robustness in scheduling, 2nd edn. Wiley, Hoboken
15. Roy B (2010) Robustness in operational research and decision aiding: a multi-faceted issue. Eur J Oper Res 200(3):629–638
16. Ourari S, Berrandjia L, Boulakhras R, Boukciat A, Hentous H (2015) Robust Approach for centralized job shop scheduling: sequential flexibility. IFAC-PapersOnLine 48(3):1960–1965
17. Herroelen W, Leus R (2005) Project scheduling under uncertainty: survey and research potentials. Eur J Oper Res 165(2):289–306
18. de Terssac G (2003) Travail d'organisation et travail de régulation. In: de Terssac G (ed) La théorie de la régulation sociale de Jean-Daniel Reynaud. Débats et prolongements. La découverte, Paris, pp 121–134

The Simulation of Extreme Situations for the Analysis of Resilience: An Original Methodology to Improve Simulation and Organizational Resilience

Cecilia De la Garza[(✉)], Pierre Le Bot, and Quentin Baudard

EDF Lab Paris-Saclay, PERICLES – FOHST,
7 Av. Gaspard Monge, 91120 Palaiseau, France
{cecilia.de-la-garza, pierre.le-bot, quentin.baudard}@edf.fr

Abstract. The purpose of this study was to study the sizing of operation teams as defined in the project concerning the reorganisation of operation teams in "extreme situations" (ES). These changes follow the Fukushima accident, which revived interest in the resilience of socio-technical systems. Through an empirical Resilience Model, accident simulation situations in ESs are analysed here, incorporating operation teams and their interactions with the crisis system. This paper presents the methodology put in place to analyse several groups involved in crisis management and discusses the results and the type of contribution, in order to characterise strengths and areas of progress in terms of organisational resilience.

Keywords: Organizational resilience · Simulation · Crisis

1 Context and Objective

Through the simulation of an accident situation considered an "Extreme Situation", the aim is to study the sizing of operation teams as defined in a reorganisation project. The aim was to identify strengths and highlight areas of progress for the operation team and on a part of the national crisis system. The notion of Extreme Situation (ES) emerged after the accident at Fukushima in order to pose the question of an organization's response to cumulative events beyond the design basis of the plant and of the organization itself. The characteristics of an ES result, a priori, from an initiating event such as an earthquake or a flood of unusual intensity: (i) the degradation of the environment can go as far as an isolation of the site preventing the emergency teams from arriving. (ii) The conditions of intervention in the field can become very difficult. (iii) People on site may have been injured or contaminated by the initiating event or its consequences. (iv) Internal and External means of communications may have been lost. (v) An accidental situation can occur on several reactors simultaneously, involving the loss of internal and/or external electrical sources, and/or loss of cooling means. In these Fukushima-type ESs, operation teams are placed under "penalising" conditions. Nevertheless, if the teams are alone on the site, there exists a national level crisis

S. Bagnara et al. (Eds.): IEA 2018, AISC 821, pp. 13–22, 2019.
https://doi.org/10.1007/978-3-319-96080-7_2

organisation that will be deployed (see Fig. 1). As a result, while focusing on operation teams, the study analyses interactions with the outside world, in order to characterise the resilience of the crisis system as a whole. In a crisis situation, resilience is the result of the collective operation of various entities, including the operation teams, the NCP and the National Technical Support Team (NTST), who must cooperate and coordinate themselves efficiently.

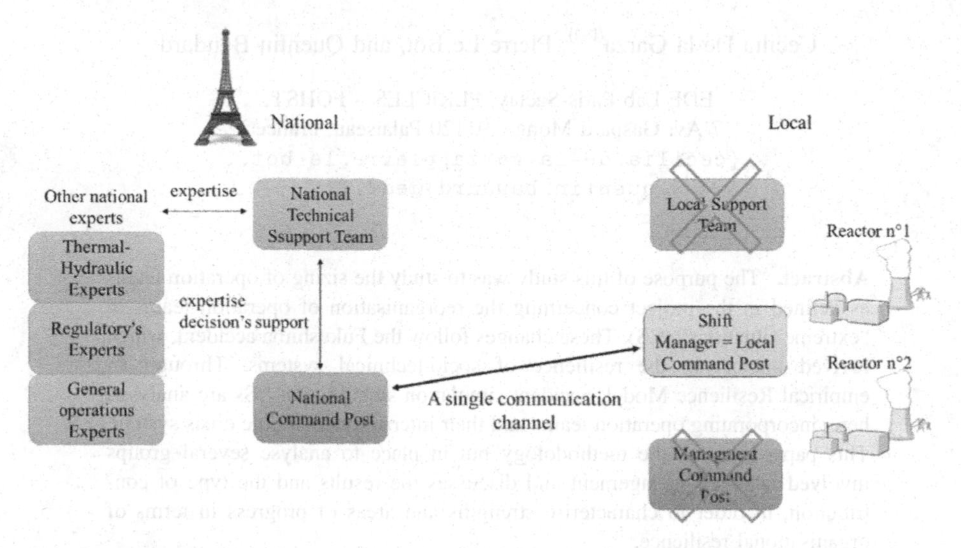

Fig. 1. Representation of crisis organisation in ES [1]: absence of the Local Support Team and the Management Command Post

2 An Empirical Resilience Model "Serving" the System Studied

The accident at the Fukushima Daiichi power plant in 2011 drastically renewed interest in resilience. Yet our approach, which was developed several years ago, is based on the Model of Resilience in Situation (MRS) for the organisation of the management of crises that may take place on EDF nuclear sites [2, 5]. The MRS was created empirically according to observations of operations in simulated accident conditions. The two main characteristics of this model are as follows:

- A dynamic description of the operation of the socio-technical system studied and the necessary use of two opposing rationalities, namely the anticipation of the rationality (technical, organisational, procedural) on the one hand, and the adaptability of the "flexible" rationality resulting above all from the collective operation of on-call and operation teams, their experience and their expertise on the other hand.
- Resilience "takes place in situation". The dynamic operation of the team is based on the succession of rupture/stability phases, where the system implements rules to deal with the situation, and stabilisation phases, during which the system applies its rules.

The organisational anticipation process is proactive, in other words *a priori* with respect to an accident situation. It provides sufficient rules and resources to recognise a situation, interact with it and enable robust and productive operation. However, any claim that all situations can be anticipated in detail is a myth. Anticipation is both a simplification of a future reality, and a category of situations with similar characteristics. It is managed upstream by national engineering divisions or on production sites.

The adaptation process during crisis management takes place reactively in a situation or just before a situation interacting with the means provided through anticipation. It is the implementation of these means in a situation experienced and only to manage this particular situation that may present specific needs. The groups involved in the management of a situation support this process by establishing, when necessary, ad hoc organisation and execution rules in real time and using them themselves. Adaptation does not therefore take place ex nihilo and reuses what has already been anticipated. For example, stakeholders will modify or complete an imprecise procedure that is impossible to apply in a particular situation. They will adapt the knowledge acquired during training and will develop new knowledge. This is a common process in "ordinary" situations. However, in extraordinary situations, in other words, when faced with an unexpected situation during the management of a critical situation, stakeholders are also expected to adapt by using their own experience and relying on support from the groups involved in crisis management.

In terms of safety, the question of where to focus efforts comes up: is it better to anticipate or rather to rely on adaptation in real time? Our studies show that both are essential. The question is then how to reconcile these two aspects of resilience that seem a priori paradoxical? How can we make the various crisis teams understand that the processes of anticipation and adaptation are equally important? And how can we identify them during observations of simulations, crisis exercises or a posteriori based on event analyses? In the following section, the methodological approach implemented to identify resilience factors during crisis management simulations will be presented, and/or unfavourable elements related to organisational changes. This stage is a preliminary stage that will subsequently enable some elements to be provided for reflection, in order to strengthen resilience in the field of crisis management.

3 Methodological Approach: Organisation of "Extreme Situation" Simulation Situations

3.1 A Multidisciplinary Approach Involving Numerous Stakeholders

Since the resilience model is an empirical model, this stage is based on the analysis of the actual activity of the teams involved in crisis management during full-scale simulations. With this is mind, we developed a multidisciplinary approach combining ergonomics, human reliability and safety. The aim is to combine analyses of cognitive ergonomics, such as decision-making or problem solving, analyses of the collective operation of teams in terms of cooperation and coordination, analyses of teams' performances with respect to the operations carried out on the reactor and their ability to deal with a complex situation [3].

From the methodological point of view, two areas are distinguished: a scientific area and a logistics area. The simulation scenarios chosen were based on the Fukushima accident: earthquake, site isolated for 24 h, multi-plant unit accident such as the "total loss of internal and external power sources", absence of the LST and the MCP (cf. Fig. 1) and the safety engineer in the control room. A test of this type had never been organised in France; it was necessary to think about what had to be simulated, the level of representativeness, the underlying goals and the underlying type of data collection and processing, against the backdrop of the resilience model.

From the logistical point of view, the organisation of this type of simulation required the implementation of an upstream working group involving stakeholders at the national level (engineering divisions, stakeholders from the crisis organisation branch, human factor consultants, business line coordination, training) and at the local level (production sites, operations department, Internal Emergency Plan managers, training centres, trainers).

The working group organised and prepared the various stages required for the performance of such a test for a period of six to eight months. In addition, the same process was required for each of the tests carried out.

From the scientific point of view, human factor specialists (ergonomics and human reliability) managed the preparation of "extreme situation" tests. Thus, in total, the organisation, preparation and performance of an "extreme situation" test, taking into account two damaged reactors, required the involvement of between 60 and 70 people.

3.2 Extreme Situation Test Preparation Stages

The various stages will be presented here, with the emphasis on the scientific aspect for this paper.

Composition of the working group. This working group included representatives from the R&D division and national representatives with skills in crisis management, the human factor, safety and operations. In addition, representatives from the site and the training centre where the test would be conducted were required, in order to plan operation teams, trainers and simulator usage slots.

Establishment of a schedule and identification of the various tasks. This part was managed by the operator for the preparation of scenarios, the recruitment of the operation teams and the national crisis teams who participated in the test, the organisation of the use of simulators with the training centre, the preparation of ad hoc documents for the test, etc.

Establishment of the test protocol in relation to the goals of the study. Even though this task is the responsibility of the R&D division, it was developed with all the members of the WG. This means:

- Defining the assumptions and themes to be tested. Here, the study focuses on the evolution of the operation teams, which have changed, in relation to the envisaged design changes and associated procedures for the specific management of the internal and external power source loss scenario, as well as for crisis management by the various team members.

- Developing scenarios and specifications to prepare scenarios and any necessary adaptations on the simulator. This concerns specific scenarios that involve two parallel simulators.
- Preparing and carrying out a technical validation phase of the scenario on the simulator. This phase requires the participation of trainers and other representatives of the WG. It purpose is to ensure that the scenario runs on the simulator and that the procedures are suitable.
- Preparing and carrying out a practice test phase on the simulator with the trainers and two operation teams other than those who will participate in the final test. The goal is to test the entire simulation system and to make sure that everything works, to test the interactions between the two simulators, the times planned, the operational actions and the impact on the installation, etc. Adjustments may still be necessary before the final test.

Decision concerning methodological choices in terms of data collection and analysis. Two types of data collection were promoted: in situ observations and collective debriefings, which will be presented in more detail later.

Selection of the operation teams participating in the practice and final tests. Even though it was not possible to select a specific operation team, some criteria were formulated, such as having teams with experienced operators and others with less experience, so as to have a minimum level of representativeness of the operations population.

Identification of information/training requirements for the various participants in the test. Since this is a situation that is in the process of being modified and for which we wanted to test the changes, specifically the organisational changes, it was necessary to provide at least some information and preparatory elements to the operation teams upstream. Depending on the case in point, one half-day to one and a half days were required to present the new missions, technical developments and associated procedures.

3.3 The Simulation System

The study system adapted to the existing crisis system, in order to be representative of a real crisis situation in ESs (see Fig. 1). It is characterised by five observation points corresponding to five separate working groups.

- On site, two full-scale operations simulators in nuclear production site training centres. One operation team manages each of the simulators for five hours (points 1 and 2). Two teams of field operators act "as though they were performing" the actions locally (point 3).
- At the national level: a National Crisis Technical Team was partially staffed (point 4) and, depending on the case in point, representatives from the National Division Command Post took part in the tests (point 5).

To meet the objectives of the study, it was necessary to be able to recreate and observe inter-group and intra-group interactions, as well as the collective operation characterised by cooperation, coordination of actions and collaboration. However, coverage of this network of stakeholders geographically located in separate locations, working on distinct objects and with constraints specific to each group, raises methodological questions for data collection and then for processing. Ergonomics, human reliability and safety observers were spread across the posts of these five centres. There were seven observers per control room on the simulator, two for field operators, two with the NTST and two with representatives of the NMCP. In total, this represented the mobilisation of twenty observers.

4 Focus on the Control Room and the Operation Teams

4.1 Working Assumptions

This paper will focus on the "operation teams" working groups and the missions of the various members. Each operation team is made up of two operators, a supervisor, a shift supervisor and a Monitoring Operator (MO), as well as Field Operators, who perform local actions. The shift supervisor carries out the "Local Command Post" mission within the crisis organisation and is the only contact person on the site able to interface with the National Management Command Post (NMCP) in extreme situations. In the event of the prolonged absence of the Safety Engineer, the MO carries out the mission of monitoring and checking the fit between the current situation and the operations strategy adopted. He also prepares the crises messages for the LCP.

These are penalising situations for the operation teams, since the site loses the means of communication, and only a satellite phone can be used to establish a single outgoing link to communicate with the national crisis team.

Following the earthquake, internal and external field inspections must be carried out by two field operators, in order to produce a review of the state of the installations. During this phase, no-one else may go into the field, in order to ensure their safety. The external field inspection is time-delayed during the first hour. However, in a situation with a loss of power sources, equipment control from the control room becomes unavailable, and it is not possible to read the indicators displaying information on the physical state of the plant unit. The only means of operating and checking the installation is to go directly to the equipment in the field.

Specifically, for these groups, five working assumptions guided the organisation of "extreme situation" simulations and their subsequent analysis. These are listed below:

- The new distribution of missions within the operation team makes it possible to manage the consequences of an accident in all situations: when the monitoring operator is required (i) and when it is not.
- The team is able to manage local actions and "field operator" resources safely.
- The operation team ensures the transmission of the relevant information from the two control rooms to the National Division Command Post in the ES context.
- The dynamics of the interactions enable the organisation to ensure both the characteristic robustness and adaptability of resilience, in order to maintain control of the accident situation, regardless of its severity.
- The tools available enable the consequences of an accident to be managed (procedures, communication tools, etc.).

4.2 The Data Collection System

For observation, a common grid was used by all observers, in order to guide chronological in situ note-taking on the actions performed, the decisions taken, communication and the difficulties encountered. The goal was to be able to reconstruct parallel chronologies of each of the groups and their interactions a posteriori. The simulations were filmed, in order to be able to closely analyse certain sequences.

At the end of the five hours of simulation, a post-simulation debriefing is carried out on the collective operation of the team, the missions, the developments, the difficulties, the favourable points and the interactions with other working groups. This debriefing is carried out in parallel by HF experts per working group, to enable each working group to communicate in depth on the various themes covered for two or three hours. During the post debriefing phase, all the observers then carry out a review, in order to share the notable events of each group. These debriefings were recorded to facilitate post hoc thematic analysis.

4.3 The Data Processing System: From Notable Events to the Identification of Favourable Elements for Resilience

Between 2014 and 2016, seven ES-type tests were carried out with variations: five in a multi-plant unit accident situation, including one with two reactors using different technologies, and two with a thermohydraulic accident in a single plant unit but in a cumulative situation with another event (fire or flood).

Multidisciplinary analysis was supposed to identify the favourable factors and difficulties for the organisational resilience of the socio-technical system in ESs. The common basis of this analysis is as follows:

- The reconstructed chronologies based on the notes taken by each observer for a final assembly of a chronology of each of the control rooms and a chronology of the interactions between the Site and the National level.
- The identification of Notable Events (NE). An NE is an event that reports an action, a “non-action”, a decision taken, a collective or individual initiative, an observation that may improve the reliability of the organisation of the operation team and/or the crisis team, operations, their robustness, or that facilitates sensemaking, or on the contrary that may make the socio-technical system less reliable and that may weaken safety barriers, thereby hindering sensemaking. The NE is either detected during in situ observation or later during the group debriefing, or during the reconstruction of the overall chronology. NEs are elements used in the analysis of the resilience of the socio-technical system. This concerns, for example, an unexpected collective or individual decision taken when dealing with an unexpected situation, which turns out to be suitable, or a reconfiguration of the team to divide the tasks differently, in other words, a supervisor who decides to take charge of the management of local actions, their prioritisation in connection with the available field operators and the state of the installation. Furthermore, by contrast, there may be NEs that could weaken resilience, such as the non-application of a part of the instruction due to misunderstanding on the part of the operator, or focusing on a problem and not identifying a change in situation.

NEs are then grouped per group: control room team, field operators, NTST and NMCP. An initial level of common analysis is carried out through work themes that enable the following to be highlighted: technical NEs relating to the result of operations on the process, expertise NEs relating to the support provided to operations, or organisational and operational NEs of the group.

A second, more detailed level of analysis is carried out by skill. Here, the ergonomics part, made up of four types of analyses used to support the initial assumptions, will above all be described:

- Cognitive analyses interacting with the operation of the team, in other words, decision-making, problem solving, understanding of the situation, diagnosis and prognosis.
- An analysis of collective operation through cooperation, coordination and collaboration.
- The construction of control room activity timetables to visualise the network of stakeholders involved in crisis management, interactions between stakeholders in the control room and with external working groups.
- The thematic analysis of debriefings based on the notable events, the assumptions explored and the specific elements reported by the teams.

5 Results: A Robust Crisis Socio-Technical System that Nonetheless Requires Adaptation Capabilities

The results show that the sizing of the partial crisis system in ESs is not called into question, although some points can be reinforced from the point of view of organisational resilience. Some examples of NEs illustrate this below, by targeting operation teams.

(1) Due to the loss of power sources, from the beginning of the accident, compliance with the instructions required a large number of works sheets to be produced by field operators. It quickly became apparent that there was a need to prioritise and re-prioritise local action sheets, depending on the evolution of the state of the reactor, since it was not possible to go into the field before having carried out a field inspection, and furthermore, field operator resources are limited. Three types of difficulty were observed.

- Local action prioritisation is an activity with which the teams are not familiar and which involves significant cognitive requirements in terms of understanding of the current situation, anticipation of the impact of a local action on the reactor, etc.
- Activity that may have an impact on the workload of the team and on its collective operation: one of the members may be "monopolised" by this prioritisation activity at the beginning of the accident scenario (approximately 2 h), and may therefore not be available to the team.
- Failure to prioritise or to prioritise properly may have an impact on the operation of the installation: a local action performed late may exacerbate the situation.

(2) The monitoring and checking mission, as well as the task of writing crises messages, assigned to the Monitoring Operator (MO) enabled the line of defence to be maintained in the absence of the safety engineer and the LST. However, the situations are different in 900 MW control rooms, which are side by side and for which only one MO applies the monitoring and checking instruction, and in 1300 MW control rooms. For the latter, because the control rooms are located far apart, there is one MO per control room. In both cases, the LCP is relieved of this checking duty, which enables it to concentrate on its other missions. In both cases, collective operation processes that made the organisation more reliable were observed. For example, the MO helped operators search for information or solve a problem, for example; the MO identified the priority actions to be carried out in the control room. However, in both cases, it was observed that coordination between the MO and the LCP was not always smooth, either because the MO, who has to manage two reactors, has a heavy workload at times, which in this case delayed the transmission of information to the LCP, which was awaiting the reactor status review in order to be able to carry out a telephone review with the NMCP. In the other configuration, the fact that there are two MOs implies that the LCP is required to communicate and coordinate with two separate people, which is more time-consuming.

(3) From the point of view of communication between control rooms and the national level, it appears that due to a single communication channel possible in ESs, and a single contact on site, the LCP, communication turns out to be difficult to manage. The LCP must not be over-solicited, but it is still necessary to understand what is happening and to take into account its requests and provide support. In addition, from the national level point of view, the support team must both manage the multi-plant unit aspect and replace the local crisis team that it was not possible to staff on the site in view of the context. In some cases, this was difficult for the support team, which, for example, made proposals for action that the site did not understand.

6 Discussion

We adapted our multidisciplinary approach to observe in situ the interactions between the national level and the site, and between the national entities, as well as between the control room and field operators. The MRS was enriched by the contributions of cognitive ergonomics and was extended to the various groups involved in crisis management, with a view to taking into account collective operation in a network.

Thus, based on the actual requirements highlighted during the simulations for each of the groups and missions, both in the control room and at the national level, it is possible to propose areas for improvement. Among other things, these are materialised through new training systems [1] and work tools suited to crisis conditions. For example, to make the management of local actions and field operator actions more reliable in ESs, a Local Action Monitoring System prototype was designed as part of an agile process [4]. This is a panel used to display the actions pending, to prioritise them and then to see those which are in progress and who is carrying them out, and finally to display successes and failures. Specifically designed for the supervisor, this support system also facilitates the management of the safety of field operators who go out to perform actions. It also enables information to be shared within the operation team and facilitates a view of the status of the reactor at a given time. However, the incorporation of this tool within operation teams requires the parallel implementation of training in both the tool and in the activity of local action prioritisation. This is all the more justified since this system could be used in other contexts, such as plant unit outage.

Other organisational reflection is underway, for example in connection with coordination and collaboration between the LCP and the MO, with a view to facilitating their interactions. The distribution of tasks between these two stakeholders or new operating and information transmission practices could be tested.

The purpose of simulating crisis situations in ESs was mainly to study the new distribution of missions within the operation team. Simulation is known to be a technique used to test concepts, tools and organisations during design and/or modification processes. However, one of the lessons learned through this study is that the organisation of this type of simulation remains expensive and time-consuming, given the number of resources mobilised. While this may be justified in some cases, it appears necessary to think of lighter simulation methods, using techniques such as storytelling and serious games [1]. Furthermore, these lighter systems could be used in two ways: to test technical, procedural and organisational changes and/or to train the groups concerned by these changes.

References

1. Alengry J, Falzon P, De la Garza C, Le Bot P (2018) What is “training to cope with crisis management situation”? A proposal of a reflexive training device for the National Technical Support Team. In: Creativity in practice, 20th congress of the international ergonomics association, 26–30 August 2018, Florence, Italy
2. Bringaud V, Le Bot P (2018) Lessons learned from crisis simulation for the shift manager in Extreme situation. In: Creativity in practice, 20th congress of the international ergonomics association, 26–30 August 2018, Florence, Italy
3. De la Garza C, Pesme H, Le Bot P (2013) Interest of combining two human factor approaches for the evaluation of a future teamwork. In: Proceedings of the european safety and reliability, ESREL 2013 conference, Amsterdam, Netherlands
4. De la Garza C, Le Bot P, Baudard Q, Alengry J, Gaillard-Lecanu E (2016) D'un «document» à un «dispositif» de suivi des actions en local dans le nucléaire. Congrès ErgoIA, Juillet, Biarritz
5. Le Bot P, Pesme H (2010) The Model of Resilience in Situation (MRS) as an idealistic organization of at-risks systems to be ultrasafe. In: PSAM 2010 - 10th international conference on probabilistic safety assessment & management, Seattle, Washington, USA, 7–11 June 2010

The Need to Present Actual Costs After an Ergonomic Intervention

Rafael Bezerra Vieira[1(✉)], Mario Cesar Vidal[1], José Roberto Mafra[1], Guilherme Bezerra Bastos[2], Rodrigo Arcuri[1], and Luiz Ricardo Moreira[1]

[1] Universidade Federal do Rio de Janeiro, Rio de Janeiro, Brazil
rbvieira84@gmail.com, mcrvidal@gmail.com, arcuri.rodrigo@gmail.com, mafra@facc.ufrj.br, luizricardo@ergonomia.ufrj.br

[2] Universidade Veiga de Almeida, Rio de Janeiro, Brazil
guilherme.bbastos6@gmail.com

Abstract. The importance of costs in Ergonomics has been demonstrated through several studies over the last years. Generally the studies tend to be biased to financially justify the ergonomic intervention, using estimated initial data. This paper aims to demonstrate that there is a need for a new cost verification at the end of the intervention, in order to present the actual values to the top management of the company, making the technical report more complete and useful to the several company's departments, and specifically the accounting and financial ones. The applied methodology was adapted from Lahiri et al. (2005) and Mafra (2004) in a case study in the food industry. The estimate presented by the team at the beginning of the work was R$ 30,380 but at the end it turned out amounting to R$ 36,479, an increase of 13%. This variation, in addition to being significant on its own, is relevant for several departments of the company, since it facilitates other decisions, as well as allowing the recognition of the values in the accounting and financial statements.

Keywords: Ergonomics · Cost · Accounting

1 Introduction

The lack of ergonomics in business can silently turn into high costs as consequences of absenteeism, decreased efficiency, and decrease in product quality, among others. In many situations, executives believe that such costs are not important enough to justify investments made in Ergonomics. However, more and more studies have demonstrated that investing in ergonomics usually brings positive financial returns for companies, often in very short periods.

Mafra (2004) demonstrates that ergonomic intervention can improve sales by making the company more competitive, and many of the investments pay themselves in less than two years. Thus, one could say that after revealing such benefits, executives would choose to perform an ergonomic analysis on the company looking for improvements.

S. Bagnara et al. (Eds.): IEA 2018, AISC 821, pp. 23–27, 2019.
https://doi.org/10.1007/978-3-319-96080-7_3

Lahiri et al. (2005) argue that one of the major objectives of the economic analysis of an intervention is to prove that the investment to be realized can bring great profits to the company, demonstrating that the intervention should be of interest to the company, once it can solve problems and may bring greater profitability.

This study argues that the absence of ergonomics can cause financial losses for companies, and aims to show that the cost exposure should be an initial part of the ergonomic intervention, as well as the final part, initially presenting the estimated costs, and at the end, the actual costs. In this way, the ergonomic report may be used in addition to its initial purposes, as it will be subject to appreciation of other areas such as financial and accounting. Some examples of loss of productivity and costs of absence of ergonomics can be cited:

- Lemstra (2016) coordinated a survey in a hospital in Lisbon and found that nurses had symptoms of injuries resulting in absenteeism and loss of productivity, generating a total cost of $ 64,575.44;
- Magagnotti et al. (2016) followed 204 workers at the Saskatchewan company in Canada and identified that there were claims for injury, recovery time and costs, implementing a company-wide injury prevention program after one year reduced incidences by 58% and cost was reduced from $ 114,149.07 to $ 56,528.14 per year;
- Falck et al. (2014) developed an ergonomic costing model, following two weeks a Swedish automotive company and during that time sought to quality deficiencies. The authors found an error cost of approximately $ 86,000, and concluded that finding and correcting errors in the plan reduces internal costs, being 12.2 times cheaper than errors found later, and this ratio is 9.2 when using market costs.

These studies have in common the fact of obtaining the costs in the initial moment of the intervention in order to justify the same, demonstrating that the expected return is generally well above the expectation of return, when compared with other investments.

It should be noted that the determination of these values is not easy to obtain, as shown by Pereira da Silva et al. (2014), who talks about the types of measurement difficulty: information cost, multifactorial aspects and simplifications, absence lack of measurement methods. Another point to be highlighted is that there is still scarce literature published on the measurement of costs in Ergonomics. Examples such as existing biomechanical measurement methodologies could, quite easily, adopt standard cost formulas as demonstrated by Trask et al. (2012).

This work uses as a case an ergonomic intervention performed in the food industry where a model adapted from Lahiri et al. (2005) and Mafra (2004) was applied, divided in two parts. For the first costing (direct and indirect costs of the ergonomic intervention) variations and modifications in the methodology of Lahiri et al. (2005) were presented. For the second costing (expected effects after ergonomic intervention), a variation of the methodology presented by Mafra (2004) was used. This approach made possible to identify the payback (return on investment), the main points of losses in the process and how much the company stopped losing after the intervention.

The models were applied before the intervention, with the purpose of financially justifying the same. After the intervention, it was verified that a new iteration model application could bring benefits to the management of the company, since it would present the real values related to the ergonomic intervention, facilitating several areas of the company and the decision making of the top managers, since it would disclose the real financial and equity variations.

This study differs from Lahiri et al. (2005) and other studies on intervention costs, since it seeks to demonstrate the importance of obtaining the actual values acquired after the ergonomic intervention. Real values are essential for the dissemination of ergonomic intervention results, making it clear to managers, executives and other stakeholders that the investment made has brought concrete results for the company.

2 Methodology and Case

The case was carried out by the Complex Systems Ergonomics Research Unit (GENTE) laboratory team of COPPE/UFRJ in a pre-pasta factory for cakes. Seeking to achieve its production goals, the plant was working at full power, but still could not reach its monthly goal. Ergonomic analysis was performed due to the large number of falls and slips in the feed sector of the mixer. Based on this, an analysis was started in the process of displacement of the cart of distribution of the sector of products. The cart helps in the displacement of the wheat that stays in the floor corresponding to the mixer until the industrial sieves where they are inserted in the process. The analysis disclosed a series of problems, and three impacts were considered more relevant: (a) loss of raw material; (b) quality and inadequacy of the floor; and (c) difficulty to handle the cart.

There was constant loss of raw material by the often spilled wheat material, due to itinerary of the cart which had irregularities and holes, and was conducted on a slippery floor, which in turn was due to inadequate coating. The cart needed constant maintenance, but we noticed that there was no such activity. Furthermore, the wheels were well worn and it did not have a good position to load. Another reason for the loss of raw material was the loss during the feeding of the silos, which was due to incompatibility with the cart.

After detecting the above problems, we pointed out some recommendations for local development and needed acquisitions. The recommendations and their estimated and actual costs are shown in Table 2.

Initially, estimates were made of current losses and possible profits in the future. After the project was completed and all changes were implemented, a new valuation was carried out to obtain the actual values.

In order to collect the data, it was important the participation of the sectors involved. This collection was carried out through questionnaires, interviews with collaborators and field surveys. At the early stages an ergonomic evaluation was performed, looking for problems in the work environment. Such as inadequate lighting, extreme temperatures, repetitive physical effort and maneuvers, inadequate postures, long working hours, etc.

After the ergonomic evaluation, the direct and indirect operating costs were estimated and the data collected regarding the costs were divided between the sections presented in Table 1, whose categories were based (with adaptations) on the methodologies of Lahiri et al. (2005), Mafra (2004) and Rickards and Putnam (2012).

Table 1. Direct and indirect costs

			Estimated value	Real value
Direct costs	Operational	Loss of raw material	R$ 2,812	R$ 5,132
		Product withheld	R$ 9,375	R$ 7,571
		Waste management	R$ 1,125	R$ 1,459
	Personal	Medical license	R$ 600	R$ 600
Indirect costs	Tax and inspection	Social security	R$ 4,167	R$ 4,167
		Health risk	R$ 834	R$ 1,117
		Social rights	R$ 2,500	R$ 2,500
Total			R$ 21,413	R$ 22,546

Table 2. Implementation costs

		Estimated value	Real value
Implementation costs	Engineering project	R$ 2,000	R$ 3,000
	Installation of retractable rings	R$ 6,900	R$ 7,024
	Regular alignment of cart	R$ 2,400	R$ 2,000
	Improvement of the handle of the stand	R$ 300	R$ 650
	Change of rotation	R$ 2,080	R$ 2,140
	Ground reform	R$ 6,000	R$ 7,238
	Introduction of docking grooves	R$ 10,700	R$ 14,427
Total		R$ 30,380	R$ 36,479

Subsequently, the costs of implementing the recommended improvements to solve the problems were evaluated. Costs were divided between engineering design, manufacturing and installation, new operating costs, installation of retractable rings, periodic alignment of the cart, improvement of the handle of the trolley, change of castors, remodeling of the floor and introduction of grooves.

At the end of the ergonomic intervention, a new data collection was performed, seeking the actual values for direct and indirect costs and implementation costs. The cost concerning the absence of ergonomics and the payback (time of return on investment) were then calculated.

The cost of the absence of ergonomics was calculated using all direct and indirect costs. While in the estimate made at the early stages of the project it amounted to R$ 21,413 per month, the real value found after the intervention was R$ 22,546 per month. The payback was calculated on top of the costs of absence of ergonomics and the monthly investment costs and the estimated time of return on investment was of 1 month and 22 days. The real value obtained after the completion of the intervention was 1 month and 19 days.

In the case presented, absolute values may not seem significant at first, but represent around 13% of implementation costs. These differences should be demonstrated to stakeholders in the company, especially to top manageers and investors, as they may represent considerable amounts, of which these parties are not aware.

3 Conclusion

In recent years, some papers have succeeded in demonstrating that ergonomics should be presented as a competitive differential for top management. Companies have diverse responsibilities, including the ones along their shareholders, so they will choose based on maximizing their earnings. Thus, several studies followed this bias in order to bring ergonomists this tool, allowing greater insertion of their work.

Thus, this work demonstrates that there is still more to be done, and achieves the same goal. It is worth noting that there is a need for reassessment after the ergonomic intervention in order to obtain the actual values of the intervention, since the estimated values may deviate from the real values, with small differences and with more significant differences. In this way, the identification of costs in ergonomics will justify the ergonomic intervention for the administration, as well as give technical support to generate reliable accounting-financial information for external stakeholders, such as investors. In this way, ergonomics costs can fulfill the initial role of presenting ergonomics as a competitive advantage for management, as well as providing useful information to other users, such as external users (banks, investors, etc.).

References

Mafra JRD (2004) Economia Da Ergonomia: Metodologia De Custeio Baseado No Modelo Operante. Federal University of Rio de Janeiro, Rio de Janeiro

Lahiri S, Gold J, Levenstein C (2005) Estimation of net-costs for prevention of occupational low back pain: three case studies from the US. Am J Ind Med 48(6):530–541

Lemstra ME (2016) Occupational management in the workplace and impact on injury claims, duration, and cost: a prospective longitudinal cohort. Disponível em: https://www.dovepress.com/occupational-management-in-the-workplace-and-impact-on-injury-claims-d-peer-reviewed-article-RMHP. Acesso em 19 Jan 2017

Magagnotti N et al (2016) A new device for reducing winching cost and worker effort in steep terrain operations. Scand J Forest Res 31(6):602–610

Falck A-C, Örtengren R, Rosenqvist M (2014) Assembly failures and action cost in relation to complexity level and assembly ergonomics in manual assembly (part 2). Int J Ind Ergon 44 (3):455–459

Pereira Da Silva M et al (2014) Difficulties in quantifying financial losses that could be reduced by ergonomic solutions. Hum Factors Ergon Manuf Serv Ind 24(4):415–427

Trask C et al (2012) Data collection costs in industrial environments for three occupational posture exposure assessment methods. BMC Med Res Methodol 12:89

Rickards J, Putnam C (2012) A pre-intervention benefit-cost methodology to justify investments in workplace health. Int J Workplace Health Manag 5(3):210–219

Ergonomics/Human Factors Education in United Kingdom

Sue Hignett(✉) and Diane Gyi

Loughborough Design School, Loughborough University, Loughborough, UK
S.M.Hignett@lboro.ac.uk

Abstract. This paper presents a summary of the Ergonomics and Human Factors (EHF) professional accreditation process in the UK. EHF education can be accredited by the Chartered Institute of Ergonomics and Human Factors (CIEHF) as qualifying courses and as short (training) courses. A framework is used as professional competencies (5 units) with expected levels of proficiency to support career development through membership grades (student, graduate, registered, fellow). An example of education is given with the 5 postgraduate programmes (MSc, Postgraduate Diploma, Postgraduate Certificate) at Loughborough University: Ergonomics and Human Factors, Human Factors in Transport. Human Factors for Inclusive Design, Ergonomics in Health and Community Care, and Human Factors and Ergonomics for Patient Safety. Finally, an opportunity is offered to explore competency with an affiliate discipline (Unser Experience) in the context of usability testing for medical devices.

Keywords: Education · Competency · Accreditation

1 Introduction

The Chartered Institute of Ergonomics & Human Factors (CIEHF) is both the professional membership body and regulator for Ergonomics and Human Factors (EHF) in the UK. Professional behaviour in EHF is guided by a Code of Professional Conduct [1] and managed by the Professional Affairs Board (PAB). The PAB leads the establishment, promotion and maintenance of high standards of professional knowledge, practice and conduct in EHF in accordance with the Charter, Byelaws, General and Council Regulations, and under the policy direction of the CIEHF Council.

To address the complexity of the accreditation function, and develop a cross-cutting system, the CIEHF professional competencies [2] were reviewed and revised in 2016. An extensive project was undertaken to both simplify systems (reducing from 9 units as IEA [3] to 5 competencies) and provide a foundation for widening the reach of CIEHF accreditation to support interest from affiliate disciplines.

There are 5 universities (Fig. 1) offering qualifying courses at postgraduate (MSc) level. These can be studied full-time, part-time (usually over 3 years) and are available I [4] n 3 formats: distance learning, week block modules, weekly lectures (over 11–15 weeks).

S. Bagnara et al. (Eds.): IEA 2018, AISC 821, pp. 28–35, 2019.
https://doi.org/10.1007/978-3-319-96080-7_4

- Heriot Watt University
  - MSc Human Factors
- Loughborough University
  - MSc Ergonomics and Human Factors,
  - MSc Human Factors in Transport
  - MSc Human Factors for Inclusive Design
  - MSc Ergonomics in Health and Community Care
  - MSc Human Factors and Ergonomics for Patient Safety
- University of Derby
  - MSc Health Ergonomics through online learning
  - MSc Ergonomics (Human Factors) through online learning
  - MSc Ergonomics and Organisational Behaviour through online learning
  - MSc Behaviour Change
- University College London
  - MSc Human-Computer Interaction with Ergonomics,
- University of Nottingham
  - MSc Human Factors & Ergonomics,
  - MSc Applied Ergonomics by distance learning

Fig. 1. Postgraduate education in UK for EHF qualifying courses

2 Professional Competency

A checklist of the CIEHF professional competencies (Table 1) has been developed as both an applicant guide to expected competencies for qualification as an EHF specialist and also for assessment of applications. It is used as a cross-cutting assessment platform across CIEHF professional activities for accreditation of degree courses, short training courses, membership applications, continuing professional development etc.

CIEHF-accredited 'Qualifying' degree courses are assessed against the professional competency checklist with accreditation renewal every 5 years (as a new application). The Degree courses provide learning for every listed competency and successful students (graduates) should have at least an awareness of every competency (proficiency level 1–3). As new Ergonomists (Graduate) career progresses and experience increases, the level of proficiency will increase through the 6 levels of competency. The level of proficiency in each competency will depend on experience, seniority and individual career paths as no-one is expected to become 'Expert' in all competencies.

0. **Unaware**: No knowledge or understanding of this competency.
1. **Aware**: Knowledge or an understanding of basic techniques and concepts particular competency for a particular competency.
2. **Novice**: Limited experience gained in a classroom and/or as a trainee on-the-job for a particular competency; expected to need help; can understand and discuss terminology, concepts, principles and issues.

Table 1. Professional competencies (UK)

1. Ergonomics/Human Factors (E/HF) principles
1 Ability to identify and apply methods of analysis, evaluation and validation with respect to human interfaces for tasks, activities and environments.
1.1 Understands the role and application of E/HF principles in optimising system performance and wellbeing across all ages and capabilities.
1.2 Demonstrates ability to enhance health, safety, comfort, quality of life, attitudes, motivation, usability, effectiveness and efficiency.
1.3 Demonstrates ability to identify potential and existing high risk tasks, activities and environments.
2. Ergonomics/Human Factors (E/HF) theory and practice
2.1 Understands theoretical and practice bases for analysis of human interactions.
2.1a Demonstrates use of E/HF theories, methods and tools for analysis of systems (including process), tasks, workload (physical and mental) including mental models, communication and anthropometry.
2.2 Understands the theoretical and practice bases for (re)design of human interfaces (physical and mental).
2.2a Understands the influence of such factors as a person's body size, skill, cognitive abilities, age, sensory capacity, general health and experience.
2.2b Demonstrates ability to integrate E/HF principles and concepts into systems, interface and product design including requirements development and validation.
2.2c Evaluates user needs for safety, efficiency, reliability, ease of use.
2.2d Determines the match and the interaction between human characteristics, abilities, capacities and motivations, and the system(s), organisation, planned or existing environment, products used, equipment, work systems, machines and tasks
2.2e Understands the management of E/HF risks, including priorities and mitigations; potential benefits and costs of E/HF solutions; short and long term goals relevant to defined problems.
2.2f Can apply relevant legislation, codes of practice, standards (government and industry).
2.2 g Determines whether the interface or interaction is amenable to E/HF intervention.
2.3 Understands the theoretical and practice bases for data collection and analysis relating to E/HF.
2.3a Understands the type of quantitative and qualitative data required for E/HF appraisal and design; selects and validates the proposed collection/analysis methods and tools.
2.3b Understands and can apply the basics of experimental design and statistics.
2.3c Understands and can apply the basics of qualitative study design and analysis including knowledge elicitation, interviews, document analysis, and observation.
2.3d Demonstrates ability to seek and obtain relevant ethical approval for E/HF data collection and analysis.
3. Human capabilities and limitations
3.1 Understands the theoretical and practice bases for E/HF relating to Physical capabilities and limitations.
3.1a Demonstrates a working knowledge of anatomy, functional anatomy, anthropometry, physiology, pathophysiology, and environmental sciences as they apply to E/HF practice.
3.1b Can apply knowledge of biomechanics, anthropometry, motor control, energy, forces applied as they relate to stresses and strains produced in the human body.

(*continued*)

Table 1. (*continued*)

3.1c Understands the effects of the environment (including acoustic, thermal, visual, vibration) and individual sensory response (sight, hearing, touch, taste, smell) on human health and performance.
3.2 Understands the theoretical and practice bases for E/HF relating to Psychological and Social capabilities and limitations.
3.2a Understands theoretical concepts and principles of social and psychological sciences relevant to E/HF.
3.2b Recognises psychological characteristics and responses and how these affect health, human performance, attitudes, perception; stress; human reliability and error
3.2c Can apply knowledge of Human Information Processing (including situation awareness, memory, decision making)
3.2d Demonstrates a knowledge of systems theory including socio-technical systems and culture (e.g. organisational and safety culture).
3.2e Understands the principles of group functioning, motivation, engagement and participation.
3.2f Understands the principles of organisational management including individual, group (team) and organisational change techniques, including training and work structuring.
4. Design and development of systems including products, tasks, jobs, organisations and environments
4.1 Understands the theoretical and practice bases for E/HF relating to design and development of systems.
4.1a Understands basic engineering (technology) concepts, with a focus on design solutions and contextual operation of technologies.
4.1b Demonstrates an understanding of the principles of E/HF and human-machine interface technology including hardware, software, internet and network based technologies and social media.
4.1c Understands the requirements for safety systems, the concepts of risk, risk assessment and risk management.
4.2 Utilises a systems approach to the human-aspects of the specification, design, assessment and acceptance of products, services and human factors interventions.
4.2a Applies E/HF principles to design of systems (and services), products, job aids, controls, displays, instrumentation and other aspects of tasks and activities.
4.2b Understands the iterative nature of design development including simulation and computer modelling.
4.2c Considers the options for achieving a balance between human and technological, task and environment to achieve optimal system.
4.2d Selects appropriate forms of E/HF solutions and recommendations, based on theoretical knowledge and practice and develops a comprehensive, integrated and prioritised approach.
5. Professional skills and implementation
5.1 Understands role of E/HF in change strategies.
5.1a Provides design specifications and guidelines for technological, organisational and E/HF design or redesign of the work process, the activity and the environment which match the findings of E/HF analysis.

(*continued*)

Table 1. (*continued*)

5.1b Develops strategies to introduce a new design to achieve a healthy and safe human interaction.
5.1c Recognises the safety hierarchy, application of primary and secondary controls and the order of introducing controls.
5.1d Recommends personnel selection where appropriate as part of a balanced solution to the defined problem.
5.1e Interacts effectively to clients at all levels of personnel.
5.2 Develops appropriate recommendations for education and training in relation to E/HF principles.
5.2a Understands current concepts of education and training relevant to application of E/HF principles.
5.2b Implements effective education and training programmes relevant to understanding the introduction of E/HF measures.
5.3 Supervises the application and evaluation of the E/HF plan.
5.3a Implements appropriate design or modifications.
5.3b Incorporates methods to allow continuous improvement.
5.3c Selects appropriate criteria for evaluation.
5.3d Produces clear, concise, accurate and meaningful records and reports.
5.4 Shows a commitment to ethical practice and high standards of performance and acts in accordance with legal requirements.
5.4a Behaves in a manner consistent with accepted codes and standards of professional behaviour.
5.4b Recognises the scope of personal ability for E/HF analysis and when it is necessary to consult and collaborate with different professional experts.
5.4c Demonstrates commitment to ongoing professional development by maintaining skill set and an awareness of wider E/HF practice.

3. **Intermediate**: Can successfully complete tasks independently in a particular competency; can understand and discuss the application and implications of changes to processes, policies, and procedures in this area - may need help from an expert; shows awareness of how a narrowly focused task can draw upon knowledge crossing a variety of different areas; can demonstrate the appropriate use of different techniques and methods in the application of Human Factors research or consultation.
4. **Advanced**: Can perform actions associated with this competency without assistance and recognised (by employer) as the go-to person; participates in senior level discussions regarding this competency; assists in the development of reference and resource materials in this competency; can bring together disparate theories and techniques or the application of novel solutions to complex problems; demonstrates use and application of multiple tools and techniques to more complex projects that require Human Factors integration.

5. **Expert**: Known as an expert or recognised authority in this area and can provide guidance, troubleshoot and answer questions related to this area of expertise with consistent excellence in applying this competency across multiple projects and/or organisations; able to create new applications for and/or lead the development of reference and resource materials for this competency (Table 2).

Table 2. Proficiency Levels

Application for Grade of membership	Expected experience (years in practice)	Indicative proficiency level
Student member	Not applicable	0–1
Graduate member	On graduation	1–3
Registered member	Minimum 3 years	2–4
Fellow	10 years	3–5

3 Loughborough University

At Loughborough University the EHF MSc suite of programmes [5] is described as:

- Examining how best to ensure a good fit between people, the things they do, the objects they use, and the environments in which they work, travel and play in.
- All MSc programmes are accredited by the CIEHF (affiliated with the International Ergonomics Association).
- Teaching is provided by world leading academics, industry experts and specialists in each area (Chartered Ergonomists/Human Factors Specialists).

All the postgraduate programmes can be studied full time or part time and are offered as:

- MSc (2000 h study)
- Postgraduate Diploma (1200 h study)
- Postgraduate Certificate (600 h study)

As well as a generic Ergonomics and Human Factors programme, the 5 streams allow specialisation in Inclusive Design, Transport, Health and Community Care, and Patient Safety. The programmes were changed from weekly lectures to one week blocks of teaching in 2012, allowing learning over a one-week intensive period of teaching at the University. This has greatly increased the diversity of the student cohort in terms of age range, work experience, and professional background and resulted in an increase of 50% in admissions over previous years. In addition, it has greatly improved access for part-time students across Europe giving the flexibility to continue to work alongside academic studies. The annual intake is approximately 30 new students each year, with a total student cohort of over 65 students.

Loughborough University also offered an undergraduate programme (B.Sc). This had perennial problems with direct recruitment from school applications, but has been popular as a transfer course (e.g. after 1 year of engineering). However increasing academic pressures has led to repeated initiatives to improve recruitment; this has included 'B.Sc. Psychology with Ergonomics' (now discontinued), 'B.Sc. Design Ergonomics' (final new intake in 2016) and most recently 'B.Sc. User-Centred Design' [6] (new in 2017). The employability of all graduates and postgraduates is high with many alumni achieving influential positions in global companies.

4 Challenges and Opportunities

In an attempt to support and ensure competence in EHF provision for UK plc., and engage with a broader audience, the CIEHF is exploring how to support accreditation enquiries from user-centred design and user experience programmes. This discussion is particularly timely with the increasing number of EHF graduates being employed in healthcare, with a focus on the usability testing of medical devices.

In 2000 the FDA issued guidance on Medical Device Use Safety (incorporating Human Factors Engineering into Risk Management), revised in 2016 with clearer definitions for Human Factors and Usability [7]:

- Human Factors are '*the application of knowledge about human capabilities (physical, sensory, emotional and intellectual) and limitations to the design and development of tools, devices, systems, environments and organisations (ANSI/AAMI HE75:2009)*' [8]
- Usability is '*Characteristic of the User Interface that establishes effectiveness, efficiency, ease of user learning and user satisfaction (ISO/IEC 62366:2007)* [9].

However, these definitions have been merged in the UK as '*Human Factors and Usability Engineering – Guidance for Medical Devices Including Drug-device Combination Products*' where the term Human Factors is used '*to encompass other terms such as ergonomics and usability*' [10].

However there are significant differences in competency expectation (and processes) in the disciplines of EHF and User Experience (UX). Furniss et al. contrasted professional competency requirements between EHF and UX [11]. They identified that EHF had clearly defined (and established) competencies (IEA and UK) with a lack of 'definitive set of competencies' in UK [12]. They suggested two possible reasons including the age of the discipline (UX is relatively new compared to EHF) and '*the diverse mix of skills and roles that make up the community, making it challenging to identify an agreed set of competencies*'.

The CIEHF has started conversations with academic course providers of User-Centred Design and UX programmes to explore how the professional competencies could be used for affiliate disciplines.

References

1. CIEHF charter documents, professional code of conduct (page 20). https://www.ergonomics.org.uk/Public/About_Us/CIEHF_Documents/Public/About_Us/CIEHF_Documents.aspx?hkey=8df03a4a-ab8a-482d-8a50-99c4a052f0c7. Accessed 11 Apr 2018
2. CIEHF professional competencies. https://www.ergonomics.org.uk/Public/Membership/Registered_Member/Public/Membership/Registered_Member.aspx?hkey=32fc9cb9-6d12-45fd-a3cb-8063e4c256f4. Accessed 13 Apr 2018
3. IEA, 2001. International ergonomics association. Core competencies. www.iea.cc/project/PSE%20Full%20Version%20of%20Core%20Competencies%20in%20Ergonomics%20Units%20Elements%20and%20Performance%20Criteria%20October%202001.pdf. Accessed 11 Apr 2018
4. https://www.ergonomics.org.uk/Public/Careers_Jobs_CPD/Degree_Courses.aspx. Accessed 11 Apr 2018
5. http://www.lboro.ac.uk/study/postgraduate/masters-degrees/a-z/ergonomics-human-factors/. Accessed 11 Apr 2018
6. https://www.ergonomics.org.uk/Public/Careers_Jobs_CPD/Degree_Courses.aspx. Accessed 11 Apr 2018
7. https://www.fda.gov/downloads/medicaldevices/deviceregulationandguidance/humanfactors/ucm320905.pdf. Accessed 13 Apr 2018
8. ANSI/AAMI HE75:2009(R) (2013). Human factors engineering—Design of medical devices. www.aami.org/productspublications/ProductDetail.aspx?ItemNumber=926#sthash.z3tHlcCS.dpuf. Accessed 13 Apr 2018
9. IEC (2007) ISO 62366-1. Medical devices – Part 1: Application of usability engineering to medical devices. International Organization for Standardization, Geneva
10. MHRA (2017) Human factors and usability engineering – guidance for medical devices including drug-device combination products. https://assets.publishing.service.gov.uk/government/uploads/system/uploads/attachment_data/file/645862/HumanFactors_Medical-Devices_v1.0.pdf. Accessed 13 Apr 2018
11. Furniss D, Curzon P, Blandford A (2018) Exploring organisational competences in human factors and UX project work: managing careers, project tactics and organisational strategy. Ergonomics 61(6):739–761
12. Gray CM, Toombs AL, Gross S (2015) Flow of competence in UX design practice. In: Proceedings of the 33rd annual ACM conference on human factors in computing systems. ACM, pp 3285–3294

Difficulties of the Modern Work Style in Office Environment

Fruzsina Pataki-Bittó(✉)

Department of Ergonomics and Psychology,
Budapest University of Technology and Economics,
Budapest 1117, Hungary
bittofruzsina@erg.bme.hu

Abstract. This Hungarian research is investigating the personal experiences of communication and collaboration at work, and aims to identify the related factors that influence the job satisfaction and the performance of the employees. Information was collected from eleven depth interviews in order to explore and compass the topic in details and understand the connections between the problems of the respondents and their work environment (work environment refers to both physical and communication environment). Therefore respondents were chosen from different office environments, varied by the size of the offices (from three workstations in a room to open-plan office) and by the type of the organizations (multinational and Hungarian companies, state-owned institutions). The interviews covered the topics of work environment, work conditions, work style, working habits (e.g. preferred hours to work on cognitive load tasks), and the personal factors that influence work efficiency. Regarding interruption overload, the different types of interruptions were discussed during the interviews to investigate the personal strategies used in order to focus on work.

The result of the research shows that the lack of functional spaces is a common issue regarding the physical environment, but there are other problems related directly to communication channels and forms. The preference of personal communication and the importance of social interactions at work were also highlighted even though respondents realized that personal inquires often keep them from effective work.

Keywords: Modern office environment · Collaboration · Office interruptions

1 Introduction

The new ways of working refers not only to the style of working but the working environment as well. Collaboration has the leading role in modern offices where the various types of communication tools assist the idea of working together. Physical office environments are also designed to support collaboration by increased level of visual and audible access (openness) and decreased level of privacy.

Several scientific studies have attempted to call the attention to the difficulties of performing mental load tasks in modern office environment. Several research investigates the effects of the different types of office layout on the individual work performance, while others focus on the effects of the office interruptions.

S. Bagnara et al. (Eds.): IEA 2018, AISC 821, pp. 36–45, 2019.
https://doi.org/10.1007/978-3-319-96080-7_5

The following literature review presents a wide range of aspects associated with work efficiency in modern office environment.

2 Related Literature

2.1 Office Environment

Considering office environments, most research examines the behaviour and the performance of the employees in cell offices in comparison with the different kinds of open-plan offices. Literature differentiates small, medium-size and large open-plan offices by the size of the space, but same size open-plan offices can be still diverse by the level of separation among individual workstations and among the blocks of workstations. Therefore office types are often studied by three characteristics, which are 'architectural privacy', 'visual access' and 'physical proximity' [1].

A conceptual model by De Croon et al. [2] outlines two groups of parameters: the office concept and the work conditions, which influence the short-term physiological and psychological reactions and the long-term health and performance of the office workers.

In the model, office concept consists of **office location** (conventional or telework), **office layout** (openness and distance between workplaces) and **office use** (fixed or desk sharing). Office concept effects work conditions, which covers **job demands** (cognitive workload and working hours) and **job resources** (communication, work autonomy, privacy, interpersonal relations) [2].

After reviewing 49 related literature by the aspects of office lay-out, office location and office use, De Croon et al. concluded that there is a strong evidence that open-plan offices reduce the job satisfaction of the employees and found limited evidence that it intensifies cognitive workload and worsens interpersonal relations. The authors offered a remark that to obtain a full picture of the effects of office environments, other aspects should also be considered, which might influence office workers' health and performance, for example: quality of furniture, IT facilities, lighting and air quality, colours and materials [2].

A unique study underlined the role of the doors in the office environments: Nichols et al. [3] describes the doors as "mediators of time and attention", and as a medium that can express the interruptibility of the employees. It can be an advantage of cell offices that the office worker can control the environment and block out distractions by closing the door, but due to communication technologies distractions come through other channels [4].

Seddigh et al. [4] investigated the relationship between the office environment and the effects of interruptions during an immediate free recall memory test. The participants completed the test in normal working conditions and in quiet conditions (other than busy working hours). Researchers found that although memory performance was better in smaller open-plan offices compared to larger open-plan offices, the cognitive performance was more intensively influenced by the number of interruptions (human and virtual) than the number of employees in the office. Further analyses revealed that employees working in cell offices had higher drop in performance caused by fewer distractions. The researchers presumed that people working in more busy office environments develop coping strategies against the high level of distractions while people in private offices don't feel the necessity of doing so. These results indicate that cell offices might not be as advantageous to conduct cognitively demanding tasks as previously thought [4].

2.2 Interruption Management

The more and more frequent use of open-plan offices draw the attention of researchers on the effects of office interruptions. The interruptions were examined by types and characteristics, but researches included the topic of the practices to avoid unnecessary interruptions and to reduce the negative effects. Interruption management techniques according to Janssen et al. [5] can be categorized by two strategies: 1. Stop unnecessary interruptions from occurring and 2. Change the timing of interruptions. Physical office environment plays an important role in interruption management, as both the level of separation and the possibility to use quiet spaces (such as focus room or quiet room) can offer a solution for concentration demand work. In addition to office layout solutions, organizations can set up communicational guidelines and flexible work arrangements which support employees to perform their cognitive tasks. There are computer-based solutions as well developed to handle the timing of interruptions, such as software settings, and intelligent notification systems.

Some researchers use the expression **availability** management instead of interruption management [6, 7]. This expression is slightly different, because it suggests that the office worker accepts that interruption exists and does not want to change it, but for his own interest he controls interruptions by his own availability.

Wiberg and Whittaker [7] tested a technology that could help users to manage their availability during working on mental load tasks. Researchers found that although it would be reasonable to manage availability effectively, users feel a **social obligation** to deal with the interruption as soon as possible. Regarding phone calls users take them not only to avoid causing inconvenience to others but also to avoid future commitments which they have to keep in mind [7].

2.3 Cultural Differences

Previous research presented that **cultural differences** have an impact on the approach to collaboration and on the office environment preferences as well. Rothe et al. [8] examined the prioritisation of the aspects of the physical work environment comparing Dutch and Finnish employees' opinion. Researchers found differences by the *opportunities to concentrate* and the level of *privacy* – which are more important for Finnish employees while *locality of spaces*, *openness* and *transparency* are more important for Dutch office workers [8].

Perlow and Weeks [9] revealed that **helping behaviour** shows significant differences by cultures. Helping behaviour means whether employees are willing to help anyone who asks for help or only to those from whom they expect help in the future. The difference between the American and Indian employees was interpreted as the difference in the framing of helping: at the American site the act of helping is considered as an "unwanted interruption", while at the Indian site it is considered more like an "opportunity for skill development" [9]. Helping behaviour influences the overall work behaviour, including the methodology to deal with interruptions.

Cultural differences can be explained based on the work-related cultural dimensions defined by Hofstede. Hofstede and his colleagues [10] explore in their study how the values in the workplace are influenced by the following dimensions: power distance, uncertainty avoidance, masculinity versus femininity, individualism versus collectivism, long term orientation versus short term normative orientation, and indulgence versus restraint. The approach of collaboration is closely related to the level of **collectivism**, while the approach to the new ways of working is connected the most to the level of **uncertainty avoidance**. Since our research is collecting information from Hungarian office workers, here we review the characteristics of the Hungarian culture by these two Hofstede dimensions. According to the collected data of Hofstede et al. Hungary is an individualist society, which means that instead of belonging to groups people rather look after themselves only. Hungarian people reached a high score for avoiding uncertainty as well, which involves an emotional need for rules and an inner urge to be busy and work hard. The result of our research might reflect these characteristics, but it is important to note that the scores of dimensions apply to countries and not individuals [11].

3 Qualitative Research

3.1 Objectives

To sum up, there is a wide range of aspects that influence the work performance (both individual and group work) in modern office environment. This study aims to uncover the aspects (including physical and communicational environment, work conditions and social attributes) that Hungarian employees consider significant regarding their personal work performance. It is believed that their personal judgement is significant if the organization intends to improve not only productivity, but the level of job satisfaction as well.

The objectives of the paper are: (1) to explore the key factors that influence the employees' satisfaction of work environment; (2) to understand how work environment affects individual work performance; (3) to explore the approach to collaboration and new ways of working of the Hungarian office workers.

3.2 Research Method

Semi-structured interview technique was chosen to allow respondents to raise aspects of work environment that has not been considered previously.

At the introduction the interviewer summarized the overall aim of the interview: to explore the factors of the work environment which influence the employees' work performance. Interviewees were asked to express their personal opinion, share both positive and negative thoughts and experiences, and speak out their problems regarding the topics of the interview.

The framework of the in-depth interviews is based on the presented literature and includes the following topics (raised by open questions):

I. General questions: Position, type of work, work practices (cognitivity, level of collaboration, place and time of individual work)

II. Evaluation of:

1. Office environment (size of the office, desk configurations, functional spaces)
2. Communication (communication channels used and preferred)
3. Focus work (importance, possibilities and working habits)
4. Work arrangements (working hours, flexibility, opportunity to telework)

III. Factors that influence the most the daily work performance

IV. Ideal work conditions

After the introduction (I.) the semi-structured interview was designed to discuss the actual work environment from different angles (II.). The last two parts of the interview (III. and IV.) meant to make the interviewee summarize and prioritize the earlier mentioned aspects and draw a conclusion which of those factors truly matter(s) to him/her. The last question of the ideal work conditions was designed to lead interviewees to describe an office environment that solves their problems and results in high-level job satisfaction.

3.3 Participants

Our aim was to explore the opinion of people working in different office environments, work conditions and organizational culture. 11 interviewees were selected by their position, the type of the organization (multinational company, Hungarian company or state-owned institution), the field of business and the size of the office. Among the 7 male and 4 female respondents there were economists, engineers, an agronomist, a designer, a lawyer and a programmer. The respondents were between the age of 22 and 45. Interviews were conducted in 2017 and lasted from 35 to 75 min each, depending on the respondents (Table 1).

Table 1. Participants by selection criterias

	Position	Type of organization	Size of the office
Number of participants in categories	7 Employee	3 State-owned institution	1 Open-plan office
	3 Manager	6 Multinational company	4 Separated open-plan
	1 Executive	2 Hungarian company	3 Medium office (4–8p)
			3 Small office (1–3p)

4 Research Result

In this section we present a model that illustrates the connections of the major themes discussed during the interviews (Fig. 1). Themes discussed formed two main groups: (1) factors that respondents think are closely related to work performance and (2) factors that respondents do not link directly to work performance but have relevance

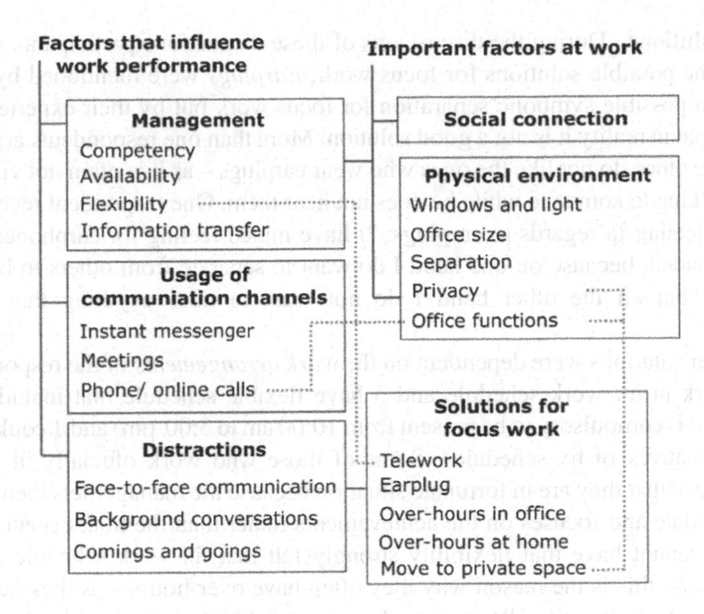

Fig. 1. Connections of the major themes

and contribute to job satisfaction. One topic, the role of management was mentioned in both context.

The factors that influence work performance consists of three main categories: 1. distractions, 2. usage of communication channels and 3. management.

Ten from eleven interviewees claimed that they face difficulties when tasks demand concentration, though the extent of the problem varied by the type of work.

Distractions were mentioned most frequently as the factor that influences work performance. A respondent started the topic with a sentence that describes the situation: "Basically I can't concentrate on work in the office, because something always happens." Among the different kinds of distractions, *face-to-face communication* was considered the most obvious factor that affects work efficiency. Interviewees emphasised that face-to-face communication is the most effective way of communication and continuous communication is necessary for those whose work is connected. The people questioned choose face-to-face communication whenever it is possible, but they expressed awareness that during concentration demand work it is a serious setback factor.

Regarding *background communication* (colleagues' conversations) respondents agreed that it is almost impossible to ignore the speech of other, especially when words are audible. This type of distraction was stated as the most problematic symptom of the office environment (in other than a private office). Nevertheless participants added that it is often advantageous to hear other's work-related conversations because these often contain important information and contribute to successful collaboration.

The third factor that causes distractions was mentioned in connection with office layout: *comings and goings*. All of the interviewees who work in open-plan offices or medium-size office rooms made a remark that although office noise is not a problem, they find it disturbing when people are walking behind or next to the workstation.

Possible Solutions. During the discussions of these distractions participants started to talk about the possible solutions for focus work. *Earplugs* were mentioned by 7 interviewees as a possible symbolic separation for focus work but by their experience they concluded that in reality it is not a good solution. More than one respondents argued that people in the office do not like the ones who wear earplugs – as it is often not visible and they start talking to someone while he does not hear them. One respondent revealed and ambivalent feeling in regards to earplugs: "I have mixed feeling for earphones. I often wear it one-sided, because on one hand I do want to separate from others to be able to concentrate, but on the other hand I do not want to miss anything that happens around me."

The other solutions were dependent on the *work arrangements* of the respondents: 5 of them work in fix work schedule and 5 have flexible schedule that includes "core hours" (e.g. it is compulsory to be present from 10:00 am to 3:00 pm) and 1 could choose from 3 alternatives of fix schedules. Some of those who work officially in fix work schedule added that they are in fortunate situation because the manager lets them work in flexible schedule and focuses on the achievements rather than the hours spent at work. Those who cannot have that flexibility strongly felt that fix work schedule is unreasonable because this is the reason why they often have over-hours – as they have to go earlier or have to wait until colleagues go home to be able to perform tasks that demand concentration. The same strategy was mentioned by those who work in flexible schedule, but they expressed less negative feelings regarding the early and late working hours.

Most of the respondents are allowed to work from home only in exceptional cases. Two respondents (from multinational companies) are allowed to work from home on regular basis, and both of them are happy to do it 1 day a week. All except one who do not have that opportunity or the boss is not in favour of telework, found this situation unreasonable and expressed their wish: "Why can't we work from home if our task is the kind that we could do at home?". But none of the interviewees would work more than 2 days a week from home, which had two reasons: (1) the lack of self-discipline to work at home, (2) the feeling of being out of something.

Two respondents working at multinational companies mentioned that sometimes they *reserve a small meeting room* for themselves when they have a lot of work that demands concentration. Some other respondents raised the topic of focus rooms as an ideal solution for concentration demand tasks, that they heard about but have not tried.

Communication Channels. Switching off or mute phone, email or instant messenger was not considered as a possible strategy to create ideal environment for focus work. One of the respondents uses email rule to control the number of notifications, but respondents agreed that availability is so important that they cannot switch communication tools off. To be always available is more like a social obligation, than an order from the management, because employees do not want to cause delay for the colleagues whose work depends on them. Regarding the different communication channels, respondents felt that instant messenger and regular meetings influence the most work performance. Instant messenger is necessary for urgent messages (as urgent messages cannot be selected from the mass of emails), but also considered as a communication tool that can decrease the negative effects of face-to-face interruption as the timing of the interruption can be negotiated. Those respondents who have no regular meetings feel the lack of it in regards to coordination and information flow, but those who have more than one a day claimed that meetings steal valuable time. Table 2 presents the often repeated problems with the communication channels.

Table 2. Frequent problems with the communication channels

Instant messenger	It is often misapplied (too long messages, too complex or not urgent questions) Colleagues and partners use different types of instant messengers which is inconvenient and causes confusion The high level of fun chat can be disturbing
Online calls	If there is no separated place for calls: it is the most disturbing factor for the colleagues, and the lively environment can be disturbing for the caller as well
Email	Email is often used for urgent messages (instead of instant messenger)
Meetings	Too many and too long meetings often 'steal' the most efficient part of the day Participants often have to be present during the whole meeting even if they are not involved in most of the topics Managers are often not available due to sequential meetings

Management. The manager or teamleader was often mentioned during the interviews. Managers were blamed for making employees work more difficult (1) if their knowledge of the specific field was not deep enough, (2) if they are often not available when their decision is needed (3) if they are not giving freedom to the employees both in the subject of work and in work schedule and (4) if they keep back information or do not support information flow. By these parameters managers were illustrated as key factors that influence overall job satisfaction.

Physical Environment. Contrary to our expectations, interviewees did not think that *office size* has a crucial importance regarding office environment. There were objections to open-plan offices without separation, but most interviewees found open-plan offices acceptable if groups are separated within the office space. Those who have worked in both open-plan and small office rooms highlighted that it does not matter how many people are in the office, because people can adjust to the office size by learning how to behave in each type. The prevalent opinion was that the most important is "to sit next to those you are working with." The ideal office size was defined by most of the respondents as 8–10 people in one office or in one well separated open-plan office section, and 3 of the respondents prefer small offices.

The position and the *privacy* of the workstations appeared to be the most important features in regards to the physical environment. Seven respondents stressed that *natural light and window sight* are vital aspects: "Now that I'm sitting on the brighter side of the office, I'm happier." Regarding the position of the desk several interviewees described themselves as "lucky" to have a workstation situated in the corner or next to the walls and windows because that offers them more privacy. All respondents except two felt strongly about having an own dedicated workstation as they like to keep their personal objects on the desk. One respondent gave a suggestive explanation: "the desk reflects the personality of it's owner and the personal gadgets make the office colourful", while another underlined the feeling of uncertainty in case of shared desk system.

The topic of *office functions* appeared in three context: in regard to the usage of communication channels, focus work and social connections. Respondents realized the importance of phone rooms as those could prevent interruptions, but the lack of phone rooms appeared to be a common problem. Respondents who prefer to block out the environmental noises completely while concentrating stated that there is a need for more small meeting rooms. Finally, respondents underlined the role of the kitchen, the coffee corners and all the spaces and time that are for socializing because these are playing an important role in information flow.

Summarizing the interviews there were values repeated frequently: **flexibility and freedom** (to feel trust from the management regarding working hours and time management), **social connections** (to have fellowship and to be a member of a team), **privacy** (to have a dedicated workstation with personal objects) and **empathy/thoughtfulness** (not to disturb colleagues).

5 Conclusions

In this sections we present the conclusions by the objectives of this paper. We discovered key factors that influence the Hungarian employees' satisfaction of work environment from which the most emphasised were the followings: the position of the workstation (in regard to privacy and natural light) and the spaces that allow social interactions and separated spaces for private and work-related phone calls. Additionally the necessity of flexible working hours was underlined which supports work-life balance and supports performing focus work as well.

To the research question how work environment affects individual work performance the answer is complex. Most of the work (even if it is considered individual work) needs support, information from the colleagues – which generates continuous communication. In case of tasks that require concentration employees try to avoid the normal working environment – as the conversations nearby and the personal inquiries disturb most of the employees who try to focus on work. Two types of solutions appeared during the interviews: (1) improve the office environment with functional spaces: (a) dedicated spaces for calls and for small meetings – to reduce the level of distractions; and (b) create separated rooms for focus work – to block out environmental noise and human interaction, or (2) make the most of the flexible work arrangements and telework in order to perform mental load tasks in quiet environment.

By discovering the approach to collaboration (and new ways of working) it was clear that the participants of the research find collaboration natural and they accept it with all its inconveniences. Nevertheless most of them are seeking solutions for focus work. One or two days/week telework would be acceptable, but respondents highlighted their social needs which leads to the preference of work in the office. Strong feelings for dedicated desks and the fear of shared desk system was expressed which corresponds with the high level of uncertainty index of the Hungarian culture.

We find it important to note that according to the interviews multinational companies are dealing better with the challenges of the office environment than Hungarian organizations and state owned institutions. We believe that the result of this qualitative research provides support for the management of the Hungarian organizations to find a way to improve work environment and increase job satisfaction.

References

1. Zerella S, Von Treuer K, Albrecht S (2017) The influence of office layout features on employee perception of organizational culture. J Environ Psychol 54:1–10
2. De Croon E, Sluiter J, Kuijer PP (2005) The effect of office concepts on worker health and performance: a systematic review of the literature. Ergonomics 48(2):119–134
3. Nichols J, Wobbrock JO, Gergle D, Forlizzi J (2002) Mediator and medium: doors as interruption gateways and aesthetic displays. In: 4th conference on designing interactive systems: processes, practices, methods, and techniques, proceedings paper, pp 379–386
4. Seddigh A, Stenfors C, Berntsson E, Bååth R (2015) The association between office design and performance on demanding cognitive tasks. J Environ Psychol 42:172–181
5. Janssen CP, Gould SJJ, Li SYW, Brumby DP, Cox AL (2015) Integrating knowledge of multitasking and interruptions across different perspectives and research methods. Int J Hum Comput Stud 79:1–5
6. Scholl J, Hasvold P, Henriksen E, Ellingsen G (2007) Managing communication availability and interruptions: a study of mobile communication in an oncology department. In: LaMarca A, Langheinrich M, Truong KN (eds) Pervasive Computing 2007. LNCS 4480, pp 234–250. Springer, Heidelberg
7. Wiberg M, Whittaker S (2005) Managing availability: supporting lightweight negotiations to handle interruptions. ACM Trans Comput Hum Interact 12(4):356–387
8. Rothe PM, Beijer M (2011) Most important aspects of the work environment. In: Proceedings of the 10th EuroFM research symposium, pp 1–12, Vienna, Austria
9. Perlow L, Weeks J (2002) Who's helping whom? Layers of culture and workplace behavior. J Organ Behav 23:345–361
10. Hofstede G, Hofstede GJ, Minkov M (2010) Cultures and Organizations: Software of the Mind. McGraw Hill Professional, New York (2010)
11. Hofstede Insights Homepage. https://www.hofstede-insights.com/country-comparison/hungary. Accessed 19 Feb 2018

Development of Behavior Markers for Emergency Response Training

Masaru Hikono[1(✉)], Yuko Matsui[1], Mari Iwasaki[2], and Miduho Morita[2]

[1] Institute of Nuclear Safety System, Inc., 64-Sata, Mihama, Fukui 919-1205, Japan
hikono.masaru@inss.co.jp
[2] Inter Quest Inc., 22-5, Entomi-cho, Shogoin, Sakyo-ku, Kyoto 606-8323, Japan

Abstract. Regarding the Great East Japan Earthquake of 2011, various reports have stated that many human factor-related challenges were involved in the emergency response at the Fukushima Daiichi Nuclear Power Plant. In consideration of this, we developed an emergency response exercise curriculum (*taikan* exercise) for site leaders that can be completed quickly on-site. This exercise requires players to react in real time to various interruptions made by controllers. Since the players' behaviors cannot be observed in real time from outside, however, observers must replay hours of video recordings, so the results usually cannot be reviewed immediately after the exercise.

Therefore, this study aimed to develop behavior markers that can be easily observed in real time from outside. Four researchers (including one with on-site work experience) watched videos of the exercise (same scenario) conducted by eight teams and wrote down the players' behaviors which they recognized as good non-technical skills (NTSs). The recorded behaviors were organized by the four researchers and the extracted behavior markers were grouped into six categories. Lower-level items (elements) of each category were also indicated. From results of a previous study it was argued that behavior observation and evaluation by observers has several constraints, but developing markers tailored to the workplace under such constraints would enable the provision of tools for better observation. In the future for improvement, the identified behavior markers will need to be tested in actual emergency exercises.

Keywords: Emergency training · Behavior markers · Non-technical skills

1 Introduction

Enhancing the safety of socio-technological systems is a challenge common to many areas. To achieve this goal, merely improving the safety of mechanical factors is not sufficient, and in the area of nuclear power, the importance of human factors and safety culture was pointed out in 2005 [1], drawing on the lessons of the Three Mile Island and Chernobyl accidents. Proposals on the human factors of managers who take command in emergency response centers have also been included in various accident investigation reports (from the government [2], an electric power company [3], the Atomic Energy

S. Bagnara et al. (Eds.): IEA 2018, AISC 821, pp. 46–56, 2019.
https://doi.org/10.1007/978-3-319-96080-7_6

Society of Japan [4], and the Institute of Nuclear Power Operations [5]) on the Fukushima Daiichi Nuclear Power Plant accident (hereinafter "Fukushima accident"). However, the safety measures implemented after the Fukushima accident focused mainly on hardware (facilities), raising concerns in plant-hosting communities that believe "people and training are what truly count" [6]. Safety cannot be improved in a well-balanced manner without addressing the software (human) aspects such as workers and their organizations.

Hikono et al. [7] attempted to extract lessons mainly on the human aspects required of site managers from interviews [8] with the plant manager at the time of the Fukushima accident. The skills related to human aspects are referred to as soft skills, CRM (crew resource management) skills, and non-technical skills (NTSs), and are receiving attention. According to Flin [9], NTSs are defined as skills such as the ability to quickly grasp circumstances, communicate effectively, and display leadership, as opposed to technical skills (expertise and skills directly related to a specific job), and which should be possessed by site staff to avoid human error and ensure safety. Improving NTSs and thereby responding swiftly and accurately to major disasters rarely experienced in daily work tasks cannot be achieved merely by preparing manuals and conducting desktop tasks; it is essential to develop disaster response capabilities by regularly repeating practical exercises such as Table-Top exercises through improvised Role-Playing [10]. Therefore, Hikono et al. [11] developed an exercise curriculum that can be conducted at on-site in emergency response centers and meeting rooms (hereinafter "*taikan* exercise") and they have begun to test it with on-site managers. As a result of categorizing good behaviors of participants on recordings of the *taikan* exercise recognized by observers as exercising NTSs [12], many NTS-related behaviors described by Flin [9] desired of site managers in an emergency situation, such as communication, grasp of circumstances, and organizational management, were identified.

Meanwhile, as the *taikan* exercise has been conducted repeatedly, the demand for identifying specific behaviors sought in exercise participants and quantitatively recording participants' behaviors has increased. Generally, Table-Top exercises are evaluated through various methods including checklists, achievement ratio, required time, tests, self-evaluation, observer's evaluation, and post-exercise discussions [10]. However, it is difficult to set evaluation criteria in advance for Table-top exercises, as they intentionally include situations that are impossible to handle due to constraints in human resources, time, and equipment, for participants to experience decision making under time pressure, information confusion, and communication errors. In sectors with high safety requirements, such as commercial aviation [13], rail transportation [14], nuclear power generation [15], and healthcare [16], an evaluation system based on behavior observation has been recommended. However, Flin [9] indicated that behavior observation also has shortfalls, such as the inability to cover all situations, low possibility of observing rare behaviors, and limits in the capabilities of the observers themselves. In the initially developed *taikan* exercise, several researchers [12] gathered after the exercise to watch hours of recorded video and to compare it with the checklist created based on the lessons of the Yoshida Testimony [8]. Thus, as it was not possible to observe the behaviors of participants in real time, the observation results could not be presented to the participants in the post-exercise meeting (where participants review their own behavior during the exercise with the help of a facilitator to identify the good points in terms of NTSs and the points requiring improvement).

Thus, this study aims to clearly define a minimum set of basic actions required of participants in a *taikan* exercise session and to develop behavior markers by which observers will observe the participants. The post-exercise meeting is positioned as an important occasion for the participants themselves to realize the importance of NTSs and has been appreciated by participants as a forum to share good practices of other participants and identify areas for self-improvement. Being able to provide the observers' input would increase the awareness of participants and enable them to recognize the significance of the exercise.

2 *Taikan* Exercise: Outline and Characteristics

The curriculum of the *taikan* exercise, for which behavior markers were developed, is outlined below. The participants were plant managers on holiday shifts who were requested to conduct an initial response to an incident that occurred late at night on a holiday. The participants were divided into groups of four or six members, and each member was assigned the role[1] of general commander, unit commander (one or two persons), contact person (one or two persons), or on-site coordinator. After being briefed on the exercise rules and the background to the initial incident to which they had to respond, the participants moved to an on-site emergency response center (or meeting room) where they responded to the initial incident for approximately 40 min. They were also required to handle various problems (interruptions) presented to them by controllers (all with on-site experience) by telephone from a separate room. The interruptions were designed based on the lessons extracted by Hikono et al. [7] from an interview with a former plant manager, and interruptions were made 13 times during the 40-min exercise. The emergency response center was recorded on audio and video throughout the exercise. A 60-min meeting was held after the exercise. The total time for the exercise and post-exercise meeting was approximately two hours.

3 Basic Actions Required of Participants During the Exercise

In drafting behavior markers, a minimum set of basic actions that should be incorporated in the exercise were considered. The basic actions attach importance to the process from the introduction of interruptions to decision making, and consisted of actions to increase resources (human resources, time margin, and information) or to create rules, norms, workspaces, or atmosphere (Table 1) based on the good practices

[1] The participants were assigned four roles, namely general commander (overall supervision and command), unit commander (commander of each reactor), contact person (in charge of each reactor, contact with parties in and outside the plant), and on-site coordinator (in command of the team that responds to an emergency on the front). Groups consist of four or six members (with two unit commanders and contact persons) depending on the number of the units.

extracted by Matsui et al. [12]. Basic actions were selected based on the following perspectives.

The action:

- is related to NTSs (meeting the purpose of the exercise);
- is in line with workplace rules and operational circumstances (high frequency);
- has low scenario dependence (can be observed under different scenarios); and
- is readily observable from outside (can be used by many observers).

The participants were informed of these basic actions in advance and were requested to think actively about the challenge that they were facing, and also to think for themselves.

Table 1. Basic actions required in the exercise

Basic action	Good practice observed in the exercise
1. Building a team	In a team with a less experienced commander, experienced members were seated close to the commander
2. Clarifying goals	Commanders clearly stated their decisions, allowing members to share goals and act quickly
3. Briefing	The commander announced, "Briefing starts now" to get everyone's attention and ended the briefing with a clear "That's all."
4. Reciting	Important information such as number of people was recited in telephone conversations. Commanders also repeated the information when receiving a report
5. Setting priorities	The commander ordered a certain task to be put off due to low priority, so that the time could be spent on another task
6. Drawing attention	Important matters were reported after drawing the members' attention by clearly saying, "This is important."
7. Asking if unclear	The circumstances of the other party were confirmed by asking, "Is your situation such and such?"
8. Grasping other members' circumstances	The commander encouraged less experienced members to "Ask if anything is unclear."
9. Voluntarily collecting information	When grasping the situations of members, various locations were contacted to find out how many workers of what job types were in each place
10. Having alternatives	In response to a phone call about a missing tool, the decision was made to take an action other than that regularly taken

4 Developing Behavior Markers

Behavior markers were created for use by observers to observe the basic activities required of the participants in the exercise. Audio and video data on the exercise sessions held in FY2017 for seven teams was used. For each of the ten basic actions, four analysts (one person with site experience, one psychologist, and two researchers of

Table 2. Category 1: Building a team

Behavior markers (element)	Good practice observed in the exercise	Basic action	NTS
E1 Supporting subordinates having difficulties	The commander at the headquarters encouraged less experienced members to "Ask if anything is unclear."	Grasping other members' circumstances	Relationship-building under stress
E2 Motivating subordinates	A senior member offered to help a subordinate having difficulty, which motivated the subordinate to act on his/her own initiative	Grasping other members' circumstances	Relationship-building under stress
E3 Checking subordinates' actions	The commander overheard the unit commander speaking on the phone and advised the subordinates to speak as they act	Grasping other members' circumstances	Relationship-building under stress
E4 Issuing specific orders to subordinates	To a member looking unconfident, the commander offered advice on what to communicate before making a phone call	Grasping other members' circumstances	Relationship-building under stress
E5 Seating, location	The clock was moved to a more visible place	Team-building	Organizational management
E6 Division of roles/assisting role	The division of roles was checked	Team-building	Organizational management
E7 Helping other members	Members actively offered to help with tasks that were not theirs	Grasping other members' circumstances	Organizational management
E8 Sharing an understanding of circumstances	Members shared each other's circumstances by looking at the event sequence chart	Team-building	Organizational management
E9 Enabling subordinates to speak out	The commander asked others if they had any concerns and encouraged them to share their thoughts	Voluntarily collecting information	Organizational management

Table 3. Category 2: Obtaining information/grasping circumstances

Behavior markers (element)	Good practice observed in the exercise	Basic action	NTS
E1 Voluntarily seeking detailed information	After receiving a report on the number of people on-site, a member voluntarily took action to learn how many workers for what job types were in each place	Voluntarily collecting information	Communication
E2 Asking to obtain details	A member asked, "Which tunnel?" to ensure the accuracy of information	Asking back	Communication
E3 Checking the white board	Members frequently checked the information on the white board	Voluntarily collecting information	Grasp of actual circumstances
E4 Checking information source	Finding that a member reporting a phone call was in a rush, the commander asked, "Who was the phone call from?"	Voluntarily collecting information	Communication
E5 Reciting numbers and names	Important information such as number of people was recited on the phone. The commander also repeated such information when receiving reports	Reciting	Communication
E6 Reciting circumstances and key points	Other parties' messages were recited in a phone conversation	Reciting	Communication

educational methods) recorded the elements of specific actions of participants observed during the exercise. The good practices identified were grouped based on the ten basic actions and then grouped further by situations corresponding to the exercise scenario to obtain 43 behavior markers (elements) in six situations[2] (categories) (Tables 2, 3, 4, 5, 6 and 7). Prior to analysis, the analysts had studied the ten basic actions. Also, on the day of the exercise session, they observed the exercise on a monitor in a different room and listened to the post-exercise meeting discussions.

[2] Category 1, Building a team (Table 2); Category 2, Obtaining information/grasping circumstances (Table 3); Category 3, Minding others (Table 4); Category 4, Having alternatives (Table 5); Category 5, Communicating information and intentions (Table 6); and Category 6, Briefing (Table 7).

Table 4. Category 3: Minding others.

Behavior markers (element)	Good practice observed in the exercise	Basic action	NTS
E1 Taking a pause	When a subordinate was asked for instructions but could not decide, the commander instructed him/her to hang up the phone for now	Grasping other members' circumstances	Relationship-building under stress
E2 Asking simple questions	When a caller did not identify him/herself, a member frankly asked, "Who are you?"	Grasping other members' circumstances	Relationship-building under stress
E3 Mentioning the other person's name	A member intentionally stated the other person's name when asking, "Where are you now, Yoshida-san?"	Grasping other members' circumstances	Relationship-building under stress
E4 Showing consideration	When ordering members to evacuate, the member added the phrase "Be very careful."	Grasping other members' circumstances	Relationship-building under stress
E5 Giving a role to play	A member waited for the other party to calm down before asking the tide level	Grasping other members' circumstances	Relationship-building under stress
E6 Calming someone down	A member waited for the other party to calm down before asking the tide level. (Same as the above)	Grasping other members' circumstances	Relationship-building under stress

Table 5. Category 4: Having alternatives

Behavior markers (element)	Good practice observed in the exercise	Basic action	NTS
E1 Assuming the worst	The on-site coordinator connected a vague piece of information ("Someone who should be here is not here") with the possibility of an injured person and called for everyone's attention	Having alternatives	Assessment of circumstances
E2 Giving thought to consequences	When requesting a rescue helicopter or ordering someone to be sent out, associated risks were considered simultaneously and specifically	Having alternatives	Assessment of circumstances
E3 Considering risks	Instructions were given to not let people on the premises, considering the possibility of contamination	Having alternatives	Assessment of circumstances
E4 Thinking not only by the rules	After hearing the number of people, instructions were given to "ask for some ID showing name, DOB, and position" and to "assign someone to that task."	Having alternatives	Assessment of circumstances

(continued)

Table 5. (*continued*)

Behavior markers (element)	Good practice observed in the exercise	Basic action	NTS
E5 Thinking widely and flexibly	Asked for technical assistance from another party and resolved the issue using other instruments	Having alternatives	Assessment of circumstances
E6 Changing plans flexibly	In response to a phone call about "not having tools," the decision was made to "stop working on low priority matters."	Having alternatives	Assessment of circumstances
E7 Using other resources	Managed to dispatch people using the information that a contractor instrumentation team was on-site	Having alternatives	Organizational management
E8 Delegating to external parties	Issued instructions to delegate some authority to headquarters due to the lack of capacity at the power station	Having alternatives	Delegation of authority

Table 6. Category 5: Communicating information and intentions

Behavior markers (element)	Good practice observed in the exercise	Basic action	NTS
E1 Communicating priorities	The commander clearly stated, "Do not assign people to another task as our current priority is stabilizing the plant."	Setting priorities	Decision-making under stress
E2 Using specific words to draw attention	Drew everyone's attention using words such as "Announcement!" and "Emergency!" before starting to speak	Drawing attention	Communication
E3 Communicating information promptly	The headquarters did not need to make inquiries because they were updated in a timely manner	Grasping other members' circumstances	Communication
E4 Indicating the reason when issuing instructions	The commander said, "I have already expressed my feelings, but am issuing the instructions again because they may not have been understood."	Grasping other members' circumstances	Communication
E5 Announcing name and role	When answering the phone, a member announced his/her name and role clearly	N/A	Communication
E6 Speaking comprehensibly	A member spoke with a moderate tone of voice (not too loud) and in an easily comprehensible manner	N/A	Communication

(*continued*)

Table 6. (*continued*)

Behavior markers (element)	Good practice observed in the exercise	Basic action	NTS
E7 Writing clearly on the board	Information was written clearly in neat handwriting on the white board	N/A	Communication
E8 Communicating decision criteria	The commander made the decision to put off a task because it had low priority	Setting priorities	Decision-making under stress
E9 Communicating goals and prospects	A member receiving instructions clearly stated what he/she was going to do and what he/she wanted the other party to do	Setting clear goals	Relationship-building under stress

Table 7. Category 6: Briefing

Behavior markers (element)	Good practice observed in the exercise	Basic action	NTS
E1 Drawing attention before starting	The commander said, "Stop what you're doing" to draw everyone's attention before starting the briefing	Briefing	Communication
E2 Making proposals to the commander	Briefing began with the unit commander's words that "Information has become confusing; let's set priorities."	Briefing	Communication
E3 Making the end clear	The commander said, "Briefing starts now" for everyone to switch gears and the briefing was ended with a clear "That's all."	Briefing	Organizational management
E4 Wrapping up in about a minute	The commander said, "Briefing starts now" for everyone to switch gears and the briefing was ended with a clear "That's all."	Briefing	Organizational management
E5 Focusing on the briefing	The rule not to pick up the phone during a briefing was rigorously observed	Briefing	Organizational management

5 Challenges and Conclusion

Behavior markers were grouped into six situational categories: (1) building a team, (2) obtaining information/grasping circumstances, (3) minding others, (4) having alternatives, (5) communicating information and intentions, and (6) briefing. Furthermore, for each situational category, elements were extracted from actually observed behaviors. Flin [9] points out that there are constraints regarding the evaluation of behavior monitoring by the observers; but hereafter by applying the developed markers to actual exercise sessions, an effective measurement method can be developed that links those behavior markers with the NTSs required of management in an emergency, as obtained from the lessons of the Yoshida Testimony [8].

An emergency can profoundly shake the people who face it as well as their perceptions, and it may never become manageable by improving a certain set of skills. From this perspective, there should be essential skills that are yet to be identified. The categories and elements identified at this time must also be tested and honed through repeated exercise sessions, by capturing and adding newly discovered viewpoints. Another challenge is developing an environment to ensure that observers can perform marking at the same level for multiple scenarios and, irrespective of a change in observers, for each exercise session. Furthermore, several levels of behavior markers should also be prepared depending on the proficiency of participants. For the future, it is necessary to continue to observe the behavior markers extracted from actual exercise sessions and present the observation results in post-exercise meetings, to provide a richer training curriculum for participants.

References

1. Frischknecht A (2005) A changing world: challenges to nuclear operators and regulators. In: Itoigawa N, Wilpert B, Fahlbruch B (eds) Emerging demands for the safety of nuclear power operations - challenge and response. CRC Press, Boca Raton, pp 5–15
2. Investigation Committee on the Accident at the Fukushima Nuclear Power Stations (2012) Final report. Media land, Tokyo. (in Japanese)
3. TEPCO (2012) Fukushima nuclear accident analysis report. http://www.tepco.co.jp
4. Atomic Energy Society of Japan (2014) Investigation committee on the nuclear accident at the Fukushima Daiichi NPP final report. Maruzen, Tokyo. (in Japanese)
5. Institute of Nuclear Power Operations (2011) INPO 11-005 special report on the nuclear accident at the Fukushima Daiichi Nuclear Power Station
6. Fukui-Shinbun Co., Ltd. (2016) Has the safety of the power plant increased? Fukui-Shinbun, 19 January 2016 morning edition, p 1. http://www.fukuishimbun.co.jp. (in Japanese)
7. Hikono M, Sakuda H, Matsui Y, Goto M, Kanayama M (2016) Learning non-technical skill lessons from testimony given in the investigation of the nuclear accident at the Fukushima Nuclear Power Stations. J Inst Nucl Saf Syst 23:153–159. (in Japanese)
8. Cabinet Secretariat (2014) Interview records of the investigation committee on the accident at the Fukushima Nuclear Power Stations. http://www.cas.go.jp. (in Japanese)
9. Flin R, O'Connor P, Crichton M (2008) Safety at the sharp end. Ashgate, Farnham
10. Workshop of table-top exercise (2011) Guidebook of the table-top exercise. Naigai Publishing Co. Ltd., Tokyo. (in Japanese)
11. Hikono M, Matsui Y, Kanayama M (2017) Development of emergency response training program for on-site commanders (1). In: Proceedings of international congress on advances in nuclear power plants: 17382 (CDROM)
12. Matsui Y, Hikono M, Iwasaki M, Morita M (2017) Development of emergency response training program for on-site commanders (2). In: Proceedings of international congress on advances in nuclear power plants: 17437 (CDROM)

13. Flin R (2010) Chapter 6 - CRM (non-technical) skills - applications for and beyond the flight deck. In: Kanki BG, Helmreich RL, Anca J (eds) Crew resource management, 2nd edn. Academic Press, San Diego, pp 181–202. https://doi.org/10.1016/b978-0-12-374946-8.10006-8
14. Madigan R, Golightly D, Madders R (2015) Rail industry requirements around non-technical skills. In: Sharples S, Shorrock S, Waterson P (eds) Contemporary ergonomics and human factors. International conference on ergonomics & human factors 2015. CRC Press, Taylor & Francis Group, Daventry, pp 474–481
15. Yim HB, Kim AR, Seong PH (2013) Development of a quantitative evaluation method for non-technical skills preparedness of operation teams in nuclear power plants to deal with emergency conditions. Nucl Eng Des 255:212–225. https://doi.org/10.1016/j.nucengdes.2012.09.027
16. Bracco F, Masini M, Tonetti GD, Brogioni F, Amidani A, Monichino S, Maltoni A, Dato A, Grattarola C, Cordone M, Torre G, Launo C, Chiorri1 C, Celleno D (2017) Adaptation of non-technical skills behavioural markers for delivery room simulation. BMC Pregnancy Childbirth 17:89. https://doi.org/10.1186/s12884-017-1274-z

Taking to the Skies: Developing a Dedicated MSc Course in Aviation Human Factors

Alex W. Stedmon[1(✉)], Rebecca Grant[1], Don Harris[1], Stephen Legg[2], Steve Scott[1], Dale Richards[1], John Huddlestone[1], and James Blundell[1]

[1] Centre for Mobility and Transport, Coventry University, Coventry, UK
ab5840@coventry.ac.uk

[2] Centre for Ergonomics, Occupational Health and Safety, School of Public Health, College of Health, Massey University, Palmerston North, New Zealand

Abstract. From flight-crew to cabin-crew, air-traffic controllers to aircraft engineers - the 'human factor' is vital to the safe and efficient operation of all aspects of the aviation industry. Over the past three decades a better understanding of Human Factors has resulted in significant safety benefits, forming a cornerstone of every aviation safety management programme. With this in mind, an innovative approach to teaching Ergonomics and Human Factors (E/HF) at Coventry University has created a unique MSc in Aviation Human Factors. From the outset this specialist part-time MSc has been designed to attract a wide range of aviation professionals (e.g. aviation engineers, flight and cabin crew, safety personnel, or air traffic controllers) from civil or military organisations in the UK or overseas. The course therefore provides a niche and bespoke learning experience for its students, and from a user requirements perspective this necessitated a particular pedagogic approach.

Keywords: Aviation human factors · Distance learning · Ergonomics
Human factors · Bespoke learning

1 Introduction

It is estimated that up to 75% of all aircraft accidents have a human factors element to them. As our understanding of the impact that human behaviour on system and individual performance has increased, the importance of the field of human factors within the aviation industry has grown significantly. The aviation industry continues to evolve and develop, it is likely there will be a growing need for human factors professionals to work within all aspects of the industry to apply the latest knowledge on human performance to help improve safety and efficiency in aviation operations.

In order to meet the demand of professionals wanting to develop and enhance their human factors knowledge and skills, the Human Systems Integration Group at Coventry University have designed and established an MSc in Aviation Human Factors to cater for industry requirements for operational staff. As a result, this MSc is delivered as a fully distance learning course targeting industry professionals who wish to study

S. Bagnara et al. (Eds.): IEA 2018, AISC 821, pp. 57–61, 2019.
https://doi.org/10.1007/978-3-319-96080-7_7

part time, alongside their current employment. Whilst the course team are based in the United Kingdom, the course is open and accessible to students around the world.

2 Course Design

Students are taught by core members of the course team who have achieved professional standing as Chartered Ergonomists, Chartered Psychologists, Occupational Psychologists and Fellows of the Chartered Institute of Ergonomics and Human Factors.

The academic staff have decades of relevant research and industry experience in aviation human factors (as well as other safety critical industries), both in the UK and overseas, that they share through their research-informed teaching activities.

The course team's experience includes non-technical skills in aviation, accident investigation, the human factors of unmanned aerial vehicles, aircraft cabin and passenger safety, human factors issues within aircraft ground handling and air traffic control, design and certification of flight deck interfaces, innovative future flight deck designs, the use of new technologies by aviation personnel, and training analysis, design and delivery in the aviation and military training domains.

As a distance learning course, the MSc has been designed to attract UK, European and International students, many of which bring with them experience from working in a range of countries and aviation-based organisations. To date the course has attracted delegate from major international airline companies, air traffic providers, manufacturers, and various armed forces. This provides a very rich environment for students to learn from each other through various discussion opportunities throughout the course. Studying by distance learning also allows students to progress their careers in industry while simultaneously being able to apply the knowledge and principles gained on the course and/or enhancing their current skills in the job they are doing.

A unique feature of this course, and one that was built on a key user requirement, is that while access to the internet is required, course materials have been developed to allow study to be undertaken anywhere in the world – even on board an aircraft! The portability of the study materials allows students to study them at a time and place that fits in around their busy work schedule.

The course is delivered via the 'Moodle' Virtual Learning Environment (VLE) but designed to be downloaded in full to provide greater flexibility where required. Course materials have been specifically designed to complement and enhance the existing knowledge of aviation professionals rather than being a general E/HF course.

Students are provided with high quality online distributed module materials providing written topic summaries; links via remote access to journal articles; book chapters and e-books; links to videos; topic exercises; asynchronous bulletin-board based discussion groups; recommended reading lists; links to external webpages. As the course is delivered through distance learning, these methods form the guided study content of the modules.

In addition to the formal course materials, students also have access to one-to-one guidance, tutor and supervisor support (via email, online video conferencing, telephone and in person at the University if desired).

3 Course Structure

The MSc Human Factors in Aviation is designed as a single intake course where students start each September on a part-time basis. The part-time model provides notional learning hours of between 20 to 25 h per week. A typical module is 150 notional learning hours studied over seven weeks, of which approximately one-third is delivered via guided module material and two-thirds via self-directed independent study, to develop independent interests beyond the guided material. The self-directed study time is likely to be divided between extended module reading, assignment preparation and independent research.

The course can be completed in a minimum of two years via a modular approach and it is expected that most students will complete the course in this time.

The course includes content providing the theoretical basis for the study of human factors in aviation, followed by further content demonstrating how these concepts are applied in aspects of aviation operations. Throughout the modules students are encouraged to apply the theoretical material to real world issues within the aviation industry, as human factors practitioners would.

The course consists of eight mandatory 15 credit modules covering the main areas of human factors in aviation. Students also undertake a 60 credit industry-related human factors research thesis, demonstrating the skills and knowledge gained during the course and allowing them to develop their ability to design, plan, execute and communicate an applied research project in the field.

Conducted over two years the full 180 credits required for the MSc qualification is further divided over six semesters as follows:

- **Human Information Processing in Aviation** (Year 1/Semester 1) - introduces the basic concepts of human information processing, and its application in the applied aeronautical context. Students also become familiar with the concepts of workload and situation awareness, associated measurement techniques and their importance in all aspects of system design and evaluation.
- **Decision Making and Error in Aviation** (Year 1/Semester 1) - examines the major schools of thought in the study of decision making and the processes for improving aviation decision making. Students then consider the nature of error, its root causes and methods by which it may be mitigated.
- **Flight Deck Design** (Year 1/Semester 2) - provides an introduction to the issues relating to human factors in flight deck design, including displays and controls, the use of automation, and human-computer interaction on the commercial flight deck.
- **Selection, Stresses and Stressors** (Year 1/Semester 2) - introduces the psychology of individual differences, focusing on personality and cognitive ability, as well as the measurement of these concepts. The course then examine the areas of psychological stress and environmental stressors within an aviation environment, alongside the effects of such stressors on human performance.
- **Training and Simulation** (Year 1/Semester 3) - an overview of the training life-cycle, training methods, media and the evaluation of training is discussed. The course then examines the role of simulation in an aviation training context.

- **Critical Literature Review** (Year 1/Semester 3): this provides students with an opportunity to develop their skills in critically evaluating scientific literature through review of a contemporary topic within the aviation domain. Previous reviews have included just culture, decision making in complex naturalistic environments, psychological stress in aviation, motion platforms in flight simulation, enhancing crew resource management training, non-compliance with procedure, fatigue in aviation, barriers to incident reporting and perceptions towards unmanned aerial systems.
- **Crew Resource Management** (Year 2/Semester 1): introduces the principles of crew/team resource management and looks at the literature behind non-technical skills, such as decision making, communication, leadership/team working, situation awareness and fatigue management. Students also learn about the process of identifying, training and assessing non-technical skills in the applied aeronautical context.
- **Human Factors and Safety Management** (Year 2/Semester 1): provides an overview of the organisation-wide safety management function where the emphasis is on risk management and avoiding (or managing) error across operations. Students also consider approaches used to investigate the human factors aspects of aviation accidents and incidents.
- **Human Factors Research Thesis** (Year2/Semesters 2 and 3): this module provides an opportunity for students to conduct their own industry focused research. In the past, students have undertaken research projects into decision making processes and error detection in air traffic control, passenger safety information, safety operators deviating from procedure, perceptions towards training and training feedback systems, the use of new technology and training for flight crew training, public attitudes to laser attacks on aircraft and reporting culture and barriers to effective incident reporting.

Alongside these modules a major advantage of this MSc is that human factors research methods and data analysis are integrated within the first eight modules. This allows a more focused approach and a demonstration of each methodological area as students progress through and build their confidence throughout the course. From a user requirements perspective, many students may not be familiar with aspects of design, measurement and analysis techniques and rather than having a single methods module that students could easily fail, this approach allows them to learn in a more developmental manner.

The MSc is assessed solely on coursework including applied written assessments, formal reports and essays and presentations. Modules include between one and three coursework elements for assessment.

4 Learning Outcomes and Career Prospects

In addition to the knowledge and application of the areas of human factors addressed, students develop a wide range of transferable professional skills during their studies, including the ability to produce written reports, engage in self-directed study, design, plan and execute a piece of applied human factors research and demonstrate a critical awareness of professional, legal, social and ethical issues.

More specifically, on successful completion of the MSc in Aviation Human Factors, students should be able to:

- Critically apply knowledge of the principles of Human Factors through the systematic investigation of complex issues to produce safe, efficient and cost-effective solutions in the aviation/aerospace industry
- Apply Human Factors theories to practical case studies in which aviation-related scenarios are analysed
- Critically evaluate research findings and theories in the area of Aviation Human Factors
- Successfully complete an independently conducted research project in a subject area of Human Factors in Aviation
- Demonstrate the skills and qualities required of a Human Factors professional (e.g. demonstration of ethical and legal considerations; critical reflection and personal responsibility) in the safety-critical aviation environment.

Since the MSc has begun, Coventry University has received the prestigious 'Gold Award' in the 2017 Teaching Excellence Framework (TEF) and the course itself received 100% overall course quality student satisfaction in the Postgraduate Taught Experience Survey (PTES) in 2017. This MSc is also approved as an accredited course leading to certification as European Human Factors Specialist in Aviation/Aviation Psychologist for appropriately qualified members of the European Association Aviation Psychology (EAAP).

Aviation is a global industry, and the discipline of human factors is recognised internationally. Potential destinations for human factors professionals within the aviation discipline include airlines and aircraft operators, air traffic control services, engineering organisations, aerospace manufacturers, safety consultancies and research organisations.

5 Conclusion

This paper provides an overview of the development of this focused and specialised course and the approach taken in developing a dedicated human factors learning experience for a dedicated user group. From a user requirements perspective the distance learning and integrated methods approach work well for the intended user group. Furthermore, traditionally, E/HF teaching at Coventry University has taken place in separate departments and while this is still the case, this MSc provides a basis for exploring a more integrated approach across other areas in a distance learning format that is open and accessible to industry professionals worldwide.

Work Activity Analysis to Support Technological Aid Supply in Vocational Training for Adolescents with Learning Difficulties

Marie Laberge[1,2,3](✉), Aurélie Tondoux[2], Marie-Michèle Girard[1,2], Fanny Camiré Tremblay[1,2], and Arnaud Blanchard[2]

[1] University of Montreal, Montreal, Canada
marie.laberge@umontreal.ca
[2] CHU Ste-Justine Research Centre, Montreal, Canada
[3] CINBIOSE Research Centre, Montreal, Canada

Abstract. Adolescents with special needs are at risk of leaving school without a qualifying certification. This pilot study aims to develop an innovative intervention based on using technological aids (TA) to help students enrolled in a semiskilled vocational training program developing work skills. The intervention consists of successive steps leading to work activity analysis, relevant means identification, and realization of pedagogical activities supported by TA. Eight teachers and fifteen students were recruited for a multiple case study. Teachers first attended a training workshop, and were then asked to apply the intervention with students who need help in their traineeship. Multiple data were collected and triangulated (interviews, log books, meetings, in situ observations, etc.). In total, the teachers implemented 46 pedagogical activities with their students. Most of the time, teachers were the principal users (initiated the activity, manipulated the device, programmed the application, and determined the use conditions). Concerning the student participation, they were actively involved at various level in most activity as well, but they did not always handle the technological devices themselves. The perceived value of the activities was generally well rated by teachers.

Keywords: Vocational training · Learning difficulties
Work disability prevention

1 Introduction

Adolescents with learning difficulties are at risk of leaving school without a qualifying certification. In Quebec (Canada), the *Work-Oriented Training Path* (WOTP) enables these young people to develop their employability by offering pre-work traineeships in actual companies. The aim of this pilot study was to develop an innovative intervention based on using technological aids to help students enrolled in the WOTP programs developing work skills.

S. Bagnara et al. (Eds.): IEA 2018, AISC 821, pp. 62–66, 2019.
https://doi.org/10.1007/978-3-319-96080-7_8

2 Intervention Development Process

The research team developed an intervention process which would facilitate the planning and use of technological aids, in order to support vocational training for disabled adolescents enrolled in a pre-work traineeship. The intervention was developed considering the findings of two previous pilot studies (Laberge et al. 2015; Laberge 2017; Laberge et al. 2018) and theoretical models in the fields of education and ergonomics, using a socio-constructivist and didactic approach to professional development. This approach was developed from recent empirical research and theoretical models used in the fields of teaching didactics, constructive ergonomics, and technology-enhanced learning. The approach consists of successive steps leading to work activity analysis, relevant means identification, and realization of pedagogical activities supported by technological aids (TA): (1) needs assessment based on work activity analysis tools; (2) selection of the best supportive mean (not necessarily a TA, but if so, continue to step; (3) identification of the appropriate technological function (e.g. memory compensation aid); (4) finding an appropriate product (specific device and application); (5) elaboration of the pedagogical scenario (context of use) considering work activity and workplace context; (6) implementation; (7) integration/transfer; and (8) subjective evaluation of effectiveness by teachers, students and workplace people (this last one was reported by teachers).

Teachers first attended a training workshop and were then asked to apply a method to conduct workplace analysis and technological aid allocation for two students. They were free to choose any applications and to determine the conditions of use (at work, either while working or off the job, at school, both, at home; various levels of student engagement in the manipulation of technological aids). Every six weeks, a collective meeting was organized to share the various experimentations.

3 Data Collection

Eight teachers and fifteen students who were enrolled in the WOTP participated in a multiple case study. Teachers participated in two semi-structured interviews (beginning and end of the experimentation year). Students participated in a semi-structured interview at the end of the experimentation year. Every two months, the content of meetings was recorded and analyzed. Additional data was collected during the course of the study (log books, questionnaires, interviews, in-situ observations). Data triangulation led to an understanding of utilization barriers and facilitators. Figure 1 illustrates the intervention and the data collection steps.

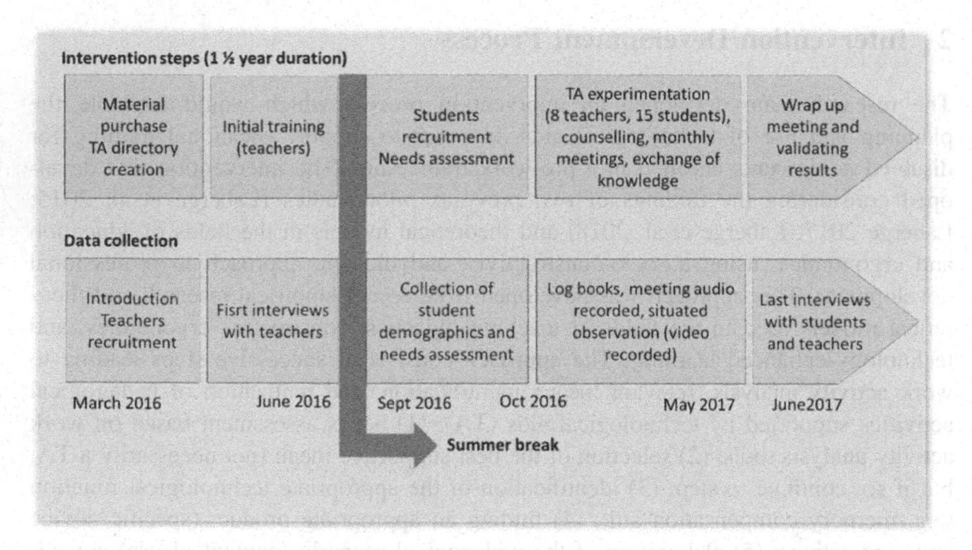

Fig. 1. Intervention and data collection steps

4 Findings

At the start of the experiment, teachers targeted various needs, mostly associated with individual difficulties, such as the ability to see one's inherent mistakes and apply corrections, memorization of useful information, problem with executive functions, motor learning, and diverse social abilities. The needs were then refined by applying a work activity analysis steps, assisted by ergonomists. This approach led the teachers to propose 46 pedagogical activities with their students, all assisted by a TA, at different extend. With the help of work activity analysis techniques, the technological supporting functions had targeted individual difficulties, but also workplace and environmental problems, such as product variability (memory function), poor signage for spatial orientation (spatial orientation), task complexity (video self-confrontation, demonstration), etc. Figure 2 illustrates the extent of pedagogical intentions relying on these activities. Table 1 presents the principal technological functions and applications used.

Among the 46 activities, most of the time teachers were the principal users of the technological means (manipulated the device, programmed the application, and determined the use conditions). Concerning the student participation, they were actively involved at various level, but did not always handle the technological devices themselves; often, teachers gave them a final artifact to use if needed (e.g. looking a printed task or photo to remember products or operations, consulted a final product on a technological device), which required minimal or no manipulation of the devices at the workplace (n = 20). In many cases, students were taught to use by oneself relatively simple applications or to participate in programming more complex applications (n = 22). In few cases (n = 4), students were thoroughly engaged in the use and programming of relatively sophisticated applications. Finally, teachers were asked to answer the question "how did you find this activity in terms of helpfulness for trainees?"

Table 1. Supporting functions and applications used (n = 46 pedagogical activities)

Techno supporting functions	N	Applications (exemples)
Memory function	20	Picstitch, Skitch, Shadow Puppet
Video self-confrontation	12	Photos, Videos, Imovie
Time/spatial orientation	3	Alarm, Google Maps, Clock
Demonstration	2	My video Coach, Explain Everything
Emotion control	1	Google Keep, Videos
Shift calculator	1	Time sheet
Communication aids	1	Dictaphone, google translate
Multiple functions	6	Clock + Notes

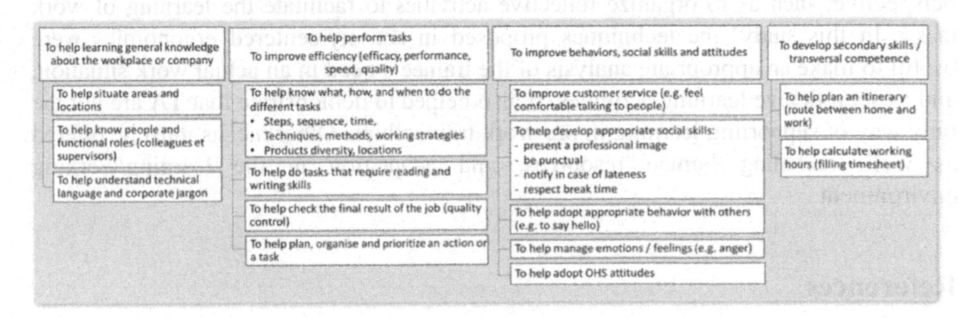

Fig. 2. Extent of pedagogical intentions

Table 2. Student's TA engagement and perceived value of activities (n = 46) rated by teachers

Level of trainee's engagement in manipulating TA	Perceived effectiveness of the activity by teachers (N)			
	+	+/−	−	N/A
No or low manipulation of TA (e.g. student received an artifact like photo or written instructions developed with a TA; student consulted info in an application set for him/her)	14	1	4	1
Autonomous use of a simple or common application or some involvement in programing a specialized application (e.g. help editing photos on Pic Stich)	15		6	1
Programing alone a specialized application (e.g. collects photos and reconstitutes a sequence in a recall task assistant, such as Shadow Puppet)	3		1	

+: the activity was judged effective to help trainee; +/−: mixed opinion; −: the activity was judged ineffective; N/A: teacher did not express an opinion

The perceived value of the activities was generally well rated by teachers (n = 32) (see Table 2). A "low or no manipulation of TA" does not necessarily mean a low engagement in pedagogical activity, and not even a weak perceived effectiveness. In fact, even if the teacher was the principal user of a TA, the students were often thoroughly engaged in the preparation of materiel (simulating tasks, verbalizing operations, etc.), which was also considered very useful.

5 Conclusion

At the beginning of the study, we thought that teachers would use technological aids mainly to compensate for disability (e.g. program a memory aid for a memory impairment), but finally, we found that teachers also used technologies in a didactic perspective, such as to organize reflective activities to facilitate the learning of work tasks. In this study, the techniques proposed in activity-centered ergonomics were useful to make an appropriate analysis of the trainee's need in an actual work situation, and to set reflexive learning activities. It also helped to demonstrate that TA are not the only way of supporting learning in pre-work traineeships. Other means must be offered as well, including human resources, and adaptation in the learning/working environment.

References

Laberge M, Tondoux A, Thonnon J, Béland J (2015) Le recours aux aides technologiques pour favoriser l'insertion professionnelle de jeunes adultes en situation de handicap, une utopie? In: 50th International conference of the Société d'Ergonomie de Langue Française, Paris, 23–25 September 2015

Laberge M, Tondoux A, Charland G (2017) Programmation de recherche sur le recours aux TIC pour soutenir l'insertion professionnelle des adolescents éprouvant des difficultés d'apprentissage lors de la transition école – vie active. In: Annual meeting of the CRWDP Québec group, CIRRIS, Québec, 15 November 2017

Laberge M, Melançon S, Martel-Octeau N, Tondoux A (2018) Approche pour intégrer des aides technologiques à l'apprentissage en stage d'insertion professionnelle auprès d'adolescents présentant des difficultés d'apprentissage. In: 43rd International conference of the Institut des troubles d'apprentissage « les troubles d'apprentissage et défis dans le monde de demain », Montréal, Westin Hotel, 21–23 March 2018

Embedding Human Factors and Ergonomics in MSc Organizational Psychology

Matthew C. Davis(✉)

Leeds University Business School, University of Leeds, Leeds LS2 9JT, UK
M.Davis@Leeds.ac.uk

Abstract. This paper reflects on the author's experience of embedding human factors and macro ergonomics content as part of an MSc Organizational Psychology program within a business school. The pedagogical underpinning of the author's Systems Thinking and Consulting Practice module is explained and key features that have been employed to engage students from a variety of backgrounds discussed. The key challenges encountered, including practical, institutional and disciplinarily issues are outlined. The paper concludes with suggestions for positioning human factors and ergonomics teaching to appeal to a broad range of students.

Keywords: Socio-Technical Systems Thinking · Education
Organizational Psychology

1 Introduction

This paper reflects on my own experience of embedding human factors and macro ergonomics content as part of an MSc Organizational Psychology program within a business school. The topics are covered as part of a 15 credits module (Systems Thinking and Consulting Practice) designed to introduce students to Socio-Technical Systems Thinking (STST) as an overarching framework for considering human-technical work systems.

The opportunity for students to encounter human factors and ergonomics topics and methods within organizational psychology/occupational psychology masters programs in the UK has diminished. Finding ways to present the content in a way that engages an increasingly diverse student profile on such programs, in addition to the need to structure modules to draw a larger cohort to ensure operational viability within a business school environment presents challenges. I will discuss the approach I have taken to tackle such a scenario, the practical challenges encountered and the potential I see for broadening the appeal of human factors and ergonomics to students from other disciplines.

S. Bagnara et al. (Eds.): IEA 2018, AISC 821, pp. 67–71, 2019.
https://doi.org/10.1007/978-3-319-96080-7_9

2 Systems Thinking and Consulting Practice Module

2.1 Approach and Pedagogy

The module reflects the research interests and projects of myself and colleagues, using STST as a consistent lens through which to consider the various topics. Reflecting the core STST philosophy on which it draws [1–4], the module is designed to be multi-disciplinary, drawing students from both the MSc Organizational Psychology in addition to Engineering, Geography and Physics programs. This diverse student cohort is designed to promote cross-disciplinary knowledge sharing and to demonstrate the value that different skill and knowledge sets can bring to the discussion of complex problems. The module is structured around traditional lecture delivery, accompanied by small group seminars to discuss practical case studies and scenarios in depth. The module attempts to engage students in the topics and demonstrate the relevance to contemporary business. This is supported through incorporating interactive sessions run by human factors and applied psychology consultants, sharing their experience working on real projects and as part of multi-disciplinary consulting teams. In addition, a group assignment requires students to work in multi-disciplinary teams to analyze the BP Deepwater Horizon oil spill using a human factors framework e.g., Accimap [5], STS Hexagon [6], to generate practical recommendations and to present their findings in a consultancy style.

The practical case study application and consultancy orientated approach to the module was instrumental in building interest from program leaders in other departments in the university. The structure of the module enables demonstration of transferable skills relating to: research methods; analysis techniques and frameworks; cross-disciplinary working; synthesis of technical and academic resources, and; translation of findings for business audiences. The transferable skills are valuable to other social science and technical disciplines who need to demonstrate program level learning and skills outcomes to their accreditors. Furthermore, the nature of human factors and ergonomics as discipline areas help to ensure that students from both technical and more behavioral backgrounds find terminology and approaches that are familiar to them, helping to anchor them as they approach more novel material.

2.2 Key Research Topics Used to Engage Students

The material covered on the module reflect my own and colleague's interests. The lecture sessions are designed to focus on a specific project or case study that the lecturer has undertaken. The majority of projects relate to industrially funded or supported work, helping to reinforce the business relevant nature of the human factors and macro ergonomics topics to which they relate. The wide variety of application areas help to illustrate the cross-cutting relevance of socio-technical and ergonomics ideas. The key topics that the module is structured around include:

- System failure analysis and prediction using socio-technical frameworks and techniques [7]. This includes coverage of traditional accident analysis topics [5], in addition to consideration of business system failures [6];

- The design of physical workspace, in particular examining the challenges of designing contemporary open plan offices that support knowledge work [8];
- Technology and software design and the management of attendant change, in particular considering the application of socio-technical principles [3, 4];
- The design of tele-health solutions and approaches for modeling and evaluating competing scenarios [9];
- The application of macro ergonomics approaches, behavior change and socio-technical principles to promote environmental sustainability [10–12];
- Approaches to crowd management and the modelling of crowd behavior in routine and emergency situations [7, 13];
- Information acquisition and knowledge sharing behaviors, with particular consideration to the context of design engineers and their social networks [14].

3 Practical Challenges

A number of practical challenges in designing and delivering a human factors and ergonomics module that runs across a series of programs and faculties have been encountered. One particular challenge concerned the difficulties that arose due to students arriving with varying knowledge bases from different disciplines and backgrounds, with differing expectations regarding teaching styles and assessment formats. Whilst a key aim of the module is to introduce students to cross-disciplinary working and to equip them with the toolset to engage in this effectively, this can be more difficult for some students when the educational environment is significantly different to that which they are used to.

The institutional environment within UK business schools means that there is pressure to ensure that modules are attractive to greater numbers of students. The popularity of business degree programs, particularly at post-graduate level, has driven a focus on efficiencies of teaching. With finite staff time and higher student numbers to accommodate, there is a danger of modules that are seen as niche (i.e., attracting small numbers of students) being cut. This is particularly relevant to human factors and ergonomics modules which are likely to be run as part of specialist organizational psychology, or related behavioral programs, that attract smaller student numbers than general management programs. This context challenges us to consider how we increase the number of students on such modules, maintaining the core human factors and ergonomics knowledge base whilst simultaneously making the material accessible to a variety of disciplines.

Furthermore, a specific threat is posed to the coverage of human factors and ergonomics teaching within UK MSc Organizational Psychology programs. The effective downgrading of human factors and ergonomics within the British Psychological Society's QOcc Psych (Qualification in Occupational Psychology) Stage 1 [15] and the concomitant reduction in emphasis within accredited courses poses an existential threat to the understanding of human factors and ergonomics within the organizational psychology profession. Whilst the changes to the QOcc Psych are welcome in that they allow institutions to increase coverage of topic areas that they are specialists in, it also provides cover for program teams to abdicate teaching topics that they consider difficult to resource.

4 Positioning Human Factors and Ergonomics to Appeal to Students Across Disciplines

My experience over the past 10 years of teaching human factors and ergonomics to psychology students has reinforced to me both the need to, and value in, teaching multi-disciplinary groups. As previously discussed, I believe that there is a pressing need to broaden the appeal of human factors and ergonomics, to ensure financial viability of modules within business school environments. I also believe that the pedagogical and practical value of teaching human factors and ergonomics ideas to students from diverse backgrounds make this beneficial to students also.

Designing and redesigning the Systems Thinking and Consulting Practice module to improve and respond to student feedback has yielded insights into student perceptions. This experience leads me to make a number of suggestions regarding how human factors and ergonomics may be positioned to attract interest from a broad range of student groups and in particular to appeal to students from business/social science disciplines. These suggestions include:

- Emphasizing transferable analytic skills and methods;
- Demonstrating application across a range of domains and problems;
- Including business case studies and very practical applications;
- Incorporating multi-disciplinary working;
- Building in practitioner interaction and industrial speakers;
- Making the inclusion within organizational psychology programs a positive point of differentiation.

5 Conclusion

Human factors and ergonomics contributes greatly to the knowledge base of organizational psychologists and remains an important part of their training. The institutional and broader disciplinary environment poses challenges to traditional ergonomics modules on MSc Organizational Psychology programs. I present a research centred and skills based approach to teaching human factors and ergonomics, an approach that emphasizes and makes a virtue of the multi-disciplinary profile of the student cohort. Human factors and ergonomics provides a valuable toolkit and mind-set for approaching a diverse range of problems for students from many backgrounds. This broad applicability offers a route to both extending the impact of human factors and ergonomics, in addition to supporting the future viability of modules and programs.

References

1. Trist EL, Bamforth, KW (1951) Some social and psychological consequences of the longwall method of coal-getting: an examination of the psychological situation and defences of a work group in relation to the social structure and technological content of the work system. Hum Relat 4(1):3–38. https://doi.org/10.1177/001872675100400101
2. Cherns A (1987) Principles of sociotechnical design revisited. Hum Relat 40(3):153–161. https://doi.org/10.1177/001872678704000303
3. Clegg CW (2000) Sociotechnical principles for system design. Appl Ergon 31(5):463–477. https://doi.org/10.1016/s0003-6870(00)00009-0
4. Mumford E (1995) Effective systems design and requirements analysis: the ethics approach. Macmillan, Basingstoke
5. Salmon PM, Cornelissen M, Trotter M (2012) Systems-based accident analysis methods: a comparison of Accimap, HFACS, and STAMP. Saf Sci 50(4):1158–1170. http://dx.doi.org/10.1016/j.ssci.2011.11.009
6. Davis MC, Challenger R, Jayewardene DNW, Clegg CW (2014) Advancing socio-technical systems thinking: a call for bravery. Appl Ergon 45(2):171–180. http://dx.doi.org/10.1016/j.apergo.2013.02.009
7. Clegg CW, Robinson MA, Davis MC, Bolton L, Pieniazek R, McKay A (2017) Applying organizational psychology as a design science: a method for predicting malfunctions in socio-technical systems (premists). Des Sci 3:1–31. https://doi.org/10.1017/dsj.2017.4
8. Davis MC, Leach DJ, Clegg CW (2011) The physical environment of the office: Contemporary and emerging issues. In: Hodgkinson GP, Ford JK (eds) International review of industrial and organizational psychology, vol 26. Wiley, Chichester, pp 193–235
9. Hughes HPN, Clegg CW, Bolton LE, Machon LC (2017) Systems scenarios: a tool for facilitating the socio-technical design of work systems. Ergonomics 60(10):1319–1335. https://doi.org/10.1080/00140139.2017.1288272
10. Christina S, Dainty A, Daniels K, Tregaskis O, Waterson P (2017) Shut the fridge door! HRM alignment, job redesign and energy performance. Hum Res Manag J 27(3):382–402. https://doi.org/10.1111/1748-8583.12144
11. Haslam R, Waterson P (2013) Ergonomics and sustainability. Ergonomics 56(3):343–347. https://doi.org/10.1080/00140139.2013.786555
12. Davis MC, Coan P (2015) Organizational change. In: Barling J, Robertson JL (eds) The psychology of green organizations. Oxford University Press, New York, pp 244–274
13. Challenger R, Clegg CW, Robinson MA (2009) Understanding crowd behaviours: Supporting evidence. Crown, London
14. Hughes HPN (2017) The role of advice networks in the design and development of jobs. Ph. D thesis, University of Leeds
15. Fletcher C, McDowall A (2014) Were 8 great? Doing more with less! An introduction to the new OP curriculum. In: Paper presented at the British Psychological Society's Division of Occupational Psychology Conference, Brighton, UK

A Chair Assessment Model for Organizational Benefit, Safety and Asset Management

Alison Heller-Ono(✉)

Worksite International, Inc., 170 17th Street, Pacific Grove, CA 93950, USA
alisonh@worksiteinternational.com

Abstract. Extensive research has been performed on ergonomic chair design and the impact chair design has on seated posture and the musculoskeletal system of office workers. The research has significantly advanced the science of sitting and the design of office ergonomic chairs. BIFMA and ANSI criteria have identified guidelines for manufactures and consumers to better understand the features an office ergonomic chair should possess. There is a gap in the literature, however regarding how to assess the ongoing performance of an ergonomic chair after it is placed into the workplace. Not as it applies to the fit to the end-user, but the quality and competency of the chair to remain in use in the workplace. Once an ergonomic office chair is purchased in the workplace, it often remains in circulation far beyond its acceptable life cycle and warranty. As a result, chairs that are old, worn, outdated and inoperable continue to be used by office workers. These older chairs often present additional ergonomic and safety risk factors exposing employees to unnecessary musculoskeletal stress and strain resulting in injury exposure claims for the employer.

This paper introduces an assessment methodology using predictive analytics to evaluate the quality and competency of an office ergonomic chair over time. Rather than relying solely on an employee's subjective, biased opinion of chair quality; instead an objective, measurable rating scale is used to determine chair status. The Chair Assessment System (CAS) and tool provides an overall score indicating whether the chair should remain in use, be repaired or removed from circulation in a timely manner.

Keywords: Ergonomic task chair · Safety · Life cycle · Predictive analytics Chair rating system

1 Introduction

1.1 The Ergonomic Task Chair

The ergonomic task chair is the most important "work tool" determining a worker's seated productivity in conjunction with the computer (and workstation). It can be said the chair is foundational to good seated workstation ergonomics. Yet, the office task chair is routinely misunderstood, undervalued and probably the least appreciated asset employers purchase.

Until now, there has never been a way to inventory and measure chair quality and competency for ongoing use in the workplace. The goal of the chair assessment system

S. Bagnara et al. (Eds.): IEA 2018, AISC 821, pp. 72–79, 2019.
https://doi.org/10.1007/978-3-319-96080-7_10

(CAS) model is to help employers manage their chair assets as a system by providing an objective measure to determine whether to keep, repair or replace chairs and then fit employees for ongoing safety, comfort and productivity.

2 An Unmanaged Asset

Few employers, if any, use an organizational or systems approach to managing chairs as an asset in the workplace. In addition, employers do not recognize or track when employee chairs are at the end of their life cycle (Fig. 1) keeping them far too long, exposing themselves and their employees to increased liability, reduced productivity and increased risk for seated musculoskeletal disorders leading to workers' compensation claims.

The problem is a combination of both employer and employee lack of awareness and understanding of the value and importance of ergonomic chairs in the workplace. Starting with how to select chairs for the workforce and demonstrable willingness to invest in quality chairs as an important asset that contributes to employee health and productivity.

Missing is an objective methodology regarding how to determine whether to keep, repair or replace and fit office task chairs once they are in the workplace. The tests in ANSI/BIFMA X5.1 - 2017 Office Chairs standard are intended to assess the performance of new products only. They are not intended to assess a product that has been in use. Essentially, there is no way to objectively assess a chair through its lifecycle to identify proactively when to repair it before the warranty ends or when to remove it from circulation before catastrophic failure.

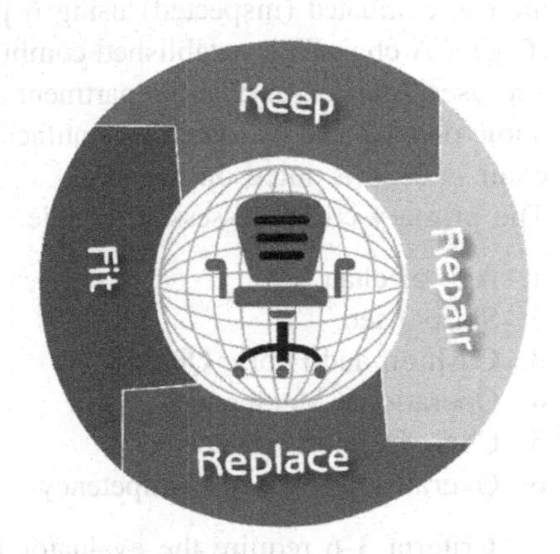

Fig. 1. The chair life cycle model developed by Worksite International, Inc. includes keep, repair, replace and fit as the cycle.

3 Gap in Ergonomic Guidelines for Chairs in the Workplace

The furniture industry has developed widely accepted ergonomic guidelines for new chairs. Most notable are the BIFMA G1 - 2013 Ergonomics Guideline - Ultimate Test for Fit and the newly released ANSI/BIFMA X5.1 - 2017 Office Chairs. These guidelines are intended to standardize on chair safety and design, so they fit most individuals. The standard defines specific tests, the laboratory equipment that may be used, the conditions of tests, and the minimum acceptance levels to be used in evaluating general-purpose office chairs. However, there is no system in place to evaluate chairs once they are brought into and used over time in the workplace.

4 Development of a Chair Assessment System (CAS)

Research was conducted over a 6-year period from 2010–2016 to develop a chair assessment system (CAS) model to coincide with the ergonomic chair life cycle (keep, repair, replace and fit). The CAS consists of an Excel data table and an Excel assessment tool along with a chair fitting form to offer predictive analytics for chair decision makers.

After the initial idea was developed and applied in a large organization, additional trials were performed over the last 2 years to test, validate and improve the CAS Excel tool design and functionality.

4.1 Chair Assessment Tool (CAT) Methodology

Chair Assessment Criteria. To utilize the CAS, each appropriate chair in the work area is evaluated (inspected) using 6 primary criteria on the Chair Assessment Tool (CAT). A chair ID is established combined with other identification to include name of end user, location of chair, department and date of assessment. The chair manufacturer, name or chair model, date of manufacturing or shipping date and number of shifts the chair is used is noted on the CAT.
The primary criteria assessed include:

1. Age of chair
2. Shifts used
3. Cushion and Fabric Quality
4. Operational Mechanics
5. Chair Comfort (perceived)
6. Overall Quality and Competency

Criteria 3–6 require the evaluator to rate the chair based on a three-point rating scale of good, fair or poor following inspection and observation of each criteria as shown in Table 1. Whenever possible, the end user participates in selecting the chair perceived comfort rating. Criteria 3–6 offers descriptive terms to select from that best describes the condition of the chair at the time. The information is entered onto the Chair Assessment Tool (CAT), which is then input to the Chair Assessment System Excel sheet for calculation using a proprietary algorithm to determine the score of the chair.

For the final score, the algorithm is translated to three responses:

1. Keep the chair
2. Repair the chair (or refurbish)
3. Replace the chair (then fit for a new chair)

Each chair score is color coded on the CAS Excel datasheet. A dashboard is created automatically for further sorting of the data to give the employer a better sense of the quality and competency of their chairs by manufacturer and by location/department.

Table 1. Chair assessment rating criteria

Rating area	Description
Cushion/Fabric quality	**Good Condition** — Good cushion comfort, foam is supportive, fabric or mesh in good condition **Fair Condition** — Fabric intact, no stains, beginning to show signs of wear, foam is compressing **Poor Condition** — Well worn, thin and stained, torn fabric, foam has collapsed; can feel seat bottom
Operational mechanics	**Good Condition** — All features work correctly, casters roll easily, cylinder adjusts properly; holds position effectively **Fair Condition** — Only some features work, chair needs maintenance or repairs; cylinder sticks (hard to raise, lower), casters need cleaning; some parts still under warranty **Poor Condition** — Features do not adjust correctly or stay in place; casters are worn/broken; cylinder failure (no longer holds position); chair is unsafe; warranty expired
Chair comfort (Perceived)	**Good/Very Comfortable** — Good chair comfort; very satisfied with chair fit and features; always supportive **Fair/Comfortable** — Fair chair comfort; satisfied with chair fit; not sure of all features; requires adjustment for support **Poor/Uncomfortable** — Poor chair comfort; not satisfied; does not adjust well for fit; causes discomfort
Overall quality and competency	**Good Condition** — Good cushion and fabric quality, all features work and the employee is comfortable to very comfortable **Fail Condition** — Fair cushion and fabric quality, only some features are functioning; warranty nearly expired; employee is comfortable **Poor Condition** — Cushion and fabric are worn and torn, features do not function; warranty expired; employee is uncomfortable

Chair Evaluator and Evaluation. For this study, the author, a Certified Professional Ergonomist evaluated all the chairs noted in the example dashboard. However, the Chair Assessment System is designed to be used by anyone who understands the basic features and functions of an ergonomic chair, all of which can easily be taught to the evaluator as needed. In the workplace, users of the CAS may be members of an ergonomics team or in-house evaluator, risk manager, safety manager, purchasing or facilities manager or technician, supervisor or even the end user/employee. The Chair Assessment Tool is easy to use and easy to understand provided the evaluator has a basic knowledge of ergonomic chair functionality.

To evaluate one ergonomic chair takes approximately 5–6 min. The more chairs evaluated, the more efficient the evaluator becomes in rating chairs for quality and competency. The evaluator needs to assess the bottom of the chair seat to look for any manufacturer labels for type of chair, model number and shipping/manufacturing date. At the same time, observe the levers and knobs necessary to adjust the chair in upright position. A quick visual inspection of the fabric and external moldings of the chair should be noted for quality, wear and tear. The remaining assessments are made while seated on the chair including assessing cushion quality; operational mechanics by adjusting all mechanical parts to assure they are operational and perceived comfort which is best provided by the end user if present during the assessment. Otherwise, the evaluator should rate the chair comfort. Finally, an overall rating is given taking all factors into consideration.

5 Outcomes

As an example, a small company participated in using the CAS. Select chairs were evaluated using the criteria described above to inventory and assess chair quality and competency. The database is provided to the employer to determine which chairs can remain in operation, which need to be repaired (while under warranty) and which should be removed due to risk of failure or harm to the user.

The CAS Excel dashboard (Fig. 2) is presented below as an example of an organization with 21 employees using a variety of chairs. Data is entered from the Chair Assessment Tool and then imported into the CAS Excel database to track chair inventory, quality and competency of each chair assessed.

An overall total score is calculated and is shown in the appropriate color, red, yellow or green. Chairs in poor condition based on the categories described in Table 1 are shown in red. Chairs in fair condition that would benefit with some degree of repair or refurbishing and are likely still under warranty are shown in yellow. New chairs and chairs continuing to be in good condition overall with little concern are shown in green.

The CAS database is also translated into a color-coded dashboard (Fig. 2) to act as a visual interface that provides at-a-glance views into key measures relevant to the type of chairs the employer has, the location of the chairs and the quality and sustainability of the chairs. The CAS dashboard provides visualization to help focus attention on key trends, comparisons, and exceptions regarding the chair asset management program.

The CAS dashboard identifies at least twelve chair trends for anyone in the organization who needs to be aware of and manage the chair fleet program. This might include safety and risk managers, EH&S, facility managers, purchasing managers, budget managers, ergonomics teams and others in the organization with an interest in understanding the state of chair asset management relative to employee health, safety and productivity.

The CAS dashboard shown in Fig. 2 identifies trends in:

- The number of chairs by manufacturer/model type
- Assessment totals and the number to keep, repair and replace
- Assessment totals by department/location
- Age of chair by manufacturer/model
- Cushion/Fabric quality by manufacture/model and by department/location
- Operational mechanics quality by manufacture/model and by department/location
- Comfort (perceived) by manufacture/model and by department/location
- Overall quality by manufacture/model and by department/location

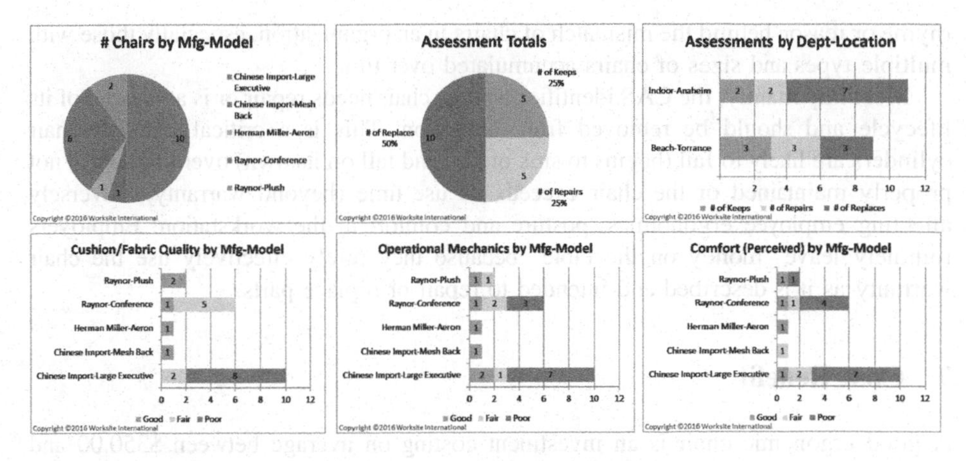

Fig. 2. The Chair Assessment System dashboard (excerpt) helps to focus attention on key trends, comparisons and exceptions in the chair fleet. (Color figure online)

6 Discussion

Using the example company, the employer can specifically identify chair concerns by person, by department and by location which chairs are holding up the best, which are perceived as most comfortable, which require repair and or replacement and many other trends, comparisons and exceptions in their chair program. Chairs scored in the green are acceptable for ongoing use. Chairs scored in yellow indicate they need some degree of repair under warranty (as applicable). Chairs scored in the red, allow the employer to identify the chair for removal, to effectively budget for replacement and perform a chair fitting with the employee.

The Chair Assessment System easily removes bias from a process often riddled with favoritism or some degree of workplace social hierarchy where certain employees receive new chairs because of status in the organization, a work injury or because their neighbor received a chair. While it may be appropriate to provide new chairs in some circumstances, many times, the current chair simply is not adjusted correctly or set for best fit. Inventorying and assessing the chair for comfort and fit is part of the chair asset management process.

The CAS helps to identify based on objective, measurable evidence whether to keep, repair or replace a chair and then fit for a new one because it has reached the end of its lifecycle or offers an inadequate fit. Through the inventory process, employees learn about the features and functions of their chairs, how to adjust for fit and comfort as well as determine if the chair is a poor fit for them.

Furthermore, this system helps employers to see which chairs are working well for the organization by looking at the perceived comfort of the ratings; further helping to assign chairs more effectively, rather than the "willy-nilly" approach employers typically apply. For example, the petite female in the large, oversized chair and the large male or obese employee in a significantly undersized chair. Typically, there is no rhyme or reason behind the mismatch of chairs in an organization, especially those with multiple types and sizes of chairs accumulated over time.

Most importantly, the CAS identifies when a chair needs repair or is at the end of its lifecycle and should be removed from operation. This is a critical issue as chair cylinders are likely to fail (begins to sink or rise and fall on its own) over time when not properly maintained or the chair exceeds its use time (beyond warranty), adversely affecting employee ergonomics, posture and comfort at the workstation. Employers routinely leave "money on the table" because they rarely effectively use the chair warranty as it is described and intended to repair or replace parts.

7 Cost Benefit

A good ergonomic chair is an investment costing on average between $350.00 and $850.00 and expected to last approximately 10 years or more. Investment in the Chair Assessment System costs literally pennies per chair compared to investing in a new chair, especially when it is not necessary. Employers routinely discard chairs perceived as broken or a poor fit simply because they don't understand how to adjust or use the chair properly; fail to use the warranty or select incorrect chairs for employees.

It is estimated utilizing the chair warranty has an expected cost saving of at least 50% over buying a new chair. By repairing an existing chair as part of a preventive maintenance program, it extends the life of a good quality chair another 3–5 years that would have otherwise been determined to be replaced without the CAS data.

It does take time to evaluate and document the status of each chair. As stated earlier, a chair assessment takes approximately 6 min using the CAT followed by brief data entry into the CAS Excel sheet. This time is well spent given the outcome.

In regard to the cost-benefit, the chair assessment system is an affordable, asset management strategy designed to optimize operational chair performance, minimize whole life costs and support an organization's corporate health and safety goals.

8 Conclusion

Based on numerous trials in the workplace, feedback from industry leaders and practitioners, the chair assessment system provides practical and informative, predictive analytical data in a simple and easy to use format allowing employers to better understand chairs as a system. The CAS shows employers it's worth the time to inventory and assess task chairs for safety, health and productivity impact.

The Chair Assessment System is an effective way to assure employee seated work health and chair satisfaction through the life cycle of chair use. Using an inventory and asset management system to measure task chair quality and competency helps employers and practitioners determine the most effective chairs in the workplace, which need to be repaired and which should be replaced to minimize risk, liability and exposure to seated work discomfort. By doing so, thousands of dollars can be saved by reducing work injuries and improving employee health and productivity.

For more information about the Chair Assessment System described in this paper, please contact https://www.worksiteinternational.com/chair-assessment-system

References

1. Rivers, TB (2017) Why workplace leaders should use predictive analytics. iOffice blog, 26 April 2017
2. Robertson M, Amick B et al (2007) The effects of an office ergonomics training and chair intervention on worker knowledge, behavior and musculoskeletal risk. Appl Ergonomics 40 (2009):124–135
3. Colombini D, Occhipinti E et al (1993) Criteria for the ergonomic evaluation of work chairs. Med Law 84:274–285
4. Groenesteijn L, Vink P, et al (2009) Effects of differences in office chair controls, seat and backrest angle design in relation to tasks. Appl Ergonomics 40(3):362–70
5. Core Working Group Members of APPA, Federal Facilities Council, Holder, IFMA and NASFA. Asset Lifecycle Model for Total Cost of Ownership Management: Framework, Glossary and Definitions
6. Watkins G (2013) Effective asset management for facilities managers. Service Works Group
7. BIFMA G1-2013 Ergonomics guideline - Ultimate test for fit
8. ANSI/BIFMA X5.1-2017 Office Chairs

Managerial Simulation: A Tool for Devising Management Organization

Laurent Van Belleghem[1,2](✉)

[1] Conservatoire National des Arts et Métiers, Paris, France
[2] Realwork, Paris, France
laurent.vanbelleghem@realwork.fr

Abstract. In ergonomics of activity, simulation of activity is a methodology for assisting in the development of new work situations. It involves asking workers to "play out" their work activity in a new prescriptive context that is represented on a simulation support. Managerial simulation consists in reproducing this device and focusing on the work of managers. This means building a simulation support capable of representing the rule and management resource system that is to be devised, and asking managers to simulate their own activity within this new prescriptive context. Such an approach provides an opportunity for collective and informed reflection on the development of managerial logics of action. To illustrate this approach, we present a mission which aimed to reorganise the management system in a packaging workshop. An initial in-depth activity analysis allowed us to characterise the work situation of the foremen. We show how the organisation of their responsibilities in the form of geographical zones (islands) impedes their transversal activities relating to the workshop as a whole. We then adopted a managerial simulation approach to build an organisation scenario that replaced the island organisation with a distribution of responsibilities allocated on a weekly basis to each manager. The simulation of managerial activity, on a suitable support, allowed us to validate the principles used and to facilitate the development of new, pertinent and efficient managerial logics.

Keywords: Activity · Organisational simulation · Managerial simulation

1 Introduction

Management is an activity. The ergonomics of activity is therefore a legitimate field through which to study and attempt to improve management performance. Such improvement is useful not only for managers themselves, but also for their colleagues. Managers must organise the productive capacity of their teams, help with their development and protect them from negative work effects. They therefore require the resources with which to achieve these objectives.

Manager evolution is traditionally seen in terms of the evolution of managers themselves, through various types of training (in leadership, communication, etc.) and team-building (sports outings, game areas, etc.). This approach has more to do with

S. Bagnara et al. (Eds.): IEA 2018, AISC 821, pp. 80–86, 2019.
https://doi.org/10.1007/978-3-319-96080-7_11

adapting managers to their jobs than with the general ergonomic principle of "fitting the job to the worker". From an ergonomic standpoint, it is managers' work situations that must be made to evolve, by acting on their determinants: objectives to be met, zones of action, tasks, resources, delegation of authority, etc. These elements are defined for specific populations and constitute the organisation of management. If the job of management is to be improved, then the organisation of management has to become a design object.

For ergonomists, helping to devise a management organisation takes place in two stages:

1. Characterising managers' work situations by carrying out an in-depth analysis of their activities. Such an analysis must make it possible to identify not only the determinants of the situation upon which to act, but also the logics of action that the managers have developed and implemented (strategies, cooperation, mutual aid, etc.).
2. Engaging in a project to evolve the management organisation that:
   - aims to act on the managers' work situations (and not on the managers themselves);
   - facilitates the implementation of pertinent and efficient managerial logics of action.

Managerial simulation methodology can be used to achieve this (Van Belleghem and Guerry 2016). In ergonomics of activity, simulation of activity is a methodology for assisting in the design of new work situations (Barcellini et al. 2014). It consists in asking workers to "play out" their work activity in a new prescriptive context, represented on a simulation support. In projects that are primarily technology or space-oriented, one might use a scale model, for example (Broberg et al. 2011; Braatz and Paravizo 2017). In projects that are essentially organisation-oriented, the simulation support must be able to represent the rule system that is being devised. In such cases we talk about organisational simulation (Van Belleghem 2012, 2017; Daniellou et al. 2014).

Managerial simulation involves reproducing this device by focusing on managerial activity. This means building a simulation support that is capable of representing the rule and management resource that is being devised, and asking managers to "play out" their work activity within this new prescriptive context. This approach provides an opportunity for collective and informed reflection on the development of managerial logics of action.

The mission described below, which took place in a cosmetics packaging workshop, is an example of this.

2 Example: Management Reorganisation in a Packaging Workshop

2.1 Project Context: Management Organised into Islands

The workshop bottles cosmetics and then packages them ready to be sent out to stores. There are a dozen automated and manual lines grouped together in islands. Each island is assigned to one of the four foremen. The islands are currently hierarchically and functionally independent. The foremen are in fact "island managers", each being assumed to be autonomous and multicompetent on his island. The result is that the islands are separated from one another, partitioned. Whilst we did observe strategies of mutual aid and cooperation between foremen, this depended to a large extent on the strength of their personal relationships and on the department's history.

Current management organisation methods are no longer considered to be sufficiently suited to the evolution of the department, and it was for this reason that the packaging manager had decided to contact ergonomists. As part of a participative approach, the mission was designed to assist with the management reorganisation project, with a view to:

- decompartmentalising the foremen's zones (currently divided into islands),
- allowing a stronger managerial presence within the teams,
- basing the new organisation on the strengths of each individual foreman.

First and foremost, the mission used activity analysis methodology to examine the work done by the foremen and to identify the characteristics of their work situation. It then employed a methodology for scripting and simulating the new management organisation. It concluded with the creation of written specifications for this new organisation.

2.2 The Job of Management: An Impeded Activity

Analysis suggests that the hypothesis of the islands operating in a partitioned and separate manner is not entirely accurate. In particular, we find strategies of cooperation between foremen, of both intra and inter-island mutual aid between personnel, of resource adjustments at workshop level (assignments, planning, etc.), of transversal regulation, of assistance to teams, of distribution of transversal roles, of performance of transversal analyses and projects, etc.

But these strategies are often informal (the workshop's prioritisation objectives are not formalised, for example), heterogeneous (variable relationships of mutual aid between foremen, for example), are not always visible, known or recognised, equitable (for example, the extent to which information is shared depends on personal relationships), are unplanned or not organised (such as when no replacement has been planned for a foreman's vacation).

There is thus a need and a perceptible desire for transversality in the managerial activity, but these are impeded by the existing management organisation, due to its history and structure. For example, the islands are configured in such a way that the lines of a given island are often at some distance from one another, which limits opportunities for "natural" cooperation between personnel working on the lines. Managers then find it hard to organise cooperation within their islands, or to support inter-island cooperation when it goes beyond their zone. In doing away with the notion of islands within an appropriate organisation should, to a large extent, help to limit or even definitively remove these impediments.

Moreover, analyses of the managers' work show that they are often "distracted" by their teams who need them to come and resolve everyday issues. Their workdays are thus split between important moments in the workshop and time spent in the offices organising progress in production. De facto, there is little or no time available to get to grips with jobs commensurate with their skill sets. Furthermore, because the foremen have similar rationales for coming into the workshop, it is not uncommon to observe periods when several foremen are in the workshop at the same time, alternating with periods when no foremen are present (Fig. 1). This observation reveals a paradox: although the foremen are "distracted" by the workshop, the latter gains no real benefit from managerial presence.

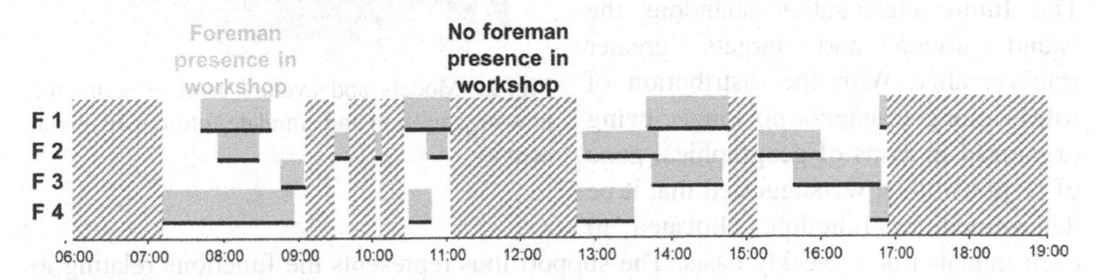

Fig. 1. Superimposition of periods during which foremen are present in the workshop. This figure shows that the time foremen spend in the workshop is concentrated within certain periods, leaving considerable periods of time when none are present.

It was on the basis of these observations, and in order to resolve their main constraints, that the participative development of a new managerial organisation was begun.

2.3 The Recommended Managerial Simulation Device

Our approach was based on a device that simulated the managers' work. The methodological tool was made up of two supports. The first made it possible to represent the future management organisation in the form of scenarios, while the second, which provided a small-scale model of the workshop, allowed managers to simulate their future activity over the course of full shifts, using avatars to represent themselves (Fig. 2). This support system thus enabled both a collective reflection on the type of organisation to develop, and an easier projection into the new organisation. The simulations also provided the opportunity to explore new managerial logics of action that were coherent with the proposed management organisation scenarios. Logics of action do not refer here to the application of prescribed rules or procedures (even if they are often in line with the latter); they relate to a more or less systematic structuring of action, built through the experience of the actors concerned, in terms of their appreciations of the situation, their personal characteristics, their know-how, experience, appetence, values, etc. On this occasion it was particularly a question of facilitating the development of individual logics of action that would build on each foreman's individual strengths, whilst at the same time enhancing their complementarity within the new management device.

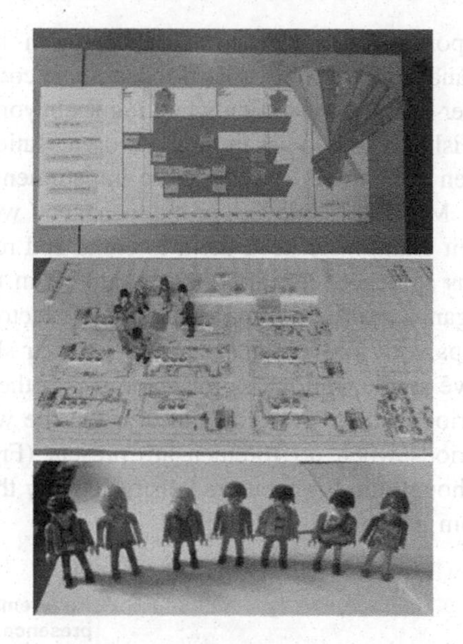

Fig. 2. Models and avatars used to script the new organisation and simulate future managerial activity

2.4 The Chosen Management Organisation Scenario

The future organisation abandons the island notion and targets greater transversality. With the distribution of roles among managers no longer being organised in terms of geographical zone of responsibility, we suggested that it be determined by function, allocated to each manager on a weekly basis. The support thus represents the functions relating to the different tasks that the managers complete over a 24-h timeline (00:00 to 24:00), thus making it possible to represent the various possible shifts (Fig. 3).

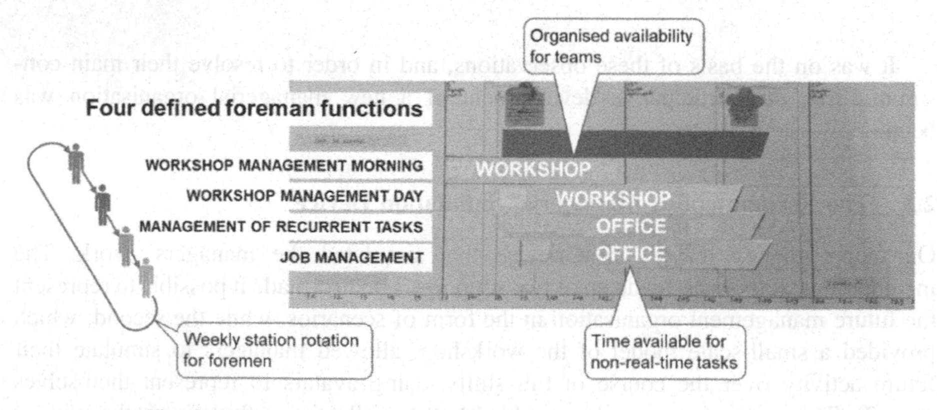

Fig. 3. The scenario chosen for the new management organisation

Four foremen functions were defined:

- Two workshop management functions (morning and day) where two foremen are freed from all administrative office tasks and provide support for the operational teams and for real-time management (unforeseen events, decision-making, etc.) for the workshop as a whole.

- Two office functions, where two foremen, freed from all workshop management tasks, deal with administrative tasks that require continuity and concentration. The first of these two functions relates to the management of recurring tasks (assignments, audits, debriefings, etc.). For each foreman, the second function consists in authorising and monitoring actions relating to the dossiers for which he is responsible, in line with his appetence and skill sets (actual or desired). Such dossiers relate in particular to the effective implementation of a process of constant improvement and to the development of personnel.

As designed, the new management organisation made it possible not only to remove the impediments that the island organisation had created for managers, but also to resolve the previously identified paradox, thus:

- giving managers time to explore the dossiers that suit their individual appetence and skills, thus allowing them to make the most of their strengths,
- providing stronger managerial support in the workshop and throughout the day.

After several months of implementation and a certain number of adjustments, this new organisation demonstrated its effectiveness by encouraging improved cohesion between managers and developing greater coherency between the responsibilities of all concerned.

3 Conclusion

Managerial simulation is more than just an additional tool in the ergonomist's toolbox. It is the mark of a broader field for ergonomic action, relating to ever-wider determinants and to increasingly diversified strategic actors. The study of managers' work is a relatively new area for ergonomists. This interest is rooted in the idea that managers must be able to help reconcile performance with the attention they pay to their teams in terms of health, subjectivity and development. Yet as far as day-to-day work is concerned, this activity is often hindered by the characteristics of their own work situations. Henceforth, addressing managers' work situations should prove beneficial to all of their colleagues.

It should be possible to extend this notion to include decision-makers (directors, presidents, etc.) whose activity, like that of managers, is determined by a set of rules (governing company operation and its decision-making and management bodies – supervisory board, board of directors, executive committee, employee representative committees, etc.). Decision-makers therefore only possess limited powers, in the sense that they are themselves part of the system of governance that they helped to create. We make the hypothesis that these systems would be better informed if they were allowed to more systematically script the forms that they might take and if they could simulate the decision-making activity of the collaborators involved. It is a case of turning the governance system into a design object. We suggest calling this emerging approach *governance simulation* (Van Belleghem 2016).

In other words, it is indeed the ergonomist's ability to act on work situations that can be augmented by action on managerial and decision-making activities. This orientation needs to be operationalized as part of the ergonomists' aim to make managerial and decision-making action an object of analysis and research in its own right. It must also be informed: this is the challenge for managerial and governance simulations.

References

Barcellini F, Van Belleghem L, Daniellou F (2014) Design projects as opportunities for the development of activities. In: Falzon P (ed) Constructive ergonomics. Taylor and Francis, New York, pp 150–163. https://bit.ly/2viBOBV

Braatz D, Paravizo E (2017) Articulating intermediary objects in the design process to enable ergonomics intervention. In: Proceedings of 12th international symposium on human factors in organizational design and management (ODAM), Banf, Canada

Broberg O, Andersen V, Seim R (2011) Participatory ergonomics in design processes: the role of boundary objects. Appl Ergon 42(3):464–472. https://doi.org/10.1016/j.apergo.2010.09.006

Daniellou F, Legal S, Promé M (2014) Organisational simulation: anticipating the ability of an organisation to cope with daily operations and incidents. In: Broberg O, Fallentin N, Hasle P, Jensen PL, Kabel A, Larsen ME, Weller T (eds) Proceedings of 11th international symposium on human factors in organizational design and management

Van Belleghem L (2012) Simulation organisationnelle: innovation ergonomique pour innovation sociale. In: Dessaigne MF, Pueyo V, Béguin P (coord) Innovation et travail. Sens et valeurs du changement. Actes du 47ème congrès de la Société d'Ergonomie de Langue Française, Lyon, France, 5, 6 et 7 septembre 2012. Editions du Gerra. https://bit.ly/2IUWCkI

Van Belleghem L (2017) What are the design requirements for an organisational simulation support? In: Proceedings of ACE-ODAM 2017, Banff, Canada. https://bit.ly/2JQVxf9

Van Belleghem L, Guerry M-H (2016) De la simulation organisationnelle à la simulation managériale. Communication présentée au 51ème Congrès de la Société d'Ergonomie de Langue Française, Marseille, France. https://bit.ly/2jCFFmE

Van Belleghem L (2016) La simulation de l'activité en conception ergonomique: acquis et perspectives. Activité 15(1):1–22. http://journals.openedition.org/activites/3129

Impact of New Technology on Job Design, Skill Profiling and Assessing the Physical, Cognitive and Psychosocial Impacts on the Workforce

David Caple(✉)

David Caple & Associates Pty Ltd, Melbourne, Australia
david@caple.com.au

Abstract. The introduction of new technology into the workplace has been associated with the reduction of safety risks where hazardous tasks have been eliminated. For example, the use of robotics and mechanical aids in manufacturing and warehouse logistics eliminates repetitive manual handling tasks. The business case for investing in this technology is based on a reduction in labour costs, including reduced injuries as well as an improvement in capability, productivity and the quality of pallet stacking. The reduction in manual labour jobs is associated with an increase in jobs requiring skills in STEM (Science, Technology, Engineering and Mathematics). Workers with these skills are required to install, program, maintain and trouble shoot problems with the technology to ensure sustained reliability. There are more semi-skilled jobs required to operate the materials handling equipment as the product is moved through the logistics process. The cognitive and psychosocial skills required for these semi-skilled jobs include the NTS (Non-technical Skills). These include skills in situational awareness; teamwork; decision-making and task management (Kodate et al. 2012). This research originated in medicine and has been applied to many industry sectors including manufacturing and mining.

In the office sector the use of hand held technology and high quality wifi has enabled the nature of work to change significantly. No longer are workers limited to sit at a workstation to access their computer. They are now able to access their handheld device in multiple settings within and outside their office workplace. This results in a major shift in the physical, cognitive and psychosocial demands of their work. They are able to stand up and walk to collaboration areas, breakout spaces and a range of meeting areas to conduct their work with colleagues.

The role of the ergonomist is to assess the impact of technology and work systems on the workers. This requires a broad set of technical and communication skills. Their role is to provide guidance on the physical, cognitive and psychosocial requirements to ensure the health and wellbeing of the workers. This role requires close engagement with members of the multidisciplinary teams who are designing and implementing the technology. It is evident that these new ways of working requires a change management strategy to ensure that the workers understand the objectives of the changes and the impact on their roles.

The long-term impact of this technology and changing ways of working require further research to assess the physical and psychological impact on the workers.

S. Bagnara et al. (Eds.): IEA 2018, AISC 821, pp. 87–91, 2019.
https://doi.org/10.1007/978-3-319-96080-7_12

Keywords: Technology · Skills · Physical · Cognitive · Psychosocial Workplaces

1 Objective

The impact of technology on the reduction of injury risk in the workplace is dependent on the overall impact on the jobs that exist within a supply chain context. This review is based on the use of new technology in manufacturing and white-collar industries and looks at the associated reduction of injury risk to workers.

2 Conceptual Framework

The introduction of robotics in manufacturing has become cost effective in many medium to large sized companies to ensure quality and consistent productivity on repetitive tasks. This includes robots used in the cutting of materials and the assembly of products in industries including automotive and food manufacturing. One of the advantages of using robots in logistics is the capacity of the robot to accurately pack cartons of goods in specific pallet configurations and to heights that maximize the logistics transportation in trucks to customers. However when the goods arrive at the customer the packaging needs to be safely handled and stored within the capacity of the distribution centre or store. For example, in the supermarket industry in Australia many of the incoming goods on pallets from large suppliers of soft drinks can be delivered in pallet loads greater than 2400 mm high. These pallets are stacked at the factory using robots. However, to minimise the manual handling risks for the supermarket staff a maximum height of 1400 mm to 1800 mm is needed for the pallet, depending on the size and weight of the cartons. Hence the ability to fully optimize the cost benefit of the investment in the robots is restricted within the supply chain context to ensure a safe system of work for all workers. Within the office workplace the increased use of handheld technology and the improved reliability of Wi-Fi has provided options for new ways of working. Previously workstations and personal computers were the focus to a workstation where the majority of office work was done. Now computer access can enable the user to work in multiple settings both within and outside their office. This introduces a new era of physical, cognitive and psychosocial challenges in defining safe and productive work.

This review has considered the impact of new technologies in a range of industry sectors and the associated need to assess the skill profiles as well as the ergonomics requirements for the worker. This includes the physical, cognitive and psychosocial aspect of the work design.

3 Technical Approach

This project reviews the experiences of workers in the manufacturing and office work environments to assess the impacts of the new technologies on their ways of working. Site visits were made to multiple work locations in Australia across the supermarket supply chain. This included the factories where the goods were manufactured as well as the agriculture areas responsible for growing the food. It included the logistics and warehousing operations where the company temporarily stored the incoming goods awaiting distribution to stores. The process of ordering, picking and palletising of store orders was reviewed, together with the transportation of goods to store, the pallet decanting, and finally, the stacking of goods onto shelves for customers to purchase.

The project also reviewed the impact of new ways of working in 10 large private companies across the banking, insurance and government sectors. This included a review of the impact of moving away from allocated workstations and using laptops and handheld devices in multiple work settings in and around the office complex.

4 Proof of Findings

The business case for investing in new technology such as robotics in manufacturing is based on a reduction in labour costs, including reduced injuries to workers, as well as an improvement in productivity and the quality of pallet stacking. The capacity of the robots in industry to eliminate repetitive work has been observed in many areas including the manufacturing of whitegoods, automotive, food products and confectionery as well as clothing and furniture. The reduction in manual labour jobs in those industries where the new technology has been introduced has been associated with an increase in jobs requiring skills in STEM (Science, Technology, Engineering and Mathematics). These skills require a more technology-focused cohort of workers, who need to demonstrate more complex skills to install, program, maintain and trouble shoot problems associated with the technology, ensuring sustained reliability. The cognitive and psychosocial skills required for these more technical jobs also include requirements for NTS (Non-technical Skills). These include skills in situational awareness; teamwork; decision-making and task and time management (Kodate et al. 2012). This research originated in medicine and has been applied to many industry sectors including manufacturing and mining where complex physical and cognitive demands are evident.

There are also more semi-skilled jobs required to operate the materials handling equipment in these industries as the product is moved through the logistics process. The skills of the workers need to include a functional understanding of the technology as well as problem solving and troubleshooting to understand how the technology is to be operated safely. The ergonomic requirements for this work involves a greater requirement for cognitive and psychosocial skills and less demand on the physical capacity to sustain repetitive production work.

In the office sector the use of hand held technology and high quality Wi-Fi has enabled the nature of work to change significantly. The design of offices is no longer restricted to an allocation of a workstation to each worker but to provide a range of work settings for focussed and collaborative work. Workers are not limited to having to sit at a workstation to access their computer, information or undertake their allocated tasks. Workers are now able to access their laptop or handheld device in multiple settings within and outside their office workplace. This results in a major shift in the physical, cognitive and psychosocial demands of their work. They are able to stand up and walk to collaboration areas, breakout spaces and a range of meeting areas to conduct their work with colleagues. This provides for greater physical and cognitive variety in their work practices.

The design of these systems of work requires the ergonomist to work as part of multidisciplinary teams to ensure that the work settings suit the proposed work practices for the workers. This starts in the concept development stage when the work organizational structures are resolved and identify the neighbourhoods or areas where groups of workers will be positioned. The detailed design of the individual work settings for the proposed activities are then developed in collaboration with the architect and designers, in consultation with the workers. This provides an opportunity for the physical ergonomics requirements to be incorporated into the design. The layout of the neighbourhoods with the "kit of parts" of work settings is then developed. These are assessed to ensure that the cognitive and psychosocial requirements for the work are integrated into the area designs. The use of the new technology enables a range of work settings that are for short and longer term use. For example casual settings with lounge chairs and couches are provided for casual use of technology for around thirty minutes at a time. Booths with tables and bench couches are available for use for periods of around one hour and focus settings with sit to stand adjustable workstation are available for work that requires longer than four hours use. As the worker takes their technology with them, between these settings, they can maintain their productivity whilst changing the ergonomics demands of their workplace.

It is evident that these new ways of working requires a change management strategy to ensure that the workers understand the objectives of the changes and the impact on their roles. Prior to relocation of workers to these new workplaces it was found that pilot sites, with examples of the new ways of working, be established. Groups of workers should be provided with the opportunity to work in and explore these settings for a few weeks. This is to enable them to experience the use of these settings and to have a holistic understanding of the physical, cognitive and psychosocial impacts on their work styles. Without the opportunity to trial and appreciate the different ways of working the workers in the new offices are more likely to try to replicate their previous ways of working.

5 Impact of Findings

It is evident that the impact of new technology on workers involves changes to the physical, cognitive and psychosocial demands of their work. The business case to invest in the new technology is primarily related to improving productivity, reducing long-term costs and injuries, as well as improving process or product quality. The ergonomics impact of new technology on the worker is not seen as a primary driver for the investment. However the sustainability of the new technology is dependent on the workers embracing the new technology and utilizing it to provide a safe system of work.

Within the context of introducing new technology in manufacturing on supply chain businesses such as supermarkets, system designers require an "end to end" understanding if the various steps in the supply chain and how the product is handled. The potential advantages of using robots during the manufacturing of products is limited in the supply chain if the downstream businesses are not capable of handling the palletised goods.

In offices the introduction of new ways of working with new technologies and new work settings is also limited if the workers are not aware of the physical, cognitive and psychosocial benefits of the new work settings. This also needs adjustment to management and supervisory practices to accept and encourage the use of the dynamic and diverse work place options. The potential for them to merely replicate their previous ways of working is evident if the change management process into the new work settings is not well managed.

The long-term impact of new technology and changing ways of working require further research to assess the physical and psychological impact on the workers and identify optimum change management practices to ensure the transition and outcome is efficient and effective.

Reference

Kodate N, Ross A, Anderson J, Flin R (2012) Non-Technical Skills (NTS) for enhancing patient safety: achievements and future directions. Jpn J Qual Saf Health Care 7(4):360–370

Working Times Profiles of Hypermarket Workers

F. Cravo[1], T. P. Cotrim[1,2](✉), and J. D. Carvalhais[1,2]

[1] Ergonomics Laboratory, Faculdade de Motricidade Humana, Universidade de Lisboa, Estrada da Costa, 1499-002 Cruz Quebrada, Portugal
filipecravo10@gmail.com, {tcotrim,jcarvalhais}@fmh.ulisboa.pt

[2] CIAUD, Faculdade de Arquitetura, Universidade de Lisboa, Rua Sá Nogueira, Polo Universitário, Alto da Ajuda, 1349-055 Lisboa, Portugal

Abstract. Increasing liberalization of the sector, especially in opening hours with extended days and Sundays opening augmented working times in unsocial hours and irregular schedules with consequences for workers health and safety. This study aims at characterizing the profile of hypermarket workers based on working time in order to alert for the consequences of the unadjusted work time organization with respect to sleep, sleepiness, satisfaction and work-family conflict. It was based on a cross-sectional questionnaire applied between November 2016 and January 2017. The sample consisted of 289 workers (response rate of 70.3%). Our sample showed a majority of women (70.5%), a mean age of 38 years old, with 21.7% above the 45 years old. Most of the workers had a 40 h full-time working period (70.1%). The shifts' organization is based mostly on different fixed shifts (77.8%). In the early morning and night shifts the median duration of hours of sleep was lower, with 5.5 h of sleep for those in shifts starting from 4:00 h to 5:00 h and 6 h for those in shifts starting from 6:00 h to 7:30 h or from 22:00 h to 24:00 h and the perception of adverse levels of work-family conflict was higher. For the night shift sleepiness was more frequent/very frequent (24.3%). Those starting working at 7–7:30 h and 22–24 h were the groups with higher frequency of dissatisfaction with working schedules.

Keywords: Working time · Hypermarket workers · Sleep

1 Introduction

The retail sector accounts for a considerable share of the EU economy [1]. This sector consists of a variety of shop size and formats, but in this study we were focused on a hypermarket, located in the city centre. Increasing liberalization of the sector, especially in opening hours with extended days and Sundays opening augmented working times in unsocial hours [1] and irregular schedules. These workers are part of the great percentage of the population in urban economies that is required to work outside the regular 08:00 h–17:00 h working day [2]. This requirement is met by nonstandard

S. Bagnara et al. (Eds.): IEA 2018, AISC 821, pp. 92–100, 2019.
https://doi.org/10.1007/978-3-319-96080-7_13

working schedules. These schedules include shift work (6 pm–6 am), extended working hours (>8 h), extended working weeks (>5 days), on-call duties, weekend work, and combinations thereof [3]. The increase in shiftwork has led to greater flexibility in work schedules, the ability to provide goods and services throughout the day and night, but with considerable negative effects on sleep loss, health and productivity [2, 4].

This study aims at characterizing the profile of hypermarket workers based on working time in order to alert for the consequences of the unadjusted working time organization with respect to sleep, sleepiness, satisfaction and work-family conflict.

2 Methodology

This exploratory and descriptive study was based on a cross-sectional questionnaire applied between November 2016 and January 2017.

2.1 Participants

The population consisted of 411 workers. The sample included 289 workers, corresponding to a response rate of 70.3%.

The workers were contacted personally, and informed about the study design and its objectives. The questionnaire was applied further to an informed consent explaining that participation in the study was voluntary and anonymous.

2.2 Questionnaire

The questionnaire was developed according to the study objectives. It integrated the socio-demographic characteristics and the variables related to working time, such as sleep duration, satisfaction with the schedules, sleepiness and work-family conflict.

Sleep duration was measured by a question about the number of hours slept after the shift based on the concepts used by the Standard Shiftwork Index (SSI) [5] and the Survey of Shiftwork (SOS) [6].

Satisfaction with the shift system was evaluated adopting the 5 point rating scales used in the Questionnaire REQUEST [7]. The satisfaction scale ranged from “very dissatisfied” to “very satisfied”.

Sleepiness was characterized, by asking participants to report the frequency of sleepiness at work during the shift from “very seldom or never” to “very often or continuously”. The item used a 5-point scale based on the study of Härmä et al. (2002) [8].

The Portuguese medium version of the Copenhagen Psychosocial Questionnaire II (COPSOQ II) [9] was used to assess the work-family conflict.

3 Results

3.1 Socio-demographic Characteristics

Our sample showed an average age of 38 years (sd = 10), with 39.5% below the 35 years old. There were a higher percentage of women (70.5%), married (47.4%), non-smokers (68.5%) and 63.5% did not practice exercise regularly (Table 1).

Table 1. Socio-demographic characteristics.

Socio-demographic characteristics		N	%
Age groups	<25	39	13.6
	26–35	74	25.9
	36–45	111	38.8
	46–55	48	16.8
	56–65	14	4.9
	Total	286	100
Gender	Female	203	70.5
	Male	85	29.5
	Total	288	100
Civil Status	Single	113	39.1
	Married/Partnership	137	47.4
	Divorced	38	13.1
	Widowed	1	0.3
	Total	289	100
Regular exercise practice	No	183	63.5
	Yes	105	36.5
	Total	288	100
Smoking habits	Yes	62	21.7
	No	196	68.5
	Ex-Smoker	28	9.8
	Total	286	100

Regarding the circadian type, 67.8% were morning and 14.8% evening type (Table 2). The workers belonging to the evening type were the younger group with a mean age of 33 years (sd = 11) and those with morning type were the oldest (mean = 40; sd = 8).

3.2 Working Time, Sleep Characteristics and Sleepiness

Concerning the duration of work, 70.1% worked 40 h/week. The different schedules were mostly fixed with 13.6% starting working from 4:00 h to 6:30 h, 14.9% from 7:00 h to 7:30 h, 15.6% from 8:00 h to 9:30 h, 7.3% from 10:00 h to 13:00 h, 13.2% from 15:00 h to 16:00 h, 7.3% from 17:00 h to 18:00 h, 5.9% from 22:00 h to 24:00 h and 22.2% with rotating shifts (Table 3).

Table 2. Circadian type and age by circadian type.

Circadian type			Age (yrs)		
	N	%	Min-max	M	SD
Morning	192	67.8	18–60	40.3	8.2
Intermediate	49	17.3	18–61	34.1	11.7
Evening	42	14.8	18–61	33.0	11.4
Total	283	100	18–61	38.0	9.9

Table 3. Working time characteristics.

Working time		N	%
Weekly work duration (in hours)	<20 h	3	1.0
	20 h	29	10.0
	30 h	45	15.6
	40 h	202	70.1
	>40 h	7	2.4
	Total	288	100
Schedules (starting time)	4 h–5 h	6	2.1
	6 h–6:30 h	33	11.5
	7 h–7:30 h	43	14.9
	8 h–9:30 h	45	15.6
	10 h–11:30 h	17	5.9
	13 h	4	1.4
	15 h–16 h	38	13.2
	17 h–18 h	21	7.3
	22 h–24 h	17	5.9
	Rotating	64	22.2
	Total	288	100

When analyzing the median age by schedule (starting time), the youngest group worked mainly in the afternoon and night shifts (Fig. 1).

The sleep duration average was 6.6 h (sd = 1.3), with a range from 3 h to 12 h, globally. After the shift, the median of hours of sleep was lower for the early morning and night shifts: 5.5 h of sleep for those shifts starting from 4:00 h to 5:00 h in the morning and 6 h for the shifts starting from 6:00 h to 7:30 h in the morning or from 22:00 h to 24:00 h (Fig. 2).

The perception of sleepiness was not high, but appeared to be more frequent/very frequent among those on night shift (24.3%) and starting the shift at 13 h (15.8%) (Fig. 3).

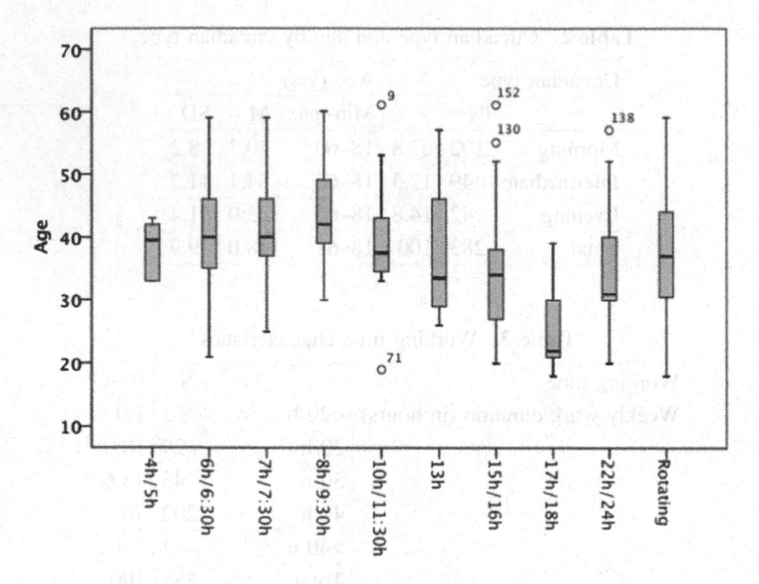

Fig. 1. Age distribution by shift starting time.

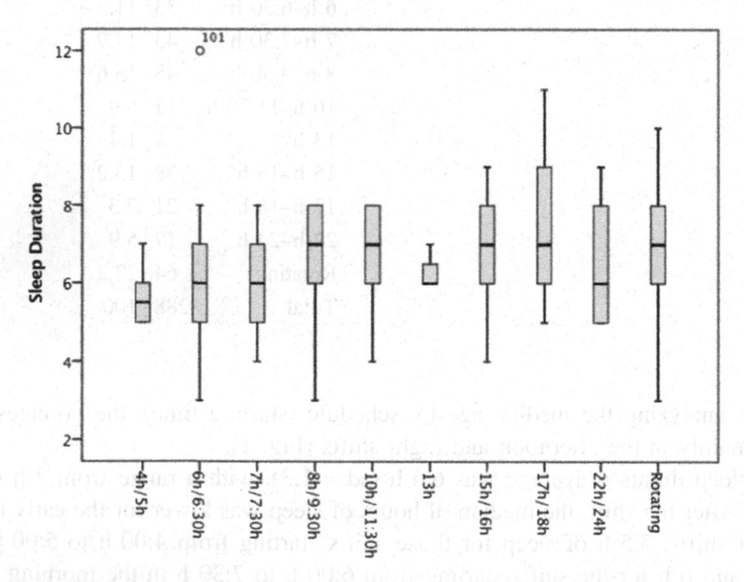

Fig. 2. Distribution of sleep duration in hours by shift starting time.

3.3 Satisfaction with Working Schedules and Work-Family Conflict

Generally, almost half of the workers were satisfied/very satisfied (46.7%) with the working schedules; over one-third was neutral (34.6%). By schedule, those starting working at 7–7:30 h and 22–24 h are the groups with higher frequency of dissatisfaction (respectively, 23.8% and 29.4%) (Tables 4 and 5).

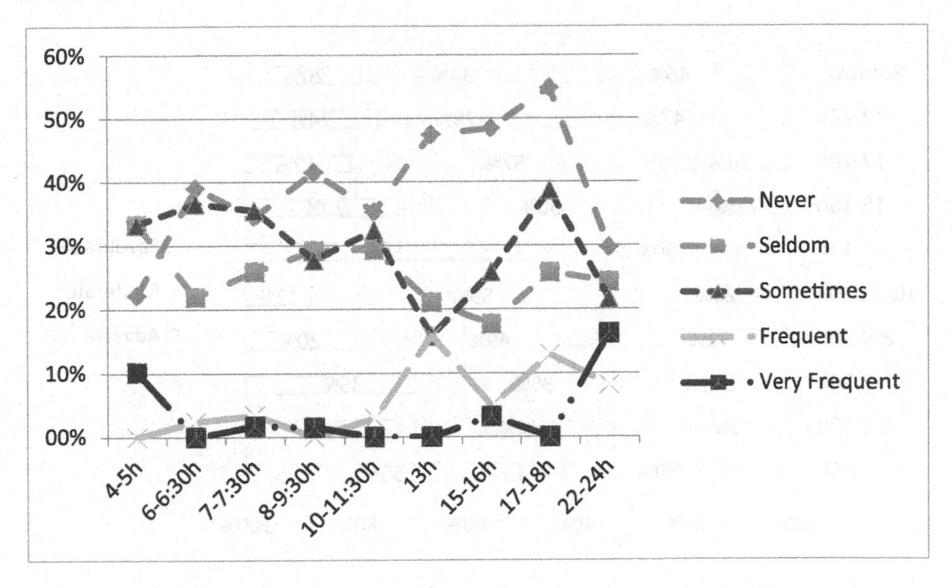

Fig. 3. Distribution of sleepiness by starting time.

Table 4. Satisfaction with the working schedules in the morning period.

Satisfaction with working schedules	Total		4 h–5 h		6 h–6:30 h		7 h–7:30 h		8 h–9:30 h		10 h–11:30 h		13 h	
	N	%	N	%	N	%	N	%	N	%	N	%	N	%
Very satisfied	42	14.5	0	0	4	12.1	4	9.5	15	33.3	2	11.8	0	0
Satisfied	93	32.2	0	0	15	45.5	17	40.5	11	24.4	6	35.3	1	25
Neutral	100	34.6	5	83.3	13	39.4	11	26.2	12	26.7	6	35.3	1	25
Dissatisfied	33	11.4	1	16.7	1	3.0	8	19.0	3	6.7	1	6.7	1	25
Very dissatisfied	21	7.3	0	0	0	0	2	4.8	4	8.9	2	11.8	1	25
Total	289	100	6	100	33	100	42	100	45	100	17	100	4	100

Table 5. Satisfaction with the working schedules in the afternoon period.

Satisfaction with working schedules	Total		15 h–16 h		17 h–18 h		22 h–24 h		Rotating	
	N	%	N	%	N	%	N	%	N	%
Very satisfied	42	14.5	3	8.1	4	19.0	2	11.8	6	9.5
Satisfied	93	32.2	10	27.0	6	28.6	5	29.4	20	31.7
Neutral	100	34.6	16	43.2	8	38.1	5	29.4	23	36.5
Dissatisfied	33	11.4	4	10.8	2	9.5	2	11.8	10	15.9
Very dissatisfied	21	7.3	4	10.8	1	4.8	3	17.6	4	6.3
Total	289	100	37	100	33	100	42	100	45	100

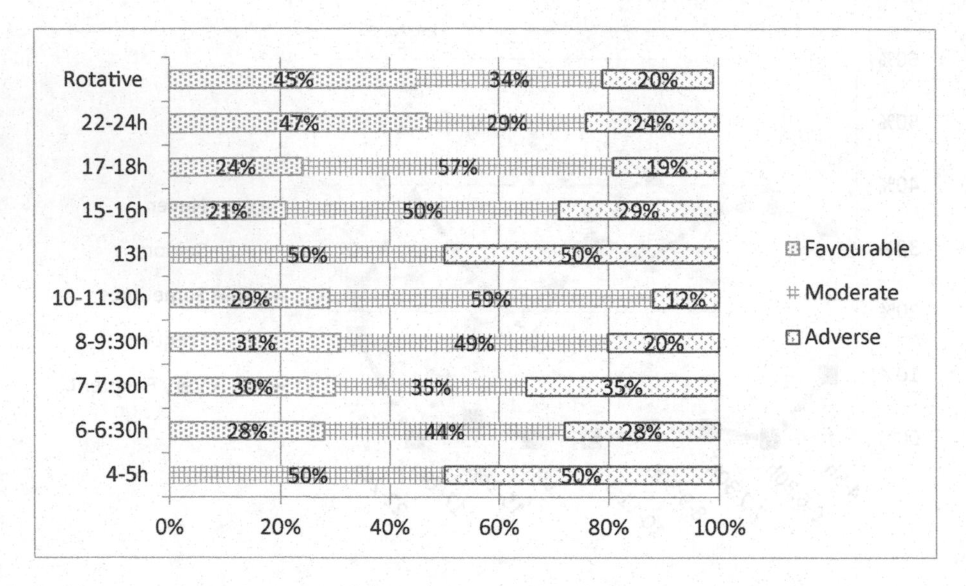

Fig. 4. Distribution of work-family conflict by starting time.

The perception of work-family conflict as an adverse factor showed higher percentages among those with early morning and early afternoon shifts (Fig. 4).

4 Discussion

Our sample showed a majority of female workers (70.5%), a mean age of 38 years old, with 21.7% above the 45 years old, what is in line with the European retail sector [1]. In 2010, the retail sector displayed a moderate ageing profile, with a share of employed aged over 50 of 20.9%, especially among women (20.7%) [1].

With respect to weekly worked hours, most of the workers had a 40 h full-time working period (70.1%), but part-time work is also common, with 15.6% working 30 h per week and 11% 20 h or less. In 2010, the European retail sector showed a share of part time workers in commerce of 22.2%, used to allow businesses to match staffing to peak footfall days and hours [1]. Part-time workers are more likely to have a second job what will influence their working time and sleep patterns.

The shifts' organization is based mostly on different permanent shifts, corresponding to 77.8% of the workers with a range of schedules varying from early morning shifts, until night shifts. This distribution is slightly different (with a lower percentage of rotating shifts) from the national results in 2010, when Portugal presented almost a third of retail workers (30.2%) working in rotating shifts [1]. The early morning shifts, starting at 4:00–6:30 h, are performed by 13.6% of the workers, while the night shifts, starting at 22:00–24:00 h, by 5.9%. It is in the early morning and night shifts that the median duration of hours of sleep was lower. Our results are similar to those found in other studies [4, 10] with 5.5 h of sleep for those in shifts starting from 4:00 h to 5:00 h in the morning and 6 h for those in shifts starting from 6:00 h to

7:30 h in the morning or from 22:00 h to 24:00 h. The perception of sleepiness is usual in night work or early morning work [4, 11]. In our study, the night shift sleepiness was more frequent/very frequent (24.3%), in spite of the youngest groups working mainly in the afternoon and night shifts.

By schedule, those starting working at 7–7:30 h and 22–24 h were the groups with higher frequency of dissatisfaction with working schedules (respectively, 23.8% and 29.4%). Some studies point that the level of satisfaction reflects on how well the shift workers are coping with the schedules and may impact on the level of sleepiness [12].

Early morning permanent shifts presented also a higher percentage of workers perceiving adverse levels of work-family conflict. The interference of the working time with the rhythms of social life is a well-known problem common to shift workers but also to early morning or afternoon permanent shifts [13].

One of the main limitations of the study was the low number of participants in some of the fixed shifts.

5 Conclusions

Liberalization in opening hours has been introduced in Portugal since 2011 with an allowed opening period from 6 am to 12 pm that impacted the working times. Our study showed a wide range of schedules varying from early morning shifts, until night shifts, with 77.8% of the workers in permanent shifts. The early morning and night shifts had a higher impact on the reduction of median sleep duration and sleepiness was perceived as frequent/very frequent (24.3%) for the night shift. In the early morning shifts the perception of work-family conflict was higher.

These results should alert the organizational managers for the risks related to some schedules that determine a shortening in sleep duration. Chronic short sleep duration is commonly related to health and safety problems [4] that must be addressed by a working time management program.

References

1. Eurofound. (2012) Working conditions in the retail sector, Dublin
2. Rajaratnam SM, Arendt J (2001) Health in a 24-h society. Lancet 358(9286):999–1005
3. Merkus SL, Holte KA, Huysmans MA, van Mechelem W, van der Beek AJ (2015) Nonstandard working schedules and health: the systematic search for a comprehensive model. BMC Public Health 15(1084):1–15
4. Kecklund G, Axelsson J (2016) Health consequences of shift work and insufficient sleep. BMJ 355(i5210)
5. Barton J, Spelten E, Totterdell P, Smith L, Folkard S, Costa G (1995) The standard shiftwork index: a battery of questionnaires for assessing shiftwork-related problems. Work Stress 9(1):4–30
6. Folkard S, Spelten E, Totterdell P, Barton J, Smith L (1995) The use of survey measures to assess circadian variations in alertness. Sleep 18(5):355–361
7. Ryan B, Wilson JR, Sharples S, Morrisroe G, Clarke T (2009) Developing a rail ergonomics questionnaire (REQUEST). Appl Ergon 40(2):216–229

8. Härmä M, Sallinen M, Ranta R, Mutanen P, Müller K (2002) The effect of an irregular shift system on sleepiness at work in train drivers and railway traffic controllers. J Sleep Res 11:141–151
9. Silva C, Amaral V, Pereira A, Bem-Haja P, Pereira A, Rodrigues V, Cotrim T, Silvério J, Nossa P (2012) Copenhagen Psychosocial Questionnaire - COPSOQ - Portugal e países africanos de língua oficial portuguesa. In: Fernandes da Silva C. (ed.) 1st edn. Análise Exacta. Universidade de Aveiro
10. Pilcher JJ, Lambert BJ, Hufcutt AI (2000) Differential effects of permanent and rotating shifts on self-reported sleep length: a meta-analytic review. Sleep 23:155–163
11. Knauth P (1996) Designing better shift systems. Appl Ergon 27(1):39–44
12. Axelsson J, Åkerstedt T, Kecklund G, Lowden A (2004) Tolerance to shift work - how does it relate to sleep and wakefulness? Int Arch Occup Environ Health 77(2):121–129
13. Demerouti E, Geurts SE, Bakker AB, Euwema M (2004) The impact of shiftwork on work-home conflict, job attitudes and health. Ergonomics 47(9):987–1002

Challenges of Telework in Brazil: A Sociotechnical Analysis

Lígia de Godoy[(✉)] and Marcelo Gitirana Gomes Ferreira

Universidade do Estado de Santa Catarina – UDESC, Florianópolis, Brazil
{ligiadegodoy,marcelo.gitirana}@gmail.com

Abstract. Amongst several transformations of our society, new paradigms of work emerge, also driven by advances in technology, as in the case of telework. The fact that it is still a recent modality in Brazil raises some concerns about human factors, which is aggravated by the lack of specific legislation and recent changes in labor laws. The variety of teleworking factors makes the subject complex and results in sparse and sometimes contradictory studies. Sociotechnical systems, objects of study of organizational ergonomics, encompass aspects that can be summarized in the technical, personnel, organizational and environmental subsystems. This study aims to identify telework challenges in Brazil, through the analysis of the current scenery, based on a sociotechnical approach. The antecedents and outcomes of telework in Brazil were raised through a multidisciplinary bibliographical research, being analyzed through a sociotechnical based teleworking framework. Common points were identified in the publications, mainly in relation to the telework outcomes for the workers, being the most discussed subject among the authors. Although reflecting research from other countries, aspects arising from the external environment can be noticed. The research has shown a lack of publications that attest to the antecedents and outcomes of telework, as well as research done under a specific bias. This demonstrates the need for sociotechnical studies of telework in Brazil, integrating aspects of all subsystems. In future studies, this can be done empirically, being guided by sociotechnics, using the sociotechnical based teleworking framework.

Keywords: Telework · Brazil · Sociotechnical systems

1 Introduction

As a result of many technological, economic and social transformations of our society, new paradigms of work emerge. The main impetus for the emergence of these new paradigms came mainly in the 1990s, when, according to Kumar [1], the nature of work was changing with incredible speed. The democratization of the access to microcomputers and the ease of distance communication in real time, through the internet, were determining factors for the evolution of the paradigms of work and, in particular, for the consolidation of telework. Telework, a work carried out remotely with the support of information and communication technologies, has become popular throughout the world, being a growing modality also in Brazil. The fact that it is a recent modality has

S. Bagnara et al. (Eds.): IEA 2018, AISC 821, pp. 101–111, 2019.
https://doi.org/10.1007/978-3-319-96080-7_14

been raising discussions about the implications of teleworking for workers, in relation to human factors. In Brazil, especially, the lack of specific legislation and recent changes in labor laws aggravate concerns regarding teleworkers and should be considered in the study of telework. It is also perceived, because it is a complex subject, often sparse studies, conducted by a specific and sometimes contradictory bias. Sociotechnics aims to integrate the various aspects related to work, studying the so-called sociotechnical subsystems, namely: personnel, technical, organizational and environmental. The aim of this research is, therefore, to identify telework challenges in Brazil, through the analysis of the current scenery and the main factors that act as antecedents or outcomes of teleworking, based on a sociotechnical analysis. The present study was carried out through a bibliographical review of telework in Brazil, permeating the four basic sociotechnical subsystems, guided by a sociotechnical based teleworking framework.

2 Organizational Ergonomics and Sociotechnical Systems

Telework, as a new paradigm of work, is one of the objects of study of Organizational Ergonomics, which is related to the optimization of sociotechnical systems, encompassing organizational, political and process structures [2]. According to Hendrick [3], the sociotechnical systems theory sees the organization as an open system with permeable boundaries. The theory points to four main subsystems: the personnel subsystem; the technological subsystem; the external environment or environmental subsystem and the work system design or organizational structure subsystem.

Hendrick [3] indicates three main characteristics of the personnel subsystem that must be analyzed: the degree of professionalism; demographic characteristics and psychosocial aspects of the workforce. The author reinforces the interaction between the personnel subsystem and the technological subsystem. While the technological subsystem designs and defines the tasks that must be performed, the personnel subsystem must be designed to describe the ways in which these tasks should be performed. Also on the technological subsystem, Hendrick [3] states that technology has been defined in several distinctly different ways:

(1) By the mode of production, or production technology; (2) by the action individuals perform on an object to change it, or knowledge-based technology; (3) by the way it reduces uncertainty, or strategy for reducing uncertainty; and (4) by the degrees of automation, workflow rigidity, and quantitative specificity of evaluation of work activities, or workflow integration. (p. 45)

The organizational structure subsystem, or work system design, is usually conceptualized in three main dimensions: complexity, formalization, and centralization. Complexity indicates the degree of differentiation and integration of the work system structure, and formalization indicates the degree to which jobs or tasks are standardized. Centralization is related to the way the formal decision-making in the work system occurs. As an open system, organizations must have the ability to adapt to the external environment. Some relevant aspects of the external environment that can positively or negatively influence the sociotechnical subsystems are: socioeconomic aspects; educational aspects; political aspects, cultural aspects and legal aspects.

Hendrick [3] states that the four sociotechnical subsystems are mutually interdependent, and that a change in one of them affects all others. The authors reinforce the influence of the environmental subsystem, or external environment, influencing the other subsystems through causal events, which the author calls "joint causation". Another concept brought by Hendrick [3] is that of "joint optimization", which seeks to articulate the subsystems, so that none of them is maximized individually.

3 Telework

According to the International Labor Office [4], telework can be defined as a: "form of work in which (a) work is performed in a location remote from central offices or production facilities, thus separating the worker from personal contact with co-workers there; and (b) new technology enables this separation by facilitating communication."

The International Labor Office [4] also mentions the difficulty in conceptualizing telework, since it is usually studied from the perspective of several areas, besides the constant emergence of new modalities of work. Because it is a complex concept, and fragmented in several fields, there are several variables to consider in the conceptualization of telework. Nicklin et al. [5] presented a framework that categorizes telework in the following aspects: proportion: part to full time; location: fixed to mobile; schedule: fixed to varied; collaboration: low to high; synchrony: serial to concurrent; autonomy: low to high.

The evolution of information and communication technologies has also enabled the practice of geographically dispersed work teams, that is, distributed in different locations, often countries. The dispersed work is, according to Corso et al. [6], the result of a convergence of two factors: spatial dispersion and contractual dispersion. The authors point out that, in addition to the workers in the dispersed model must present some characteristics in their profile, such as autonomy, traditional management systems do not meet their needs, representing, therefore, a challenge for the organization. In a bibliographic review, Hertel, Geister and Konradt [7] suggest some principles for the management of virtual or dispersed teams, namely: clear goals and team functions; careful implementation of efficient communication and collaboration processes; ongoing support for team awareness, with sufficient feedback and information on the work situation of each member; creating experiences of interdependence among the team; development of trainings to prepare the team for specific challenges of the virtual work team.

4 Telework in Brazil

According to IBGE demographic data, raised in the last census in 2010, approximately 23% of Brazilians employed in Brazil work in the same place where they live, representing around 20 million people [8]. The Home Office Brazil survey [9], conducted by SAP Consulting with the institutional support of SOBRATT (Brazilian Society of

Telework and Tele-activities), collected data on telework from the study of 325 companies, in the period of October 2015 to March 2016. Of the companies surveyed, 68% already had the practice of teleworking in several modalities, and 37% adopted home-office telework, however, only 61% of these companies adopted the practice formally. Among the companies studied, 93% were private companies and 7% were public. Despite the small number of public companies surveyed, there is a rapid expansion of teleworking in the Brazilian public administration. In the judiciary, the practice is regulated by the CNJ - National Council of Justice, through Resolution 227/2016. After almost a year of the regulation of the practice by the judiciary, it is estimated that up to June 2017, almost 2% of the employees of the labor sphere had already joined the modality [10].

One of the reasons for concerns regarding teleworkers in Brazil is the lack of specific legislation that regulates teleworking. There is in article 6 of the Consolidation of Labor Laws - CLT an article, updated in 2011, which states that there is no distinction between the work performed in the employer's establishment, the one performed at the employee's home and the one performed at a distance, if the employment relationship assumptions are characterized [11]. However, Law 13.467 of 2017 [12] amended the CLT text and began to include teleworkers in Article 62 of the CLT, which regulates workers who are not subject to working hours control. In practice, this has removed from workers the right to overtime pays, which are bonuses they receive for working beyond set times, as well as other work-related amounts. In addition, some specific provisions on teleworking have been included. The text indicates that issues regarding the activities to be performed and the responsibility for the acquisition and maintenance of equipment and infrastructure should be regulated by individual contracts of work. Likewise, the text provides for the instruction of the worker as to the precautions to be taken to avoid accidents and diseases arising from work, registered from a term of responsibility signed by the worker [12], but not disposing of ways of evaluation and control of these recommendations.

5 Method

To analyze the antecedents and outcomes of telework in Brazil, empirical publications attesting to these factors were sought. First, the Scopus and Web of Science databases were searched. As these databases returned few results, due to the need for the surveys to be Brazilian, then searches were made at the Scielo and Google Scholar databases. Articles published in journals and doctoral theses were included, excluding those that contained only bibliographic reviews. The research was not restricted by the study area, being found articles of several areas, such as administration, ergonomics or psychology. Only the articles that were accessible by the university were included, filtering them to analyze if they were suitable for research, that is, if they allowed to evaluate the antecedents and/or outcomes of telework. In addition to the results of the bibliographical survey, information about Brazilian legislation and sociodemographic data were searched, in order to raise characteristics of the external environment that may influence the telework outcomes. To present the research results and facilitate the discussion, a framework adapted from Bélanger, Watson-Manheim and Swan [13] was used. The framework (Fig. 1) indicates the antecedents, divided between the personnel, technical and organizational structure subsystems, and the outcomes of telework, analyzed according to three levels, individual, group and organizational. The framework was adapted to indicate possible external environment interference at all levels. It is important to emphasize that the framework was adapted and used for the presentation of the research results, seeking to show a perspective of the current telework scenery in Brazil for discussion purposes, but it cannot be used to identify causal relationships between specific antecedents and outcomes, since the information presented comes from different organizations.

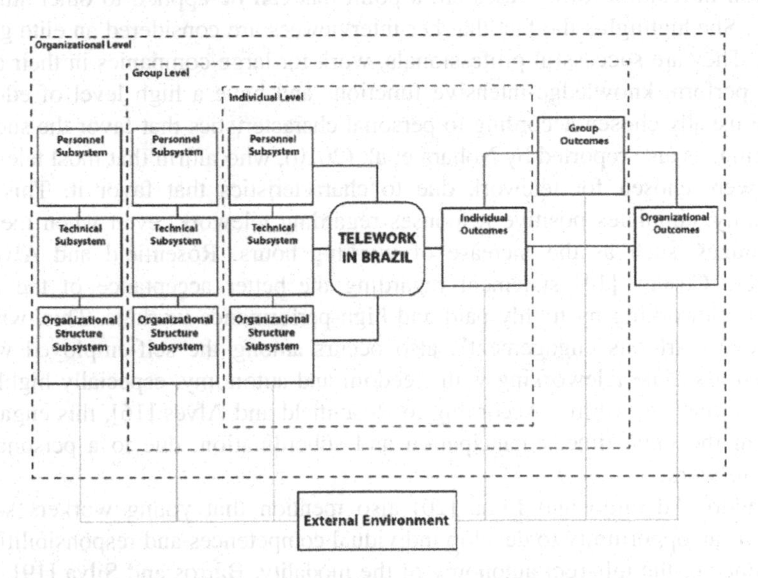

Fig. 1. Sociotechnical based teleworking framework (Source: Adapted from Bélanger, Watson-Manheim and Swan, 2013)

6 Telework in Brazil: Results and Discussion

It is noticed, among the Brazilian publications, a greater tendency to approach teleworking from the perspective of the personnel subsystem, followed by the organizational subsystem. Costa [14] sought to contribute to the understanding of telework, with a focus on workers perceptions, as well as Nohara et al. [15], with home-office workers who perform tasks characterized by the author as knowledge intensive. Rosenfield and Alves [16] conducted research through interviews with teleworkers in Lisbon (Portugal) and three cities in Brazil, in the following modalities: wage earners; self-employed or independent workers and call center attendants, making a comparison between wage earners and self-employed workers, the same comparison made by Rafalski and Andrade [17]. Boonen [18], Barros and Silva [19] and Aderaldo, Aderaldo and Lima [20] conducted case studies of a single company, the last two studies being directed at Brazilian subsidiaries of multinational companies. Aderaldo, Aderaldo and Lima [20] analyze the perception of teleworkers in relation to telework as a practice of human resources, being the only work that focused on the research with interns. Alves [21] studies the technologies as mediators of the changes in the work, interviewing Brazilian and Portuguese teleworkers. Mello [22] seeks to identify aspects of home-office teleworking in call center and contact center companies through an organizational approach. The authors Pereira Junior and Caetano [23] bring a counterpoint between the perceptions of teleworkers and conventional workers on teleworking.

The workers studied by Costa [14] perceive flexibility and autonomy in the organization of work as the greatest benefits of teleworking. The author makes a reservation in relation to her research, a point that can be applied to other studies on telework. She highlights the fact that her interviewees are considered an elite group of workers. They are successful professionals, work for large companies in their areas of activity, perform knowledge-intensive functions and have a high level of education. They are usually chosen according to personal characteristics that favor the success of teleworking, as also reported by Nohara et al. (2010), who affirm that most teleworkers studied were chosen for telework due to characteristics that favor it. This profile occasionally stimulates positive responses regarding telework, even when mentioned disadvantages, such as the increase of working hours. Rosenfield and Alves [16] corroborate Costa's [14] statement regarding the better acceptance of the adverse effects of teleworking by highly paid and high-performance workers. This, which the authors call "virtuous engagement", also occurs among the self-employed workers, who often associate teleworking with freedom and autonomy, especially highlighting freedom in the use of time. According to Rosenfield and Alves [16], this engagement reflects, at the same time, emancipation and subordination, due to a personal commitment to work.

Aderaldo, Aderaldo and Lima [20] also mention that young workers see teleworking as an opportunity to develop individual competences and responsibilities, as a consequence of the inherent autonomy of the modality. Barros and Silva [19], as well as Pereira Junior and Caetano [23], point out among the main advantages reported by teleworkers: better perceived quality of life, especially in relation to the family; autonomy and flexibility in work organization. Barros and Silva [19] consider that, because they live in the two largest Brazilian metropolises, a good part of the interviewees gives importance to the fact that they do not need to go to the organization when teleworking.

Among the publications compiled, flexibility and autonomy are perceived as the great benefits of teleworking for the worker. Rosenfield and Alves [16], however, put the issue of autonomy as a paradox: while the worker has the autonomy to manage his own time, he has no control over it, since working time is determined by the volume of work and by the deadlines demanded. The authors consider telework as a more flexible than autonomous modality, since it is subordinated to the demands of organizations, in the case of wage earners, or of the market, in the case of self-employed workers. Alves [21] corroborates this view by mentioning a supposed transfer of control, where control, which would usually be centralized and external to the worker, becomes the self-control of the worker, so that he can meet the stipulated deadlines. According to the author, productivity is now linked to delivery of the service in the stipulated deadline, no longer to working time.

Alves [21] also indicates a decrease in deadlines because of the power of information and communication technologies to accelerate tasks, bringing simultaneity and instantaneousness in communications. Rosenfield and Alves [16] talk about the strategy adopted by the organizations to "stretch" the time, virtually, to intensify the work and reduce the teams. Consequently, an issue recurrently brought by teleworkers, both in Brazil and in other countries, is the amount of hours worked, which in Brazil can be aggravated by the latest changes in the labor law. In Boonen's [18] survey, about 97% of teleworkers claimed to work more than eight hours a day, which among traditional workers, those working in the organization environment, accounted for only 57%. Among teleworkers, 49% also stated that they work more than 13 h a day, which does not appear among traditional workers, probably due to labor legislation. The excess hours also appeared in the studies of Costa [14] and Alves [21], which verified the excess hours worked between the two modalities of work studied, wage earners and self-employed teleworkers.

Costa [14] also mentions other reported disadvantages, such as cost transfer, personal and professional isolation and the invasion of family space by work. Interviewees in this study mention some conflicting situations, such as physical presence and, at the same time, non-availability to family members, self-control needed and pressure to present results. Among other common outcomes, the inseparability between space/time of work and of nonwork is a recurrent question in the research of Rosenfield and Alves [16], being related to a series of adversities reported. Alves [21] also points out that the workers themselves make strategies to overcome this inseparability, by creating artificial controls of working hours and space, physically delimiting the work environment and dressing formally, as if they were in the organization's environment, for example. Barros and Silva [19] also point out as a disadvantage the need for self-control on the part of the worker, already mentioned by other authors as a crucial aspect.

Focusing on organizational issues, Mello [22] points out the main reasons for the companies to adopt telework, according to the managers: cost reduction; increased productivity; compliance with Law 8.312, which determines quotas to be filled by people with disabilities, with the consequence of strengthening the company's social responsibility program. As a telework important antecedent, the author also points out the acquisition by companies of telecommunications technologies that allow the monitoring and management of activities. Because Mello's research [22] deals with an activity with quite singular characteristics, which are call center and contact center services, some aspects are specific to these activities. For example, the author points out, as barriers, the difficulty of finding people with disabilities qualified for teleworking and the resistance of client companies to home-office teleworking because of information security. Some managers studied by Mello [22] point out, as restrictive forces, the infrastructure of workers in the residential environment, which may be deficient, also loss of control of managers and lack of specific labor legislation. As an effective contribution to the company, managers mention measured productivity, cost reduction, pollutant reduction and social inclusion.

The antecedents and outcomes from the publications were organized with the support of the presented framework, through which it is possible to visualize a scenery of telework implications in Brazil (Fig. 2). The aspects that were mentioned isolated, but which are considered relevant, were placed in gray and the outcomes considered negative were presented on a gray background.

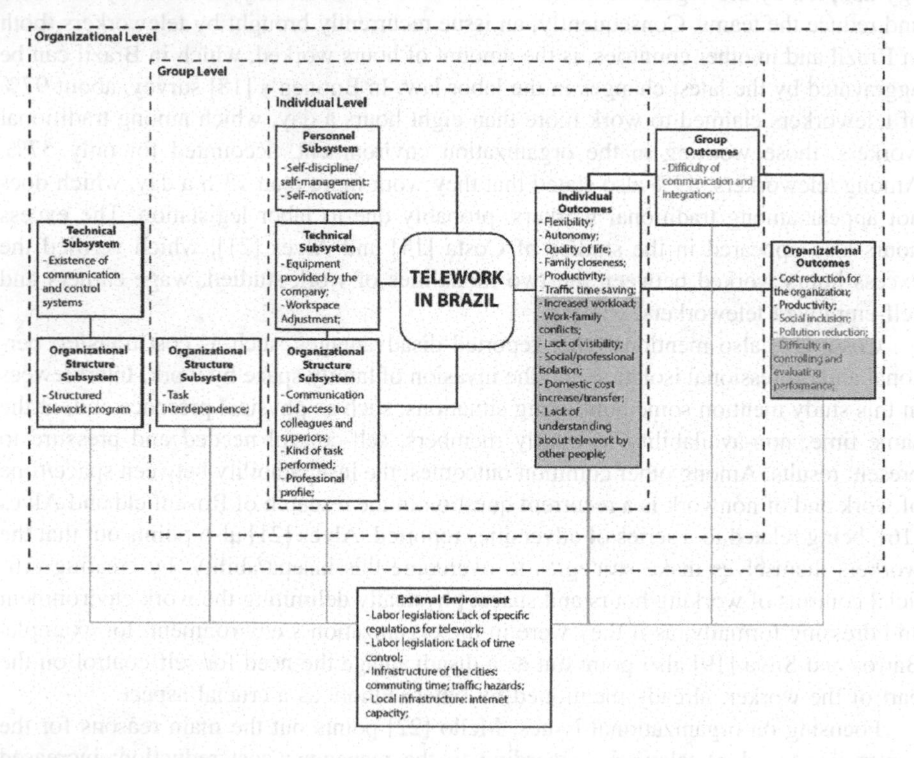

Fig. 2. Telework in Brazil (Source: The authors)

Among the selected publications, only one studies telework from the organizational point of view, although some of them mention aspects of the organization. Most publications highlight a specific subsystem, not analyzing the interrelationship between the subsystems, which could help identify the causes of the generated outcomes, as well as result in improvements in the work system. The analysis of telework from a specific perspective, the organization or the teleworker, can generate some inconsistencies as identified in the research in relation to the cost. The transfer of costs in teleworking appears from two different ways in the outcomes, in different publications: from the organizational point of view, the authors indicate the reduction of costs as a teleworking advantage [17, 22], and from the point of view of the workers, it is pointed out a possible transfer of costs to the worker [14], referring to fixed costs, which can be understood as those that would be reduced in the organization.

In general, individual outcomes reflect research on teleworking in other countries [24], with some specificities related to the external environment, that is, to Brazil's own characteristics. In Brazil, telework represents challenges for workers, as the modality demands personal characteristics that support its inherent flexibility, as well as the balance between personal life and work. In relation to the latter, the responsibility for the adequacy of the work environment can also be pointed out as a challenge when equipment and financial aid are not provided by the organization. For the organization, it is possible to point out the need of adapting the management and control models to this new modality and its specificities, such as the performance of activities at distance and the flexibility of schedules. The organization must also do its best to overcome the individualization of the teleworker, seeking integration in the work teams. Another challenge that should be given attention by the organization is the possible personal and professional isolation of the worker, besides the feeling of lack of visibility. The organization must therefore provide support to avoid such outcomes. The influence of the external environment also poses challenges, especially in relation to legislation. Considering the legal dispensation of control of working hours, it is also up to the organization to manage the work of these workers to avoid excess hours worked. Also in relation to the external environment, the infrastructure of cities can be considered a challenge, in relation to the quality of the internet, for example, which can influence the performance of telework.

7 Conclusions

The present study proposed to identify telework challenges in Brazil, through the analysis of the current scenery, based on a sociotechnical perspective. For this, a multidisciplinary bibliographic review was carried out, and the analysis was guided by a framework adapted from other authors. The framework was used to synthesize the antecedents, related to the personnel, technical and organization structure subsystem, and the outcomes of telework, according to three levels, individual, group and organizational. The research returned few scientific publications that attest to the antecedents and outcomes of telework in Brazil, besides they often analyze teleworking from a specific bias. In the selected publications, the authors, in general, presented common points, mainly in relation to the outcomes perceived by the teleworkers. Although some aspects reported also appear in researches of other countries, the influence of environmental factors, or of the environmental subsystem, was perceived in both the antecedents and the outcomes analyzed. The compilation of antecedents and outcomes from the publications allowed the identification of possible challenges of telework practice in Brazil, both for workers and for the organization. The scarcity of publications that attest to the implications of telework in a judicious way, coupled with the lack of specific legislation on the modality, raise a possible demand for research on the subject. Future studies could analyze telework from a sociotechnical perspective, in Brazil, in an empirical way, also being guided by the sociotechnical based teleworking framework.

Acknowledgments. The authors gratefully acknowledge UDESC and CAPES for the financial support that made this research possible.

References

1. Kumar K (1997) Da sociedade pós-industrial à pós-moderna: Novas teorias sobre o mundo contemporâneo. Jorge Zahar, Rio de Janeiro
2. IEA: Definition and domains of ergonomics. https://www.iea.cc/whats/. Accessed 02 Oct 2017
3. Hendrick HW (2002) Macroergonomic methods: assessing work system structure. In: Hendrick HW, Kleiner BM (eds) Macroergonomics: theory, methods, and applications. Lawrence Erlbaum Associates, Mahwah, pp 45–66
4. ILO: Challenges and opportunities of teleworking for workers and employers in the ICTS and financial services sectors. International Labour Office, Geneva (2016)
5. Nicklin JM, Cerasoli CP, Dydyn KL (2016) Telecommuting: What? Why? When? and How? In: Lee J (ed) The impact of ICT on work. Springer, Singapore, pp 41–70
6. Corso M, Martini A, Pellegrini L, Massa S, Testa S (2006) Managing dispersed workers: the new challenge in knowledge management. Technovation 26:583–594
7. Hertel G, Geister S, Konradt U (2005) Managing virtual teams: a review of current empirical research. Hum Resour Manag Rev 15:69–95. https://doi.org/10.1016/j.hrmr.2005.01.002
8. IBGE: Censo Demográfico. http://www.ibge.gov.br/home/estatistica/populacao/censo2010/default.shtm. Accessed 02 Oct 2017
9. SOBRATT: Pesquisa HOME OFFICE BRASIL 2016: Teletrabalho e Home Office. http://www.sobratt.org.br/index.php/11-e-12052016-estudo-home-office-brasil-apresenta-o-cenario-atual-da-pratica-no-pais-sap-consultoria. Accessed 02 Oct 2017
10. CNJ: Normatizado há um ano, teletrabalho agrada tribunais e servidores. http://www.cnj.jus.br/noticias/cnj/84854-normatizado-ha-um-ano-teletrabalho-agrada-tribunais-e-servidores. Accessed 15 Oct 2017
11. Brasil: Lei no 5.452 (1943) Consolidação das Leis do Trabalho. Rio de Janeiro
12. Brasil: Lei no 13.467 (2017) Altera a consolidação das leis do trabalho (CLT). Brasília
13. Bélanger F, Watson-Manheim MB, Swan BR (2013) A multi-level socio-technical systems telecommuting framework. Behav Inf Technol 32:1257–1279
14. Costa ISA (2007) Teletrabalho: subjugação e construção de subjetividades. Rev Adm Pública 41:19. https://doi.org/10.1590/s0034-76122007000100007
15. Nohara JJ, Acevedo CR, Ribeiro AF, Silva MM (2010) O teletrabalho na percepção dos teletrabalhadores. RAI–Rev Adm e Inovação 7:150–170
16. Rosenfield CL, Alves DA (2011) Autonomia e trabalho informacional: O teletrabalho. Dados - Rev Ciências Sociais 54:207–233 (2011). https://doi.org/10.1590/s0011-52582011000100006
17. Rafalski JC, Andrade AL (2015) Home-office: aspectos exploratórios do trabalho a partir de casa. Temas em Psicol 23:431–441. https://doi.org/10.9788/tp2015.2-14
18. Boonen EM (2002) As várias faces do teletrabalho. E G Econ e Gestão 2/3:106–127
19. Barros AM, Silva JRG (2010) Percepções dos indivíduos sobre as consequências do teletrabalho na configuração home-office: estudo de caso na Shell Brasil. Cad. EBAPE.BR. 8:71–91. https://doi.org/10.1590/s1679-39512010000100006
20. Aderaldo, I.L., Aderaldo, C.V.L., Lima, A.C.: Aspectos críticos do teletrabalho em uma companhia multinacional. Cad. EBAPE.BR. 15:511–533 (2017)
21. Alves DA (2009) Tecnologias como mediadores das mudanças sociotécnicas no teletrabalho. TOMO 15:143–165
22. Mello AAA (2011) O uso do teletrabalho nas empresas de call center e contact center multiclientes atuantes no Brasil: estudo para identificar as forças propulsoras, restritivas e contribuições reconhecidas, USP

23. Pereira Junior E, Caetano MES (2009) Implicações do teletrabalho: um estudo sobre a percepção dos trabalhadores de uma região metropolitana. Rev Psicol Organ e Trab 9(2):22–31
24. Gajendran RS, Harrison DA (2007) The good, the bad, and the unknown about telecommuting: meta-analysis of psychological mediators and individual consequences. J Appl Psychol 92:1524–1541. https://doi.org/10.1037/0021-9010.92.6.1524

Participatory Macroergonomics Study – Schools and Kindergartens as Shared Workplaces

Päivi Kekkonen[✉] and Arto Reiman

Well-Being at Work and Productivity Research Team,
Industrial Engineering and Management, University of Oulu, Oulu, Finland
paivi.kekkonen@oulu.fi

Abstract. Currently, many service organizations encounter challenges that set new requirements for management: individual employees face changes to worksites, job tasks, and work communities while there is a simultaneous decrease in recruitment and increase in the average age of employees. Both physical and psychosocial burdens caused by these factors can lower the work ability and productivity of the employees. The aim of this study was to find solutions for the management of these load factors in workplaces where stakeholders from different subdivisions inside the municipal organization work together. The concept of a shared workplace, which is common in industry, was contemplated to find successful ways to manage work ability and productivity. The case organization in this study was a municipal business unit providing meal and cleaning services to target workplaces, namely two kindergartens and four schools. The objective of the study was to find practical solutions for observed challenges related to work environment and practices at the target workplaces. Study materials were comprised of Occupational Safety and Health documents and statistics, interviews, and observations of work activities. Root cause analysis, by applying the 5*Why-methodology, was carried out to find ultimate causes for the work ability challenges. Practical solutions for the challenges were sought at participatory development sessions. Based on the results, a generalizable model for the management of load factors at shared workplaces in the public sector was proposed.

Keywords: Participatory development · Shared workplace · Work ability

1 Introduction

1.1 Background

Public organizations face various challenges in the present economic situation in Europe. One main challenge is related to work life itself. Working life is in a continuous process of change and jobs and work tasks are constantly examined and restructured for optimization purposes. While the average age of employees is increasing, there is a simultaneous decrease in recruitment due to budgetary reasons. The common goal of the municipal processes being able to provide a sense of well-being at different

S. Bagnara et al. (Eds.): IEA 2018, AISC 821, pp. 112–121, 2019.
https://doi.org/10.1007/978-3-319-96080-7_15

levels may become unclear. These factors cause both physical and psychosocial burdens that can lower an employee's work ability and productivity (Reiman et al. 2017).

In this article, we focus on a participatory development process in which both macroergonomic and microergonomic development needs (Hendrick and Kleiner 2001) were identified inside a municipal organization. More specifically, our study targeted workplaces where stakeholders from different subdivisions inside the municipal organization work together. In industry, such workplaces can be identified as shared workplaces (Väyrynen et al. 2012). This study was carried out as part of the "Work ability management at a shared workplace" project which sought solutions for the management of harmful load factors at work. An important point in the project was to examine if the concept of a shared workplace could be useful for finding means to successfully manage work ability and productivity at municipal workplaces.

1.2 General Concepts

Finnish legislation defines the shared workplace as a workplace where one employer exercises main authority and other employees of several employers or self-employed individuals work simultaneously or successively so that their work can affect other workers' safety and health. In this case, the employers and self-employed individuals are required by mutual cooperation to make sure that their actions do not endanger the health and safety of any employee at the workplace (Occupational Safety and Health Act 738/2002).

During recent decades many organizations have increasingly outsourced many operations, especially various support operations, in order to focus on their core business (Nenonen 2012). This has increased the occurrence of both shared workplaces and situations that are in many ways similar to shared workplaces.

According to many studies, allowing employees to participate in the development of their own work increases both well-being and productivity at work (van Eerd et al. 2010). The concept of participatory design and development covers all design and development activities where the users of the systems or products are systematically involved in and actively provide input into the design and development process. For example, common design tools are different kinds of models and prototypes, analysis methods, check-up lists and various discussion, ideation and group work techniques (Langford and McDonagh 2003). Participatory design aims to extract and utilize the tacit knowledge possessed by the users or employees (Spinuzzi 2005).

1.3 Objectives

The case organization in this study was a municipal business unit that provided meal and cleaning services. The case organization and its' work ability management processes supplemented with economic analyses have been reported earlier by Reiman et al. (2017). In their research, further studies for workplace level were issued as a topic for future research. In this study, we respond to this research challenge by providing an in-depth study of two kindergartens and four schools to whom this business unit provided its services. In our study, the subject workplaces were considered shared workplaces where different actors interacted to provide services aimed at educating children and youngsters. Our special interest was to find microergonomic Occupational Safety and Health (OSH) development needs and to provide a macroergonomic development process in which solutions for these challenges are provided.

A prerequisite for our study was to include all stakeholder groups working at the target workplaces in the development process. Even though the business unit providing the meal and cleaning services was our initial observation unit, we expanded our study to cover all relevant actors at these workplaces. This included the employees, management, and OSH actors from all sectors. By engaging these stakeholder groups in collaborative work, management practices were brought forward to further spread best practices, healthier work environments, and improved work communities. The principles of continuous improvement were also present throughout the participatory process. Furthermore, macroergonomics were emphasized as cooperation processes and organizational practices inside the shared workplaces and inside the municipal organization were assessed. A wider goal of the study was to formulate a generalizable model for the management of harmful load factors at shared workplaces.

The research questions of the study were the following:

1. What kind of physical and psychosocial load factors occur at shared workplaces of a municipal organization from the viewpoint of meal and cleaning service employees?
2. What kind of practical solutions can be found to reduce the load factors?
3. What kind of a management model could be used to manage work ability at such shared workplaces?

2 Methods

The materials for this study were gathered from six workplaces (two kindergartens and four schools). The kindergartens and schools that acted as target workplaces were considered shared workplaces although the researchers identified that these workplaces did not necessarily fulfill all the requirements for a shared workplace according to legislation since all the employees at these workplaces worked under one employer (municipality). Figure 1 presents the actors in one of the schools. The students and their guardians are in the center surrounded by various actors, including the employees of the school and the meal and cleaning services organization. Some of the actors work permanently at the school, while some carry out work tasks on the school premises irregularly or are important stakeholders in relation to the operations carried out at the school but primarily work elsewhere. The actors in the outer ring formed the group of stakeholders that were engaged in the participatory development process depicted in the following discussion.

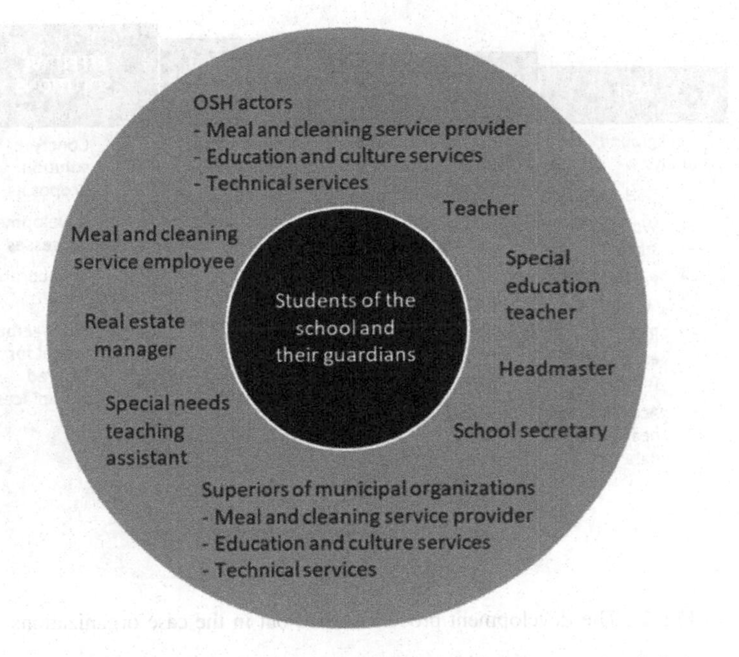

Fig. 1. The actors in one of the schools that acted as a target workplace.

The kindergartens and schools acting as target workplaces were chosen so that they would form a representative sample of workplaces where the meal and cleaning service employees of the municipal case organization work. The target worksites could be characterized by the following properties (the acronyms for the target workplaces are used in the Results section):

- K1: Kindergarten operating in an old building.
- K2: Kindergarten operating in a new building.
- S1: School with a small number of students operating in a new building.
- S2: School with a large number of students operating in a new building together with many other operations, such as library, kindergarten, and youth work.
- S3: School with a large number of students operating in an old listed building.
- S4: School with a large number of students operating in an old building that was currently under renovation.

The participatory development process that was carried out in the study is presented in Fig. 2. After choosing the target workplaces and identifying the relevant stakeholder groups in each case, the development process was carried out in each target workplace over a nine-month period.

PAST	PRESENT	PARTICIPATORY DEVELOPMENT	EFFECTIVE SOLUTIONS
Background analysis: • OSH statistics • Workplace surveys • Personnel surveys • Risk assessment reports • Accident and near-miss data	Current situation at workplaces: • Observations • Interviews	• Root cause analyses • Participatory workshops	• Concrete solution proposals • Development processes • Collaboration models • Management model for "shared workplaces"

Fig. 2. The development process carried out in the case organizations.

In the background (PAST) and current status (PRESENT) analysis stages, multiple data sources were utilized. First, the past situation at the target workplaces was analyzed through workplace specific OSH statistics, accident and near-miss data, risk assessment reports, personnel surveys, and workplace survey reports. Second, the present work environments and practices were observed at the target workplaces. The observations were performed by following the cleaning and meal service personnel during their daily work routines for approximately 0.5 workdays per workplace. This was carried out by the researchers, in most cases in cooperation with the OSH representative of the case organization. Personnel were also directly interviewed during the observations. Based on the analyses described above, researchers made tentative proposals for the most significant development targets for every workplace. The development targets were categorized by their origins into physical and psychosocial categories. Based on the past and present data, visualized descriptions of the problems were formed. These descriptions contained both written and visual information about the observed individual challenges, as well as their causes and possible solutions.

In the participatory development stage, the researchers followed the premises of Lean–oriented 5*Why–methodology to identify the root causes for the existing development targets (Ohno 1988). Basically, the question "Why?" was expressed several times for each identified development challenge. Participatory development sessions (workshops) comprised of personnel from all groups at the target workplaces, as well as other actors relevant to the operations at the workplace, were organized to identify development solutions for the identified challenges. The goal of the workshops was to engage all relevant stakeholder groups for each workplace in the development of the work. At each workplace, representatives of the employee groups, employers, and OSH personnel took part in the workshops and, at some workplaces, representatives of stakeholders functioning in the same building, such as library staff, youth workers, or caretakers, were present. Participatory development sessions resulted in practical ideas and sharing of best practices.

By working out concrete and practical solutions to the observed challenges, utilizing participatory design and the cooperation of the various stakeholders the development process aimed at designing better operational models and work practices, as well as creating safer and healthier work environments and communities. In addition, the idea of continuous improvement and engaging different stakeholders in the development of shared workplaces was illustrated in the development process. As a result, to contribute to organizational management processes, a simplified model for identifying and managing OSH problems at shared workplaces in the public sector was created for future use.

3 Results

The past and present phases carried out at the target workplaces provided material for the root cause analysis and following workshop phases. As an example, a visualization of the root cause analysis for the musculoskeletal disorders of the meal service employee working at a kindergarten is presented in Fig. 3. Similar analyses were carried out on the cleaning service employees, and also with problems related to psychosocial load factors. The participatory development sessions focused on finding solutions, especially to these root causes, that had been identified by the 5*Why – process.

The development challenges identified at target workplaces were divided into 17 separate themes. The number of practical challenges related to each theme in each kindergarten (K1–K2) and school (S1–S4), as well as the total number of solutions suggested in the participatory workshops, are listed in Table 1. The solutions suggested in the workshops included both new practices and good practices already in use at some workplaces that could be spread to other worksites by the different stakeholders taking part in the workshops. In addition to development challenges listed in Table 1, wider subjects for development that were often common for several worksites were recognized during the process. These included challenges in identifying and engaging all stakeholder groups in the initial foundational planning process of premises, promoting communality between the actors of the shared workplace, the demand and supply of knowledge related to ergonomics among the employees, and recognizing the mutual interest of the community in addition to the interest of individual actors or organizations.

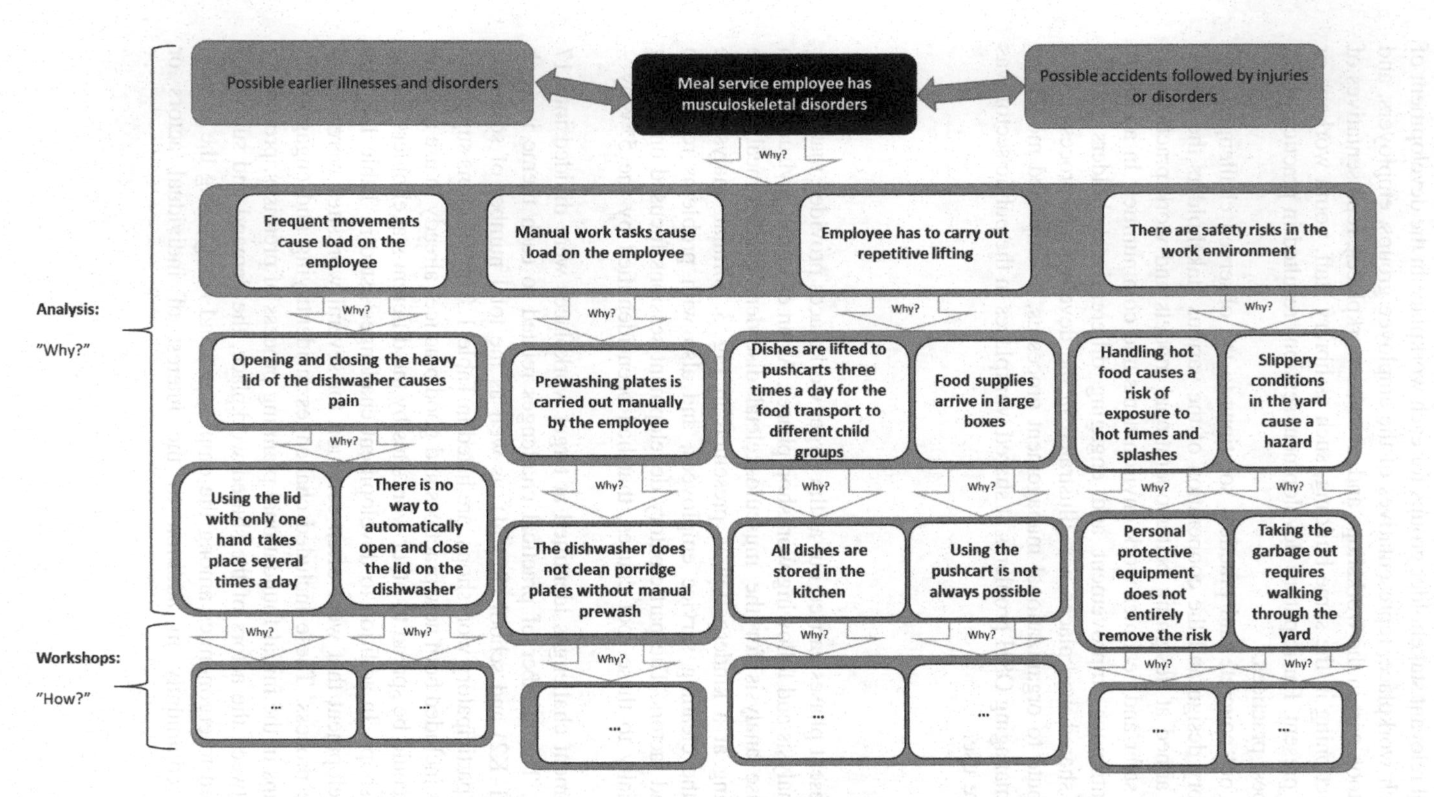

Fig. 3. As an example, the results of the root cause analysis conducted on musculoskeletal disorders caused by physical load factors in the work of the meal service employee working at a kindergarten.

Table 1. Number of development challenges identified at each worksite, along with new solutions and already identified good practices suggested in the workshops.

Development challenge	K1	K2	S1	S2	S3	S4	Solutions (N)
Premises and materials	–	–	2	2	6	1	5
Furnishings	3	–	1	2	4	2	8
Lifting	3	1	1	2	4	1	11
Work postures	2	1	3	2	2	2	7
Tools	2	1	–	2	4	1	4
Machinery	1	–	2	3	1	1	4
Tidiness of premises	–	–	–	2	2	3	2
Waste management	–	–	1	2	1	2	4
Ventilation	–	–	2	1	1	1	1
Hearing protection	–	–	1	1	1	1	3
Danger of slipping/tripping	1	1	1	2	4	1	6
Accident risk	2	–	–	4	4	2	8
Work clothing	–	–	2	1	1	2	1
Cooperation and flow of information	1	–	1	–	1	–	11
Interruptions and changes	1	1	2	1	4	2	–
Adequacy of time	1	1	3	3	3	1	5
Simultaneous actions	1	–	1	4	–	2	6

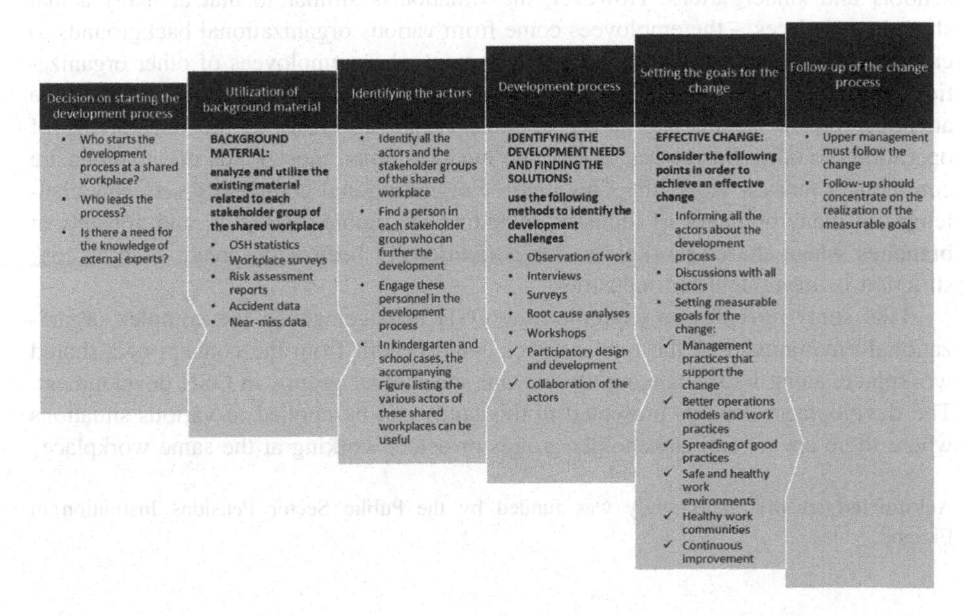

Fig. 4. The model for the management of harmful load factors at shared workplaces in the municipal sector.

4 Discussion and Conclusions

This study aimed at identifying physical and psychosocial load factors in the work of meal and cleaning service employees working at kindergartens and schools, as well as finding practical solutions to observed challenges in their work. The identified load factors were divided into 17 development challenge themes and solutions were found to almost all of these challenges in workshops that engaged all relevant stakeholder groups of these shared workplaces.

Furthermore, a generalizable model for the management of harmful load factors at shared workplaces in the municipal sector was formulated during the study. A model based on the analysis of background material related to the current OSH situation in the organization, observations carried out at the workplace, and participatory development workshops that engaged all relevant stakeholder groups was developed and determined to be functional. In addition to solutions to individual development challenges, the process also provided information on the management of larger subjects for development, which can be generalized to other similar municipal workplaces. The model is presented in Fig. 4.

In this research, the concept of a shared workplace was applied to a situation in the municipal sector, where a separate municipal business unit provided services to municipal organizations. This situation does not entirely fulfill the legislative definition of a shared workplace, since the municipality acts as an employer for both the employees of the business unit providing meal and cleaning services, as well as the schools and kindergartens. However, the situation is similar to that at many actual shared workplaces – the employees come from various organizational backgrounds to carry out their own tasks in a work environment where employees of other organizations simultaneously carry out their own work, in such a manner that their actions have an effect on the health and safety of other employees. Due to the outsourcing of operations to other businesses or separate business units, these kinds of situations are common in today's work life. This kind of organizational complexity sets new challenges at many branches of industry, including the municipal sector, and not only at branches where shared workplaces have traditionally been widespread, such as construction or manufacturing industries.

This study provides an example how OSH management in the complex organizational environment of the public sector could benefit from the concept of a shared workplace, along with engaging all relevant stakeholder groups in OSH development. The development process presented in this study can be applied to various situations where there are several stakeholder groups or actors working at the same workplace.

Acknowledgements. This study was funded by the Public Sector Pensions Institution in Finland.

References

Hendrick HW, Kleiner BM (2001) Macroergonomics – an introduction to work system design. Human Factors and Ergonomics Society, Santa Monica

Langford J, McDonagh D (eds) (2003) Focus groups: supporting effective product development. Taylor & Francis, London

Nenonen S (2012) Implementation of safety management in outsourced services in the manufacturing industry. Tampere University of Technology, Publication 1023, Tampere

Occupational Safety and Health Act (738/2002), Chap 6: Special situations of organising work. Ministry of social affairs and health. Unofficial translation. http://www.finlex.fi/en/laki/kaannokset/2002/en20020738.pdf. Accessed 4 Apr 2018

Ohno T (1988) Toyota production system: beyond large-scale production. Productivity Press, Portland

Reiman A, Ahonen G, Juvonen-Posti P, Heusala T, Takala E-P, Joensuu M (2017) Economic impacts of workplace disability management in a public enterprise. Int J Publ Sect Perform Manage 3(3):297–310

Spinuzzi C (2005) The methodology of participatory design. Tech Commun 52(2):163–174

van Eerd D, Cole D, Irvin E, Mahood Q, Keown K, Theberge N, Village J, St. Vincent M, Cullen K (2010) Process and implementation of participatory ergonomic interventions: a systematic review. Ergonomics 53(10):1153–1166

Väyrynen S, Koivupalo M, Latva-Ranta J (2012) A 15-year development path of actions towards an integrated management system: description, evaluation and safety effects within the process industry network in Finland. Int J Strateg Eng Asset Manage 1(1):3–32

Activity Analysis as a Method for Accompanying Industrial Craft Companies to Internal Changes and Market Challenges

Mariachiara Pacquola(✉) and Patrizia Magnoler(✉)

TINTEC Department Member, Macerata University, Macerata, Italy
cpacquola@hotmail.com, p.magnoler@unimc.it

Abstract. The competition that exists in today's markets increasingly calls for more flexibility and continuous changing of operating schemes depending on the type of products in question. The worker's creativity constitutes a resource that constantly regulates the actions to be carried out during the productive process, and that sometimes also endangers the psychophysical sustainability of the worker her/himself. This paper presents a research project which, by using Activity Analysis, Professional Didactics and Neo-Functional methodologies and tools, engaged the workers in a training process of self-development in collaboration with their colleagues and the company management, allowing the change of failed schemes of action, improving the wellness of the worker and the performance of the company.

Keywords: Activity analysis · Professional Didactics · Workplace learning

1 Introduction

The footwear craftsman was considered a figure with having full and complete autonomy when designing and creating an artefact, able to conceive and respect the intrinsic reasoning to the artefact functioning (Simondon 1989). His cognitive image (Ochanine 1978) allowed him to simultaneously have a general and detailed vision of the product when preparing for the manufacturing process. In this way he could control and balance the effect of the different variables that would regularly present themselves in a constantly different relationship with one another, from product to product. The introduction of a taylorist organisation for mass production of shoes in Italy in the 1950s led to the breaking down of the production process into specialised operations, automatically nullifying the system view of the process, the perception of the result and of its quality. The control was exteriorised, assigned to "responsible" production figures. The division in phases of the manufacturing process led to the de-pauperisation of the worker's power to act (Mendel 2002) and the effectiveness of the gesture as well as the pursuit of adopting the same actions in similar situations became the focus of the

Mariachiara Pacquola is the author of Sects. 3 and 4.
Patrizia Magnoler is the author of Sects. 1 and 2.

S. Bagnara et al. (Eds.): IEA 2018, AISC 821, pp. 122–132, 2019.
https://doi.org/10.1007/978-3-319-96080-7_16

process, in order to optimise the production time. Each micro-problem presented by the situation requires the mobilisation of a capacitative agency of the worker (Moretti 2013), a dose of creativity when it comes to inventing solutions, both for protecting the quality of the product and also for ensuring the necessary conditions of security and protection for the individual health (Dejours 2003). Tackling the continuous string of unexpected events and changes, requires the development of creative strategies, postures and gestures, in order to ensure the successful outcome of the specific action: at the same time, creativity, proved essential in the production phase, sometimes causes the development of both physical and psychological problems that could also potentially lead to mistakes.

In this paper, we present a pathway of research inspired by the theories of Professional Didactics (Pastré 2011) and Activity Analysis (Barbier and Durand 2017) performed in a luxury footwear manufacturing Small Medium Enterprise (SME). The working conditions were found to be regulated externally by a rigid taylorist work organisation, but work solutions were self-regulated by workers in order to keep up with the rhythm and with the task to be accomplished. Based on research results, it was possible to develop a training programme that could be used to support the workers in breaking their habitual operating processes and their views of the action that caused them physical disturbance and lowered their performances, and to help them to acquire new professional gestures sustained by a different self-perception and self-design.

2 Identifying the Status Quo

The study was carried out by a multi-disciplinary team with skills in research, professional training and organisational process analysis. The initial reconnaissance performed in the enterprise presented the following organisational structure: 50 operators spread along an assembly line which organises the various phases of footwear production; 3 managers who carry out continuous monitoring. When the production chain is working at maximum capacity, the assembly line is loaded with 750 pairs of shoes per day and this means that the worker has approximately 1 min to perform his/her own specific action, by taking into account his/her first diagnosis made on each semi-finished product; each day, many different shoes models can appear, requiring action schemes to be continuously redesigned (Vergnaud 1996)[1]. Faced with a similar diversity of situations in relation to the available time, the operators react by developing personal interpretations of the gesture, attempting to limit their physical and psychological disturbances, protecting their health and, at the same time, assuring that a good result is achieved, in aesthetic, functional and technical terms. The operational diversity among workers, associated with the evolving requests of brand's designers, increasingly focused on manufacturing products with new materials and shapes, automatically

[1] According to Vergnaud, the scheme is an invariant organisation of the action for a class of specific situations. It is composed of four components: an objective, the rules of action, the operating invariants and the theorems in progress (concepts deemed true by the party that helps direct and guide the action).

multiply the probability that errors as well as temporal and qualitative misalignment will occur during the production process.

The study began by analysing the work organisation, and subsequently focused on a particular block of operations of the assembly line, in which operators join the top part of the shoe (last and upper) to the lower half parts (heel and sole). The workstations entailed in this part of assembly process (micro-line) are those responsible for marking, roughing, gluing and pressing.

The training intervention requested by the SME was based on 3 improvement objectives: to increase the ability of workers to learn other professional gestures, in order to make those transferrable to other stages of the micro-line; to reduce the number of errors when shoes leave the micro-line (the purpose of the experience discussed herein); to identify and train operators capable of undertaking the responsibility to monitor the progress of the work.

The second phase aimed at understanding the state-of-art was developed by using the techniques envisaged by activity analysis[2]: the researcher video-recorded and subsequently analysed each operation together with the operator who performed it (simple self-confrontation of Theureau 2006) and among those who execute the same professional gesture (crossed self-confrontation of Clot, 2008)[3]. A more in-depth study was performed by reconstructing the course of action (Theureau 2006; Pacquola et al. 2013), accompanying the operator to explicit the information on which s/he bases his diagnosis during the assembly process (I perceive that…), the development of the decision s/he takes (then I will do….), the realisation (what I am doing…) and the immediate reflection on what has been done (assessment of the adequacy of the action). In particular, the activity analysis on the first phase of the micro-line, the marking phase, highlighted the lack of a reference standard for aligning the production of the three people assigned to this specific phase: the different practices adopted did not always generate the expected quality production. The demand for quantity/quality of the task, coupled with the speed of the assembly line, led the operators to resiliently develop creative, self-adjusted solutions, even if these were not sufficient to ensure a constant product quality or the worker wellness.

Based on the activity analysis, the following operational working models emerged. S., with 13 years of experience, self-trained, has in turn trained C., who has been working in the company for 6 months, adopting a simple apprenticeship model, (observation and explanation). For both women, the action scheme performed was as follows: take a last and the relative sole (weight 400 to 500 g) from the production chain cart; combine sole and last; turn them round; keep them pressed tightly together supporting them by laying on the stomach; mark with a technical pen the edge of sole and last, interrupting the line in correspondence with the fingers of the other hand, which at the same time keep the last-sole pressed. The stress of the muscles of the left

[2] Activity analysis is the focus of the Activity Clinic, a process designed to favour the development of the power to act by the subject and of a professional group in various situations.

[3] The aim of these techniques is to show how the action is performed and the reasons behind it. It allows the worker to also consider other possible ways of performing the same action. While in simple self-confrontation, there is only one worker who analyses their own action, in cross-self-confrontation, several workers are present.

arm, folded and blocked against the body to support the weight and to improve the operator's grip on the last-sole, was linked to the compromised, contracted movements of the right hand, which was prevented from making fine, accurate marks; finally, the lack of an adequate distance between the gaze and the right arm and hand, led the operator to draw a blurry, non-continuous line that went beyond the tolerance margins of the edges of the sole (2 mm). This kind of action caused two fundamental problems: a) it increased the risk of mistakes, which were not immediately detected by the operators of the following workstations, b) it also caused evident physical problems. C. has had to undergo surgery on the left carpal tunnel; S. reported a chronic state of stress, inflammation and pain in her left arm, loss of sensitivity in her left hand, circulation problems in her arms and gastrointestinal problems. To protect herself from the pressure of the shoe on her stomach to be marked, S. wore a special apron with a specific type of padding, which she had designed and made herself.

Z., with 8 years-experience in another company, knowed how to perform the marking operation using the correct professional tool, the ast peg, but did not do it correctly: she turns her chest rather than turning the last anchored to the pin with her hand (Fig. 1).

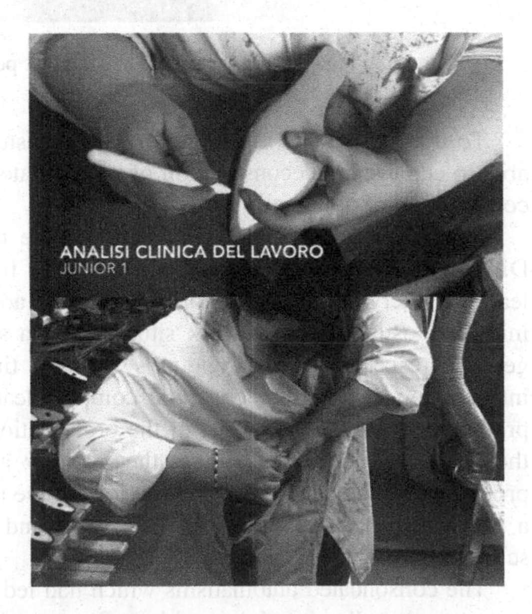

Fig. 1. Operating models of S. (left) and C. (top right) and Z. (bottom right).

3 The Training Intervention

To identify a model of action that would serve as reference for aligning the marking practices, according to the advice of the researcher-trainer, the company sought an external expert who, when video-recorded as he performed the operation on the ast peg, showed the correct marking gesture to make. The simple self-confrontation of the expert displayed and easily communicated the rules of the action and the series of movements to be adopted to improve both the quality of the shoe and the working wellness: take one last at a time; turn it upside down (180°); place it on to the rotating tip of the ast peg using the hole provided; take the corresponding sole, lay it on to the last, align the edges of sole and last and pressing down evenly with one hand to hold it still; using a technical pen, mark the edge of the sole on the upper, first on one side of the last and then, by pivoting it on the mobile point of the ast peg and turning it 180°, on the other side, without letting the sole slip off the last (Fig. 2).

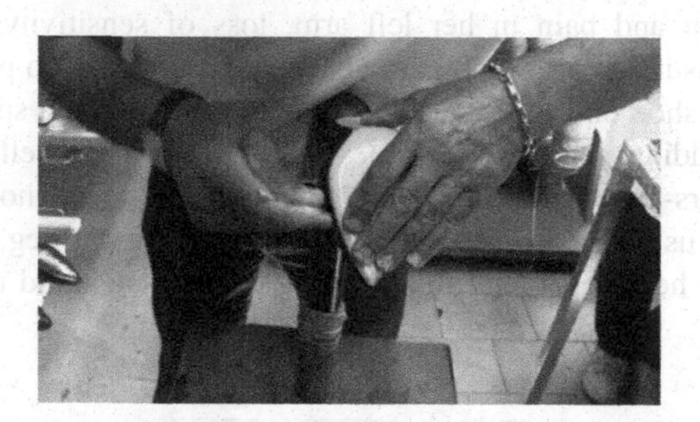

Fig. 2. The correct use of the ast peg in the marking phase

To train on a more functional marking gesture, the trainer had to design devices able to achieve the deconstruction of consolidated action schemes, and the intentional construction of new ones.

In designing the training intervention, the theoretical references of Professional Didactics were fundamental[4]. These provide frameworks and tools for developing learning interventions in workplace situations adopting a pedagogical progression: an initial exposure to simple work situations in a simulated environment (acquiring the gesture, respecting the individual's learning time, using simple models); then an introduction to progressively more complex learning situations by gradually adding production variables (elements of the organisation regarding the task itself, deadlines, the speed of the assembly line, the quantity and complexity of shoe models, the production peaks), and finally, the actions in the real workplace. The goal was to reach a different conceptualisation of the action, and then to the building of new action schemes replacing the previous ones.

The consolidated automatisms which had led to postural imbalances and muscular compensation allowing the repetition of gestures for long periods but causing physical problems, had to be deconstructed by re-awakening the workers' sensitivity to perceptions, movements and sight-hand coordination. In order to allow the worker to "unlearn" a professional gesture and to activate a professional process of self-design, it

[4] Professional Didactics is an inter-disciplinary theoretical perspective that revisits the beliefs of Developmental Psychology (Vergnaud 1996), Didactics of the Disciplines (Brousseau 1998) and of French ergonomic Psychology (Ombredane and Faverge 1955; Leplat 1997). It is founded on the theories of Piaget and uses the analysis of the action as a basis for designing training devices.

was necessary to identify which capabilities (Nussbaum 2001) – abilities gained through growth and with an appropriate external educational influence along the stages of development- had been altered by the harmful performance. The analysis of the professional gesture within the theoretical framework of Neo-Functionalism (Rispoli 2016) helped to identify the basic Functions behind the automatic action and to design adequate interventions for re-integrating the perceptive and motor skills. By observing the worker, interviewing her/him and filling out a form on the quality and quantity-based aspects of the task (functional diagram), the researcher-trainer assessed the basic modalities used by the worker in her/his approach to the working activity, such as her/his sensory and perceptive capacitance, strength, consistency and control. The alteration level was assessed based on the hyper/hypotrophy, the lack of connection, the rigidity of the functions in the four psycho-corporeal dimensions: the postural one (wideness of the movements, the mobility of the postures); the physiological one (vagotonic or sympathicotonic respiration, pain thresholds…), the emotional one (anxiety, fear, calmness…), the cognitive one (conceptual capacity, awareness, developmental imagination…) (Pacquola et al. 2013).

Given the diversity of the cases identified during the activity analysis, three different types of training intervention have been designed.

3.1 Perfecting the Professional Gesture (Z.)

The training intervention with Z. (Fig. 3) was based on a session in which the new professional gesture was simulated in a classroom situation, driven by a simultaneous simple self-confrontation on the expert's video. Later, a crossed self-confrontation between the expert and Z. on the Z.'s video took place. Finally, the acquired performance was assessed at the workplace.

At the end of the intervention, which lasted a total of 2 h in a period of time spanning two weeks, Z. succeeded in integrating the plan of action with a distal rotation movement of the ast peg pin (new arm-hand coordination) maintaining the posture of the neck-shoulders-back straight. The expert, the SME's managers and trainers positively assessed the learning result and engaged Z. as a tutor in the process of reinforcing the skills of the other two students in the production workplace, identifying her as a future senior of the micro-line.

3.2 Learning a Professional Gesture (C.)

The training intervention with C. envisaged an assessment of her usual performance and a group session (Z., S., C.) designed to teach C. and S. about the new working station layout (equipped with the ast peg) and the new professional gesture. Subsequently a simple self-confrontation was organised to favour her awareness of her own operating model, confronted with the expert one. The rapid re-conceptualisation and acquisition of autonomy in accomplishing the activity (after a total of 3 h of training) enabled the operator to proceed to the phase of consolidation in a production place, directly on the assembly line, accompanied by the tutor (Fig. 4).

Fig. 3. The training intervention of Z.

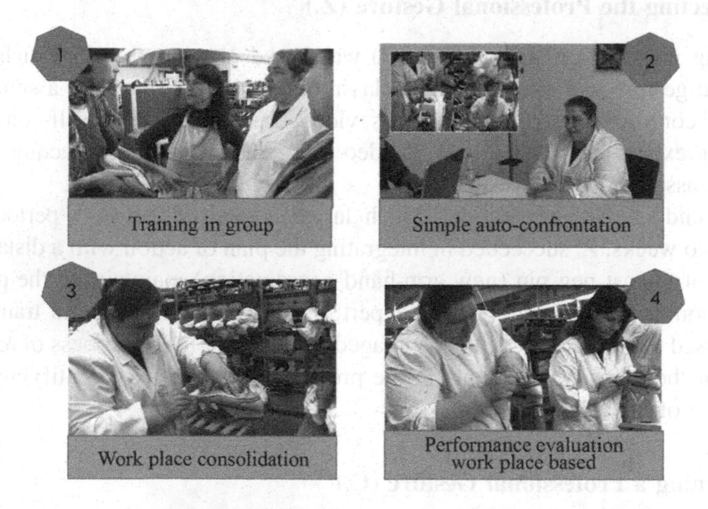

Fig. 4. The training intervention of C.

3.3 Unlearning a Professional Gesture (S.)

S. was experiencing a situation of chronic work-related stress, worsened by the results of a number of traumas caused by a car accident. At the end of each day she had to do an hour of stretching to relieve the pain in several parts of her body. She was showing signs of burnout, lack of interest in her work and a feeling of inefficiency, for which she compensated with hyperactivity and by accelerating her movements.

During the group session, when acquiring the new method, S. expressed the fear of losing her grip of the shoe because she felt that her hand was too small to press down hard enough on the ast peg with enough force to hold it still. The new action of pressing down with her hand spread out was also painful for her (Fig. 5). The habitual gesture of placing the shoe on her stomach and holding it with her arm and hands reassured her that she was not likely to make any mistakes and allowed her to control her anxiety.

To design of experiences integrating perceptive and motor skills for S. implied the reconstruction of several functions of the hand that the deeply-seated automatism of the gesture had reduced, and at the same time required to support the development of new self-perceptions and self-beliefs. The tasks of the trainer therefore focused, in parallel: on extending the range of possible movements of the hand, perceived as small because it was usually contracted; on the perception of the strength effectively required to perform the new gesture; and on the worker's belief that it was possible to overcome a situation of discomfort. Firstly, the trainer guided the worker through the learning experience, by taking the action out of the assembly line context and, in a separate environment, asking the worker to imitate the gesture (Figs. 6 and 7) leading by example.

Fig. 5. The arm, shoulder and neck of S. are still hypertonic and the hand is contracted and closed.

Fig. 6. Contracting-Leaning Range. S. gets her shoulder and arm used to lean her weight on the supported hand.

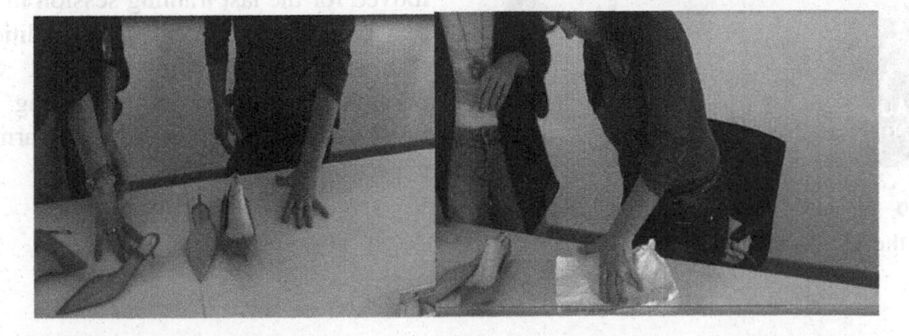

Fig. 7. Closing-Stretching Range. S. gets her hand used to making stretching movements, by smoothing out a piece of paper.

Once she had learned the new movements, she was able to integrate them into a sequence of actions, trying out the professional gesture on an ast peg simulated by the trainer's hand: while S. learns to modulate the pressure of the hand in holding the sole still on the last, the trainer, who perceives the force effectively exerted by the worker, guides her, by providing constant feedback and monitoring the process and the result (Fig. 8).

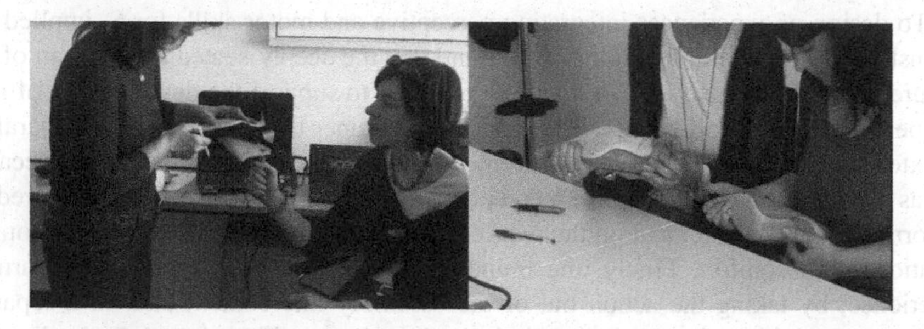

Fig. 8. Holding-Pressing Range. S. simulates the gestures needed to use the ast peg correctly.

3.4 Building a New Skill and Launching a New Working Practice

In this training phase, S. was accompanied in the simulation of the new professional gesture with the actual working tool, the ast peg, but the activity was still staged outside the production context, so that she could focus on the effectiveness of the gesture. By combining practical activity with a simple self-confrontation, the trainer encouraged her to think about her learning, helping her to be aware of the differences between the movements performed before and after the Functional intervention, to consolidate a new representation of herself and her action (Fig. 9).

Fig. 9. S. comments on her action before and after the Functional intervention

The cycle of training-Functional intervention- simulation on the ast peg-self-confrontation was replicated twice for a total of 6 h of training, within a period of two months. Once S. consolidated the new gestures, the trainer moved for the last training session in the real production context and conditions (Bernstein 1997). Only then, it was possible to proceed in performing the final assessment of the learning experience.

3.5 Redesigning the Ergonomics of the Workplace as a Team

The acquisition of the new professional gesture activated a true devolution, not only of the operators, but of an entire professional group. The appropriation of the new gesture and its conceptualisation allowed Z., C, and S. to express their user requirements and the required functions of the tool. In this way, a design group was formed, composed by the quality manager, the trainers and the technical experts to set out the specifications of the new ast peg (with a height regulation option, a wide enough base on which comfortably lay down the pens and the components of the shoe) and to redefine the workstation layout in order to enable the workers to move more easily. The indications provided by the operators and by the group of experts allowed the manufacturing technician to create the product required, perfectly in line with the operators' needs and suitable for its intended purpose.

4 Conclusions

The combination of research and training enabled us to identify the relationship created between innovation in the work analysis methodologies, training devices design increasingly tailored on individual needs, and improved organisational and work wellness. By accurately analysing workers in action, we were able to identify the gestures and the finalities underlying the actions schemes; then, starting from organisational and working representations of workers and management, it was possible to design training actions that allow the work to be transformed and improved. The ability to resolve difficult situations with DIY methods, an unreported resource used to solve day-to-day problems in SMEs, was identified as research object and reinterpreted to introduce new and different forms of individual and group self-regulation. The role of the trainer thus becomes that of mobilizing internal and external knowledge of the enterprise by renewing and encouraging the sharing of new representations of work and work organization. The result of this process has an impact on production, knowledge, action and well-being (Kaneklin and Scaratti 2012).

References

Alastra V, Kaneklin C, Scaratti G (2012) La formazione situata. Repertori di pratica. Franco Angeli, Milan

Barbier JM, Durand M (2017) Encyclopédie d'analyse des activités. Ed. Puf

Bernstein NA (1996) On dexterity ans its development. In: Latash ML, Turvey MT (eds) Dextirity

Dejours C (2009) Travail vivant, 2 tomes. Payot, Paris

Mendel G (2002) L'acte. In Barus-Michel J, Enriquez E, Lévy A (dir publ) Vocabulaire de psychosociologie. Octarès, Toulouse, pp 27–37

Nussbaum M (2001) Giustizia sociale e dignità umana. Da individui a persone. Il Mulino, Bologna

Ochanine DA (1978) Le rôle des images opératives dans la régulation des activités de travail. Psychologie et éducation 2:63–72

Pastré, P. La didactique professionelle: approche antropologique du développement chez les adultes. Puf, Paris (2011)
Simondon G (1989) L'individuation psychique et collective. Aubier, Paris
Pacquola Mc, Pacquola B, Rizzi G (2013) Transferer les savoirs d'experience: un chantier italien dans le secteur des chaussures. Revue Travail et apprentissage, pp 41–64
Rispoli L.: Il corpo in psicoterapia oggi. Neo-Funzionalismo e sistemi integrati. Ed. Franco Angeli (2016)
Theureau J (2006) Le cours d'action, méthode développée. Octarès, Toulouse
Vergnaud G (1996) Au fond de l'action, la conceptualisation. In: Barbier JM (dir) Savoir théoriques et savoir d'action. PUF, Paris, pp 275–292

The Effectiveness of Labour Inspectors in Improving the Prevention of Psychosocial Risks at Work

Rafaël Weissbrodt(✉), Marc Arial, and Maggie Graf

State Secretariat for Economic Affairs, Department of Labour, Work and Health Research Unit, Bern, Switzerland
grweissbrodt@bluewin.ch, marc.arial@seco.admin.ch, maggie.graf@fastmail.com

Abstract. Psychosocial risks at work are preventable through ergonomic design of the work organisation, working environment and tools. Taking appropriate action to tackle these issues belongs to an employer's duty to protect health and safety at work under Swiss law and the law of many countries. Labour inspectors are responsible for checking workplace compliance with such regulations. This study assessed the outcomes of a Swiss national labour inspection campaign focused on psychosocial risks. A quasi-experimental research design was used, with two groups of workplaces (one inspected and the other, not) and two questionnaire surveys with an interval of one year. Inspected firms improved their management of health and safety and reported more competence for the management of psychosocial issues compared to the control group and their pre-inspection levels. They also demonstrated an increased willingness to take action on prevention issues. To a lesser extent, they were more likely to implement specific psychosocial risk management measures, particularly those aimed at individual support for stressed employees. However, inspection visits did not lead to increased employee participation in the prevention of workplace risks, improvements in work organisation, working schedules or staffing levels. Based on these results, it was recommended that the labour inspectors more strongly emphasise a prevention approach grounded in the assessment and improvement of job design, content and organisation.

Keywords: Labour inspection · Psychosocial risks · Stress · Outcome Evaluation

1 Introduction

Psychosocial risks at work may develop from many causes but most could be avoided through appropriate design of the work organisation, workplaces and tools. Laws exist in most European countries that require the ergonomic design of work, specifically in relation to both physical and mental loads. Labour inspectors are required to check that companies comply with these laws, but little research exists on company responses to labour inspectorate interventions regarding psychosocial risks. These risks are not readily observable and, therefore, require an audit of company practice by the

S. Bagnara et al. (Eds.): IEA 2018, AISC 821, pp. 133–139, 2019.
https://doi.org/10.1007/978-3-319-96080-7_17

inspectors, rather than walk-through inspections. The literature identifies many hindrances to inspectors' intervention in dealing with these issues. Their complexity also increases the pressure on limited inspectorate resources [1–4]. However, studies on this subject are still rare and come mainly from Nordic or Anglo-Saxon countries.

This paper reports on the results of a study on an inspection campaign conducted in Switzerland by the cantonal labour inspectorates and the State Secretariat for Economic Affairs (SECO). Between 2014 and 2018, the labour inspectors conducted their inspection visits with a special focus on psychosocial risk prevention. In the first years, an evaluation study was conducted to assess the effectiveness of this campaign on company practices. The aim was to assist the inspectors to develop appropriate working methods for this task.

In this context, the study aimed to answer the following questions: What operating methods do Swiss labour inspectors use to integrate psychosocial risks into their enforcement visits, and with what effects? We deal with these questions using an activity-centred ergonomics approach as a conceptual framework to analyse the psychosocial risk prevention practices [5]. We further postulate that this approach can contribute to the evaluation of occupational health public policies, by identifying the strategies [6] used by labour inspectors in their interactions with companies. Research on front-line civil servants (also known as "street-level bureaucrats") [7] has shown that they make use of the autonomy they have, in order to adapt their services according to circumstances and interactions with their "clients"; in so doing, they influence, sometimes unexpectedly, the implementation of public policies. This finding presents analogies with ergonomic theories on how operators regulate their activity [8]. To our knowledge, these links have not been addressed in the literature to date.

2 Research Field and Methods

2.1 Research Field

Switzerland has 30 cantonal labour inspectorates, employing around 140 inspectors for around 400,000 companies. The cantons are responsible for controlling the enforcement of the occupational health and safety laws in enterprises without high safety risks, in the secondary and particularly the tertiary sectors. The SECO coordinates their activities. After being trained in the enforcement of psychosocial risk prevention measures, inspectors were required to incorporate these aspects into their inspection visits to raise awareness in enterprises about the processes and measures that employers should implement to protect their employees.

2.2 Sample Design and Data Collection

The evaluation of the campaign was based on a mixed methodology, combining questionnaire surveys, activity observations and individual and group interviews.

Using questionnaire surveys, comparisons were made between an intervention group of 185 companies audited by inspectors and a control group of 161 similar companies, not inspected during the study period. As the inspectors were free to choose the companies for inspection (they generally concentrate their efforts in known problem sectors and branches), a quasi-experimental study resulted. The surveys were conducted either by telephone interview or online (as desired by the company) before the inspector visited the company (pre-test) and then one year later. The control companies were selected using stratified random sampling, based on company size, branch and region. They were asked to respond to the same battery of questions in the same way twice with an interval between them of one year.

The questionnaire consisted of 120 questions, focusing on occupational health and safety (OHS) management, worker participation, respondents' perceptions of psychosocial risk factors and psychosocial risk management. It was addressed to the person "most familiar with OHS issues in the company"; in most cases, this was the employer or a senior manager. Responses to open-ended questions and spontaneous comments were transcribed.

In parallel, for each company included in the quantitative study, we sent a questionnaire to the inspector concerned, in order to collect his or her impressions following the visit. It contained questions about how he or she perceived the interlocutor in the enterprise, his or her inspection style, and the nature of any actions he or she had required.

In order to gather qualitative data, we accompanied inspectors on some of the enterprise visits. Participation was voluntary; we sought to reflect some diversity in the initial training of inspectors, their seniority and the type of company audited. We made paper-and-pencil observations, reflecting as faithfully as possible the exchanges between the inspector and his or her interlocutors. To limit study interference, we did not record the dialogues. The observations were discussed individually with the inspector concerned. We invited the inspector to comment on extracts from our notes and his or her visit report, and to explain his or her way of proceeding. In addition, an interview guide was used to gather information on how the inspector more generally dealt with psychosocial risks. Notes were taken during the interviews. Finally, a workshop was organised with some 20 inspectors. Each participant cited at least two measures that they had prescribed to a company; all of the material was discussed in sub-groups and collated for analysis by the researchers.

2.3 Data Analysis

Employer perceptions and practices were analysed using factor analysis. Nine scales were constructed to measure staff exposure to psychosocial risks, as perceived by employers, as well as the situation of companies with regard to OHS management, employee participation and psychosocial risk prevention measures. The factor analyses led to a distinction between two types of preventive measures. The first category mainly deals with the organisation of psychosocial safety management. It includes measures such as company charters or regulations, role definition, procedures in the event of conflicts, harassment, violence, etc. The second category consists of changes to work organisation, design and environment. The effects of the inspection visits on the evolution of the scores were measured by means of multiple regression analyses, taking into account the size of the company and its business sector.

The replies to the questionnaire sent to the inspectors were analysed by means of descriptive statistics and bivariate association tests. The qualitative material gathered during observations and interviews was the subject of thematic coding and analysed with descriptive statistics, based on a general "activity regulation" framework [8]. Its dimensions were as follows: conditions external to the inspector (constraints and resources), internal conditions (skills and objectives), inspection strategies, consequences for the task (effects of the inspection on the company) and for the inspector (job strain). The coding was done using the RQDA software [9], and the statistical analyses with Stata [10].

3 Results

3.1 Sample Description

For the quantitative study, 692 companies were contacted; 404 agreed to participate in the first survey and 346 in the second. They came, in decreasing order, from the service sector, trade, hotels and restaurants, industry and construction, the health and social sector, administration and education. The sample reflected the distribution of workplaces inspected during the study period and the area of responsibility of the cantonal inspectorates. It is not representative of the structure of the Swiss economy, which includes a higher proportion of small enterprises and more enterprises in other sectors.

70 inspectors announced at least one planned visit to the researchers; the total number of announced visits was 287. The inspectors responded in 275 cases to the questionnaire sent to them; 2 were excluded due to missing values.

The observations and interviews concerned 9 inspectors (by original trade 6 technicians or engineers, an occupational health nurse, a psychologist and a lawyer) with between 1.5 and 37 years as an inspector. We accompanied them on 11 company visits, including 3 factories, 2 hotels, a bank, an educational institution, a home for the elderly, a police force, a dental laboratory and a drugstore.

The remainder of the chapter summarises the results of the company survey and then details the inspectors' strategies.

3.2 Outcomes of Inspection Visits

The analyses revealed a significant effect ($p < .05$) of inspections on several dimensions. The inspected enterprises improved their systematic OHS management score (average increase of 1.6 to 2.0 points, depending on the branch and size, on a scale from 0 to 12). They also reported more competence in the management of psychosocial issues compared to the pre-inspection level (+1.4 to 2.3 points on a 12-point scale). There was an increased willingness to take action, but this result was statistically significant only in tertiary sector companies (+ about 1 point on a 12-point scale). Overall, the effects were more pronounced in tertiary enterprises with more than 100 employees, then in smaller tertiary enterprises, and in secondary sector enterprises regardless of their size. Some improvements were also noted in the control group, but they were significantly less than those in the experimental group.

On the other hand, inspection visits did not lead to increased employee participation in the prevention of workplace risks, nor to improvements in work organisation, working schedules or staffing levels ("changes to the organisation, design and environment at work"). An impact on the "organisation of psychosocial safety management" was only found in tertiary sector companies with more than 100 employees (+1.4 points on a 12-point scale). The inspections had a significant effect on the following single measures: the establishment of a charter or regulation regarding psychosocial risks, the clarification of roles in managing these risks, the adaptation of the tasks of some employees to prevent stress, the offer of confidential advice in the event of a problem, and the implementation of a conflict resolution procedure. The inspections also led to a statistically significant increase in the recording of working time, in comparison with non-inspected workplaces.

3.3 Inspection Strategies

Observations and interviews showed that the inspectors must deal with a number of constraints linked to their professional context: sometimes difficult relations with the social partner representatives, lack of time to carry out in-depth analyses, scarcity of exchanges between colleagues, absence of specific provisions on psychosocial risks on which to base their requirements, etc. These concerns were more evident when inspectors addressed psychosocial risks than when they checked the physical work environment. Other difficulties were related to relationships with interlocutors within companies, company characteristics (low levels of safety and health knowledge, economic difficulties, small size) and the difficulty of involving staff representatives in the inspection process.

According to the questionnaire completed by the inspectors, psychosocial risks were addressed in 96% of visits and the inspectors required measures against these risks in 51% of cases. Most often, they were comfortable addressing psychosocial risks (median score of 87 on a scale of 0 to 100). They considered their interlocutors to be generally sincere (85/100), to be not very defensive (13/100), to have enough decision-making power to be able to take measures (85/100) but as having only moderately good knowledge of psychosocial risks (67/100). Most inspectors adopted a very motivation oriented style of inspection (85/100), and more rarely a directive one (30/100). The measures they required were most often formal (77/100), such as company declarations or charters, procedures, etc.; more rarely, they were practical (41/100) or related to the content or organisation of work (55/100). Overall, the inspectors reported having formulated their requirements or recommendations in a general (77/100) rather than detailed and precise manner (61/100).

The inspectors accompanied in the field particularly checked the processes, documents and measures in place in the companies. They also asked about working conditions (working hours, social climate, organisation, etc.). When they required the employers to take action, they focused on formalising a process or offering support to staff. None of these inspectors required a detailed risk assessment to be undertaken. They tried to motivate their interlocutors by emphasising the company's vulnerability, arguing for the business case, and detailing health issues. Legal arguments were rarely used. The inspectors were flexible, provided advice and highlighted actions already taken by the company rather than shortcomings. They adapted to the person they were speaking to by giving examples and asking open-ended questions. They tried to create a connection, explaining the reason or process of the inspection visit, and avoiding irritating their interlocutors. Most of the time, they took a long-term view of their visit, either because they had already intervened in the company or planned to return, or by offering support when needed.

4 Discussion

To our knowledge, this is the first study showing positive outcomes of labour inspection visits on the management of psychosocial risks. These results should encourage the authorities to sustain their efforts in this area.

The inspectors addressed psychosocial risks in a motivating manner, with more emphasis on employee support and procedures in the event of problems, than prevention of their root causes. This approach could reinforce the pre-existing tendency of many employers to see psychosocial risks as an individual problem and not as the result of poor work organisation [11]; such a tendency also appeared in the spontaneous comments of many survey respondents. Our observations support some results from other publications [1, 3]. The "wicked nature" of psychosocial risks [2], as well as social, political and economic constraints, may explain the adoption of the cautious strategy of the labour inspectors. This strategy itself contributes to explaining the effects of inspections, as measured in this study: The increase in the level of companies' self-reported competency and willingness to take action probably results from the motivational approach adopted by the inspectors. The more frequent implementation of measures for the organisation of psychosocial safety management also reflects the type of requirements formulated by inspectors. Worker participation and changes to work organisation, design and environment, on which inspection visits did not have a significant effect, were relatively seldom addressed by inspectors. These relationships between inspection styles and contents, contextual constraints, and public policy outcomes are consistent with Lipsky's works on street-level bureaucrats [7].

Based on these results, it was recommended that the inspectors are specifically trained to emphasise a prevention approach grounded in the assessment and improvement of job design, content and organisation. The connections between these factors and psychosocial risks are poorly understood by companies. Inspectors should more systematically require employers to integrate these dimensions into their risk assessments. Moreover, worker participation in prevention and inspection processes should be given greater value, *e.g.* by providing in-depth training for inspectors to conduct individual and group interviews with employees. As time constraints may limit the effectiveness of this measure, its effectiveness should be studied. Finally, it would be useful to further encourage exchanges of practices and collaborative work between labour inspectors, for instance through visits in pairs and expert networks.

Acknowledgements. This study was funded by the State Secretariat for Economic Affairs and the Federal Coordination Commission for Occupational Safety, Switzerland. We are very grateful to every company and labour inspector who took part in it. We also thank Prof. David Giauque

(University of Lausanne) and our colleagues Tarek Ben Jemia, Marc Huber, Samuel Iff, Stephanie Lauterburg Spori, Pascal Richoz, Margot Vanis and Christine Villaret D'Anna who assisted with the study.

References

1. Bruhn A, Frick K (2011) Why it was so difficult to develop new methods to inspect work organization and psychosocial risks in Sweden. Saf Sci 49(4):575–581
2. Jespersen AH, Hasle P, Nielsen KT (2016) The wicked character of psychosocial risks: implications for regulation. Nordic J Working Life Stud 6(3):23–42
3. Quinlan M (2007) Organisational restructuring/downsizing, OHS regulation and worker health and wellbeing. Int J Law Psychiatry 30(4):385–399
4. Weissbrodt R, Giauque D (2017) Labour inspections and the prevention of psychosocial risks at work: a realist synthesis. Saf Sci 100:110–124
5. Petit J, Dugué B (2012) Psychosocial risks: acting upon the organisation by ergonomic intervention. Work 41:4843–4847
6. Faye H, Falzon P (2009) Strategies of performance self-monitoring in automotive production. Appl Ergon 40(5):915–921
7. Lipsky, M.: Street-level bureaucracy. Dilemmas of the individual in public services. Russel Sage Foundation, New York (1980/2010)
8. Leplat J (2006) The concept of regulation in activity analysis (La notion de régulation dans l'analyse de l'activité). Perspectives interdisciplinaires sur le travail et la santé 8(1) (2006). http://journals.openedition.org/pistes/3101. Accessed 10 Apr 2018
9. Huang R (2016) RQDA: R-based qualitative data analysis
10. StataCorp (2015) Stata statistical software: release 14. StataCorp LP, College Station
11. Weissbrodt R, Arial M, Graf M, Ben Jemia T, Villaret D'Anna C, Giauque D (2018) Preventing psychosocial risks: a study of employers' perceptions and practices (Prévenir les risques psychosociaux : une étude des perceptions et des pratiques des employeurs). Relations Industrielles/Ind Relat 73(1):174–203

"To Him Who Has, More Will Be Given…"– A Realist Review of the OHSAS18001 Standard of OHS Management

Christian Uhrenholdt Madsen[1(✉)], Marie Louise Kirkegaard[2], Peter Hasle[1], and Johnny Dyreborg[2]

[1] Department of Materials and Production, Aalborg University, AC Meyers Vænge 15, 2450 Copenhagen, Denmark
cum@mp.aau.dk

[2] National Research Center for the Working Environment, Lersø Parkalle 105, 2100 Copenhagen, Denmark

Abstract. The OHSAS18001 standard is now the most widely adopted management systems for occupational health and safety worldwide. The standard is intended to support companies in attaining a higher health and safety standard. However, there is limited knowledge on how this standard in fact is working in practice and thus can improve health and safety at work.

In order to investigate how the OHSAS18001 standard is working in practice, we identified the main mechanisms assumed to be actively involved in the successful implementation and management of the standard, by using a framework inspired by a realist methodology. In line with this methodology, we assessed how the context of the adopting organizations impinges on the identified mechanisms and synthesized the findings into useful knowledge for practitioners and fellow researchers alike.

The starting point for the analytical process is the program theories that we identified in the standard and supplementary materials from key stakeholders. Thus we analyze how key stakeholders and policymakers expect the standard or program theory to work when it is implemented in an organizational setting. The three program theories (PT) we identified are: An 'operational' PT, a 'compliance' PT, and an 'institutional' PT.

Then we compared these 'assumed' program theories to how the OHSAS18001 actually worked in real-life settings. We identified four so-called context-mechanism-outcome configurations by reviewing available empirical studies and by extracting knowledge from them. These CMO-configurations are: 'Integration', 'learning', 'motivation' and 'translation'. This analytical approach means that our paper provides both i -depth understanding of the assumed program theories behind the OHSAS18001standard and understanding of the actual mechanisms of certified management systems in occupational health and safety management in various context presented by the included implementation studies.

Keywords: OHS · OHSAS18001 · Realist review

S. Bagnara et al. (Eds.): IEA 2018, AISC 821, pp. 140–149, 2019.
https://doi.org/10.1007/978-3-319-96080-7_18

1 Introduction

Organizations increasingly seek to resolve and mitigate occupational health and safety concerns with the use of voluntary and certified occupational health and safety management systems. The OHSAS18001standard, which was first published in 1999 by British Standards, furthered this tendency, and is by now the most widely adopted management systems for occupational health and safety worldwide. The OHSAS18001 standard is becoming more and more important in the governance of health and safety in companies, and thus it becomes paramount to understand both how these new management systems are intended to work, but also importantly, how they actually 'work' within organizations.

There are already a number of empirical studies on the use of OHSAS18001, generally divided into two streams of research: One stream of research has sought/has tried to measure effects of the OHSAS18001 on organizational performance in general and on organizational safety outcomes in particular [1–3]. These studies point to a positive effect of the certification process on organizational safety performance. Lo and colleagues [2] report a decline in compliance failures from US-based manufacturers [2]. Abad et al. not only describe a positive effect of the OHSAS-certification but also demonstrate how this effect increases with the number of certification years, leading them to conclude a learning effect in the Spanish organizations in their study [4]. However, studies of this kind rarely describe the processes and social mechanisms inside the organizations that may lead to safety outcomes in the end.

A second stream of research has tried to uncover the internal organizational dynamics related to the implementation and utilization of OHSAS18001 (cf. [5–7, 9, 10]). This research mainly utilizes case studies based on qualitative methods to uncover the social interactions and structurations related to the organizational adoption of the standard. These types of studies are often specific and bound to the particularities of the cases without providing much details and prescriptions for future implementations in other contexts. The integration of these two streams of research is so far not developed, and knowledge about the organizational mechanisms related to positive OHS outcomes is therefore lacking. We intend to fill this research gap by reviewing the existing studies on OHSAS18001 in order to integrate knowledge that is rich on contextual details, but still provides valuable insights into the generalizable mechanisms and processes for future implementations.

To achieve this goal, we have conducted a review inspired by a realist approach [11] of the use of OHSAS18001 and the social mechanisms it activates within organizations. In this way, we have utilized both qualitative and quantitative studies in our synthesis. We have researched the use of the OHSAS18001 across a range of different albeit comparable organizational and regulatory contexts in different industries and countries across OECD-member states. Thereby, we have developed suggestions for the 'demi-regular' [12] mechanisms and causal potentials inherent in the OHSAS18001, and how they interact in various contexts.

2 Methodology

In contrast to other review-methodologies, the realist approach compares and analyzes the CMO-configurations (context-mechanism-outcome (cf. [13]) that appear in empirical studies of both quantitative and qualitative studies, and thereby provide a deeper insight into the interplay between actual mechanisms and contexts related to the successful implementation of OHSAS18001. Our review followed the realist review procedure [11] by identifying the intended program-theories of policy makers and standard publishers [11] first, and then by gradually refining the theories through our iterative review and through the comparison of available and relevant empirical data. In other words, we compare how the voluntary and certified OHSMs are 'supposed to operate' (i.e. the program theories) [11] and how they actually operate (i.e. the CMO-configurations) when put under researchers' magnifying glasses. By doing so, our review adds valuable knowledge to the discussion on how and when OHSAS18001 works or fails to work. In the first step, we investigate how the OHSAS18001 is supposed to work by unpacking policymakers and stakeholders assumptions about how the OHSAS18001 would 'operate' – the so-called 'rough' program theories [12, 14]. We selected a number of key documents, which included the OHSAS 18001 standard and the accompanying 'Guidelines for implementation' (OHSAS18002)[1]. As suggested by Wong and collegues [11], we also consulted public communication and promotional material from the original publishers, British Standard, which we found on their website. Finally, as an illustrative example of materials from more distant stakeholders, we used material from the Danish government's public sources (i.e. legal documents, debates from parliament and public statements from the ministry of occupation) about the reasons for incentivizing the use of the standard among Danish companies from 2005 and onwards. Together these key documents provided us with a somewhat clear overview of the expectations about the outcomes of the implementation of OHSAS18001 from a range of stakeholders that promotes OHSAS18001 as a way of securing safe and healthy workplaces. These key documents were then used to identify possible rough program theories. We scanned all these sources for descriptions of intended outcomes (i.e. what organizations gain from certification). Then we searched for described resources in the text that would lead to said outcome (continuous improvement demands, demands about documentation etc.). Finally, we described the mechanisms that would have to be in place for this resource-outcome constellation to 'work', and that is known from other empirical areas within mainstream organizational and management literature (e.g. legitimacy).

In step two, we identified the CMO-configurations [14]. First, we designed an exhaustive list of concepts, terms and synonyms that would help us to find the most relevant studies. We included studies that covered the three following aspects:

1. An actual example of a OHSAS18001 ('management system' terms)
2. The 'processual elements of a OHSAS18001' ('processual terms'), and
3. Sources that had descriptions of OHS concepts and issues ('OHS terms').

[1] Both found in 'Arbejdsmiljøledelsessystemer Danish Standard, 2.ed. 2010'.

The studies were hereafter carefully read by the reviewers and classified into the 'active mechanisms' (e.g. translation) involved in the study, contextual elements that affected the outcome of the mechanism (e.g. how the existing organizational OHS approach becomes the blueprint for implementation), and finally into outcomes (e.g. OHSAS18001 is implemented in a formalistic and compliance focused manner).

3 Findings

3.1 The Program Theories

We identified three program theories [14] on the basis of the key documents mentioned above. Each program theory has its own expected positive outcome for organizations (See Table 1). First, we identified an 'operational' program theory (PT1) that emphasizes operational gains to the safety management system of the organization. In PT1, the standard is expected to provide the organizations with systematic tools such as PDCA[2]-procedures, continuous improvement or the possibility to integrate the OHSAS18001 with other key management systems within the organizations. When the tools are used as stipulated in the standard, the assumption is that these tools will increase the maturity level of the OHS management system, ensure a systematic and lasting approach, and increase the focus on OHS issues within the organization. A second program theory (PT2) was identified that emphasizes 'compliance' as the most important outcome. PT2 emphasizes the resources in the standard that help the organization to streamline and comply with increasingly complex and diverging regulatory frameworks, which in particular was salient for organizations operating in multiple countries. Finally, an 'institutional' program theory (PT3) was identified. PT3 assumes that the institutional legitimacy public image is improved, and that the company thereby gains the advantages and resources that accompany these institutional advantages such as better stakeholder relations. These outcomes are achieved by using proper documentation systems and mandatory stakeholder communication – both mandatory within the OHSAS18001 standard.

3.2 The CMO-Configurations in OHSAS18001 Implementations and Management

The included OHSAS18001 studies showed four different CMO configurations. They are described in Table 2, and in further detail below. Each CMO-configuration is described in terms of what resources in the OHSAS18001 are activated, the active social mechanisms, the contextual elements, the outcomes, if they relate to the assumed program theories described above.

Multiple included studies indicate that organizations that have compatible systems already in place have an easier time implementing and managing the OHSAS18001 standard as well [2, 5, 8, 17]. Together they describe a CMO-configuration which we have dubbed 'integration'. In short, this means that the main organizing principles in

[2] Plan-do-check-act.

Table 1. Program theories of OHSAS18001

Program theory (PT)	Resources within standard	Mechanism	Outcome	Examples from OHSAS18001 and OHSAS18002[a]
'Operational' program theory (PT1)	- PDCA approach - Compatible with other CMS (e.g. ISO9001 or ISO14001) - Demands of continuous improvement - Demands of management and employee participation	- Systematization - Specialization - Continuous improvement processes - Maturing management system	- Improved OHS performance - Preventive and proactive approach (e.g. registration of near-misses) - Increased awareness and commitment from management and employees - Lower accidents and health risks within organization	*"A system of this kind enables an organization to develop an OH&S policy, establish objectives and processes to achieve the policy commitments, take action as needed to improve its performance..."* [15]
'Compliance' program theory (PT2)	- Structured approach to compliance - Demands about formulation and running update of official organizational OHS-policy	- Formalization - Meta-regulation	- Always ensured compliance with all legal requirements across geographical and regulative boundaries	*"To meet its policy commitments the organization should have a structured approach to ensure that the other requirements can be identified, evaluated for applicability, accessed, communicated, and kept up to date"* [15]
'Institutional' program theory (PT3)	- Demands for documentation - Demands about stakeholder communication	- 'License to operate' - Legitimacy through adoption of institutionalized practice	- Better reputation - Reputational gains and advantages	*"Demonstration of successful implementation of this OHSAS standard can be used by an*

(*continued*)

Table 1. (*continued*)

Program theory (PT)	Resources within standard	Mechanism	Outcome	Examples from OHSAS18001 and OHSAS18002[a]
	- Certificate 'to show'		- Improved relations with key stakeholders	*organization to assure interested parties that an appropriate OH&S management system is in place"* [15]

[a]Examples from OHSAS18001 and OHSAS18002 reproduced with permission from Dansk Standards Foundation:

the standard work best, if the implementing organizations have a context characterized by similar principles already in place. Lo and colleagues [2] found a clear effect of certification in their study of large American manufacturing companies. Furthermore, they show that organizations characterized by 'tightly coupled' production systems have even bigger safety advantages from certification. This is supported by a Danish research project [5, 8, 10] which demonstrates that organizations used to operate with systematic and formalized routines more easily implement and operate certification schemes. Finally, two studies show that organizations that already operate OHSAS18001's 'next-of-kin' management systems (ISO14001 and ISO9001) also experience greater safety and processual advantages [9, 17]. Overall, the studies indicate that previous experience with implementing standards provides an important mechanism and context in order to explain the operational effects of the OHSAS18001 certification.

The CMO-configuration 'learning' likewise shows that existing structures and resources determine whether implementation of OHSAS18001 leads to effective organizational learning and continuous improvement processes. Two of the included studies support this mechanism [9, 10]. The standard has a clear learning component in its 'continuous improvement' demands and processes. It is also this mechanism that can be seen as the logical driver behind the maturity model envisioned in PT1 – the operational program theory. Abad and colleagues (2013) found that the effects increase with the number of years with a certification program (OHSAS18001). This indicates that a learning process takes place after the implementation of OHSAS18001. However, as both Granerud and Rocha [10] and Silva et al. [9] show, implementation of the OHSAS18001 standard is by no means a guarantee for a higher-level OHS learning effect.

Table 2. CMO configurations of OHSAS18001

Type of CMO?	Resource(s)	Mechanism(s)	Context(s)	Outcome(s)	Related program theory	Source(s)
Integration	Plan-Do-Check-Act Compatibility with other CMS	Integration	Existence of similar organizational-, production- or management systems	Improved OHS performance	Operational	[2, 5, 8, 17]
Learning	Continuous improvement	Organizational learning	Existing structures and culture for learning (e.g. other continuous improvement frameworks, such as TQM, LEAN etc.)	Continuous improvement of OHS efforts (e.g. learning from accidents)	Operational	[9, 10]
Motivation	Certificate Obligatory management involvement in OHS efforts Continuous Improvement	Commitment	Motivation for certification process	Successful implementation	Institutional Operational Compliance	[4–6, 8, 18]
Translation	Internal audits	Translation and adaptation	Existing OHS approach	OHSAS18001 is tailored to already existing approaches OHS management becomes 'auditable'	Institutional Operational	[5–8]

Finally, five of the included studies show that the organizational motivation for seeking certification is a rather important factor in the operational success of the intervention. In this CMO-configuration, the motivation is the primary contextual element that defines successful implementation and the commitment to the OHSAS18001 in larger parts of the organization. Motivation is the active mechanism. The resources that is supplied from the certification process is both the potential of improvement frameworks to deal with OHS issues [4, 18] as well as the promise of social legitimacy that comes with having a certification [5, 6, 8]. We know from Bevilaqua and colleagues' study (2016) that the motivational factors for implementing the management system are quite important in determining the success of an OHSAS18001 certification in terms of actually improving OHS factors at the organizational level. These motivational factors can take the form of desires to comply with rules and regulation, with displaying social responsibility, and with improving important OHS issues within the organizations [18]. The same overall mechanism is identified in the studies by Rocha and Granerud [6], Kristensen [8], and Hohnen and Hasle [5]. To a various degree, they demonstrate that many of the case organizations see the certificate as a social and institutional necessity, and as something an organization like them should have. This external legitimacy-seeking behavior in many cases turns into the main driver of commitment, which the internal actors interested in the implementation of OHSAS 18001 subsequently use to their advantage in order to further the implementation process.

A fourth CMO-configuration centers on 'translation'. The translation configuration shows how the certification is carried into organizations by carriers who interpret and translate the certification process to fit a new context. To varying degrees in the four included studies [5–8], the configuration describes how the new policy is tailored to fit the internal context of the organization and the known processes, and furthermore how the translated standard becomes a part of internal positioning and hierarchical processes and is made a part of the political processes and discussions in the organizations. The translation mechanism thus explains how the systems and organizational hierarchies which the carriers are embedded in influence the implementation of the OHSAS18001 certification, and thus partly relates to the other CMO-configurations described above.

4 Discussion

The CMO-configurations each shows, how the existing contexts in terms of existing structures ('integration'), capabilities ('learning') or incentives ('motivation') influence how the resources and mechanisms inherent to the intervention end up shaping OSH efforts. This is not surprising. Researchers within organization studies and operations research have long pointed to the fact that organizations are best suited to absorb what is recognizable from neighboring fields [19, 20]. Furthermore, it is posited that advantages often, although not always (cf. [21]), are building on top of existing capabilities in a cumulative process, not replacing them in a trade-off [22]. These findings bring us to the title of our paper, which we borrowed from The New Testament "…for to him who has, more will be given…" (Matthew 13: 11–12).

However, the fact that organizations build on already existing knowledge, systems and structures can also potentially have somewhat unintended consequences for the implementation of the standardization, as is also pointed out by some of our included studies. First of all, the fact that the certification is integrated into already existing procedures also means that organizations are prone to detect and assess risks and health issues that fit into the existing data and systems within the organizations. The occupational health and safety becomes 'auditable' as pointed out by Hohnen and Hasle [5]. This means that so-called 'wild' OHS issues such as psychosocial factors tend to be somewhat overlooked in the OHSAS processes in these companies. The same result is also shown in one of the municipal cases described by Jespersen and colleagues [7]. However, we do not know the extent of this, because none of the included studies that measure effects on OHS [2, 4, 17], actually take psychosocial issues into specific consideration. And furthermore while it is clear that organizational motivational factors for implementation are important for successful implementation, there is also a danger that this can lead to symbolic and decoupled adaptations within the organizations, meaning that the system is only implemented to satisfy external stakeholders (e.g. regulators, customers, partners). In [6], there is one case example of this. We tentatively suggest that organizations that implement the system because of external pressure, but then recognize key features and practices in the OHSAS18001 certification from their already implemented systems will have a higher likelihood of avoiding symbolic adaptation. This is also hinted at in two included sources [5, 8].

Together, the four CMO-configurations explain specific paths to success in the OHSAS18001 certification process. What our findings tentatively point out, is that the OHSAS18001 certification will not work as a 'silver bullet' or panacea for every organization with OHS issues, if there is not any existing capabilities, structures or routines to build on. Furthermore, the organizational motivation seems to play a role in determining positive outcomes, as well as if there is a will to implement, and whether the organizations in question see the certification as something positive for the organization. Finally, multiple sources describe how the certification process is driven by actors and their interpretations of the policy implications which again helps determining the final implemented system of certified OHS management.

References

1. Pagell M, Klassen R, Johnston D, Shevchenko A, Sharma S (2015) Are safety and operational effectiveness contradictory requirements: the roles of routines and relational coordination. J Oper Manag 36:1–14. https://doi.org/10.1016/j.jom.2015.02.002
2. Lo CKY, Pagell M, Fan D, Wiengarten F, Yeung ACL (2014) OHSAS 18001 certification and operating performance: the role of complexity and coupling. J Oper Manag 32(5):268–280. https://doi.org/10.1016/j.jom.2014.04.004
3. Fernández-Muñiz B, Montes-Peón JM, Vázquez-Ordás CJ (2012) Occupational risk management under the OHSAS 18001 standard: analysis of perceptions and attitudes of certified firms. J Clean Prod 24:36–47. https://doi.org/10.1016/j.jclepro.2011.11.008

4. Abad J, Lafuente E, Vilajosana J (2013) An assessment of the OHSAS 18001 certification process: objective drivers and consequences on safety performance and labour productivity. Saf Sci 60:47–56. https://doi.org/10.1016/j.ssci.2013.06.011
5. Hohnen P, Hasle P (2011) Making work environment auditable–a 'critical case' study of certified occupational health and safety management systems in Denmark. Saf Sci 49(7):1022–1029. https://doi.org/10.1016/j.ssci.2010.12.005
6. Rocha RS, Granerud L (2011) The search for legitimacy and organizational change: the agency of subordinated actors. Scand J Manag 27(3):261–272
7. Jespersen AH, Hohnen P, Hasle P (2016) Internal audits of psychosocial risks at workplaces with certified OHS management systems. Saf Sci 84:201–209. https://doi.org/10.1016/j.ssci.2015.12.013
8. Kristensen PH (2011) Managing OHS: a route to a new negotiating order in high-performance work organizations? Saf Sci 49(7):964–973. https://doi.org/10.1016/j.ssci.2011.02.001
9. Silva SA, Carvalho H, Oliveira MJ, Fialho T, Soares CG, Jacinto C (2017) Organizational practices for learning with work accidents throughout their information cycle. Saf Sci 99(A, SI):102–114. https://doi.org/10.1016/j.ssci.2016.12.016
10. Granerud RL, Rocha RS (2011) Organisational learning and continuous improvement of health and safety in certified manufacturers. Saf Sci 49(7):1030–1039. https://doi.org/10.1016/j.ssci.2011.01.009
11. Wong G, Greenhalgh T, Westhorp G, Buckingham J, Pawson R (2013) RAMESES publication standards: realist syntheses. BMC Med 11(1):21. https://doi.org/10.1186/1741-7015-11-21
12. Pawson R (2006) Evidence-based policy: a realist perspective. SAGE Publications, Thousand Oaks
13. Dalkin SM, Greenhalgh J, Jones D, Cunningham B, Lhussier M (2015) What's in a mechanism? Development of a key concept in realist evaluation. Implement Sci 10(1):49. https://doi.org/10.1186/s13012-015-0237-x
14. Wong G, Westhorp G, Pawson R, Greenhalgh T (2013) Realist synthesis. RAMESES training materials, July 2013
15. Arbejdsmiljøledelsessystemer (2010) 2nd ed. Dansk Standard
16. BSI (2017) BS OHSAS 18001 features and benefit. BSI Group
17. Wiengarten F, Humphreys P, Onofrei G, Fynes B (2017) The adoption of multiple certification standards: perceived performance implications of quality, environmental and health & safety certifications. Prod Plan Control 28(2):131–141. https://doi.org/10.1080/09537287.2016.1239847
18. Bevilacqua M, Ciarapica FE, De Sanctis I (2016) How to successfully implement OHSAS 18001: the Italian case. J Loss Prev Process Ind 44:31–43. https://doi.org/10.1016/j.jlp.2016.08.004
19. Cohen WM, Levinthal DA (1990) Absorptive capacity: a new perspective on learning and innovation. Adm Sci Q 35(1):128–152. https://doi.org/10.2307/2393553
20. Zahra SA, George G (2002) Absorptive capacity: a review, reconceptualization, and extension. Acad Manag Rev 27(2):185–203. https://doi.org/10.5465/AMR.2002.6587995
21. Schroeder RG, Shah R, Xiaosong Peng D (2011) The cumulative capability 'sand cone' model revisited: a new perspective for manufacturing strategy. Int J Prod Res 49(16):4879–4901. https://doi.org/10.1080/00207543.2010.509116
22. Ferdows K, De Meyer A (1990) Lasting improvements in manufacturing performance: in search of a new theory. J Oper Manag 9(2):168–184. https://doi.org/10.1016/0272-6963(90)90094-T

Creativity and Performance: A Case Study in a Highly Regulated and Constrained Domain

Nathalie de Beler[1](✉), Christelle Casse[2], and Alain Noizet[3]

[1] EDF-R&D, 92140 Clamart, France
nathalie.de-beler@edf.fr
[2] Laboratoire PACTE, 38040 Grenoble, France
[3] Société LIGERON SONOVISION, 69003 Lyon, France

Abstract. The outage of an energy production plant for maintenance is a complex domain submitted to numerous stakes and constraints: productivity, safety, security and environmental protection. Outage projects have to manage a set of events of various natures: technical, organizational or human hazards and changes in regulation. Obviously, the Socio-Organizational and Human (SOH) dimensions are essential to overtake these difficulties, and to reach and maintain the performance of the outage projects at an acceptable cost (human, social, financial). For several years, the outage projects did not reach the expected performance. Concluding that the organization would have reached its limits, the operator launched a 3 years R&D project in 2015. This project aimed at enlightening the mechanisms and SOH characteristics that contribute to reach and maintain the targeted performance. First, the R&D project developed an "effective organization" model for an outage. Second, an approach were designed to perform a diagnosis of the organization, using the model as a framework for analysis. This paper first describes the approach and the main lessons learnt from the field studies, focusing on collective creativity that is needed to manage hazard events even in a highly regulated and constrained domain. Then, the paper will present lessons from applying the approach to a production plant. Finally the paper will outline the limitations of both model and approach and the issues to be addressed to transmit the approach to the operator.

Keywords: Resilience · Reliability · Energy · Creativity · Performance

1 Context: An Outage of Production for Maintenance

The outage for maintenance of an energy production plant is a complex domain submitted to numerous stakes and constraints: productivity, safety, security and environmental protection. Its main purposes consist of the replacement of parts of the resource of production, as well as test running, regulatory controls and maintenance activities on the equipment. The company –composed of 19 production units- adopted since the middle of the nineties an organizational structure called "outage project", cross-functional to the various departments of each unit.

S. Bagnara et al. (Eds.): IEA 2018, AISC 821, pp. 150–159, 2019.
https://doi.org/10.1007/978-3-319-96080-7_19

The project's organization main features rely on a national high level requirement report, which describes the process of preparation, realization and collection of experience feedback. The organization is supported by a dedicated structure which mobilizes from about fifty to one hundred people. They are almost 100% dedicated on this outage project from the preparation phase until the experience feedback phase.

Outage projects have to manage a set of events of various natures: technical, organizational or human hazards and changes in regulation. Obviously, the Socio-Organizational and Human (SOH) dimensions are essential to overtake these difficulties, to reach and maintain the performance of the outage projects at an acceptable cost (human, social, financial).

For several years, the outage projects did not reach the expected performance. The various analyses realized by the operator revealed recurring difficulties, such as Non-quality of maintenance, insufficient preparation of maintenance activities or deficiency in the resolution of technical hazards. Concluding that the organization would have reached its limits, the operator launched a 3 years R&D project in 2015, aiming at enlightening the mechanisms and the SOH characteristics that contribute to reach and maintain the targeted performance.

2 The Development Process of the Model: A Combination of State of the Art and Field Studies

The R&D study adopted a classical approach in human and social sciences, combining a state of the art of the characteristics of the organizations in risk areas, and field studies during several outages projects.

2.1 State of the Art: The Concepts of Robust, Reliable or Resilient Organizations

The literature review retained three paradigms of industrial security: the robust, the reliable and the resilient organization.

The robustness paradigm aims at designing a system in a rational way to optimize its performance, and then intends to obtain the conformity of the system to the established specifications. The organizational reliability paradigm identifies some social and organizational factors that allow systems to control their endogenous risks, caused by the complexity of interdependent technical processes (Roberts 1993; Rochlin 2001). Organizational resilience is an ability to deal with unknown situations. Three approaches were distinguished. The first one is the resilience engineering (Hollnagel et al. 2006; 2010) that relies on four "functions": anticipation, monitoring, reaction and learning. A second approach (Woods 2010) insists on reconfiguration capabilities through the mechanisms of information circulation in networks, shared values between stakeholders and an appropriate situation awareness. A third approach considers the resilience of organizations as the result of social dynamics. They allow adaptation (Christmann et al. 2014; Tillement et al. 2009) meaning specific modes of organization specific to the context. They also allow improvisation in response to unknown situations (Weick 1993; 1995).

The capitalization of many outage project studies from the sight of the human and social sciences as well as a continuous collaboration with two production units in the first phase of our research, enabled us to build a model of an efficient organization for outage projects. Concretely, several "back and forth" were made between the state of the art and the field studies, to develop an operational model based on the concepts of organizational robustness, reliability and resilience.

2.2 A Model of a Performant Organization for Outage Projects

In a traditional approach, performance analysis verifies that the organization has the appropriate means to achieve its objectives (relevance), to produce the expected results (effectiveness) at the right time and at the lowest cost (efficiency). In the outages project's context, an organization will be qualified as efficient if it succeeds in managing the variety of situations encountered -more or less disturbed, unforeseen or unprecedented- without degrading the targeted performance, and maintaining it durably. The model of the organization proposed is based on a characterization of the situations encountered during outage projects. Thus, in the course of their implementation, outage projects must manage different situations that are characterized by two main axes (See Fig. 1): the degree of understanding of the situation by the actors (x-axis), ranging from "usual situations, consistent with expectations" to "unprecedented situations". And the existence and the availability of a response to the situation (y-axis): the solution can indeed be planned and available or, on the contrary, must be built possibly by (re) combination of existing solutions (i.e., transfer of resource, change in sequence of activities or using innovations).

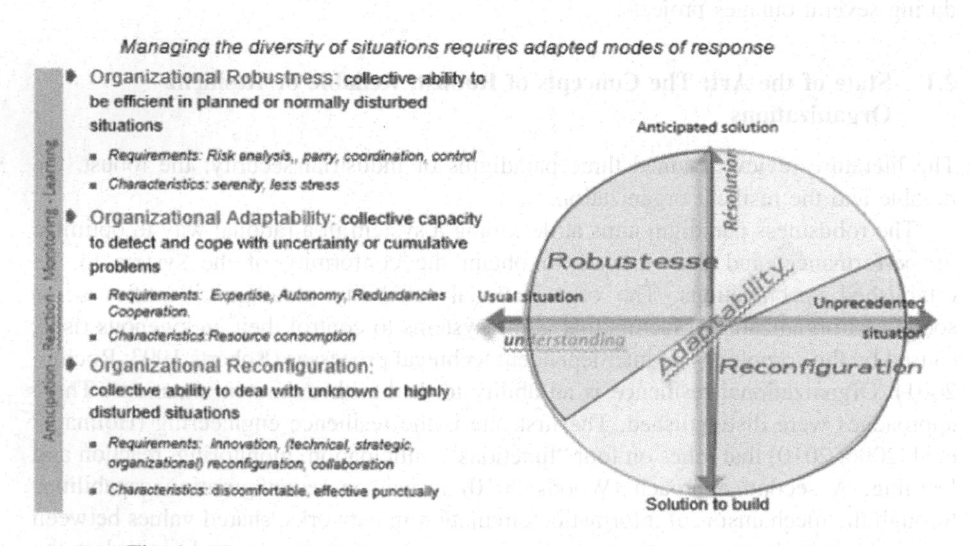

Fig. 1. Three modes of governance, to manage the diversity of situations

These two axes determine four sets of situations:

The north-west quarter represents the most frequently encountered situations, known or "normally disturbed". The situation is foreseen, the potential hazards are well-known and the solution(s) to deal with them are also prepared. If the event occurs, the prepared solution is implemented. In the south-west quarter, the situation is recognized, understood and the diagnosis quite easy to make. This is usually a combination of problems, often in cascade, making the usual solutions inapplicable in this context. It is therefore necessary to reconfigure the organization or to build a new solution by adapting the usual responses. In the North-East quarter, the actors have difficulties to characterize the situation which presents a problem whose diagnosis is laborious, long or partial because getting reliable and up-to-date information on the situation is difficult. Indeed, the situation is new for those who have to manage it, but a similar problem has often already been encountered in other projects or other units. There is therefore an answer that could be accommodated to the current situation. The South-East quarter brings together unknown situations, even unprecedented and dimensioning, for which a new solution must be built. The consequences of the unusual situations are significant on the project outages issues. A new solution must be built, requiring innovations (technical, material, strategic, organizational). These situations are the rarest.

The general hypothesis of the model is the existence of three organizational functioning or modes of governance, each of them being performant to manage some of the situations described. We call them "operating regime".

- The first one is the "Robustness regime", a collective ability to be efficient in planned or normally disturbed situations. This regime is based on the anticipation of potential hazards and the preparation of appropriate responses (programs, procedures, and resources), the effective application of these preparations, and the control of the results. The performance of this regime is high for ordinary or normally disturbed situations, in particular because it is economic at cognitive level, and because it facilitates collective activity by sharing the global situation awareness of the outage project.
- The second one is the "Adaptability regime", a collective capacity to detect and cope with uncertainty, excesses and hazards that cannot be handled by the robustness regime. It makes it possible to cope with situations that are unclear or unknown for those who confront them, but which are an accumulation of known problems, often cascading, finally leading to complex and singular situations. Adaptability overcomes the limits of the rigorous preparation and application mechanisms of the robustness regime. There are several key mechanisms of adaptability such as collective resolution of complex problems, adaptation of processes and rules and anticipation of margins, reserves and resources.
- The third one is the "Reconfiguration regime" which is a collective ability to deal with unknown or highly disturbed situations as they cannot be managed with robustness or adaptability mechanisms. These disturbing situations require punctually a reorganization and technical, strategic or organizational innovations. Then, capacity of creativity is necessary to deal with unknown situations. For example, reconfigurations of the strategy may be an arbitration between long term and short

term issues, which can go as far as abandoning certain initial objectives (duration or budget, program change, etc.).

The general hypothesis is that performance of the organization relies on its capability to process each regime and to move from one to another according to the situation evolution.

3 A Case Study: Lessons Learned from Applying the Approach to a Production Plant

3.1 Methodology

The model presented in the previous chapter was tested during a field study. This study focused on an outage project of an energy production plant. It covered about 11 months from preparation of the outage, through its realization and until the experience feedback phase. The SOH data collection was based on observations by immersion within the outage project/power plant staff and interviews. Observations were based on a "sampling" approach of different situations by studying either the organization of a specific service or position or by following an event/incident through the scope of different actors of the situation. Interviews allowed to gather information on the organization's explicit and implicit mechanisms during the preparation phase and were a tool for situation/incident debriefing during the realization phase. Eventually, cross section data and seven specific events were analyzed through the model developed in 2.2.

To go further, a second field study was led in the same production plant one year after. It covered a period of 6 months during the phases of realization and feedback of the outage project. This second study aimed to analyze the learning dynamics in the outage project at 3 levels: individual, collective and organizational and to link this analysis with the performant organization model. The data collection was based on 3 periods of observations by immersion and interviews. The observations aimed to collect data on collective learning processes in the outage project management team and in the technical teams, particularly when they were facing hazards, events and new situations. The interviews were led to collect data on individual trajectories, learning strategies and tools of the outage project managers and maintenance technicians as well as the organizational and collective devices that have supported the learning processes.

3.2 Dynamics and Mobilization of the Different Organizational Regimes

The results of the first field study confirmed the relevance of the model elaborated. In fact, typologies of situations and organizational regimes concepts were proven helpful to characterize, understand and analyze the outage organization and its way of dealing with long term issues and short term incident or hazards. Furthermore, the field study helped to bring to light that an organization would change its organizational regime according to the context. It appears at this stage that, while one regime seems to be predominant, the organization would occasionally switch to the two other regimes to handle the variety of situations encountered. Specifically, depending of the novelty of the situation it was facing and of its degree of preparation, the organization could often switch from its main robustness regime to an adaptability regime. This mechanism could be observed at a very local level (i.e. a maintenance team) or at a much bigger scale (i.e. whole outage project).

In some cases, the switching process between organizational regimes could occur naturally in the collective or be requested by the outage project management. Most of the time, it would be illustrated by a change of strategy to handle an incident: creation of a short term task force, brainstorming session in order to enhance creativity in problem solving, mobilization of various expert profiles from different staff (outage project, power plant, contractors or sometimes national experts…).

The flexibility allowing to change organizational regimes seems to rely on specific characteristics of the organization. In that field study, the outage project tended to prepare the outage in the most robust way in order to easily deal with classic problems and have more time and resources to face the unknown. This was achieved thanks to the prevision or temporal and resources margins, as well as specific training of the staff in order to facilitate collaboration and collective decision making skills. Typically, the "robust" organization observed during the field study would get ready for potential incident with specific risk analysis, preparation of solution through several different strategies formalized in a document. Yet, they would avoid the bias of believing that every incident is predictable and also get ready to face unknown situations by training the staff to work spontaneously in short term task forces and facilitate communication and creativity. Eventually, it appears that an organization's ability to switch to different regimes while facing a situation relies on specific "SOH dimensions" that are listed below.

3.3 The Main SOH Dimensions of the Performance

The analysis led to the conclusion that **some specific SOH dimensions were particularly important to ensure the ability to change organizational regimes** and more generally were keys for the performance of the organization. Those SOH dimensions so far identified are:

- **The capacity to prepare and manage time and resources margins** in order to absorb the human/temporal cost of switching to an adaptability or reconfiguration regime while facing a problem.
- **Having a direct and non-ambiguous communication about the margins status** for outage project staff, power plant staff and contractors. A clear communication on the margins seems to contribute to the mutual trust and the effective collaborations between stakeholders. Those elements are particularly needed under the adaptability and reconfiguration modes.
- **Developing a non-technical competence management** by preparing and animating the collectives (through role playing, training, simulation…) and establish a **participative leadership.** This seems to contribute to the needed collective dynamics under adaptability or reconfiguration regimes.
- **The management ability to keep alive the idea that « not everything is predictable »** among the staff. By doing so, the organization avoid the bias of a robust

focused preparation that would not be flexible enough to face unprecedented situations.

- **The ability to make the existing "robust" organization evolve** according to the hazards and events encountered during the previous outage project. This is typically done by capitalizing the experience and using it to adapt procedure, staff organization, or meetings planning accordingly.
- **The importance of individual and collective experience of the outage project staff**, power plant staff and contractor's staff. The collective experience helps the development of collaboration and problem solving skills necessary to handle hazards, incidents and uncertainties during an outage.

3.4 The Necessary Learning Process at Individual, Collective and Organizational Levels

The learning processes based on experience at individual, collective and organizational levels plays an essential role in promoting the necessary creativity to manage the unprecedented or very disturbed situations.

The second study which focused on the learning processes of the project management team and the maintenance teams showed that each operating regime requires specific skills that are supported by specific organizational arrangements:

- The **'Robustness regime'** requires rigor and anticipation. Robustness requires learnings during the preparation phase of the outage project to explore the potential hazards and events and after the realization phase to get feedback and lessons from the experience to be reusable for the next projects. A first condition is a good structuring of the outage project preparation phase. Robustness depends on the upstream anticipation processes of hazards, involving the whole concerned outage project actors in order to well define collectively the parries and actions to put in place in case of occurrence. It also requires a good formalization of procedures and decision-making processes in case of events. The robustness is also based on a good 'learning from experience' system that analyses all the past incidents to get lessons for future outage projects.

The 'Adaptability regime' requires a high level of expertise in the different areas represented in outage projects; a good level of autonomy and accountability of the actors in their area of expertise; cooperation and confidence in each other; a constant vigilance in maintaining a global situation awareness, a monitoring of the consumption of margins, reserves and "social climate". This is why the outage project team's composition is a crucial point, in connection with the forward-looking management of jobs. Outage teams must be composed of experienced technicians and managers (used to face complex and variable situations) mixed with novices or less experienced people who need to acquire these skills. Outage teams also need to integrate people with complementary backgrounds and expertise in the different core areas of maintenance and operation. The managerial competences of team's composition and forward-looking management of jobs are therefore a determining factor in building teams' adaptability. Other factors are the shared experience on the field of the outage project actors and common benchmarks. Those factors enhance trust, mutual understanding and dialogue that are essential for cooperation in unforeseen situations. Psychological security and collective awareness of the situation are also important for outage team's members to develop the capability to be creative.

- The 'Reconfiguration regime' is based on the same type of arrangements and learning as the 'Adaptability regime' but exacerbated. New situations require creative skills and 'do-it-yourself' skills at collective and cross-functional levels that rely on the experience and personal skills of the actors, but above all on the collective skills of sharing diagnoses and knowledge. These individual and collective creative skills are important to analyze a complex situation from different points of view with different expertise to build an original solution in a constrained time. They contribute to building the capacity for reconfiguration. As adaptability, reconfiguration requires the managers to anticipate the composition of the outage project teams integrating complementary 'pillars of experience' in different areas of plant operation and maintenance as well as the most novice to transmit skills. The setup of collective reflexive devices from the upstream phase of the outage project is also an asset for developing the reconfiguration capacity. In the energy production plant where we led this study several reflexive devices were set up to prepare and train the managers and technicians to solve complex situations: different working groups focused on anticipating potential hazards to construct parries and prepare alternative scenarios, collective simulations of events and complex situations completed by a collective debriefing meeting drove by an engineer and planning simulations. Our study shows that these devices contribute at improving the managers and technicians' problem solving skills, and also at developing trust and common knowledge between the outage project actors. Expertise, collective training, trust and common knowledge allow the outage teams to be resilient and collectively able to face uncertainty and to invent new strategies and ways of doing.

In addition to the competencies related to different regimes, other organizational conditions have impacts on learning processes in the outage project of energy production plant that are summarized below:

- The resources in terms of relational networks and information systems knowledge, mobilized and shared by the actors of the outage project when facing new problems depend on their background and the positions that they previously held in the company. This analysis reinforces the importance of reflection about forward-looking jobs management and outage team' composition in the long term.
- The training of newcomers in the outage project management team coming from different positions in the company has to be adapted regarding their background. It is more efficient when they are accompanied by a senior during their first experience in project and when they also have had at least few weeks of experience with outage maintenance teams on the field.
- The management of interfaces between trades and between project managers and technicians is an issue for the effectiveness of outage projects and the quality of the collective work. However, it appears that the management of these interfaces is regularly set aside during the preparation phase. Firstly, because it is planned at the end of the phase in the formal organization and secondly, because it involves

complex tuning processes that are 'naturally' fended off by the actors to the realization phase when the management of conflicts or incompatibilities is unavoidable. As a result, the interfaces between trades and between project managers and technicians are under-worked. The collective devices on simulations and hazards anticipation can be a solution to better manage interfaces if they are planned from upstream of the phase of preparation and involve different trade teams and both project managers and technicians.

The learning processes are one of the main levers for developing adaptability, creativity and resilience in the outage project. The management of the outage team's composition, the looking-forward management of jobs, the adaptive and practical training and the reflexive collective devices contribute at supporting learning dynamics for improving the performance of the outage projects.

4 Conclusion

Having confronted the concepts of robustness, adaptability and reconfiguration to a variety of actors involved in maintenance outages in several production units has led to promising results in terms of usefulness and usability of the developed model. The model being a simplified but not simplistic representation of reality, the transmission of concepts requires a method that facilitates both the understanding of the complexity and its appropriation by the greatest number (at least all along the managerial line). Experience has shown that the appropriation of the concepts is facilitated when the explanation is based on real examples. Then, the model stimulates the reflection of the stakeholders on their own practices, the strengths and weaknesses of their organisation, thus contributing to the identification of working areas regarding the SOH dimensions.

The practical experience brought some pre-requisite and conditions for a successful application of the approach. It requires a fundamental step of diagnosis of the current functioning of the outage organization. This diagnosis is currently processed in another production plant. It is based on an "image reflection" realized by SOH experts: 19 interviews were led with actors of the outage project management team. This macroscopic study aims at identifying the main strengths and weaknesses of the organization in the light of the model developed. The first objective is to obtain a diagnosis shared by the diversity of actors within the outage project organization. The second objective is to select some events –3 of them- which will be the subject of a more in-depth retrospective analysis. This phase is currently underway. The expectation is to review each event towards the model in order to identify in which of the 4 sets of situations it took place, which of the operating regime was processed and to determine if it was the appropriate operating regime. It would then be possible to point out the SOH dimensions that should be object of more attention or recommendations to improve the efficiency of the organization.

The next step will be to consolidate the approach and ensure its sustainable applicability in the production units. A challenge to overcome is the reduced availability of operational actors to participate in interviews and working sessions. One of the conditions of success will therefore be the managerial commitment which will itself be based on its conviction of the interest and the contribution of such an approach to build a more robust and resilient organization.

References

Hollnagel E, Woods DD, Leveson NG (2006) Resilience engineering: concepts and precepts. Ashgate Publishing, Hampshire

Hollnagel E, Paries J, Woods DD, Wreathall J (eds) (2010) Resilience engineering in practice: a guidebook. Ashgate, Farnham

Le Bot P (2011) Suivre la prescription et prendre l'initiative pour être sûr: la résilience en situation. IMDR-Les entretiens du risque (29–30 November 2011)

Roberts K (1993) New challenges to understandings organizations. Macmillan Publishing Company, New York

Rochlin GI (2001) Les organisations à "haute fiabilité": bilan et perspectives de recherche. In: Bourrier M (ed) Organiser la fiabilité, Paris, L'Harmattan, Collection risques collectifs et situations de crise

Tillement S, Cholez C, Reverdy T (2009) Assessing organizational resilience: an interactionist approach. Management 12(4):230–265

Weick K (1993) The collapse of sensemaking in organizations: the mann gulch disaster. Adm Sci Q 38:628–652

Weick KE (1995) Sensemaking in organizations. Sage Publications, London

Woods DD (2010) Resilience and the ability to anticipate. In: Hollnagel E, Paries J, Woods DD Wreathall J (eds) Resilience engineering in practice: a guidebook. Ashgate, Farnham

Effect of Shift Work on Health and Performance of the Workers – Comparison Between Turkey and the Czech Republic

Marek Bures[1](✉), Andrea Machova[1], and Ali Altunpinar[2]

[1] University of West Bohemia, Univerzitni 8, 30614 Pilsen, Czech Republic
buresm@rti.zcu.cz

[2] Gaziantep University, Üniversite Bulvarı, 27310 Şehitkamil – Gaziantep, Turkey

Abstract. The aim of this study was to map out how shift work is perceived by workers in two shift regimes and two national groups. The research was carried out by means of a questionnaire survey of the Czech manufacturing workers and the Turkish airport ground personal. The impact of the shift work was studied especially from the physical, mental, social and health aspects with connection to family status. The 55 Czech male workers, 49 Turkish male workers and 60 Turkish female workers have participated in the survey. The dependence between family status and responses of all three groups together was examined and confirmed. Also the differences in responses between the sexes of Turkish respondents were analyzed and only partially confirmed in the area of sleeping habits and mental aspects. Last but not least the effect of shift work on workers performance and scrap rate was analyses only on the sample of the manufacturing workers.

Keywords: Shift work · Physical and psychical condition · Family status Performance · Scrap rate

1 Introduction

Shift work is broadly defined as scheduled work that is completed outside the parameters of the traditional day shift e.g. from 9 a.m. to 5 p.m. [1]. Shift workers are than defined as a workers who change their working schedule and don't follow the biological rhythm of sleeping during the night [2]. The medical interest for harmfulness of shift work started between the two world wars and still is very actual. The context of shift work in the early 21st century is changing rapidly, and in comparison with previous centuries, those involved in or required to work shift work are now spread over many different business sectors. At present, approximately one in five workers around Europe and in United States work on shift basis [2].

Shift work and especially night work can have a negative impact on health and well-being of the workers. The ongoing researches usually see the negative impact of the shift work in these areas.

S. Bagnara et al. (Eds.): IEA 2018, AISC 821, pp. 160–172, 2019.
https://doi.org/10.1007/978-3-319-96080-7_20

A. The physical aspects – Due to disturbance of normal circadian cycle shift work negatively influence the sleeping routines, eating habits and have further consequences in fatigue, digestive problems, obesity or cardiovascular functions.
B. The mental aspects – Disturbance of normal sleeping routine leads also to many neuro-psychical problems like anxiety, depression, hysteria or exaggerated sensitivity.
C. The social aspects – Difficulties in maintaining the usual relationships both at family and social level, with consequent negative influences on marital relations, care of children and social contacts. This aspect can have a specific adverse effects by females due to their family roles.
D. The performance aspects – Different work performance and efficiency as well as consequent errors and accidents.

In fact, the effects of such stress condition can vary widely among the shift workers in relation to many intervening variables concerning both individual factors (e.g. age, personality traits, physiological characteristics), as well as working situations (e.g. workloads, shift schedules) and social conditions (e.g. number and age of children, housing, commuting) [3]. In the following chapters some studies which focused on specific aspects are mentioned.

1.1 The Physical Aspects

One of the most researched areas in the shift work literature is the impact on the physical health of the shift worker. Cardiovascular disease has been studied more comprehensively than other shift work-related disorders. Much of this research has been undertaken in Scandinavia, where a team of researchers has been examining the potential effects of shift work on the cardiovascular system of shift workers for over two decades. This extensive research programme has come to no definitive conclusions however [4, 5] as well as the research performed by [6].

Meal times are important synchronizers of the human life. They have both physiological and social contents: therefore, they represent a crucial point of the shift worker's life. The digestive disorders, which are often complained of by shift workers, are certainly favoured by the derangement of normal eating habits, particularly on night shift. Although the calorie intake remains substantially unaltered, the quality of food eaten by shift workers changes: on night shift they usually have quick meals, consisting of pre-packed food, and increase the intake of 'pep' drinks, such as coffee, wine and tea [3]. Abnormal eating behaviour was positively associated with shift work in a recent study conducted on nurses [7]. Increased body mass index was associated with night and rotating shift nurses, but not with permanent day shift nurses in another study [8].

1.2 The Mental Aspects

The mental aspects are usually connected with low amount of sleep or sleep–wake disturbances which results in neurological problems. Such conditions in the long run can not only give rise to permanent and severe disturbances of sleep but may also be implicated in troubles of the nervous system, such as chronic fatigue, changes in behaviour patterns, persistent anxiety or depression. 10. Jamal [9] suggests that characteristics of the shift schedule (e.g. working a rotating or night shift) are associated with a higher likelihood of health problems including trouble getting to sleep and headaches. The study performed by Geiger-Brown et al. [10] explored the relationship between demanding scheduling variables and mental health indicators of depression, anxiety and somatization among 473 US female nursing assistants working in nursing homes. Working two or more double-shifts per month was associated with increased risk for all mental health indicators, and working 6–7 days per week was associated with depression and somatization.

1.3 The Social Aspects

Shift work has been also studied in regard to social aspects in several studies. For instance Demerouti et al. [11] investigated the impact of shift work characteristics on work-family conflict, job attitudes, and health perceptions in a sample of military police. Not surprisingly, respondents working non-day or weekend shifts reported significantly greater work-family conflict compared to respondents working day shifts. Another research examining the impact of work schedules and preventative measures at work on work-family conflict was published by [12]. The authors found that different work schedules had a differential impact on work-family conflict.

Women shift workers may face even more stressful conditions in relation to the irregular working schedules and their additional domestic duties, especially if they are married and with children. Domestic circumstances have been primarily examined in terms of impact on sleep for shift workers.

The addition of children to a shift working household meant more sleepiness and more domestic disruption particularly for the female shift workers [13].

Several studies examined the reactions and feelings of the partners of shift workers. In one study, 53% of participants were unhappy or very unhappy with their partners' shift work, and a third of all respondents had tried to persuade their partners to change their working hours [14].

1.4 The Performance Aspects

The circadian fall in psycho-physical performance at night, in association with sleep deficit and stronger feeling of fatigue, decreases the work efficiency of the night worker and increases the possibility of errors and accidents. However, the studies concerning work accidents among shift workers are quite controversial: some investigations have reported more accidents on night shifts, others on day shifts, while others report accidents as less frequent, but more serious on night shifts [3].

Further field research is then necessary on this important aspect, also in relation to the recent introduction of new technologies which require more alertness and vigilance, and are then more vulnerable to errors than are manual work activities.

1.5 The Type of Industry

The current research studies can be also divided by field of industry they are focused on. Shift work is quite common in service industry and manufacturing due to economic reasons like increasing the productivity and decreasing production costs. Chou and Hsieh [15] for instance focused on the impact of shift work on sleeping quality and job performance in semiconductor manufacturing company. The same type of industry with focus on females childbearing and birth weight in different work schedules was examined in the study of [16]. A long term 10-year observational study which investigated the risk of mental health among shift and daily workers was performed by [17]. The ability to fall asleep in connection with environmental and somatic factors in the shift workers from manufacturing industry was exploited in [18]. However not much of the researches from manufacturing industry focused on work performance and the rate of scraps in shift work.

Some service industries like supermarkets or transport also assume activities 24/7 nearly all days of the year. Several studies like [19] focused on air traffic controllers which needs to work in rapid shift rotation and are also exposed to huge working stress as they are responsible for people's lives. Not many studies focused also on the airport ground staff like [20].

Several studies were performed on police officers or military police. Rajaratnam et al. [21] estimated that shift-work disorder, defined as excessive sleepiness and insomnia, was present in 14.5% of police officers who worked night shifts. Another research which resulted of poor sleeping and resting habits by police officers was published by [22]. Previously mentioned study by [11] focused on sample of military police.

Numerous studies have been also focused on mapping the influence of shift work in healthcare especially by nurses where the adverse effects may have fatal consequences on patients' health. Just only like an example we can name [7, 12, 13, 23].

Not many researchers focused on relative impact of multiple shift work features on outcomes in different national settings like [24, 25].

Based on the previous research studies, the following hypothesis were stated: H1: There will be a difference in perception of the shift work between sexes. H2: There will be a difference in perception of the shift work between nationalities. H3: There will be a difference in perception of the shift work according to the family status. H4: There will be a difference in worker performance on different shifts.

2 Methodology

2.1 Participants

There were three types of shift workers groups that were monitored and analyzed for further differences. There were several differences; sex, nationality, type of industry and finally shift regime. The first group consists of 55 Czech male workers from manufacturing industry. The mean age of the participants was 38 years (SD = 10.5) and the age range was 18–51 years. Because few females worked in this factory, only male participants were included. The shift system consists of three shifts; from 06:00 a.m.

to 02:00 p.m., from 02:00 p.m. to 10:00 p.m. and from 10:00 p.m. to 06:00 a.m. Each worker works the morning shift from 06:00 a.m. to 02:00 p.m. for one week (five working days). After that continues one week afternoon shift from 02:00 p.m. to 10:00 p.m. and last one week on night shift from 10:00 p.m. to 06:00 a.m. The Saturdays and Sundays are off. The state holidays are also a reason for day off. This shift system can be marked as slowly rotating shift work.

The second and third group consists of 49 Turkish male workers and 60 Turkish female workers from airport service in concrete ground personal. The mean age of the Turkish male participants was 28 years (SD = 3.3) and the age range was 23–40 years. The mean age of the Turkish female participants was 29 years (SD = 3.4) and the age range was 24–38 years. The shift system here also consists of three shifts; from 07:00 a.m. to 03:00 p.m., from 03:00 p.m. to 11:00 p.m. and from 11:00 p.m. to 07:00 a.m. Every worker must start 8-h shift from 07:00 a.m. and have to work two days in this shift. After that, continues two days from 03:00 p.m. shift and finally two days from 11:00 p.m. shift to get two days off. This system proceeds 24/7 uninterruptedly in every day of year. This shift system can be marked as rapidly rotating shift work.

2.2 Procedures

The assessment was done by a questionnaire survey. In order not to discourage the participants the questionnaire contained only several questions. The questions focused on individual perception of following areas:

1. Do you mind working in a changeable shift system?
2. Do you think that the shift work affects making the time for yourself and your loved ones?
3. Do you think that the shift work influences you physically? (eating habits, visage)
4. Do you think that the shift work influences you psychologically and mentally? (anger, sensuality overreacting, immediate response, anhedonia etc.)
5. Do think that shift work influences your health? (getting sick easily, having a chronic disease, using drug permanently)
6. Do you think that the shift work influences your sleeping routine?

Beside the questions the information about sex, age and family status were collected. The five point Likert scale was used for possible answers, ranging from 1 (never affect) to 5 (always affect). The questionnaires have been distributed to employees on one joined meeting where the purpose of the study and the form of the questionnaire was explained. The employees were asked to fill in the survey in their leisure time in order to avoid them feeling under pressure or stress and to consult answers between themselves. The deadline for completing the questionnaire was set to five working days. After the five day deadline all questionnaires were collected at the personnel department where they were picked up for further assessment.

The rate of return in Turkey was almost 100%. The rate of return in Czech Republic was 73%. All questionnaires were checked for errors and the questionnaires that showed confusion or misstatement with completion have been excluded.

In the case of Czech Republic (manufacturing company) there was also the possibility of evaluate the effect of shift work on workers performance. For this purpose the data on productivity and scrap were collected. It was dropped from the original intention to assess workplace accidents due to incomplete data. The data were collected for one month, during morning, afternoon and night shift. In total 3600 reports were received for individual products. In order to prepare the final evaluation, the reports were reduced to twenty reports representing each working day and each shift in the company, which were subsequently multiplied by a coefficient to avoid sensitive data being exploited.

3 Results

First evaluation which was done was categorization according to the family status. There were three groups of family status: in a relationship, married and single. The Czech males representation was 33% in relationship, 49% married and 18% single. The Turkish males representation was 43% in relationship, 35% married and 22% single and the Turkish females representation was 39% in relationship, 28% married and 33% single. We can say that the group's proportion was almost equal except for Czech workers here the single group was represented as least.

As a next step the histograms for individual questions and answers were generated. On the following pictures you can see the results (see Figs. 1, 2 and 3.). Before statistical analysis, only from these graphs we can see some major differences. For example the Turks (males and females) don't mind working in the shift system as much as the Czechs despite the fact that they work in rapidly rotating shift work which is more difficult to adapt to than the slowly rotating shift work. Another obvious difference is between males and females as the sleeping routine is heavily affected by Turkish females. This is probably due to different approaches to sleep by males and females and the fact that females sleep longer in general.

Because of insufficient amount of responses from different age groups the statistical evaluation was performed only for gender and family status variables. There were three possible relations examined: the relationship between family status and answers for each group separately, the relationship between family status and answers for all groups together regardless of nationality and the relationship between sex of Turkish respondents and answers.

To confirm/reject the hypothesis about whether the two variables are independent/dependent (whether the answers depend on the family status), the Chi-Squared Distribution of the good match was performed. In order to use this test the answers 1 and 2 had to be merged. In the case of monitoring the relationship between family status and the answers for each group separately the condition was violated, so we could not safely say whether there is a statistically significant dependence between these variables for lack of data (there are only five singles in Czech nationality). Based on this result other test was performed where the dependence between family status and responses of all three groups together was examined. In this case, it turned out that it depends on the family stratus. The test rejected the zero hypothesis of independence at level of significance $\alpha = 0.05$ for answers to the 2nd, 3rd and 4th questions. So answers for those questions depend on family status. In the case of level of significance $\alpha = 0.1$, we rejected the zero hypothesis for all questions, which means that family status and

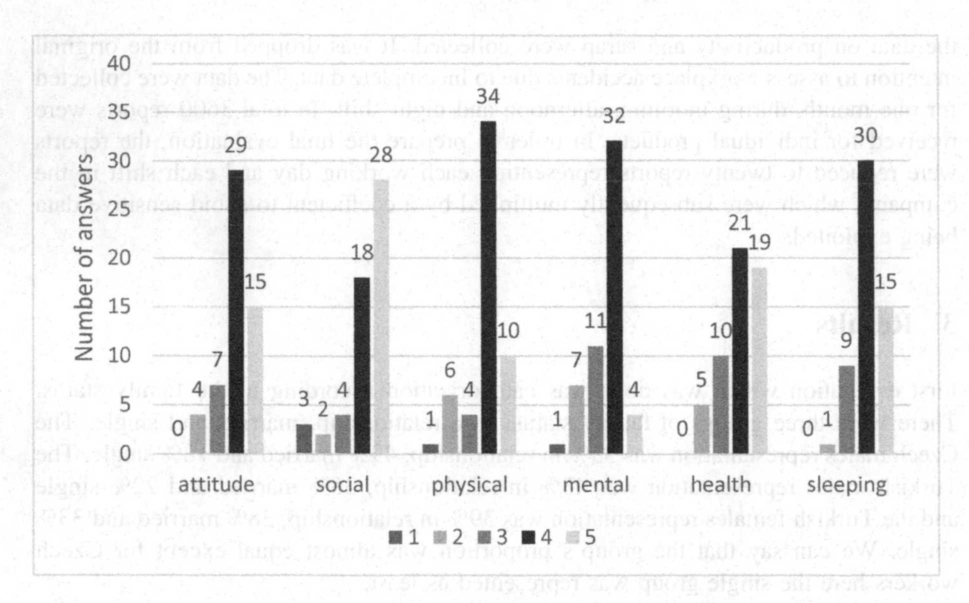

Fig. 1. Histogram of answers for Czech males.

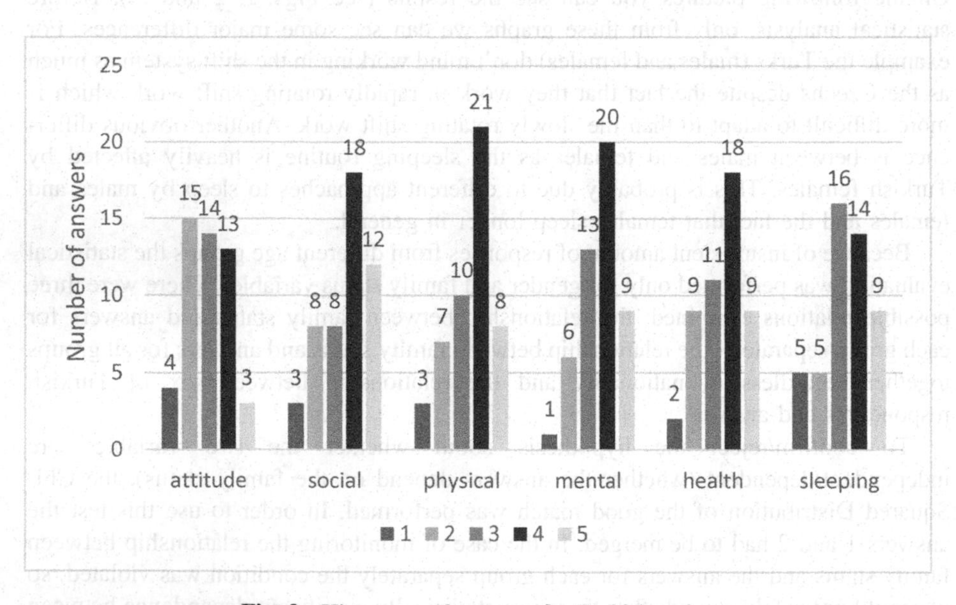

Fig. 2. Histogram of answers for Turkish males.

responses from all groups (Turkey men/women and Czech men) were dependent. Thus the H3 hypothesis was confirmed.

The last test, which was carried out, focused on differences in responses between the sexes of Turkish respondents. The results showed that men and women from Turkey have a different opinion on the sleeping and mental aspects on level of

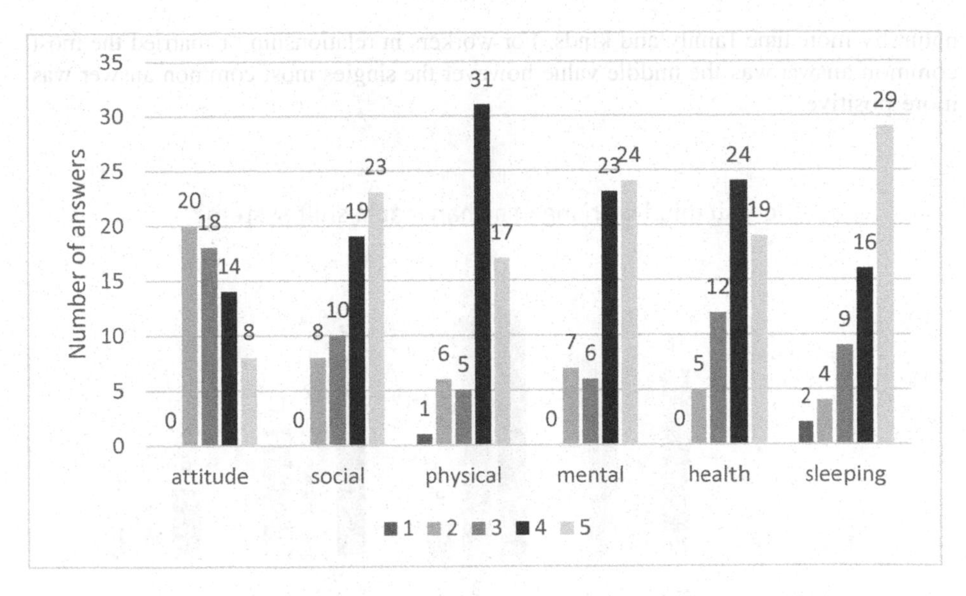

Fig. 3. Histogram of answers for Turkish females.

significance α = 0.05, mainly due to the large differences on the scale 5). The hypothesis H1 was only partially confirmed as the Turkish females concerned more about the shift work influence than males but only in the area of sleeping habits and mental aspects.

4 Discussion

According to several authors, about 20% of all workers have to leave shift work in a very short time because of serious troubles; on the other hand, only 10% of all workers do not complain about shift work during their working life, while the remaining 70% withstand shift work with different levels of intolerance [3]. In our research the only acceptance of shift work was stated by Turkish males. On contrary the Czech males were rather strict in non-acceptance of shift work. So It seems that this fact is rather dependant on external factors (like company culture, shift schedules etc.) than on sex or nationality.

Although there has been some evidence presented suggesting that the impact of shift characteristics on some aspects of worker well-being may differ by nation [25]. Evaluation of data in [24] did not provide evidence of cross-national differences in the magnitude or direction of relationships between shift characteristics and indicators of off-shift quality of life. The same result was obtained within our study. The nationality didn't make such a difference on the shift work perception thus the H2 hypothesis was not confirmed.

What was confirmed was that there are statistically significant differences between different family status groups. From the graph on Fig. 4. We can see that the singles cope better with the shift work than married people or in relationship who needs naturally more time family and kinds. For workers in relationship or married the most common answer was the middle value however the singles most common answer was more positive.

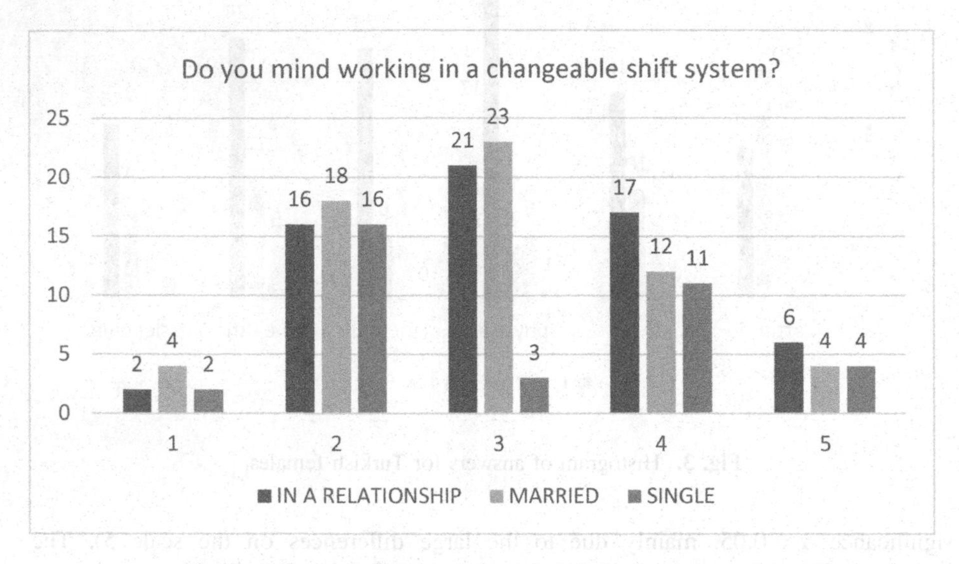

Fig. 4. Influence of family status on shift system acceptance.

As described previously we have been also able to monitor the productivity and scrap rate in the Czech manufacturing company. The results described in the following graphs (Figs. 5 and 6) show obvious differences. The data represents 20 working days. Regarding the productivity (Fig. 5) there's probably no surprise that the night shift had the lowest values but quite surprisingly the afternoon shift reached the maximums in several products. The same results can be seen also on the scrap rate (Fig. 6). As presumed the highest scrap rate was observed on night shift oscillating between average rates and the values above but surprisingly the lowest values were observer not at the morning shift but in the afternoon shift. The above outputs confirmed the H4 hypothesis.

5 Conclusion

Shift work is still very actual in these days and therefore we targeted our research on this area. The aim of this research was to compare the effects of different shift regimes and to find out if nationality of respondents has an influence on the shift work acceptance. A structured questionnaire was used to assess the impact of the shift work on attitude to shift work, social, psychological, mental and health aspects and, last but not least, sleeping routine. Four hypotheses were stated at the beginning of the research which were confirmed or disproved in the following manner.

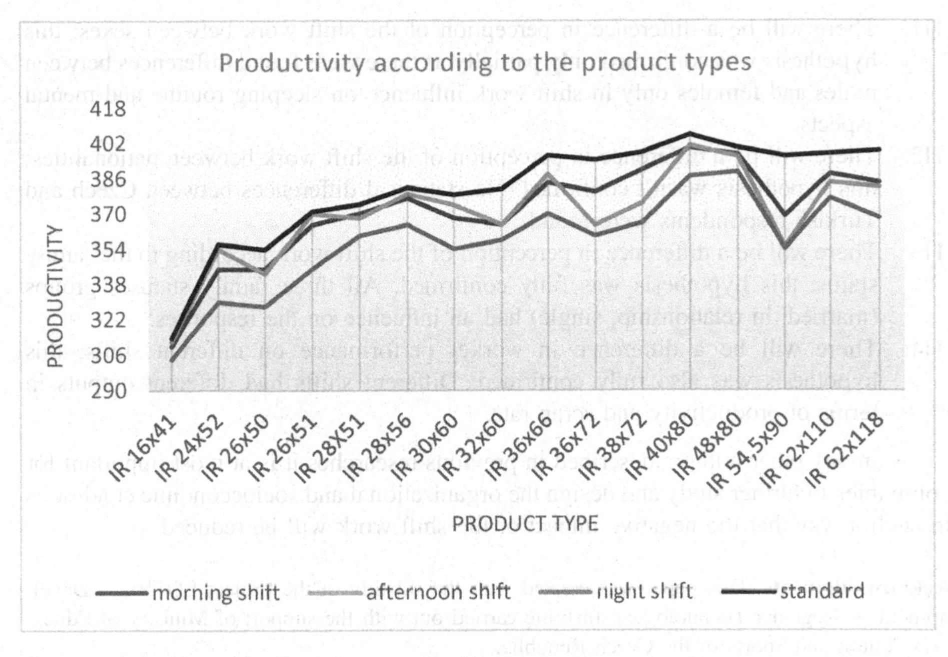

Fig. 5. Difference in productivity according to the shift type.

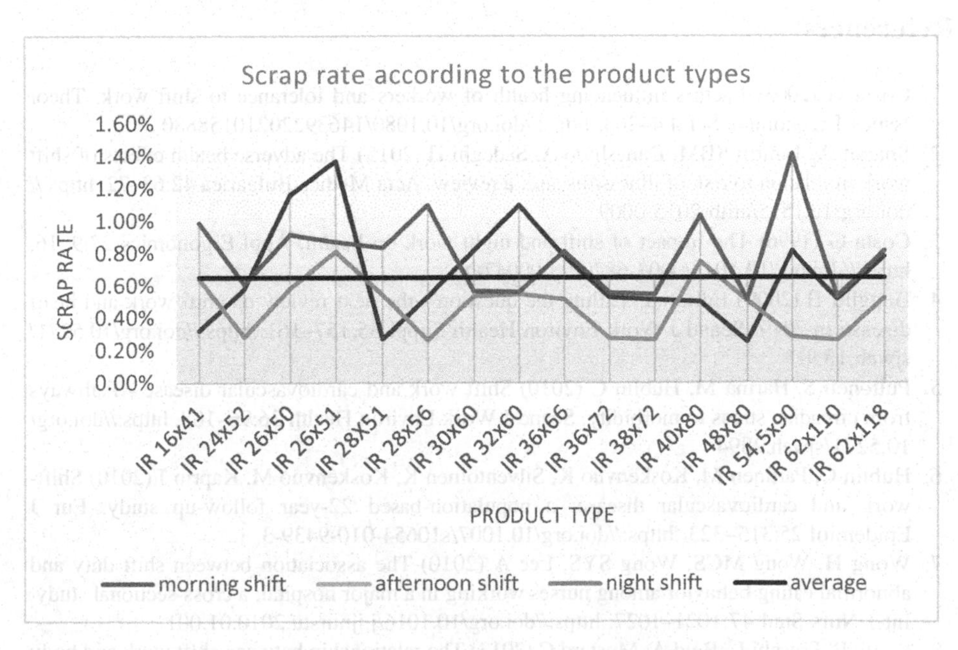

Fig. 6. Difference in scrap rate according to the shift type.

H1: There will be a difference in perception of the shift work between sexes; this hypothesis was confirmed only partially as there were found differences between males and females only in shift work influence on sleeping routine and mental aspects.

H2: There will be a difference in perception of the shift work between nationalities; this hypothesis wasn't confirmed. No statistical differences between Czech and Turkish respondents were found.

H3: There will be a difference in perception of the shift work according to the family status; this hypothesis was fully confirmed. All three family statuses groups (married, in relationship, single) had an influence on the responses.

H4: There will be a difference in worker performance on different shifts; this hypothesis was also fully confirmed. Different shifts had different outputs in terms of productivity and scrap rate.

As many other authors described in previous researches it is at most important for companies to further study and design the organizational and socioeconomic conditions in such a way that the negative impact of the shift work will be reduced.

Acknowledgment. This paper was created with the subsidy of the project LO1502 – Development of Regional Technological Institute carried out with the support of Ministry of Education, Youth and Sports of the Czech Republic.

References

1. Costa G (2003) Factors influencing health of workers and tolerance to shift work. Theor Issues Ergonomics Sci 4:4–263. https://doi.org/10.1080/14639220210158880
2. Shariat A, Tamrin SBM, Daneshjoo A, Sadeghi H (2015) The adverse health effects of shift work in relation to risk of illness/disease: a review. Acta Medica Bulgarica 42:63–72. https://doi.org/10.1515/amb-2015-0009
3. Costa G (1996) The impact of shift and night work on health. Appl Ergonomics 27:9–16. https://doi.org/10.1016/0003-6870(95)00047-X
4. Bøggild H (2009) Editorial: settling the question - the next review on shift work and heart disease in 2019. Scand J Work Environ Health Suppl 35:157–161. https://doi.org/10.5271/sjweh.1330
5. Puttonen S, Härmä M, Hublin C (2010) Shift work and cardiovascular disease - Pathways from circadian stress to morbidity. Scand J Work Environ Health 36:96–108. https://doi.org/10.5271/sjweh.2894
6. Hublin C, Partinen M, Koskenvuo K, Silventoinen K, Koskenvuo M, Kaprio J (2010) Shift-work and cardiovascular disease: a population-based 22-year follow-up study. Eur J Epidemiol 25:315–323. https://doi.org/10.1007/s10654-010-9439-3
7. Wong H, Wong MCS, Wong SYS, Lee A (2010) The association between shift duty and abnormal eating behavior among nurses working in a major hospital: a cross-sectional study. Int J Nurs Stud 47:1021–1027. https://doi.org/10.1016/j.ijnurstu.2010.01.001
8. Smith P, Fritschi L, Reid A, Mustard C (2013) The relationship between shift work and body mass index among Canadian nurses. Appl Nurs Res 26:24–31. https://doi.org/10.1016/j.apnr.2012.10.001

9. Jamal M (2004) Burnout, stress and health of employees on non-standard work schedules: a study of Canadian workers. Stress Health 20:113–119. https://doi.org/10.1002/smi.1012
10. Geiger-Brown J, Muntaner C, Lipscomb J, Trinkoff A (2004) Demanding work schedules and mental health in nursing assistants working in nursing homes. Work Stress 18:292–304. https://doi.org/10.1080/02678370412331320044
11. Demerouti E, Geurts SAE, Bakker AB, Euwema M (2004) The impact of shiftwork on work-home conflict, job attitudes and health. Ergonomics 47:987–1002. https://doi.org/10.1080/00140130410001670408
12. Camerino D, Sandri M, Sartori S, Conway PM, Campanini P, Costa G (2010) Shiftwork, work-family conflict among Italian nurses, and prevention efficacy. Chronobiol Int 27:1105–1123. https://doi.org/10.3109/07420528.2010.490072
13. Lushington W, Lushington K, Dawson D (1997) The perceived social and domestic consequences of shiftwork for female shiftworkers (nurses) and their partners. J Occup Health Safety Aust N Z 13:461–469
14. Matheson A, O'Brien L, Reid J-A (2014) The impact of shiftwork on health: a literature review. J Clin Nurs 23:3309–3320. https://doi.org/10.1111/jocn.12524
15. Chou AWM, Hsieh C-L (2010) The impact of shift work implementation on sleeping quality and job performance: a case study of semi-conductor manufacturing company. In: Presented at the 40th international conference on computers and industrial engineering: soft computing techniques for advanced manufacturing and service systems, CIE40 2010
16. Lin Y-C, Chen M-H, Hsieh C-J, Chen P-C (2011) Effect of rotating shift work on childbearing and birth weight: a study of women working in a semiconductor manufacturing factory. World J Pediatr 7:129–135. https://doi.org/10.1007/s12519-011-0265-9
17. Norder G, Roelen CA, Bültmann U, van der Klink JJ (2015) Shift work and mental health sickness absence: a 10-year observational cohort study among male production workers. Scand J Work Environ Health 41:413–416. https://doi.org/10.5271/sjweh.3501
18. Taniyama Y, Nakamura A, Yamauchi T, Takeuchi S, Kuroda Y (2015) Shift-work disorder and sleep-related environmental factors in the manufacturing industry. J UOEH 37:1–10
19. Sonati J, de Martino M, Vilarta R, Maciel É, Moreira E, Sanchez F, de Martino G, Sonati R (2015) Quality of life, health, and sleep of air traffic controllers with different shift systems. Aerosp Med Hum Perform 86:895–900. https://doi.org/10.3357/amhp.4325.2015
20. Bellier S, Briet M, Chaix S, Colin J, Collet R, Fau-Prudhomot P, Monel C, Picou S, Robineau B, Rolland C, Sanchez-Bréchot M-L (2017) Effects of shifts in work hours for airport ground staff. Archives des Maladies Professionnelles et de l'Environnement. 78:137–146. https://doi.org/10.1016/j.admp.2016.07.002
21. Rajaratnam SMW, Barger LK, Lockley SW, Shea SA, Wang W, Landrigan CP, O'Brien CS, Qadri S, Sullivan JP, Cade BE, Epstein LJ, White DP, Czeisler CA (2011) Sleep disorders, health, and safety in police officers. JAMA, J Am Med Assoc 306:2567–2578. https://doi.org/10.1001/jama.2011.1851
22. Wang X-S, Armstrong MEG, Cairns BJ, Key TJ, Travis RC (2011) Shift work and chronic disease: The epidemiological evidence. Occup Med 61:78–89. https://doi.org/10.1093/occmed/kqr001
23. Yarmohammadi H, Pourmohammadi A, Sohrabi Y, Eskandari S, Poursadeghiyan M, Biglari H, Ebrahimi MH (2016) Work shift and its effect on nurses' health and welfare. Soc Sci (Pak) 11:2337–2341
24. Barnes-Farrell JL, Davies-Schrils K, McGonagle A, Walsh B, Milia LD, Fischer FM,

Hobbs BB, Kaliterna L, Tepas D (2008) What aspects of shiftwork influence off-shift well-being of healthcare workers? Appl Ergonomics 39:589–596. https://doi.org/10.1016/j.apergo.2008.02.019
25. Tepas DI, Barnes-Farrell JL, Bobko N, Fischer FM, Iskra-Golec I, Kaliterna L (2004) The impact of night work on subjective reports of well-being: an exploratory study of health care workers from five nations. Rev Saude Publica 38:26–31

Ergonomic Maturity Model: A Practical Macroergonomic Tool

Yordán Rodríguez Ruíz[1(✉)] and Elizabeth Pérez Mergarejo[2]

[1] National School of Public Health, Universidad de Antioquia, Cl. 62 #52-59 Medellín, Colombia
yordan.rodriguez@udea.edu.co

[2] Industrial Engineering School, Universidad Pontificia Bolivariana, Cq. 1 #70-01 Medellín, Colombia
elizabeth.perezme@upb.edu.co

Abstract. This paper presents the Ergonomic Maturity Model, a new macroergonomic tool that allows organizations to assess their capacity to introduce, apply and integrate ergonomics into their processes. The model establishes five levels of maturity: Ignorance, Understanding, Experimentation, Regular Use and Innovation, the latter being the desired state. The maturity assessment with the model is supported by an assessment matrix, which describes the organization's behavior at each maturity level through four macro factors and their corresponding factors: (1) Culture (acceptance and teamwork); (2) Integration (strategic alignment, management, commitment and resources); (3) Performers (knowledge and skills, person in charge and compensation); and (4) Surveillance (indicators, information systems and risk assessment). The Ergonomic Maturity Model is a practical tool, which from a systemic perspective, allows the identification of barriers, opportunities and strategies for improvement during the development and integration of ergonomics in the organization. Finally, the results of the application of the Ergonomic Maturity Model in two Colombian organizations are presented.

Keywords: Systems ergonomics · Macro-ergonomic assessment
Organizational ergonomics

1 Introduction

For several decades, ergonomics has been regarded as a profession and a scientific discipline based on a systems and philosophy approach [1, 2]. In North America, system ergonomics has become popular as macroergonomics [2] and aims to ensure that work systems are fully harmonized with their socio-technical characteristics. In terms of general systems theory, a fully harmonized and compatible system can result in synergistic improvements in several organizational criteria: health, safety, comfort, productivity, quality of products and services, job satisfaction and quality of working life [3].

Despite the macroergonomic approach is not a new perspective, its application in Latin American countries is incipient and even unknown to many. This poses a challenge to all researchers and practitioners of ergonomics in the region, since this perspective

S. Bagnara et al. (Eds.): IEA 2018, AISC 821, pp. 173–185, 2019.
https://doi.org/10.1007/978-3-319-96080-7_21

makes it possible to make the design, development, intervention and implementation of ergonomics more successful [4].

For the application of the principles promoted by macroergonomics a set of methods, tools and techniques have been developed [5, 6]. These methods have been critically reviewed by several researchers in order to define new strategies aimed at increasing the effectiveness of their application in real contexts and in correspondence with new theories and trends [5, 7]. Among several ideas discussed, it is mentioned that in the development of new macro-ergonomic methods a balance must be achieved between their generality, validity and usefulness [7].

A recently published critical analysis of maturity models related to the security culture [8] shows how has the interest in the use of these models grown in recent years. On the other hand, reporting on the development and application of maturity models in ergonomics is limited [9–12].

This paper presents the Ergonomic Maturity Model, a macroergonomic tool for assessing the ability of an organization to introduce, develop, and apply ergonomics, using a set of influential factors as a reference [10]. This model is an easy-to-use tool that allows you to visualize the current status of a company and strategize for improvement. This model, together with other available tools, is intended to contribute to the understanding and application of the principles of macro-ergonomics in organizations, extending beyond the workplace, the impact that ergonomics has today in many Latin American countries. Also, the results of the application of the model in the maturity assessment in ergonomics in two Colombian organizations are presented.

2 Development of Ergonomic Maturity Model

The design of the Ergonomic Maturity Model is based on the approach proposed by the maturity models for the improvement of processes and on the principles of macro-ergonomics. This was designed in two stages: planning and design, taking as a reference the development of other models [13, 14].

Planning stage:

1. Team building.
2. Bibliographic review of available maturity models [15], identifying the main limitations and advantages presented. It should be noted, as a result of the review, that there is a paucity of available maturity models in the field of ergonomics [9, 11]. The models that contributed most to the development of the proposed model were: Process and Enterprise Maturity Model [16]; Business Process Maturity Model [17]; the model proposed in this paper Assessing Ergonomics Maturity Level [9]; and the Ergonomic Maturity Model [11].
3. Definition of the objective and the name of the model to be developed.

Design stage:

1. Definition of maturity levels of the model. The behavior of each level is described in general terms, based mainly on the literature studied.

2. Definition of the macro factors and factors of the model taking into account the literature studied and the opinion of the experts.
3. Preparation of the model assessment matrix defining the behavior of macro factors and their factors for each of the five levels of maturity model.
4. Tools development (checklists and surveys), which together with other participatory techniques (workshops, interviews, focus groups, etc.), allow the collection of information to assess the maturity of the organization.
5. Design of a procedure for the application of the Ergonomic Maturity Model in organizations [12].

3 Ergonomic Maturity Model

The aim of the Ergonomic Maturity Model is to assess the ability of an organization to introduce, develop and apply ergonomics. This model [10, 12] allows the maturity of the organization to be classified into five possible levels (see Fig. 1). The classification is gradual, from level 1: Ignorance to level 5: Innovation, the highest level that can be reached, the latter being the desired state. An overview of each model maturity level is presented below.

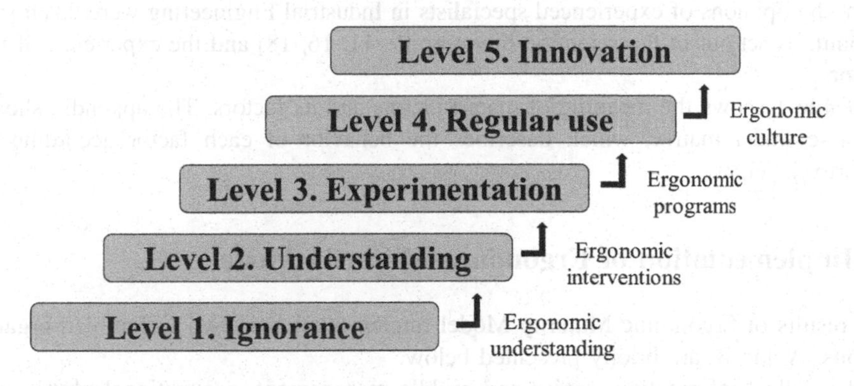

Fig. 1. Ergonomic maturity levels.

Level 1: Ignorance. At this level, organizations do not know what ergonomics is. They do not understand why they have problems related to the human factor in their production processes/services and ergonomics is not considered as a source of solution.
Level 2: Understanding. At this level, organizations recognize that they have problems associated with the human factor. They are concerned about the increase in the number of injuries and accidents at work, and are paying compensation to their employees for this. They look for quick solutions to their problems and begin to understand how ergonomics can help them.

Level 3: Experimentation. At this level, the profitability of ergonomics is recognized primarily for reducing injuries. Small ergonomic interventions are performed to experience their effectiveness. Some positive results are obtained as a result of the interventions, but in isolation. The use of ergonomics is limited due to limited experience.
Level 4: Regular use. At this level the organization regularly uses ergonomics to prevent injuries, accidents and improve performance. Ergonomics programs are developed under the leadership of ergonomics committees. The role of the ergonomist is recognized. Good practices related to ergonomics are replicated and disseminated.
Level 5: Innovation. At this level, ergonomics has been harmoniously integrated with the organization's processes. Ergonomics is part of the organization's culture. Ergonomics programs are designed and implemented with the participation of all the employees involved. Surveillance systems are used to predict and monitor the performance of ergonomic indicators. The generation of innovative proposals is promoted and rewarded.

To determine the level of maturity, the Ergonomic Maturity Model evaluates four macro factors in the company: culture, integration, performers and surveillance. These macro factors are deployed in 12 factors: acceptance, teamwork, strategic alignment, management, commitment, resources, knowledge and skills, person in charge, compensation, indicators, information systems, risk assessment [10, 12]. In order to define them, the opinions of experienced specialists in Industrial Engineering were taken into account, as set out in the scientific literature [9, 11, 16, 18] and the experience of the authors.

Table 1 shows the meaning of macro factors and its factors. The appendix shows the assessment matrix, which describes the behavior of each factor according to maturity level.

4 Implementation of Ergonomic Maturity Model

The results of Ergonomic Maturity Model implementation in two Colombian organizations, A and B, are briefly presented below.

To collect information, senior and middle management, occupational physicians, occupational safety and health officers (SHO) and operational workers were interviewed.

The evaluation was based on the assessment matrix (see Table 4 in Appendix). Surveys, interviews and group surveys were applied to identify the level of maturity of each Ergonomic Maturity Model factor. The maturity level of each macro factor was determined, given the lower level reached by its corresponding factors. Finally, the level of maturity of the organization was determined, given the lower level reached by the Ergonomic Maturity Model macro factors. What supports this procedure is the fact that, if an organization has some factor at a lower level than the rest, it will not be able to enjoy all the benefits of being at the higher level of maturity, always having some element that prevents it from moving forward.

Table 1. Macro factors and factors of the Ergonomic Maturity Model.

Culture: Disposition and way of working of the organization for the use of ergonomics	
Acceptance	Scale that people in the organization accept ergonomics as a useful tool for solving problems and improving the performance of their processes
Teamwork	How teamwork is used to solve ergonomic problems in the organization
Integration: The degree that management structures and policies prevailing in the organization condition the integration of ergonomics with the organization's processes	
Strategic alignment	How the implementation of ergonomics in the organization contributes to the fulfillment of strategic objectives
Management	The way that ergonomics is planned, executed and controlled to achieve the objectives
Commitment	Commitment of senior management to the development of ergonomics in the organization
Resources	Ability of human and financial resources for the development of ergonomics
Performers: Individuals (internal and external) who perform ergonomics in the organization	
Knowledge and skills	Set of knowledge and skills of the executors to implement and develop ergonomics
Person in charge	Individuals or group of people responsible for ergonomics in the organization
Compensation	Moral and material incentives for good practices related to ergonomics
Surveillance: How ergonomic information is collected, analyzed, interpreted and used in the organization	
Indicators	Type and nature of the indicators defined in the organization related to ergonomics. How the indicators are used
Information systems	Technologies used in the organization to collect, analyze, interpret and communicate information related to ergonomics
Risk assessment	Comprehensive process of hazard identification, risk analysis, and risk evaluation

Table 2 shows the results obtained in the assessment for both organizations.

Both organizations were classified at maturity level 1: Ignorance, although there are differences in the behavior of the factors of organization A and B (see Table 2).

Organization A was classified in level 1 of maturity, due to the teamwork factor, which, being in a lower level, will not allow, in practice, the development of the rest of the elements. For organization B, it is clear that many more factors need to be worked on to improve in order to advance at least to maturity level 2.

Table 2. Maturity assessment using Ergonomic Maturity Model for organizations A and B.

Macro factors	Factors	A	B
Culture	Acceptance (CA)	3	2
	Teamwork (CT)	1	2
Integration	Strategic alignment (IA)	2	1
	Management (IM)	2	1
	Commitment (IC)	3	2
	Resources (IR)	4	1
Performers	Knowledge and skills (PK)	2	1
	Person in charge (PP)	3	1
	Compensation (PC)	4	1
Surveillance	Indicators (SI)	2	1
	Information systems (SS)	3	1
	Risk assessment (SR)	3	1
Ergonomic maturity level		1	1

As part of the organizations' improvement plans, strategies were developed to surpass the barriers that prevent the transition to maturity level 2. This approach facilitates the gradual maturing of the organization [17, 19]. These strategies were given a level of priority for implementation, starting with the most immature and inconsistent factors. Table 3 shows the priorities for organization A.

Table 3. Priorities in the factors to be improved for organization A.

Macro factors	Factors	Maturity level 1	2	3	4	5
Culture	Acceptance (CA)			■		
	Teamwork (CT)	■				
Integration	Strategic alignment (IA)		■			
	Management (IM)		■			
	Commitment (IC)			■		
	Resources (IR)				■	
Performers	Knowledge and skills (PK)		■			
	Person in charge (PP)			■		
	Compensation (PC)				■	
Surveillance	Indicators (SI)		■			
	Information systems (SS)			■		
	Risk assessment (SR)			■		

+ Priority -

5 Conclusions and Future Work

The Ergonomic Maturity Model is a macroergonomic tool available to ergonomic professionals. It is expected its use under a systemic perspective will contribute to broadening the limited scope (human-machine interaction) of ergonomics that is still prevalent in most Latin American organizations. The Ergonomic Maturity Model has been useful in identifying barriers and opportunities, with respect to the introduction, development and application of ergonomics and in developing actions aimed at integrating ergonomics into the organization's processes.

The usefulness of the Ergonomic Maturity Model could be extended beyond the organization boundaries, as the knowledge gained during assessment can be collected and used by government institutions, ministries, insurance organizations, universities, unions, professional associations, etc. in establishing policies and strategies to develop ergonomics in a sector, region or country. In this way, self-knowledge would be created, which has proven to be applicable and to work in the real conditions of a specific context; leaving behind the historical and classical approach of adapting models created in first-world countries to the Latin American context which is quite different.

The ergonomic maturity of an organization will be influenced by the legal, cultural, economic, political, geographical and historical context in which it is located. This may impact on the relative weight of the factors measured in the Ergonomic Maturity Model to determine the maturity level of the organization. For this reason, in order to increase the flexibility of the Ergonomic Maturity Model, a procedure is being developed that allows the model factors to be weighted according to the context in which it is applied.

Another aspect that is being developed is the creation of an ergonomic maturity index, which will enable the maturity levels obtained to be quantitative differentiated between various organizations, even when they have been classified at the same maturity level.

Appendix

Table 4. The evaluation matrix of Ergonomic Maturity Model.

		L 1. Ignorance	L 2. Understanding	L 3. Experimentation	L 4. Regular use	L 5. Innovation
Culture	**Acceptance**	It is not known about Ergonomics and how it can help solve problems	The existence of Ergonomics problems in the company is recognized and a positive image of Ergonomics is beginning to be created	The need for the ergonomic use to reduce injuries is accepted and the benefits of its use are recognized	Full acceptation of the need and use of ergonomics in reducing injuries and improving process performance (efficiency and effectiveness)	There is an ergonomic culture and ergonomics is included in the organization's decisions
	Teamwork	Teamwork is not used to solve ergonomic problems	Occasionally, a group is in charge of the analysis of Ergonomics problems informally	A group is formally in charge of Ergonomics problem analysis and solution development	Ergonomics committees train company staff and lead the development of Ergonomics programs	As part of the organization´s culture, teamwork is employed for the analysis and solution of Ergonomics problems
Integration	**Strategic alignment**	The application of Ergonomics is not associated with the strategic objectives of the company	The actions related to Ergonomics are isolated and are not related to the strategic objectives of the company	Ergonomics projects are developed that may or may not contribute to the company's strategic objectives	The objectives of the Ergonomics programs are bonded to the strategic objectives of the company	Ergonomics is taken into account in shaping the company's strategic objectives

(continued)

Table 4. (*continued*)

	Management	It is not known that Ergonomics can be an aspect to manage in the company	The individual changes you make are not planned or controlled (not managed)	Efforts are aimed mainly at redesigning, rebuilding and repairing workplaces and systems	Ergonomics is integrated into the company's existing prevention programs	Ergonomics management is integrated into the company's management system
	Commitment	Senior Management does not support initiatives related to Ergonomics. They're skeptical	Senior management is interested in Ergonomics as an aid to solving company problems	Senior Management supports the development of attractive Ergonomics projects to solve the most urgent problems	Senior Management recognizes the need for the use of Ergonomics to reduce injuries, reduce costs and increase productivity. Supports the development of Ergonomics programs	The Senior and Intermediate Management supports and promotes the development and application of Ergonomics
	Resources	No resources are assigned to Ergonomics	Some resources (human and financial) are assigned to training in Ergonomics	Resources its assigned only to those Ergonomics projects that are expected to succeed or solve the most urgent problems	Resources are assigned for Ergonomics programs whose activities are duly justified	The company plans and assigns the necessary resources for the application and development of Ergonomics

(*continued*)

Table 4. (*continued*)

		L 1. Ignorance	L 2. Understanding	L 3. Experimentation	L 4. Regular use	L 5. Innovation
Performers	**Knowledge and skills**	They do not have any knowledge or skills and do not employ personnel specialized in solving ergonomic problems	There is no knowledge or skills, but they identify expert staff who can help them solve their problems	Companies has people trained in specific topics of Ergonomics and rely on specialized external personnel to execute projects in the area	The company has a specialized group of accredited people. Occasionally, they rely on specialized external personnel to carry out specific activities	The company has qualified, trained and experienced personnel in the identification and solution of Ergonomics problems and the development of innovations for improvement
	Person in charge	People are unaware of the role of the ergonomist and there is no one in charge of ergonomics in the company	The role of the ergonomist is recognized and a person who could assume that role is identified	Ergonomics projects has people in charge appointed for the duration of the project	Existence of an Ergonomics Committee that is responsible for all activities related to the ergonomics	Ergonomics responsibilities are established at all levels of the company
	Compensation	Good practices related to Ergonomics are not encouraged	Is recognized the contribution of good practices related to Ergonomics	Good practices achieved in Ergonomics projects are morally encouraged	The good practices achieved in the Ergonomics programs are morally and materially encouraged	Emphasis on the moral and material stimulation of good practices and innovative solutions related to Ergonomics

(*continued*)

Table 4. (*continued*)

Surveillance	**Indicators**	No indicators are defined or used to collect information related to workers' health and performance	There are registries related to work injuries that reflect the signs and signals of existing ergonomic problems	Some indicators related to the activities of Ergonomics are defined to monitor diseases and measure the results (cost-benefit) of projects	Indicators of safety and health, performance, cost and quality bonded to ergonomics are defined and used to detect and monitor problems	Proactive indicators are defined, integrated with the processes and aligned with the company's strategic objectives
	Information systems	IT isn't used for ergonomics	Identification of IT that can be used for Ergonomics activities	Information is gathered from existing databases and some communication channels or IT are used to report on the achievements of the implementation of Ergonomics projects	IT are implemented, together with those existing in the company, provide useful information for the performance of the Ergonomics programs and the publication of results	IT (database software management) are integrated to support management and development of Ergonomics in the company
	Risk assessment	Relevance of risk assessment is neither made nor recognized	Assessment of isolated risk are carried out with the support of external personnel	Risk assessments are limited to regulatory compliance and are conducted only in positions and activities that present frequent problems	Systematic risk assessment is done in order to control and minimize risk	Risk assessment is carried out integrated and systematically with the company's processes. It is taken into account in the implementation of new projects

References

1. Dul J, Bruder R, Buckle P, Carayon P, Falzon P, Marras WS, Wilson JR, van der Doelen B (2012) A strategy for human factors/ergonomics: developing the discipline and profession. Ergonomics 55:377–395. https://doi.org/10.1080/00140139.2012.661087
2. Wilson JR (2014) Fundamentals of systems ergonomics/human factors. Appl Ergon 45:5–13. https://doi.org/10.1016/j.apergo.2013.03.021
3. Hendrick HW, Kleiner B (2002) Macroergonomics: theory, methods, and applications, 1st edn. CRC Press, Boca Raton
4. Kleiner BM (2008) Macroergonomics: work system analysis and design. Hum Fact 50:461–467. https://doi.org/10.1518/001872008X288501
5. Waterson P, Robertson MM, Cooke NJ, Militello L, Roth E, Stanton NA (2015) Defining the methodological challenges and opportunities for an effective science of sociotechnical systems and safety. Ergonomics 58:565–599. https://doi.org/10.1080/00140139.2015.1015622
6. Stanton NA, Hedge A, Brookhuis K, Salas E, Hendrick HW (2005) Handbook of human factors and ergonomics methods, 1st edn. CRC Press, Boca Raton
7. Waterson P, Robertson MM, Carayon P, Hoonakker P, Holden R, Hettinger L, Robertson MM, Waterson P (2014) Macroergonomics and sociotechnical methods current and future directions. Proc Hum Fact Ergon Soc Annu Meet 58:1536–1540. https://doi.org/10.1177/1541931214581320
8. Goncalves Filho AP, Waterson P (2018) Maturity models and safety culture: a critical review. Saf Sci 105:192–211. https://doi.org/10.1016/j.ssci.2018.02.017
9. Gibson SL (2000) Assessing ergonomics maturity level. Proc Hum Fact Ergon Soc Annu Meet 44:578–579. https://doi.org/10.1177/154193120004402846
10. Rodríguez Ruíz Y, Pérez Mergarejo E, Montero Martínez R (2012) Modelo de Madurez de Ergonomía para Empresas (MMEE). El Hombre y la Máquina 40:22–30
11. Vidal MC, Guizze CL, Bonfatti RJ, Silva e Santos M (2012) Ergonomic sustainability based on the ergonomic maturity level measurement. Work 2721–2729. https://doi.org/10.3233/wor-2012-0516-2721
12. Rodríguez Ruíz Y, Pérez Mergarejo E (2016) Diagnóstico macroergonómico de organizaciones colombianas con el Modelo de madurez de Ergonomía. Ciencias de la Salud 14:11–25. https://doi.org/10.12804/revsalud14.especial.2016.01
13. De Bruin T, Freeze R, Kaulkarni U, Rosemann M (2005) Understanding the main phases of developing a maturity assessment model. In: Campbell B, Underwood J, Bunker D (eds) Faculty of Science and Technology. Australasian Chapter of the Association for Information Systems, Sydney, pp 8–19
14. Zeb J, Froese T, Vanier D (2011) Development and testing of a process maturity model in the domain of infrastructure management. In: Proceedings of the CIB W78-W102 2011: international conference, Sophia Antipolis, France, 26–28 October 2011
15. Pérez Mergarejo E, Pérez Vergara I, Rodríguez Ruíz Y (2014) Maturity models and the suitability of its application in small and medium enterprises. Ingeniería Ind 35:184–198
16. Hammer M (2007) The process audit. Harv Bus Rev 85:111–119, 122–123 (142)

17. Fisher DM (2004) The business process maturity model a practical approach for identifying opportunities for optimization. Bus Process Trends 9:11–15
18. Alexander DC, Orr GB (2003) Success factors for industrial ergonomics programsin. In: Karwowski W, Marras WS (eds) The occupational ergonomics handbook. Univ. Lynn, Boca Raton, pp 1561–1573
19. Röglinger M, Pöppelbuß J, Becker J (2012) Maturity models in business process management. Bus Process Manag J 18:328–346. https://doi.org/10.1108/14637151211225225

Fragility of Tertiary Ergonomics/Human Factors Programmes

Stephen J. Legg[1(✉)] and Alex W. Stedmon[2]

[1] Centre for Ergonomics, Occupational Health and Safety, School of Health Sciences, College of Health, Massey University, Palmerston North, New Zealand
sjlegg2@gmail.com

[2] Centre for Mobility and Transport, Coventry University, Coventry, UK

Abstract. For many years Massey University in New Zealand ran a postgraduate diploma and masters degree in Ergonomics/Human Factors (EHF) that were small, successful, comprehensive, internationally recognized, linked directly to professional certification and innovative. However in 2016 these qualifications were closed. This occurred soon after some other recent prominent university EHF course closures, worldwide. The demise of the Massey EHF programme was largely due to governmental restriction in tertiary academic budgets and poor understanding of the holistic nature of EHF amongst senior tertiary decision makers. This paper describes the history and composition of the Massey EHF programme. It outlines the reasons for its fragility and closure and indicates that it could easily be revived. The paper concludes with consideration of a way forward and the potential for evolution of EHF at Massey, by being more widely integrated into a broader range of academic undergraduate courses. A logical and co-operative way forward for a small country would be to integrate undergraduate teaching of EHF into wider teaching programmes across all of its universities, not just one, and to complement this with a single cross-university national masters degree.

Keywords: Future · University · Education · Integration · Undergraduate Postgraduate · Masters

1 Introduction

In recent years there have been prominent worldwide closures of Ergonomics/Human Factors (EHF) university programmes (Vitalis et al. 1999; Legg and Stedmon 2017). The EHF postgraduate programme at Massey University in New Zealand is one such case. As an exemplar, this paper describes the former Massey EHF postgraduate and masters programme, considers the reasons for its fragility and closure and its potential for evolution.

S. Bagnara et al. (Eds.): IEA 2018, AISC 821, pp. 186–191, 2019.
https://doi.org/10.1007/978-3-319-96080-7_22

2 The Massey University EHF Postgraduate Programme

Since the 1960s, and until very recently, Massey University has been the sole tertiary institution in New Zealand to have offered ergonomics and human factors modules, courses and qualifications. This started with a single introductory Ergonomics module integrated into a Graduate Diploma of Health and Safety and in 1986 became a stand-alone Graduate Diploma in Ergonomics. In the 2000 the University developed and ran a Postgraduate Diploma and Masters degree in EHF. In practice, the postgraduate diploma was the 'first year' of a two-year masters degree. The structure of the master of ergonomics degree is shown in Table 1. It provided a comprehensive postgraduate ergonomics education in New Zealand that was internationally recognised. The mode of study was flexible part-time/full-time distance learning (extramural) (Legg and Moore 1999), similar to those described by Watson and Horberry (2003) and Richardson and Baird (2008). The two qualifications exemplified interdisciplinary, cross-college/faculty co-operation and multi-campus delivery, initially being in the Faculty of Technology, then the College of Sciences. Later it came under the auspices of the College of Business.

The postgraduate diploma in ergonomics (PGDipErg) was usually completed over a number of years of part-time study. It contained five taught 'papers': work capacity and performance; ergonomics analysis; micro/macro ergonomics; people, technology and design, and; advanced research methods. Each paper was composed of a number of discrete modules aligned with and covering all of the educational components for national and international professional certification.

The master of ergonomics (MErg) degree built on the postgraduate diploma. It was characterized by provision for four alternative routes to graduation: (i) a masters thesis - to encourage research in ergonomics leading to a doctorate, (ii) a research report and two taught papers – a research focused alternative for students who wished to supplement their thesis with taught papers in an area allied to their research field, (iii) a research report and an ergonomics professional practice paper - specifically designed in recognition of the supervised professional training components of the national and international professional certification schemes and involving supervised professional training in the practice of ergonomics under joint academic and external industrial supervision, and (iv) an ergonomics professional practice paper and two taught papers.

All of the papers were offered in the distance (extramural) mode and included block 'contact' workshops of between one and five days, to supplement the coursework that was assigned in hard copy or web based 'study guides' that were sent out at the beginning of the academic year. This approach optimized contact between the teaching staff and the students yet retained the advantages of the distance learning. There was provision for students to take ergonomics papers at some overseas universities and have these credited towards their coursework, or to undertake their research projects at overseas universities.

The two qualifications encouraged the development of increased research activity in the field of ergonomics and provided all of the educational and much of the supervised professional practice components required for national and international certification (Legg 2008). Students graduating with the Master of Ergonomics had

Table 1. Massey University Master of Ergonomics degree until 2016.

Paper Title
Year 1 (120 Credits required)
Work Capacity and Performance (15 credits)
Ergonomics Analysis (30 credits)
Micro-macro Ergonomics (30 credits)
People, Technology and Design (15 credits)
Advanced Research Methods (30 credits)
Year 2 (120 Credits required)
Alternative i)
Masters thesis (120 credits)
Alternative ii)
Research Report (60 credits)
Other relevant papers (60 Credits)
Alternative iii)
Research Report (60 credits)
EHF Professional Practice (60 Credits)
Alternative iv)
EHF Professional Practice (60 Credits)
Other relevant papers (60 Credits)
Other relevant papers: Special topic in EHF (30 Credits); Advanced Occupational Health and Safety (30 Credits); Organisational behaviour (30 Credits); Human Factors in Professional Aviation (30 Credits)

demonstrable competence in the educational, professional practice components of the requirements for professional certification as a New Zealand Ergonomist. This included the following topics: ergonomics principles; human characteristics; work analysis and measurement; people and technology; ergonomics applications; professional issues; professional practice and research. In order for graduating students to become fully certified, they required some additional supervised practice and unsupervised work experience in independent practice.

3 Fragility of the Massey EHF Programme

In 2010, national governmentally driven tertiary budget restrictions and consequent academic realignment resulted in a recommendation for closure of Massey's EHF programme. However, in 2014 a newly formed College of Health at Massey aimed to revitalize the EHF programme by integrating it into a suite of health-related qualifications. Surprisingly, the new College failed to advertise or promote EHF as a unique or useful qualification. Unsurprisingly, this led to formal closure of the postgraduate EHF programme in 2016 due to 'unprofitability' and 'low student demand'. There was a university wide 'push' to delete all small modules because they were unprofitable, regardless of their national and industrial relevance. However, some of the modules were retained as an EHF option/specialization in revised Masters of Public Health and Health Sciences qualifications. Most recently however, due to non-replacement of key retiring staff, these options will no longer be offered.

4 Strategic Review of EHF at Massey

In 2015 a strategic review of EHF at Massey explored a wide range of viewpoints about EHF from academics, administrators and managers via interviews and focus groups (Stedmon 2015). There was widespread goodwill and empathetic support for EHF, particularly amongst academics. The university contained numerous academic staff who also used EHF in 'other' discipline areas e.g. aviation human factors, industrial/organizational psychology, management systems, human resource management, industrial design, product development, rehabilitation and sport science. However, there were wide differences and misunderstandings about the nature of EHF. There were also widespread systemic 'organisational' barriers within the university (a 'silo culture') that limited EHF integration across disciplines. University management was driven by a need to pigeon-hole EHF within a specific single wider discipline area. This made it difficult for EHF modules to be included in a wide range of relevant disciplines e.g. engineering/technology, design, psychology. The review recommended stronger effort to redress widespread misunderstandings about the nature of EHF, to promote EHF teaching through integrated modes (internally and distance-learning; undergraduate courses feeding postgraduate programmes) with clear relevance to a wider range of disciplines (e.g. science, health, public health, physiotherapy, occupational therapy, sport, psychology, design, technology, engineering, business, management, defence, security and aviation) and industry.

5 Potential for Evolution

In 2016 the College of Health introduced two new undergraduate modules in EHF (Work and Health and Healthy Workplace Design) These, together with an existing module (EHF: work, performance, health and design), which was a core component of a Graduate Diploma in Health and Safety, formed an EHF Minor in a revised Bachelor of Health Sciences. Unfortunately, however, in 2017 the College decided to exclude all

Minors from its undergraduate programmes. This left no undergraduate routes to EHF learning at Massey except via the Graduate Diploma in Health and Safety. Thus, at present, and despite the positivity of the strategic EHF review, the potential for further evolution of EHF learning at Massey is severely limited. The way forward would be to integrate EHF undergraduate modules into courses such as science, health, public health, physiotherapy, occupational therapy, sport, psychology, design, technology, engineering, business, management, defence, security and aviation.

For a small country, like New Zealand, it would be counterproductive for there to be competition for small niche academic programmes, such as EHF, between universities. A logical and co-operative way forward would be to integrate undergraduate teaching of EHF into wider teaching programmes across all of the universities, not just one, and to complement this with a single cross-university national masters degree. The cross-university model for teaching small but important niche disciplines such as EHF may be difficult to achieve in the competitive economically driven tertiary environment. It might however be attractive to both young and mature web-savvy students, produce internationally certified ergonomists and could even be cost effective for the Universities.

6 Conclusions

In conclusion, ergonomics education at Massey University was originally comprehensive, internationally recognized, linked directly to professional certification and innovative. It became fragile due to budgetary restrictions, misunderstandings about the nature of EHF, low student demand, unprofitability, a university-wide systemic competitive 'silo' culture, and a lack of both course promotion and institutional support. The potential for evolution was identified as teaching EHF in multiple modes (internal and distance-learning) in undergraduate modules, where it is clearly relevant to a wider range of disciplines, and feeding a revived postgraduate EHF programme in a practice-focused context relevant to industry. A logical and co-operative way forward for a small country would be to integrate undergraduate teaching of EHF into wider programmes across all of its universities, not just one, and to complement this with a single cross-university national masters degree in EHF.

References

Legg S (2008) Tertiary education and professional certification in ergonomics: an international exemplar in New Zealand. In: OHSIG Conference, 11–12 September 2008

Legg SJ, Moore D (1999) Ergonomics education and professional certification in a small country: developments in New Zealand. In: Proceedings of the international ergonomics association and hellenic ergonomics society symposium: strengths & weaknesses, threats & opportunities of ergonomics in front of 2000, Santorini, Greece, pp 101–109, 31 August–1 September 1999

Legg SJ, Stedmon A (2017) Tertiary ergonomics/human factors programmes – who cares, what next? In: Proceedings of the 48th annual conference of the association of Canadian ergonomists and the 12th international symposium on human factors in organisational design and management: organising for high performance, 31 July–3 August 2017, Banff, British Columbia, Canada, pp 201–203

Richardson M, Baird A (2008) Teaching ergonomics at a distance: a virtual workplace. In: Bust PD (ed) Contemporary ergonomics. CRC Press, pp 34–39

Stedmon AW (2015) Human factors/ergonomics at Massey University: perspectives, perceptions and potential for developing a unique teaching, learning and research capability within New Zealand. Massey University Research Fund Report, December 2015

Vitalis A, Walker R, Legg SJ (1999) Prion dromo ergonomia? Wherefore ergonomics? In: Proceedings of the international ergonomics association and hellenic ergonomics society symposium: strengths & weaknesses, threats & opportunities of ergonomics in front of 2000, Santorini, Greece, 31 August–1 September 1999, pp 53–58

Watson M, Horberry T (2003) The value of an online distance education program in human factors. In: McCabe PT (ed) Contemporary Ergonomics. Taylor and Francis, London, pp 475–480

An Exploratory Study of the Issues of a Professionalization to the Industrial Safety in a French Railway Company

Audrey Marquet[1,3](✉), Vincent Boccara[2], Stella Duvenci-Langa[3], and Catherine Delgoulet[1]

[1] LATI, Université Paris Descartes, 92100 Boulogne-Billancourt, France
audrey.marquet@etu.parisdescartes.fr
[2] LIMSI-CNRS, Université Paris Sud, Université Paris-Saclay, 91400 Orsay, France
[3] SNCF, 93210 Saint-Denis, France

Abstract. This communication aims to explore the leverages and impediments of a professionalization to an integrated safety culture. The two objectives of the study are: (1) to understand the way the strategic actors perceive the safety model and the role of the managerial line and, the operational staff, and (2) to identify the professionalization initiatives. Nine semi-structured interviews were conducted with HOF (Human and Organizational Factors), safety, and training experts. The thematic analysis was conducted with MAXQDA® software for qualitative data processing. The results show that the safety model is mainly perceived as "rule-based" and professionalization refers for the most part to training systems. This model affects the forms of professionalization of the operational staff, the missions assigned to the managerial line and the operational staff, as well as the professionalization initiatives via compliance with the prescriptions (procedures, standards). These results underline the gap to be filled between the company's desire to establish an integrated safety culture and the current experts' perception of its operationalization in the field.

Keywords: Safety · Professionalization · Rail industry

1 Professionalization as Proceeds and Lever of the Safety Culture

This thesis is conducted in the framework of a French industrial partnership (SNCF, FONCSI, Paris Descartes University, CNRS). It focuses on the evaluation of professionalization to the safety culture in a railway company (SNCF). This company is part of an organizational trend promoting a culture of "integrated safety" considered as a guarantee of enhanced reliability of technical social systems [1]. The professionalization process, proceeds of the safety culture, is seen as one of the levers for supporting the migration to this culture, conducive to organization and efficient action in work situations. The approaches to evaluation would be an entrance to understand the existing and endow the organization of a process of reflection on itself. For it, it is a

S. Bagnara et al. (Eds.): IEA 2018, AISC 821, pp. 192–198, 2019.
https://doi.org/10.1007/978-3-319-96080-7_23

question of evaluating the relevance and efficiency of the professionalization mechanisms implemented in training and at work [1]. In this context, the first exploratory study of this thesis aims to identify how experts in safety and its professionalization nowadays perceive current practices.

The general theoretical framework of the thesis joins constructive ergonomics [2, 3] and professional didactics [4], which are complementary while mobilizing the concept of "activity" [5]. For the constructive ergonomics, the work situations have, in particular, to integrate from their conception the acquisition and the development of the skills of the operators [2]. The professional didactics focuses in understanding the vocational skills' development processes and the characteristics of learning situation, both in vocational training and in the work [6].

The professionalization to the safety is considered here as the process by which an individual the necessary acquisitions and competencies recognized by a professional institution or a professional body [7]. It is thus becoming a professional. The professionals are attached by a code of ethical and technical standards. In specific cases, they exercise a judgement and do not always rely on application of general rules [8]. It is a constant concern of companies with major risks, and defines itself by four characteristics [9, 10]: (1) various objects/goals concerning the acquisition of knowledges and know-how through a training, for the exercise of their profession; (2) a process which participates in the development of an identity and a professional culture [11]; (3) a training where intervenes important changes and where can be set up the specific courses which favors the professionalization; (4) courses which are a set of means arranged in compliance with a plan, and are a place of mediation between two contexts: the conception and the use of courses [12].

The characteristics of the professionalization refer both to knowledge and know-how to trades and their identity. A narrow link exists between professionalization and culture of the company. Safety professionalization can be considered as a product of the "safety culture" (a set of characteristics and a set of attitudes shared by the individuals of an organization concerning the control of the risks) [13], which would be expressed in a professionalization program. It can also be considered as a driving force behind its deployment and effective development.

The safety model underlying this culture is then a strategic element that can be found in several forms in the literature. Three main orientations of the safety model can be distinguished: (1) "rule-based" [15] grounded in the formulation and the strict respect of rules elaborated from exogenous norms; (2) "managed" [16] related to the capacity of all the operators to take initiatives to be able to cope with variabilities and unforeseen events of the real work; (3) "integrated" [1] which does not limit itself to the regulatory and procedural dimensions, but considers jointly the necessary adjustments in real work situations [1]. The safety model chosen by a company should therefore influence the role of the managerial line and the operational staff in the safety professionalization, as well as the safety professionalization initiatives. Safety professionalization relates to training, but also to the daily work practice and to the activities favoring reflexivity (exchange times and spaces for exchanges, discussions around the practice, etc.) [17]. Given that, we propose to distinguish two main types of professionalization initiatives (a) those whose primary goal is the learning of the safety dimensions of trades (the training courses set up and financed by the company);

(b) those whose primary goal is production but underlies a reflexive dimension favorable to the development of safety skills [18]. The workplace will highlight the quality of the production, while training is the place where the personal development and the learning are prioritized [19].

In this perspective, this study explored three objectives: (1) understand the way the experts in safety and its professionalization (i.e. HOF, Safety, and Training experts) perceive (a) the safety model within the three institutions of the company; (b) the role of the managerial line and the operational staff regarding the current safety model and safety professionalization process; (2) identify the professionalization initiatives (a) according to their first aim of learning or production orientation; (b) according to the formal and informal situations which support their operationalization.

2 Investigative Methodology

2.1 Field of Investigation and Participants

In order to collect a diversity of viewpoints, the field of investigations concerns the three establishments of the company: infrastructure management, steering/support functions, railway undertaking.

The interviews were conducted with 10 experts with a strong experience and knowledge in their respective domain: four in Human and Organizational Factors (HOF), three in safety departments and three in vocational training. They were chosen for three main reasons: (1) HOF experts contribute or contributed to the policy of transformation of safety culture and training; (2) safety experts (a) translate in rule or procedure frameworks the orientations regarding safety program decided by the steering company, and (b) make animations for the safety with the workers and with their managerial staff in production situations; (3) training experts have among their primary objectives the professionalization to the safety of the workers.

2.2 Collect and Process Data

The data collection was made from semi-structured interviews [20] led to leave an interview guide covering three topics: (1) the perceived model(s) of safety mobilized in production and in training, (2) the role of the managerial line and of operational staff concerning the safety and the professionalization to the safety, (3) the learning situations in training and in production. These topics were systematically considered regarding the available resources and identified obstacles by interviewees. Interviews were conducted in a place assuring the confidentiality of exchanges. Eight of them were individual. They were recorded then transcribed. The ninth was conducted simultaneously with two experts. The counting from the point of view of the experts relate to nine interviews. It was based on a note taking (interviewees not having agreed to be recorded). The nine interviews last on average 83 min (standard deviation = 38.53; Min = 48; Max = 180). The corpus of each interview was on average 3,839 words (SD = 1,726.78; Min = 1,360; Max = 6,464). A thematic analysis [21] was conducted with MAXQDA® software for qualitative data processing [22].

The schema of coding contained two general categories: (1) the professionalization to the safety with three subcategories: (a) the role of the managerial staff (director of establishment, director of operational unit, proximity manager), (b) the role of operational staff, (c) the professionalization initiatives; (2) the model(s) of safety including three subcategories: (a) "rule-based" safety, (b) "managed" safety, (c) "integrated" safety. These subcategories were systematically considered regarding the available resources and identified obstacles by the interviewees. Then, we analyze the frequency and the contents of arguments in all categories.

3 Results: A Safety Culture Centered on the Prescription

3.1 A Dominant "Rules-Based" Institutions Safety Model

For seven out of nine experts the current model of safety in the company was "rule-based" (2 HOF, 2 Safety, 3 Training), and for two it was "integrated" (1 Safety, 1 HOF).

The experts defining the model of safety spread as "rule-based" mention four main characteristics: (1) be accountable in his national authority of safety by reports and monitoring of indicators related to safety (2/7, Training and HOF); (2) use prescription (rule and procedure frameworks) to define the work of the managerial line and the operational staff (1/7, Safety); (3) deploy prescriptions (referential) which were conveyed by a double mission of the managerial line on: (a) the management by the writing of directives to deploy (1/7, HOF), (b) the "monitoring" (e.g. control of the execution of the procedures; 1/7, Safety); (4) apply the prescriptions for the operational staff (1/9, Safety). No resource had been mentioned, but two of the seven experts identify two obstacles concerning the application of this model (2/9, Training and HOF): (1) an omnipresent regulation in the work due to the culture of the company (1/9, Training), (2) a partial use of the tools which support this model (1/9, HOF).

Two experts (HOF, Safety) defined the current model of safety as "integrated", and mentioned two characteristics: (1) adapt the prescriptions according to the context for the operational staff through four missions of application of the "monitoring" (2/9, HOF and Safety), as for example the adaptation of the procedures (1/9, HOF), (2) combine aspects "rule-based" and "managed" by a double mission of the managerial line through the "monitoring" (e.g. the assessment of the operational staff and their accompaniment; 1/9, HOF). The HOF expert identified one resource with this model: the climate of the company through the confidence between the managerial line and their operational staff to favor the feedback about events with which they were confronted. The safety expert believed that the weight of the rules can be an obstacle to the integrated-safety implementation because of the time that operational staff must devote to the rules.

3.2 Safety Missions Characterized by the Respect of the Prescription

The missions of the managerial line and the operational staff were essentially perceived as centered on the deployment and the compliance with the safety prescription.

In that perspective, all experts identified four missions relating to the safety professionalization which vary according to the hierarchical level: (1) the management and deployment of a safety strategy for the Director of Establishment (DE) and the Director of the Operational Unit (DOU) (4/9, Training, 2 Safety, HOF); (2) the management by the managerial line (4/9, 2 Safety, Training and HOF) of the unit (e.g. budget, human resources) and the management of the interfaces (e.g. computer station) by the operational staff (1/9, 1 Safety); (3) the use of "monitoring" by the DOU and the Proximity Manager (PM) to monitor compliance with the prescription and support operational staff (6/9, 2 Safety, 2 Training, 2 Training, HOF); (4) application of the "monitoring" (procedures) by the operational staff for the compliance with the prescription (5/9, 2 Safety, 2 Training, HOF).

Two resources were identified by experts (2/9, Safety, HOF): (1) a prescription (referential) on whom the managerial line relies (1/9, Safety), (2) capacity of the operational staff to progress (1/9, HOF). The experts mentioned thirteen obstacles among which organizational eleven (7/9, 2 HOF, 2 Safety, 3 Training) due to the complexity of the system (1/9, Safety) and to the organization of the units (e.g. a vast sector which does not allow the DE to see PM, or numerous reorganizations; 7/9, 2 FOH, 2 Safety, 3 Training); two cultural, on a strong presence of the prescription and due to the culture of the company (2/9, Training, Safety).

Regarding professionalization to the safety, the nine experts identified three missions in training which vary according to the hierarchical level: (1) "monitoring" by the PM supporting the operational staff (e.g. contributes to their training; 2/9, Safety); (2) human resources management by the DOU (1/9, 1 Safety); (3) training's design by the PM (1/9, 1 Safety).

For the work in production, the nine experts identified two missions: (1) "monitoring" (4/9, 1 FOH, 1 Training, 2 Safety) concern the DE, the DOU, and PM who ensuring the respect for the prescription (e.g. controls); (2) human resources management on the track of training and skills by PM (3/9, 2 Safety, Training). One expert identified an organizational resource which concerns an accompaniment of the managers by the Steering (1/9, 1 Training). Two organizational obstacles were concerned: (1) a lack of availability of the operational staff (1/9, 1 Safety), (2) a lack of time of PM to professionalize their subordinates (1/9, 1 HOF).

3.3 Professionalization Initiatives as a Compliance with the Prescription

The nine experts identified ten professionalization initiatives based on the respect for the prescription: eight were in the primary aim of learning, and two in the primary aim of production.

The first group of initiatives was directed on: (1) the respect for the prescription (training habilitation, local, continuous, at the post of work, awareness of occupational hazards, and educational toolkit; 5/9, 3 Safety, Training and HOF); (2) the understanding of the system (training habilitation and local; 2/9, Training and Safety). No element has been mentioned on training in the analysis of events. These initiatives grouped three advantages: (1) ride-along (2/9, 2 Safety), (2) an accompaniment of operational staff (1/9, Training), (3) exchanges between peers (1/9, 1 HOF). Four limits were identified: (1) a lack of human resources (trainer and operational staff; 4/9,

Training, 2 HOF, Safety), a lack of financial and material (e.g. low budget and tools in training local; 1/9, Safety), (2) a lack of practice of their job (e.g.: traffic manager, station agent, signaler) (3/9, HOF, Safety, Training), (3) a content centered on the prescriptions (2/9, HOF, Safety), (4) a gap between the prescription and the real work (1/9, 1 Safety).

The second group of professionalization initiatives was characterized by: (1) the application of the prescription of the company (i.e. Computer-based monitoring; 1/9, 1 Safety), (2) a traceability of the accidents (i.e. the experience feedback; 1/9, 1 FOH) No limits were mentioned, but two experts (1 Safety and 1 FOH) identify three advantages which concerned the identification of a non-compliance with the prescriptions.

4 Discussion - Conclusion

The objectives of the study were: (1) understand the way the strategic actors in safety and its professionalization perceive the safety model, and the role of the managerial line and the operational staff; (2) identify the professionalization initiatives.

The different expertise gathered in this first exploratory study show three main patterns: (1) according to the majority of experts the safety model is perceived as "rule-based" [15]. This model is not adapted because the omnipresent regulation in the work is time-consuming for the operational staff; (2) regarding professionalization and production to the safety, the role of the managerial line and operational staff essentially concerns on the deployment and the compliance to the prescription, and less on the professionalization; (3) the majority of professionalization initiatives identified are in vocational training. The professionalization initiatives in production area allow verifying the application of prescription knowledge concerning the operational staff.

Thus, there is still a gap between the objective of the company regarding culture of safety – go out of the "rule-based" safety [15] in moving towards an "integrated" safety [1] – and the perception of what currently operates in the production situations and professionalization situations. This lack of adequacy can lead to difficulties in daily work and training, and harm the production of safe rail operations [23].

However, these interviews do not allow to understand how this gap is played out in the daily lives of the company's employees and therefore what the levers would be to reduce it. This is the next step of the present research: (1) from the point of view of work: what are the trade-offs on the part of the managerial line and operational staff to work and professionalize? (2) from the point of view of professionalization: how are currently mobilized the systems with the primary aim of learning or production; what are the challenges of redesigning these trade initiatives?

The investigation of these questions, via the analysis of the activity (operational staff and their managers, and trainers/trainees within two trades) in real situations, will aim in a third stage to define an evaluation approach for such systems capable of supporting an integrated safety culture.

Acknowledgment. The authors thank the experts from the railway sector who agreed to participate in this preliminary study, the SNCF company and the French Foundation for a Culture of Industrial Safety (FonCSI) who support this research project.

References

1. Johansen JP, Almklov PG, Mohammad AB (2016) What can possibly go wrong? Anticipatory work in space operations. Cogn Technol Work 18(2):333–350
2. Falzon P (2013) Constructive ergonomics. CRC Press, New York
3. Guérin F, Laville A, Daniellou F, Durraffourg J, Kerguelen A (2007) Understanding and transformingwork. The practice of ergonomics. ANACT, Lyon
4. Filliettaz L, Billett S (2015) Conceptualising and connecting francophone perspectives on learning through and for work. In: Filliettaz L, Billett S (eds) Francophone perspectives of learning through work. Springer, New York, pp 19–48
5. Daniellou F, Rabardel P (2005) Activity-oriented approaches to ergonomics: some traditions and communities. Theor Issues Ergon Sci 6(5):353–357
6. Lenoir Y, Habboub E (2011) Professional didactics and teacher education: contributions and questions raised. Educ Sci Soc 11–40
7. Boccara V (2017) Training design oriented by works analysis. In: Bieder G, Gilbert C, Journé B, Laroche H (eds) Beyond safety training. Springer, New York, pp 117–126
8. Hayes J (2014) The role of professionals in managing technological hazards. The Montara blowout. In: Lockie S, Sonnenfeld DA, Fisher DR (eds) Routledge international handbook of social and environmental change. Routledge, London & New York
9. Bourdoncle R (1993) La professionnalisation des enseignants: les limites d'un mythe. Revue Française de Pédagogie 105:83–114
10. Wittorski R (2008) La professionnalisation. Savoirs 17(2):9–36
11. Hugues EC (1958) Men at their work. The free Presse, Glenoce
12. Bourdet JF, Leroux P (2009) Dispositifs de formation en ligne: de leur analyse à leur appropriation. Distance et Savoir 7(1):11–29
13. IAEA (1991) Safety culture. Safety series No. 75-INSAG-4. IAEA, Vienna
14. Dekker S (2003) Failure to adapt or adaptations that fail: contrasting models on procedures and safety. Appl Ergon 34:233–238
15. Hollnagel E (2004) Barrier and accident prevention. Ashgate, Aldershot
16. Reason J (1998) Achieving a safe culture: theory and practice. Work Stress 12(3293):293–306
17. Gilbert C (2015) La Sécurité: Une affaire de "professionnels"? Tribunes de la Sécurité industrielle 2015(06):1–6
18. Schön DA (1983) The reflective practitioner: how practioners think in action. Temple Smith, London
19. Beckers J (2007) Compétences et identité professionnelles. L'enseignement et autres métiers de l'interaction humaine. De Boeck Université, Bruxelles
20. Fylan F (2005) Semi-structured interviewing. In: Miles J, Gilbert P (eds) A handbook of research methods clinical and health Psychology. Oxford University Press, New York, pp 65–78
21. Anderson R (2007) Thematic Content Analysis (TCA) descriptive presentation of qualitative data. http://www.wellknowingconsulting.org/publications/pdfs/ThematicContentAnalysis.pdf
22. MAXQDA. https://www.maxda.com. Accessed 20 Apr 2018
23. Gouldner AW (1954) Patterns of industrial bureaucracy. The Free Press, New York

The Contribution of Creativity of Action to Safety: The Key Role of Requisite Imagination

Justine Arnoud(✉)

Université Paris-Est, IRG (EA 2354), UPEC, UPEM, 94000 Créteil, France
justine.arnoud@u-pec.fr

Abstract. The aim of this paper is to show that requisite imagination by designers and more broadly by all workers – the ability to wonder and imagine key aspects of the future we are planning/designing – can be considered as a manifestation of creative action in organization and should be encouraged. We conducted an intervention in a nuclear power plant facing with a new nationwide policy seeking to extend the lifespan of nuclear facilities through important alterations. The first alteration made turned out badly, due to poor anticipation and a focus on technical effects, whilst organizational aspects were largely excluded. We decided to facilitate the conduct of a collective inquiry with the stakeholders involved with these ongoing evolutions, to discuss organizational design and stimulate requisite imagination. We discuss our main finding from three perspectives: theoretical, methodological and practical.

Keywords: Required imagination · Creativity in practice · Systems safety
Collective inquiry · Nuclear industry

1 Introduction

The inclusion of human factors engineering in the design of safer sociotechnical systems first began, with the advent of Human Reliability Analysis methods, with a shift from technical to human failures as possible causes to system failures and accidents (Swain and Guttman 1983). Subsequent authors went on to argue that failures of the human component of these systems (i.e. "unsafe acts") might in some cases be best understood in terms of "latent conditions" – system characteristics which only lead to failure in specific conditions, and which are often the result of organizational decisions (Reason 1990). This crucial shift led to two major consequences: (1) a shift away from the view of human agents as an agent of unreliability to an agent of reliability, and (2) the acknowledgement that organizational factors might contribute to system reliability, and not just to possible failures (Weick and Sutcliffe 2001).

Creativity is often described as the ability to produce work that is both novel and suited to task characteristics (Amabile 1996). In the context of ensuring systems safety, this is generally understood in one of two ways: (1) the ability for workers to adapt to situations where the resources provided by the organization to ensure system safety are ill-suited to the situation, and to improvise novel and more appropriate solutions; and

S. Bagnara et al. (Eds.): IEA 2018, AISC 821, pp. 199–213, 2019.
https://doi.org/10.1007/978-3-319-96080-7_24

(2) the ability for systems designers and other stakeholders involved in organizational design to exhibit "requisite imagination", i.e. the ability to anticipate what might go wrong and propose means to prevent such failures or mitigate their effects (Adamski and Westrum 2003). The existence of these kinds of resources to ensure systems safety is sometimes referred to as a distinction between *constrained*, or *regulated*, safety on the one hand and *managed* safety on the other (Morel et al. 2008; Nascimento et al. 2014).

In this paper, we argue that this anticipation of potential situations of system operation can be considered as a manifestation of creative action in organizations and requires collective inquiries. We conducted an intervention within a French public electrics company whose goal was to anticipate potentially adverse effects of evolutions in domain regulations concerning systems safety, notably in the face of a nationwide policy seeking to extend the lifespan of nuclear facilities. In order to anticipate future problems and propose possible solutions, various stakeholders in the nuclear facility we studied underwent a process of inquiry. However, this inquiry was mostly of a technical nature, whilst organizational aspects were largely excluded. To remedy this, we opened up a discussion space to gather the stakeholders involved with these ongoing evolutions, to discuss organizational design whilst stimulating requisite imagination. Exchanges uttered in this discussion space were transcribed and subjected to an analysis of the dynamics of narration (Lorino 2005), making it possible to identify the narratives constructed around past, present and future situations of system operation. We discuss the implications of these findings from three perspectives: (1) theoretical, concerning the relationship between creativity and reliability at a *meso* level, (2) methodological, focusing on the use of such inquiries to develop required imagination; and (3) practical, for stakeholders in the nuclear industry who must find creative solutions to deal with a complex landscape of evolving requirements.

2 Creativity in Practice in the Context of Ensuring Systems Safety

Creativity is often described as the generation of ideas that are both useful and original (Amabile 1996; George 2007). In the context of work, creativity is defined as the ability to produce work that is both novel and suited to task characteristics (Amabile 1983; Sternberg and Lubart 1999). Much has been written on what creativity is and where it happens, but little is known about how it is developed in practice (Stierand 2015). Recent works investigate creativity as a concrete, collective and everyday form of practice, inherent in daily work (Carlsen et al. 2014) with a focus on a comprehensive understanding on creative action in organizations (Arjaliès et al. 2013). This focus appears as an opportunity for the ergonomics discipline and opens a new area: designing work for enhancing creativity (Dul and Ceylan 2010). In the field of systems safety and reliability, the role of creativity remains underexplored and source of debate. This is generally understood in one of two ways: the ability for workers to adapt to situations – safety is derived from the beneficial creativity of workers' action (2.1), the ability for systems designers and other stakeholders to exhibit "requisite imagination", i.e. the ability to imagine what might go wrong and propose means (Adamski and Westrum 2003) – safety depends on attention to organizational design (Bourrier 2005)

(2.2). After describing these two ways, the central thesis in this paper is the claim that the anticipation of potential situations of system operation by designers but also workers can be considered as a manifestation of creative action in organization. This anticipation requires a process of collective inquiries.

2.1 A Comprehensive Understanding of Creative Action in Organizations

Employees at any level of the organization are a potential resource for creativity (Dul and Neumann 2009): in daily work, "there will always be situations that are either not covered by the rules or in which the rules are locally inapplicable" (Reason et al. 1998, p. 297). The activity is adaptive, it constantly manages the variability of reality; hence, activity is necessarily creative. For systems safety, it is a major shift away from the view of human agents as agents of unreliability, to agents of reliability with cognitive sense-making skills and strategies (Norros et al. 1986; Weick and Sutcliffe 2001). Therefore, the understanding of situation by the operators and the meaning they attach to the information or events that occurs proves an essential factor to the reliability of systems (Grosdeva and de Montmollin 1994). Creativity thus appears as practices unfolding in everyday work (Carlsen et al. 2014; Kurtsberg and Amabile 2001; Sternberg and Lubart 1999). A call for adopting a comprehensive understanding of creative action in organizations is formulated (Arjaliès et al. 2013). The reflection of Joas in his book "The creativity of action" (1996), and American pragmatism, especially the work of Dewey (1938); Peirce (1903/1998), offer a fruitful way of understanding creative practice and answering this call.

In his book, Joas (1996) proposes a model that emphasizes the creative character of human action and asserts that there is a creative dimension to all human action often marginalized in the history of action theory. He offers an understanding of processes of collective action and shows "both the meaning and the necessity of taking the creative character of human action in consideration" (p. 6). By focusing on collective action, rather than purely individual action, "*it becomes much easier to proceed to issues such as the emergence, reproduction and transformation of social order*" (p.199). In ergonomics, studies on the collective dimension of work have drawn strong interest from the 1980s onwards. A new area is opening to take into account creativity in the collective work (Dul and Ceylan 2010) that exists as a potential in even the most mundane, everyday plodding actions of social practice (Kilpinen 1998). But how does creativity arise?

Creativity arises as a response to uncertain and unanticipated situations that call out changeful actions (Arjaliès et al. 2013). Uncertain and unanticipated situations call for a process of collective inquiry. The concept of inquiry was developed by pragmatist authors (Dewey 1938; Peirce 1903/1998) and closely interlaces creative thought, logical reasoning and experimental action to make sense of and transform situations Lorino et al. (2011). In pragmatist thinking, the interruption of habitual action is a key point as it is this that initiates the cycle of inquiry. Peirce's notion of abduction appears as a key element to creativity: "*abduction is the process of forming an explanatory hypothesis. It is the only logical operation which introduces any new idea*" (Peirce 1903, p. 216). For pragmatists, action consists not in the pursuit of clear-cut goals or in the application of norms, and creativity is not the overcoming of obstacles along these prescribed routes: "anchoring creativity in action allows the pragmatists to conceive of creativity precisely as the liberation of the capacity for new actions" (Joas 1996, p. 133). This view offers a way for understanding creativity in practice and offers an opportunity for ergonomics to take into consideration this concept.

For Bourrier (2005), it is important to suggest that safety is derived from the beneficial creativity of agents but it is also important to take attention to design, especially the organizational (re)design because "*when organizations are redesigned, the routes to organisational reliability are also modified*" (p. 102). The ability for systems designers and other stakeholders to exhibit "requisite imagination" (Adamski and Westrum 2003) is critical to the design for complex man-machine systems.

2.2 Requisite Imagination: Imagination and Wonder as a Manifestation of Creative Action

Requisite imagination is the ability to imagine key aspects in the future we are planning (Westrum 1991). Most important, "*it involves anticipating what might go wrong and how to test for problems when the design is developed*" (Adamski and Westrum 2003, p. 195). For these authors, requisite imagination is an "art" to reflect on the design and acknowledge potential problems. The failure to use requisite imagination opens the door to incidents, accidents or major catastrophes. If we look into the analysis of the Fukushima disaster from Hollnagel and Fujita (2013), the catastrophe stems from the "inadequate engineering anticipation or risk assessment during the design in combination with inadequate response capabilities" (p. 13). Designers have to take the time to look deeply into the design, look forward and consider longer-range consequences, eliminate as many "side effects" as possible (Adamski and Westrum 2003; Westrum 1991). What does "imagination" mean and what are the possible links with creativity?

Wonder is often examined as an invitation to imagination (Carlsen and Sandelands 2015; Weick 1989) and seems to be at the origin of acts of creativity (Glaveanu 2017). Glaveanu (2017) has taken up both sides of the phenomenon – the surprise and the receptivity to what is new, *wondering at*, and the active search for meaning, *wondering about* – while specifying the lived experience "*whereby the person becomes (more or less suddenly) aware of an expanded field of possibility for thought and/or action and engages (more or less actively) in exploring this field*" (Glaveanu 2017, p. 2). This definition grounds it firmly within the complex dynamic between the actual (what is here) and the possible (what is not here or not yet here), a dynamic at the core of creativity even if the "what is" (the realm of the *actual*) is more considered than the "what is not (yet)" (the realm of the *possible*) (Glaveanu 2017). Concerning the reliability and safety issues, the concept of "requisite imagination" allows taking into account "what is not (yet)" and more precisely, "what might go wrong".

Our main thesis is the claim that requisite imagination by designers (and more broadly by all workers) – the ability to wonder and imagine key aspects of the future we are planning/designing – can be considered as a manifestation of creative action in organization through collective inquiries.

3 Extending the Lifespan of Nuclear Facilities: A Call for Imagination

3.1 Research Context and Objectives

The present study is a part of a 2-year research project conducted in a nuclear power plant that wished to understand and develop the collective dimension of work, especially in the operations department which features an abundance of meanings, revealing a particularly complex collective activity: "*to operate a facility means to coordinate the interventions (either simultaneous or occurring at different times) of specialists from various fields. This intervention is essential for the smooth operation of the facilities*" (Girin and Journé 1997, p. 3, our translation). Over the course of these 2 years, researchers were present on the field at regular but impromptu times - November 2014, March 2015, June 2015, December 2015 and February 2016 – for durations of 7 to 10 days.

We decided to apprehend the complexity of collective activity through the observation of collective inquiries which seek the order of an explanation for disordered and enigmatic facts. We focus on "*abnormally disturbed situations*", i.e. sufficiently disturbing situations in the eyes of the operators to call for a work of definition and interpretation of the situation, to call for a collective inquiry (Journé and Raulet-Croset 2007; 2008). Among these situations, we observed inquiries triggered by interruption of habitual action caused by a nationwide policy seeking to extend the lifespan of nuclear facilities. This policy has resulted in new – technically and temporally – large-scale "alterations". More broadly, the project wanted to associate renovation, replacement of equipment, integration of post-Fukushima measures, and improvement of the safety level of nuclear power plants. The nuclear power plant studied was one of the first plants to begin the alterations; the others were waiting for feedbacks.

"Alterations" come from engineering reflections (*Contracting*) which confides a list of work to actors within the production site (*Project management*) in charge of managing the alterations (understanding, adaptation to the site, preparation …). The project management works in relation with the *Operations department* who will make the alterations with the professionals concerned (boilermaking, plumbing, electricity, etc.). The *Methods department* is invited to facilitate links and exchanges between the project management and the operation department in order to ensure that the work is carried out safely and as quickly as possible.

3.2 Technical Nature of Inquiries Focused on "What Is" and Tensions Among Various Actors

From our periods of immersion, we were able to observe questions and construction of a problem around the first alteration, realized with delay and unexpected consequences. This first alteration generically represents a category of unintelligible situations; it reveals the failure of the stories by which the actors are used to working. These actors were invited to reconstruct a new intelligibility of the collective activity: "*The usual course of activity is disrupted, raising precognitive questions (...) the second step defines a problem (...) the inquiry is an exploratory interaction with the world rather than a problem-solving procedure, since the problem is not given.*" (Lorino et al., p. 778).

We gathered facts around the first alteration made: "PNPP 2311", the number designating a material that was to undergo a alteration. This first alteration, carried out later than intended by the organization (in December 2015 instead of June 2015), led to environmental incidents. Attempts to construct the problem have been observed but carried out separately, each of the actors having a tendency to see the problem from the point of view of their own activity and to blame the actors of the other activities, the project management or the operations department: "*The alterations are not prepared, it is not normal, why do not we see the project management teams? They bypass the organization!*" (Operation department, 12/2/2015); "*Why does the operations department block our activities the day before whereas we explained everything to them beforehand?*" (Project management, 12/3/2015). The Methods department tried to facilitate the exchanges but encountered difficulties; a file with important documents was created for the operation department but didn't seem to be used.

No collective inquiry actually took place prior to the first alterations, as the disruption with the activities carried out until now was probably underestimated. Once the difficulties were known around the first alteration, we observed inquiries during daily meetings associating the various actors involved in these alterations. But these investigations were primarily technical and mainly focused on the past event and its consequences. Here the findings are closely akin to Girin and Journé's reflection (1997): team members in a nuclear power plant tend to overestimate the technical aspect of their function (on the specialist model) in relation to its organizational dimension.

Anxiety and fears about the next coming alterations remained perceptible: "*The tension is strong and all the first alterations have gone very badly ... It is all the alterations coming from the national which are often not very clear (...) with PNPP 2311 we observed bad consequences... These alterations have not been well prepared or many more will happen, we are not super confident ... in addition we must succeed a large volume of alterations*"(Chief in operation department, 12/15/2015). Each of the actors had the vague feeling of "being able to do better together" but the only technical inquiries centered on the alterations already realized ("what is") did not allow to exchange and plan about "what might go wrong" and " how to prevent it ".

To remedy this, we opened up a discussion space to gather the stakeholders present on the nuclear site (project management, operations and methods department) involved with this ongoing evolutions to discuss organizational design, whist stimulating requisite imagination.

3.3 A Community of Inquiry to Stimulate Requisite Imagination

We decided to work with the various actors to imagine these new alterations and to further anticipate their effects. The difficulties observed around the first alteration constituted a starting point for the organizational (re)design in which we wanted to initiate. Through these ongoing evolutions, organizations can be redesigned and the routes to organizational reliability can also be modified (Bourrier 2005). A recent paper demonstrates how work debate spaces contributes to the development of an integrated safety culture (Rocha et al. 2015). In line with this work, we want here to show the interest of debate spaces for facilitating a collective inquiry to stimulate the requisite imagination, i.e. to go from a reflection on "what is" to "what is not yet" and "what might go wrong" and "how to prevent it". As Vygotsky observes, studying and transforming activity go hand in hand: "to study something means to study it in the process of change (in order to) discover its nature, its essence, for it is only in movement that a body shows what it is" (Vygotsky 1978, p. 65). So we asked different actors to draw together the "Alterations" process by specifying in turn the work done until now, to go back over the difficulties and to project in the next coming.

This exchange took place on February 16th, 2016 at the nuclear power plant and brought together 7 volunteers including two people from the Operations department, two people from the Methods department, and three people from Project management.

In order to stimulate "requisite imagination", actors and researchers drew the process together and engaged in reflective conversations in which they jointly re-narrate and made sense of the experience by drawing on past narrations, present emplotments, and by considering future possibilities (Cunliffe et al. 2004). Sharing wonder can thus be targeting in collective inquiry (Carlsen and Sandelands 2015) where researcher and practitioners alike look at empirical samples, designed a process plan (Lorino et al. 2011) to co-produce useful knowledge. Here the "designed plan" was not pre-built by the researchers but to be built during the meeting with the practitioners.

3.4 Methods of Data Analysis

The meeting lasted approximately three hours and was recorded and transcribed (445 verbal utterances). Exchanges uttered in this discussion space were subjected to a first analysis focused on the various moments of potentials for wonder and imagination that triggers inquiry (Carlsen and Sandelands 2015). Then exchanges were subjected to an analysis of the dynamics of narration (Lorino 2005) making it possible to identify the narratives constructed around past, present and future situations of system operation. Lorino (2005) suggests analyzing narrative practice as a form of situated action, oriented towards the abductive creation of meaning: "*the narrator tells his story to listeners who, far from being passive, participate in fact in the construction of the narrative by the reception that they reserve for it, by the expectations that the narrator anticipates from them or observes over the course of the story (…) the reconstruction of the events relies on the testimonies and the analyzes of a multiplicity of actors*" (p. 202, our translation). The interest is then focused on "what are the stories used for and what do they tell us?"

More precisely, we chose first to examine conversational turn-taking on certain themes and "situations": actors not only address each other, but also have to deal with situations. In a turn-taking conversational process, "*I cannot know what I have said till the other speaker answers me. I cannot know what I have done till the practical effects*

Table 1. Variables, codes and target definitions for data analysis

Variables	Codes	Definition
"What is?" Narratives around the past and actual situation	Subjects Roles Narratives of PNPP2311	Real work and difficulties Effects on others: surprise? Receptivity? … How is PNPP231 addressed?
"What might go wrong and how to prevent it" Narratives around the future	Subjects Roles Propositions/Experimentation	Anticipated future events Fears, threats/hopes, dreams … Imagined actions and experiments
What are the stories used for and what do they tell us?	Subjects Roles	Technical/organizational Illustrate/Argue/Convince… Links with the future?

of my act become perceivable and intelligible" (Lorino 2014, p. 454). After this first general reading dedicated to the identification of the main themes of the discussion, second coding defined sub-variables (Table 1):

– "What is": narratives around the actual situation;
– "What might go wrong" and "how to prevent it": the future situations of system operation. What narratives concern the future?
– What are the stories used for and what do they tell us?

4 Main Findings

4.1 Narratives Around Past and Present Situations: An Opportunity to Show Work in the Process and State Difficulties

Taking turns, the stakeholders described the process in terms of the work they carried out at each stage, in a chronological manner. Here, the description of the activity was intended for other stakeholders, who interacted at various times to obtain more information, to add new information, or to express some form of surprise. On two occasions, the stakeholders in Project management and in the Methods department mentioned a type of work which they consider today to be "invisible", and expressed satisfaction with being able to bring this work to the attention of other stakeholders. Table 2 below identifies key interruptions in the meeting, conveying surprise and wonder, and leading to debate between participants.

These interruptions are related to the organizational level and, for some time, do not mention the purely technical level. Here, the issues relate to falling behind, to not use artifacts, to inefficient meetings, etc. In their description of the process, the stakeholders question it and take their own collective activity as an object of inquiry.

Table 2. Main interruptions: surprises

When?	Interruptions	Consequences
During the description of the project management activity, it is mentioned that stakeholders often work in "suboptimal conditions", in part because of the fact that engineering reports arrive late	Made by the Operations department, expressing surprise: "why are the conditions suboptimal? Why do these reports arrive late?"	Explanations and a new understanding expressed by the participant from the Operations department: "So, okay, we work in two incompatible organizations!" Discussion ensues
During the description of the activities of the methods department…	Made by the Operations department: "The files you prepare for us, we don't use these at all" Methods Department: "Is that true? No, don't say that!""	Reflection on: how to better support the work of the Operations department, how to make all the information concerning alterations at the disposition… Work habits are questioned by Operations Department: "We never used to work like that before … It really needs to all come together". Later on, the role of the files is further questioned by Project Management, and fears are expressed (concerning multiple avenues of information communication)
During the description of the activity of project management, one of the participants mentions weekly meeting with the schedule department	The head of the project management department interrupts and seems surprised: "I don't know anything about that". The other stakeholder adds: "But it doesn't work"	A discussion ensues concerning the scheduling of alterations

4.2 What Might Go Wrong and How to Prevent It? Actors Imagine the Future and Anticipate Potential Problems

On six separate occasions, the collected materials produced in the meetings exhibited instances of "what might go wrong":

1. Origin and contents of "what might go wrong": when describing the activity of the methods department, the stakeholders anticipated a risk: "*at the moment, files have to arrive four months before the work, but it can sometimes arrive in an emergency. We haven't been confronted to that yet, but it could happen.*" Stakeholders anticipated a suboptimal mode of work related to the need to carry out emergency alterations, and to the need to implement several alterations at the same time. What

are the anticipated consequences of these elements? What difficulties will emerge from the need to manage these alterations in a hurry? Is there a risk of making errors? Of blocking the process? How to prevent this? There is some ongoing reflection concerning upcoming evolutions in work organization (engineering), with the implementation of a Lean production approach. This approach should make it easier for the files to arrive in a more regular manner, taking into account the constraints related to preparation in the nuclear power plant facilities.

2. Origin and contents of "what might go wrong": During the description of the activity of project management, it is acknowledged that there are failings of anticipation and that at the present time, it may be impossible to reconcile the alterations with everyday activities and fortuitous events: *"it's difficult for us to make the links"*. A discussion begins on the topic of how to "do better" in the future, and avert consequences in terms of risk and delays.

How to prevent it? The group quickly mentions projections, listing the stakeholders who are in a position to analyze these interactions and interrelations.

3. Origin and contents of "what might go wrong": while describing the activities of the Project management department, the stakeholders go back to mentioning the creation of a file which would allow making the main documents – concerning the alterations that need to be made – available to the Operations department. Project management then expresses some fear concerning the potential risks such files may lead to, and how these documents should be passed on and updated.

How to prevent it? Following these discussions, the stakeholders from the Methods department suggest interacting with the Operations department when a change is made, in order to add meaning to the documents and to update them when necessary.

4. Origin and contents of "what might go wrong": During the account of an alteration carried out during unit outage with a dedicated worker, stakeholders mention the fact that it will be difficult to find time and means to do this, particularly during in-operation periods. Here, the participants anticipate the fact that Project management will likely be unable to be constantly present, and that those from the Operations department will be unable to be dedicated to carrying out only one type of alteration.

How to prevent it? The methods department reminds the group that its role is to work ahead of such issues, and that this will be made possible by the expected arrival of resources in the group within the near future.

5. Origin and contents of "what might go wrong": when discussing issues surrounding PNPP2311, the Project management team mentions the following: *"this situation could happen again. Documents had been prepared to deal with it, but the meaning of these documents was unknown..."* A reminder is made of the problems encountered.

How to prevent it? The stakeholders reflect on the possibilities for avoiding this kind of situation, engaging in a broader questioning of the specific nature of the alterations that need to be made and the difficulties related to *"having to adapt to the existing organization, to its modes of operation, etc.*", in particular the separation between activities related to preparation and to implementation, resulting in a loss of information; the lack of visibility of the Logistics function, although it has a part to play in making these alterations (e.g. preparing a working site, setting up scaffolding, etc.). Specific examples are used to highlight an unsuitable organization. The stakeholders attempt to project themselves into the future, and wish to continue to convey these difficulties in order for decisions to be made.

6. Origin and contents. When describing the activity of the methods department, stakeholders mention the ongoing issue of training agents. The current training provided at the National level is deemed unsatisfactory by all, and a reflection is undertaken to "go beyond" the current system.

How to prevent it? Project management wishes to train and inform agents concerning alterations to the system in order to give meaning to these alterations and prevent their potential effects, notably in terms of nuisance. A suggestion is made for Project management to interact more closely with the organizational/HF department onsite.

Through these questions concerning the future and through their joint reflection, stakeholders imagine (often negative) scenarios, as well as solutions to prevent either existing failings from persisting, or new failings from appearing or from remaining unanticipated. When the stakeholders paint the picture of the inefficiencies of the process today, they also project means to improve it and rethink their ways of working with each other.

4.3 The Story of the "2 Wires": What Is It for and What Does It Tell Us About the Past, Present and Future? a Collective Search for Meaning

On several occasions, stakeholders spontaneously mention "histories": the analysis made below (Table 3) of one of these stories shows that these past stories bring about questions concerning the future and trigger inquiries into "how things should be done" and "how people should work better together", in order to implement the upcoming alterations that need to be made to extend the lifespan of nuclear facilities.

Through this account of an agent being confronted to an unusual situation of "cutting the two wires", the group questions the way in which alterations are made. They disrupt normal courses of action and confront practitioners with the need to question the frameworks of understanding in their collective activity. Technically, these interactions show that the stakeholders have exchanged on the potential effects of this cut, particularly in terms of a sudden rise in temperature that needs to be anticipated, etc. Here, a reflection is initiated concerning how to work better together and to make sure that the alterations make sense and can be carried out safely, and in a timely fashion. The inquiry triggered by this account calls for primarily organizational reflections. How can the workers build a shared meaning for these alterations, how can they question their work habits, how can they design a suitable organization, etc.

Table 3. The story of the "2 wire": a collective search for meaning

Project management 1 – there was a worker who had been around for awhile, and he said this: "in twenty years of experience, it's the first time I've ever seen people cutting the two wires! Usually we only cut the one". So he didn't know what was going to happen in terms of temperature. My feeling is, when you say you're going to cut the two wires for the water cooling system, it makes people freak out. (…). It's just that there's going to be a time when people have to be reassured. Except that sometimes, to reassure them, you need to spend half an hour to an hour to do it	Narratives around a past « story » : this is the narrative from a participant in the Operations department confronted with an unusual situation which requires "cutting the two wires". The narrative reveals the need to reassure the workers. This alteration is unusual. The risk is for operations to enter a state of standstill
Operations department 1 – That's true. I think meeting people physically is really important, because you see a mode of operations arrive just like that and you may not be familiar with it. So the fact of explaining what's going to be done, how it's going to be done, etc. It makes it possible to anticipate several kinds of trouble, and to simplify the representation you have of a problem. You may not have enough time to spend half an hour to an hour fixing an alteration	How to prevent standstill? Though physical interactions
Project Management 1 – The thing is we have to work. When we collect a mode of operations, when professionals collect a mode of operation from the Operations department, they have to be able to explain in what context it is they're working. When the Operations department ask them why they need to cut both wires like that, there's no answer	How to prevent it? Working ahead of issues so that the people involved are able to explain the nature and meaning of the alterations
An account of an alteration carried out in good conditions during unit outage *Participants are working closely together, there are active exchanges between the Operations department and project management, there is time available to read the file, etc. Discussions are held*	Past story: an alteration was carried out correctly during unit outage. What are the conditions for this success?

(*continued*)

Table 3. (*continued*)

Methods department - So here, typically, with that type of thing, we'll be there to do it instead of the Operations department (…). If ever the team has a question, we'll be there to provide support, so that they don't need to do this work of preliminary research and all this reading	How can this be carried out? The role of the methods department
Project management – Yeah, because we won't be able to do this for all the files, all of the time	Expected impossibility for all the conditions of success to be met in the future
Methods department – that's the point of working ahead of the problem	Need to anticipate, to work ahead of the problem

5 Discussion

We discuss the implications of these findings from three perspectives:

– A theoretical perspective concerning the relationship between creativity and reliability/safety at a *meso* level (processes). This paper offers a comprehensive understanding of creative action in organizations through collective inquiries triggered by "inhabitual/doubtful situations" where elements "do not hang together" (Dewey 1938). The new alterations are doubtful, a new process "disturbed, troubled, ambiguous, confused, full of conflicting tendencies, obscure, etc." (Dewey 1938, p. 105) which modify routes to organizational reliability and involve rethinking organizational design. The first alteration question usual practices; stakeholders are invited to imagine a new organizational design by imaging what might go wrong and how to prevent it. This imagination is built here from the first experience that showed that the extension of the lifespan of nuclear facilities disrupted operations and habits. Surprises can be observed, the effects and "side effects" of these alterations have been minimized from the beginning. There is a complex dynamic between the *actual* (what is here, what are the problems) and the *possible* (what can last if nothing is done, what might go wrong in the future), a dynamic at the core of creativity (Glaveanu 2017). Actors are invited to wonder and imagine key aspects of the future, not only technical aspects but organizational one. As Arjaliès et al. (2013) note, creativity arises as a response to uncertain and unanticipated situations that call out changeful actions. Creativity appears as the liberation of the capacity for new actions.
– A methodological perspective focusing on the role of the formation of a pluralistic community of inquiry (Dewey 1983 [1916]) to develop requisite imagination: the fine art of anticipating what might go wrong. The formation of a community of inquiry is here an essential issue to make emerge, in an explicit, reasoned, debatable and transformable way, the collective activity that otherwise the dispersed and isolated actors undergo as a series of constraints (Lorino 2007). The methods developed in ergonomics around participatory approaches (Wilson 1991) and work

debate spaces (Rocha et al. 2014) help us to foster discussion in this community and to create an opportunity to develop imagination and wonder around a new process which requires new joint intervention of multiple actors.

- A practical perspective for stakeholders in the nuclear industry who must find creative solutions to deal with a complex landscape of evolving requirements; as Hollnagel and Fujita (2013) underline, "*nuclear power plants are complicated, not only in how they function during everyday or exceptional conditions but also during their whole life cycle*". And what is going on when this life cycle gets longer? Issues are high… It does not imply that any reform must be blocked, but only "*that reform should be implemented once there is sufficient understanding of how a given system behaves*" (Bourrier 2005). Moreover, this understanding should be completed by imagining key aspects of any future reform…

References

Adamski AJ, Westrum R (2003) Requisite imagination: the fine art of anticipating what might go wrong. In: Hollnagel E (ed) Handbook of cognitive task design. Lawrence Erlbaum Associates, Mahwah, pp 193–220

Amabile TM (1996) Creativity in context. Westview Press, Boulder

Amabile TM (1983) The social psychology of creativity: a componential conceptualization. J Pers Soc Psychol 45(2):357–377

Arjaliès DL, Lorino P, Simpson B (2013) Understanding organisational creativity: insights from pragmatism. In: Kelemen M, Rumens N (eds) American pragmatism and organization: issued and controversies. Gower Publishing, Burnington, pp 131–145

Bourrier M (2005) The contribution of organizational design to safety. Eur Manag J 23(1):98–104

Carlsen A, Rudningen G, Mortensen TF (2014) Playing the cards: using collaborative artifacts with thin categories to make research co-generative. J Manag Inquiry 23:294–313

Carlsen A, Sandelands L (2015) First passion: wonder in organizational inquiry. Manag Learn 46:373–390

Cunliffe AL, Luhman JT, Boje DM (2004) Narrative temporality: implications for organizational research. Organ Stud 25(2):261–286

Dewey J (1938) Logic: The theory of inquiry. Holt, New York

Dul J, Ceylan C (2010) Work environments for employee creativity. Ergonomics 54(1):12–20

Dul J, Neuman WP (2009) Ergonomics contribution to company strategies. Appl Ergon 40 (4):745–752

George JM (2007) Creativity in organizations. Acad Manag Ann 1:439–477

Girin J, Journé B (1997) La conduite d'une centrale nucléaire au quotidien. Les vertus méconnues du facteur humain, Exposé à l'Ecole de Paris du Management, France, Paris

Glaveanu VP (2017) Creativity and wonder. J Creative Behav Spec Issue Creativity Paradox Multi Perspect 1–7

Grosdeva T, de Montmollin M (1994) Reasoning and knowledge of nuclear power plant operators in case of accidents. Appl Ergon 25(5):305–309

Hollnagel E, Fujita Y (2013) The Fukushima disaster – systemic failures as the lack of resilience. Nucl Eng Technol 45(1):13–20

Joas H (1996) The creativity of action. Polity Press, Cambridge

Journé B, Raulet-Croset N (2007) Of organizations and situations: a pragmatist view of organizing through the process of inquiry. In: 23rd EGOS Colloquium, 5–7 July, Vienna
Journé B, Raulet-Croset N (2008) Le concept de situation: contribution à l'analyse de l'activité managériale en contextes d'ambiguïté et d'incertitude. M@n@gement 1(11):27–55
Kilpinen E (1998) Creativity is coming. Acta Sociol 41:173–179
Kurtzberg TR, Amabile TM (2001) From Guilford to creative synergy: opening the black box of team level creativity. Creativity Res J 13:285–294
Lorino P (2005) Contrôle de gestion et mise en intrigue de l'action collective. Revue Française de Gestion 31(159):189–212
Lorino P (2007) Communautés d'enquête et création de connaissances dans l'organisation : le modèle de processus en gestion. Ann des Télécommunications 62(7–8):753–771
Lorino P (2014) From the analysis of verbal data to the analysis of organizations: organizing as a dialogical process. Integr Psychol Behav Sci 48(4):453–461
Lorino P, Tricard B, Clot Y (2011) Research methods for non-representational approaches to organizational complexity: the dialogical mediated inquiry. Organ Stud 32(6):769–801
Morel G, Amalberti R, Chauvin C (2008) Articulating the differences between safety and resilience: the decision-making process of professional sea-fishing skippers. Hum Factors J Hum Factors Ergon Soc 50(1):1–16
Nascimento A, Cuvelier L, Mollo V, DiCioccio A, Falzon P (2014) Constructing safety: from the normative to the adaptive view. In: Falzon P (ed) Constructive ergonomics. CRC Press, Boca Raton, pp 95–109
Norros L, Sammatti P (1986) Nuclear power plant operator errors during simulator training, Research report No. 446, Technical Research Centre of Finland, Espoo
Peirce CS (1903/1998) The essential Peirce. Indiana University Press, Bloomington
Reason J (1990) Human error. Cambridge University Press, Cambridge
Reason J, Parker D, Lawton R (1998) Organizational controls and safety: the varieties of rule-related behaviour. J Occup Organ Psychol 71:289–304
Rocha R, Mollo V, Daniellou F (2015) Work debate space: a tool for developing a participatory safety management. Appl Ergon 46:107–114
Sternberg RJ, Lubart TI (1999) The concept of creativity Prospects and paradigms. In: Sternberg RJ (ed) Handbook of creativity. Cambridge University Press, Cambridge, pp 3–15
Stierand M (2015) Developing creativity in practice: Explorations with world-renowned chefs. Manag Learn 46(5):598–617
Swain AD, Guttman HE (1983) Handbook of human reliability analysis with emphasis on nuclear power plant applications. NRC, Washington
Vygotsky LS (1978) Mind in society: The development of higher psychological processes. Harvard University Press, Cambridge
Weick KE (1989) Theory construction as disciplined imagination. Acad Manag Rev 14:516–531
Weick KE, Sutcliffe KM (2001) Managing the unexpected: assuring high performance in an age of complexity, 1st edn. Jossey-Bass, San Francisco
Westrum R (1991) Technologies and society: the shaping of people and things. Wadsworth, Belmont
Wilson JR (1991) Design decision groups e a participative process for developing workplaces. In: Noro K, Imada A (eds) Participatory ergonomics. Taylor and Francis, London

The Association Between Safety Climate and Musculoskeletal Symptoms in the U.S. Logging Industry

Elise Lagerstrom and John Rosecrance(✉)

Colorado State University, Fort Collins, CO 80523, USA
John.Rosecrance@colostate.edu

Abstract. The purpose of this research was to assess the association between safety climate and musculoskeletal symptoms (MSS) among workers and management in the logging industry. The Nordic Safety Climate Questionnaire, NOSACQ-50 and modified Standardized Nordic Questionnaire were administered to 743 loggers. Five safety climate dimension scores were assessed. The disparity between management's views of their own safety priority versus workers' views of management's safety priority indicates a need to focus on safety interventions in those specific dimensions. The relationship between these measures can be used to identify possible areas and opportunities for future interventions.

Keywords: Logging · Safety · Safety climate · Musculoskeletal symptoms

1 Introduction

The construct of safety climate has been developing since 1980 in response to the need for leading, rather than lagging, indicators of occupational safety performance, including the prevention of occupational injuries and incidents (Zohar 1980). Zohar (1980) developed one of the first measures of safety climate, which was designed to discriminate between companies with high and low accident rates by measuring different dimensions of organizational climate (Zohar 1980). Based on the compilation of ideas and research regarding safety climate and culture, Zhang et al. (2002), defined safety climate as: "the perceived state of safety at a particular place at a particular time, making the definition relatively unstable, and subject to change depending on the features of the current environment or prevailing conditions" Zhang et al. (2002).

Safety climate researchers have determined that it was important to measure safety climate before beginning an intervention to ensure an adequate climate for change (Neal et al. 2000). While studying the link between organizational climate and safety climate, researchers indicated that interventions aimed at improving the safety of an organization would be more successful if they occurred in a positive climate (Neal et al. 2000). The authors of the same study found that safety climate had an effect on worker motivation and compliance, which was important for determining safe work behavior and performance (Neal et al. 2000). Assessment of safety climate is a critical and

S. Bagnara et al. (Eds.): IEA 2018, AISC 821, pp. 214–219, 2019.
https://doi.org/10.1007/978-3-319-96080-7_25

underutilized tool for occupational groups with a history of and/or high risk for worker injury.

Inherent dangers of the logging industry are induced by environmental conditions, heavy machinery, manual labor, and can vary based upon season, regional logging practices, and terrain (Fig. 1). For example, in the Southeastern United States, logging is highly mechanized, the terrain flat, and the weather mild in comparison to logging in the Pacific Northwest or Intermountain regions of Montana and Idaho, where logging is characterized by harsh conditions, steep terrain, severe weather, and remote work locations (Lagerstrom et al. 2017; United States Department of Labor (OSHA) 2017).

Fig. 1. Professional logger operating heavy equipment (loader/processer) in the mountains of Montana, USA.

There is a scarcity of published studies that specifically assess safety climate and musculoskeletal symptoms (MSS) in the logging industry. The purpose of this cross-sectional study was to quantify safety climate and prevalence of MSS in the logging industry of the intermountain states of Montana and Idaho. The secondary aim was to investigate the association between MSS and five dimensions of safety climate.

2 Methods

Surveys were administered to loggers participating in a required emergency first-aid training workshop. The training sessions are held annually at different locations across the state of Montana. The surveys consisted of three questionnaires: demographic questionnaire, a MSS questionnaire, and a safety climate questionnaire. Participation in the survey was voluntary and anonymous. Compensation was not provided to participants. All workers in attendance at the workshops were eligible for participation.

Demographic information collected as a part of the survey included age, gender, and education level. Job information collected by the survey included logging system type, supervisory status (leader v. worker), whether the logger was an accredited logging professional (designation requiring continuing education on safe and environmental logging practices) and years spent employed in the logging industry.

To determine the presence of musculoskeletal symptoms, a modification of the Standardized Nordic Questionnaire (SNQ) was administered. The modified questionnaire included three questions in reference to nine anatomical regions of the body as follows: (1) "During the last 12 months have you had a job-related ache, pain, or discomfort?"; (2) "During the last 12 months has this ache, pain, or discomfort prevented you from doing your day's work?"; (3) "During the last 12 months have you seen a physician or physical therapist for this pain, ache or discomfort?" For each question, participants checked either yes or no for each of the nine anatomical region.

To assess safety climate, we used a modification of the English translation of NOSACQ-50 that was modified for this study to fit within survey time and space requirements. The modifications included using five of the seven dimensions: (1) management safety priority and ability, (2) workers' safety commitment, (3) workers' safety priority and risk non-acceptance, (4) peer safety communication, learning, and trust in safety ability, and (5) workers' trust in efficacy of safety systems. The 38 items that pertained to the five dimensions were answered with a Likert scale (1–4) ranging from strongly disagree (1), to strongly agree (4). Multiple studies have validated this measure in various industries, countries, and languages. The NOSACQ-50 was developed by a team of Nordic researchers trying to determine reasons why different occupational groups have higher accident and injury rates than other groups performing the same work (Kines et al. 2011). The definition of safety climate is a measure of "a workgroup members' shared perceptions of management and workgroup safety related policies, procedures, and practices" (Kines et al. 2011).

2.1 Statistical Analysis

Means, standard deviations, and frequency statistics were calculated for all demographic variables. The continuous variable, years of experience in the logging industry, was transformed into a categorical variable by decades of experience. Two binary variables were created based on the results of the modified SNQ, to identify workers who experienced MSS in any anatomical area (Yes/No), or missed work due to MSS in the past 12 months (Yes/No).

Dimension scores for NOSACQ-50 were analyzed and interpreted in accordance with published guidelines. A score for each dimension was calculated. Scores for negatively worded items were reversed when calculating mean dimension scores. Safety climate dimension scores were analyzed separately for leaders (owners/supervisors) and workers.

T-tests were performed to determine if there was a significant difference in safety climate dimension scores based on leader-worker status and MSS status, i.e. whether the respondent had experienced any MSS (Yes/No), or had missed work due to MSS (Yes/No).

A categorical response variable was created for safety climate scores corresponding to recommended levels published by the National Research Centre for the Working Environment in Denmark as soft guidelines for interpretation (Kines et al. 2011). Safety climate dimension scores above 3.30, on the scale of 1–4, indicate that the safety climate level of the workplace is good, dimension scores from 3 to 3.30 correspond to a fairly good safety climate, and scores below 3.00 correspond to fairly low or low safety climate dimension scores.

Multinomial logistic regression was performed to determine which demographic, workplace, and injury variables were associated with the categorical interpretation of the safety climate dimension scores. Separate logistic regression models were run for each of the five safety climate dimensions. Variables in the model included logging system type, supervisory status, if the worker was certified as an accredited logging professional (ALP), education level, age, years of experience in logging, and whether the respondent had reported any MSS.

Data analysis was performed using SAS 9.4 (SAS Institute Inc, 2012). Significance was based upon $p < 0.05$. The study protocol and consent was approved by the university's Research Integrity and Compliance Review Office.

3 Results

One thousand fifty-nine workers attended the training workshops and 743 surveys were returned for an overall response rate of 70.2%. Mean age of respondents was approximately 46 (SD: 13.67); mean number of years employed in the logging industry was 22 (SD: 14.11). The mean number of hours worked per week in logging was 47.1 (SD = 15.5), while the mean number of months worked each year in logging was 9.2 (SD: 2.6). Ninety-four percent of participants identified as male.

Most workers indicated that their primary logging system type as mechanical (84%), with 16% of respondents identifying that they primarily use a conventional logging system (chainsaw).

Overall, 48% of the respondents reported experiencing musculoskeletal symptoms due to their work in the past year, and 6% of the respondents reported missing work in the past year due to MSS (Table 1). The anatomical area with the highest 12-month period prevalence of MSS for all loggers was the low back (38.1%), followed by the shoulders (27.6%), neck (24.8%) and the knees (24.7%).

The results of the safety climate survey indicated that when all responses were considered, the dimension of "Management safety priority and ability" had the highest mean overall score (3.40), followed by the dimension "Workers' safety commitment" (3.39), "Workers' trust in the efficacy of safety systems" (3.34), and "Peer safety communication, learning and trust in safety ability" (3.34). The dimension "Workers' safety priority and risk non-acceptance" had the lowest mean score (3.10) and, when interpreted, was the only dimension found to not fall into the "good" category. Across all five dimensions assessed, leaders had higher dimension scores survey than workers. In dimensions "Management safety priority and ability" and "Workers' safety commitment", the difference between leader ratings and worker ratings was significant ($p < 0.05$).

Table 1. Percentage of workers reporting musculoskeletal symptoms (one-year period prevalence).

Anatomical region	Percent (n = 649)
Neck	24.8%
Upper back	17.8%
Lower back	38.1%
Shoulders	27.6%
Elbows	14.5%
Wrist/Hands	21.0%
Hip/Thighs	17.1%
Knees	24.7%
Feet	13.8%
Symptoms in any area	48.1%
Missed work due to symptoms in any area	6.0%
MSS Score (Mean number of MSS categories reported)	1.8

Leaders who experienced MSS had a significantly lower score on the safety climate dimension "Workers' safety priority and risk non-acceptance" ($p = 0.05$) than leaders who did not experience MSS. Workers who reported MSS had significantly lower scores on the dimension "Management safety priority and ability" ($p = 0.03$), and dimension "Workers' safety priority and risk non-acceptance" ($p = 0.013$), in comparison to workers who did not report MSS. No significant differences in safety climate dimension scores were found with leaders nor workers who did or did not miss work due to MSS.

4 Discussion

This study conducted a quantitative evaluation of the determinants of safety climate within the logging industry, and provided a baseline measure of the safety perceptions for this population. While not significant in the regression model, the disparity between leaders' and workers' safety climate scores is of interest. Across all five dimensions, leaders (owners/supervisors) of logging companies had higher safety climate scores than workers, and in two dimensions, when workers were compared directly to leaders, workers had significantly lower responses. The significant differences were found in dimensions one (management safety priority and ability) and four (workers' safety commitment).

In the safety climate dimension of "workers' safety priority and risk non-acceptance", workers who reported work-related MSS were nearly three times more likely to be assigned to the low category of safety climate than the high category, meaning, loggers who experienced MSS in the past year were more likely to have low safety priorities and accept risks in the workplace than loggers who did not report MSS.

As indicated in the results of the study, the authors provide a quantitative evaluation of the current safety climate in the logging industry. The data and subsequent results obtained during this study provide a baseline measure of both musculoskeletal symptoms and safety climate, which can be used as a standard of comparison after the application of safety interventions.

This research indicated that injury prevention efforts in the logging industry should focus on sustaining the relatively high level of safety climate within the logging industry. This can be accomplished efficiently by specifically targeting the mismatch between leaders (owners/supervisors) versus worker's safety perceptions, and the association between MSS and safety climate.

Quantifying the current safety climate of the logging population and investigating determinants of safety climate is needed to identify possible areas and opportunities for future interventions. Measuring leading indicators, such as safety climate, is a step toward proactive injury surveillance and control.

References

Kines P, Lappalainen J, Mikkelsen KL, Olsen E, Pousette A, Tharaldsen J, Törner M (2011) Nordic Safety Climate Questionnaire (NOSACQ-50): a new tool for diagnosing occupational safety climate. Int. J. Ind. Ergonomics 41(6):634–646

Lagerstrom E, Magzamen S, Rosecrance J (2017) A mixed methods analysis of logging injuries in Montana and Idaho. Am J Ind Med 60(12):1077–1087

Neal A, Griffin MA, Hart PM (2000) The impact of organizational climate on safety climate and individual behavior. Safety Sci 34(1):99–109

United States Department of Labor OSHA (2017). Safety and health topics: logging. https://www.osha.gov/SLTC/logging/

Zhang H, Wiegmann DA, Von Thaden TL, Sharma G, Mitchell AA (2002) Safety culture: a concept in chaos? In: Paper presented at the proceedings of the human factors and ergonomics society annual meeting

Zohar D (1980) Safety climate in industrial organizations: theoretical and applied implications. J Appl Psychol 65(1):96

Quality of Work Life and Differences in Demographic Characteristics Among Managerial Staff in Algerian Tertiary Sector

Bouhafs Mebarki[1(✉)], Mohammed El Amine Ahmed Fouatih[1,2], and Mohamed Mokdad[1,2,3]

[1] Laboratory of Ergonomics and Prevention of Risks, University of Oran2, Oran, Algeria
mebarkibouhafs@gmail.com
[2] Tlemcen University, Tlemcen, Algeria
[3] Bahrain University, Zallaq, Bahrain

Abstract. The aim of the present paper is to measure the level of quality of work life (QWL) and to study the differences in demographic characteristics (gender, age, work experience and socio-professional category) among managerial staff in two public settings from tertiary sector based in Oran - Algeria.

A total sample of 252 managerial staff members participated in the study. Data were collected using a questionnaire technique.

The analysis of the results showed that the level of QWL was medium, and there were no statistically significant differences in QWL between demographic characteristics categories (1) gender, (2) age, (3) work experience in these two organizations and (4) socio-professional categories. The results are discussed in the light of previous researches.

Finally, the study concluded that the QWL needs more attention from management levels in the public sector in Algeria.

Keywords: Quality of work life (QWL) · Demographic characteristics Managerial staff

1 Introduction

During the last two decades Quality of Work Life (QWL) has gained the attention of the research community, for it is "becoming an imperative issue to achieve the goals of the organization in every sector" [1]. There is a consensus in the research literature on the importance of QWL as it is a prerequisite to increase employees' productivity and wellbeing. "As a result, high QWL organizations may enjoy better sustainable efficiency, productivity and profitability" [2].

As a research issue, QWL has been defined in a variety of ways [3], and the term QWL includes quality of work and employment quality [4]. Walton [5], one of the early researchers of QWL, asserted that the concept suggested comprehensiveness and was broader than the aims of the unionization movement, labor laws, or equal employment

S. Bagnara et al. (Eds.): IEA 2018, AISC 821, pp. 220–232, 2019.
https://doi.org/10.1007/978-3-319-96080-7_26

struggles". While, Sirgy et al. [6] define QWL "as employee satisfaction with a variety of needs through resources, activities, and outcomes stemming from participation in the workplace". Hence, Sirgy et al. [7] shifted the conceptualization of QWL dimensions from the traditional Walton's approach who proposed eight major conceptual categories relating to QWL dimensions [8], to the need-hierarchy theory [9]. Sirgy et al. [7] identified seven dimensions of QWL. These are: (1) health and safety needs, (2) economic and family needs, (3) social needs, (4) esteem needs, (5) actualization needs, (6) knowledge needs, and (7) aesthetic needs.

These conceptual categories of QWL dimensions has gained a consensus among researcher community [10–14], as they are important to both employees and management, who should find the appropriate ways to meet the perceived needs of employees.

Many dimensions of QWL were thoroughly investigated; their effects on QWL were studied, while studies of the effect of demographic characteristics (gender, age, work experience and Socio-professional categories) on QWL have conflicting results [1, 15–17]. Thus, the question needs to be thoroughly investigated, particularly in different cultural contexts, and among different socio professional categories.

Measuring the level of QWL was a challenge to many researchers as it embodies many dimensions [1, 12, 15] and it enhances cultural and organizational ingredients, in different sectors of occupational activity. QWL "is an umbrella term which includes many concepts. QWL means the sum total of values, both materials and non-materials, attained by the worker throughout his life" [1].

Pioneers of QWL studies found the industrial sector to be a good breeding ground for their research activities. In a later stage, the tertiary sector aroused the interest of studies [1, 18], as it encompasses a broad spectrum of professional activities, private and public sector. According to Martins et al. [18] public sector does not invest enough in QWL. Previous studies [19] have shown that the organizational culture in public sector is a decisive factor in the QWL.

In the Algerian case, the situation is still ambiguous, especially in the services (tertiary) sector of activity, which is economically considered as the most important sector. According to the Algerian National Office of Statistics [20] the tertiary sector contributes nearly 48% of GDP and employs nearly 60% of the labor force. The share of services in GDP has increased recently, well ahead of agriculture (13% of GDP and employs 10.8% of the labor force) and industry (39% of GDP and employs almost a third of the workforce). The tertiary sector encompasses a large spectrum of activities, among which: transport, distribution, sale of goods and the provision of services.

The aim of the present paper is to measure the level of quality of work life (QWL) and to study the relationship between QWL and some demographic characteristics (gender, age, work experience, and socio-professional category) among managerial staff in two public settings from tertiary sector based in Oran - Algeria.

Based on the above literature review, the dimensions of the QWL adopted in this study can be categorized in seven categories as seen in Fig. 1.

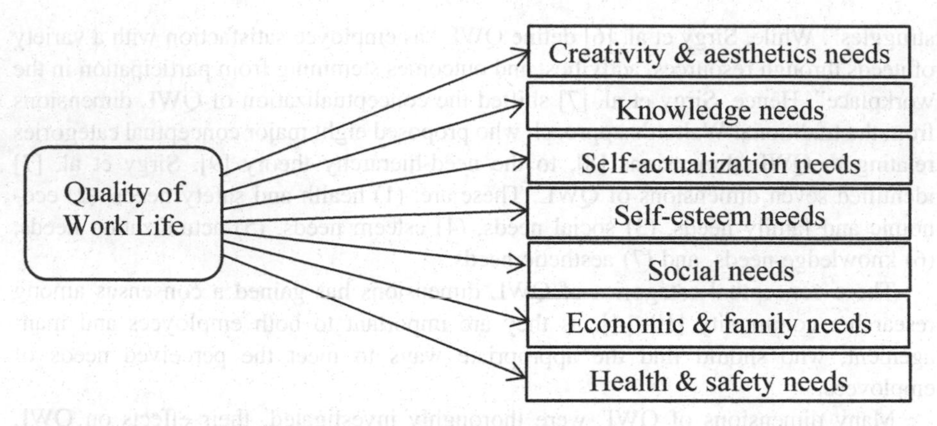

Fig. 1. The QWL dimensions.

2 Research Framework and Hypotheses

To examine the relationship between QWL and demographic characteristics of managerial staff, a research framework was developed for the purposes of the present study, based on Bolhari's et al. [21] study, as illustrated in Fig. 2.

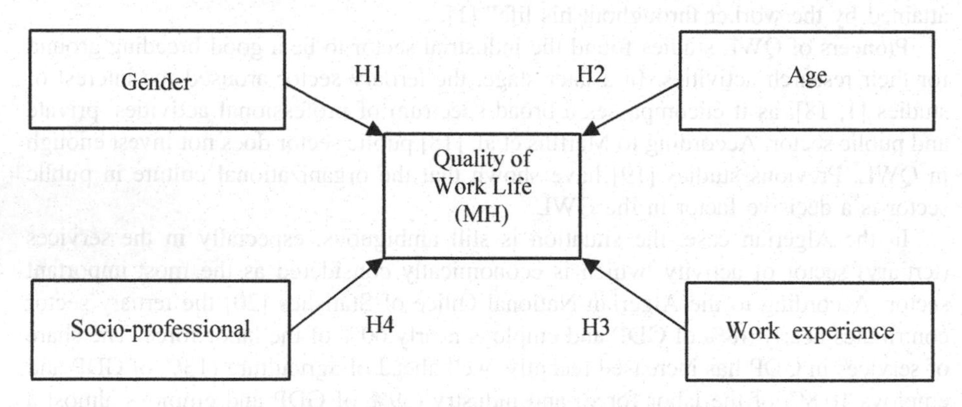

Fig. 2. Research framework (adopted from Bolhari et al. [21]).

The main hypothesis (MH) of this study was formulated as follows: The level of QWL among managerial staff of the present sample is low.

While relationships of demographic characteristics with QWL were formulated through the following hypotheses:

H1: There are no statistically significant differences between males and females in QWL.

H2: There are no statistically significant differences between age groups in QWL.

H3: There are no statistically significant differences between work experience categories in QWL.
H4: There are no statistically significant differences between socio-professional categories in QWL.

3 Method

3.1 Sample

A random sample of 252 managerial staff members from two service companies based in Oran - Algeria participated in the study, during the period from November 2015 to March 2016. Their demographic characteristics are shown in the Table 1.

Table 1. Demographic characteristics of the sample (n = 252).

Characteristics		No. of respondents	(%)
Gender	Male	138	54.8
	Female	114	45.2
Age	21–30	64	25.4
	31–40	127	50.4
	41–50	43	17.1
	≥ 51	18	7.1
Work experience in the organization	<5	92	36.5
	5–9	78	31.0
	≥ 10	82	32.5
Socio-professional categories	A senior executive	31	12.3
	Middle class manager	186	73.8
	Supervisor	35	13.9

Participants filled in the questionnaires, in the presence of researchers, during approximately a half-hour session, depending on their availability.

3.2 Tool

To measure the seven dimensions of the QWL, a structured questionnaire was designed, inspired from previous studies [7, 10, 12, 14]. It consisted of two sections; the first section dealt with demographic characteristics of the sample, namely: gender, age, Work experience in the organization, Socio-professional category. The second section, consisted of 75 items, dealt with the following dimensions of the QWL:

1. Health & safety needs consisted of 14 items
2. Economic & family needs consisted of 11 items
3. Social needs consisted of 10 items
4. Self-esteem needs consisted of 10 items

5. Self-actualization needs consisted of 10 items
6. Knowledge needs consisted of 10 items
7. Creativity & aesthetics needs consisted of 10 items

Respondents were asked to rate their level of agreement on each statement on a five-point Likert scale, from "1" as "strongly disagree" to "5" as "strongly agree". Hence, the level of QWL was determined from mean value of the respondents' attitudes towards items of the questionnaire as follows (Table 2):

Table 2. Determining values of the level of QWL.

Meaning	Very high level	High level	Medium level	Low level	Very low level
Mean value (total of items)	375–315.75	315–255.75	255–195.75	195–135.75	135–75

The reliability of the questionnaire was assessed on a sample of 100 respondents by Cronbach's alpha, which assumes a range from r = 0 to 1, with r = 0.7 or greater considered as sufficiently reliable; results are illustrated in Table 3.

Table 3. Reliability test for QWL questionnaire.

QWL dimensions	Number of items	Reliability (Cronbach's α)
Health & safety needs	14	0.668
Economic & family needs	11	0.755
Social needs	10	0.633
Self-esteem needs	10	0.807
Self-actualization needs	10	0.855
Knowledge needs	10	0.834
Creativity & aesthetics needs	10	0.878
Total	**75**	**0.946**

While, the internal consistency validity was assessed on the same sample of 100 respondents using Pearson's coefficient of correlation between the value of each QWL dimension and the overall value of the questionnaire, as follows (Table 4):

3.3 Data Analysis and Research Findings

The data were analysed using the Statistical Package for the Social Sciences (SPSS), version 20.0.

Table 4. Validity test for QWL questionnaire.

QWL dimensions	Pearson's (r)
Health & safety needs	0.710**
Economic & family needs	0.771**
Social needs	0.730**
Self-esteem needs	0.854**
Self-actualization needs	0.871**
Knowledge needs	0.746**
Creativity & aesthetics needs	0.824**

**Significant at the 0.01 level

4 Results

Results of the study are presented along the following subsections:

4.1 Level of QWL Among Managerial Staff

Results of the evaluation of the level QWL among participants in the study are presented in Table 5.

Table 5. Shows the level of QWL among managerial staff (n = 252).

Indicators	Mean	SD	Min. score	Max. score	Theoretical mean	Level of QWL
Scores of QWL	253.61	46.775	106	375	225	Medium level

As shown in Table 5, the mean total score of QWL was 253.61 (SD = 46.775), this value is greater than the theoretical mean (225), which means that respondents attitudes towards QWL dimensions were at a medium level.

4.2 Gender Differences in QWL

Results on gender differences in QWL are presented in Table 6.

Table 6. Shows the difference between males (n = 132) and females (n = 114) in QWL.

Indicators	Gender	n	Mean	SD	Degrees of freedom	T test value	Sig. or P-value	Level of significance
Scores of QWL	Male	138	257.71	48.992	250	1.536	0.126	n.s.
	Female	114	248.64	43.640				

Tabulated T = 1.972, df = 250, α = 0.05

4.3 Age Differences in QWL

Results on age differences in QWL are presented in Table 7.

Table 7. Shows the QWL in different age groups.

Test	Sum of squares	Degrees of freedom	Mean square	F test value	Sig. or P-value	Level of significance
Between groups	5570.161	3	1856.720	0.847	0.469	n.s.
Within groups	543597.946	248	2191.927			
Total	549168.107	251				

Tabulated F = 2.63, df = 3, 248, α = 0.05

4.4 Work Experience Differences and QWL

Results on work experience differences in QWL are presented in Table 8.

Table 8. Shows the QWL in different work experience years.

Test	Sum of squares	Degrees of freedom	Mean square	F test value	Sig. or P-value	Level of significance
Between groups	1178.402	2	589.201	0.268	0.765	n.s.
Within groups	547989.705	249	2200.762			
Total	549168.107	251				

Tabulated F = 3.02, df = 2, 249, α = 0.05

4.5 Differences of Socio-Professional Categories in QWL

Results of the differences between socio-professional categories in QWL are presented in Table 9.

Table 9. Shows the QWL in different socio-professional categories.

Test	Sum of squares	Degrees of freedom	Mean square	F test value	Sig. or P-value	Level of significance
Between groups	352.038	2	176.019	0.080	0.923	n.s.
Within groups	548816.069	249	2204.081			
Total	549168.107	251				

Tabulated F = 3.02, df = 2, 249, α = 0.05

5 Discussion

5.1 Level of Quality of Work Life

As shown in Table 5, the mean total score of quality of work life was 253.61 (SD = 46.775), this value is greater than the theoretical mean (225), which means that respondents attitudes towards QWL dimensions were at a medium level.

Although, comparisons with other studies might be misleading, for the diverse forms of work organization, and the marked differences between countries in their characteristic work systems, their occupational legislations and degrees of technical and economic development are key determinants of QWL, we try to compare the results of the present study with findings of similar studies.

To the best of our knowledge, the only study, which, evaluated the level of QWL in Algeria, is that of Boukhemkhem [22] where the results revealed a low/unfavorable level of QWL among both male and female university employees.

The experience of job insecurity is known to be a source of psychological stress, as unemployment itself, and have clear implications for employee welfare [23]. Thus, the medium level of QWL among managerial staff in our sample can be explained, firstly, by the fact that, managerial staff do not experience job insecurity, as both enterprises of the present study have the monopole of gas and water transportation and distribution, and, have no threat of any sort of competition. Secondly, the managerial regime, which can be described as inclusive regime, where trade unions have their say in daily life of the organization. As has been pointed out by Gallie [24]: "The quality of work, it is suggested, will be better in inclusive regimes where trade unions have high levels of participation in national decision making, than in dualistic regimes where they protect only core employees or in Liberal market regimes where regulation is generally very weak".

In some similar context of developing countries, Bolhari's et al. [21] findings in Iran, on a sample of Information Technology Staff, and that of Salah [25] in Saudi Arabia among university teaching staff, were similar to the results of the present study. While, Eslamian et al. [26] study on nursing staff in emergency departments, revealed a low level of quality of work life, associated with workplace violence. Rastegari's et al. [27] study on a similar population showed a moderate to low level of QWL, associated with moderate nurses' task performance. Differences between the results of the present study, and previous research findings, can be explained by the difference in socio economic contexts, in which these studies were conducted.

The results of the present study, there for, advocate for new management strategies to enhance the level of QWL among managerial staff in the two setting under study. The scope of the study was limited to public economic service sector; future researches may undertake studies in organizations of the civil service sector, like the public health service and education, which are nowadays knowing many unrest movements in Algeria (strakes, turnover, etc.).

5.2 Gender Differences in QWL

As shown in Table 6, there were no significant differences in terms of QWL that are attributable to gender ($P = 0.126 > \alpha = 0.05$).

In a similar study on QWL among University employees in Algeria, Boukhem-khem [22] found no significant difference in the level of QWL between male and female employees. Thus, the results of the present study agree with the findings of previous researches. [17, 21, 22, 28], in terms of gender effect on QWL, in other terms, male and female employees are experiencing the same level of Quality of work life. However, differs from that of Tabasum et al. [29], on employees of private commercial banks, where, they showed male employee's perception of QWL differs from the female employees.

This result was not expected, as the general belief word-wide, especially in developing countries is that most women at work are exerting themselves in combining work and home responsibilities, and at the same time, aspire towards self-actualization in their career, as Rani and Kritika [30] pointed out. Particularly in traditional societies, as advocated by feminist movements. In the case of the present study, managerial staff, of both gender categories were issued from the same university education levels, and belonging to middle class backgrounds, working under the same work legislation rules, particularly gender equality aspects, which are applied under the control of strong union movements.

5.3 Age Differences in QWL

As shown in Table 7, there were no significant differences in terms of QWL that are attributable to age ($P = 0.469 > \alpha = 0.05$). This result confirms previous research findings, where age variable had no significant influence on Quality of Work Life among specific categories of employees, like university teaching staff [31, 32]. However, differ from results of other studies on Information Technology Staffs [21], university teachers [33] where there were significant differences among employees belonging to different age categories in their perception towards QWL.

The reason behind these conflicting results have to be further investigated, as age factor of an employee is synonym to his/her work experience and career perspectives.

5.4 Work Experience Differences in QWL

As shown in Table 8, there were no significant differences in terms of QWL, that are attributable to work experience ($P = 0.765 > \alpha = 0.05$). This same finding is on line with Salah's [25] results on faculty teaching staff, and Xhakollari's [34] results on mental health workers. Nevertheless, findings of the present study on the effect of work experience on QWL differ from Bolhari's et al. [21] study on Information Technology Staffs, Indumathy and Kamalraj [17] study on textile industry workers, Tabasum et al. [15] on faculty members of private universities, and Aarthy and Nandhini [31] on engineering faculty members.

As for age, work experience ought to be deeply investigated, for the content behind the term "work experience", which is not only the number of years spent in one organization, but the term embraces other variables, like, education qualification, skills level, job content, job opportunities, career growth and development, employment traits and personal characteristics of each employee.

5.5 Differences Between Socio-Professional Categories in QWL

As shown in Table 9, there were no significant differences in terms of QWL, that are attributable to socio professional categories ($P = 0.923 > \alpha = 0.05$). This result, can be explained by the fact that, the three professional categories of the present study (a) senior executives, (b) middle class managers and (c) supervisors are belonging to the managerial class, which work under similar conditions.

Although, the result confirms previous findings [35, 36], we point out differences with findings of other studies [21, 37] which might be attributed to environmental, organizational culture and climate, activity sector and size of the setting, as these factors differ from one study to another.

Regardless of their age, gender, work experience and socio professional category, managerial staff of the present study showed the same level of QWL. The demographic variables have no effect on QWL, as an independent variable. A possible explanation of this result, may reside in the type of organizational culture or management methods, work conditions, rules and work procedures, which, equally apply for all members of the sample in the same way. As, these factors are known to be important ingredients of QWL.

Results of research works, on the relationship between Quality of Work Life and demographic characteristics of employees, are conflicting. In their analysis, of the literature on the subject Yadav and Khanna [1] pointed out that 6 out 25 literatures said that there was no relationship between gender and QWL, age affected the QWL according to 4 out of 25 literature works, whereas, experience gave a positive relation with QWL in 4 literatures.

The main methodological drawback, we noticed on previous research work, is that most of the studies on QWL and demographic characteristics of the populations from which study samples were drawn, did not clearly describe the socioeconomic, cultural and organizational contexts of the samples, as these contexts are key determinants of employees attitudes towards QWL.

The current study was confined, only to the managerial Staff, in two Algerian public service companies. Further studies on other socio professional categories, in both public and private sectors, may throw more light on different issues of QWL, in a large spectrum of industries and among different working populations.

6 Conclusions

The study revealed a medium level of quality of work life among managerial staff, and no significant differences in the QWL that are attributable to demographic variables were obvious. These are in agreement with some previous studies, while differences with other studies in some issues of the QWL, were also noticed. Hence, our findings should be treated with some reserve, as organizational socio cultural contexts are known to influence QWL dimensions.

Although, the results of the present study are a useful tool for elaborating new management strategies and programs to enhance the level of QWL and promote a better QWL, other salient variables which were not in the scope of this study, should be included in such strategies. In addition, more complex interactions of QWL with demographic characteristics, and other variables, among managerial staff should be examined.

The comparison between QWL levels in different professional categories seems to be a promised theme for future research, as it gives a general view on the health state of the organization in terms of QWL of its employees.

References

1. Yadav R, Khanna A (2014) Literature review on quality of work life and their dimensions. IOSR J Hum Soc Sci (IOSR-JHSS) 19(9):71–80
2. Mosadeghrad AM (2013) Quality of working life: an antecedent to employee turnover intention. Int J Health Policy Manag 1(1):43–50
3. Ilgan A, Ata A, Zepeda SJ, Ozu-Cengiz O (2014) Validity and reliability study of Quality of School Work Life (QSWL) scale. Int J Hum Sci 11(2):114–137
4. Carayon P, Sainfort F, Smith MJ (1999) Macroergonomics and total quality management: how to improve quality of working life? Int J Occup Saf Ergon 5(2):303–334
5. Walton RE (1973) Quality of work life: what is it? Sloan Manag Rev 11–21
6. Sirgy MJ, Reilly NP, Wu J, Efraty D (2008) A work-life identity model of well-being: towards a research agenda linking Quality-of-Work-Life (QWL) programs with Quality of Life (QOL). Appl Res Qual Life 3(3):181–202
7. Sirgy MJ, Efraty D, Siegel P, Lee DJ (2001) A new measure of Quality of Work Life (QWL) based on need satisfaction and spillover theories. Soc Indic Res 55(3):241–302
8. Walton RE (1975) Criteria quality in work life. In: Davis LE, Cherns RL (eds) The quality of working life: problems, prospects, and the state of art, vol 1. Free Press, New York
9. Maslow AH (1970) Motivation and personality. Harper & Row, New York
10. Lee DJ, Singhapakdi A, Sirgy MJ (2007) Further validation of a need-based Quality-of-work-life (QWL) measure: evidence from marketing practitioners. Appl Res Qual Life 2(4):273–287
11. Mortazavi S, Yazdi SVS, Amini A (2012) The role of the psychological capital on quality of work life and organization performance. Interdiscip J Contemp Res Bus (IJCRB) 4(2):206–217
12. Marta JKM, Singhapakdi A, Lee DJ, Sirgy MJ, Koonmee K, Virakul B (2013) Perceptions about ethics institutionalization and quality of work life: Thai versus American marketing managers. J Bus Res 66(3):381–389
13. Elamparuthi D, Jambulingam S (2014) Need satisfaction and quality of work life in Chennai automobile industry. Int J Econ Commer Manag 2(3):1–7
14. Viljoen A, Kruger S, Saayman M (2014) Understanding the role that Quality of Work Life of food and beverage employees plays in perceived service delivery and productivity. South Afr Bus Rev 18(1):27–52
15. Tabassum A, Rahman T, Jahan K (2012) An evaluation of the quality of work life: a study of the faculty members of private universities in Bangladesh. ABAC J 32(3):36–57

16. Sandhya Nair GS (2013) A study on the effect of Quality of Work Life (QWL) on Organisational Citizenship Behaviour (OCB) – With special reference to college teachers is Thrissur District, Kerala. Integr Rev J Manag 6(1):34–46
17. Indumathy R, Kamalraj S (2012) A study on quality of work life among workers with special reference to textile industry in Tirupur District – a textile hub. Int J Multidiscip Res 2(4):265–281
18. Martins JC, Pereira MG, Pinheiro AG (2013) The contemporary acceleration as demarcation of styles and quality of working life among healthcare professionals. Lusoph J Cult Stud 1(2):322–337
19. James G (1992) Quality of working life and total quality management. Int J Manpow 13(1):41–58
20. ONS (2014) Statistical yearbook of Algeria, Edition 2014, Résultats 2010/2012, vol n° 30. http://www.ons.dz/IMG/pdf/AnRes10-12No30.pdf. Accessed 10 Nov 2017
21. Bolhari A, Rezaeean A, Bolhari J, Bairamzadeh S, Soltan AA (2011) The relationship between quality of work life and demographic characteristics of information technology staffs. In: Proceedings of international conference on computer communication and management, CSIT, vol 5, pp 374–378
22. Boukhemkhem D (2015) Quality of work life: theoretical concepts and evaluation Case study of employees of Jijel University. Journal of Economie & Société 11(11):131–162. https://www.asjp.cerist.dz/en/article/1975
23. Burchell B (2011) A temporal comparison of the effects of unemployment and job insecurity on wellbeing, sociological research, 16 September 2011. http://www.socresonline.org.uk/16/1/9.html
24. Gallie D (2017) The quality of work in a changing labour market. Soc Policy Admin 51(2):226–243. https://doi.org/10.1111/spol.12285
25. Salah AA (2013) Quality of work life among faculty of education teaching staff. J Arab Stud Educ Psychol 2(39):158-189. [in Arabic]
26. Eslamian J, Akbarpoor AA, Hoseini SA (2015) Quality of work life and its association with workplace violence of the nurses in emergency departments. Iran J Nurs Midwifery Res 20(1):56–62. https://www.ncbi.nlm.nih.gov/pmc/articles/PMC4325414/
27. Rastegari M, Khani A, Ghalriz P, Eslamian J (2010) Evaluation of quality of working life and its association with job performance of the nurses. Iran J Nurs Midwifery Res 15(4):224–228. https://www.ncbi.nlm.nih.gov/pmc/articles/PMC3203281/
28. Jnaneswar K (2016) A study on the level of quality of work life experienced by the employees of public sector units in Kerala. J Inst Public Enterp 6(1/2):51–61
29. Tabassum A, Rahman T, Jahan K (2011) Quality of work life among male and female employees of private commercial banks in Bangladesh. Int J Econ Manag 5(1):266–282
30. Rani AK (2013) Quality of work life of female employees in Universities of Haryana. https://editorialexpress.com/cgi-bin/conference/download.cgi?db_name=IAFFE2013&paper_id=38&file_type=slides
31. Aarthy M, Nandhini M (2016) A study on quality of work life among the engineering college faculty members in Coimbatore District. Int J Manag Res Rev 6(8):1051–1057
32. Arunkumar B, Saminathan R (2017) Work life balance among women teachers of Self Financing Colleges (SFC) in Thanjavur District, Tamilnadu. J Hum Soc Sci (IOSR-JHSS) 22(7):48–55. www.iosrjournals.org
33. Rao T, Arora RS, Vashisht AK (2013) Quality of work life: a study of Jammu University teachers. J Strat Hum Resour Manag 2(1):20–25

34. Xhakollari L (2013) Quality of work life and mental health professionals in Albania. Mediter J Soc Sci 4(1):529–534
35. Rajasekar D (2017) An empirical research on quality of work life among executive level employees in shipping industry. In: 4th IHRC, ELK Asia Pacific journals
36. Balaji R (2013) A study on quality of work life among employees. Int J Innov Res Sci Eng Technol 2(2):470–473
37. Johnsrud LK (2006) Studied on quality of faculty work life: the University of Hawaii

How Holistic Approaches of Activity Analysis in Ergonomics Renew Training Design?

Christine Vidal-Gomel[1(✉)], Vincent Boccara[2], Dominique Cau-Bareille[3], and Catherine Delgoulet[4]

[1] Université de Nantes, CREN, Chemin de la Censive du Tertre, BP 81227, 44312 Nantes Cedex 3, France
christine.vidal-gomel@univ-nantes.fr
[2] Université Paris Sud XI, LIMSI, Rue Von Neumann, Bat 508, 91403 Orsay Cedex, France
boccara@limsi.fr
[3] Université Lyon 2, IETL, 86 rue pasteur, 69007 Lyon, France
dominique.cau-bareille@univ-lyon2.fr
[4] Université Paris Descartes, LATI, 71, avenue Edouard Vaillant, 92774 Boulogne-Billancourt Cedex, France
catherine.delgoulet@parisdescartes.fr

Abstract. This communication presents the benefit of holistic approaches of work analysis in ergonomics and professional didactics in the field of occupational training. The research-intervention processes discussed have the originality aiming to conduct jointly learning questions, conditions of training and prevention of occupational risks. Its specificities are presented from three research-interventions in the field of aeronautics industry, agriculture and education. We stress questions that are emerging, such as the design of digital tools for training, the importance of subjective commitment at work and the identity transformations and we precise how the holistic approach advocated, based on a model of operator development that articulates mastery of tasks and work and the construction of health and safety over the long term, may be used to treat these questions.

Keywords: Work analysis · Holistic approach · Health · Safety Training design

In line with previous symposia on Ergonomics work analysis and training (EAWT), since Rabardel, Teiger, Laville et al. [1], the ergonomics research-intervention processes defended in the field of occupational training had the originality aiming to conduct jointly learning questions, conditions of training and prevention of occupational risks. In previous symposia a situated approach of ergonomic work analysis had always been the reference. Teiger and Montreuil [2] specified the need to reconstitute "in great details, with the help of operators on how the later carry out their tasks […]. It then becomes possible to identify constraints on an activity and the resulting consequences in particular for health, reliability and work". The specificity of this approach is to be centered on operator activity in situation with an objective of constraints and

S. Bagnara et al. (Eds.): IEA 2018, AISC 821, pp. 233–241, 2019.
https://doi.org/10.1007/978-3-319-96080-7_27

resources identification - i.e. at the levels of tasks characteristics, social and organizational context [2, 3].

A holistic approach (or systemic approach) is defended. In ergonomics, it designates the point of view that is adopted to jointly take into account a set of dimensions of human activity (biological, psychological, cognitive, social) and that also seeks to highlight a set of determinants of operators' activity, depending on the problem addressed [4]. It is neither a question of dealing with all the dimensions of the activity and all its determinants, nor of taking into account only some of them on a recurrent basis. The scope of a holistic approach can only be relevant for the purpose of the research: "When defining problems and formulating solutions, system boundaries are defined, and the focus of HFE [human factors/ergonomics] can be on specific aspects of people [...], on specific aspects of the environment [...], or on a specific level (e.g. micro), but the broader context of the human within the environment is always taken into consideration ('contextualisation')" [5, p. 379]. In line with this position Lundberg [6], for example, summarizes various researches on situation awareness (SA), which were conducted either by taking SA as a process, or as a state or from a systemic point of view. His synthesis and analysis lead him to propose a holistic model, relevant to deal with resilience issue. In ergonomics, a holistic approach focuses on two "related outcomes: performance and well-being" [5].

The originality of the research-interventions presented in this communication is to point out different level of analysis that are required, the combinaisons of different approaches - i.e. ergonomics and professional didactics, the extension of the actors included in the research-intervention process and an evolution of the objects of the analysis. Furthermore, all the studies choose to triangulate the methods used [7]. This triangulation consisted in combining several different methods of data collection, in order to: (1) reduce the biases inherent to each method and thus increase the reliability and validity of the empirical data gathered; (2) to provide a qualitative richness and a better understanding of the phenomenon studied; (3) to facilitate understanding of processes which cannot be directly observed, such as conceptualisation in action [8]. This approach thus combines various classic ergonomic methods: e.g. field observations [9], several kind of interviews [10, 11], and a documentary collection [9]. Three studies, conducted in France in the field of aeronautics industry, agriculture and education are presented to stress these specificities. Their research objects were more or less broad, which is why their exposure will go from the circumscribed point to the broadest one. The first study focused on the design of a virtual training environment (VTE) for a vocational training (training of assembly line operators); the second aimed at designing a training program for a specific professional operation (late gathering of bovine embryos) in a real work situation and the third dealt with the place given to continuing training (by the employer and the employees themselves) in the process of retraining secondary school teachers. Each of them proposed an original way of developing a holistic approach to the analysis of activity in the field of vocational training, by mobilizing expanded theoretical and methodological resources to conduct the analysis. In conclusion, we will discuss the required evolutions of the work analysis scope to reach the objectives of health and security at work shared in this symposium.

1 Early Design of Virtual Reality Environment for Training in Aeronautics Industry: From Work Analysis to Works Analysis

The first study took place in a multidisciplinary research project[1] bringing together a manufacturing company and one of its subsidiary sites, computer engineers, automation engineers, designers and ergonomists [12]. It was an upstream design project for a VTE for the assemblers in the aeronautics industry. Ergonomists were initially to contribute to the needs analysis (by the analysis of the work to be trained by the demonstrator) and to its early evaluation. However, the first investigations conducted in real work and training situations prompted the ergonomists to support the idea that this device not only aimed at training for a specific trade (which required analysis) but also had to be designed as one of several pedagogical tools for training (which also required analysis from the point of view of the trainees and trainers). Under these circumstances, classical approaches often reduce users to learners alone and the self-assessment of the device in use [13]. The aim of the ergonomist contribution was then to propose an alternative approach that focus on both technological and user-centred design approaches throughout the project duration. To do so, a two-phasc approach was adopted: Phase 1 was devoted to the "analysis of the actual works", Phase 2 aimed at the "multidimensional evaluation of the uses of demonstrator designed". We present here only the first phase.

We developed a mean of analysing "the works" of trainers, trainees and production operators. The originality of this stance lay more particularly in the fact that the training situation was considered as a full working situation. This assumption was based on three complementary arguments [14]. Trainers are workers caught within socio-economic issues of viability, of quality, of the performance levels of the training tools that they design, coordinate and/or facilitate; their activity is subject to constraints and must meet requirements just as much as that of production operators. (2) The trainees are employees of a company; their work can therefore be taken to be "an apprenticeship" which has to be linked to the context in which it takes place. (3) These training works (of both trainers and trainees) challenge and must orient the way the training is designed, in the same way as the more "obvious" work of production operators. The objective therefore aimed to examine the place of work analysis, its scope and the actors involved, so that ergonomists can help to design a training tool (definition of objectives, training devices and situations, properties of the virtual environment and of the manipulated objects, tool functionalities, etc.). Without going into the details of all of the analyses performed, "work" analysis is considered as a mean of identifying the multiple horizons of the training situations to be defined: "training for what?", "training how?", "what device(s)?", "for what purposes?" [15], considering working conditions for training to be learning conditions for trainees [16].

---

[1] The "Natural Interactions, Knowledge, Immersive system for Training in Aeronautics" (NIKITA) research project, funded by the Agence National pour la Recherche (ANR) (http://www.emissive.fr/nikita/).

Analyses conducted in this phase produced and sustained a frame of reference of situations [17] synthesizing benchmarks for design which fulfil the role of intermediary objects within the design process [18]. It integrated: (1) organizational and situational factors in production and training, (2) potential sources of accidents and defects in production, (3) tasks, processes, rules and tools, and (4) elements relating to pragmatic conceptualization and to the organisers of the journeymen activity. It was designed with the dual objective of an operational tool for: (1) training course design managers (trainers and the training center director); (2) the VTE designers, with a view to proposing "potential development situations" in and with the future VTE [19]. More specifically, it was a resource and a tool with which to orient the choice of the material interface (commands, technical device, etc.) and the software interface with a set of training scenario, potentially mobilizing virtual agents and satisfying the different modes of use depending on the type of real training situations envisaged (in the classroom vs. in the workshop, individual vs. group, with a trainer vs. without a trainer, etc.). The suggested frame of reference thus offers a set of elements which may be used and combined in various ways, in order to simulate training situations in line with the questions asked during the design process and as they evolved. It also helped to guide the functional analysis and the development of specifications with and for the design team, taking into account the utility, the usability and the learning aspect of the future system [14].

This holistic approach thus contributed to broadening the point of view often centred on just one category of users in the design processes, both in the phase dedicated to the analysis of needs and in the phase devoted to the evaluation of the demonstrator. Works analysis with production line operators, learners and trainers made it possible to review the intentions of the project, identify ways of improving the VTE and the conditions of its operational deployment.

2 From Training to the Task Mastery to the Work Context as an Obstacle of the Learning Process

The second study mixed ergonomics and professional didactics methods [8, 20]. This research-intervention was carried out at the request of an agronomy research centre and was aimed at improving a situation of skills transfer between an animal handler expert in the late gathering of bovine embryos and two of its experienced colleagues who were novices in this operation. The training took place in working situations: the two animal handlers carried out the gathering operation with the help and guidance of the expert. The operation is carried out without visual control and only from haptic sensations on which the actors must concentrate to understand the progression of the tools in the animal's genital tract, the difficulties encountered and how to overcome them. However, these skills are tacit, which makes their learning and the expert guidance difficult. Self- and cross-confrontation interviews have been carried out to enable actors to become aware of the skills to be acquired and transmitted, and to encourage collective debates on different practices and difficulties [21]. However, given the conflicts between the actors and more generally their state of irritation, at times their ill-being, this training porgram has been put in difficulty. An hypothesis of psychosocial risk factors (PRF) started to emerge [22]. We investigated it through deeper interviews with animal handlers, researchers requesting the gathering operations and their management. We identified different determinants of operators work situations as PRF [23]: instability linked to the fear of closing the center, which get poor results, in a context of restructuring of the establishment and linked to the turnover of the local management in recent years; the impossibility to do a quality work. Many changes have created a situation where the herd management and the task of animal selection for experimentations that is done upstream of the gathering of embryos is impeded. These tasks partly determine their results. The identification of this psychosocial risks context in which this training took place, the sharing of this diagnosis with the operators allowed to create a more serene situation, at least temporarily, and to make the training succeed.

This research-intervention was initially centered on the realization of the gathering of embryos task. The context of psychosocial risks led to a complementary study, changing the scale for data collection and analysis. If activity is always at the center, this time it is its social and organisational determinants that are the object of the intervention. The holistic approach therefore consists here in highlighting the determinants of the activity in a work situation which constitute obstacles to the development of a training course in situation and thus to the learning of the actors. This study thus points out an articulation between the contributions of professional didactics which emphasizes the characteristics of tasks and the learning of their control in situation and ergonomics which is interested in a broader set of determinants of the operators activity.

3 From Training to the Activity Systems to Understand of the Development

The third study that we are presenting is concerned with the french national education, in which continuing training is mostly limited to the presentation of ongoing reforms, in very short sessions, aimed at conforming teachers to reforms. All the work of the decryption and the understanding of the new prescriptions and the development of new lessons is often done alone at home, rarely in collaboration with colleagues. The recent development of e-learning for primary education teachers (called M@gistère) is symptomatic of this trend of outsourcing training outside of working hours, which is considered by the institution as an expense that must be reduced at all costs by relegating it as the personal work of the teacher. These professionals are therefore in a process of self-training occurring mostly in their private space, participating in an invisibilisation of the training and of the work required. This leads to the confiscation of personnal time on other spheres of life.

This configuration escapes the traditional training activity analysis models used in ergonomics; which led us to mobilize the System of Activities model [24]. Coming from Francophone psychosociology, it allows us to identify relationships between different spheres of life (professional, social, family) and to analyze the resulting trade-offs and consequences [25]. Work activity is part of a global system in which regulations are played out, often invisible or subject to little analysis; even more so in the field of education where a large part of the professional activity (preparations, corrections, research, etc.) is carried out at home, blurring the boundaries between work and personal time.

Understanding teachers' relation to their training, self-training in this case, imposes to decompartmentalize the analysis of the activity from taking into account only their workplace to extend it to other spaces of life in order to grasp the interdependencies between professional and personal life. We need to take under consideration the fact that any increase in constraints at a given point in the system of activities (e.g. overwork at home when a reform is put in place) mobilizes regulations that go beyond the other subsystems (sphere of personal and social life); constituting what Curie and Hajjar [26] call a "systemic determinism of activities" (p. 46); undermining the overall balance of the system. Resources competition or conflicting requirements can create conflicts between different subsystems that can lead to burnout and health problems.

This analysis model was also particularly relevant in a research that we carried out in the agricultural education system [27]. We were interested in the procedures of accompaniment of the disciplinary retraining of "Accounting and Office Technologies" teachers after the disappearance of the "Rural Services" sector. This study was conducted to respond to the request of a trade union organization which took note of the weakening or even the suffering of the teachers concerned. Our analysis reveals that this reform forced 57 teachers to engage in a double professional transition: a transition from a teaching point of view, involving a phase of learning and sometimes training to the discipline of reconversion, to invest in a new process of professionalization - and - a transition from an identity point of view because of their requalification in a new education classification validated by an inspection led by their hierarchy. In the absence of sufficiently structured training procedures and devices enabling them to acquire the basics of their new teaching disciplines, they had to engage in a very cumbersome self-training process over a period of several years, invading all areas of life. Such an endeavor was a source of identity fragility, burnout and psychosocial risks for all those concerned.

4 Broader Points of View on the Analysis of the Activity

These three studies follow the objectives of previous symposia by focusing on the learning, training, health and safety of men and women at work. However, new questions are emerging, such as the design of digital tools for training or the importance of subjective commitment at work and the identity transformations. In the first case, the aim is to provide designers with information based on the results of the analysis of the activity. Although this is not new for ergonomists, the originality here is twofold. First, the scope of the activity analysis was not limited to the work activities to be trained, but rather aimed to take into account the work activities of trainers and trainees from the very beginning of the design of the future VTE. Secondly, the approach developed aimed at understanding work situations rather than only the tasks that must be mastering. The notion of situation here has a triple meaning [28]: (1) it is a psychological construct of what needs to be done, (2) it is an effective coupling between the actor and the environment at the time the task is performed, (3) it is a given environment (including material and objective). The challenge was then to hold together these three meanings of the notion of situation in activity analysis to guide the design of the future VTE at the artifactual level, the professional knowledge to take into account as well as the future conditions of use in training for the trainers and the trained. Thirdly, the centration on "situation" was oriented by the revelation of their potentialities for vocational training in relation to the characteristics of the trainees actual skills. The holistic approach is redimensioned from a focus on the task to a consideration of the situation as psychological construct, coupling and environment, and integrating into the training design process all actors who may be involved in learning process.

Subjective engagement at work is a recent object for ergonomists [22], as are questions about identity transitions. If this leads to the mobilization of knowledge from related disciplines (psychology, psychosociology in particular), this knowledge feeds the understanding of the work activity and its effects. In the second study, the holistic view adopted is to understand how the issues and difficulties of everyday work impede making workplace training a protected space for learning. While the learning process can benefit from taking place in situations, which gives it a highly "ecological" character, it also constitutes a de facto exposure to occupational risks from which it is difficult to protect apprentices. The third study shows the consequences of training designed with a limited model of learning and professional development. It underlines the need to explore the spheres of life and their interactions in order to understand how health is constructed at work, and then leads to taking the analysis of activity out of its usual sphere to try to understand what is played out in the private domain.

The holistic approach advocated in this communication is based on a model of operator development that articulates mastery of tasks and work, and the construction of health and safety over the long term [29]. It draws on existing knowledge in ergonomics but less used for vocational training (for example on design processes) but also on related disciplines and theoretical frameworks (professional didactics, psychology, psychosociology).

References

1. Rabardel P, Teiger C, Laville A, Rey P, Desnoyer L (1991) Ergonomics work analysis and training. In: Queinnec Y, Daniellou F (eds) Design for everyone. Taylor and Francis, London, pp 1738–1740
2. Teiger C, Montreuil M (1996) The foundations and contributions of ergonomics work analysis in training programmes. Saf Sci 3(2–3):81–95
3. Cloutier E, Ledoux E, Fournier P-S (2012) Knowledge transmission in light of recent transformations in the workplace. Relations industrielles/Ind Relat 67(2):304–324
4. Falzon P La (1998) construction des connaissances en ergonomie: éléments d'épistémologie. In: Dessaigne M-F, Gaillard I (eds) Des Évolutions En Ergonomie. Octarès, Toulouse
5. Dul J, Bruder R, Buckle P, Carayon P, Falzon P, Marras WS, Wilson JR, van der Doelen B (2012) A strategy for human factors/ergonomics: developing the discipline and profession. Ergonomics 55(4):377–395. https://doi.org/10.1080/00140139.2012.661087
6. Lundberg J (2015) Situation awareness systems, states and processes: a holistic framework. Theoret Issues Ergon Sci 16(5):447–473. https://doi.org/10.1080/1463922X.2015.1008601
7. Jick TD (1979) Mixing qualitative and quantitative methods: triangulation in action. Adm Sci Q 24(4):602–611

8. Tourmen C, Holgado O, Métral J, Mayen P, Olry P (2017) The Piagetian Schème: a framework to study professional learning through conceptualization. Vocations Learn 10 (3):343–364
9. Guérin F, Laville A, Daniellou F, Duraffourg J, Kerguelen A (2007) Understanding and transforming work. The practice of ergonomics. ANACT, Lyon
10. Flick U, Von Kardoff E, Steinke I (2004) A companion to qualitative research. Sage, London
11. Mollo V, Falzon P (2004) Auto-and allo-confrontation as tools for reflective activities. Appl Ergon 35(6):531–540
12. Delgoulet C, Boccara V, Carpentier K, Lourdeaux D (2015) How designing a virtual environment for professional training from an activity framework? Dialog between ergonomists and computer scientists. In: Proceedings 19th triennal congress of the IEA, 9–14 August 2015, Melbourne (Australia). http://ergonomics.uq.edu.au/iea/proceedings/Index_files/papers/325.pdf. Accessed 16 Apr 2018
13. Sonderegger A, Sauer J (2009) The influence of design aesthetics in usability testing: effects on user performance and perceived usability. Appl Ergon 41(3):403–410
14. Boccara V, Delgoulet C (2015) Works analysis in training design. How ergonomics helps to orientate upstream design of virtual training environments. Activités 12(2):137–158 http://activites.revues.org/1109. Accessed 16 Apr 2018
15. Olry P, Vidal-Gomel C (2011) Conception de formation professionnelle continue: tensions croisées et apports de l'ergonomie, de la didactique professionnelle et des pratiques d'ingénierie. Activités 8(2):115–149 https://journals.openedition.org/activites/2604. Accessed 16 Apr 2018
16. Ouellet S, Vézina N Activity analysis and workplace training: an ergonomic perspective. In: Filliettaz L, Billett S (eds) Francophone perspectives of learning through work. Springer, Cham, pp 241–256
17. Mayen P, Métral J-F, Tourmen C (2010) Les référentiels en formation: enjeux, légitimité, contenu et usage. Recherche et formation 64:31–46
18. Vinck D (2009) De l'objet intermédiaire à l'objet-frontière. Revue d'anthropologie des connaissances 3(1):51–72
19. Mayen P (2015) Vocational didactics: work, learning, and conceptualization. In: Filliettaz L, Billett S (eds) Francophone perspectives of learning through work. Springer, Cham, pp 201–219
20. Vidal-Gomel C (2016) Formation en situation de travail et troubles psychosociaux. Travail Apprentissage 18:25–47
21. Mollo V, Nascimento A (2014) Reflexive practices and the development of individuals, collectives and organizations. In: Falzon P (ed) Constructive ergonomics. CRC Press, Boca Raton, pp 205–220
22. Van Bellghem L, De Gasparo S, Gaillard I (2014) The development of the psychosocial dimension of work. In: Falzon P (ed) Constructive ergonomics. CRC Press, Boca Raton, pp 33–47
23. Gollac M, Bodier M (eds) (2011) Mesurer les facteurs psychosociaux de risque au travail pour les maîtriser. DARES, Ministère du travail et de l'emploi, Paris. http://travail-emploi.gouv.fr/IMG/pdf/rapport_SRPST_definitif_rectifie_11_05_10.pdf. Accessed 16 Apr 2018
24. Curie J (2002) Parcours professionnels et interdépendances des domaines de vie. Education Permanente 150:23–32
25. Almudever B, Le Blanc A, Hajjar V (2013) Construction du sens du travail et processus de personnalisation: l'étude du transfert d'acquis d'expériences et des dynamiques de projet. In: Baubion-Broye A, Dupuy R, Prêteur Y (eds) Penser la socialisation en psychologie. Erès, Toulouse, pp 171–185

26. Curie J, Hajjar V (1987) Vie de travail, vie hors travail: la vie en temps partagé. In: Lévy-Leboyer C, Spérandio J-C (eds) Traité de psychologie du travail. PUF, Paris, pp 37–55
27. Cau-Bareille D (2016) Se reconvertir au gré des réformes de l'Enseignement Agricole: quels enjeux… pour quelles actions syndicales? Rapport de recherche du laboratoire de recherches Education Cultures et Politique, Lyon
28. Mayen P (2012) Les situations professionnelles: un point de vue de didactique professionnelle. Phronesis 1(1):59–67
29. Delgoulet C, Vidal-Gomel C (2014) The développement of skills: a condition for the construction of health and performance at work. In: Falzon P (ed) Constructive ergonomics. CRC Press, Boca Raton, pp 3–17

Collaborative Design Methods and Macroergonomics as Organizational Tools for Distance Education's Design Teams

Cid Boechat(✉) and Claudia Mont'Alvão

PPGDesign, Pontifical Catholic University of Rio de Janeiro, Rio de Janeiro, Brazil
cidboechat@gmail.com, cmontalvao@puc-rio.br

Abstract. This paper discusses the possibility of using collaborative methodologies of Design and Ergonomics, such as Participatory Design, Design Thinking, Management Design and Macroergonomics, among others, as tools for solving organizational challenges faced by multidisciplinary teams that produce educational material. Distance Education can present, due to its intrinsic characteristics and cultural issues, difficulties of adoption and implementation by students and professionals. On top of that, new technologies have changed social and educational paradigms, requiring the educational institutions to form multidisciplinary teams and to invest in equipment to keep up with the development of information and communication technologies (ICTs). Therefore, this paper contextualizes the process of educational content creation, based on current technological and pedagogical demands. It also points out the importance of choosing management options that help the multidisciplinary team to reach its full potential and produce high quality material. For that, this paper addresses Design and Ergonomics as collaborative tools to achieve the management guidelines required to improve the multidisciplinary processes. Finally, it proposes a case study to ascertain the applicability of the theoretical concepts raised.

Keywords: Distance Education · Macroergonomics · Ergodesign

1 Introduction

Distance education has intrinsic characteristics that can make its adoption and implementation more difficult for students, teachers and professionals. Bejerano (2008) reminds us that, in this type of education, students loose opportunities for academic and social integration. They also need more self-discipline and greater initiative to access, learn, and understand the didactic material. Specially, those who arrive at distance higher education may become easily discouraged when faced with particular academic challenges. On the other hand, some teachers miss the interaction with students inside and outside the classroom. In distance education, teachers need more organization and more time ahead of class to translate the content taught in a traditional classroom to other medium.

There is another complicating factor: new technologies. While it is challenging and costly to keep up with their evolution, educational institutions cannot ignore them, as

S. Bagnara et al. (Eds.): IEA 2018, AISC 821, pp. 242–251, 2019.
https://doi.org/10.1007/978-3-319-96080-7_28

these new technologies have changed social and educational paradigms. Innovations have made large-scale Open and Distance Education a more complex challenge. The world has seen devices such as smartphones, e-books and tablets increase the need to form and train multidisciplinary teams for didactic material production. Adapting educational content to different medium requires specific language and visual resources. It also needs specialized professionals, who work in partnership with teachers and authors (see Oliveira 2013), such as content managers, instructional designers, graphic designers, illustrators, web interface designers, programmers and tutors, among others. It would not be surprising if all these factors affected the proper functioning of work teams and their internal processes, leading to management and organizational problems.

In the scenario outlined so far, this paper proposes a reflection based on the difficulties that Distance Education faces and the new kinds of interaction between teachers, students and technicians. It addresses the need for a new relationship between these stakeholders and the educational material. If all these changes have brought organizational problems, how can we seek solutions through the collaborative methodologies of Design and Ergonomics that apply to work systems?

The paper also relates the differences and similarities between the traditional Management methodologies and new approaches, such as Design Thinking. For these approaches, the designer acts as someone who facilitates collaborative processes, using daily and tacit experiences of those involved in problematic work systems. Something Boland Jr. and Collopy (2004) called a 'design attitude' for problem solving.

Finally, to verify the practical applicability of these methodologies as tools to solve Distance Education's organizational problems, this paper proposes a future case study in a department of educational material production.

2 Designing Educational Material

The use of information and communication technologies to overcome the gap between teachers and students is a Distance Education characteristic since its inception, from textbooks, broadcasting and tele-education, to recent technologies such as video streaming, Augmented Reality, Interactive Digital TV and Virtual Learning Environments. According to STRIDE (2005), Distance Education is the field of educational attempt in which there is physical separation between the student and the teacher, and a technological medium replaces the interpersonal communication of conventional education.

In addressing the current state of Distance Education, the role of new ICTs (Information and Communication Technologies) and innovations cannot be ignored. They act instrumentally in the teacher-student relationship and on the educational and cultural paradigms. When used appropriately, ICTs can expand access and strengthen the relevance of education. They can also raise educational quality by making teaching and learning more engaging and connected to real life. However, they demand a more complex process that involves curriculum and pedagogy, institutional readiness and teacher competencies (see Tinio 2003).

As computers, mobile devices and the Internet got popular, they modified the gap between students and teachers. Teixeira and Weschenfelder (2013) state that we can speak of Distance Education before and after the Internet and the technologies associated with it. Before the Internet, there was a 'one-to-many' (radio, TV) or 'one-to-one' (correspondence) communication. Via the Internet, there are three communication possibilities: 'one-to-many', 'one-to-one' and, above all, 'many-to-many'. Internet is different from other media used before for Distance Education: it is a carrier of data that is left to receiving hosts to interpretation. It is a 'rather a force which impacts all previous forms and redefines their limits under a new paradigm' (see Samans n.d.).

Gbenoba and Dahunsi (2014) say that customized and self-instructional materials are fundamental to the success of institutions in Open and Distance Learning (ODL). It could be said that the fast pace of technological development and its adoption by students made the production of educational materials more complex. In the past, leaders of educational institutions used to choose the kind of educational technology and media their students would use to learn. Nowadays, leaders seek to adapt educational material to technologies students already possess, forcing institutions to invest in innovation. Therefore, students' access and practicability can be as influential as educational effectiveness and production costs when choosing the media for delivering the educational material.

However, educational materials are not the only issue involved in this scenario. There are also virtual environments for material availability, communication and interaction with students, the so-called Learning Management Systems (LMS). The website edutechnica.com has been tracking and trending the use of LMS in the US, Australia, Canada, and the United Kingdom since 2013. In the United States, only 6,9% of higher education institutions with more than 500 students did not use a LMS in 2016 (see Kroner 2016). These Learning Management Systems allow multi-functional Virtual Learning Environments (VLE) creation for blended and distance learning. Educating in VLEs requires greater dedication of teachers and more support of the technical-pedagogical team. This increases preparation time for teachers and professionals. Courses that use VLE need a more regular follow-up of the entire process. Bejerano (2008) says that faculty needs more time and organization ahead class to set up distance education courses.

This scenario sometimes ends up demanding the formation of multidisciplinary technical teams to produce the educational material and to maintain the VLEs. Freeman (2005) agrees, saying that, in general, distance-learning materials are not produced by an individual, but rather by small teams that may include curriculum and instructional designers, tutorial support and print or web designers. According to him, there are four principal ways of producing educational materials:

(1) by an instructional designer who is the content-provider and the writer;
(2) by an instructional designer who commissions freelance content-providers to write the materials;
(3) by an instructional designer who converts text provided by a faculty; or
(4) by a team of people, including content-providers, instructional designers and specialists such as audio and video producers.

Moreira (2009) recalls that, in a small-scale distance-learning project, professionals still manage to accumulate functions. On the other hand, larger projects, involving the production and management of different media, demand the composition of a larger team with different roles.

3 The Multidisciplinary Teams that Design the Educational Material

As Mallmann and Catapan (2007) remind us, multidisciplinarity permeates the entire conception and development of Distance Education, since the relationship between the stakeholders is multidimensional. According to Moreira (2009), in general one can divide the material production team into three sub-teams:

(1) authors, content specialists and pedagogues (who develop content, gather materials, organize dynamics, develop research, promote pedagogical discussions);
(2) instructional designers (multidisciplinary professionals who adapt the content to different medias); and the
(3) media and art team (who work on graphic design, animations, illustrations, navigability).

What would be the best way to manage a multidisciplinary team that splits into several, but needs to function as one? Which management techniques could bring consistent results, considering the didactic characteristics required? The correct administration of educational material production process is crucial for distance education.

Technological, political, economic, scientific and cultural changes steadily modifies educational production processes. Innovation demands a more integrated management of processes and people. Like a network, these processes must reflect new ways of learning in a connected society. According to Otto Peters, the mass consumption of distance learning material could reduce education to an industrial 'assembly line', as the material is mass-produced with an industrial division of labor (see Jarvis et al. 2008). However, Búrigo et al. (2016) point out that multidisciplinary Distance Education can no longer use production models close to Fordism. Mass production, with task divisions and separate units, allows little communication between teams, in particular between those who produce and those who manage.

Dialogue and active participation of stakeholders are essential attributes in multidisciplinary distance education teams (see Mallmann and Catapan 2007). The production of didactic material requires that teammates go beyond their tasks. They need to understand the production process as a whole, need to exchange information with their co-workers of other fields of knowledge, thus producing their full potential and enhancing the material. To achieve this goal, the institution must know the personal and professional profile of each teammate. It is necessary to understand those people and build a scenario where they can produce and dialogue, bringing gains for the whole team and for the produced material.

Distance education specificities require tools to produce a suitable material. Tools that take into account personal profiles, the external scenario and organizational characteristics. Tools that address the stakeholders' daily difficulties and ask them how to improve their work processes. What this paper asks is: can these answers, these tools, come through collaborative methodologies of Design and Organizational Ergonomics, such as Macroergonomics?

4 Design and Ergonomics as Organizational Collaborative Tools

Management and organizational needs are topics addressed by several areas of knowledge for over a century. What is the difference that Design methodologies can bring? What results these methodologies can obtain that other approaches have not achieved? According to Buchanan (2015), a fundamental principle of Design makes the difference: the focus on the experience of all those served by an organization. The concern should not only be about profitability, but also providing significant improvement in people's lives.

With that mindset, designers realized that organizations as well could be treated as a project. Benefiting, that way, from Design methodologies and a "design way of thinking" that would make scarcely noticed elements stand out. That would generate different results if compared with traditional management methodologies. One can relate this approach to the concept of Design Thinking (see Brown 2009): Design ceases to be only operational to become strategic, providing ways to integrate teams, transforming what is multidisciplinarity in interdisciplinarity. Buchanan (2015) says this new Design role, which focuses on the functioning of systems, seeks the organizing idea or principle that operates behind systems, organizations, and environments - behind mutual interactions.

In the same way, Organizational Design and Management have been present in the scope of Ergonomics for some time. In the 1980s, the Human Factors Ergonomics Society set up a committee to analyze modern trends for the next twenty years, until 2000. Organizational design and management of work systems were topics examined (Hendrick and Kleiner 2001). The Committee's conclusion was that the complexity of the changes would require a greater presence of Ergonomic Science in Organizational Design and Management (ODAM). At the same time, they identified the need for Ergonomics to absorb ODAM concepts, adopting a broader view of their performance. That way, Ergonomics would go beyond the operator interacting with an interface paradigm (see Kleiner and Hendrick 2008).

Macroergonomics is the domain that addresses the design of work and focuses on the relationship between Ergonomics and Organizational Design and Management (Hendrick 2005). While 'traditional' human factors focused on the individual employee at the workplace level, Macroergonomics' approach embraces a broader view of the work system, including environmental topics such as social and economic issues (see Carayon et al. 2012). According to Hendrick and Kleiner (2002), Macroergonomics focuses on the human-organizational interface (HOI). It also approaches human, technological and contextual factors that influence the design and analysis of work systems. It takes into account environmental components, work design, cognitive and physical ergonomics (see Hoeft and Fitzhugh 2013). For example, the macroergonomical work system proposed by Smith and Carayon-Sainfort (1989) is based on systems theory and work complexity. It positions the employee at the center of the work system, while he is surrounded by his needs and physical, cognitive and psychosocial characteristics that can influence his interactions with the work system. The model includes, that way, three domains of specialization of Human Factors: physical, cognitive and organizational (Carayon et al. 2012).

Watson et al. (2009) remind us that Macroergonomics has produced satisfactory results in developing work systems using analytical methods, with a perspective that identifies and fixes interface incompatibilities between human capabilities, work design and technologies. From this, specific methodologies have emerged, such as Macroergonomic Analysis of Structure (MAS), which focus on the context of socio-technical work systems, specifically the relationship between Personnel (individuals who perform tasks), Technological (tools, methods, software and equipment), Environment (internal and external to the company) and Organizational (structures and processes) subsystems.

In addition, Macroergonomics uses and adapts other tools and methodologies for its scopes, such as questionnaires, semi-structured interviews, field research and focus groups. It is also possible to simulate work systems in the laboratory, where the ergonomist manipulates sociotechnical variables, or the system itself, in search of answers (Hendrick 2005).

The professionals who produce the educational material work in an environment pressured by constant changes in technological innovations and communication tools. They have to deal with various pedagogical specificities and dialogue with co-workers who come from backgrounds different from their own. They need a management system that understands their job of adapting educational content to different media. It is a technical task, but also creative, which may not work in the most common production models. According to Getty (1989), the macroergonomical perspective, applied to evaluation of an organization design, can lead stakeholders to 'avoid problems or develop improved approaches to operate and organize teaming arrangements'. However, Watson et al. (2009) state that, although Education is a complex system of work, it has been little explored by Macroergonomics' methods and theories. According to them, in the educational field, Ergonomics still focuses on biomechanical issues.

Whatever methodology is chosen to help solve multidisciplinary problems, it is important that they be participatory. People of all hierarchical levels are essential resources for an organization's success (see Carayon et al. 2012). In addition, the best work systems are those designed using the needs and knowledge of workers in their management. Reflecting on the work of Wilson and Haines (1997), Carayon et al. (2012) also say that management processes that include employee participation into 'production, problem solving, designing, and/or opinion sharing activities' benefit both the organization and the individuals.

The main gain from employees' participation is that they know the symptoms of problems they face every day. In addition, they can help by choosing the changing tool that best fits their profile and context (see Hendrick 2005).

The users' involvement in the process is fundamental in the new Management theories based on Design (Buchanan 2015). One can explain this importance by relating it to the Participatory Design, in which the users are the best source of information and answers to the question addressed by the designer, since they experience the object of study on a daily basis. A participatory design project should be based on the appreciation of tacit knowledge. Designers should interact with the users until they outsource this knowledge, and, having this information, use it as a basis to create and evaluate possible solutions (Silva 2012).

Brown Jr. (2005), analyzing the work of Brown (2002) and Hendrick (1996), say that participatory practices are some of the most important methodologies in Ergonomics and product design. According to the authors, in the literature, one can find similar concepts and terms, such as Participation, Employee Involvement, Participatory Ergonomics and Participatory Management. They are approaches that include the worker and are concerned with the design and ergonomic analysis.

Dul and Ceylan (2010), addressing Shalley et al. (2004), say that all employees in an organization can produce new and potentially useful ideas not only for products or services, but also work procedures, improvements and solutions to everyday problems.

The stakeholders' participation is active even in scenarios where the proposed change tool is not of employee preference. They are more likely, when included in the process, to support changes in the work system, even if their own preferred approach is not adopted (Hendrick 2005).

In the multidisciplinary context of Distance Education, methodologies that include the stakeholders are chosen for a reason: the service provided is educational and has much specificity. Therefore, it is important to consider not only the daily experiences, but also the pedagogical needs in the foundation of organizational proposals. For this reason, the intense participation of stakeholders is essential, so that they contribute to their theoretical pedagogical knowledge, daily practical experience and tacit knowledge. This range of information centered on the user is the differential that Design brings to the Management process (see Buchanan 2015).

It is important to emphasize that participatory methodologies can suffer resistance from adoption by the organizations. Brown Jr. (2005), when analyzing Participatory Ergonomics, say that two disadvantages of the methodology could be the increase in costs and the need for commitment of the hierarchically higher managers. The greater the involvement required by a change tool, the greater the commitment of managers. It is necessary to adopt, permanently, an organizational philosophy that promotes active participation and involvement.

5 Conclusion and Future Perspectives

The multidisciplinary scenario of distance education has to deal with interaction issues between people from different backgrounds and fields of knowledge. It also deals with a constant pressure of external and internal factors, such as task characteristics, legislation, and technology, among others. This would require not only a humanistic focus, centered on individuals and their explicit and implicit needs, but, also, to focus on the setting and the systemic interrelations around these people.

This paper understands that an 'Organizational Ergodesign' approach, gathering the Design humanistic characteristics and the Macroergonomic systemic and contextual approach would bring the necessary balance to the analysis of work systems in the distance education scenario. Authors such as Moraes (2013), Yap et al. (1997) and Wilcox (1998) say that the complexity of modern systems narrows the differences between Ergonomics and Design. Ergodesign is the integration of those two disciplines, where its focus is humanistic and creative as much as ergonomic and systematic. That way, an organizational ergodesign approach should not be taken as a mere set of methods and techniques, but as a different take on work systems that could help to guide the relationship between human beings, organizations and the complex world surrounding them.

To test these concepts in practice, this paper proposes, in the future, a field research in a distance higher education institution. The goal is to observe, in practice, the applicability of the participatory view of Design methodologies and Macroergonomics on the management of multidisciplinary content production. The institution must have its own educational material production team, able to gather instructional design, graphic and web design, creation of the web support material (Moodle, for example) and construction of learning objects (animations, illustrations, videos, among others).

A case study is proposed because, according to Yin (2014), it can deal with a wide variety of evidence - documents, artifacts, interviews, and observations. Bonoma (1985) also says that a case study is a very useful method when the phenomenon is broad and complex, and in cases where the phenomenon cannot be studied outside the context where it naturally occurs.

As previously seen in this paper, Design and Ergonomics have several methodologies that can be applied in the management of organizational systems. They also have several methods that focus on the stakeholders' inclusion in their analysis processes. A practical research would be relevant as an opportunity to evaluate what methods, proposals and tools can work better in the multidisciplinary scenario of Distance Education found by the designer or ergonomist.

With the field research results collected and analyzed, a proposal for procedural change and work culture in the team would be built collectively with the stakeholders, according to the chosen methodologies. In the future, the results could be deepened and compared in other researches, to get more parameters on the use of Design methods and Macroergonomics as tools of change and construction of organizational settings in multidisciplinary Distance Education environments.

Another unfolding is possible. Several studies, such as Dul and Ceylan (2010), say that the influence of the physical environment (furniture, lighting, acclimatization) is as important in employee performance as socio-organizational factors (company policies, incentives, tasks). Subsequent research can focus on these issues by applying them in the context of Distance Education.

References

Bejerano AR (2008) Face-to-face or online instruction? Face-to-face is better. Commun Curr 3(3):1–3

Boland RJ Jr, Collopy F (2004) Managing as designing. Stanford University Press, Stanford

Bonoma TV (1985) Case research in marketing: opportunities, problems, and a process. J Mark Res 22(2):199–208

Brown O Jr (2005) Participatory Ergonomics (PE). In: Stanton N, Salas E, Hendrick H, Hedge A, Brookhuis K (eds) Handbook of human factors and ergonomics methods. CRC Press, Boca Raton, pp 81-1–81-7

Brown T (2009) Change by design: how design thinking transforms organizations and inspires innovation. Harper Business, New York

Buchanan R (2015) Worlds in the making: design, management, and the reform of organizational culture. She Ji J Des Econ Innov 1(1):5–21

Búrigo C, Cerny R, Teixeira G, Marcelino L (2016) A Gestão Colaborativa no processo formativo da EAD. Revista Gestão Universitária na América Latina - GUAL 9(1):165–176

Carayon P, Hoonakker P, Smith MJ (2012) Human factors in organizational design and management. In: Salvendy G (ed) Handbook of human factors and ergonomics. Wiley, Hoboken, pp 534–552

Dul J, Ceylan C (2010) Work environments for employee creativity. Ergonomics 54(1):1–25

Freeman R (2005) Creating learning materials for open and distance learning: a handbook for authors and instructional designers. Commonwealth of Learning, Vancouver

Gbenoba F, Dahunsi O (2014) Instructional materials development in ODL: achievements, prospects and challenges. J Educ Soc Res 4(7):138–143

Getty RL (1989) The macroergonomical challenge of industrial teaming arrangements' organizational structure. In: Proceedings of the human factors and ergonomics society annual meeting, vol 33(13). Sage, London, pp 836–840

Hendrick HW, Kleiner B (2001) Macroergonomics: an introduction to work system design. Human Factors and Ergonomics Society, Santa Monica (CA)

Hendrick HW, Kleiner B (2002) Macroergonomics: theory, methods, and applications. CRC Press, New Jersey

Hendrick HW (2005) Macroergonomic methods. In: Stanton N, Salas E, Hendrick H, Hedge A, Brookhuis K (eds) Handbook of human factors and ergonomics methods. CRC Press, Boca Raton, pp 81-1–81-7

Hoeft RM, Fitzhugh SL (2013) Applying a macroergonomic approach to the design and analysis of business software. In: Proceedings of the human factors and ergonomics society annual meeting, vol 57(1). Sage, London, pp 409– 413

Jarvis P, Griffin C, Holford J (2008) The theory & practice of learning. Routledge Falmer, London

Kleiner BM, Hendrick HW (2008) Human factors in organizational design and management of industrial plants. Int J Technol Hum Interact 4(1):113–127

Kroner G (2016) 4th Annual LMS data update, 3 October 2016. http://edutechnica.com/2016/10/03/4th-annual-lms-data-update/. Accessed 7 Mar 2018

Mallmann EM, Catapan AH (2007) Materiais Didáticos em Educação a Distância: gestão e mediação pedagógica. Linhas 8(2):63–75

Moraes A (2013) Ergonomia, ergodesign e usabilidade: algumas histórias, precursores: divergências e convergências. Ergodesign e HCI 1(1):1–9

Moreira M (2009) A composição e o funcionamento da equipe de produção. In: Litto F, Formiga M (eds) EDUCAÇÂO A DISTÂNCIA: O ESTADO DA ARTE – VOLUME 1. Pearson, São Paulo, pp 370–378

Oliveira D (2013) O uso do vídeo em EAD: desafios no processo de ensino aprendizagem. Revista Cesuca Virtual: Conhecimento sem Fronteiras 1(1):1–15

Samans J: The impact of web-based technology on Distance Education in the United States. http://www.nyu.edu/classes/keefer/waoe/samans.html. Accessed 7 Mar 2017

Silva NAN (2012) Abordagens Participativas para o Design: Metodologias e plataformas sociotécnicas como suporte ao design interdisciplinar e aberto a participação. PUC-SP, São Paulo

Smith MJ, Sainfort PC (1989) A balance theory of job design for stress reduction. Int J Ind Ergon 4(1):67–79

STRIDE (2005) Development and revision of self-learning materials. IGNOU, New Delhi

Teixeira OAF, Weschenfelder GV (2013) Evolução do EAD e as novas mídias. Revista Cesuca Virtual: Conhecimento sem Fronteiras 1(1):1–21

Tinio VL (2003) ICT in education: an e-primer. United Nations Development Programme, New York

Watson J, Smith TJ, Kraemer S, Halverson R, Woodcock A (2009) Macroergonomics in education: on your mark, set, GO! In: Proceedings of the human factors and ergonomics society annual meeting, vol 53(1). Sage, London, pp 1042–1046

Wilcox S (1998) Human factors in the industrial designers' society of America. In: Human factors and ergonomics society bulletin, vol 41(10). HFES

Yap L, Vitallis T, Legg S (1997) Ergodesign: from description to transformation. In: Proceedings of the 13th triennial congress of the international ergonomics association, vol 2. Finnish Institute of Occupational Health, Helsinki, pp 320–322

Yin RK (2014) Case study research: design and methods. Sage, London

A Cooperative and Transversal Methodology to Improve the Making of Prescriptions in a High Reliability Organization: A Constructive Ergonomics Approach

C. Thomas[1(✉)], F. Barcellini[1], Y. Quatrain[2], B. Ricard[2], and P. Falzon[1]

[1] Conservatoire National des Arts et Métiers, Paris, France
camille.thomas.m@gmail.com
[2] Electricité de France Research and Development, Chatou, France

Abstract. This communication presents an action-research conducted in a French Electrical Power Generation company addressing the making of prescriptions, i.e. the documentation design process intended to ensure the quality of prescriptive documents. This action-research has two goals: fostering the development of a cooperative and transversal collective work and improving the documentation design process. In this way, a methodology was set up based on the principles of the constructive ergonomics approach, i.e. allowing the development of both the individuals and of the organization. 3 one-day workshops were conducted, implying a set of methods (explanation of a case, design of scenarios of organizational change, and organizational simulations) and involving 7 participants who represent the diversity of the actors implied in the documentation process.

The results show that the methodology allows the participants to better comprehend the process and the role of the other actors. This refers to constructive dimension because the development of new elements of knowledge and the construction of a shared frame of reference are prerequisite to develop cooperative work. The highlighting of difficulties leads the participants to design scenarios and organizational simulations. This allows them to think about the organization of their work and lead them to co-constructed organizational solutions taking into account the constraints of all of them.

Now, the main issue is the identification of the implementation conditions of the organizational solutions in order to foster their effective setting up.

Keywords: Constructive approach · Documentation process design

1 Introduction

This communication presents an action-research conducted in a French Electrical Power Generation company addressing the making of prescriptions, i.e. the documentation design process intended to ensure the quality of prescriptive documents.

S. Bagnara et al. (Eds.): IEA 2018, AISC 821, pp. 252–258, 2019.
https://doi.org/10.1007/978-3-319-96080-7_29

1.1 Context and Field Research

The documentation design process under study is complex, mainly because the process

- Organizes the design of specific prescriptive documents which aim at ensuring the safety of a High Reliability Organization; thus the organization of the process need numerous steps of control and validation;
- Deals with a variety of prescriptive documents: depending on the objective concerned (more or less strategic or operational), there are 4 levels of abstraction of documents. Designing a very operational document is based on the frame defined by a more strategic document, which itself is designed from a more strategic document. For each level of abstraction, different types of document are designed according to the environment (normal, accidental), the material concerned, the type of plant, etc.
- Implies numerous actors of the company (around 1300 people) and that is necessary to coordinate;
- Implies very diverse actors: they are distributed in numerous and very different entities of the organization, they are geographically distant (distributed throughout the country), they do not consider the same requirements and do not have the same constraints ("real time" nuclear operations, relations with an external control body, the financial viability of the nuclear operation, design requirements of the material, etc.), and the documents they design or modify vary.

In this context, the documentation design process requires a joint collective activity, i.e. a set of local activities coordinated in order to produce a result [1, 2]. In this case, producing prescriptive documents regarding power plants operation. This joint collective activity is cooperative and transversal: activities have to be coordinated locally or in various entities of the company.

1.2 Research Issues

The action-research we conducted has two goals. On one hand, fostering the development of a cooperative and transversal collective work. On another hand, improving the documentation design process.

In order to support the cooperative and transversal work, a methodology was set up based on the principles of the constructive ergonomics approach [3], i.e. allowing

- individuals to acquire or develop know-how, knowledge and skills;
- the organization to develop reflective processes that are open to the workers' capacities for innovation.

Thus, the goal is to enable the actors of the process to discuss and redesign the organization of the documentation process. The objectives of the action-research are both productive and constructive. The productive issues deal with the definition of organizational scenarios and simulations to improve the organization of the documentation process. The constructive issues are relative to the development of the cooperation in order to improve the cooperative work, necessary to perform the documentation process.

2 Methodology

2.1 Population Concerned

Among the numerous and diverse actors implied in the documentation design process, the study focused on a given type of prescriptive documents, those which are used to operate power plant in a normal environment, and thus on the actors who write and coordinate the production of these documents.

A sample of 7 from 7 different entities took part to the study, representing the diversity of the actors implied in the documentation process.

2.2 Description of the Methodology and Data Collection

3 one-day workshops involving 7 participants (technicians, engineers) were conducted, recorded and transcribed. A set of methods were set up during the workshops and lead to:

- Explanation of a case (selected by the participants among a set of cases collected from interviews performed during the diagnosis step) to reveal the point of view and the constraints of each actor;
- Design of scenarios of organizational change in order to tackle the identified difficulties and develop the suggested solutions;
- Organizational simulations [4] to test the scenarios. This was done using a shared representation of the documentation process (actors, input, outputs, performed actions);

2.3 Data Processing

The processing of the qualitative data (transcriptions of the 6 workshops) addresses different aspects. It aims at identifying:

- On one hand, what the individuals know or do not know about the organization of the process and the role of other actors: what do they explain about their work? Which difficulties do they express? What do they need to understand about the organization of the process and the role of other actors?
- On another hand, the potential changes to the process organization: on which organizational aspects do the actors think it is possible to act considering their level in the hierarchy? Which changes are considered necessary? Which steps of the process are impacted by the changes? Which are the expected gains?

3 Results

The workshops allowed the participants (who belonged to different entities) to meet. Discussions during the workshops lead the participants to explain their activities and their difficulties, to better understand the constraints of others and the global organization of the process.

3.1 A Better Comprehension of the Process and of the Role of the Other Actors: A Constructive Issue

The qualitative analysis of interactions during work group shows that the discussions between the different actors allow the participants to have a better comprehension of the work of the other actors, of the documentation process, and of the organization and strategy of the company.

Improving the comprehension of the work of the other actors means a better understanding of:

- The impact of the work of an actor on the work of another actor;
- The origin of an input necessary to design a document;
- The field reality that actors have to consider when writing prescriptive documents;
- The repartition of tasks between different actors;
- The role of an actor in relation to another;
- And more generally the work performed by the other actors.

Improving the comprehension of the documentation process implies:

- The possibility to define a common language to define and name some steps of the process;
- A better comprehension of the process as a whole: the chain of the different steps and where each actor takes part in;
- But also a better comprehension of the sub-processes which are specific to the entities;
- A better understanding of the prescribed process;
- A better visibility about the gap between the prescribed process and the actual process;
- The understanding of the influence of an external actor on the organization of the process.

Improving the comprehension of the organization and strategy of the company is reflected by:

- The discussion of pros and cons of the distributed design of the documents;
- The understanding of the impacts and stakes of strategic decisions (about the organization of the process, about the strategy of recruitment and of training);
- A better comprehension of the hierarchical organization in plants.

Sharing knowledge about actual activities and about the process leads participants to construct a shared representation of the process and of the role of each actor. The construction of this shared representation is part of the constructive dimension of the action-research: the actors develop new elements of knowledge and construct a shared frame of reference, which is a prerequisite to develop cooperative work [5].

3.2 Emergence of Difficulties, Design of Scenarios and Organizational Simulations: Productive and Constructive Issues

Expressed Difficulties. The difficulties expressed by the work group relate to 3 dimensions: the prescribed organization of the work, the actual work – in contrast with the work prescribed by the organization [6] - and the representation that actors not involved in the documentation process have about the work of the actors of this process.

About the prescribed work organization. Two examples illustrate this dimension.

- The distributed organization of documentation design: In order to tackle the workload, within deadlines, a national entity need to develop a strategy, consisting in sharing the workload with local entities (located in plants). But this has consequences: the duration of the production is longer, it needs more coordination, there may be conflicts of interest with the local entity which pursues other objectives, it increases workload in the local entity.
- The lack of visibility about the process: each actor knows "his/her part" of the process, i.e. the steps of the process in which he/she takes part. The lack of visibility about the rest of the process can constitute a difficulty to understand the global operation of the process.

About real work. Three examples can illustrate this second dimension.

- Considering the requirements of the field reality: writers highlighted the importance of taking into account the requirements of the field reality in order to avoid: maladjusted designed documents to their use, difficulties using the procedure and greater workload.
- The delay of production and the evolution of documents: the end of the production and the evolution of the national level documents sets the beginning of the production of the local level documents. Depending on the production time of national documents, the coordination between the writers of the national and the local level have need to be different to write and to review the national and local documents.
- The quality of the documents: the validation of a document by a power plant simulator is not sufficient to ensure its quality; the simulator does not allow the anticipation of all possible situations. To address this problem, the writers need the support of a network of experts.

About the representation that the actors not involved in the documentation process have about the work of the actors of this process. According to the actors implied in the documentation process, there is a lack of recognition about their work. This can be due to an unclear view of what is done by the involved actors for those who are not involved and an unclear view of what is at stake in the documentation design process. According to the actors, the lack of recognition resulting of this implies different consequences: a lack of attractiveness for this type of jobs, and a lack of training for actors.

Scenarios and Organizational Simulations. The enhanced comprehension of the process and of the role of the other actors, and the highlighted difficulties allowed participants to focus the reflection about one issue: the reinforcement of the exchanges between the different actors of the process about the technical aspects of the documents. From this issue:

- Scenarios were designed to develop alternative processes;
- Organizational simulations allowed scenarios to be tested.

Participants proposed 14 scenarios. These 14 scenarios constitute concrete and potential ways to improve the documentation process. In this paper, we present the processing of one of them and then the organizational simulation that was conducted.

In order to help participants writing the script, the participants were asked to answer 3 questions. Table 1 shows the answers to the three questions for the finally selected scenario.

Table 1. Scriptwriting questions and responses.

Question	Scriptwriting response
What do I want to change in the process?	Reinforce the context simulation of documents
Which solution do I suggest?	Testing the procedures (on the field, by simulator) before their implementation
How do I explain or justify this suggestion?	(1) Avoiding the need to set up a specific process intended to review the documentation (2) Designing a document well-adapted to his use.

After this, the workgroup played this scenario by an organizational simulation.

The first organizational simulation resulted in the addition of a new step allowing procedures to be tested and in precisions regarding this step: which actors? Which timelines? which possible outcomes?

The second organizational simulation considered that the outcome of the first simulation demonstrated a need for changing the input triggering the design of procedures. Another new step was then added.

The simulations allow the work group to think about the organization of their work and lead them to co-constructed organizational solutions taking into account the constraints of all of them. This refers to the constructive dimension of the approach, because the collective construction of an alternative organization of the process:

- allows participants to develop new knowledge about the process and the work of the other actors;
- allows participants to develop their ability to reflect about work organisation;
- leads to a more relevant organization, improved by the reflective methodology.

4 Conclusion and Perspectives

The results of this approach are productive and constructive. These 2 dimensions are present and interact at the same time throughout the approach.

On the one hand, the productive results deal with what the participants visibly produce: the scenarios and the simulations.

On the other hand, the constructive results deal with the constructive dimension of the approach, i.e. a shared representation of the global process and of the work and needs of all actors, the development of a transversal cooperative work, the development of the possibility to collectively think and collectively act (by suggesting organisational scenarios and by considering the constraints of each of them). All this contributes to developing cooperation.

The methodology we set up is unusual in this company because:

- It is co-constructed with the participants: the methodology is adapted and discussed several times with a sample of participants;
- It is real work-centred;
- It is a bottom-up approach (as opposed to the present top-down approach);
- It is transversal (while the entities usually perform their work in isolation);
- It deals with documentation design (a domain not so frequently addressed).

Now, the issue consists in the identification of the implementation conditions of the organizational solutions in order to foster their effective setting up.

A boarder perspective is to identify the necessary conditions to implement and sustain such a constructive approach of organizational design.

References

1. Lorino P (2009) Concevoir l'activité collective conjointe: l'enquête dialogique. Etude de cas sur la sécurité dans l'industrie du bâtiment. Activités 6(1):87–110
2. Thomas C et al (2015) The production of prescriptive documents in safety-critical organizations: an exploratory diagnosis. In: 19th triennial congress of the IEA, Melbourne 9–14 August
3. Falzon P (2014) Constructive ergonomics. CRC Press, Boca Raton
4. Barcellini F, Van Belleghem L (2014) Organizational simulation: issues for ergonomics and for teaching of ergonomics' action. In: Broberg O, Fallentin N, Hasle P, Jensen PL, Kabel A, Larsen ME, Weller T (eds) Proceedings of human factors in organizational design and management – XI and nordic ergonomics society annual conference–46
5. Caroly S, Barcellini F (2014) The development of collective activity. In: Falzon P (ed) Constructive ergonomics. CRC Press, Boca Raton, pp 19–34
6. Daniellou F, Rabardel P (2005) Activity-oriented approaches to ergonomics: some traditions and communities. Theor Issues Ergon Sci 6(5):353–357

Management of Psychosocial Work Environment: Outline of a Multidisciplinary Preventive Intervention in a Large North Italian Municipality

Angela Carta[1,4(✉)], Paola Manfredi[2,4], and Stefano Porru[3,4]

[1] Department of Medical and Surgical Specialties, Radiological Sciences and Public Health, University of Brescia, Brescia, Italy
angela.carta@unibs.it
[2] Department of Clinical and Experimental Science, University of Brescia, Brescia, Italy
paola.manfredi@unibs.it
[3] Department of Diagnostic and Public Health, University of Verona, Verona, Italy
stefano.porru@univr.it
[4] University Research Center - Integrated Models for Prevention and Protection in Environmental and Occupational Health - Mistral, University of Brescia, Brescia, Italy
mistral@unibs.it

Abstract. There is a need of evidence-based multidisciplinary intervention studies to prevent work-related stress in municipality workforces. This study was carried out with the aim of improving management of psychosocial work environment in a large municipality. A steering committee of two occupational physicians and a psychologist designed the intervention, consisting of: preliminary meetings with management; comprehensive data collection (risk assessment, health surveillance, assessments from psychologists, educators, mediators; training activities; collection of sentinel events, data from focus groups); qualitative and quantitative data analysis to identify critical points. The 1730 workers are employed in 7 areas, 32 sectors and 74 services. Psychosocial risk assessment was carried out by questionnaires developed by national and international bodies. Risk assessment identified critical areas and sectors and needs update. Low response rate and non-comparability of data were critical issues. Health surveillance often reported stress related problems, but no structured intervention or integration with other municipality functions were noted. Internal transfer and mobility were critical issues too. Psychologists and consultants evaluated many cases and often suggested operative solutions. Focus groups were few, with no clear participation rate. Based on these data, the committee designed an intervention entailing: (a) formal municipality policy for work-related stress; (b) a multidisciplinary working group, managed by a psychologist, involving occupational physicians and consultants to evaluate single cases and general issues, as well as to elaborate, adopt and monitor specific interventions; (c) use of evidence based risk assessment and health surveillance tools; (d) implementation of good practice in transfer and mobility management.

S. Bagnara et al. (Eds.): IEA 2018, AISC 821, pp. 259–266, 2019.
https://doi.org/10.1007/978-3-319-96080-7_30

Keywords: Psychosocial work environment · Municipality workforce
Multidisciplinary intervention

1 Introduction

According to the European Survey of Enterprises on New and Emerging Risks [4] mental health complaints, such as stress, depression or anxiety, are the second most frequently reported work-related health problem, following musculoskeletal disorders. Work related psychosocial factors have been shown as major contributors to mental health problems and "psychosocial risk" has become a familiar term within the occupational health management domain. This term refers to health risks arising from work-related "psychosocial hazards", which have been defined as "aspects of the design and management of work and its social and organizational contexts that have the potential for causing psychological or physical harm" [3, 8, 10]. Two of the most prevalent and costly of the occupational health problems influenced by psychosocial hazard are musculoskeletal and mental health disorders (MSD and MHD) in most industrialized countries [12]. Psychosocial risk factors, especially high demands and low control at the workplace, have been identified as additional risk factors for MSD: main etiologic mechanisms of MSD and physical and psychosocial risk factors seem to be particularly based on the relationships between biomechanical load and corresponding pathophysiologic alterations of tissues, and stress-induced alterations of the neurohormonal system, pain perception, hippocampal neurogenesis and gene expression, respectively [9, 11].

Over the last decade, a number of significant developments toward the prevention and management of psychosocial risks have been achieved at the international, regional and national level. These include both regulatory approaches, such as ILO conventions, European Union Directives and national legislation, as well as "non-binding/voluntary" approaches, which may take the form of specifications, guidance, social partner agreements and standards. Despite these efforts, there is a lack of evidence on intervention efficacy to reduce psychosocial risk factors. A systematic review overview of systematic reviews of workplace-based interventions for employees' mental health graded evidence of effects of workplace-based interventions as low or very low [1]. Overall, the analysis of the literature show that there is a need of evidence-based multidisciplinary intervention studies to prevent work-related stress in workforces [12].

Aims of this study are:

- development of an intervention to prevent the effects of work-related stress in the employees of a large municipality.
- evaluation of the effectiveness of the intervention, through the use of appropriate indicators.

2 Methods

A steering committee (SC) of two occupational physicians and a psychologist designed the intervention, consisting of: preliminary meetings with management; comprehensive data collection (risk assessment, health surveillance, assessments from psychologists, educators, mediators; training activities; collection of sentinel events, data from focus groups); qualitative and quantitative data analysis to identify critical points.

The preliminary meetings had the objective to acquire knowledge of the organization and to reconstruct every activity aimed at managing psychosocial risk or concerning the promotion of workers' wellbeing. Municipality management identified Human Resource Office (HRO) as the internal project contact referent. On the basis of preliminary information a comprehensive data collection focused on psychosocial risk factors and well-being was perform over five years (2012–2017). Data collection included risk assessment documents, annual reports on health surveillance, supervisory organ reports, consultants' reports on focus groups, educational interventions and conflict mediation, mobility and turn-over, disciplinary trials and claims on municipality services. In order to improve data collection, semi-structured data forms were developed and when no information was available, motivation had to be reported.

HRO had complete retrospective data collection in 6 months. Data were analyzed by each member of SC and results were discussed in collective working sessions, in order to point out critical aspects and to propose solutions.

3 Results

The 1730 workers were employed in 7 areas, 32 sectors and 74 services. The organization underwent 4 major renovations over the last 5 years and 3 changes of general manager.

3.1 Risk Assessment

The evaluation of April 2015 was analyzed. The assessment was performed with the methodology (check list) of the National Network for the prevention of psychosocial workplaces [6] with the integration of subjective data, collected online through the Occupational Stress Questionnaire (OSQ) [2]. The following critical issues emerged:

- no sector/area was overall "High" risk (but 11/37 medium risk); this appeared in contrast with the results emerging from the analysis of subjectivities through the OSQ.
- no moments of coordination with the MCs appeared, nor was the role of the RLS been explained, if it had been exercised at all
- the evaluation did not appear representative of all sectors, areas or situations.
- the comparison with the evaluation carried out in 2012 was not clear.
- as for the OSQ, the percentage (51%) of completed questionnaires did not make this assessment representative. The statistical elaborations did not produce helpful

insights. In any case, the numerical data showed values all beyond the OSQ (medium-high) benchmark, with a deterioration compared to 2012.
– no indicators of effectiveness were foreseen or adopted.

3.2 Counselor for Organizational Well-Being Report

The document analyzed presented an interesting case - series. Essentially, problems of mobility, discrimination, motivation, climate, interpersonal relationships, privacy were addressed.

Excellent system proposals, with wellness table and case monitoring was presented. A possible intervention certainly appeared to be the comparison and coordination with SC activities' program.

3.3 Health Surveillance Reports

All the health reports of the last 4–5 years were been examined. Overall, among the critical issues observed, the risk of stress appeared to be unbalanced and underappreciated. There were no analyses in relation to health variables (for example, accidents at work, absenteeism, cases of anxiety/depression, turnover, age, overtime, etc.). In some cases, formulations of the fitness for work could have be more operational.

3.4 Psychologist Report

Some important issues emerged through the analysis of the activities of the consulting psychologist, such as:

– the number of the assisted population was relevant; a clear needs for targeted interventions emerged;
– the reasons for the requested interventions were clear, but it appeared very likely that there were unexplained, hidden motivations;
– there was a need for coordination with occupational physician and its activation through well-designed procedures
– a certain spontaneity was evident, when solutions were searched or adopted.

3.5 Mobility

This theme is of considerable importance for management and operations.

The analysis of the data showed a high number of mobility requests, which underline the need to improve their management. As far as acceptance of mobility request is concerned, the level of communication was limited and this led to situations of obvious discomfort. Also, motivation of request was often not clearly reported, even if burn out and "environmental incompatibility" were adduced by the workers.

3.6 Education Program

The area is strategic, but so far there was insufficient focus on the topic of work-related stress. Educational programs design was not clear and efficacy evaluation data were not available.

It is necessary to acquire data on the real training needs, that should be selected and processed. The training should aim at the management/organizational positions, supervisors and workers. Trainers should work in harmony with the present project.

3.7 Focus Groups

The analysis of the main critical issues that emerged from the focus groups and possible improvement proposals, as indicated by the answers to some questions of the OSQ contained in the return document, made it possible to identify the following main problems: vertical communication, horizontal communication, training, internal conflict, turnover, computerization, work organization, conflict with users and security problems, workload, sources of environmental stress.

The data of the document were read taking into consideration the number of participants with respect to the group's composition and the actions already carried out by external consultants in the last 5 years.

A first general observation regards the number of sectors: 14 high risk and 16 medium-high risk. In these sectors, the recurrent themes concerned the vertical and horizontal communication of training and work organization. The interventions made on the educational side by the consultants (psychologist-educator) had in fact affected, in the three-year period 2014–2016, only the high and medium-high stress risk sectors. In particular, the SIA Sector (Services for Children) was the subject of a total of 20 interventions, the CPC (Culture and Promotion of the City) 12 interventions, the SERS (Social Services) 10 (with probable underestimation consequent to the refusal of the worker to contact to the consultant on proposal of the competent doctor).

In general, it was emphasized that the representativeness of the answers to the questionnaire was variable according to the sector. In the sectors with the highest risk of stress, there was a greater participation and this data seemed to indicate both the worker's interest and the trust in the possibility of identifying and implementing possible solutions.

Another relevant observation was the dispersion of data (both with respect to the response range and with reference to the standard deviation). This could be indicative, presumably, of a discrepancy in the representation of work problems and this would require a preliminary work of sharing the same and priorities.

Another limitation regarding the interpretation of the data coming from the focuses, was the limited number of both meetings and people involved. Moreover, in the return document the data relating to the number/role of the participants or the detail was not made explicit and available of group work.

Key elements, detectable from the analysis, seemed to be the transversal lack of an overall vision of the organization of the entity and a difficulty in perceiving the meaning of one's work, with consequent demotivation.

In some sectors this interpretation was explicit and related to ongoing restructuring and outsourcing of some services.

Overall, there are few problems with the workload or sources of environmental stress, reinforcing the hypothesis that both the organization of work, which is also linked to the management of horizontal and vertical communication, was the cornerstone of this work sickness.

The training request was transversal to the various sectors. While acknowledging that there is an objective utility in seeking an update in one's work, the more so in rapidly changing sectors, the impossibility of differentiating the various sectors with respect to this training need, suggests the presence of other needs that find expression generic and improper in the training request. It can be assumed that rather than a real technical formative lack, there is rather the perception of a lacking capacity to face problematic situations.

In some situations (5 sectors) there was also an express reference to the education of superiors. It will be a matter of assessing how much the expectations are full of "saving" values, as much as it is in the effective power of the managers and how much the sense of powerlessness is more or less founded with respect to the responsibility of each worker.

In situations of high risk, the presence of problems related to horizontal communication is highlighted and aspects of internal conflict were also present.

3.8 Authority Reports

The Authority recognizes that the Municipality operated beyond the "minimum", the positivity deriving from having carried out training meetings and the risk assessment were highlighted, although noting that the consultation of the laborer representatives took place during the programming phase only. He also declared that the corrective actions were "modest, punctual and insufficient", that did not give rise to improvements or even produced worsening.

4 Discussion

This study enabled to highlight some critical points on work related stress assessment and management within a large Municipality. The multidisciplinary SC therefore proposes some solutions and interventions, as follows:

(a) Outline a general organization chart for the assessment and management of work-related stress
(b) Adopt a general policy at the Municipality regarding work - related stress
(c) share the project among those who should and could collaborate in a multidisciplinary manner and based on specific mandates and competencies (e.g., General Manager/Auditor, Human Resources, Administrative management of personnel, Safety and prevention Responsibles, Managers, Supervisors, workers' representatives, Advisory committees, Counselor for organizational wellness);
(d) Identification of priorities (areas, sectors, people, issues);

(e) Establishment of a Support and Consultancy Unit, dedicated to Work-related Stress. The SC proposes the activation of a Center, named Research Center for Mental Health and Wellness (*Italian Acronym: Ri.Be.S Centro di Ricerca per il benessere e la salute*) with dedicated space and time and a commitment to evaluate single cases and general issues, as well as to elaborate, adopt and monitor specific interventions, using evidence based indicators;
(f) Update risk assessments and health surveillance, by means of updated, evidence based tools and standardized methods [6], as well as adopting strategies to improve participation and response rate. Involvement of area managers, workers and their representatives is recognized as a critical issue
(g) concrete adoption and monitoring of operational procedures and good practices, within the framework of a multidisciplinary vision, concerning, for example: sentinel events, aggressions, disciplinary measures and reward systems, analysis and management of accidents at work, absenteeism, job suitability, communication and counseling
(h) consult and involve worker representatives in all the successive phases of the project
(i) define a dedicated educational program with a focus on management, initially through a process of assessing training needs and including an assessment of effectiveness. A specific area of training intervention, verified the permanence of a fragmented representation of the working entity as emerged from the analysis, could be addressed in parallel to managers and workers. In particular, it could be proposed a group training activity aimed at gathering congruence and/or differences with respect to work activities, aims, tasks, priorities and methods, in an ergonomic perspective
(j) Improve the collection of health surveillance data, through the activity of occupational physician, by means of questionnaires to evaluate subjectivity
(k) The activities on work - related stress should be seen and performed within the more general view of the Corporate Social Responsibility, aiming, in perspective, on a more friendly work organization, that looks with confidence and positivity to the evaluation and management of certainly delicate and complex topics such as aging, staff mobility, reconciliation of working hours, extra-work life-family, commuting, disability.

5 Conclusion

Despite psychosocial factors in workplaces are recognized as one of the main problems for workers in industrialized countries, evidence of the effectiveness of interventions is modest. The approach described in this applied research tries to establish preventive strategies, in order to manage psychosocial factors and implement intervention of effectiveness. The overall proposal was well received by the management and will be carried out over the next two years. The intervention should allow to assess whether the proposed integrated multidisciplinary approach will be effective in achieving an improvement of work related stress over the short and long term.

Acknowledgement. This research has been funded by Brescia Municipality.

References

1. Dalsbø TK, Dahm KT, Austvoll-Dahlgren A, Knapstad M, Gundersen M, Reinar LM (2013) Workplace-based intervention for employees' mental health. Accessed 21 May 2018. https://www.ncbi.nlm.nih.gov/books/NBK464819/pdf/Bookshelf_NBK464819.pdf
2. Elo AL, Leppanen A, Lindstrom K (1992) OSQ occupational stress questionnaire: users' instruction. Review/Institute of Occupational Health 19. Helsinki (1992)
3. EU-OSHA (2012). Management of psychosocial risk at work: an analysis of the finding of the European Survey of Enterprises on new and emerging risks. European Agency for Safety and Health at Work, Luxemburg
4. European Agency for Safety and Health at Work (2014) Second European survey of enterprises on new and emerging risks (ESENER-2) overview report: managing safety and health at work. https://osha.europa.eu/it/tools-and-publications/publications/second-european-survey-enterprises-new-and-emerging-risks-esener
5. Hassard J, Teoh K, Cox T (2018) Job satisfaction: evidence for impact on reducing psychosocial risks. https://oshwiki.eu/wiki/Job_satisfaction:_evidence_for_impact_on_reducing_psychosocial_risks. Accessed 21 May 2018
6. INAIL (2011) Valutazione e gestione del rischio da stress lavoro-correlato. Manuale ad uso delle aziende in attuazione del D.Lgs 81/08 e s.m.i
7. INAIL (2017) La metodologia per la valutazione e gestione del rischio stress lavoro-correlato. Manuale ad Uso delle aziende in attuazione del D.Lgs 81/2008 e s.m.i
8. Jain A, Leka S, Zwetsloot G (2011) Corporate social responsibility and psychosocial risk management in Europe. J Bus Ethics 101(4):619–633. https://doi.org/10.1007/s10551-011-0742-z
9. Kraatz S, Lang J, Kraus T, Munster E, Ochsmann E (2013) The incremental effect of psychosocial workplace factors on the development of neck and shoulder disorders: a systematic review of longitudinal studies. Int Arch Occ Env Health 86(4):375–395
10. Leka S, Cox T (2008) PRIMA EF: guidance on the European framework for psychosocial risk management. http://www.prima-ef.org/uploads/1/1/0/2/11022736/prima-ef_ebook.pdf
11. McFarlane AC (2007) Stress-related musculoskeletal pain. Gen Musculoskeletal Conditions 21(3):549–565
12. Montano D (2014) Upper body and lower limbs musculoskeletal symptoms and health inequalities in Europe: an analysis of cross-sectional data. BMC Muscoloskeletal Disord 15:285 (2014). https://www.ncbi.nlm.nih.gov/pmc/articles/PMC4153890/pdf/12891_2014_Article_2230.pdf. Accessed 21 May 2018
13. Oakman J, Macdonald W, Bartram T, Keegel T, Kinsman N (2018) Workplace risk management practices to prevent musculoskeletal and mental health disorders: what are the gaps? Saf Sci 101:220–230

Event Cross-Simulation: A Tool for Structuring Debates on Work, Enhancing Collective Learning and Improving Safety Management

Christelle Casse(✉)

Laboratoire PACTE/IEP Pacte/IEP, BP 48, 38040 Grenoble Cedex 9, France
christelle.casse@umrpacte.fr

Abstract. In recent years, various studies have shown the importance of instituting Work Debate Spaces (WDS) within companies to improve safety management, team performance and team members' health within organizations [12, 16]. For years, ergonomists have been using simulation in various projects as a method for eliciting workers' actual activity in the discussions [19]. The aim of the paper is to show how a cross-simulation method implemented in WDS can contribute to foster collective learning and improve professional practices and procedures in a high risk organization. This paper will present in a first part the context and the theoretical framework of the action research and the cross-simulation' methodological device built by the researchers. In a second part, it will develop the results of an experiment carried out by using the cross-simulation device in WDS in a road-tunnel operating company. We will conclude presenting the conditions and limits for both the cross-simulation method and WDS to be effective.

Keywords: Simulation · Learning from experience · High risk organization

1 Introduction

In recent years, various studies have shown the importance of instituting Work Debate Spaces (WDS) within companies to improve safety management, team performance and team members, health within organizations [12, 16]. Research in ergonomics, psychology and management [4, 7–9, 14] highlights the need for operators as managers, to set up collective regulation spaces to discuss concrete problems of daily work, to elaborate together acceptable procedures and practices to manage safety. WDS appear as means to deal with operational problems, and to develop individual and collective apprenticeships. The results from the literature show that they play a positive role in the performance of organizations in terms of production and safety, and for the health of workers. They contribute to the construction of efficient work teams, that support for individuals in their daily activity [5].

However, several conditions are necessary for the debate to take place and lead to concrete changes in the rules of organization [9, 18]. Effective discussion forums on work involve, in particular, the formal objective of solving the difficulties of daily

S. Bagnara et al. (Eds.): IEA 2018, AISC 821, pp. 267–275, 2019.
https://doi.org/10.1007/978-3-319-96080-7_31

work, times for collective exchange, animators trained in discussion facilitation and methods to elicit workers' activity such as work simulation. For years, ergonomists have been using simulation in various projects as a method for eliciting workers' actual activity in the discussions. "*Simulation aims to help employees to look forward into their future work situations by getting them to "play" their probable future activity on the basis of action scenarios represented by intermediary objects* [3, 19]. Simulation appears to be a methodology of great potential for fuelling discussion on work and enhancing learning from experience because it fosters articulation of past experience and future situations. Our question is which conditions and methodology are needed to do that?

The aim of the paper is to show how a cross-simulation method implemented in WDS can contribute to foster collective learning and improve professional practices and procedures in a high risk organization.

2 Context: An Action-Research in a Road-Tunnel Operating Company

The present study is part of a 3-years action-research project conducted in a road tunnel operating company to implement a new approach of operating feedback system based on WDS for better collective safety management [6]. Road tunnels are increasingly sophisticated infrastructures. They comprise a civil engineering structure and a wide array of safety equipment and facilities, designed to ensure the safety of tunnel users (lighting system, ventilation system for fires, fire network, extinguishers, emergency call points, emergency exits, etc.). As well as these items of safety equipment, tunnels are also fitted with cameras, various sensors, and an Automatic Incident Detection (AID) system. These provide continuous information on equipment status and on any problems or anomalies that may occur. The system is managed by a Control Centre supervisor, working in conjunction with patrol teams (who are on-hand to intervene in the event of an incident). These two professions interact constantly in the interest of day-to-day equipment and user safety. The tunnel operation process involves four key responsibilities: traffic management, user safety, maintenance and repair works. The supervisor is responsible for equipment and traffic monitoring and incident management. Patrollers respond during incidents to protect user safety, check the condition of traffic lanes and conduct minor maintenance work on the equipment. Our research focuses on a heavy-traffic urban infrastructure comprising 10 km of road, including three tunnels (respectively 200, 300 and 1,200 meters in length). Formal event feedback meeting are organized in the company since 2006. These meetings focus on "significant" events, as defined in tunnel regulations (i.e. accidents with injured casualties or fatalities, fires, any incident leading to non-planed tunnel closure). The debrief meetings provide an opportunity for the professionals to develop mutual knowledge and exchange on their event management practices. However, these meetings mainly concern the hierarchy of operations (team leaders, managers). In most cases, they lead to the definition of new rules that modify existing event management procedures, but they do not make it possible to address the difficulties of daily routine and situations without immediate consequences but which can be complex to manage for the operators in the field.

The challenges of action research are the improvement of risk management by setting up WDS for supervisors and patrollers to discuss their practices and the real difficulties encountered on the ground dealing with critical situations.

3 Methods: Work Debate Spaces Based on Cross-Simulation

We organized WDS with all the CC supervisors and patrollers. In order to structure the debates, we designed an original method of collective simulation based on past event scenarios. The purpose of this method was to help the operators to put themselves in a situation in order to foster the confrontation of the different strategies and practices and to stimulate debate.

3.1 Choice of Situations and Construction of Scenarios

The choice of events to simulate situations was made by the researcher and the CC team leader who was involved in the entire process of the action-research. The events were selected for their pedagogical dimension, according to three main criteria: their generic character, the variety of contexts, the singularity of each situation. The event scenarios submitted to the simulation groups were built by the CC team leader. Information about the dates and operators of the event was neutralized to make them "anonymous". Simulation supports were constructed from photos, situation maps and context information to help professionals to project themselves into the situation. Information forms were also created to collect each professional's decisions and actions at each stage of the simulation.

3.2 Conduct of Cross-Simulation Sessions

The session that we discuss in this article concerned all the CC supervisors and patrollers spread over three meetings made a few days apart. The simulated events are two fire starts. Both events involve specific operating modes and the mobilization of different procedures. *Scenario 1 - Fire outside a tunnel:* The scenario takes place at night, at an hour of light traffic outside the tunnels, on the side of the road in an area bordering the territorial area of the operating company; the straps to access this area are operated by another company. *Scenario 2 - Fire inside a tunnel:* This second, more "classic" scenario concerns a tunnel fire starting at rush hour. It falls into the category of 'significant events' in the road-tunnel regulation.

Each meeting was organized in the same format. The team leader prepared and facilitated the meeting, at which the ergonomist attended. Three inter-trade groups were formed consisting of one or two CC supervisor(s) and two or three patrollers (more or less experienced). For each meeting, four steps were followed. Step 1. - The team leader presented the event scenario with a photo and the geographical location. He gave the instructions for the collection of information and distributed the information forms.

Step 2. - Each group performs the simulation and discusses; a representative from each trade take notes of decisions and actions on the information form. Step 3. - After all the groups have finished the simulation, the team leader asks one of them to present its analysis of the situation, strategy, decisions and actions. Step 4. - The other two groups commented on the strategy presented, giving their own analysis of the situation and their actions. The team leader regularly intervened to question one or the other about the motivations of their decisions or actions and to discuss the proposed strategies, particularly in terms of security. At the end of the session, the debates were synthesized by the team leader, recalling the points of divergence that had emerged and the important rules that had been shared, debated or built during the session. Of the three meetings of this cross-simulation session, the nine groups productions were compared.

3.3 Data Analysis

The analysis of the collected data focused on three dimensions (Rocha, 2014): the themes addressed during the debates, the content and the function of the exchanges between the participants in particular in terms of contribution of knowledge, discussion and elaboration of practices and procedures, and finally the collective dynamics of the debates. The analysis is essentially qualitative, the discussions in the groups being singular according to the interactions between the trades on the selected event.

4 Event Cross-Simulation: A Tool for Structuring Sebates on Work, Enhancing Collective Learning and Improving Safety Management

The cross-simulation session conducted from the two event scenarios opened up debates on event management practices. It allows the professionals to share and transmit knowledge and create new collective procedures.

4.1 The Diversity of Knowledge and Strategies Implemented in the Field

The main contribution of the WDS based on the cross-simulation is to highlight for the operators of the two trades the diversity of the analyzes and the strategies. For example, the simulation for scenario 1 (outdoor fire), conducted by the nine groups, resulted in nine different strategy proposals. However, different strategies emerged regarding actions to close access and manage traffic to the fire zone. Access actions: Some groups decided to close the access ramps to the fire zone although they were not part of their area of responsibility, while others argued that they did not have to do it. Traffic management: Some groups did not completely close the traffic at the level of the fire, as it took place at night in an area with little traffic. They closed the lane closest to the burning vehicle and leaved the furthest route in circulation in order to avoid blocking users at least until the arrival of firefighters, who can then ask to completely close the traffic lanes. The other groups decided to block the traffic to the fire zone and arranged a diversion.

For this case of fire, the interpretation of the context and the safety procedures was very different from one group to another. The analysis of the various strategies indicated that they were more related to the fire experience and the knowledge of the work than to the job. Two main practices were questioned and discussed collectively: Is it necessary and allowed to close access ramps that are not part of the territorial area of the company in case of fire, particularly in this context of low traffic? Is it acceptable not to totally stop the traffic when there is a car fire beside the road? in what conditions? The answers to these questions conditioned different way of intervention.

4.2 Novice Training and Procedural Recall

Fire management simulations allowed novices and all-rounders, who are less present on the ground, to review event management procedures. They also allowed the team leader to assess the level of integration of the procedures by the patrollers and to identify the rules that are less known to accompany them in their training. The group also had a training role, the more experienced operators spontaneously transmitted their knowledge of rules and procedures to novices. For example, for scenario 2, a collective choice was made as to the situation of the patrollers, so that they were inside the tunnel at the time of the fire, which is a rare case (in general they come from outside) for which the rules of intervention are specific. This made it possible to pass on to the novices this particular way of operation. Experienced patrollers have all proposed the same strategy of action in this case, which consisted in a first time to create a 'mobile traffic jam' so that users still in the tunnel could slow down and stop gradually, and in a second time to accompany the users in the evacuation by foot to the emergency exits if necessary. A young patroller questioned about his strategy by the team leader did not know where the patrollers should be in such a fire situation. An experienced patroller gave him the principles of field dispatch procedure. We then witness a horizontal transmission which, in addition to allowing the young patroller to learn and the experienced to put forward the experience acquired, had the advantage of reinforcing the social links between the elders who have a 'fixed status' in the company and the novices who have all a 'all-rounders' status. The team leader only had to validate the knowledge passed on.

4.3 The Development of Common Rules and Procedures Adapted to Changing Contexts

The debates that emerged from the cross-simulation sessions resulted in the development of a common rule regarding the strategy to be implemented in the particular contexts involving the perimeter of responsibility of the operating actors. For example, in the first scenario (Fire outside the tunnel) following the team leader's reminder regarding the priority of user safety, the common rule defined in this type of context was to close all the ramps and all the lanes in the zone of the fire.

A second example of elaboration of rules emerging from the debates concerns Scenario 2 (Tunnel Fire), which takes place in a context of significant congestion of traffic due to work in progress. This scenario was chosen because it highlighted recurring intervention constraints for patrollers. Cross-simulation made it possible to compare how the different teams deal with these constraints: a novice offers innovative itineraries, which valorized him and made it possible to approach a new scenario. Patrollers and supervisors discussed, in particular, a practice authorized only in exceptional circumstances because it involves a danger, it was the counter-directional intervention in the tunnel. Some teams were asking for it systematically when they were blocked. Others did not use it because they followed the principle of exceptional practice. After discussion, given the evolution of the traffic and the recurrence of the blockade situations of the patrollers, which had a certain cost for the company, the team leader validated the evolution of the rule by authorizing the practice of intervention in the opposite direction to manage an event in case of traffic jam. This new rule resulting from the debates between the operators in the field was then validated by the management and the operations manager.

5 Discussion and Conclusion

5.1 Contribution of Cross-Event Simulations

Our study brings out several levels of contribution from cross-simulation sessions to the improvement of WDS. Simulations based on past real events offer the opportunity to revisit these events in a more open constructive perspective than traditional post-event debriefing, to draw potentially broader lessons. Cross-simulations make it possible to confront operators with several learning situations in a short time. This is of interest for training and preparing novices especially for managing rare and diverse situations, but also for developing ongoing training of the most experienced operators. Interest is reinforced by the fact that the transmission of knowledge is as much horizontal, between peers, as vertical. This confirms the skill development and collective support function that is advanced in the literature [16]. These methods make it possible to work on a wide range of situations and in particular on critical situations that do not fall into the category of 'major events' and for which the procedures are not very formalized and the operating modes vary. The operators can build and update the collective action rules to better control the risks associated with the variability of these situations on the ground. The WDS that are opened from the simulation sessions also have the advantage of encouraging the establishment of a bottom-up process for improving the procedures, thanks to the relay of the team leader facilitator who carries validated proposals at the operational level in the debate and decision-making bodies of senior management. Thus, the WDS have favored on the one hand learning from experience and analysis of the past, and on the other hand the construction of common experience, the definition of frameworks of action (the procedures) and the development of collective and individual skills for the present and the future. It thus contributes to strengthening cross-trades activity and the teams' ability to manage risks and safety.

5.2 Engineering of Work Debate Spaces Based on Cross-Simulation

In this section, we develop concrete methods for implementing WDS based on event cross-simulation.

The Learning Potential of the Cross-Simulation Method
Simulation of work is a traditional method of ergonomists, particularly developed in spatial design approaches [1]. But it is also increasingly used to drive organizational change, or for training purposes, especially in risky contexts [2]. By revealing the differences between the interpretations of the situation and between the strategies of action of the workers, the cross-simulation generates *'socio-cognitive' conflicts* which foster learning [13]. Based on groups that mixed instituted teams, the cross-simulation method skirts the effects of cleavage between teams. This cross-approach of simulation is therefore particularly structuring for the debate and the collective learning processes. The limits of this method, however, lie in the preparation of the simulation and debating sessions and in the training of managers to animate the debates in a constructive framework.

Preparation as a Decisive Step
The objective of the approach is ultimately to set up an autonomous and sustainable device of WDS piloted by the internal actors. The achievement of this objective is based on the process initiated by the stakeholder and the actors of the company well in advance of the debating sessions, which proceeds from several important stages, from the work analysis in the field, to the diagnosis and the choice of situations to be simulated, plus the preparatory stage for designing the tools and the modalities of the simulation and the debates. The autonomy of the device depends on the involvement of the company's stakeholders in each of these stages, so that they take ownership of the approach and master the tools, especially the local managers. The temporal organization (a long-term approach), the choice of actors (experienced and multi-skilled) and the content of the exploratory phase (work analysis in the field) are essential for the smooth running of the process and its sustainability.

Train Local Managers to Prepare and Facilitate WDS
Managers must be trained and prepared to be able to pilot the simulations and facilitate the WDS. Cooperation within a multi-professional group is not spontaneous, even less when the collective work is led by a line manager. The actors must feel in a context of psychological security [10], that is to say that they must have confidence in the fact that they will not be punished in case of error. The quality of the animation is based on the formalization of clear rules for the participants and the establishment of a dynamic of constructive dialogue, based on listening, respect for the diversity of points of view and the ability to arbitrate between the people. For this, managers must incorporate participative management principles. The facilitation of work debates also requires the acquisition of action-oriented interview techniques, such as the *explicitation interview* [19, 20], for which training are required. Finally, it is important to train line managers in the construction of simulation scenarios, based on past situations. They must both be able to extract the key elements for understanding the context, and build or prepare the tools so that the actors can project themselves better in the situation. The training of managers is a decisive condition for the smooth running of exchanges and the production of concrete results, as well as the commitment of operators over time.

Organizing the Treatment of Emerging Issues at the Top Level
For the WDS to work, it is important to organize the shift from emerging issues at operational level simulations to higher levels in the organization. The WDS make it possible to elicit the activity of ground, but it also summons that of the management, revealing in particular the insufficiencies or inconsistencies of the procedures in certain situations and the latent imbalance between the situations framed by the procedures of urgency and other more non-framed daily situations. This is why it is useful after the WDS at the operational level, to organize the relay at the level of the top-management to be able to deal with outstanding issues which need decisions, according to a principle of subsidiarity [17]. During the action-research, we set up at the end of each debate session a restitution time in the presence of a senior executive, so that he/she can provide answers to emerging questions and in any case he/she guarantees to the participants that they will be treated.

References

1. Barcellini F, Van Belleghem L (2014) Organizational simulation: issues for ergonomics and for teaching of ergonomics' action. In: Proceedings of 11th international symposium on human factors in organizational design and management (ODAM), pp 885–890
2. Barcellini F, Van Belleghem L, Daniellou F (2014) Design projects as opportunities for the development of activities. In: Falzon P (ed) Constructive ergonomics, pp 150–163
3. Boujut JF, Blanco E (2003) Intermediary objects as a means to foster co-operation in engineering design. Comput Support Coop Work (CSCW) 12(2):205–219
4. Caroly S (2010) Activité collective et réélaboration des règles: des enjeux pour la santé au travail. Doctoral dissertation, Université Victor Segalen-Bordeaux II
5. Caroly S, Barcellini F (2014) The development of collective activity. In Falzon P (ed) Constructive ergonomics, p 19
6. Casse C (2015) Concevoir un dispositif de retour d'expérience intégrant l'activité réflexive collective: un enjeu de sécurité dans les tunnels routiers. Doctoral dissertation, Grenoble Alpes
7. Daniellou F, Simard M, Boissières I (2011) Human and organizational factors of safety: a state of the art. Number 2011-01 of the Cahiers de la Sécurité Industrielle. Institute for an Industrial Safety Culture, Toulouse, France. http://www.FonCSI.org/cahiers/
8. Detchessahar M (2001) Quand discuter c'est produire. Revue française de gestion 32–43
9. Detchessahar M (2013) Faire face aux risques psycho-sociaux: quelques éléments d'un management par la discussion. Négociations 1:57–80
10. Edmondson A (1999) Psychological safety and learning behavior in work teams. Adm Sci Q 44(2):350–383
11. Haines HM, Wilson JR (1998) Development of a frame work for participatory ergonomics. Research report. Health and Safety Executive, p 72
12. Hendry J, Seidl D (2003) The structure and significance of strategic episodes - social systems theory and the routine practices of strategic change. J Manag Stud 40(1):175–196
13. Johnson DW, Johnson RT (2009) Energizing learning: the instructional power of conflict. Educ Res 38(1):37–51
14. Lorino P, Tricard B, Clot Y (2011) Research methods for non-representational approaches to organizational complexity: the dialogical mediated inquiry. Organ Stud 32(6):769–801

15. Rocha R (2014) Du silence organisationnel au développement du débat structuré sur le travail: les effets sur la sécurité et sur l'organisation. Doctoral dissertation, Bordeaux
16. Rocha R, Mollo V, Daniellou F (2015) Work debate spaces: a tool for developing a participatory safety management. Appl Ergon 46:107–114
17. Rocha R, Mollo V, Daniellou F (2017) The structured debate on work as an approach to developing an enabling environment. @ctivités 14(2)
18. Van Belleghem L (2015) Eliciting activity: a pathway to real work. In: Proceedings 19th triennial congress of the IEA, August 2015, vol 9, p 14
19. Van Belleghem L (2016) Eliciting activity: a method of analysis at the service of discussion. Le travail humain 79(3):285–305
20. Vermersch P (2006) Les fonctions des questions. Expliciter 65:1–6

The Nature of the Firm – A Social Cybernetic Analysis

Thomas J. Smith[1] and Robert Henning[2(✉)]

[1] University of Minnesota, Minneapolis, MN, USA
smith293@umn.edu
[2] University of Connecticut, Storrs, CT, USA
robert.henning@uconn.edu

Abstract. In 1937, Ronald Coase published 'The Nature of the Firm' [1], addressing the question of why firms exist. He concluded that firms emerge to reduce costs of transactions. A 'transaction' is defined both as the action of conducting business, as well as an interaction between people. Both senses of the term prompt the present social cybernetic analysis of the nature of the firm. Social cybernetics focuses upon the reciprocal feedback control and feedforward interactions between two or more individuals in a group or organizational setting, a process termed *social tracking* that involves dynamic linking of the social behavior of two people, of multi-person teams of people, or of larger groups of individuals engaged in intra- or inter-institutional transactions.

From the perspective of social cybernetics, the potential for continued market success of a firm thus is equated with the degree to which the fidelity of social tracking among transactional participants is developed, maintained and refined through organizational design and management.

Keywords: Transactions · Social cybernetics
Organizational design and management

1 Introduction

'The Nature of the Firm,' published by Coase in 1937 [1], offers an explanation of why firms exist. A firm in modern economic theory is an organization which transforms inputs into outputs [2, p. 5]. In his analysis, Coase addresses the question of why the economy features a number of business firms instead of consisting exclusively of a multitude of independent, self-employed people who contract with one another. His conclusion is that there are a number of transaction costs involved in using the market, such that the cost of obtaining a good or service via the market actually exceeds the price of the good. It is a host of other costs, including search and information costs, bargaining costs, keeping trade secrets, and policing and enforcement costs, that can all potentially add to the cost of procuring something from another party. This suggests that firms will arise in order to internalize the production of goods and services required to deliver a product, thereby reducing the costs of transactions tied to production [3].

A 'transaction' is defined both as the action of conducting business, as well as an interaction between people. Both senses of the term prompt the perspective provided in

S. Bagnara et al. (Eds.): IEA 2018, AISC 821, pp. 276–285, 2019.
https://doi.org/10.1007/978-3-319-96080-7_32

this report, which offers a social cybernetic interpretation of transactional exchanges that underlie the nature of the firm. The present focus is on firms in the U.S.—it is not assumed that the analysis necessarily applies to firms in other countries.

1.1 Social Cybernetic Analysis of Transactional Behavior

The most comprehensive behavioral theory of how people interact and exert control over themselves and the world around them is that of social cybernetics [4–6, 7, Chap. 8], focusing upon the reciprocal feedback and feedforward control interactions between two or more individuals in a group or organizational setting, a process termed *social tracking.* Social tracking is conceived as a dynamic linking of the social behavior of two or more people which is often goal-oriented as it usually is in the workplace.

The cybernetic model of social behavior as a social tracking process is illustrated in Fig. 1. Social tracking is based on the specialized, coordinate motor behavioral and physiological responses which a given individual in a social group initiates in order to track and thereby control sensory feedback generated by the motor behavior of others in the group. Movements of one person generate sensory input to a second, who, in controlling this input as sensory feedback, generates input of a compliant sort back to the first person, and so on. During group social tracking, one individual generates sensory feedback which all other group members track in a compliant manner through their behavioral-physiological sensory feedback control mechanisms. In this manner, the group as a whole establishes a system of reciprocal social tracking relationships, in which the social partners become engaged in mutual exchange and control of sensory feedback to establish a yoked, behavioral-physiological, feedback-integrated system.

As suggested by the feedback parameters and control characteristics depicted in Fig. 1, social tracking typically involves many modes, variations, and conditions of mutual sensory feedback control. This social cybernetic model can be applied generally to interpret and analyze the systems properties of the entire spectrum of social behavior, encompassing work, verbal and nonverbal communication, language, predation, courtship and mating, artistic expression, parent-child bonding, education and training, and organizational and institutional behavior. It can involve parallel control (that is, symmetric or matched) or series-linked control (asymmetric and occurring in a serial manner via multiple people). It may involve positive, negative, compensatory, complementary, differential, integrative, and/or transformed types of social feedback control of activity by the interacting persons. All of these varied social tracking modes and conditions may also occur between groups, between a group and an individual, and between organized groups and institutions.

In comparison to social interaction between two individuals, group social tracking embodies more complex and ramified patterns and modes of sensory feedback control. Figure 1 illustrates several distinct types of group social tracking relationships, namely: (1) individual-group; (2) intergroup; (3) intragroup; (4) mediated group; (5) inter-institutional; (6) intra-institutional; and (7) group-institutional interactions. A general systems analysis of the human factors design of group and institutional organizations suggests that this limited number of social interactive groupings describes the large majority of social interactions in firms, as well as in society generally [4]. Indeed,

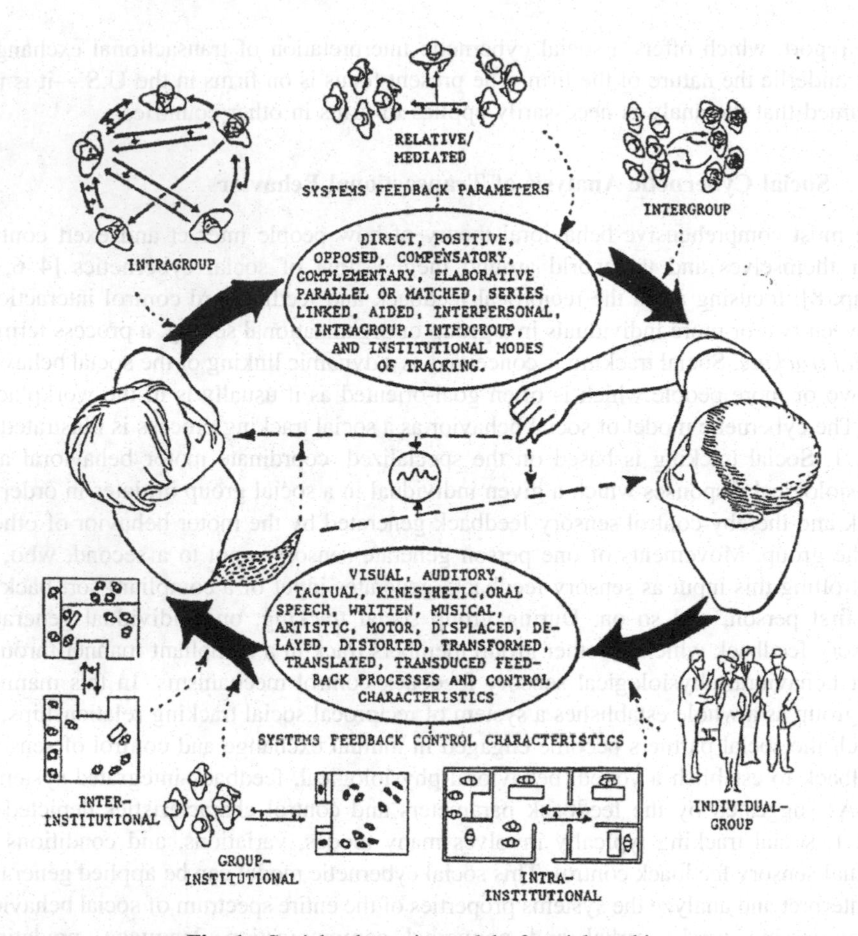

Fig. 1. Social cybernetic model of social tracking.

Coase himself adopts a cybernetic perspective with his suggestion that economic systems are self-organizing [1, p. 387].

From the perspectives of social cybernetics therefore, the market success of a firm is dependent on the degree to which social tracking fidelity among transactional participants is developed, maintained and refined in the service of organizational development and management of the firm through application of macro-ergonomics.

2 Social Tracking – A Paradigm for Enhancing Existing Transactions and Meeting Future Challenges

Social cybernetic principles can be used in three important ways to understand how firms establish and maintain the high levels of transactional efficiency necessary to survive and remain competitive in the face of changing economic, technological, demographic, cultural, and political environments. To begin with, application of social cybernetic principles can be used to rationalize why macro-ergonomic interventions are so important. For example, close examination of an existing transactional process may reveal that the feedback control behavior necessary to manage one aspect of a transaction has become burdensome due to delayed (non-compliant) feedback following a reorganization that has reduced transactional efficiency. Once feedback control aspects of this transaction are identified as problematic, new protocols and technologies can be considered for restoring prior levels of transactional efficiency.

Firms also need to act proactively to maintain and even improve transactional efficiency, and this has distinct advantages over relying on compensatory approaches which only occur following a degradation in transactional efficiency. For example, if management learns that new computer-mediated communication systems offer a means to boost transactional efficiency, adoption of these technologies is definitely worth considering. However, deliberating over which system is the best fit for the firm requires a means to predict whether a new communication system will function in the context of the firm and deliver on the promise of improved transactional efficiency. Such deliberation by management involves feedforward control of behavior that relies much more on existing social tracking relationships than is normally appreciated. As revealed in laboratory research on driving behavior, feedforward control actions in the form of proactive behaviors were found to be yoked to continuous steering behaviors [8]. Scaling up this relationship to an organizational level, a firm's capacity to engage in proactive steering behaviors to benefit transactional efficiency will similarly depend on yoking these steering behaviors to the realities of ongoing social tracking relationships within the firm. A firm's capacity via management to successfully engage in proactive steering behavior regarding transactional efficiency cannot magically emerge from a vacuum. It will depend on the extent that the firm's management is able to yoke their steering behaviors to existing social tracking relationships which form the backbone of all ongoing internal transactions among employees. Put another way, a firm's capacity to capitalize on its internal strengths and engage in proactive steering behaviors will to a large extent depend on management's ability to leverage ongoing social tracking relationships within the firm.

As a case in point, a recent study in commercial aviation reported that the effectiveness of a multi-team system under simulated emergency conditions depended on team members functioning as boundary spanners who then communicated critical information between the teams [9]. When seeking a long-term solution to supporting these types of boundary-spanning transactions, an examination of the various modes of social tracking found that the existing communication protocols within teams would represent a good starting point. Any new communication protocols would need to consider the macro-ergonomic design factors necessary to support both feedback and feedforward control during social tracking, such as minimizing response delays in series-linked communications and supporting the parallel mode of social tracking during shared decision making.

Social cybernetic research on organizational learning conducted by Haims and Carayon [10] corroborates the importance of adopting macro-ergonomic designs that promote action-feedback relationships during social tracking, between individuals and teams and across all levels of a firm. Formation of these action-feedback relationships was found to be a prerequisite to organizational learning and development, and this therefore represents the third way that social cybernetic principles can be used to understand how a firm can both establish and maintain transactional efficiency over the long term in the face of changing environments. To this end, there is growing evidence in favor of implementing participatory programs that offer a structured process of employee engagement whereby employees at all levels of the firm are able to collaborate on intervention design efforts, and with line-level employees able to assume a central role in both initiating and steering the course of organizational change efforts [11]. The parallel mode of social tracking is at the heart of these participatory initiatives because it is integral to any shared decision-making process, and this explains why promoting yoked social tracking relationships in these initiatives can provide ancillary benefits beyond improved transactional efficiency, such as also improving the organizational health of the firm [12].

3 Types of Transactions Involved in the Management of a Firm

This section introduces selected examples of the types of transactions that the manager of a firm might be expected to engage in. Coase's postulate — firms exist to reduce the cost of transactions — provides a rationale for this analysis. That is, to explore the implications of the social cybernetic model of firm transactions, it is desirable to first understand the different types of transactions that we are talking about. Nevertheless, the analysis offered here cannot be considered definitive. Indeed, there is every reason to believe that, given the likelihood that the organizational design of each different firm is context specific (i.e., varies in relation to the social, cultural, economic, etc., environment in which the firm operates [7, Chap. 9], the transactional environment of different firms also is highly context specific. Indeed, in his 1937 article, Coase notes that, "nothing could be more diverse than the actual transactions which take place in our modern world" [1, p. 396]. This analysis therefore focuses on the transactional features of one small firm, and on selected examples from large firms, on the assumption that at least some of the observed relationships may be found in other firms sharing similar organizational designs and contexts.

3.1 Small Business Transactions—Examples for One Firm

To gain some insight into the nature of transactions, and the concomitant social cybernetic challenges, confronting a present-day small firm, in early November, 2017, Smith interviewed the owner of a small silicon rubber-molding company located in a suburb of the Twin Cities of Minneapolis and St. Paul in Minnesota. This small firm was selected for analysis on the assumption that the owner should be responsible for the entire range of transactions affecting company operations. At the time, the employee workforce comprised one product manager, three inspectors (including quality inspection), and 6–8 molders.

This focus on a small business is aligned with the original emphasis of Coase [1, p. 392], who notes that, "the operation of a market costs something, and by forming an organization (*i.e.*, a firm) and allowing some authority (an "entrepreneur") to direct the resources, certain marketing costs are saved. The entrepreneur has to carry out his function at less cost, taking into account the fact that he may get factors of production at a lower price than the market transactions which he supersedes." It seems reasonable to assume that, with this thesis (the crux of his report), Coase is equating the idea of an entrepreneur with the organizer of a small, rather than a large, business.

Six types of owner-mediated transactions were identified with the interview, involving social interactions (SIs) by the owner with: (1) the 10–12 employees (18 face-to-face SIs/hour); (2) 100 customers (once/day to once/year SIs, mostly by phone); (3) 24 suppliers (12 SIs/week, by phone); (4) regulators (5 face-to-face SIs/year); (5) major equipment maintenance (1 face-to-face SIs/year); and (6) the landlord (1 face-to-face/month). Based on these results, a rough estimate in the range 150–175 SIs per 8-h workday by the owner can be calculated, most involving face-to-face SIs with employees.

The social cybernetic model in Fig. 1, coupled with the above results, underscore the considerable social behavioral skills required of the owner to engage in the transactions required for successful operation of the business. In particular, face-to-face transactions involve social tracking on the part of participants that requires control of both visual and auditory (speech) feedback, likely featuring both parallel and series-linked modes of social tracking either occurring alternately or in combination. Phone transactions involve asymmetric auditory social tracking because of required turn-taking. The social-behavioral skills necessary to successfully engage in these social tracking behaviors must be exercised multiple times per hour by the owner. As outlined in Sect. 2 above, the social behavioral demands of these transactions are amplified if the owner interacts with two or more respondents simultaneously. The social tracking demands on the owner remind us that social tracking is the most difficult type of behavioral skill to be mastered by the maturing individual. It is typically not fully refined until the late teen or early adulthood years, and for some individuals it is not ever effectively mastered [13, pp. 503–504].

The interview findings support the conclusion of Coase, noted above, that a firm consists of the system of relationships that comes into existence when the allocation of resources is dependent on an entrepreneur. Yet it is also clear that "the system of relationships" in even a small firm, can be highly complex, demanding refined social behavioral skills on the part of the participants in each transaction, a social tracking complexity that Coase's broad analysis does not address. Moreover, it is clear that in the case of this small business, and possibly with many small businesses, the burden of managing transactional relationships falls primarily on the owner.

3.2 Modes of Transactions in a Large Firm—One Example

The question of why firms of various sizes exist represents a recurrent theme in the original 1937 report of Coase. His key conclusions [1, p. 396] are that an increase in the number of transactions as the size of a firm grows results in: (1) a drop in organizing costs; (2) a slower rate of increase in these costs with more transactions; (3) fewer mistakes by the entrepreneur, plus a smaller rate of increase in mistakes; and (4) a smaller rise in the supply price of factors of production. From these perspectives, we may expect both the magnitude and the modes of transactions to increase with larger relative to smaller firms, as this section documents with a few selected examples. One key consideration in this regard for large relative to small firms is that senior management of large firms may delegate responsibility for managing transactions to subordinate individuals in the firm.

Wells Fargo, one of the largest banks in the U.S., offers a prime example of the types of transactions involved with the management of a large firm. The context of this analysis is a notorious scandal that has engulfed the bank in recent years, involving [14]: (1) opening several million unauthorized retail customer accounts; (2) charging hundreds of thousands of customers for unauthorized protection insurance for their auto loans; and (3) firing thousands of employees for improper sales practices. As a result of these transgressions, the U.S. Federal Reserve Board imposed strong sanctions and, as well, identified a series of performance failures by the bank's Board of Directors that needed to be redressed. Recommendations regarding the latter embody an ordered set of transactions involving various modes of social tracking: (1) communication of corporate values to all company employees; (2) auditing of system operations through SIs with relevant company personnel; (3) candid conversations among key personnel about integrity issues; (4) enterprise-wide review of risk analysis and mitigation policies and procedures; (5) SIs with employees, customers, and suppliers regarding company values and their application; (6) analysis of disputes and litigation confronting the company; (7) review of the impact of company culture and decision-making on operational performance; (8) in-depth analysis of the signals that the board conveys to company employees through its actions; (9) review of the nature and validity of performance reviews at all levels; and (10) promotion of a company culture of integrity and honesty. The common theme spanning all of these recommendations is that of feedforward control for continuous improvement. Presumably, some or all of these transactional strategies also may apply to other large firms.

3.3 Time Marches on — the Nature of the Firm Beyond Coase

In the years since Coase published his seminal report [1], a number of new transactional challenges for firms have emerged that did not exist or were of minor importance 80 years ago. Selected examples include: (1) automation of work (based on computer-mediated transactions) has been implicated both in improving or reducing productivity [18] as well as with the degradation of the quality of work [19]; (2) the internet enables computer-mediated transactions worldwide, yet because of cyberattacks (a malignant form of such transactions), Samuelson [20] ranks the internet as the greatest threat to the U.S. economy, and therefore the viability of U.S. firms; (3) one factor associated with the drop in U.S. productivity growth observed over the past decade is worker distraction attributable to interaction with social media, another popular mode of computer-mediated transaction [21]; (4) the 2010 Supreme Court's Citizens United decision opened the gates for virtually unrestricted corporate political campaign giving, an heretofore entire ecosystem of firm transactions that has attracted both intense support as well as controversy [22]; and (5) the most prominent example of a new challenge for firms is the emergence of a globalized economy, that for firms in most countries of the world (and certainly in all developed ones) requires understanding of state-dependent language, culture, regulation, and political differences for purposes of engaging in cross-border transactions, likely associated with increased cost per transaction compared to within-state transactions. New hierarchies of transactions are also emerging that undermine control by management over the transactions that define a firm. It is the associated sociotechnical systems challenges that can define the role of macro-ergonomics interventions going forward.

4 Implications

What does the foregoing analysis contribute to the original insights of Coase regarding the nature of the firm and its health and survival? One answer is suggested by Coase himself—as summarized above (Sect. 3.2), he notes that as the number of transactions increase as the size of a firm increases, both costs and entrepreneur errors tend to decrease. However, he also notes that, "as a firm gets larger…the costs of organizing additional transactions within the firm may rise," a relationship termed by economists, "diminishing returns to management" [1, pp. 394–395]. The interpretation of this pattern from a social cybernetic perspective is that social interactions during a given transaction impose inherently demanding behavioral control challenges (Fig. 1), especially if multi-person teams are involved (Sect. 2). However, the diminishing returns relationship suggests an amplification of difficulty in managing the behavioral demands of a growing number of complex behavioral transactions, thereby increasing the cost per transaction and reducing the economic viability of the firm.

This hypothesis underscores another putative insight of the analysis offered in this report, namely the need for further research to more carefully evaluate the behavioral parameters and dynamics of what is termed a "transaction" (a question not addressed by Coase). In terms of the cost per transaction: (1) Do different types of transactions exhibit different behavioral control demands? (2) Which modes of exchange of sensory feedback and of sensory feedback control are more effective for a given transaction? (3) To what extent do answers to these questions vary with multi-person versus two-person transactions? (4) How do possible behavioral differences in age, gender, religion, and cultural and political backgrounds influence the effectiveness of a given transaction? and (5) How can application of sociotechnical systems theory and macro-ergonomics principles contribute to the effectiveness of firms with the growing influence of information technology? In other words, the social cybernetic perspective offered here opens an entirely new domain of analysis regarding the role and the costs of transactions in defining the nature of the firm, and a means of planning interventions to address emerging challenges.

Additionally, demographic trends in the U.S. and other developed countries towards a more aging workforce have compelled greater attention among firms to engaging and satisfying workers. Two transactional strategies dependent upon social tracking between management and employees that are recommended to achieve this goal include: (1) promoting a sense of shared leadership [15] and decision-making [16]; and (2) nurturing a company culture that fosters respect for employees [17].

Finally, this analysis suggests that the customary view of macro-ergonomics —'concerned with improving productivity and the quality of work life by an integration of psychosocial, cultural, and technological factors with human-machine performance interface factors in the design of jobs, workstations, organizations, and related management systems' [23]—needs to directly address the specific transactional underpinnings of the firm itself, including modes of social tracking along with their associated feedback and feedforward control relationships. This analysis also brings a social behavioral perspective to the domain of market forces that—as Coase points out—have traditionally been assumed to represent the primary influence on the nature of the firm.

References

1. Coase RH (1937) The nature of the firm. Economica 4(16):386–405
2. Coase RH (1988) The firm the market and the law. University of Chicago Press, Chicago
3. Wikipedia. https://en.wikipedia.org/wiki/Ronald_Coase#The_Nature_of_the_Firm. Accessed 09 Mar 2018
4. Smith KU (1974) Industrial social cybernetics. University of Wisconsin Behavioral Cybernetics Laboratory, Madison
5. Smith TJ, Henning RH, Smith KU (1994) Sources of performance variability. In: Salvendy G, Karwowski W (eds) Design of work and development of personnel in advanced manufacturing, Chap. 11. Wiley, New York, pp 273–330
6. Smith TJ, Henning RA, Smith KU (1995) Performance of hybrid automated systems - a social cybernetic analysis. Int J Hum Factors Manuf 5(1):29–51
7. Smith TJ, Wade MG, Henning R, Fisher T (2015) Variability in human performance. CRC Press, Boca Raton
8. Smith TJ, Smith KU (1987) Feedback-control mechanisms of human behavior. In: Salvendy G (ed) Handbook of human factors. Wiley, New York, pp 251–293
9. Bienefeld N, Grote G (2014) Shared Leadership in multiteam systems. How cockpit and cabin crews lead each other to safety. Hum Factors 56(2):270–286
10. Haims M, Carayon P (1998) Theory and practice for the implementation of "in-house" continuous improvement participatory ergonomics programs. Appl Ergon 29(6):461–472
11. Robertson MM, Henning RA, Warren N, Nobrega S, Dove-Steinkamp M, Tibirica L, Bizarro A (2015) Participatory design of integrated safety and health interventions in the workplace: a case study using the Intervention Design and Analysis Scorecard (IDEAS) tool. Int. J. Hum Factors Ergon 3(3–4):303–326
12. Nobrega S, Kernan L, Plaku-Alakbarova B, Robertson M, Warren N, Henning R (2017) Field tests of a participatory ergonomics toolkit for Total Worker Health. Appl Ergon 60:366–379
13. Schiamberg LB, Smith KU (1982) Human development. Macmillan, New York
14. Stout J (2018) Fed's message to Wells Fargo board: you're responsible. In: Star Tribune, 26 February 2018, p D4
15. Owens E (2018) The keys to satisfied workers. In: Star Tribune, 5 March 2018, p D2
16. Smith TJ (2002) Macroergonomics of hazard management. In: Hendrick HW, Kleiner BM (eds) Macroergonomics. Theory, methods, and applications. Lawrence Erlbaum, Mahwah, pp 199–221
17. Crosby J (2018) Managers must look beyond age to keep workers engaged. In: Star Tribune, 25 February p D3
18. Brynjolfsson E (1993) The productivity paradox of information technology. Commun ACM 36(12):67–77
19. Hancock PA (2014) Automation: how much is too much? Ergonomics 57(3):449–454

20. Samuelson R (2018) What most threatens the economy? You might be surprised. St. Paul Pioneer Press, 1 March, p 12A
21. Smith TJ, Henning R (2017) The productivity paradox – a distracted working hypothesis. In: Presentation to: Work, Stress, and Health 2017, contemporary challenges and opportunities, American psychological association, the 12th international conference on occupational stress and health, Minneapolis, MN, 7–10 June 2017
22. Winkler A (2018) We the corporations: how American businesses won their civil rights. Liveright, New York
23. Human Factors and Ergonomics Macroergonomics Technical Group. http://tg.hfes.org/metg/about.html. Accessed 17 Mar 2018

Chronicle Workshops as Data Collection Method in Evaluation of National Work Environment Intervention

Kirsten Olsen(✉), Mark Lidegaard, and Stephen Legg

Centre for Ergonomics Occupational Safety and Health, School of Health Sciences, Massey University, Palmerston North 4442, New Zealand
k.b.olsen@massey.ac.nz

Abstract. When evaluating national work environment initiatives, it is important to choose methods through which it is possible to gather necessary and relevant information in a time efficient way for researchers and involved organisations. This article evaluates the usefulness of chronicle workshops as a data collection method to help assess the effectiveness of national work environment initiatives aiming to create interventions in organisations. Chronicle Workshops were used as one of three methods in case studies evaluating a national guideline on moving and handling people. Chronicle workshops were found to be an efficient method to identify specific interventions, when they occurred, who had been instrumental in implementing them, what contextual factors had influenced the intervention and factors facilitating and hindering intervention. They lacked specificity on individual strategies and why these did or did not work. Thus Chronicle workshops are good at creating an overview of implementation efforts but need to be supplemented with other methods to gain more detailed information.

Keywords: Participation · Moving and handling patients · Realist analysis

1 Introduction

This article evaluates the usefulness of chronicle workshops as a data collection method to help assess the effectiveness of national work environment initiatives aiming to create interventions in organisations. These initiatives are complex interventions [1] because organisations are influenced by an ever changing environment with multiple initiatives occurring at the same time and further because implementation in an organization is influenced by many factors on many levels with feedback loops changing the organizational context. Often evaluations take place after the national initiative has been launched. This makes it even more difficult to assess the process of implementation, what influenced the implementation and what the outcomes were. To evaluate a national initiative aimed at organizational intervention it is important to evaluate how and if the initiative reaches the target organisations, what makes the organisations decide to initiate interventions, how the intervention is implemented, what influences

S. Bagnara et al. (Eds.): IEA 2018, AISC 821, pp. 286–296, 2019.
https://doi.org/10.1007/978-3-319-96080-7_33

the implementation, to what extent the implementation is completed, what the outcome is, and what has influenced it.

This article only looks at how chronicle workshops, as a method, can help identify the implementation process in an organisation and what influenced this process- both in a positive and negative way.

The chronical workshops were used in a project that evaluated the uptake, use and effect of a national guideline for moving and handling people (MHP): "Moving and handling people: The New Zealand guidelines" [2], launched in 2012. The project consisted of three phases: (1) Evaluation of national injury claims rates and costs related to moving and handling people over an eleven years period [3]; (2) Evaluation of the awareness, uptake and use of the guidelines through a questionnaire survey to intended users [4, 5] and; (3) Evaluation of factors facilitation and hindering implementation of the guidelines in hospitals using a case study approach [6]. The chronicle workshops were used in the case studies in phase three. The aim of the third phase and the case studies was to identify what made the MHP guidelines work or not work for different organisations and under what circumstances.

The article describes what a chronicle workshop is, the purpose and use of the chronicle workshops (organizing, conduction and extracting findings), the outcome of the workshops (what types of data were extracted) and finally includes a reflection on the usefulness of the chronicle workshops in relation to the purpose.

2 Chronicle Workshops

The chronicle workshops were used in a mixed method multiple case study. A large public hospital and a small private hospital were selected as case study organisations. Each case study consisted of in-depth semi-structured interviews, a chronicle workshop and document analysis. Chronicle workshops have previously been used to create the shared history of an organization to visualise and retrieve information about the past, reviewing the history and creating a common explanation and understanding of the organization [7, 8].

2.1 Aim of the Chronicle Workshops

The aims of the chronicle workshops were to establish how the MHP guidelines and MHP related initiatives had been implemented to manage risk factors related to MHP and identify factors that had facilitated or hindered the implementation and impact over the period 2007–2017.

2.2 Organising the Chronical Workshops

The organisations were informed about the methods, including the chronicle workshop and time commitment before they agreed be part of the project. Initially three semi-structured interviews were conducted with a senior manager, a health and safety (H&S) manager and a MHP coordinator. Part of these interviews explored who it would be appropriate to invite to the chronicle workshops. The MHP coordinators contacted and invited prospective participants and informed them about the purpose. A time for the workshops that suited the people that decided to participate was identified. The intention was that the participants would cover different job roles that had influence or had been affected by the implementation, to give a diverse perspective on the implementation process [9]. We aimed to include participants to balance: common reference (same organisation, knowledge about MHP); length of service (including some employed before 2007); position in the organisation (from senior management to frontline people); expertise within MHP (people involved with implementation and people conducting MHP). The participants from the two hospitals are presented in Table 1.

Table 1. Participants at the two chronicle workshops.

Participants: Private hospital	Tenure (years)	Participants: Public hospital	Tenure (years)
Hospital general manager	2	Clinical nurse educator	15
Theatre manager	14	Physiotherapist in a ward	4
Quality development manager	10	Physiotherapist in community service	10
Contracted radiographer	10	Moving and handling advisor	3
Health and safety representative (theatre staff)	25	Safe handling representative (emergency)	7
Health and safety representative (administrative staff)	10	Safe handling representative (ward)	7
		Health and safety and safe handling representative (Neonatal)	10
		Nurse employee representative	20+

The MHP coordinators organized a suitable room with a large plain wall big enough to indicate a ten-year period with A4 sheets of paper. Each workshop lasted 4 h. Coffee, tea and food were provided.

2.3 Conducting the Chronicle Workshop

All participants sat in a half-moon facing towards the wall (see Fig. 1). The chronicle workshops were divided into two main phases: Exploration and Interpretation.

The exploration phase consisted of three sessions with separate topics: (i) What significant events have marked MHP as a priority at the hospital, and when?; (ii) Which stakeholder, entities or institutions have characterised and driven the development and implementation of MHP programmes at the hospital, and when?; (iii) What kind of initiatives and debates have arisen during the development and implementation of the MHP programme at the hospital, and when?

The facilitator presented the topic and invited clarifying questions at the beginning at each session. The participants wrote their personal inputs on 'sticky' notes, one issue per note. The notes for each session had different colures. The participants could write on as many notes as they needed within the time of five to ten minutes. The participants

Fig. 1. Introduction to the chronicle workshop. The setup of the venue.

then took turns at placing their note(s) on the wall and briefly explained what the note was about. At the end of the session, they had the opportunity to ask clarifying questions. Additional notes could also be added to the wall.

The interpretation phase consisted of two sessions: (i) plenum session,- interpretation of key trends in the collective history of MHP at the hospital, - dividing the history into chapters; (ii) group session, - identification of factors that facilitated or hindered the process of implementing MHP initiatives.

In the plenum session, the participants identified trends in MHP for different time periods, and created headings reflecting the trends for the periods. These heading were placed on the wall. A plenary discussion of the headings resulted in agreed headings for each time period. In the group session, the participants were divided into groups of people with similar background and experience. The groups analysed and interpreted the history, identified factors that facilitated or hindered the process of implementing MHP initiatives related to the different events. They wrote notes for each factor that they identified. The groups placed the notes on the wall and described each factor.

At the end of the workshop, the participants reflected on the workshop and contributed with any additional comments. Table 2 outlines the structure of the workshop.

Table 2. Structure of the chronicle workshops

'Moving and handling people safely -reviewing the moving and handling people effort/programmes from 2007–2017'
Introduction (15 min)
What significant events have marked MHP as a priority at the hospital, and when? (40 min)
Which stakeholder, entities or institutions have characterised and driven the development and implementation of MHP efforts/programmes at the hospital, and when? (40 min)
What kind of initiatives and debate have arisen during the development and implementation of the MHP programme at the hospital, and when? (40 min)
Interpretation of key trends in the history of MHP at the hospital. Dividing the history into chapters (25 min)
Reflection on factors that supported or hindered the process of implementing the MHP programme (50 min)
Evaluation and closure of the workshop (10 min)

Three researchers facilitated the workshop. One led the process. One took written notes and photographed the wall with the timeline. One helped by facilitating the process and in identifying themes when they emerged from discussions in plenum.

The outcome of the chronicle workshops were timelines on the wall, with headings and notes in different colours showing significant events (yellow notes), stakeholders driving development and implementation (blue notes), initiatives and debates during the development (orange notes), factors facilitating implementation (green notes) and factors hindering implementation (red notes) (see Fig. 2). To be able to document the chronicle workshops and their outcomes, the sessions were voice-recorded, notes were taken and the wall was photographed after each session.

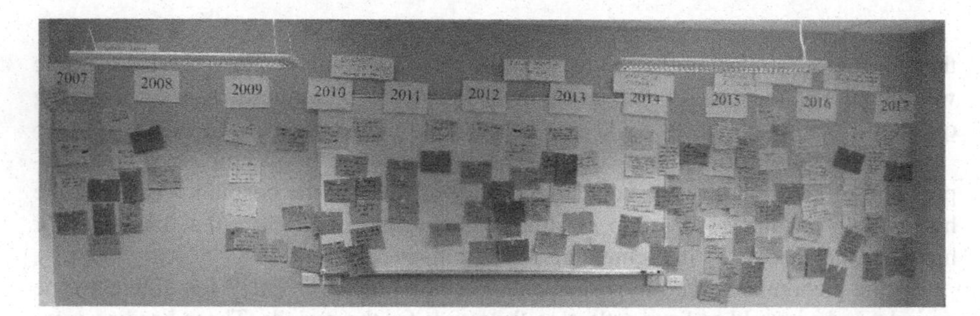

Fig. 2. The final timeline at the end of the chronicle workshop (Color figure online)

2.4 Extracting Findings – Writing the Story

The photographs of the timeline on the wall (Fig. 2) were transferred to a digital picture in Prezi. The written notes were transcribed. One researcher listened to the voice recordings, consulted the notes and digital picture and wrote the story chapter by chapter. Each chapter contained a narrative and a table summarizing significant events, stakeholders, initiatives and debates, and factors facilitating and hindering implementation of MHP. The participants were sent a draft of the story and had the opportunity to comment on it. The story was then revised. Thus, each story reflected the joint experience of the workshop participants.

3 Outcomes of the Chronicle Workshops

Outcomes of the workshops are presented in two separate sections. Section 3.1 presents results from the workshop at the private hospital, and Sect. 3.2 from the public hospital. Story chapters, significant events and related stakeholders, are presented in Tables 3 and 4. Initiatives and debates, and factors facilitating and hindering MHP implementation are illustrated by short descriptions of selected events following each table.

Table 3. Chapters of the private hospital's story about implementation of a moving and handling people programme 2007–2017, showing important events and influential stakeholders

Events	Stakeholders
2007–2009: Getting together - Period leading up to the merge	
Introduction of National MHP Programme	National H&S manager
Upgrading MHP equipment	National office; Senior management; H&S representatives (reps.)
Hospital rebuild	National office; CEO of the private hospital
Merge of two private hospitals	National office; the two CEOs
Updating local policy	H&S reps.
Cultural change	H&S reps.
2010: New team	
Creating consensus	The private hospital
2010–2012: Consolidation	
Policies- change in focus	Quality development and H&S manager
Improved organisational structure	Private hospital; H&S reps.
Continuous improvement of equipment	Private hospital; H&S reps.
2012: Wake up	
Organisational audit	Private hospital, Senior management team
2012–2015: Activation - Consolidating MHP standardisation	
Appointment of MHP coordinator	Private hospital, Senior management team
Sharing knowledge and specialist roles	National office; Senior management; H&S reps.
Reporting injuries	National office; H&S reps, MHP coordinator
Communication with accident compensation corporation	National office; Accident compensation corporation
2015–2017: Evolution	
Development of the organisation	Private hospital, CEO
Consolidation of policies and procedures	Private hospital, CEO; H&S reps.
New H&S legislation	National office & H&S manager, Private hospital
Increased openness about injuries	H&S reps
Keep focus on MHP	H&S reps, MHP coordinator

3.1 Outcomes of the Chronicle Workshop at the Private Hospital

Table 3 identifies the story chapters, events and stakeholders for the private hospital.

Debates and Initiatives, Facilitating and Hindering Factors

One event (2007–09) that influenced the private hospitals implementation of MHP programme was the merge with another local hospital that was not part of the national chain. The influential stakeholders were the national office and the two CEOs. This event was related to "updating local policy" and "culture change". For these two events, H&S representatives (reps.) were identified as influential stakeholders. Discussions about which H&S and MHP policies were relevant in the new hospital were

Table 4. Chapters of the public hospital's story about implementation of a moving and handling people programme 2007–2017, showing important events and influential stakeholders

Events	Stakeholders
2007–10: Change in focus	
Employment of MHP advisor	Public hospital
Ceiling hoists and Electrical beds	Public hospital
Changed 'On-bed transfer'	MHP advisor
Increased use of sliding sheets	MHP advisor
On ward training profession specific, Train the trainers & Ward trainers	MHP advisor; Safe handling reps., MHP/ward trainers
Triannual trainer meetings	MHP trainers
More area trainers on external courses	MHP advisor;
Introduce 'Competency sign-off'	MHP advisor; MHP trainers
Weekly staff meeting	Staff, management
2010–12: Education & accountability	
Study days for safe handling reps	MHP advisor; Safe handling reps
Nurse educators on the wards	Nurse educators; Safe handling reps
Online training - videos and E-learning	ACC; MHP advisor
Ward specific MHP programme	Safe handling reps.
H&S reps for individual work areas	Hospital board; MHP advisor; H&S reps.
H&S at monthly ward meetings	H&S manager; Safe handling reps; Central nurse manager
2012–14: Public reporting & discussion – Effect of a mine disaster	
Increased obesity & injuries - a public subject	The media; District health boards
More MHP information and guidance	MHP advisor; Safe handling reps, ACC
Recognition of impact of safe handling reps.	MHP advisor
Biannually refresher training reduced	MHP advisor
Safe handling reps at each area	Safe handling reps.
Electronic incident reporting system	Hospital board; Safe handling reps; staff
Incidents reviewed by manager, safe handling reps, and H&S department	H&S department; Clinical nurse manager; Safe handling reps.
2015: New H&S legislation and a bariatric focus	
Implementation of 'Bariatric bundle'	Public hospital; Assistant director of nursing; MHP advisor
Appointment of new MHP advisor	MHP advisor; Safe handling reps.
Introduction of competency checks	Operation officer; H&S manager; Charge nurse
2016–2017: A more holistic model	
New operation officer and H&S manager	Public hospital's board
Forming a MHP board	H&S manager; MHP advisor
Letter from OT in MAPU	Equipment advisory board, MAPU OT
Hover-matts	MHP advisor
Sliding sheets at bed heads	MHP-advisor
All new clinical staff introduced to MHP	Safe handling reps.

identified as debates related to the merge and "merge of two cultures" was identified as a facilitating factor. No hindering factors were identified. A debate identified in relation to "updating local policy" was an increased awareness of new needs related to MHP. "Change in design of the facility" (which was also mentioned as an event in Table 3: "Hospital rebuild") was identified as a facilitating factor. An initiative and debate related to "Culture change" was that staff and particularly H&S reps. focused on getting the best out of the two different cultures and changing the 'them and us' terminology. An increased emphasis on H&S reps.' work was identified as a facilitating factor. Poor communication and resistance to change were identified as hindering factors.

An important event (in the period 2012–15) was the 'appointment of MHP coordinator', where the influential stakeholders were the private hospital and the senior management team. No initiatives and debates were related to this event and no hindrance were identified, but two facilitators were identified; that the senior management team was involved and the small size of the organisation. It was not described how this event influenced MHP.

The National Office and the national H&S manager were identified as key stakeholders initiating changes following the implementation of the new health and safety legislation (2015–17). The national H&S manager organised workshops about the legislation, which informed the private hospital to assess where they did not comply with the legislation. The Management team for the Private hospital gave H&S reps. the responsibility to find solutions to the gaps that were identified. The H&S reps. were identified as key stakeholders in relation to "increased openness about injuries" (2015–17). This was related to debates about Managers' responsibility for H&S triggered by the new H&S legislation. The openness was facilitated by an increase in injuries resulting from MHP and a steadily aging workforce. The H&S reps pushed for a more systematic discussion with the senior management team about injury trends, resulting in increased awareness of MHP risks and potential issues being addressed more rapidly.

3.2 Outcomes of the Chronicle Workshop at the Public Hospital

Table 4 identifies the story chapters, events and stakeholders for the public hospital. Only the most distinct events are included because too many events were identified to present here. Some of the omitted events did not have identified related initiatives and debates or facilitating and hindering factors.

Debates and Initiatives, Facilitating and Hindering Factors

Four events: "On-ward training profession specific"; Train the trainers & ward trainers"; "Triannual trainer meetings"; "More area trainers on external courses" and; "Introduce 'Competency sign-off'" (2007–10), were driven by the MHP coordinator and supported by safe handling reps and MHP trainers. These events were related to the same initiatives and debates and were influenced by the same facilitating and hindering factors. The events focused on improving the MHP training and making it relevant to the attendees and their work, which made the training more, professional and increased staff attendance. The events were influenced by debates about who should drive safe

MHP, if staff should lead the change and the training needs of staff. Factors that facilitated the events were a general support from middle management (which was higher on wards where managers previously had worked as clinicians) and passionate staff, educators and trainers. Heavy staff workload and insufficient time and resources for training were identified as hindering factors, negatively influencing the events and debates.

The event "Weekly staff meeting" (2007–10) is different. It was the forum for debates about the trade-off between time and staff allocation and safety, and whether staff and patient safety could go hand-in-hand. These debates were influenced negatively by the work injury insurance experience rating scheme, heavy workload, insufficient time for training, poor staff attitude and buy-in to safe MHP practice, lack of vision from senior management, and that the health care sector focused on patient safety.

The events: "Letter from OTs"; "Hover-matts"; "Sliding sheets at bed heads" (2007–10) were influenced by the same debates and hindering factors. The debates were about how much equipment was needed to secure safe MHP. This debate had been present at the hospital earlier, just not mentioned earlier. A particular hindrance was a change in procurement policy that transferred the authority for procurement from the charge nurse to a central equipment advisory board. The board required a strong rationale to accompany procurement requests for it to be approved. This was a result of reduced budget and tighter regulation from the Ministry of Health. The reason they managed to introduce new equipment was that the MHP advisor was persistent and successfully managed to get approval for equipment purchase.

4 Discussion

Both chronicle workshops identified external (e.g. the new H&S legislation) and internal events (e.g. hospital rebuild) that had influenced implementation of the hospitals MHP programmes and identified contextual factors at national, industry and organisational levels that had influenced the implementation. They also identified stakeholders and influential people, but did not describe these influential people's strategies to overcome the barriers that were identified.

For example, the chronicle workshop at the private hospital identified resistance to change and the MHP advisor as an influential stakeholder. During the interviews with a Senior manager, the National H&S manager and the MHP advisor, it was described how the MHP coordinator had been influential in overcoming resistance to change and the strategy she had employed. She engaged with two of the senior managers and got them on board to support MHP. She changed staff's attitude and behaviour towards safer MHP by taking a team approach and working with staff on the floor.

The workshops did not describe in any way how the MHP guidelines had been used. Rather, this was identified through interviews with MHP advisors and H&S managers. These showed that the MHP guidelines were used to check the programme elements that the hospitals had already implemented and to support the H&S managers or MHP advisors involvement in renovation projects and new facilities. The H&S manager at the public hospital also used them to support suggested changes to MHP training.

The workshops revealed that some professions, and some individual ward managers acted as barriers for implementation. Others might have functioned as barriers but only people not participating in the workshop were mentioned as barriers.

The chronicle workshops were organised by the MHP advisors and it was difficult to get enough participants to participate. It was also difficult to ensure that all relevant types of stakeholders were present. It was easier at the smaller private hospital than for the larger public hospital. The participants at the public hospital were mostly staff rather than managers involved in the MHP programme as trainers, H&S reps or safe handling reps. It was not possible to get participants from procurement or facilities management. This was partly due to the time commitment the chronicle workshops required.

5 Conclusion

Chronicle workshops used in the present project were a good method to identify interventions and when they occurred, and who had been instrumental in the implementation. They provided good information on contextual factors, important events and factors that hindered and facilitated the implementation. However, they did not provide enough details about individual strategies and why they worked or did not work. Thus in conclusion, chronicle workshop is a good method to gain an overview of implementation effort but details need to be collected via other methods to inform how national programmes can help implementation in organisations.

References

1. Rogers PJ (2008) Using programme theory to evaluate complicated and complex aspects of interventions. Evaluation 14(1):29–48
2. Accident Compensation Corporation (2012) Moving and handling people: the New Zealand guidelines. Accident Compensation Corporation, Wellington
3. Olsen KB, Lidegaard M, Legg S (2017) Assessment of the uptake and impact of the ACC New Zealand moving and handling people guidelines (2012) - Stage 3 report - Trends in injury claims and claims cost related to moving and handling people 2005–2016. Palmerston North
4. Olsen KB, Lidegaard M, Legg S (2016) Assessment of the uptake and impact of the ACC New Zealand moving and handling people guidelines (2012). Stage 2 uptake and use - Part A: Descriptive analysis of questionnaire findings. Massey University
5. Olsen KB, Lidegaard M, Legg S (2017) Assessment of the uptake and impact of the ACC New Zealand moving and handling people guidelines (2012). Stage 2 uptake and use - Part B: Analysis of questionnaire findings stratified by role in relation to moving and handling people and by sub-sector in health care. Massey University
6. Olsen KB, Lidegaard M, Legg S (2017) Assessment of the uptake and impact of the ACC New Zealand moving and handling people guidelines (2012). Stage 4 factors facilitating and hindering implementation and impact of the MHPG and MHP programme elements: case study report. Massey University

7. Gensby U (2014) Assessing the present in perspective of the past: experiences from a chronicle workshop on company-level work disability management. Nord J Work Life Stud 4(2):85–115
8. Hansen AM, Pedersen MH (2014) Vidensproduktion, positionering og magt i historieværksteder. Tidskrift for arbejdsliv 16(3):23–37
9. Limborg HJ, Hvenegaard H (2011) The chronicle workshop. In: Rasmussen LB (ed) Facilitating change. Polyteknink Forlag, Lyngby

Current Status of Exercise Habits and Job Satisfaction of Nurses in Japan

Takumi Iwaasa[1(✉)], Motoki Mizuno[1,2], and Yuki Mizuno[3]

[1] Graduate School of Health and Sports Science, Juntendo University, Chiba, Japan
tiwaasa@gmail.com
[2] Faculty of Health and Sports Science, Juntendo University, Chiba, Japan
[3] Tokyo College of Transport Studies, Tokyo, Japan

Abstract. Nurses' duties have become more diversified and increasingly complex with Japan's ultra-aging population and advancements in medical technology. Additionally, the long working hours, the enormous workload, the chronic shortage of human resources, and night shifts place a huge physical burden on nursing staff. These factors have contributed to a job turnover rate of around 11% for nurses in Japan. The rate has been around the same for the last decade. In a bid to retain highly skilled and talented employees for the long-term and to enable them to demonstrate and build on their skills, there is a need to improve job satisfaction. This study aimed to ascertain the exercise habits of nursing staff and determine how their exercise habits relate to their overall job satisfaction.

Methods: A paper-based survey was conducted at a university hospital located in the Tokyo metropolitan area. The survey consisted of a face sheet and questions about respondents' job satisfaction and exercise habits.

Results: Of the total 659 respondents, 304 nurses responded that they had exercised more than once in the last year. There were 40 respondents who had exercised between 1 and 2 times in the last year, 62 respondents said between 6 and 11 times in the last year, and 16 respondents said more than twice a week, with only 5% of respondents saying that they had a regular exercise routine in place. When comparing the average job satisfaction score with exercise frequency, the group that exercised once a week had the highest job satisfaction score of 3.41(SD = 0.37). It was significantly higher than the job satisfaction score (3.19, SD = 0.41) for the group that did not exercise ($p < .05$).

Keywords: Exercise habit · Job satisfaction · Japanese nurse

1 Introduction

Nurses' duties have become more diversified and increasingly complex with Japan's ultra-aging population and advancements in medical technology. Additionally, as nurses are responsible for a huge variety of duties including taking care of patients' nursing care and daily recuperation, great expectations are placed on them as key members of the medical team by patients, doctors, and other medical staff. When examining the workplace, it is possible to see that there have been moves to improve

S. Bagnara et al. (Eds.): IEA 2018, AISC 821, pp. 297–303, 2019.
https://doi.org/10.1007/978-3-319-96080-7_34

the working environment for nurses such as putting a support system in place for childcare leave. However, the long working hours, the enormous workload, the chronic shortage of human resources, and night shifts place a huge physical burden on nursing staff. These factors have contributed to a job turnover rate of around 11% for nurses in Japan (Fig. 1). The rate has been around the same for the last decade, but is now climbing [1]. There is a need for retention management, in order to secure and retain skilled and experienced nursing staff [2]. In a bid to retain highly skilled and talented employees for the long-term and to enable them to demonstrate and build on their skills, there is a need to promote diverse working styles, improve the work/life balance (WLB), and enhance welfare benefits in order to improve job satisfaction.

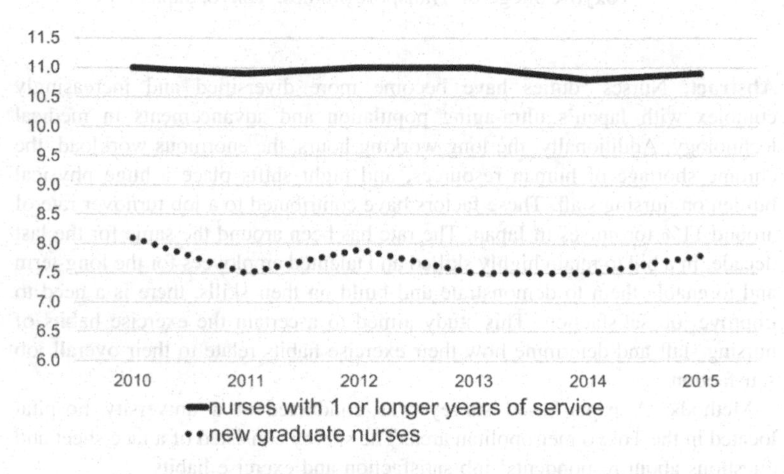

Fig. 1. Turnover rate of nurses in Japan

Health management is stealing the spotlight in corporate organizations. This means that employees' health management is viewed from a corporate management perspective, and companies are taking strategic measures to maintain and promote the health and well-being of their employees [3]. There are also many studies on the link between employees' exercise habits and their job performance. For instance, previous studies show that regular exercise is an important factor in feeling satisfied with one's work/life balance, and it is also seen to be beneficial for one's mental health [4]. However, there is very little research into the connection between the exercise habits and job satisfaction of nurses. This study aims to ascertain the exercise habits of nursing staff and determine how their exercise habits relate to their overall job satisfaction.

2 Methods

A paper-based survey was conducted at a university hospital located in the Tokyo metropolitan area. The survey consisted of a face sheet and questions about respondents' job satisfaction and exercise habits. The survey was left with respondents to complete between January 13–17, 2017. It was distributed to 692 registered nurses (excluding the directors of nursing departments) employed on a permanent basis at a university hospital.

2.1 Job Satisfaction

Interviews were conducted and 25 satisfactory and unsatisfactory factors were identified at the target hospital. From there, through discussions with nursing staff and administration specialists, 61 items related to job satisfaction were identified [5].

2.2 Exercise Habits

The survey asked respondents about exercise and sports performed in the past year, how frequently they engaged in exercise, who they exercised with, what type of exercise they did, how long (minutes) each exercise session lasted for, and the intensity of the exercise. In response to the question about who respondents exercised with, the survey asked if they exercised "alone", "with family", "with coworkers or friends", "with pets", or "with others". Regarding intensity, respondents were asked to identify if their exercise was of "high intensity", "moderate intensity", "low intensity", or "non-intensive".

3 Results

3.1 Job Satisfaction Results

A high rate of job satisfaction was identified with items such as communication with colleagues, a sense of pride in being a nurse, enjoyment from contact with patients, instructions from superiors, the frequency of nightshifts, working with superiors and colleagues who can be respected, one's salary, the atmosphere of the workplace and a culture were colleagues help each other. However, a low rate of job satisfaction was identified with factors such as trust from doctors, their own vital role, appropriate number of nursing staff, the clinical ladder, work outside of nursing duties, break rooms and hospital facilities, break times, and the amount of overtime.

3.2 Rate and Frequency of Exercise

Of the total 659 respondents, 304 nurses responded that they had exercised more than once in the last year. There were 40 respondents who had exercised between 1 and 2 times in the last year, 62 respondents said between 6 and 11 times in the last year, and 16 respondents said more than twice a week, with only 5% of respondents saying that they had a regular exercise routine in place.

Looking at the rate and breakdown of respondents' ages, the highest rate was seen in those in their 40 s with 34 of the 68 respondents engaging in some form of exercise, and the lowest rate was 30-something respondents of which 62 of the 153 said that they engaged in exercise (Fig. 2).

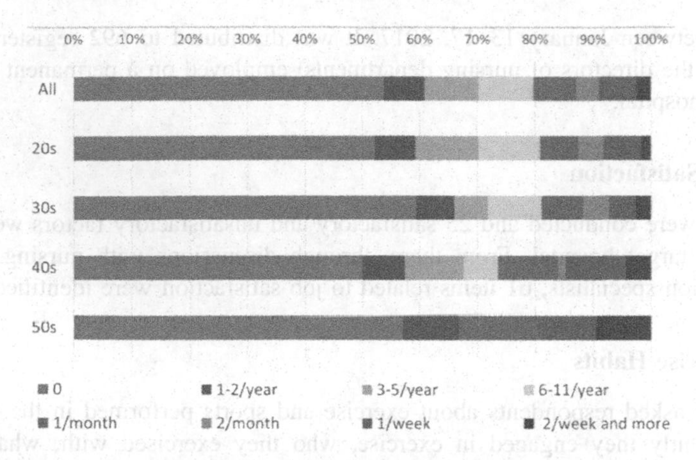

Fig. 2. Exercise frequency by age group

In terms of job titles, there was no discrepancy found among the rate of exercise by general nursing staff, charge nurses, and head nurses, however there was a difference in the frequency of exercise with general nursing staff exercising the least and head nurses exercising the most (Fig. 3).

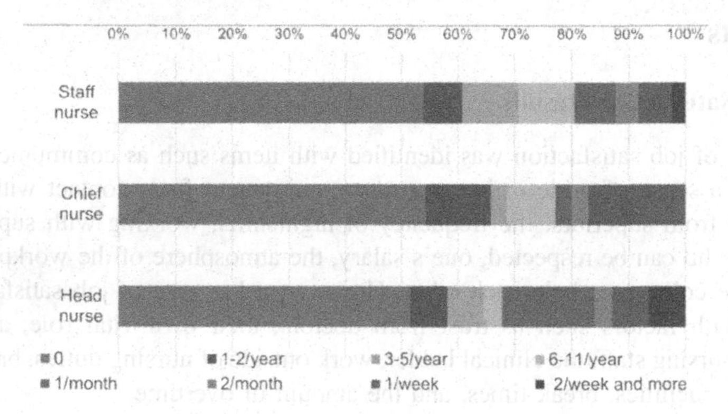

Fig. 3. Exercise frequency by position

3.3 Type of Exercise

The most common type of exercise was jogging with a 15% response rate, followed by swimming (7.8%), and walking (7.3%).

3.4 Exercise Companions

The most common response when asked who they exercise with was "alone", followed by "non-work friends" and then "work-friends".

3.5 Exercise Rate

When comparing the average job satisfaction score with exercise frequency, the group that exercised once a week had the highest job satisfaction score of 3.41($SD = 0.37$). It was significantly higher than the job satisfaction score (3.19, $SD = 0.41$) for the group that did not exercise (Fig. 4).

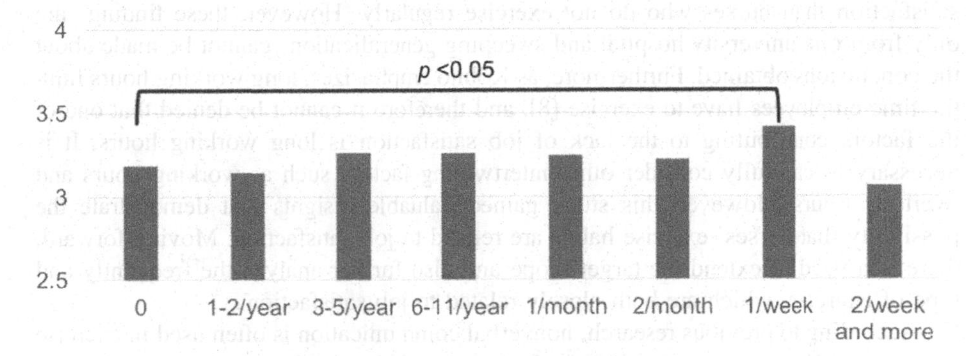

Fig. 4. Comparison of job satisfaction by exercise frequency

4 Discussion

4.1 Job Satisfaction Results

Regarding the types of exercises, the results of this survey showed a higher exercise tolerance compared to the results of the Public Survey on the Status of Sports Implementation by the Japan Sports Agency. This may be a reflection of the fact that the average age of the subjects of this survey was relatively young (under 30 years old).

4.2 Exercise Companions

There were few respondents who said that they exercised with family members, and this is a reflection on the fact that 455 (69%) of the survey respondents are not married, and 385 respondents (58.4%) live alone.

4.3 Exercise Rate

The overall exercise rate was 46.1% (304 respondents). In the national survey conducted by the Japan Sports Agency, the exercise rate was 34.5% for those in their twenties (27.8% for females and 40.8% for males). For 30-somethings, the exercise rate was 32.5% (27.7% for females, 37.2% for men). However, in this independent survey the exercise rate for those in their 20 s was 47.7% (47.2% for females and 55.6% for males), and 40.5% for those in their 30 s (41.0% for females and 35.7% for males). Compared to the national survey findings, these results are slightly lower for 30-something males and significantly higher for males and females in their 20 s and females in their 30 s.

The scope of the study would need to be expanded further in the future in order to determine if this tendency is characteristic to nurses or not.

5 Conclusion

The results of this study indicate that nurses who exercise once a week have higher job satisfaction than nurses who do not exercise regularly. However, these findings are only from one university hospital and sweeping generalizations cannot be made about the conclusions obtained. Furthermore, as Kohno emphasizes, long working hours limit the time employees have to exercise [8], and therefore it cannot be denied that one of the factors contributing to the lack of job satisfaction is long working hours. It is necessary to carefully consider other intertwining factors such as working hours and overtime hours. However, this study gained valuable insights that demonstrate the possibility that nurses' exercise habits are related to job satisfaction. Moving forward, there is a need to extend the target scope and also further analyze the frequently and type of exercise, which are both closely related to job satisfaction.

According to previous research, nonverbal communication is often used in exercise scenes, and this promotes a shift from formal communication in the workplace to informal communication, therefore not only does exercise in the workplace promote improved physical well-being and relaxation, but it has also been shown to promote sound mental health by encouraging communication [9]. Furthermore, to date there have been many reports that demonstrate that aerobic exercise such as running and cycling are effective for improving one's mood and reducing stress [10].

It is hoped that further research will be conducted into the possibility of enhancing job satisfaction by promoting regular exercise among nurses, through initiatives that encourage nurses to find more opportunities to exercise as well as find exercise companions.

References

1. Japanese Nursing Association. http://www.nurse.or.jp/up_pdf/20170404155837_f.pdf. Accessed 25 Nov 2017
2. Mizuno M, Iwaasa T, Yamada Y (2015) Current tendency of retention management of nurses at University Hospital in Japan. In: Final program 8th international conference on applied human factors and ergonomics
3. Ministry of Economy, Trade and Industry. http://www.meti.go.jp/policy/mono_info_service/healthcare/kenko_meigara.html. Accessed 25 Nov 2017
4. Motojima M, Togashi-Arakawa C, Doi T, Sakai T (2017) Satisfaction levels and factors influencing work-life balance among married female nurses. J Jpn Health Med Assoc 26(1):7–16
5. Iwaasa T, Yamada Y, Mizuno M (2017) Relationship between diversity faultlines and turnover intentions of nurses in Japan. In: International conference on applied human factors and ergonomics. Springer, Cham, pp 468–473
6. Japan Sports Agency. http://www.mext.go.jp/sports/b_menu/toukei/chousa04/sports/1381922.htm. Accessed 25 Nov 2017

7. Japan Sports Agency. http://www.mext.go.jp/sports/b_menu/houdou/29/02/__icsFiles/afieldfile/2017/02/15/1382064_001.pdf. Accessed 25 Nov 2017)
8. Kohono T (2009) Roudokosokujikan ga Undosyukan ni Ataeru Eikyo Nitsuite.In FRI Research Report No. 246. Fujitsu Research Institute, pp 1–16
9. Kawamata K (2016) The role of physical therapy in occupational health. J Jpn Phys Ther Assoc 43(Suppl 1):9–12
10. Nakahara-Gondoh Y, Nagamatsu T (2015) The fundamental study of the effect of stretching on mood and stress in female workers. Bull Phys Fitness Res Inst 113:15–18

Ergonomics Integrated into Degree Program in Health Promotion in the University of Eastern Finland

Susanna Järvelin-Pasanen(✉) and Kimmo Räsänen

Faculty of Health Sciences, Institute of Public Health and Clinical Nutrition, University of Eastern Finland, P.O. Box 1627, 70211 Kuopio, Finland
susanna.jarvelin@uef.fi

Abstract. The purpose of this paper is to describe studies in Ergonomics integrated into degree program in Health Promotion in the University of Eastern Finland (until 2010 University of Kuopio), where Ergonomics has been major subject since 1990. The studies in Human Factors and Ergonomics (HFE) have been a part of the degree program in Health Promotion since 2016, including Bachelor in Health Sciences and Master in Health Sciences. The duration of the qualification is five years (180 + 120 ECTS). The objective of the degree program is to educate specialists in Ergonomics, who can design work and work environments based on scientific knowledge. The content of the education corresponds to the qualification of the European Ergonomists set by Centre for Registration of European Ergonomist (CREE). The students are mainly adult learners living in different areas of Finland. Integrating their thesis into the ongoing development projects in working life provides a meaningful way of learning. In addition, our degree programme utilizes web based learning varying from tutored courses and blended learning to self-study and case-based courses on central topics in HFE. Web-based learning is particularly suitable in Finland due to long distances. The graduates in Ergonomics have commonly been employed in many occupations where they can utilize the knowledge from their studies. The field of Ergonomics in working life is broad, and it is impossible for one educational organization to offer comprehensive education, but in collaboration with different stakeholders we believe we have a good chance to develop the discipline in Finland.

Keywords: Education · Bachelor's degree · Master's degree · Ergonomics

1 Background

One aim in the strategy for Human Factors and Ergonomics (HFE) for the future at the University of Eastern Finland (until 2010 University of Kuopio) is to promote the education of specialists in Ergonomics since this will strengthen the application of high-quality HFE. This aim is aligned with the strategy for Human Factors and Ergonomics for the future [1].

S. Bagnara et al. (Eds.): IEA 2018, AISC 821, pp. 304–309, 2019.
https://doi.org/10.1007/978-3-319-96080-7_35

The University of Eastern Finland is the only university where Ergonomics can be studied as a major subject in Finland. Studies in Ergonomics have been available since 1990. A total of 101 students have graduated master's degree, and 13 have completed doctoral degree in the discipline of Ergonomics. The studies in Ergonomics have been a part of the degree program in Health Promotion since 2016.

The degree program in Health Promotion majoring in Ergonomics includes bachelor's and master's studies (i.e., Bachelor in Health Sciences and Master in Health Sciences). The Bachelor's degree consists a total of 180 ECTS (European Credit Transfer System) and the master's degree 120 ECTS. The duration of the qualification is five years (i.e., three years in Bachelor's degree and two years in Master's degree).

The purpose of this paper is to describe studies in Ergonomics integrated into degree program in Health Promotion in the University of Eastern Finland.

2 The Objectives of Studies in Ergonomics

The objective of the degree program in Health Promotion majoring in Ergonomics is to educate specialists in Ergonomics, who can design work and work environments based on scientific knowledge. The specific objectives in Bachelor's degree in Ergonomics is that graduates are able to:

- know the principles of interaction between work and human,
- understand the basics in physiology, psychology and sociology related to the work,
- recognize the development needs in work, working conditions and the work community,
- know the basics research, development and design methods of Ergonomics,
- understand the relationship between ergonomics, work ability and well-being at work,
- know the legislation regulating working life in Finland and in the European Union, and
- follow up the development of the Ergonomics as a discipline in Finland.

Furthermore, the specific objectives in Master's degree in Ergonomics is that graduates are able to:

- examine, analyze and develop work, working conditions and work communities,
- apply scientific knowledge and methods to promote work ability and well-being at work, and
- follow up the development of the Ergonomics as a discipline at national and international levels.

3 The Content of Curriculum in Ergonomics

The Bachelor's degree includes basic (25 ECTS) studies of Ergonomics (e.g., principles in health, safety and well-being at work), which are provided by the Open University of Eastern Finland [2, 3]. The basic studies in Ergonomics, called Ergonetti studies, is a Finnish learning program for promoting occupational wellbeing that is conducted entirely via the internet [3]. The content of Ergonetti studies follows the definitions of International Ergonomics Association (IEA) (i.e., physical ergonomics, cognitive ergonomics and organizational ergonomics) [4]. Ergonetti studies provide a functional, web-based learning environment that encourages students to develop their working environment [2]. The aim of Ergonetti studies is to enable students to learn how to improve, both in theory and in practice, health and well-being in their workplaces in a multidisciplinary manner [3]. The web (Moodle) provide a technical learning environment, and students apply the theory they learn in their workplaces. Ergonetti studies focuses on training and improving ergonomics in order to increase the work capacity of individuals within the students' own workplaces or in target jobs acquired through the studies. The Ergonetti learning program consists of five learning modules: Keys for the development of work (4 ECTS), Work environment (4 ECTS), Work community and competence (7 ECTS), Diverse strain at work (7 ECTS), and Summary (3 ECTS). The Ergonetti learning modules focus on the diverse loads and strains at work, such as physical and psychological load and strain, work and the work environment, work organization and leadership, and professional competence [3].

In addition, the Bachelor's degree includes subject studies (40 ECTS) in Ergonomics (e.g., work physiology, assessment and design methods and legislation). Further, minor studies (25 ECTS) are included. Since ergonomics is a multidisciplinary discipline, via minor studies students have a possibility to deepen their knowledge in one area in Ergonomics (e.g., work and organizational psychology, physiology, work and safety technology or studies in occupational health).

Furthermore, a total of 64 ECTS of mandatory studies in health promotion and health sciences (e.g., epidemiology, health psychology, health politics, and evidence based health care), as well as language and communication skills (10 ECTS) and statistics (12 ECTS) are required in the Bachelor degree (Fig. 1). Recommended time for graduating in Bachelor's degree is three years (i.e., 60 ECTS per year).

Curriculum of Degree Program in Health Promotion Specialization to Promotion of Ergonomics and Well-Being at Work	
Bachelor in Health Sciences (180 ECTS), 3 years	Master in Health Sciences (120 ECTS), 2 years
Basic and Subjects Studies in Ergonomics and Well-being at Work (including Bachelor´s thesis) (65 ECTS) Minor Studies (25 ECTS) Studies in Health Promotion and Health Sciences (64 ECTS) Academic Skills (i.e., Statistics, Language and Communication skills) (26 ECTS)	Advanced Studies in Ergonomics and Well-being at Work (including Master´s thesis) (70 ECTS) Minor and Optional Studies (36 ECTS) Statistics (14 ECTS)

Fig. 1. The curriculum of health promotion at the University of Eastern Finland

The Master's degree (120 ECTS) includes advanced studies in Ergonomics (70 ECTS), statistics and optional studies (50 ECTS). The advanced studies includes Master's thesis (Fig. 1). Like in the Bachelor's degree, the minor studies in master's degree give an opportunity to deepen knowledge in one area in Ergonomics. It can be the same area as in Bachelor's degree or student can choose another one. Typically, the minor studies are related to the topic of Master's thesis. Recommended time for graduating in Master's degree is two years.

The content of the HFE education in the University of Eastern Finland corresponds to the qualification of the European Ergonomists set by Centre for Registration of European Ergonomist (CREE). Based on the minimum requirements for registration of European Ergonomist (Eur.Erg.), CREE [5] requires evidence of a basic education across the ten following areas of knowledge:

1. Principles of Ergonomics
2. Populations and General Human Characteristics
3. Design of technical systems
4. Research, evaluation and investigative techniques
5. Professional issues
6. Ergonomics: Activity and/ or Work Analysis
7. Ergonomic Interventions
8. Ergonomics: physiological and physical aspects
9. Ergonomics: psychological and cognitive aspects
10. Ergonomics: social and organisational aspects

Thus, the graduated students have an opportunity to apply registration for Eur.Erg. after three years of professional experience.

The students of Ergonomics are mainly adult learners, employed, and they live in different areas of Finland. Thus, studying is influenced by several factors. For example, the geographic distance from University, work, family and the factors of coordinating them, bring challenges to studying. However, students are very committed to studying and usually graduate within the recommended time (i.e., in five years). Integrating Bachelor's and Master's thesis into the ongoing development projects in working life provides adult learners a meaningful way of learning. Students can use their previous skills in projects and focus their thesis on a theme that genuinely interests them [6].

4 After Graduating

The degree of Master of Health Sciences majoring in Ergonomics prepares graduates for the requirements of working life and acting as a specialist in Ergonomics. The graduates in Ergonomics have commonly been employed in many occupations where they can utilize the knowledge from their studies e.g. in education or in occupational health and safety. Based on the survey for graduates in Ergonomics during years 1990–2015 (n = 89, with the 30% response rate), the majority of graduates were satisfied with the content of the Ergonomics curriculum and they found the content valuable in their present occupation and working life requirements. Majority of the graduates were employed in the field of education (i.e., vocational education and training, polytechnic education or University education). Further, they were working as a well-being/occupational safety managers, project or research coordinators or inspectors in occupational safety [7].

5 The Challenges in Ergonomics Education

The expertise of Ergonomists is required increasingly in the working life due to labor and population aging, and growing standards of production and quality, as well. Further, continuous development of information and communication technology changes work organisations and organisational design, and the ways of working [1]. Thus, there is a need to increase awareness of Ergonomics and its possibilities to develop work, working conditions, and thus, increase the quality and productivity of work.

The changing working life involves new needs for Ergonomics as a discipline and to education in the field of Ergonomics. Thus, the main challenge is to develop studies that address rapid changes in requirements of modern working life. One of the strengths of our degree program is a possibility to combine theory and practice in Ergonomics. Students can implement their studies (i.e., study assignments and thesis) to real working life. In addition, our degree programme utilizes web based learning varying from tutored courses and blended learning to self-study and case-based courses on central topics in HFE. Web-based learning is particularly suitable in Finland due to scarce population and long distances. The Ergonetti studies are one example of this.

The field of Ergonomics in working life is broad, and it is impossible for one educational organization to offer comprehensive education. Thus, it is important to strengthen co-operation with other actors. This is achieved by building networks between Finnish stakeholders, for example Schools and academies of vocational education, Universities of Applied sciences and Universities that provides Ergonomics training or courses. In addition, it is important to strengthen co-operation and interaction with the companies, the research institutes (e.g., Finnish Institute of Occupational Health, National Institute for Health and Welfare), the funding agencies of working environment research, R&D departments of consumer product manufacturing companies, and government authorities (e.g., Ministry of Social Affairs and Health, Social Insurance Institution of Finland).

The challenges for the future are hard to overcome, but in collaboration with different stakeholders we believe we have a good chance to succeed.

References

1. Dul J, Bruder R, Buckle P, Carayon P, Falzon P, Marras WS, Wilson JR, van der Doelen B (2014) A strategy for human factors/ergonomics: developing the discipline and profession. Ergonomics 55(4):377–395
2. Ropponen A (2009) Experiences of learning and satisfaction with teaching of basic courses of ergonomics over internet – the Ergonetti program. Educ Inf Technol 14:81–88

3. Randelin M (2013) Sustainable well-being at work through ergonomics via the web-based learning program of Ergonetti. Publications of the University of Eastern Finland. Dissertations in Health Sciences, Number 170 (2013)
4. IEA, International Ergonomics Association Homepage. Accessed 15 Apr 2018
5. Centre for Registration of European Ergonomist (2017) Minimum requirements. https://www.eurerg.eu/the-eur-erg-title/minimum-requirements. Accessed 1 May 2018
6. Järvelin-Pasanen S (2009) "Fiksummin – ei kovemmin" – kokemuksia ergonomian opiskelijoiden alku-vaiheen ohjauksen kehittämisestä ["Smarter – not harder" – Experience from the development of tutoring of ergonomic students in their first academic year]. In: Rahkonen A, Isola M, Wennström M (eds) W5W2-pilotoinnilla kohti oppimisen kumuloitumista [With the W5 W - pilot project towards cumulating of learning]. University of Oulu, Publications of the Teaching Development Unit. Dialogies 12, 118–133 (2009)
7. Järvelin-Pasanen S, Rissanen T, Räsänen K (2016) The further employment and experiences of education among graduates in ergonomics. In: Järvelin-Pasanen S (ed) NES2016 - Ergonomics in theory and practice. Proceedings of 48th annual conference of nordic ergonomics and human factors society, Kuopio, Finland, 14–17 August 2016. Publications of the University Eastern Finland, Report and Studies in Health Sciences, No 22, pp 151–154 (2016). http://urn.fi/URN:ISBN:978-952-61-2192-5

The Dynamics of a Service Relationship: The Example of Technicians Working for Pension Schemes

Valérie Zara-Meylan(✉) and Serge Volkoff

Cnam CRTD, CEET, Gis Creapt,
29 promenade Michel Simon, 93166 Noisy le Grand, France
valerie.meylan@lecnam.net

Abstract. Relations with the general public are a key component of many working situations. This is the case for the body employing "Pension Advice Technicians" responsible for handling the case files of beneficiaries - future pensioners - and managing relations with them. In these situations, the quality of service is conventionally measured through an array of indicators used to steer the services provided and control the work carried out. However, these indicators do not measure the full range of skills used by members of staff. Our research aims to understand the work of the technicians and address its complexity in order to show how they ensure the reliability of the service they provide. Our methodology was based on an ergonomic analysis of the activity, adopting a holistic approach to the work and its challenges. It examines the dynamic management they deploy both within the working day and over longer time scales to manage the different requirements and logics which underpin the work to support beneficiaries and oversee their case files. This profession, which requires high levels of administrative rigour, also requires an ability to manage the multiple dynamics involved, including the relationships the technicians established with the beneficiaries themselves, as well as with the internal control bodies. From an ergonomics perspective, the aim is to ensure the working methods do not hinder the construction and mobilisation of these skills, but on the contrary support their development.

Keywords: Activity analysis · Job analysis and skills analysis
Errors, accuracy and reliability · Organisational design and management

1 Introduction

Relations with the general public are a key component of many working situations. This is notably the case for the people responsible for handling administrative queries from service users or beneficiaries of insurance or pension schemes. There are very high expectations of these staff members both from the general public and their employer, in particular when the latter is streamlining and reinforcing work assessment and control methods. In these situations, a classic distinction is made between the two aspects of the work: the first concerning the case files and the second the beneficiaries. This distinction also affects how the quality of service is overseen and assessed. This

S. Bagnara et al. (Eds.): IEA 2018, AISC 821, pp. 310–319, 2019.
https://doi.org/10.1007/978-3-319-96080-7_36

quality of service is mainly measured using two sources: an array of indicators used to steer the services provided and control the work carried out, which are considered to be synonymous with quality; and continuous customer satisfaction surveys which aim to measure quality from the service beneficiary's perspective.

The research in administrative sectors has focused on the one hand on the face-to-face contact with client-beneficiaries who need to be guided through the process, taking into account the constraints linked to organisation and the positioning of the social relationship with each person [1]; and on the other hand on the back-office work on the case files, the background to the service relationship [2]. This research has shown that these indicators do not measure the full range of skills used by members of staff nor all the components of the quality of their work. Furthermore, they often have unintended consequences which actually complicate the work, due to the multiplication of tasks and objectives and increasing levels of pressure.

This is the case for the agency employing "Pension Advice Technicians" responsible for handling the case files of beneficiaries – future pensioners – and managing relations with them. Our research resulted from a concern expressed by the agency management, regarding the role experience plays in the quality of pension services, given that the rate of error in the agency relating to pension payments, as measured by management indicators, has not decreased despite the commitment to meeting national targets.

In order to understand the work of the technicians and address its complexity, this research focuses on their work as a whole [3]. It examines the dynamic management they deploy both within the working day and over longer time scales to manage the different requirements and logics which underpin the work to support beneficiaries and oversee their case files.

From this point of view, their work corresponds to the management of dynamic processes, as analysed in production sectors [4, 5]. The technicians manage overlapping temporal frameworks in order to steer different processes, with the aim of meeting the targets set in terms of quality, quantity and deadlines, for all the case files in their portfolio. The challenge is to guarantee the reliability of their production, taking into account the inherent uncertainty in the situations they encounter, in conjunction with their colleagues responsible for reconstituting people's careers upstream, and the controllers responsible for checking and clearing the case files (to authorise pension payments) downstream.

Our study focuses on what we have called "managing the temporal environment" [6] i.e. how they approach the different situational requirements, in terms of temporal frameworks, in order to take action and open up possibilities for future action. According to our findings, this management is what ensures the technicians produce quality - the scope of which needs to be specified from the work perspective. From an ergonomics perspective, the objective is to relate the difficulties they face to the configuration of the temporal frameworks which do not facilitate, and sometimes even hinder, their management of their temporal environment, in order to find ways to support this management.

2 An Approach Based on an Ergonomic Analysis of the Work

The research approach involved numerous exchanges with actors at different decision-making levels, and ergonomic analyses of the technicians' work, adopting a holistic approach to the work and its challenges.

The intervention with the technicians took place over a fifteen-month period at four agencies located in different regions of France. In order to capture the work dynamics, a purpose-developed methodology was deployed, cross-referencing two forms of analysis:

- semi-structured interviews lasting around one hour, with 22 technicians, concerning the trajectory of five case files from their portfolio currently being processed, focusing on: the type of case file and its specificities, the background and any developments, relations with the beneficiary, the follow-up dynamics and expectations;
- systematic observations of the back-office work of 12 other technicians (a half-day each) recording their actions, interactions (managers, colleagues or users), movements, and the tools used (software, telephone, paper etc.).

The interpretation of the results was discussed and validated firstly during individual self-confrontation sessions [7] with the technicians concerned, then with the whole team in each agency (76 people in total, including six agency managers). These phases were indispensable both for ethical reasons and to properly understand the work and its challenges.

These analyses were part of a broader co-development approach with the agency, overseen by a steering group. They were debated with actors at different decision-making levels. The results were also supported by the feedback from discussions with several groups of managers and staff representatives.

3 The Dynamics of the Technicians' Work

3.1 The Temporal Frameworks to Manage

The temporal frameworks that need to be managed by technicians firstly relate to the agency's rules and procedures at national level, implemented in the regions, and then filtered down to the agencies. These frameworks indicate how to adjust career records, calculate pensions and control case files, according to the terms of their profession. These change constantly in line with current developments in legislation, which in turn leads to frequent changes in the instructions issued to technicians.

The case files correspond to a wide variety of situations for beneficiaries, who need to be informed and supported according to their situation and intentions for their retirement. Case-file management is a multi-faceted, long process, which takes at least three to four months. The operations required are discontinuous, with waiting periods during which the file needs to be followed up because the situation continues to change, partly because the due date for the beneficiary's first pension payment will be getting closer, and partly because the file may be completed by the employer or other pension schemes which have been asked to provide information.

Other important temporal frameworks relate to technical conditions, in particular the IT system. This constitutes a reliable system for memorising the data in all the case files and computing the sums and calculations, but it also requires a series of operations to be carried out and sets deadlines.

Furthermore, the organisational conditions of the technicians' work are also changing, with the regrouping of the agencies the teams work in and the distribution of work which aims both for certain employees to specialise on certain types of case files and for greater versatility, allowing them to also work at reception. In addition, relations with the external organisations (like other pension schemes) asked for additional information are becoming more difficult, and these organisations often tend to delay in returning the requested information.

Finally, the temporal frameworks are partly the result of the methods for assessing the skills of technicians: this assessment is done by the manager, but also via a computerised account of the number of files processed, deadlines met, the errors which are returned to them from internal control.

3.2 Organising Your Work and Organising Yourself Day-to-Day

During their working day, the technicians deploy a multitude of actions which they organise at different moments in time (see Activity chronicle 1). These moments correspond, on the one hand to the means by which these complementary case files or documents reach them (❶ case file processing actions in the diagram), and on the other hand, to the interactions with the beneficiaries which they combine with case-file processing (❷ beneficiary management actions).

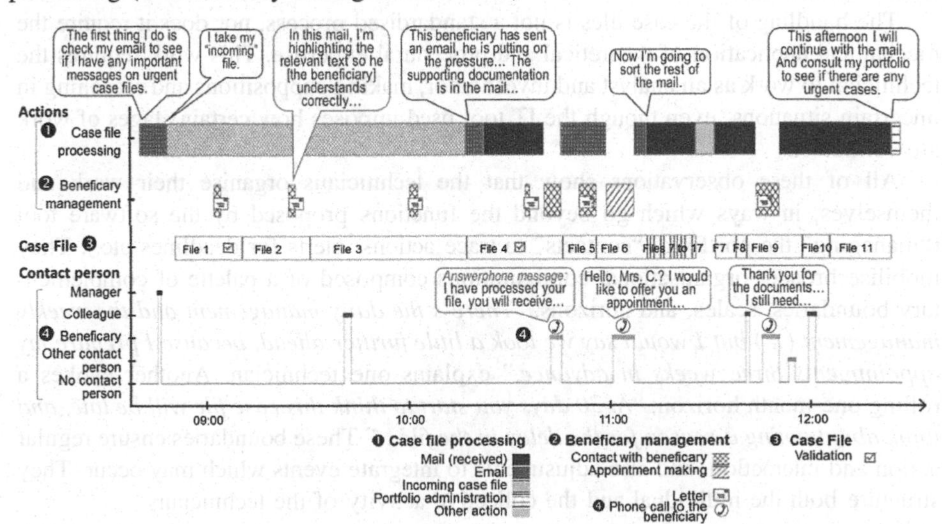

Activity chronicle 1 - Example of a pension technician's work morning: the dynamic management of case files activity, combined with beneficiary follow-up (records processed using ActoGraph® [8])

In order to initiate and sequence these moments of action, the technicians comply with a general organisation that is specific to each of them, but they all express this spontaneously, as at the start of the morning in timeline above: "*The first thing I do is check my [email], to see if I've got any important messages, regarding urgent files [...].*" In the example presented, the technician checks their email, then executes data entry operations for six files (❸Case File, Files 1 to 6), prioritises the next eleven files (Files 7 to 17) and then starts processing these (Files 7 to 11).

The majority of case files bring the technician into contact with the beneficiaries. Case files 1 to 5 and 9 require a letter to be sent by post to provide the beneficiary with information or ask them to provide further information. Other (or the same) cases require contacting the beneficiary by telephone (❹), such as case file 4 which relates to a very persistent beneficiary extremely worried about their situation, case file 5 to make an appointment at the agency, and case file 9 for which the technician follows up their written request for further information with a telephone call.

A total of 11 case files are processed on that morning (up to 21 different case files opened per half-day according to our records), of which only two were validated with a view to clearance for payment (❸ Case file validation). It should be made clear at this point that their work is not yet finished for these case files, as they need to wait for the reply to the letter sent to the beneficiary, take into account the latest information, and then ensure the files are validated by the control department. This means that all 11 case files processed by the technician that morning still require further intervention. This was a situation frequently observed, the technicians studied had 70–135 open case files in their portfolio.

The handling of the case files is not a standardised process, nor does it require the mechanical application of theoretical and procedural expertise. This work requires the technician to work as an analyst and investigator, making suppositions and reasoning in uncertain situations, even though the IT tool used imposes how certain stages of work are conducted.

All of these observations show that the technicians organise their work and themselves, in ways which go beyond the functions proposed by the software tool ("managing" the portfolio, "contacts" to trace actions, alerts for deadlines etc.). They mobilise broader organisational configurations composed of a palette of complementary boundaries, scales, and horizons. "*There's the daily management and the weekly management (...) but I would say we look a little further ahead, because I prepare my appointments three weeks in advance,*" explains one technician. Another evokes a rolling one-month horizon: "*At 30 days you start to think this case file will be late, and think about noting a reason for the delay in the file...*" These boundaries ensure regular action and interaction, and also adjustments to integrate events which may occur. They structure both the individual and the collective activity of the technicians.

3.3 Taking Decisions, Sometimes with the Risk of Rejection by Internal Control

Uncertainty and unknown factors are an integral part of working in dynamic situations. In many cases, the technicians have to take decisions which anticipate future changes. For example, on another day the technician examines a case file for which they are waiting for the reply from a former employer of the beneficiary. "*It's the end of October, for a retirement on 1st December, I am not going to wait until the end of November...,*" he explains. "*The quarters [worked for this employer] had to be recorded, which I did [manually] to ensure we met the deadline.*" Indeed, if the software detects that a request is overdue (deadlines set based on management indicators) not only is the technician penalised, but the beneficiary's pension payments will also be delayed. However, in this situation, the employer eventually completed the beneficiary's case file at the same time, so control reported an "*error for dual entry of quarters.*"

In another case, the technician received an error report from internal control regarding the supporting documentation the beneficiary had to provide on their (means tested) welfare payments. The technician closed the case file without requesting this document, as he knew it would not influence the amount due for the current year. His intention was to avoid delaying the payment of the pension. The error report from internal control was instructive according to the technician: the attention paid by internal control to this supporting documentation led to discussions within the team of technicians who are all now careful to comply with this procedure regardless of the beneficiary's level of income. However, it also increased their lack of understanding of why this procedure was implemented in the first place. Indeed, the amount due for the current year had already been established and the beneficiaries are asked each year to establish their income, so the document would be requested the following year in any case.

The analyses highlight different combinations of technical and organisational conditions which complicate or undermine the management executed by the technicians. The reports received from internal control are a particular bone of contention; the notion of *error* which characterises these reports is debatable, as it is often disconnected from the reality of the quality issues experienced both in terms of the case file and the beneficiary, as identified by the technician.

4 Case File Management

During the period of several weeks, and even several months, over which the case file is managed, a number of the findings presented in the paragraph above regarding the organisation of the working day, are also observed. Once again, the issue is how to best manage the temporal frameworks, identifying them correctly and taking into account the inherent uncertainty.

The rules in force are supposedly homogeneous. In theory, in any given situation, any technician would act the same way, for any case file presenting the same characteristics at this stage of processing. In practice, however, there are a huge range of situations which require permanent efforts to adjust between very strict instructions and the very fluid reality. This adjustment involves:

- making suppositions: interpreting the past or present behaviour of a given contact person or the beneficiary themselves, and anticipating future actions;
- reasoning: determining how urgently a request needs to be followed up on, taking into account the final deadline, the expected duration of subsequent stages in the process, the most effective order in which to execute these tasks and any possible unforeseen events;
- decision-making which often involves arbitrating between distinct objectives, all equally legitimate, but irreconcilable in the short-term: for example, the technician aims to both establish a full and accurate career record for the beneficiary but cannot allocate too much time to this. So, which of these objectives should be prioritised at each stage of the case file processing procedure?

We have already explained that case file processing is discontinuous. When re-examining the case files with the technicians, we observed that they took action on each case file at numerous different points in time, often at large intervals. We also found traces of actions taken by colleagues, who replaced the technician during their holidays, for example. Despite this, they have to keep control over the process as a whole to ensure the right calculations are made and the beneficiary does not lose their way through the process. It is also necessary to obtain a reliable representation of previous episodes and (as far as possible) of the following stages in order to plan for these.

In light of this need for consistency, when looking back over their successive actions on the same case file it becomes apparent that the systems, organisations, and people the technicians interact with constitute both resources and obstacles. The IT tool contains essential information and sophisticated processing options, but the technicians need to take into account gaps which they have difficulty understanding: some processes are missing from the software and require manual entries, system updates based on changes in regulations take months to implement etc. In the same way the legal rules and organisational principles are vital frameworks for their activity, but the underlying logic behind these is not always clear, and these rules and principles cannot possibly take into account the numerous specific circumstances encountered. Finally, the other departments within the pension scheme agency and those of its external partners, provide useful information, but the procedures for communicating between then are often rigid and the required cooperation is often lacking when the organisations are overworked, or if their practices are unclear for the technicians.

In this context, the quality of case-file management largely depends on the relationships the technicians establish with the beneficiaries - from this perspective the importance of the face-to-face appointments is paramount. During the self-confrontation sessions we conducted, the technicians explained how face-to-face appointments help to resolve a significant proportion of the difficulties they encounter, and were able to back this up with examples. These appointments notably help to overcome inertia and reduce the time spent waiting for feedback from beneficiaries, as the date of the appointment sets a de facto deadline: they have to prepare and come with the right documents or they will have attended the meeting for nothing. The discussions at these appointments also highlight technical misunderstandings on the beneficiaries' behalf. These misunderstandings are frequent as the regulations governing pensions are complicated and change frequently. It is much easier to identify these issues face-to-face rather than by letter or telephone as it allows the technician to adapt their explanation. Furthermore, some beneficiaries may be embarrassed to show some of the information relating to their career path or financial resources, yet this is required in order to calculate their pension. At a face-to-face meeting the technicians, who are used to dealing with this sort of embarrassment, can approach these aspects tactfully. Finally, the beneficiaries can take the opportunity to express their hesitations regarding the strategy to adopt at the end of their career. They may even find out at this meeting that they have more options open to them than they realised. The technician can then identify and provide them directly with the information they require to navigate these options, which saves time and ensures quality of service for the ongoing processing of the case file.

5 Conclusion

In their interaction with their working environment and through experience, the technicians have developed skills [9] which allow them to manage the multitude of demands their work places on them in the long term, with a view to fulfilling the different quality components which make up this service relationship.

The results of this research can be summarised by referring to the notion of "distance" between the technicians and the beneficiaries. This notion is often evoked in studies of service relationships but based on an acceptance focused on the expressive and emotional aspects of these relationships: the technicians are instructed to strike the right balance between indifference and empathy, and between calming the situation and upholding their dignity in conflictual situations. However, our analysis of their work suggests that it is important to strike a balance in other "distances", and this continuous adjustment is a major component of the dynamics of a service relationship.

Indeed, it would appear that the technicians also adjust these distances on other levels: spatial, cognitive and temporal. These adjustments are dependent on the technical and organisational working conditions in the agencies, the methods for assessing their results, as well as the skills they have built over their career.

The "spatial distance" relates to the communication between the technician and the beneficiary. Over the course of case-file processing, there will be multiple discussions between them. These take the form of telephone conversations (or answerphone messages), letters sent by post or emails, requiring time for delivery, waiting times and follow-up which needs to be scheduled. At various points in time, the question of whether to propose a face-to-face appointment or not will arise and we insisted on the importance of this type of exchange - which needs to be used wisely because the logistics can be complicated if the beneficiary lives far away from the agency or has transport or mobility issues, or needs to be accompanied etc. The technician therefore needs to make careful choices.

The "cognitive distance" is due to the gaps in the beneficiaries' knowledge of rules and procedures. To a certain extent the technicians try to fill these gaps, either with the beneficiary themselves or via a member of their entourage, without going into too much detail, but with the aim of ensuring they are able to understand the ongoing process.

The aforementioned issue of "emotional distance" is perhaps less present than in other client-facing professions, because most beneficiaries have had the time and opportunity to prepare their future retirement. However, the technicians have to combine a posture of support, explanation and advice, with a posture of control - reinforced by the downstream intervention of the controllers who do not have any contact with the beneficiaries. The predominance of one or other of these attitudes may vary over time, as the information in the case file is confirmed or awaits confirmation.

In conclusion we would like to turn our attention to the vital issue of "temporal distance" which we have evoked throughout this paper. The beneficiary who initiates a procedure with their pension scheme usually has a simplified vision of how the process will unfold and is not necessarily aware of the multiple factors which can slow down the processing of a case file. As the case file is processed - and even more so if the process stagnates - the technician must share with the beneficiary, as far as possible, the reasons for any delays; yet these reasons are numerous and often complex, as we have already explained. It is important that the technician is able to reduce waiting times where necessary, by entering incomplete data and correcting them later, even if this method is inherently flawed as it requires the technician to open an entire new procedure.

In order to effect these various "distance adjustments" the technicians' experience would appear to be an essential resource. This is what allows them to develop specific skills in order to manage multi-faceted and endlessly reconfigured dynamics. From an ergonomics perspective, the aim is to understand these skills in order to ensure the working methods (IT systems, workspaces, means of instructing tasks, forms of organisation, career management, etc.) do not hinder the construction and mobilisation of these skills, but on the contrary support their development.

References

Fernie S, Metcalf D (2000) (Not) hanging on the telephone: payment systems in the new sweatshops. In: Lewin D, Kaufman B (eds) Advances in industrial and labour relations, vol 9. JAI Press, Greenwich

Star SL, Strauss A (1999) Layers of silence, arenas of voice: the ecology of visible and invisible work. Comput Support Coop Work (CSCW) 8(1):9–30

Zara-Meylan V, Volkoff S (2017) Gérer l'assuré de près et de loin: la conduite des dossiers retraite Actes du séminaire Âges et Travail CREAPT [mai 2016], Travailler avec et pour un public: l'expérience des autres. Rapport de recherche CEET, vol 104. CEET, Noisy le Grand, pp 119–138. http://www.cee-recherche.fr/publications/rapport-de-recherche

De Keyser V (1995) Time in ergonomics research. Ergonomics 38(8):1639–1660

Rasmussen J (1997) Risk management in a dynamic society: a modelling problem. Saf Sci 27(2–3):183–213. https://www.sciencedirect.com/science/article/pii/S0925753597000520

Zara-Meylan V (2016) Quelles conceptions temporelles pour analyser l'activité? Une proposition issue de recherches en ergonomie dans l'horticulture -what kind of temporal concepts to analyze work activity? A proposal from a research in ergonomics in horticulture production. Activités 13(1). http://activites.revues.org/2732

Mollo V, Falzon P (2004) Auto-and allo-confrontation as tools for reflective activities. Appl Ergon 35(6):531–540. https://hal.archives-ouvertes.fr/hal-01584557/document

Boccara V, Delgoulet C, Zara-Meylan V., Barthe B, Gaillard I, Meylan S (2018) The role and positioning of observation in ergonomics approaches: a research and design project. In: Delgoulet C, Zara-Meylan, V (eds) Observations in actual work situations: is this method still a key part of ergonomics practices? IEA 2018 symposium, Florence, Italy

Volkoff S, Gaudart C (2015) Working conditions and "sustainability": converting knowledge into action, Centre d'études de l'emploi (CEE), Rapport de recherche n°92, 47 p. http://www.cee-recherche.fr/publications/rapport-de-recherche/working-conditions-and-sustainability-converting-knowledge-action

Conceptual and Practical Strategy Work to Promote Ergonomics/Human Factors in Sweden

Cecilia Österman[1(✉)], Anna-Lisa Osvalder[2], Hillevi Hemphälä[3], Jörgen Frohm[4], Susanne Glimne[5], Göran M. Hägg[6], Olle Janzon[7], Per Johan Pettersson[8], Mathias Stavervik[9], and Jane Ahlin[10]

[1] Linnaeus University, Kalmar, Sweden
styrelsen@ergonomisallskapet.se
[2] Chalmers University of Technology, Gothenburg, Sweden
[3] Lund University, Lund, Sweden
[4] Swedish Transport Administration, Borlänge, Sweden
[5] Karolinska Institutet, Stockholm, Sweden
[6] ErgoMusic, Stockholm, Sweden
[7] SSAB Europe, Borlänge, Sweden
[8] MeridentOptergo AB, Lohja, Sweden
[9] SHL Group AB, Nacka Strand, Sweden
[10] President of Swedish Ergonomics and Human Factors Society (EHSS), Linköping, Sweden

Abstract. This paper describes the results of the conceptual and practical strategy work performed by the Swedish Ergonomics and Human Factors Society (EHSS) today. The rationale of EHSS is to strengthen the quality of ergonomics/human factors knowledge and practice in Sweden and form a multidisciplinary platform across disciplines and professions for collaboration and for knowledge sharing. EHSS gathers about 350 members, representing different occupations in industry, academia and the public sector. Together, EHSS members hold knowledge and experience in physical, cognitive and organizational ergonomics and its application in working life and society.

The overall aim of this paper is to inspire related societies and stakeholders to initiate discussions about strategies and future projects that allow for collaboration and knowledge sharing. Proposedly follow the EHSS model where we have formed a multidisciplinary platform for collaboration across disciplines and professions. The activities initiated and supported by EHSS are one step towards broadening the knowledge and application of HFE in Sweden, and to comprise new arenas of specialization. By participating in the key areas in society such as teaching, standardization, product development and occupational safety and health, the work of EHSS is one piece of the puzzle to improve human activities in the future. The vision is that together, we can improve safety, efficiency and well-being for all.

Keywords: Future of ergonomics · Knowledge sharing · Design for all

S. Bagnara et al. (Eds.): IEA 2018, AISC 821, pp. 320–329, 2019.
https://doi.org/10.1007/978-3-319-96080-7_37

1 Introduction

The systems-oriented science and practice of ergonomics/human factors (E/HF) promotes a holistic approach that embrace all aspects of human activity. By understanding the physical, cognitive and social characteristics of people it is possible to design products, jobs, tasks, organizations, and environments to fit peoples physical and mental needs, abilities and limitations. The ultimate objectives of E/HF are to optimize human well-being and overall system performance for work organizations, and the broader society.

There is no shortage of knowledge available on the relevance E/HF to successful (and unsuccessful) systems, and to joint performance and well-being outcomes. Scientific literature, handbooks, and standards across domains of specializations and industries describe how system solutions that are safe, sustainable, and comfortable for the users can be developed, implemented and evaluated. The former IEA President Hal Hendrick's [1] argument that *Good Ergonomics is Good Economics*, has been demonstrated for individuals, companies and for the society [2]. Yet, as argued by Dul et al. [3], the knowledge and application of E/HF principles and methods has failed to reach its full potential. Whereas the benefit of E/HF is more easily appreciated to the users in the sharp end, the strategic importance of system design may be less accepted by decision-makers in the blunt end. There are islands of knowledge and pockets of practice that remain to be bridged.

In their position paper, Dul et al. [3] propose a joint world-wide E/HF development plan and strategies for the future, at several national and international levels to reach decision-makers and designers. By improved communication and establishment of partnerships, a '*demand development cycle*' is created which stimulate and foster research, knowledge transfer and application of high-quality E/HF.

The Swedish Ergonomics and Human Factors Society (EHSS) has attempted to meet this strategy from a national perspective, identifying relevant stakeholders and their needs, developed strategies for knowledge sharing, communication and collaboration across disciplines and professions, as well as building partnerships for research and practice, and advance E/HF education and training.

The EHSS gathers about 350 members, representing different occupations in industry, academia and the public sector. Together, EHSS members hold knowledge and experience in physical, cognitive and organizational ergonomics. The composition of the EHSS' board purposely mirrors the diversity of the members in gender and professions.

The purpose of this paper is to describe the ongoing conceptual and practical strategy work performed by EHSS of today. The work is inspired by the notion of the development of joint strategies at national and international levels to strengthen and develop the E/HF discipline and profession. Further, it is a continuance of the EHSS work presented at the IEA conference 2015 [4]. It is our intention to inspire other ergonomics and human factors societies and other stakeholders world-wide and evoke discussion about strategies, activities and possible outcomes. The overall aim of the work performed within the EHSS, entirely on a non-profit basis, is to develop and strengthen the quality of E/HF knowledge and practice in Sweden.

2 EHSS Conceptual Strategy Work

The rationale of the Swedish Ergonomics and Human Factors Society (EHSS) is to strengthen the quality of ergonomics/human factors knowledge and practice in Sweden. Specifically, the objectives are to [5]:

- create value for members;
- form a multidisciplinary platform for collaboration across disciplines and professions and for dissemination of research results, methods and tools into practice;
- strengthen the role of ergonomics/human factors in the society, including the development of easily available quality education and training.

The EHSS is a non-profit organization where the work performed by the board members and other members is entirely unpaid. Any surplus of revenues from seminar, conferences or other activities is re-invested in the organization for benefit of the members. To be able to fulfill the above objectives in a cost-effective way, with limited time, monetary and personnel resources, the EHSS has formulated a conceptual strategy. Table 1 summarizes the conceptual framework and illustrates how the objectives are linked to target group, current activities already in place, and future activities.

Table 1. Summary of the conceptual framework for EHSS strategy work.

Objective	Target group	Ongoing activities	Planned activities
Create value	EHSS members	Seminars EHSS-Newsletter Scholarship Student prize Social Media Podcast	Live stream of seminars Webinars Support for activities
Platform for collaboration and knowledge sharing	EHSS members Society	Networks - Visual ergonomics - Work load ergonomics - Voice ergonomics Research collaboration NES CREE	Network - Cognitive ergonomics
Promote E/HF in society	Society	Education Standardization Product evaluation	EHSS Award

3 Creating Value for EHSS Members

To facilitate recruiting and retention of members and make sure membership in EHSS remains valuable, numerous opportunities for professional development, learning and networking are provided.

Current and future activities to keep members updated on the latest in E/HF and potentially fill knowledge gaps that can add practical value for members include:

- **Seminars.** Breakfast seminars are arranged in Stockholm about six times a year with various current themes. It can be either members or invited guest that gives a lecture, followed by questions, discussions and breakfast. AftErgo is an after-work activity arranged about four times a year in Gothenburg. It is often sponsored by a company that presents their work or products, followed by drinks and snacks.
- **EHSS-Nytt.** An online newsletter distributed to members four times a year
- **Scholarship.** Twice a year, all members are eligible to apply for financial support of up to 10 000 SEK (about 1 000 EUR) to be able to participate in a conference, seminar or another professional development course in any of the Nordic countries
- **Student prize.** Annually, the best university student thesis in ergonomics and human factors is elected. Students and supervisors from all Swedish universities, as well as Swedish students studying abroad are eligible to nominate a thesis. The winner gets to present the thesis at the annual Nordic Ergonomics and Human Factors Society (NES) Conference.
- **Social media.** The social media practices have changed rapidly, and a major challenge for EHSS – as for many other organizations – is to bridge the generational gap [6]. For easy sharing of news and information on upcoming events, as well as social connections among members, EHSS is active on Facebook, LinkedIn and Twitter. The board of EHSS has developed an annual schedule for national and international events, and a take turn in managing the social media accounts.
- **Podcast.** This is another way of connecting members are podcasts. The first podcasts were broadcasted from the NES Conference 2015 in Norway, and featured informal interviews with keynote speakers, researcher and conference participants. It has since been a recurring feature at NES conferences and EHSS annual meetings.

To further develop membership value, the primary focus of the EHSS is now to improve the possibility for all members to participate in activities regardless of location. Up to now, member activities are primarily arranged in Stockholm and Gothenburg, the two largest cities in Sweden. With members living and working in all corners of the country, (Sweden lies between latitudes 55° and 70°N, and is geographically the fifth largest country in Europe), EHSS have decided to develop the possibility to participate online. As a first step in 2017, the seminars were recorded and put on the homepage for later view. In 2018, the aim is for all members to be able to follow some seminars in real time with Facebook Live, and further to arrange special webinars that are exclusively attended by an online audience. Members living outside Stockholm and Gothenburg can also apply for financial and logistic support from the board to arrange activities in their hometown.

4 Platform for Collaboration and Knowledge Sharing

EHSS is a forum for multidisciplinary E/HF knowledge in the widest sense, that encompass both the working life and on the private arena, and bring together professionals with diverse backgrounds and skills, as well as researchers, practitioners, labor inspectors and other officials.

Knowledge sharing can be seen as the act of making knowledge held by one individual available presented in a way so that it can be understood, absorbed, and put to practical use by other individuals [7]. For a rather small organization such as EHSS to be able to leverage its knowledge and make a difference, it is highly dependent on the members. It is the people active in the society who make sure that knowledge and skills are created, shared, and applied.

4.1 EHSS Networks

Currently, EHSS has three networks that address different domains of specialization: visual ergonomics, work load ergonomics, and voice ergonomics. The networks are open for all members, without any extra charge. The purpose of creating these networks is to facilitate networking and transfer of knowledge and practical experiences between members that work within a specific field, or who are interested in requiring deeper competencies in this area. The activities typically consist of seminars, workshops and informal meetings a few times per year. Work is going on to create new networks in other areas of interest.

Visual Ergonomics Network (SNiS). Visual ergonomics is an area that is inherently multidisciplinary. It deals with the understanding of human visual processes and the interactions between humans and other elements of a complex system [8]. Many different disciplines and professionals are needed to get a good visual environment; *e.g.* optometrists, ergonomists, lighting designers, architects and ophthalmologists.

Within the EHSS, visual ergonomics knowledge has been discussed for a long time. To facilitate collaboration between members representing these different specialist areas, a network for visual ergonomics practitioners, called SNiS (Synergonomiskt Nätverk i Sverige) was started already in 2007. Twice a year, SNiS arranges free seminars, open to all EHSS members. These seminars have contributed to increased awareness of the importance of a good visual environment.

Several members are also involved in research related to visual ergonomics. For example, a visual ergonomics risk analysis method, VERAM [9], have been developed through a cooperation between members active at Lund University, University of Gävle and KTH Royal Institute of Technology. VERAM is web based and recently, additional research funding was approved to further develop the method into an application for Android and Apple products.

Another visual ergonomics research project was initiated as a direct result of the EHSS providing opportunities for collaboration and networking. Members active at Karolinska Institutet and Linnaeus University got together and formulated the research project that investigate visual ergonomics in control room environments. Early results from this three-year project is presented at IEA 2018 [10].

Work Load Ergonomics Network (BEN). The work load ergonomics network originates in the Swedish Load Ergonomic Association (SBEF) and is now part of the EHSS. BEN is a network for EHSS members who are interested in load ergonomics and the prevention of musculoskeletal disorders. The network intends to be a forum for good cooperation for anyone interested in, affected by and working with work load ergonomic issues. The aim is to arrange two seminars per year, one in the autumn and one in the spring, to help monitor relevant research and developments, and maintain knowledge in the area of the load ergonomics. As an example, in spring 2018, a seminar was held on methods for risk assessments of physical load, to prevent musculoskeletal disorders.

Voice Ergonomics. The voice ergonomics group (REN) is the youngest network within EHSS. The is-sues addressed within the group are related to the loads on the vocal folds, which may lead to voice disorders with consequences for personal health as well as de-creased functionality at work. The core of the network consists of speech pathologists but also EHSS members from other professions have shown interest for the group. The speech pathologists have profound knowledge in the anatomy and physiology of the voice but in general little knowledge about ergonomic conditions in working life. Ergonomists and physio therapists on the other hand, generally lack knowledge regarding voice and speech. Hence, the purpose of this network is to facilitate the development and transfer of knowledge in both directions between disciplines and professions.

4.2 Research Collaboration

Several EHSS members are active researchers at universities and other research institutes and are often assisted by practitioner members in formulating and discussing research problems, getting access to work places for field studies, and circulating questionnaires. Below are some examples of projects with a strong presence of EHSS members, and where the results also have been, or are planned to be communicated back to the members at seminars and through the EHSS newsletter.

Activity-Based Offices. Several members of EHSS, especially from Chalmers University of Technology and the KTH Royal Institute of Technology collaborate in research regarding Activity-based Flexible Offices (A-FOs) [11]. A-FOs are workspace solutions without assigned workstations that provide various kinds of spaces to choose from, depending on employees' activities or preferences. The performed empirical studies have evaluated employees' satisfaction with premises, adoption of A-FOs, and consequences of working in A-FOs about 0.5 to 3 years after relocation.

Ergonomic Design for All. Though association between EHSS members from Chalmers University of Technology and the Swedish Transport Administration, a part of a research project has been initiated with focus on how spoken information in public transportation spaces, especially at railway stations, should be accessible for all people.

In short, the results show that a variety of inclusive design strategies are needed to enhance the perception of inclusivity for all people in public transportation spaces. Suitable concepts should be used for improved understanding, mobility and user experience. For example, enhance the number of visual signs and improve their readability and intelligibility, decrease the number of spoken announcements and minimize the information given in each message, focus on useful traveler in-formation during disturbances, and improve real time up-dating of the mobile applications. The project is presented in more detail at IEA 2018 [12].

MO Concept. This project is an example of how product development and scientific research merge into practical solutions for dentists, dental hygienists and surgeons.

MeridentOptergo AB started as a result of many years development if the field of special optics and a specific ergonomic need for dentists to reach a more upright working posture while working in the mouth of a patient. A prism segment in the glasses or the loupe systems interacts with a specially designed frame to angle the field of vision, which makes it possible to maintain an upright sustainable work position. The MO concept has been evaluated several times and the results show that dental personnel wearing the prismatic spectacles reduced their neck pain significantly compared with the reference group. Future work includes more evaluation in other working situations with high visual demands and sustained awkward postures [13].

Planned Future Activities. The research collaborations that has been formed so far between EHSS members, have originated through more or less informal discussions during other meetings and actives. To further stimulate collaboration and evoke innovative ideas, EHSS plan to arrange special assemblies between members who have an idea to pursue and who are looking for partners to establish a consortium with and then apply for research funding.

4.3 NES Conference

The annual Nordic Ergonomics and Human Factors Society (NES) conferences have become a meeting ground for E/HF researchers and practitioners. Every five years, the conference is held in Sweden, arranged by the EHSS. In August 2017, the NES conference was held in Lund in a cooperation between EHSS and Lund University. The theme of the conference was *Joy at Work*. During the conference, about 160 people participated to present and discus the latest research, practice, trends and conditions that can make the workplace a source of knowledge, inspiration, health and joy.

4.4 Centre for Registration of European Ergonomists (CREE)

The aim of the Centre for Registration of European Ergonomists (CREE) is to harmonize the certification practices of the ergonomics societies in Europe. The national societies have agreed on a necessary standard of knowledge and practical experience required to become a certified European Ergonomist, Eur. Erg. The certification assures that the ergonomist has a formal ergonomics education at university level, at least three years of practical work experience, and continuously develop his or her skills in ergonomics. The registration is valid for five years and the certified ergonomists are listed in the CREE directory online [14]. Currently, about 40 EHSS members are registered as Eur. Erg.

5 Strengthen the Role of E/HF in the Society

To strengthen the role of E/HF knowledge and practice in the society, EHSS are involved in various activities. Many members in the EHSS are occupational safety and health professionals, focusing on E/HF and employee performance in the working life, but this objective includes also the design of everyday systems. For example, consumer products, public spaces, grocery stores, shops and public transport should be accessible for all, regardless of age, size, gender, or physical or cognitive function abilities and limitations. Examples of EHSS outreach activities to bring about E/HF awareness to society includes work with E/HF education and training, standardization and product evaluation.

5.1 Ergonomics Education

A specific sub-goal is to influence and assist in the development of easily available quality education and training. Recently, several EHSS members have taken active part in the development of three new educations that cater for different target groups. These educations address different specializations within E/HF:

Work and technology on human terms is an online Massive Open Online Course (MOOC) in human factors engineering, ergonomics, work science and related subjects [15]. The course has been developed by Prevent (a non-profit organization owned by the Confederation of Swedish Enterprise and the trade union organizations LO and PTK), in collaboration with teaching professors at five Swedish universities. The online course is free of charge, given in English and comprise of 14 chapters and four work-place cases. The user is free to start and finish the online course independently of others and will receive a certificate after finishing the complete course.

International Masters' Programme on Technology, Work and Health at KTH Royal Institute of Technology will start for the first time in August 2018. This programme offers knowledge in proactive occupational safety and health management, and how to plan, design and analyze work environments. Two tracks of specialization are offered: Work Environment Engineering (WEE), or Human Factors and Ergonomics (HFE). The education is based on blended learning. Most of the courses offer teaching on campus every three weeks. Between the campus meetings, the students will study independently or in groups with course projects and e-learning activities. The masters' programme is described in more detail at IEA 2018 [16]

Visual Ergonomics and VERAM Risk Assessment Method. Starting in the autumn of 2018, the Faculty of Engineering at Lund University will give a new course in visual ergonomics and plan to use the VERAM method for assessment of risks in the visual environment.

5.2 Standardization

The Swedish Standardization Institute (SIS) is in charge of and coordinates standardisation in Sweden. SIS represents Sweden in the European standardisation organisation CEN and the global organisation ISO.

EHSS has representation in three working groups within the Technical Committee for ergonomics; physical ergonomics; human–system interaction; and visual ergonomics. The EHSS representatives keep the EHSS board and members informed on ongoing standardization work, and by this participation, EHSS is also furthering international development within the respective area of expertise.

5.3 Product Evaluation

EHSS performs evaluations of products to make sure they meet elementary requirements for physical, cognitive and/or organizational ergonomics. The evaluation is done by E/HF experts within the EHSS using heuristic evaluation methods that assess, for example, anthropometric design, interaction design, usability, physical load or comfort linked to the intended user group's abilities and limitations. After evaluation, the successful product is advertised on EHSS website for an annual fee.

6 Concluding Remarks

The purpose of describing the conceptual and practical work of EHSS in this paper is to inspire related societies and stakeholders. We wish to stimulate discussions about strategies and future projects that allow for collaboration between disciplines and professions, thus enabling and strengthening the quality of E/HF knowledge and practice in society. The activities initiated and supported by EHSS are one step towards developing and transferring E/HF knowledge and application in Sweden. By participating in key areas in society regarding E/HF, such as teaching, standardization, product development and occupational safety and health, the work of EHSS is one piece of the puzzle to improve human activities in the future. The mission is that together, we, product can improve safety, efficiency and well-being for all people working and living in the society.

References

1. Hendrick HW (1996) Good ergonomics is good economics. In: The human factors and ergonomics society 40th annual meeting - the presidential address
2. Rose LM, Orrenius UE, Neumann PW (2011) Work environment and the bottom line: survey of tools relating work environment to business results. Hum Factors Ergon Manuf Serv Ind 23(5):368–381
3. Dul J et al (2012) A strategy for human factors/ergonomics: developing the discipline and profession. Ergonomics 55(4):377–395
4. Ahlin J et al (2015) Strategies to develop and strengthen human factors and ergonomics knowledge among stakeholders in Sweden. In: 19th Triennial congress of the International Ergonomics Association (IEA), Melbourne 9–14 August 2015. International Ergonomics Association
5. EHSS (2018) On EHSS. http://ergonomisallskapet.se/

6. Mainsah H, Brandtzæg PB, Følstad A (2016) Bridging the generational culture gap in youth civic engagement through social media: lessons learnt from young designers in three civic organisations. J. Media Innovations 3(1):23–40
7. Ipe M (2003) Knowledge sharing in organizations: a conceptual framework. Hum Resour Dev Rev 2(4):337–359
8. Long J et al (2014) A definition of visual ergonomics. Appl. Ergon 45(126):3
9. Hämphälä H et al (2017) A risk assessment method for visual ergonomics, VERAM. In: FALF Conference: Arbetslivets utmaningar i staden och på landsbygden. SLU, Alnarp
10. Glimne S, Brautaset R, Österman C (2018) Visual ergonomics in control room environments: a case study from a Swedish paper mill. In: IEA 2018 - 20th congress international ergonomics association, Florence
11. Babapour M, Karlsson M, Osvalder A-L (2018) Appropriation of an Activity-based Flexible Office in daily work. Nordic J Working Life Stud 8(S3)
12. Nybacka M, Osvalder A-L, Inclusive design strategies to enhance inclusivity for all in public transportation - A case study on a railway station. In: Proceedings at in IEA 20th congress international ergonomics association, Florence
13. Lindegård A et al (2016) Opting to wear prismatic spectacles was associated with reduced neck pain in dental personnel: a longitudinal cohort study. BMC Musculoskelet Disord 17(1):347
14. CREE (2015) The CREE directory. https://www.eurerg.eu/the-cree-directory/
15. Lagerström G et al (2017) Development of the online course "Work and Technology on Human Terms". In: NES 2017 Joy at Work
16. Rose L, Österman C (2018) Developing an international master's programme in ergonomics at a technical university in Sweden, In: IEA 20th congress international ergonomics association, Florence

Autonomy at Work, Can (Too) High Autonomy Cause Health Complaints and Sick Leave?

Kari Anne Holte[1(✉)], Kåre Hansen[1], Lars Lyby[1,2], and Astrid Solberg[1]

[1] International Research Institute of Stavanger, Stavanger, Norway
kari.anne.holte@norceresearch.no
[2] Faculty of Social Sciences, University of Bergen, Bergen, Norway

Abstract. Autonomy is seen as a core aspect with work. High autonomy can be used in work design for alleviating negative consequences of work, as well as being a design principle for handling interdependencies. However, high autonomy is also suggested to be a burden and not necessarily beneficial. The aim of this study is therefore to explore how high autonomy may be a cause for long-term sick leave among knowledge workers. The study is designed as an explorative case study, with 5 highly educated female workers (age range 28–52) with a present or former long-term work–related sick leave (burnout, fatigue, stress, depression, MSD). The interviews followed an interview guide with open questions, where freedom was used as a proxy for autonomy. The analysis was performed by using Balance-theory. The results showed that although appreciating their autonomy in the task performance allowing for individual planning, independency, skills discretion and creativity, autonomy was an ambiguous term. One challenge was how to handle tasks with high autonomy at the individual level. Stress and health complaints arose also when high autonomy was not balanced by structures in the organizational domain. Findings are in line with contemporary studies questioning the ultimate positive associations of autonomy and high autonomy as a leverage. The results indicate that for specific groups of workers, high autonomy should be balanced with predictable organizational structures, not being a leverage in itself.

Keywords: Autonomy · Work organization
Sociotechnical system theory (STS) · Stress

1 Introduction

1.1 Autonomy as a Design Principle

High autonomy is seen as a core aspect with work being a leverage for change, alleviating negative consequences [1]. A model highly influential in this respect is the demand-control model by Karasek and Theorell [2]. The model emphasizes the psychosocial aspects of work, but the model was developed during the 70s [3], a period that still had a high amount of manual work. Last decades, economical, technological,

S. Bagnara et al. (Eds.): IEA 2018, AISC 821, pp. 330–336, 2019.
https://doi.org/10.1007/978-3-319-96080-7_38

informational and industrial changes have entailed new requirements for organizations, enforcing new employment relations and organizational forms [4]. Moreover, there has been a shift from manufacturing and manual work towards knowledge and service, changing work itself (ibid). Now high autonomy has become important in work design, for companies taking a proactive approach to handle interdependency across cultural, geographical and organizational boundaries (ibid). Grant and Parker [4] specifically address high autonomy stimulating proactivity by increasing mastery, acquire new skills and master new responsibilities (higher role-breadth self-efficacy, confidence etc.).

Recent years several researchers have pointed to negative aspects of high autonomy, being a stressful obligation having to take control over the work performance [5, 6]. A Norwegian case study identified high autonomy perceived as an ambiguity. Independency was important for the performance of their job. On the other hand, autonomy became a demand as well as entailing unfavorable consequences like unpredictability, exploitation, and establishment of informal structures [7]. Similarly, a Danish study among knowledge workers found contradictions between overcoming limits and limits as establishing predictability in the work situation [8]. These qualitative studies add to the already existing critics towards the ultimate positive understanding of autonomy.

The importance of autonomy as a design principle across several research traditions [9], call for the use of a sociotechnical system perspective, particularly the Balance-theory [10], the development of this model being an attempt to integrate knowledge about work design across different research areas [9]. Moreover, this model puts the individual at the center of the system, allowing for a bottom-up perspective, exploring the individual within its context of activity [11]. The aim of this study is by use of this model to explore how high autonomy may be a cause for work-related long-term sick leave, asking how workers with a present or former sick leave understand autonomy and how this understanding reflects the design of the surrounding work system.

2 Methods

The study is designed as a qualitative interview study, with 5 highly educated female workers (age range 28–52) with a present or former long-term work-related sick leave (burnout, fatigue, stress, depression, MSD), mainly attributed to the work situation, more specific high autonomy at work.

The interviewees were recruited trough advertising, information given in courses of stress mastery, as well as snowballing. Those that were interested to take part, were invited to contact the researchers group. When contact was established, they received both oral and written information. Before participation in the study, informed consent was required. Thereafter time and place for interviews were arranged. The interviews took place either at their home, their workplace or at the research institute.

The interviews were based on a semi-structured interview guide, covering topics like perception and understanding of autonomy, work organization, management and power relations, work home-relation and job satisfaction, stress, health and sick leave. As the word autonomy could be difficult to understand, we used freedom as a proxy for autonomy. The interview took about one hour. The interviews were recorded and thereafter transcribed verbatim and anonymized.

The Balance-theory model by Carayon [9] was used as our analytical framework. The first step in the analysis was to identify meaning units describing perceived individual autonomy. These units were labeled according to the way autonomy was described. In step two, the text was analyzed by means of the balance-theory model (i.e. the work system: task design, technology, organization, individual, and environment) [9]. For each interviewee we depicted the actual configuration showing the dynamics within the work system that described the individual workers' experience of autonomy. Afterwards aggregated pictures for similar patterns across the informants were made and described.

3 Results

In this section we will firstly describe the interviewees' understanding of autonomy. Thereafter, we will address how the strain caused by high autonomy is a reflection of the dynamics and interplay between the different domains in the work systems surrounding them.

3.1 Perception and Understanding of Their Autonomy

The five interviewees all appreciate the possibilities to work independent, influence own work day, decide tasks and work in accordance with skills and interests. However, when talking about autonomy they added additional aspects. When an interviewee was asked how she appreciated having freedom in her work, she answered like this: «*clearly, this freedom is good to have, but at the same time, freedom has in a way become my own enemy*». High autonomy is something good, but also an enemy, indicating even a personal fight between herself and the autonomy. This ambiguity is also stated by another interviewee:

«*for example, when writing this book chapter, I was told what to do, but I was free to do what I wanted, within the frames given. I perceived this as a very free and nice situation. [Do you appreciate having this freedom?] Yes, but it is something with the balance, because it can be too much and too little*».

When reflecting around a specific task, she appreciates both a certain freedom, while on the other hand having to comply to a defined framework. This is not addressed negatively, actually more the opposite. Both these interviewees therefore underline autonomy as a kind of balance. We will in the forthcoming explore what this balance could be about.

3.2 Autonomy as Interplay Between Tasks, Organization and Individual

The Balance Theory [9, 10] allows us to explore how high autonomy shapes and is shaped by the work system design. Based on the five interviewees descriptions we identify two main configurations. These are the dynamics between *task design and the individual* and the dynamics between *the task design and organization*. Both configurations act similarly on the individual, considering health and sick leave. In both configurations the work load becomes high for a longer period of time, causing high strain and health problems.

Dynamics Between Task Design and the Individual

The first configuration is dominated by the interaction between the domains **task design** and **the individual**. In this configuration informants talk about using a *variety of competencies and expertise*, as well as *difficulties in assessing task completeness*. Moreover, they talk about *unpredictability*, especially in relation to customers, entailing *lack of control* when planning in a daily basis. In combination with *personal characteristics* as *ambitions* and *high expectations* considering own job performance and quality, this add to the total work load, in the longer run being work overload. They evaluate the quality as not being good enough, it becomes impossible to assess when a task is completed and when it is satisfactory for their customers. Moreover, they also characterize themselves as personally engaged, *dedication and meaningfulness* play a huge role. One of the interviewees told us about herself: «*Eh, because you, you become so dedicated and engaged and ... perceive this as so important, that you forget ... you forget yourself, your own health and setting limits for when enough is enough. That is probably the largest challenge*».

In this configuration aspects within the organizational domain is to a little degree considered by the interviewees. Still it is a configuration where high engagement and perceived expectations is unfolded within a task design having a minimum of borders. However, **technology** is influential as technological solutions allow for working home, thus influences the work-family balance and time for recovery.

Dynamics Between the Task Design and Organization

The second configuration is dominated by the interaction between **task design** and the **organization**. High work load and stress is the cause for sick leave also in this configuration. However, when strain related to autonomy is emphasized, it is the dynamics between task design and the organizational domain that is described as challenging.

Considering **task design**, a high degree of working independent, having high autonomy to develop their own projects and *few routines* to consider is described. On the other hand, an interviewee stated this: «*it is the wrong type of freedom*». She described late incoming information, alterations and sudden changes in job content leading to *high time pressure*. In her case, *cognitive demands* are perceived as overwhelming, with a perception of *lack of competence* further entailing a *lack of perceived task completeness*.

The task design is influenced by several factors in the **organizational domain**. Two of the interviewees had to deal with regulations and political decisions, perceiving that the organization is not able to absorb and converting into *information*, *procedures*, *goals* and *explicit expectations*. Others clearly addressed management issues in other ways. One interviewee stated this: "*it is... actually it is a boundaryless organization. There is so little control and management that in a way... that could be ok, but when... when you come from outside and become socialized into this boundaryless... fellowship. Eh, this can create problems in the longer run. It is not a sustainable solution to anything*». This quote points to several aspects associated with *lack of management and leadership*. Several of the interviewees address an ambiguity considering a wish for being seen and acknowledged for the work they perform, but still wanting to retain control over their work situation. Secondly, lack of management control increase opportunities for informal leaders, that for others entail perceptions of *exploitation* and *lack of fairness*. One interviewee stated this when talking about a coordinator: «*I feel, that we cooperate in a daily basis, agreeing upon things, but still, it ends up with a result that I'm getting tasks I do not want or perceive as out of my control. I don't know why it end up this way*". An invisible management may also allow pressure and expectations not solely produced «from above» but among colleagues, allowing for *organizational (sub)cultures* to arise. One interviewee reflected around organizational routines and stated: *"Yes, yes, those defining the culture, ... and counteract plans and strategies within the organization, considering how we should work together, ... so... then the plan and the strategy loose"*. This may allow for experienced workers and seniors to establish a culture for working hard, putting pressure on younger workers. This could also be experienced as lack of accept for setting limits within own work situation, protecting for instance leisure.

Interviewees underline another sort of «blurry» character found in the relation between managers and employees especially in *project-based* organizations, becoming a source of distress and frustration. This «blurriness» show up when for instance, tasks are assigned, and projects are being developed. As the manager-employee relationship rest on notions of autonomy, openness, «horizontal dialogue» and self-management, the still intact difference between superior and subordinate, can be harder to confront and «put into words», because the «authority» in the process of assigning and developing projects is usually not unilateral but a two-way street.

4 Discussion

Among knowledge workers with an experience of work-related sick leave, autonomy is perceived as an ambiguity. When using the Balance-theory [9] as an analytical framework, two work system-configurations were identified. In one configuration personal characteristics were problematized pointing to lack of ability to cope with high autonomy. In the other configuration lack of management and leadership, structures and formal norms were problematized.

High autonomy is seen as a leverage to reduce strain and negative aspect of work [1]. Moreover, autonomy is in todays' working life a proactive approach to handle interdependency across cultural, geographical and organizational boundaries [4], and therefore to be understood as an organizational recipe and a management strategy. By using Balance-theory [9] as our analytical framework, we get indications that for health problems and sick leave associated with high autonomy, lack of structures and boundaries within the organizational domain might be the actual cause for the strain, as these groups of workers still value their autonomy and independency at task level.

The ambiguity found in this study, that means the importance of being independent while still being strained by perceiving lack of structures is in line with other studies on knowledge workers [7, 8]. Moreover, a Swedish study among highly educated young women working in male-dominated areas found that female workers struggled with overwhelming opportunities and demands as well as finding the balance between stress and recovery [12]. A mechanism for coping with this situation was setting individual boundaries as well as leaning on contextual boundaries (ibid). Balance-theory can be used for identification and elimination of negative aspects with the work situation or balancing the work system [9]. For work groups with high autonomy, attention should be given to stabilizing factors like clarity in organizational expectations and goals, a visible leadership and management and transparency regards information and decision processes.

These studies add to our critical understanding of high autonomy and question how we understand autonomy with regards work design. Specific principles of work design are suggested not being applicable in all contexts [9]. Within the sociotechnical design tradition, the emphasize of employees being involved at all levels of decision making, was elaborated in the aftermath of world war II, rebuilding the industry [13]. Moreover, the model by Karasek and Theorell was developed during the 70s [3]. The way we have understood autonomy, may be valid for industrial settings. However, the transformation seen last decade have entailed large changes in organizational forms as well as in job content, hence we may allow for several perspectives, depending on the work context.

References

1. Wu CH, Luksyte A, Parker SK (2015) Overqualification and subjective well-being at work: the moderating role of job autonomy and culture. Soc Indic Res 121(3):917–937
2. Karasek R, Theorell T (1990) Healthy work: stress, productivity, and the reconstruction of working life. Basic Books, New York
3. Theorell T, Karasek R (1996) Current issues relating to psychosocial job strain and cardiovascular disease research. J Occup Health Psychol 1(1):9–26
4. Grant AM, Parker SK (2009) 7 redesigning work design theories: the rise of relational and proactive perspectives. Acad Manag Ann 3(1):317–375
5. Kalleberg AL, Nesheim T, Olsen KM (2009) Is participation good or bad for workers?: effects of autonomy, consultation and teamwork on stress among workers in Norway. Acta Sociol 52(2):99–116
6. Hvid H, Lund H, Pejtersen J (2008) Control, flexibility and rhythms. Scand J Work Environ Health 83–90
7. Holte K, et al (manuscript) Structures and premises for perceived individual autonomy in knowledge organisations, a sociotechnical approach
8. Ekman S (2012) Authority and autonomy: paradoxes in modern knowledge work. Springer, New York
9. Carayon P (2009) The balance theory and the work system model … twenty years later. Int J Human-Comput Interact 25(5):313–327
10. Smith MJ, Sainfort PC (1989) A balance theory of job design for stress reduction. Int J Ind Ergon 4(1):67–79
11. Carayon P et al (2015) Advancing a sociotechnical systems approach to workplace safety - developing the conceptual framework. Ergonomics 58(4):548–564

12. Löve J, Hagberg M, Dellve L (2011) Balancing extensive ambition and a context overflowing with opportunities and demands: a grounded theory on stress and recovery among highly educated working young women entering male-dominated occupational areas. Int J Qual Stud Health Well-Being 6(3)
13. Mumford E (2006) The story of socio-technical design: reflections on its successes, failures and potential. Inf Syst J 16(4):317–342

Integrating Humans, Technology and Organization (HTO) in European Railway Safety Management Systems

Lena Keckland(✉), Jan Skriver, Sara Petterson, and Marcus Lavin

MTO Safety AB, PO Box 171 07, 104 62 Stockholm, Sweden
lena.kecklund@mto.se

Abstract. Legal requirements on safety culture and human factors has been part of the regulations in high hazard industries such as nuclear and aviation for several years and will now be a part of the European railway safety legislation.

In 2016, the concepts of safety culture and human factors was introduced in the re-cast European railway safety legislation (Directive EU 2016/798). The new railway safety regulations on Common Safety Method (CSM) for Safety management systems (SMS) 1169/2010 and 1158/2010 will include requirements on integration of human and organizational factors and safety culture within the railway license holders′ safety management system. Guidance on implementation will be published by the European Agency for Railways (ERA). The new regulations must be applied by the EU members states from 16 June 2019.

The concept of human and organizational factors is used to emphasize the broad range of factors influencing human performance, the interaction of such factors as well as a systemic safety view. The use of a new definition will reflect the systemic safety perspective where the interaction between humans, technology and organizations is explicitly considered. A new definition of the concept of safety culture is introduced.

This paper will give an overview of the definitions and requirements related to safety culture and the interaction between humans, technology and organization in the re-cast railway safety legislation.

Keywords: European railway safety directive · Common safety methods
Human and organizational factors · Safety culture
HTO (human-technology-organization)

1 Introduction

1.1 New Requirements in the European Railway Safety Legislation

Legal requirements on safety culture and human factors has been part of the regulations in high hazard industries such as nuclear and aviation for several years and will now be a part of the railway safety regulations. In other industries, for instance, in the nuclear sector [1], the importance of human and organizational factors as well as safety culture in the safety management system has been widely recognized.

S. Bagnara et al. (Eds.): IEA 2018, AISC 821, pp. 337–343, 2019.
https://doi.org/10.1007/978-3-319-96080-7_39

As a consequence, while elaborating the safety management system requirements regarding railway undertakings and infrastructure managers, the European Union Agency for Railways (ERA) decided to integrate safety culture, as well as human and organizational factors, into the secondary European Union legislation in the Fourth Railway package. The re-cast legislation and the new supporting guidance material includes requirements related to a systemic safety view, safety culture and the interaction of humans, technologies and organizations. MTO Safety assisted in developing the regulations in addition to providing guidance material.

In 2016, the concepts of safety culture and human factors were introduced in the re-cast Railway Safety Directive (EU 2016/798) [2] and in the revised Common safety methods for safety management systems (CSM SMS). The new railway safety regulations on CSM SMS (safety authorization) 1169/2010 and CSM SMS (safety certificate) 1158/2010 [3] includes requirements on integration and management of human and organizational factors and safety culture within the safety management system. The new legislation will be enforced in the EU member states from 16 June 2019.

1.2 Why Requirements on Safety Culture and Human and Organizational Factors Are Needed in Railway Safety Legislation

Global markets and technical developments introduces challenges to the achievement of safety and business goals in high-risk industries.

New business models require changes in business strategies, management systems, work processes, employment models and working conditions. Safety is an important part of the product delivered to clients in high-risk industries, e g railways and air traffic control. Therefore, features of the new business models present major challenges related to safety management and safety culture. Thus, new business models may change the interactions between humans, technologies and organization (the HTO system) in addition to challenging safety culture.

In order to ensure safety and safety culture in high risk industries new requirements and methods have to be developed both for both license holders and regulators. It has been suggested by the European Aviation Safety Agency (EASA) that the focal point of aviation operators and regulators should be on management systems including new forms of employment, safety culture and the governance structure of the company, e.g. subcontracting and outsourcing. This is an example of the importance of addressing human and organizational factors and safety culture. Therefore, requirements on safety culture and human and organizational factors has been introduced in the railway sector. The European Agency for Railways (ERA) has developed requirements on the safety management system of the license holders and supporting guidelines.

This paper will present the requirements related to safety culture and the interaction between humans, technology and organization in the re-cast European railway safety legislation. Also, general guidance on application of the legislation is given.

1.3 Safety Culture and Human and Organizational Factors in the Safety Management System (SMS)

The new guidance for railways safety management is based on the ISO high level structure (HLS) [4] in order to support the design of an integrated safety management system (SMS).

The ISO high level structure and the EU legal requirements consist of the following elements:

- Context of the organization
- Leadership
- Planning
- Support
- Operation
- Performance evaluation
- Improvement

The requirements and supporting guidance will require the license holders to demonstrate a systematic approach for integrating human and organizational factors within an integrated safety management system.

In order to meet these requirements the license holder must submit a strategy for human and organizational factors and a strategy for the continual improvement of safety culture. The license holders should demonstrate the use of expertise and recognized methods as a part of the application for a railways safety authorization or certificate. Furthermore, human and organizational factors to be integrated and considered for all activities and SMS elements.

2 Safety Culture and Human and Organizational Factors in European Railway Safety Management Legislation

2.1 Human and Organizational Factors

A definition and guidance has been developed for addressing human and organizational factors within European railway safety management systems.

Background. The concept of human and organizational factors is used to emphasize the broad range of factors influencing human performance as well as the systemic safety view. The new definition will reflect the systemic safety perspective where the interactions between humans, technology and organizations are considered in addition to the importance of using knowledge on human behavior in order to design safe and fit-for purpose systems [5, 7] (Fig. 1).

Definition of Human and Organizational Factors (HOF) in European Railway Safety Legislation. The definition is; "Human and organizational factors means all human performance characteristics and organizational aspects that must be considered to ensure the lifelong safety and effectiveness of a system or organization." [6].

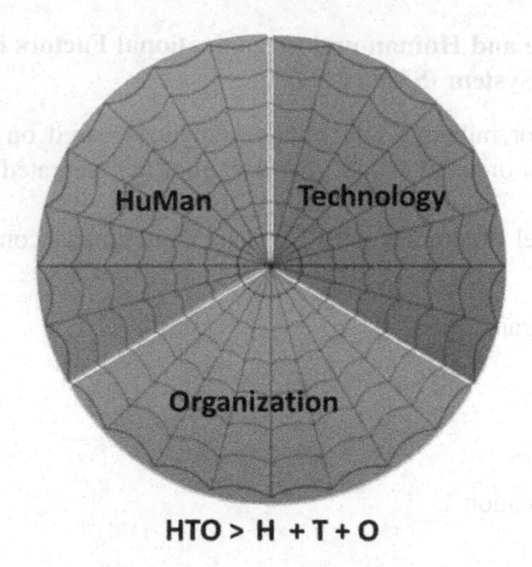

Fig. 1. Illustration of the systemic safety view and the interaction between humans, technologies and organizations.

Human and organizational factors (HOF) is a multidisciplinary field focusing on how to increase safety, enhance performance as well as increase user satisfaction. HOF is a user-centered approach, that is, the design is based upon an explicit understanding of users, tasks and environments. The starting point is always the users' capabilities and limitations and how these are influenced by and interact with the systems encountered during task performance. The goal is to identify how to best perform the task in a safe and efficient way with an emphasis on usability. HOF is used both as a proactive means of ensuring adequate design processes as well as a reactive means of identifying the key issues when something has gone wrong.

Application. The organization shall demonstrate a systematic approach to integrating human and organizational factors within the safety management system. This shall include the development of a strategy and the use of expertise and recognized methods from the field of human and organizational factors to address risks associated with the design and use of equipment, tasks, working conditions and organizational arrangements, taking into account human capabilities as well as limitations, and the influences on human performance [7].

Methods are drawn from many different fields, for example experimental psychology, industrial engineering, organizational psychology, sociology, management science, cognitive engineering, ergonomics, computer science and safety engineering [7].

HOF involves taking a systemic perspective, that is, not just looking at the human, technological and organizational factors in their own right but also emphasizing the interactions between the different factors. For example, if a train driver has been involved in a SPAD (Signals passed at danger) incident, the suggested factors to investigate (not a comprehensive list) relate to fatigue, cognitive overload, competence, etc. (Human), the technology's influence on performance, such as human-system interfaces, layout, signal placing (Technology), the organization's influence on performance, such as training, SMS, organizational priorities (Organization) as well as the interaction between the three areas such as the influence of procurement on design or management of change with the introduction of new design.

Other examples would include areas related to organizational factors affecting human performance such as staffing and work hours.

2.2 Safety Culture

A definition and guidance has been developed for safety culture within European railway safety management systems.

Background. There are many studies and reports that demonstrate the importance of safety culture for an organization's proactive protection against accidents and deviations [8–11].

There are also several definitions of safety culture ranging from the vague to all encompassing. For example, a commonly used definition is: The way we do things around here [12]. This definition provides a starting point for understanding safety culture but is imprecise and difficult to measure. At the other end of the scale a highly elaborate definition is: An organization's safety culture is the product of the values, attitudes, perceptions, competencies and patterns of behavior of individuals and groups that determine the commitment to and the efficiency of the safety management of the organization [13]. Safety culture is about people in their organizational context: the identity and behavior of employees, their managers and leaders and also that of their organization. Safety culture is defined at the group level and refers to shared values and identity among the group members.

Definition of Safety Culture in European Railway Safety Legislation. The definition is "Safety culture refers to the interaction between the requirements of the safety management system, how people make sense of them, based on their attitudes, values and beliefs and what they actually do, as seen in decisions and behaviors. A positive safety culture is characterized by a collective commitment by leaders and individuals to always act safely, in particular when confronted with competing goals." [6].

The first part of the definition is directed towards the description of the organizational culture (actual attitudes, decisions, and behaviors) and states the influence of the safety management system in the organization. The second part is more normative: if the actual organizational culture leads to a positive safety culture, it should be reflected in daily routines and arbitrations at all levels of the organization.

Application. Continual development of safety culture is an important characteristic of a high reliability organization. The new legislation emphasizes the importance of assuring continual improvement of safety culture. This includes establishing the maturity level and identify and implement appropriate measure to ensure continual development. In order to ensure continual development a safety culture strategy including regular safety culture assessments should be applied. Safety culture expertise can assist in performing independent safety culture assessments which will be important for continual improvement and learning processes.

A model for safety culture assessment can be based on the simple model of safety culture [5, 14] as presented in Fig. 2. The model contains three interdependent factors: safety management, sense making and behavior. Safety management refers to the written word, i.e. how the organization has converted regulatory requirements into policies, strategies, goals, procedures, checklists etc. essentially the safety management system. Sense making refers to the personnel's knowledge about and attitudes to the safety management systems meaning how personnel makes sense of it. Sense making can vary and even be contradictory at different organizational levels. Sense making is reflected in behavior. Behavior refers to what is actually done. This is seen in how safety is prioritized on daily basis, e.g. compliance with procedures, decision-making and procurement. The key to the model is that you can't analyse one factor on its own but must include all three factors to ensure that the whole picture is assessed.

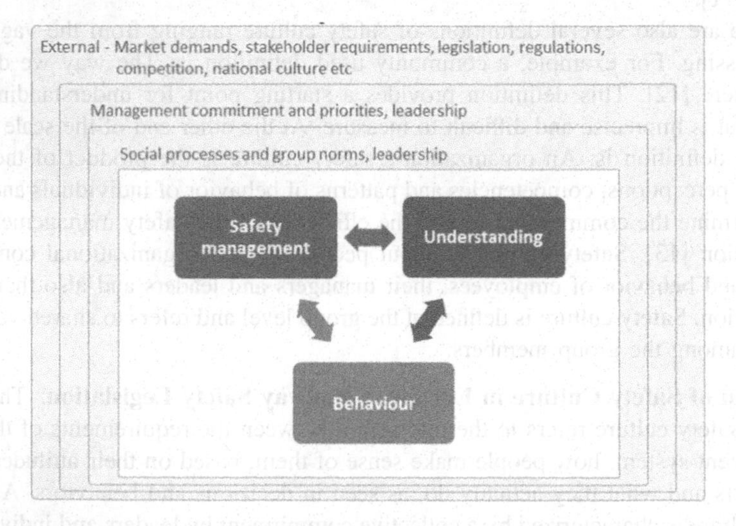

Fig. 2. Safety culture and the relation to external demands, management commitment and social processes.

3 Conclusion

In conclusion, focus on safety culture and HTO interactions is essential in order to ensure safety and reliability in railways as well as in other transportation systems. In particular, it is important when facing a rapid rate of change related to technical and organisational challenges in an interoperable and global context.

Addressing human and organisational factors and safety culture in all life-cycle stages of a complex system and using professional expertise and methodologies will be a requirement for licence holders on the European railway system. Professional application and implementation of human and organisational and safety culture expertise will also improve the business case and provide a competitive advantage for the licence holder.

Management and leadership commitment as well as knowledge and skills in human and organizational factors and safety culture will be essential to manage the challenges ahead.

Introducing the new requirements will also contribute to the achievement of the 17 sustainable development goals of the 2030 Agenda for Sustainable Development adopted by the UN [15]. The Agenda emphasizes the interaction of the core elements of economic growth, social inclusion and environmental protection where the interaction of the three elements is crucial for the well-being of individuals and societies. Furthermore, creating safe societies is imperative to achieve the sustainable development goals.

References

1. IAEA (2009) The management system for nuclear installations, IAEA Safety Standards Series, Safety Guide, No GS-G-3.5, Vienna, Austria
2. EU Directive 2016/798 of the European Parliament and of the Council of 11 May 2016 on railway safety (recast). https://eur-lex.europa.eu/legal-content/EN/TXT/. Accessed 27 May 2018
3. EU Common Safety Method for Conformity Assessment 1169/2010. Commission Regulation (EU) No 1169/2010 of 10 December 2010 on a common safety method for assessing conformity with the requirements for obtaining a railway safety authorization. https://eur-lex.europa.eu/legal-content/EN/TXT/. Accessed 27 May 2018
4. ISO/IEC Directives (2016) Part 1, consolidated ISO – Procedures specific to ISO, seventh edition. https://www.iso.org/sites/directives/2016/consolidated/index.xhtml. Accessed 27 May 2018
5. Kecklund L, Lavin M, Lindvall J (2016) Safety culture: a requirement for new business models. Lessons learned from other High Risk Industries. In: Proceeding presented at the international conference on human and organizational aspects of assuring nuclear safety–exploring 30 years of safety culture, Vienna, 22–26 February 2016
6. European Agency for Railways. Guide (2018). Safety management system requirements for safety certification and safety authorization. Draft, version 0.18 23 March 2018
7. Salvendy G (2012) Handbook of human factors and ergonomics. Wiley & Sons, New Jersey
8. Nævestad TO (2008) Safety cultural preconditions for organizational learning in high-risk organizations. J Contingencies Crisis Manag 16:154–163
9. Hudson P (2007) Implementing a safety culture in a major multi-national. Saf Sci 45:697–722
10. Rundmo T, Hale AR (2003) Managers attitudes towards safety and accident prevention. Saf Sci 41:557–574
11. Prussia GE, Brown KA, Willis PG (2003) Mental models of safety: do managers and employees see eye to eye? J Saf Resilience 34:143–156
12. Cooper MD (2000) Towards a model of safety culture. Saf Sci 36:111–136
13. ACSNI (1993) ACSNI Human Factors Study Group. Third Report: Organizing for Safety, Advisory. Committee on the Safety of Nuclear Installations, Health and Safety Commission
14. Skriver JA (2004) Simple model of safety culture. In: Waard D, Brookhuis KA, Weikert CM (eds) Human factors in design. Shaker Publishing, Maastricht
15. United Nations 2030 Agenda for Sustainable Development. Transforming our world: the 2030 Agenda for Sustainable Development http://www.un.org/ga/search/view. Accessed 27 May 2018

The Team Building for Human Resource Development in Elite Youth Soccer Players in Japan

Yasuyuki Hochi[1(✉)], Yasuyuki Yamada[2], Yukihiro Aoba[3], Tomoki Ebato[4], and Motoki Mizuno[2]

[1] Japan Women's College of Physical Education, 8-19-1 Kitakarasuyama Setagaya-ku, Tokyo 157-8565, Japan
hochi.yasuyuki@jwcpe.ac.jp
[2] Juntendo University, 1-1 Hiragagakuendai, Inzai, Chiba 270-1695, Japan
[3] Chuo Gakuin University, 451 Kujike, Abiko, Chiba 270-1163, Japan
[4] JEF United Corporation, 1-38 Kawasaki-cho Chuo-ku, Chiba, Chiba 260-0835, Japan

Abstract. Japan Football Association (JFA) carries out overall approach to National team's strengthening, youth development, coaches training. Especially, youth generation players learn about the severity of becoming professional football players, look at the reality myself and image for the future. In other words, youth players seriously seek out the problem of how they live in the real society, it is time to establish themselves (Erikson 1959). Therefore, we conducted the adaptability development program which applied the human resource development methods and the organization development methods to the youth (U-18) team belonging to the J-League (professional football league in Japan) in 2017. This program was composed of specialized tools such as TPI (Todai Personality Inventory) based on MMPI (Minnesota Multiphasic Personality Inventory).

We used Sport Self- management Skill scale (Takemura et al. 2013) and the Belief in Cooperation Scale (Nagahama et al. 2009), to examine the effects of this program from a longitudinal perspective. Furthermore, to verify the effect from a qualitatively, we collected the free descriptions about the impression of the program from participants. From the analysis, it was shown that the score of Sport Self-management Skill and Belief in Cooperation improves by this program ($p < .001$). From the text analysis, various factor about program effect were clarified such as promotion of self-understanding, clarification of self-concepts and goal setting.

Keywords: Organizational development · Measurements of effects
Collective efficacy · Intervention · Sport Self-management Skill

1 Introduction

Japan's football is based on the club system in Germany and is constructing an organizational system with the Japan Football Association (JFA) at the top. In order to compete with the world on equal levels, JFA carries out overall approach to National

S. Bagnara et al. (Eds.): IEA 2018, AISC 821, pp. 344–348, 2019.
https://doi.org/10.1007/978-3-319-96080-7_40

team's strengthening, youth development, coaches training. Especially, youth generation players learn about the severity of becoming professional football players, look at the reality myself and image for the future. In other words, youth players seriously seek out the problem of how they live in the real society, it is time to establish themselves (Erikson 1959). Therefore, we conducted the adaptability development program which applied the human resource development methods and the organization development methods to the youth (U-18) team belonging to the J-League (professional football league in Japan) in 2017. This program was composed of specialized tools such as TPI (Todai Personality Inventory) based on MMPI (Minnesota Multiphasic Personality Inventory). This program was intervened by the third-party change agent (professional facilitator).

2 Methods

2.1 Participants

This research carried out in one elite youth soccer team consisted of 35 members. Through the informed consent procedure, a total 35 members (male = 35) agreed with this study (response rate = 100%, cover rate = 100%). The mean age was 16.57 (SD = ±1.065) years old.

2.2 Team Building (TB) Program

The Team building program based on OD methods, had the theoretical background of the Transactional Analysis—focused on human relations. Then, this program had been designed by Kitamori who was structured the OD program in Japan. In this program, learning experience and many of the image replacement performed. Furthermore, this program was composed of specialized tools such as TPI (Todai Personality Inventory) based on MMPI (Minnesota Multiphasic Personality Inventory).

2.3 Measures

We used the Sport Self- management Skill scale (Takemura et al. 2013) and the Belief in Cooperation Scale (Nagahama et al. 2009), to examine the effects of this program from a longitudinal perspective. Responses to these questionnaires collected from the participants before and after the intervention of this program. Furthermore, to verify the effect from a qualitatively, we collected the free descriptions about the impression of the program from participants.

2.4 Analysis Procedure

To compare with each score of pre and post, we carried out paired t-test using statistical analysis software (IBM SPSS statistics 24) of IBM Corp. analyzed and text analysis software (Text Analytics for Surveys 4.0.1).

3 Results

The paired t-test showed that the score of subscales were higher than before the experience of TB program ($p < .001$) (Table 1). Then, the score of Sport Self-management Skill and Belief in Cooperation improves by this program.

Table 1. The result of paired t-test

	Pre		Post		Paired *t* -test					
	M	SD	M	SD	Pre-Post	SD	95%	CI	*t*	*p*
Sport self- management skill										
F1: Contribution to the team	18.54	1.36	18.66	1.47	−0.11	1.47	−0.62	0.39	−0.46	*n.s.*
F2: Thinking	15.31	1.78	16.37	1.85	−1.06	1.86	−1.70	−0.42	−3.36	***
F3: Self-recognition	17.00	1.81	17.60	1.80	−0.60	1.44	−1.09	−0.11	−2.47	**
F4: Sincere attitude	17.80	2.14	18.57	1.54	−0.77	1.85	−1.41	−0.14	−2.47	**
F5: Continuous Initiatives	13.66	2.57	13.60	3.23	0.06	1.91	−0.60	0.71	0.18	*n.s.*
F6: Achievement effort	17.74	2.03	17.83	2.08	−0.09	1.54	−0.62	0.44	−0.33	*n.s.*
F7: Improvement of issues	17.03	1.74	17.40	1.63	−0.37	1.37	−0.84	0.10	−1.60	*n.s.*
F8: Inventive ingenuity	17.31	1.68	17.71	1.27	−0.40	1.42	−0.89	0.09	−1.67	*n.s.*
Total	129.63	8.89	137.74	9.31	−8.11	6.05	−10.19	−6.04	−7.93	***
Belief in cooperation										
F1:Usefulness of cooperation	4.54	0.33	4.67	0.32	−0.13	0.26	−0.21	−0.04	−3.07	***
F2: Individual-oriented	3.00	0.51	2.60	0.60	0.40	0.50	0.23	0.57	4.72	***
F3: Reciprocity concern	1.74	0.71	1.60	0.59	0.14	0.63	−0.07	0.36	1.34	*n.s.*

***$p < .001$ **$p < .05$

Furthermore, the description on the program experience was analyzed by the KJ method and text analysis software (Text Analytics for Surveys 4.0.1). As a result, it was shown that the element corresponding to the experience effect can be classified from five elements at least “self-discovery”, “self-understanding”, “task discovery”, “introspective”, “understanding others”. Especially, descriptions related to “Problem finding” were obtained from all participants (35 people, 100%). In addition, the description on “Self-understanding” was obtained from 29 people (83%), and the description on “Reflection” was obtained from 27 (77%). Likewise, 16 people (46%) were “Self-discovery” and 6 people (17%) “Understanding others” (Table 2).

Table 2. The result of text analysis of the description

Category	Description (element data)	n	%
Problem finding	Points that should be worked out, a new task has been clarified I understood the improvement points I found out what I am missing now	35	100%
Self-understanding	I learned about the good points of myself I understood the expectation from others I knew the gap of self-cognition	29	83%
Reflection	I could deeply deal with myself Opportunity to look back on yourself Accept opinions from others	27	77%
Self-discovery	There was a new discovery I found a new one	16	46%
Understanding others	Recognize good points of fellow Opportunity to think about others	6	17%

4 Discussions and Conclusions

This study, carried out TB program to elite soccer team, was to compare the scores of each factor in the intervention before and immediately after the TB. As a result, it observed that tends to increase the score of several factors after the intervention. These results suggested that Sport Self-management Skill and Belief in Cooperation improves by this TB program. We would like to try to guess about why the difference in the score before and after the intervention of TB. The characteristic of the TB program, which was adopted in this study is to obtain the "new awareness" for the environment (self, others, and organization) in a short time. Therefore, image exchange is widely used in the TB program to promote mutual understanding and interactions of the participants. Therefore, the results of this study, it can be said to be an important discovery from the viewpoints of both practical and academic.

5 Future Work

The result of this study just compared with before and after the intervention of TB program. That had not compared to the control groups. In the future, it is necessary to longitudinal research and compared with the control groups to examine the effects.

References

French WL, Bell CH (1973) Organization Development: Behavioral Science Interventions for Organization Improvement. Prentice-Hall, Englewood Cliffs

Mizuno M, Hochi Y, Inoue M, Kaneko I, Yamada Y (2012) Construction of the integrated model for practical career support to the professional athletes. Work J Prev Assess Rehabil 41 (Suppl 1):5767–5768. IOS Press

Mizuno M, Yamada Y, Hochi Y, Honda R, Takahashi H, Shoji N, Aida H, Okada A, Mizuno Y (2015) Behavioral sensor-based organizational design and management in Japan: from the perspectives of communication channel in nursing organization. In: International ergonomics association, proceedings 19th triennial congress of the IEA, 7 pages

Myers ND, Feltz DL, Short SE (2004) Collective efficacy and team performance: a longitudinal study of collegiate football teams. Group Dyn Theory Res Pract 8(2):126–138

Shoji N, Hochi Y, Fujii H, Moriguchi H, Nakayama T, Mizuno M, Kitamura K (2014) A study of the organizational support for building resilience of the fitness club employees. In: Advances in social and organizational factors. AHFE conference, pp 425–432

Takemura R, Shimamoto K, Kato T, Sasaki M (2013) Self-management for student athletes in sports groups: development of a sport self-management skill scale. Jpn J Phys Educ Hlth Sport Sci 58:483–503

Hochi Y, Mizuno M, Kitamura M, Kitamura K (2014a) Effect of career support program "self-analysis seminar" on the self-efficacy of career decision making among Japanese university students in sport science. In: Proceedings of the 5th international conference on applied human factors and ergonomics, AHFE 2014, USB

Hochi Y, Mizuno M, Nakayama T, Kitamura K (2012) A longitudinal study on the effects of team building for university baseball team in Japan: from the view point of team-vitalization. Work J Prev Assess Rehabil 41:5762–5763

Hochi Y, Mizuno M, Nakayama T, Kitamura M, Honda R, Kitamura K (2014b) A study on the effects of team building for university baseball team in Japan: focused on morale and belief in cooperation. In: The 7th Asian-South Pacific association of sport psychology international congress in Tokyo, CD-ROM

Hochi Y, Ohsiro T, Minato S, Mizuno M (2017a) The psychological contract of employee among professional sport company: a case of Japan independent baseball league. In: 8th international conference on applied human factors and ergonomics, conference proceedings, USB

Hochi Y, Yamada Y, Mizuno M (2015) Effects of the organization development of first-year experience and developed evaluation scale in Japan: focused on the psychosocial aspects among freshman of university. In: International ergonomics association, proceedings 19th triennial congress of the IEA, 6 pages

Hochi Y, Yamada Y, Mizuno M (2017b) Team building program for enhancement of collective efficacy: the case of a university baseball team in Japan. In: advances in human factors in sports and outdoor recreation. Advances in intelligent systems and computing, vol 496. Springer, Heidelberg, pp 125–129

Ergonomics in the Age of Creative Knowledge Workers – Define, Assess, Optimize

Lorenz Prasch(✉) and Klaus Bengler

Chair of Ergonomics, Technical University of Munich, Garching, Germany
{lorenz.prasch,bengler}@tum.de

Abstract. While the performance of knowledge workers is becoming increasingly important, optimization for one of their key responsibilities is a difficult endeavor: The generation of new knowledge or products, meaning the creation of creative solutions. This is partially due to the fact that a coherent understanding of the term creativity as well as reliable assessment methods are yet to be found. After a brief overview of available definitions, and distinction and classification of measurement methods, this paper describes a new approach that can enable creativity assessment. Focusing on an outcome-centered view, a method is proposed, that describes how an environment or process can be tested towards it's potential in enhancing creative performance. By devising carefully crafted tasks, a machine can objectively evaluate a product's creative score in terms of usefulness and novelty. This enables systematic testing and comparable assessment throughout different contexts of interest, ultimately bearing the potential to unify testing across different domains. With growing datasets, it should be possible to discover additional common features of creative solutions, thus allowing further insights in the complex construct of creativity and how it is composed. This can facilitate understanding of work environments that foster creativity and shed light on an immanent human ability that is of increasing importance in the future.

Keywords: Ergonomics · Knowledge work · Creativity

1 Introduction

Ergonomics is a discipline concerned with achieving the optimal synthesis of human well-being and system-performance in the context of work [1]. However, the ever-changing concept of work as such entails a wide variety of different scenarios in which work is being carried out. From agricultural workspaces over traditional production workers up to office clerks, there are almost no workspaces and working systems that are exactly alike. Nevertheless, our domain strives to find common rules and laws to ensure aforementioned design goals.

In order to classify the different types of workers, several attempts have been made, including the popular distinction in blue and white-collar workers. Whereas blue-collar work typically refers to manual labor, white-collar workers usually perform administrative or managerial tasks. In recent years however, with faster product development cycles, the internet as a motor of change and the increased necessity for companies to

S. Bagnara et al. (Eds.): IEA 2018, AISC 821, pp. 349–357, 2019.
https://doi.org/10.1007/978-3-319-96080-7_41

foster innovation, a new type of worker has gained scientific attention: The knowledge worker, creative worker, or gold-collar worker [2–4]. This term refers to a specific type of worker whose main asset for the company is the acquisition, procession, generation as well as communication of knowledge [5].

Over the last century, the amount of knowledge work being carried out and the number of knowledge workers employed has risen continuously [6]. It is reasonable to assume that this kind of work will further increase, since digitization, industry 4.0 and artificial intelligence will advance and cover additional areas of human expertise, leaving the human worker obsolete but for his/her ability to think creative and generate novel ideas and knowledge. Many of the tasks in a knowledge worker's daily schedule cannot be described in traditional fashion and generate several new demands that have not been addressed by ergonomics so far (flexibility, creativity, …). In contrast to manual work, which has seen a rise in productivity throughout the 20th century following the principles of Frederick Winslow Taylor [7] and has been proposed to be responsible for all gains (economical as well as social) of that century, work on the productivity of knowledge workers has barely begun [8].

In the year 1999, Peter Drucker proposed making knowledge workers more productive as the central challenge for the 21st century. This, however, entails not only the question on *how* to increase productivity, but also the question *what* productivity in the context of knowledge work means. As a central business dogma states: "if we can't measure it, we can't manage it" [6]. The productivity of manual work can usually be quantified by something like units per time or more generally speaking, output divided by input. For knowledge work in contrast it is often hard to determine what exactly the desired output is and what kind of input there is to consider.

Especially in the domain of creative performance this issue leads to several uncertainties. A specific creative idea that is implemented and shows societal impact can be worth a multitude of marginally less creative ideas that do not have a comparable range. The scientific study and operationalization of creativity has a long and dreary history and has been described as a degenerating research program [9]. Despite the fact that the generation of creative ideas and the underlying processes has already been an interest of ancient poets like Aristotle and great scientific minds like Helmholtz (1821 – 1894) and Poincaré (1854 – 1912) [10], a coherent understanding has yet to be found. To this day the definition of "humankind's ultimate resource" [11] is not satisfactory [12].

2 Defining Creativity

The definition of creativity is an ambivalent endeavor. While it is encouraging that the approaches are manifold and reach from psychoanalytic, psychometric, cognitive social-psychological or in recent years neurobiological [12], thus emphasizing the importance, it is simultaneously frustrating that no clear definition has been agreed upon yet. While the ancient Greek attribution to a divine aspect and the association of creativity with mystical powers of protection and good fortune [13] are no longer in place, the distinction "what creativity is, and what it is not, hangs as the mythical albatross around the neck of scientific research on creativity" [14]. Several scholars of creativity have tried to accurately describe the construct and to identify components as well as types of creativity. Some of these attempts shall be examined in the following sections.

2.1 Types, Components and Dimensions of Creativity

Types of Creativity. There is consensus that differentiation is necessary between at least two types of creativity. There are ideas that appear to be novel in the mind of the individual having the idea (although many people might have had the same idea before) and ideas that are truly novel in historic proportions, having the potential to change an entire domain. Czikszentmihalyi separates two kinds of individuals, *personally* and *unqualifiedly* creative ones [15], Boden contrasts *psychological* (P) and *historical* (H) creativity [16] and Gardner distinguishes *little-c* and *Big-C* instances of creativity [17]. The latter model was later complemented by two additional instances, *mini-c* and *Pro-c* [18]. The general distinction however remains and is coherent despite the different approaches in classification and wording.

Components of Creativity. There is a variety of attempts describing several factors or components that influence creativity. One model and its factors in particular have reached a relatively wide consensus in the scientific community however [19]. Mel Rhodes introduced his model considering four components of creativity in 1961 [20]: The *4Ps model*. It incorporates the *person* that creates (and subsequently intellect, attitude, personality, …), the *process* involved (i.e. thinking, motivation, communication, …), the environmental *press* or interaction between surroundings and creative actors, and finally the *product* that is the result of creative work. This product has to be communicated to others by means of music, paint, words or others.

Dimensions of Creativity. When trying to differ between presence or absence of creativity (say trying to tick boxes or quantify certain aspects that are fulfilled when creativity is considered present), the view of what matters most is highly dependent on what components are emphasized. In general, there are two different approaches. The more traditional view, often found in psychology, focuses on creative *individuals*. In contrast, many economical scholars adopted the perspective of creative *ideas* or *products*. While the view of creativity as a personal trait often leads to an assessment via certain *personality characteristics* [21], the level of creativity of a certain idea or product is usually evaluated with respect to said product's *novelty*, *usefulness* and *impact* [22]. There is broad consensus that the focus chosen should be tailored to specific research interests [23], therefore we prioritize an output-centered point of view. The focus on novelty and usefulness can be found in most popular definitions of creativity, that is both explicit as well as implicit ones (e.g., [24–28]). The additional consideration of *impact* is mostly determining the type of creativity present (Big-C or little-c). The greater an ideas influence on the domain (or society in its entirety for that matter), the further up on the hierarchy of creativity it should be considered. Finally, a product's creativity is considered a continuous rather than a categorical quality [22].

2.2 A Definition of Creativity

A review of recent definitions of creativity in peer reviewed papers revealed that they are rarely consistent – if offered at all [29]. In the definitions provided, more than nine common characteristics could be identified with various frequencies of use. Depending on frequency, applicability and general consensus Plucker and colleagues propose a definition on the basis of the literature considered:

> Creativity is the interaction among *aptitude, process* and *environment* by which an individual or group produces a *perceptible product* that is both *novel* and *useful* as defined within a *social context*. [29]

Three specific factors that can influence creativity are described, namely the workers aptitude, the process used, and the environment in which work is being carried out. Furthermore, two dimensions are provided that have to be fulfilled by a product in order to be considered creative, namely *novelty* and *usefulness*. When combining the dimension usefulness and the social context in which a product has to be deemed useful, the requirement for a creative idea to be *appropriate* [23] does emerge.

3 Assessing Creativity

Similar to the definition itself, the assessment of creativity has seen a multitude of approaches from different domains interested in the construct. Depending on the focus of interest, or point of view, different priorities are chosen [12]. With focus on the person carrying out the creative task, attributes like personality or intelligence are called upon (e.g. [30, 31]). When stretching the importance of processes employed, oftentimes problem-solving is the modus operandi (e.g. [32, 33]). In emphasis of the environment in which the task is being carried out, assessment is governed by a design for creativity (e.g. [34, 35]). The general consensus however appears to be that creativity involves the generation of novel and useful products [36]. This implies that it does not necessarily have to be measured directly with the creative thinker. It suggests that the existence and/or the amount of creativity can be measured by evaluating the outcome with respect to these dimensions. Individuals producing novel and useful products would then be deemed creative in hindsight.

3.1 How Are We Doing so Far?

Historic Approach. By using ideational tests that required participants to describe objects seen in an inkblot, complete pictures, or produce long lists of words, as early as 1927 Hargreaves was able to discover a factor of intelligence that corresponded with ideational production, but was independent of general intelligence [37]. This had an enormous impact on the view of creativity as a personal factor and led to a common set of tests used to measure it, all based on the principle of *divergent thinking*: an individual's ability to explore many possible solutions to a problem in a non-linear fashion and to draw unexpected connections (e.g., [38–40]).

Current Approaches. The current approaches of assessing creativity are a little more structured and spread across a variety of metrics. Generally, it is differentiated between different types of foci, as well as corresponding categories of tests. As mentioned before, we can differentiate between the two different foci of locating creativity: the person versus the product as carrier of creativity. Additionally, three distinct types of measurement can be found in the literature: objective versus subjective ones, with the subjective category being divided into self-assessment and other assessment (see Table 1 and [12, 23] for an overview).

Table 1. Overview of creativity measurement methods. For detailed description see [12, 23].

Type	Focus	
	Person	Product
Objective	Intellect, personality, motivation Remote association Biological traits	E.g. Baron-Welsh art scale
Self-rated	Self-ratings	Recognition of creativity in own ideas
Other-rated	Expert-ratings of individuals	Expert-ratings of products

3.2 What's the Problem with That?

The biggest and most challenging problem is that, despite the lack of a definition that is commonly agreed upon, all methods that attempt to measure creative performance in an experimental setup make certain problematic assumptions about the construct. Be it the fact that experts are reliably able to evaluate different ideas or products, that self-assessment is properly reflecting one's ability to work creatively in a certain environment, or the assumption of an underlying distribution. Additionally, several of the tests commonly used have been shown to have small predictive, ecological and discriminant validity [23].

Another main concern about current methods is the fact that hardly any of them scale practically and reliably. The subjectivity of ratings makes results across different studies difficult to compare. Additionally, the excessive need for resources and materials in objective as well as subjective measurement results in a small corpus of results in total. For fundamental research, it is essential to find a reliable, easy to use, and accurate measurement tool of creative performance. This tool should relate to a certain context of interest in order to be able to properly compare and conclude from aggregated results.

To achieve this, we suggest devising a set of creative tasks that are compared with machine learning. This way the result's usefulness as well as novelty can be assessed.

3.3 So, What Exactly Do We Have to Do? Where Should the Focus Be?

According to the 4Ps model, a combination of person, process and press result in a certain product. This can be formalized as the equation

$$person \times process \times press = product \qquad [12].$$

Under these circumstances it seems adequate to assume that, if the goal is an optimal synthesis of well-being and system performance, the result or product itself is the only viable option for performance assessment. While personal factors should be taken into account (especially in terms of well-being), variations of the other two factors, process and press, seem possible in order to alter the systems output. Hence, it should be possible to optimize process as well as environment in experimental studies.

From the three dimensions in which the outcome, or product, of creative performance could be measured (novelty, usefulness or appropriateness and impact), it is apparent that for ergonomic purposes a consideration of on only the first two is sufficient. The dimension impact in this area is negligible for several reasons. First, it essentially determines only the effect of a creative idea or product; the absence of said effect does not mean that the idea itself was not creative in the first place [22]. Secondly, as impact can only be assessed in hindsight, it is not only impractical for operationalization in an empiric setting, it is also some sort of an assessment system itself. This is true at least if we agree on the fact that truly novel and useful ideas have a higher potential of having an effect in the context they comply with. Lastly, impact is mainly responsible for differentiation between Big-C and little-c creativity (or various stages hereof). However, any Big-C discovery must initially be considered a little-c discovery. Since historic novelty is impossible without personal novelty, we can confidently exclude impact as a relevant dimension for human factors. If it is possible to optimize for mundane creativity, the probability for a flash of genius is certain to increase as well.

4 Measuring Creativity Using Machine Learning

The inability of the community to agree on one test for creativity has, as described above, a multitude of well-grounded reasons. However, one of the major problems, the lack of a precise understanding of the term creativity, could indeed be overcome. Concentrating on the most agreed upon properties of creative products, *novelty* and *usefulness*, it is possible to devise tasks specifically tailored to enable differentiable results in these two dimensions. If agreed upon that creative outcomes can be compared in terms of usefulness and novelty, no further assumptions are necessary.

Depending on the domain chosen, several standardized tasks can be designed that have to fulfill only one criterion. From the product generated, it has to be possible for a machine to recognize a metric (ideally continuous instead of categorical) that describes the usefulness, or appropriateness of the solution. For example, in case of a graphical constructive task (e.g. tangram), the usefulness could be devised from the percentage of fulfillment, or how precise the shape that is provided is matched by a certain solution. This allows for an automated assessment of the dimension usefulness.

For an automated assessment of the second dimension, an algorithm could utilize unsupervised machine learning to cluster similar solutions. This would enable the tool to allocate any solution to a cluster of similar outcomes. Depending on the amount of solutions in the specific cluster, a value for novelty can be assigned. That is, the more individuals already have created a similar solution before, the less novel it is to be considered.

This enables an automated assessment of products in a certain context, or solutions to a specific, domain dependent task. A better understanding of the inherent structure of the potential solutions and the rules for their generation could help to make the algorithm more efficient; especially if the complexity of the problem to be solved increases.

Advantages. This method has several advantages over current approaches and could potentially unify creative assessment in several domains:

- Due to no human raters, results obtained are completely objective and therefore comparable across different studies
- No further assumptions besides the novel and useful paradigm are made
- The automation of assessment ensures unlimited scaling
- The underlying distribution of solutions to certain creative problems can be discovered
- Domain specific tasks can be devised in order to meet the requirement of contextual creative performance

Disadvantages. However, there is a certain set of potential drawbacks to this method. It relies heavily on the machine readability of the tasks chosen. This poses the threat of a very artificial set of tasks and therefore small ecological validity. Another factor that might impair the assessment of tasks such as tangram is the fact that motivation has been stressed as a crucial factor for creative performance [36]. In case of a suitable, but boring task, participants might not engage with their full ability to find the most creative solution they would be able to generate. Careful task design and the use of specific motivational tools however should be able to minimize these risks.

5 Summary

The complexity of scientific research on creativity is obvious. However, it is an interesting and worthy endeavor. Not only the rich history, but also the predicted rise in importance of a coherent and well-founded understanding on human production of creative products make it a field worth studying. Despite the difficulties in accurate definition as well as reliable assessment, we believe to have found a method to enable researchers to design processes as well as environments in order to foster creative performance through automated testing.

This paper proposes an approach of measurement of creative products in two dimensions, novelty and usefulness. By outlining requirements for task design on a conceptual level, we describe a powerful method to make testing for creativity possible in various contexts. Through the automation of assessment, this approach should be able to overcome most of the drawbacks of current assessment tools while remaining objective in a variety of contexts. With growing datasets, the measurement of creativity by machine learning should also provide further insides in how the construct can be understood and how it is composed. Eventually, time will tell this method's value, but from a theoretical standpoint it should be able to enhance the overall grasp on this complex matter. It should enable the community to get a deeper understanding of working conditions that foster creativity and to understand the mechanisms behind an inherent human ability that will be of increasing relevance in the upcoming years.

References

1. Dul J et al (2006) A strategy for human factors/ergonomics: developing the discipline and profession. Ergonomics 55(4):377–395
2. Kelley RE (1985) The Gold-Collar worker: harnessing the brainpower of the New Work force. Addison-Wesley, Reading
3. Bubb H (2006) A consideration of the nature of work and the consequences for the human-oriented design of production and products. Appl Ergon 37(4 spec. iss.):401–407
4. Roongrerngsuke S, Liefooghe A (2013) Attracting gold-collar workers: comparing organizational attractiveness and work-related values across generations in China, India and Thailand. Asia Pac Bus Rev 19(3):337–355
5. Drucker P (1959) Landmarks of tomorrow: a report on the new "Post-Modern" world. Harper & Row, New York
6. Ramírez YW, Nembhard DA (2004) Measuring knowledge worker productivity. J Intellect Cap 5(4):602–628
7. Taylor FW (1914) The principles of scientific management. Harper & Brothers, New York
8. Drucker PF (1999) Knowledge worker productivity: the biggest challenge. Calif Manag Rev 41(2):79–85
9. Glover JA, Ronning RR, Reynolds CR (1989) Perspectives on individual differences: handbook of creativity. Plenum Press, New York
10. Wallas G (1926) The art of thought, London
11. Toynbee A (1964) Is America neglecting her creative minority? In: Widening horizons in creativity, pp 3–9
12. Batey M (2012) The measurement of creativity: from definitional consensus to the introduction of a new heuristic framework. Creat Res J 24(1):95–104
13. Albert RS, Runco MA (1999) A history of research on creativity. Handb Creat 2(0502167):16–34
14. Prentky RA (2001) Mental illness and roots of genius. Creat Res J 13(1):95–104
15. Csikszentmihalyi M (1988) Society, culture, and person: a systems view of creativity. In: Sternberg RJ (ed) The nature of creativity. Cambridge University Press, Cambridge, pp 325–339
16. Boden MA (2003) The creative mind: myths and mechanisms, 2nd edn. Basic Books, New York
17. Gardner H (1993) Creating minds: an anatomy as seenthrough the lives of Freud, Einstein, Picasso, Stravinsky, Eliot, Graham and Gandhi. HarperCollinsPublishers, New York
18. Kaufman JC, Beghetto RA (2009) Beyond big and little: the four C model of creativity. Rev Gen Psychol 13(1):1
19. Runco MA (2004) Creativity. Annu Rev Psychol 55(1):657–687
20. Rhodes M (1961) An analysis of creativity. Phi Delta Kappan 42(7):305–310
21. Guilford JP (1959) Traits of creativity. In: Anderson HH (ed) Creativity and its cultivation. Harper, New York, pp 142–161
22. Piffer D (2012) Can creativity be measured? An attempt to clarify the notion of creativity and general directions for future research. Think Skills Creat 7(3):258–264

23. Zeng L, Proctor RW, Salvendy G (2011) Can traditional divergent thinking tests be trusted in measuring and predicting real-world creativity? Creat Res J 23(1):24–37
24. Besemer S, O'Quin K (1986) Analyzing creative products: refinement and test of a judging instrument. J Creat Behav 20(2):115–126
25. Cropley AJ (1999) Creativity and cognition: producing effective novelty. Roeper Rev 21(4):253–260
26. Sternberg RJ, Lubart TI (1999) The concept of creativity: prospects and paradigms. Handb Creat 1:3–15
27. Runco MA, Jaeger GJ (2012) The standard definition of creativity. Creat Res J 24(1):92–96
28. Corazza GE (2016) Potential originality and effectiveness: the dynamic definition of creativity. Creat Res J 28(3):258–267
29. Plucker JA, Beghetto RA, Dow GT (2004) Why isn't creativity more important to educational psychologists? Potentials, pitfalls, and future directions in creativity research. Educ Psychol 39(2):83–96
30. Eysenck HJ (1993) Creativity and personality: suggestions for a theory. Psychol Inq 4(3):147–178
31. Guilford JP (1950) Creativity. Am Psychol 5(9):444–454
32. Finke RA, Ward TB, Smith SM (1992) Creative cognition: theory, research, and applications
33. Mednick S (1962) The associative basis of the creative process. Psychol Rev 69(3):220
34. Amabile TM, Conti R, Coon H, Lazenby J, Herron M (1996) Assesing the work environment for creativity. Acad Manag J 39(5):1154–1184
35. Dul J, Ceylan C (2011) Work environments for employee creativity. Ergonomics 0139 (January):1–25
36. Mumford MD (2003) Where have we been, where are we going? Taking stock in creativity research. Creat Res J 15(2–3):107–120
37. Hargreaves HL (1927) The "Faculty" of imagination: an enquiry concerning the existence of a general "Faculty," or group factor of imagination. Br J Psychol Monogr Suppl 3(10)
38. Wallach MA, Kogan N (1965) Modes of thinking in young children
39. Guilford JP (1967) The nature of human intelligence
40. Torrance EP (1974) Torrance tests of creative thinking. Directions manual and scoring guide, verbal test booklet B. Scholastic Testing Service

Implementing Tele Presence Robots in Distance Work: Experiences and Effects on Work

Christine Ipsen(✉), Giulia Nardelli, Signe Poulsen, and Marco Ronzoni

Management Science, Technical University of Denmark, 2800 Kgs. Lyngby, Denmark
chip@dtu.dk

Abstract. As companies move toward globalization, companies use distance work to accomplish work more effectively and efficiently. A telepresence robot (TPR) is a mobile remote presence device that allows a two-way communication and interaction between a distance manager and the employees. The objective of the study was to improve the understanding of how distance workers and managers experience the use of TPR in the daily management and in which tasks the TPR is suitable to ensure employee well-being and thus performance. The data collection included three phases – before, during and after the implementation of the TPR, where we conducted 25 semi-structured individual and group interviews, on-site observations of the TPR in use and research notes. The distance manager (user) controlled the TPR from a distant site when using it in the home office. The managers were able to create a sense of proximity and via the camera feature, enable eye-contact, which the managers considered essential and beneficial for assessing the employee's feelings and well-being. The majority of the users had a positive experience regarding the TPR basic functionalities′ utilization. In all three cases the participants, both managers and employees, agreed that the TPR is most useful in planned project meetings. On the other hand, the lack of trust, problems with the technology, privacy issues and intrusive emotions affected the use of the TPR in a negative way in some cases. The TPR was not suitable for meetings where people needed to share physical documents or important meetings, i.e. private talks or decisions meetings.

Keywords: Telepresence robots · Implementation · Distance management

1 Introduction

1.1 Dispersed Work and Distance Management

Over the years, organizational changes in large traditional organizations and the development of new business opportunities across the world have dispersed workplaces and employees [5]. As companies move toward globalization and communication technologies facilitate a quicker pace of change within organizations [6], companies use distance work [1] to accomplish work more effectively and efficiently. Distance

S. Bagnara et al. (Eds.): IEA 2018, AISC 821, pp. 358–365, 2019.
https://doi.org/10.1007/978-3-319-96080-7_42

work and management can occur at different locations, from home (telework), in satellite offices (intra-organizational work), or at the customers' or clients' locations (interorganizational work) [2, 3].

Distance managers are concerned about the wellbeing of their employees working across distances so that the employees can perform [4]. Therefore, distance managers look for processes and technologies that can support the trusting relationship and create a sense of proximity across distances, which are two key elements of employee wellbeing in distance work [1, 4]. With regard to this, the dialogue is a core activity to ensure well-being, where the managers listen to their employees and acknowledge their situation and job conditions. The primary technologies are emails and real-time communication tools allowing for synchronic communication like Skype, phone, Lync, and if possible, telepresence technologies [7].

1.2 Telepresence Robots (TPRs)

Telepresence technologies, specifically, exist also in mobile versions, termed mobile remote presence (MRP) technologies, where telepresence robots (TPRs) as the Double from Double Robotics is an example of this. The combination of video conference and robotics forms the ground of the invention of telepresence robots (or remote robotic presence). Thus, TPR is a mobile remote presence device that allows a two-way communication and interaction between a distance manager (called user) driving and utilizing the TPR, and a person exposed to the interaction with the TPR (called participant).

TPRs require that two people interact through one robot - the person that controls the robot, which acts as the user, and the person exposed to and interacting with the robot (the double in this case) as part of his/her daily work. Thus, the interaction becomes two-way via the TPR. Consequently, TRPs can improve interaction across distances, however, the use requires considerations for implementation and operations as they involve human-robot interaction (HRI). HRI can be studied from a robotics-centered or human/user-centered perspective.

1.3 Objective

Despite increased attention to the use of robots in work processes, research has yet to uncover the effect of dual/two-way interaction and application of TPRs in daily work and management across distances. With this paper, we contribute to a clearer understanding of the effect on work and employees of applying and interacting across TPRs.

The exploration of the dual interaction, introduction and application of TPR in distance work allows to outline some conclusions, not yet depicted in previous research, on user experiences involved. Moreover, this study enables to discover tasks that are suitable for the TPR usage and to formulate suggestions for new users.

The objective of the study is to improve our understanding of *how distance workers and managers experience the use of TPR in the daily management and in which managerial tasks the TPR is suitable to ensure employee well-being and thus performance.*

1.4 Research Design

The study completed early 2018 with a total number of three companies. The paper presents the findings of an exploratory case study of three knowledge intensive companies (KICs), representing software development, engineering consultancies and finance and banking. We used the following criteria to select the cases:

1. Knowledge intensive companies,
2. Located in the surroundings of Copenhagen, to have quicker access to the facilities,
3. The daily manager works from another site than the his/her employees i.e. intra-organizational work,
4. Trial period of the TPR for four weeks as part of daily management,
5. "Open space" office design to allow the independent motion of the TPR.

The study builds on literature on distance management [1, 8], knowledge work [4, 9–12] and human-robot interaction [13, 14]. To investigate how distance workers experience their manager's use of TPR in their daily management and in which managerial tasks the TPR is suitable to ensure employee well-being we did an empirical investigation in knowledge intensive work following general guidelines for conducting qualitative research [15, 16].

1.5 Data Collection

We focused the data collection and analysis on the interaction between distance employees and their manager's use of the TPRs with a three-fold purpose: (1) assess positive and negative user experiences (Managers), (2) determine in which situation TPRs are useful, and (3) highlight the employees' personal feelings and emotions when communication via the TPR.

Data collection was structured in three phases – before, during and after the implementation of the TPR. The methods included semi-structured interviews before and after the trial period of the TPR, on-site observations of the TPR in use and research notes to explore the individual experiences of working with a tele-presence robot. The distance manager controlled the TPR from a distant site when using it in the home office. We conducted, eight interviews, three managers (before and after, that was the main user of the TPR) and 17 employees (mainly after), were interviewed across the three cases focusing on the users' experiences and personal emotions involved when working with the robot (Tables 1 and 2).

Table 1. Interviews with managers (User)

Company	Before implementation	After
Consultancy	1	1
Bank	1	1
Finance (IT)	1	1

Table 2. Interviews with employees (Participants)

Company	Before implementation	No. of Group-interview (after)	Total no. of participants
Consultancy	0	1	8
Bank	1	1	2
Finance (IT)	0	1	7

To assess positive and negative user experiences, determine in which managerial tasks the TPR is suitable and highlight the personal feelings and emotions involved, we observed the employee tele-presence robot interaction. In total, we observed 20 situations where the TPR was in use. When operating the TPR, the managers worked from home or other branch offices and the latter respondent group worked at the home office. We thus focus on intra-organizational distance work where distance was geography and/or time. Distance work at customers (inter-organizational distance work) is excluded from this study.

To do so, we explored the use and adoption of the TPR technology building from Venkatesh and Davis' Technology Acceptance model (2000) TAM, which focuses on the user's acceptance of the technology to ease technology adoption and thereby performance. Particularly the newest version is used: the TAM3, evolution of TAM2 [17], that embraces new determinants of perceived usefulness and perceived ease of use [18].

1.6 Data Analysis

The interviews were transcribed and coded in an open process applying Atlas.ti where the initial codes derived from the research question [19]. The initial codes were: 1) User experiences (distance manages) 2) Situations/tasks applying the TPR 3) Personal emotions regarding work and distance management applying the TPR. An inductive approach [20] was also applied to explore other factors that characterize distance management, which have not already been identified. To organize and synthesize data and develop themes, we applied a template analysis model [21].

2 Findings and Analysis

2.1 Distance Manager Experiences (User)

The cumulative results across the three case studies show that the distance managers, i.e. the user of the TPRs, were able to conduct meetings from a distance and create a sense of proximity, even if not physically there but using the robots. The Double's ability to drive and navigate eased the experience. Moreover, the TPR, via the camera feature, enabled the eye contact, which the user considered essential and beneficial for the well-being and the monitoring of colleague´s feelings. Thus, the majority of the users had in general a positive experience regarding the TPR basic functionalities´ utilization.

On the other hand, the lack of trust, problems with the technology, privacy issues and intrusive emotions affected the use of the TPR in a negative way in some cases. The users therefore suggested technical improvements of the Double and identified the best-suited situations and tasks. Finally, the managers expressed the importance of involvement all the people interested in the change, trying to integrate actively the TPR into the current working routines and activities.

2.2 Employees Experiences (Participants)

The positive employee experiences related to the technical side where the respondents appreciate the technology and the video contact as the manager became present with the TPR as the robot created the ability to see each other and not just listen to a voice. Some employees expressed that it as a positive experience to be able to communicate with both eyes and ears as it gave a live feeling, when the manager could not be on the same site as the employees. Though not physically present, the TPR reminded the employees of the presence of the manager and did not consider it disturbing that he/she moved around. The TPR was silent, facilitated a conversation across a distance and was easy to use.

However, interviewees criticized the audio, the weight and the audio quality (in some instances). Others felt monitored and controlled when the TPR was present (not in use) and the attention it got disturbed the work. Across the three cases, some employees compared the TPR with Skype and questioned the gains of shifting to TPR.

2.3 Situations and Suitable Tasks

Useful Tasks and Situations. In all three cases the participants, both managers and employees, agreed that the TPR is most useful in planned project meetings (3:3) (See Table 3). Other tasks and situations in which the TPR was useful was in group meetings and interoffice meetings (even with a questionable distance involved). Only one in three (varied across the cases), found it useful to apply the TPR as a management substitute when working from home, for both interoffice and intra-office meetings, status update, small talk, supervision tours, ideas sharing and brainstorming.

Not useful tasks and situations for TPR. Data shows, that the users only found the TPR unsuited and not useful in four tasks and situations (see Table 4). Managers from all three cases expressed frustrations regarding the TPR however, each respondent stated different tasks as problematic. The TPR was found not useful for meetings in which there was the need of sharing physical documents among people, multiple users were driving the TPR at the same time and for important meetings, as private talks and crucial decisions were taken. Additionally, the TPR was not considered useful when meetings were held outside the firm or in meetings with (potential) new clients.

2.4 Emotions

The analysis shows that the personal emotions and feelings aroused, mainly from the user (manager) side but also in some participants, mainly related to mistrust regarding

Table 3. Cumulative analysis – useful tasks and situations for TPR

Situation and task	No of cases	Final evaluation
Planned project meetings	3:3	Useful
Big group meetings (> 5)	2:3	Useful
Small group meetings (< 5)	2:3	Useful
Interoffice meetings	2:3	Useful
Work from home	2:3	Useful
Status update meetings	2:3	Useful
SCRUM meetings	1:3	Useful
Intra office meeting	1:3	Useful
Small talk	1:3	Useful
Supervision	1:3	Useful
Brainstorming	1:3	Useful

Table 4. Cumulative analysis – not useful tasks and situations for TPR

Situation and task	No of cases	Final evaluation
Important meeting	1:3	Not useful
Chair/host of a meeting	1:3	Not useful
Share documents	1:3	Not useful
Acquire or keep a client	1:3	Not useful

the TPR. That is foreseeable since the implementation of new technologies requires some adaptation time. Later the users felt confident and found the TPR easy to use when they adopted the new channels of communication. The distance managers appreciated that they could perceive facial expressions and emotions through the camera contact. These aspects led to a renewed feeling of being present in same the place, which they had experienced before with other technologies. The TPR various functionalities also resulted in less travels which relieved some stress among the managers. Among the employees, some felt a bit disturbed by the unusual presence of the TPR, sometimes even controlled (simply by having it there) and did consider the TPR as part of their current practices.

3 Discussion and Conclusion

The data shows that the initial motivation of wanting to use the TPR to improve savings sand reduce expenses due to less travelling has been met. Across the three cases, the managers also considered the current Skype solution already installed and implemented, with the costs and timings related to the implementation of a new technology as the TPR. Finally, the TPR demands extra planning to use it properly. TPR has been found beneficial in those situations in which the people that are present know well each other, with a long-lasting relation and for managerial tasks that are scheduled and familiar for the people involved, possibly recurrent over time. Conversely, the tasks found difficult for a manager are the ones that include clients or people not commonly contacted and placed outside the company. Lastly, sensitive and private speeches are reputed to be non-manageable through the TPR.

3.1 Conclusion

When implementing a new technology, it is interesting to discuss if the problems that occur are related to the maturity of the technology, is a user issue, or perhaps a combination of both but also the gains that the technology provides. Specific characteristics of manager's experiences affect the introduction and implementation of the TPR in both positive and negative ways.

On the positive side, the cumulative findings of this study show that most users (managers) found that the basic functionalities of the TPR worked satisfactorily. The feeling of being present and the usage of the camera function to see colleague's reactions are unquestionably the most useful ones among the several detected. According to the experiences from the companies usages, there are some specific managerial tasks in which the TPR can be easily used by distance managers, bringing some benefits and easing their daily practices. For example, the TPR relieve some frustration in distance managers due to the fact that some travels can be avoided or decreased (as expected), improving the perception of being actually present. Moreover, the TPR opens new communication channels that are better compared with other solutions used before, since as declared by the users, it provides eye contact considered crucial by managers.

On the other hand, some unexpected problems with the technology happened, including mistrust and privacy issues that could negatively affect the TPR usage. There are tasks (not preventively foreseen), in which it is difficult to use the TPR, provoking additional complications instead of easing the manager's journey. People are not sure to use the TPR, in situation where they are heavily dependent on the context, the people involved and the environment that can obviously vary from one situation to another. In addition, personal emotions from both the user and the participant side are crucial when implementing the TPR, because they directly affect the user's behaviours. The TPR brings some negative feelings, for instance, when dealing with a robot and not a human. Across cases, people felt disturbed and in some cases monitored and controlled, simply by having the TPR present there. Altogether, the case study compose a set of suggestions and guidelines, based on the outcome of the data analysis, to be followed when implementing a new technology as the TPR investigated in this work.

References

1. Fisher Kimball, Fisher Mareen D (2001) The distance manager. A hands-on guide to managing off-site employees and virtual teams. McGraw-Hill, New York
2. Cropper S, Huxham C, Ebers M, Smith Ring, P (2008) The Oxford handbook of inter-organizational relations. 1st edn. Oxford University Press. https://doi.org/10.1093/oxfordhb/9780199282944.001.0001

3. Verburg Robert M, Bosch-Sijtsema Petra, Vartiainen Matti (2013) Getting it done: CRITICAL success factors for project managers in virtual work settings. Int J Project Manage. https://doi.org/10.1016/j.ijproman.2012.04.005
4. Poulsen S, Ipsen C (2017) In times of change: How distance managers can ensure employees' wellbeing and organizational performance. Saf Sci 100. https://doi.org/10.1016/j.ssci.2017.05.002
5. Hinds P, Kiesler S (2002) Distributed work, vol 47. https://doi.org/10.2307/3094928
6. Bell BS, Kozlowski SWJ (2002) A typology of virtual teams: implications for effective leadership. Group Org Manage 27:14–49
7. Ipsena C, Poulsena S, Nielsena L (2015) Management across distances – how to ensure performance and employee well- being. In: Proceedings of the 19th triennial congress of the international ergonomics association. International Ergonomics Association, Melbourne. Publication date: 2015 Document Version Peer reviewed version Link back
8. Arnold J, Randall R (2016) Work psychology. Understanding human behaviour in the work place. 5th edn. Financial Times/ Prentice Hall, New York
9. Newell S, Robertson M, Scarbrough H, Swan J (2002) Managing knowledge work. Palgrave, New York
10. Newell S, Scarbrough H, Swan J (2001) From global knowledge management to internal electronic fences: Contradictory outcomes of intranet development. Brit J Manage 12:97–111. ISI: 000169785900001
11. Hislop Donald (2009) Knowledge management in organizations, vol 2. Oxford University Press, New York
12. Docherty P, Forslin J, Shani AB, Mari K (2002). Creating sustainable work systems. In: Emerging perspectives and practice. Routledge, London
13. Frennert S, Eftring H, Östlund B (2017) Case report: implication of doing research on socially assistive robots in real homes. Int J Social Robot. https://doi.org/10.1007/s12369-017-0396-9
14. Sheridan TB (2016) Human-robot interaction: status and challenges. Hum Factors 58:525–532
15. Crabtree B, Miller W (1999) Doing qualitative research, vol 2. Sage, London
16. Miles MB, Huberman AM (1994) Qualitative data analysis an expanded sourcebook, vol 2. Sage Publications, London
17. Venkatesh V, Davis FD, Smith RH, Walton SM (2000) A theoretical extension of the technology acceptance model: four longitudinal field studies. Manage Sci 46(2):162–332
18. Venkatesh V, Bala H (2008) Technology acceptance model 3 and a research agenda on interventions subject areas: Design Characteristics, Interventions, Management Support, Organizational Support, Peer Support, Technology Acceptance Model (TAM), Technology Adoption, Training, User A. Decis Sci 39:273–315
19. Miles MB, Huberman AM, Saldana J (2014) Qualitative data analysis: a methods sourcebook. Qual Data Anal. https://doi.org/10.1080/0140528790010406
20. Bryman A (2004) Social research methods, vol. 2. Oxford University Press
21. King N (1998) Template analysis. In: Casell C (ed) Qualitative methods and analysis in organizational research. A practical guide, 1st edn. SAGE Publications, London, p 118–134

Language Issues in the Activity of Interaction with the Company Players

Building Tools to Guarantee a 'Common Ground'

Marta Santos[1] and Denise Alvarez[2](✉)

[1] CPUP, Faculdade de Psicologia e de Ciências da Educação, Universidade do Porto, Porto, Portugal
marta@fpce.up.pt

[2] Programa de Pós-Graduação em Engenharia de Produção, Universidade Federal Fluminense –UFF, Niterói, Brazil
alvarezdenise@id.uff.br

Abstract. Quite often companies from different economic sectors ask for the intervention from research groups connected to the fields of Ergonomics, Production Engineering or Work Psychology, both in Brazil and in Portugal. In certain situations, those requests unfold in new demands. The attitude we adopt as consultants to answer those requests privileges a clinical approach and the activity's perspective and it demands, straight away, a positioning that relies on the actual work situations. Consequently, we value the coproduction in context, both for the meaning and for the experience brought to work. We advocate the development of a reflexive contract between the consultant and the client, given the adoption of a collaborative attitude, with constant changes, making it possible to reach a commitment that satisfies both parties.

Keywords: Ergonomics and training · Activity analysis · Tools
Common language

1 The Construction of Tools in Training Projects

1.1 Activity Analysis, Training and Transformation of the Work Situations

There is already a long tradition to reflect upon the training issues together with the work analysis, the concerns regarding health and safety and/or the development of professional competencies [1, 2]. Indeed, at first the training activities implemented according to this perspective were merely parallel activities or the continuation of other intervention projects [3]. Gradually, however, they began to earn a full status, which means the training was designed explicitly as a way to contribute to the work transformation or as a way to help the workers understand and act upon the situations, preserving their health [2].

This evolution took different forms and paces in different countries. Actually, a company's interest in and implementation of interventions led by research teams from

S. Bagnara et al. (Eds.): IEA 2018, AISC 821, pp. 366–373, 2019.
https://doi.org/10.1007/978-3-319-96080-7_43

Ergonomics, Production Engineering or Work Psychology is, both in Brazil and in Portugal, fairly new. Quite often, however, an initial request originates further research-training-action requests. The activity's point of view is an important consideration in the teams' attitude during the intervention [4], which seems to explain why the companies reorder these teams' cooperation.

1.2 The Construction of Tools to Incorporate the Activity's Point of View

To choose the activity's point of view demands, to begin with, the need to take a stand anchored in the real work situations. For that reason, in these research-training-action projects, it is possible to show the choice for a clinical approach [5]. This choice can be characterized by the observation and attentive listening to what defines each particular situation, as the team's expertise does not allow an a priori apprehension of the meaning that they give to the intervention object. This attitude requires the development of an effort so that the companies' representatives participate in the construction of an understanding of the situation that generated the request. A collaborative attitude is essential to meet a win-win commitment for both parties [6].

The thoughts led by Teiger and Laville [7] since the last decades of the 20th century on the relations established between "training-research-action" were already defining the matrix of this stance. Indeed, it was acknowledged by then that such relations are built upon the confrontation between different modalities of knowledge (the workers' knowledge, experts on the work situation, and the researchers' knowledge). It was then underlined the need to gradually master a common language on the activity, which would allow the development of a mutual understanding between researchers/trainers and workers.

According to Re [8], despite the analyses of the work situations imply different contexts and levels of complexity, they refer to a common theoretical grounding: on the one hand regarding the acceptance that an implicit knowledge and an operational strategy do exist; and on the other hand regarding the existing difficulties for the transmission of such competency. The author defends that there are constraints both for the creation of a common place and for the language underneath it. In order to deal with those constraints, she points to the assumption that the communication area among the experts committed to the work analysis shall be a "no place" that will gradually be transformed into a "common place", hence taking the shape of a "target language". This direction pushes the ergonomist, at an early stage, to manage the construction of the common models that trigger the negotiations and the system's self-regulation. That requires social dynamic management skills and negotiation skills.

The development of a common language is frequently mediated by the use of tools constructed in action as a production that can be considered tailor made, given the use of multiple means designed for specific moments in the training process and created due to the need to answer to diverse, but complementary perspectives [9].

Montreuil [10] takes a model developed to assess the implementation of a training program in ergonomics to distinguish and identify the right place for several tools and their role throughout those training sessions, setting them in the process that leads to the transformation of the work situations. According to the model, carrying out a training program in activity analysis complies 3 stages: the involvement of the company's players (using convincing tools), the training in diagnosis for the trained people (using learning tools) and, the action/assessment (using action tools), that is, the transformation of the work situation.

The "convincing tools" [10] are developed specifically to involve the players who did not take part directly in the request, so to inform and raise their awareness towards the action that will be taken. These tools play a major role in the process, as some of these players are responsible for giving permission to the trainees' initiatives, for instance, releasing them from the working hours, participating in the creation of solutions following the diagnosis, or even joining visits to locations, etc.

Regarding the "learning tools", the author highlights what she calls a "reality-based pedagogy" which tries to be a pedagogy emerging continuously from the work reality to the abstract and it aims at creating the most abstract knowledge from the analysis of familiar phenomena.

Finally, to close the classification, the author calls for the need to measure or assess the training effects, referring to the "action tools" from the ergonomics field. According to her, those are essential for the stages to be carried out in the local structures, given the continuity they provide and the initiatives they may engender to improve the work situations. It is worth mentioning, however, that the work situations do change while the diagnosis is presented by the participants (or by the pilot action group) in the training sessions and sometimes it is not easy to keep traces of the changes. It is important to think how such changes can be recorded to show the training results.

This framework reflects experiences from researches-trainings-actions in Brazilian and Portuguese companies. In this text, however, the examples and the considerations will prioritize the tools constructed specifically to induce the dialogical relationship with the companies' representatives. The goals are to understand the characteristics of these tools, their purposes, how they could be classified according to Montreuil's model [10] and how the creation of a dialogical "common place" [8] actually happens.

2 Two Examples of Tools Constructed Under the Scope of Research-Training-Intervention Projects

2.1 Tools that Call for, Convince, Impel Learning and Action: Exercise of On-the-Job Otherness in an Energy Company in Brazil

The request was made by the managers from the health sector within the operational unit of an energy multinational company. At first, the request was about the difficulties felt by the health team to relate and communicate with the company's internal clients, hence causing problems for both parties. The team running the action has then reconstructed the terms of the request. The team members were a large group of research and investigation, encompassing the manager, 2 advisors, 2 professional researchers in work analysis (working closely to the companies' representatives) and a "back office group" composed by three work psychologists who monitored the research-training-action without actually going to the field and they have helped choosing and designing the tools. The process to reformulate the request with those players enabled the design of an action plan to lead the intervention and the joint selection of the tools to be used in 3 moments:

- Moments to train the key concepts on work and health (considered a possibility to establish norms), on ergonomics and activity analysis, aiming at raising awareness among the team members involved at this stage (causing a multiplier effect later on) about the signs of otherness present at work and about the need to understand them and respect them as well;
- Moments to watch and describe the activity of one another with experiences on the field lived by health professionals working in pairs to apprehend and describe the activity; the group later on discusses and analyses the highlighted situations. This activity was conducted alternately and the participants realized the "volume" of the work activity with its variabilities, regulations, strategies, debates, conflicts and decisions;
- Dialogical device brought to implement the results achieved in the training with the group of health professionals during the company's annual strategic planning (ASP), with health experts, some of the system's users (internal clients) and company managers. Those meetings produced a spreadsheet filled with resolutions and the creation of new procedures to apply in the short, medium and long terms in different attendance locations that constitute the health sector.

The implementation of these tools during the intervention led us to two findings. The first finding states that the three stages of the model presented by Montreuil [10] may be interconnected. Throughout the intervention, given the way the process unfolded, there was a contribution for its dissemination which, on the other hand, rang a bell to other company sectors and players about the possibility to discuss all the controversies altogether, within the work environment. The training moments were crucial for the trainees to dedicate a brand new look to the work of others and it empowered them to meet the client users, fostering their ability to listen the requests from the later. Additionally, it gave them tools to analyze the work. Finally, the moments at the ASP served the dialogue between the different hierarchy levels where procedures and rules established by the company embraced a new significance conferred by the work analysis. That joint analysis was determinant for the preparation of an implementation plan based on dialogue, which, in turn, was made of activity traits. Consequently, it is clear that the three stages evoked by Montreuil [10] may intersect one another because during the intervention, the tools that were used served to convince (alert) the other company departments about the importance of the initiative, to train the people involved in the capacity to diagnose work situations and to act upon their change.

The second finding refers to the difficulties pointed out by Re [8] regarding the creation of a "common place" both in terms of language and in terms of coping. In order to face those difficulties within this intervention, one of the exercises was the construction of a model. The initial request was designed by the health sector manager, who happens to be one of the researchers as well. The manager and her advisor were the direct requesters, who would then negotiate the possibility of an intervention-training with their direct line manager. Later on, another sector consultant joined them and that was when we had the opportunity to reformulate the request. The creation of this "common ground" happened gradually along the meetings with the companies' representatives and it required both parties to compromise on language issues. The commitment between the company and the theoretical vocabulary brought by the researchers extended to the "back office group" of researchers. One of the keys to reach a "common place" was the agreement that part of the intervention should be inserted in the company's annual strategic planning (ASP). Therefore, the first two stages dedicated to train the team of experts – which consisted in moments when the researchers made the concepts explicit and in dynamics and field trips to perform practical "exercises" on work analysis – had a common horizon, the moment at ASP that represented the last stage of the intervention. That common mediator, the ASP, was crucial for the research team to adjust to the vocabulary at use in the organization and it made the participants – the health experts attending the training – comfortable, as the ASP "event" was familiar to them.

These two findings led us to a better understanding of what is at stake in the construction of a relationship between companies' representatives and ergonomist researchers. There is the possibility for a common social place that will benefit from influences from both parties and will leave traits in the organization.

2.2 Share, Analyze and Reflect with the Company's Players: A Specific Example in a Portuguese Project

A company from the retail trade sector in supermarkets requested an intervention on the relations between work and health in a specific store. This request was triggered by a sequence of significant periods of absence due to sick leave by some workers with a clear impact on the teams' management and on the work to be performed in the store.

There were two strands to the project:

- Development of training sessions with the store workers about the relations between health and work: reflection about the impact of the working conditions and organization on health; consideration about possible surveillance measures to monitor group and individual health to be implemented in store;
- Construction of an implementation plan with the company, including tools for the key players (Human resources, occupational physicians, and hygiene and safety technicians) that could be used to identify and act upon critical situations and to improve the conditions of health and wellbeing.

This second strand predicted the involvement of those interveners from a first moment to collect information on the operation in store, up until the successive feedback with systematized information on the work activity in store and the design and validation of the implementation plan.

The work sessions with the group selected by the company to follow the project (two people from human resources, the occupational physician and, every now and then, the store manager) were not limited to data collection or sharing. In fact, there was an attempt for these sessions to be formative themselves. Thereto, three purposes were to be met: (i) highlight the type of useful information to carry out a careful activity analysis (for example, collect micro demographic data about the store workers and data on work accidents, sick leaves and analyses of hazards by the safety technicians);

(ii) share key theoretical concepts on the relations between health and work; (iii) and share a systematization of elements of the store activity collected beforehand in moments where the activity of these workers was analyzed (Fig. 1).

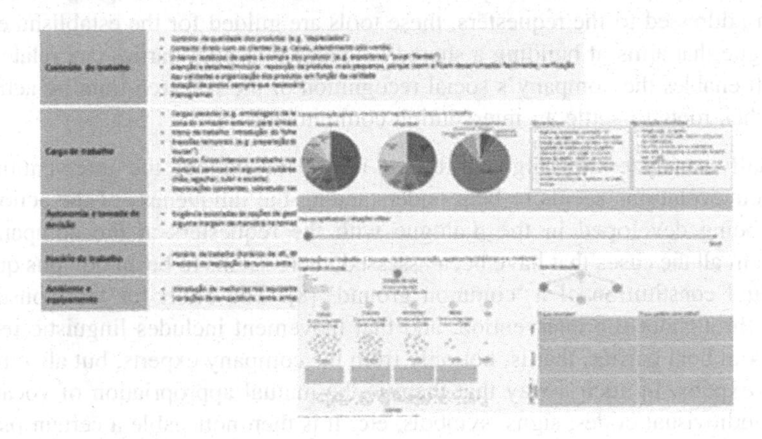

Fig. 1. Examples of tools used in the sessions with the company's interveners.

The work developed with these tools punctuated the assumptions and points of view the research team took on about the topics under analysis.

For that reason, the use of these tools seemed not to be limited to one of the uses predicted in Montreuil's model [10]. In fact, in addition to conquer the interveners (raising their awareness towards the pertinence and the impact of the undergoing analysis), it was expected to realize and understand the actual specificities of the work performance by the store workers as well as what can be generalized from this one experience. Lastly, the tools were expected to help in the definition of actions to transform the work situation. The tools developed for the requesters would then have this three-part simultaneous role: convince, learn and act.

The discussion and the reflection that followed the use of these tools would bring to the work sessions the requesters' perspectives, their concerns, the possibilities and the limitations they anticipate in the implementation of specific actions. Consequently, they also have an effect on the research team. This means those same tools will also act gradually upon the team's stand, which will partly absorb those concerns and perspectives in its speech. The groundings for a collaborative attitude are then settled, given the gradual share of a language that is reformulated every step of the way, giving a strong contribution for the mutual recognition.

3 Discussion

The considerations expressed in this text about the tools created in research-training-action situations show that:

- sometimes the tools are not meant to be used by the trainees/workers; sometimes they are addressed to other stakeholders in the process;
- in certain situations, these tools end up fulfilling more than one purpose simultaneously: they are used not only to "convince" but also to "learn" [10], for example;
- when addressed to the requesters, these tools are guided for the establishment of a dialogue that aims at building a shared language that both parties can relate to and which enables the company's social recognition of the research-training-action and launches roots to settle an intervention commitment.

Actually, analyze the language-related issues throughout the interventions and along their evolutions seems to help understanding the uniqueness of the actions that end up being developed in the dialogue with the requesters at the companies. In addition, in all the cases that have been assessed, there seems to be an obvious quest for the gradual constitution of a 'common ground' [8]. The effort for that construction happens throughout the intervention, and that movement includes linguistic reformulations from both parties, that is, not only from the company experts, but also from the research experts. In such a way that there is the mutual appropriation of vocabulary, diverse audiovisual codes, signs, symbols, etc. It is then noticeable a certain plasticity in the choice of the tools to be used during the interventions to guarantee they fit not only the contexts, that correspond to the training-intervention objects, but also the evolutions of the common understanding.

Acknowledgements. This work was funded by the Center for Psychology at the University of Porto, Portuguese Science Foundation (FCT UID/PSI/00050/2013) and EU ERDF through COMPETE 2020 program (POCI-01-0145-FEDER-007294).

References

1. Lacomblez M, Bellemare M, Chatigny C, Delgoulet C, Re A, Trudel L, Vasconcelos R (2007) Ergonomic analysis of work activity and training: basic paradigm, evolutions and challenges. In: Pikaar R, Konongsveld E, Settels P (eds) Meeting diversity in ergonomics. Elsevier Ltd., Amsterdam, pp 129–142
2. Delgoulet C, Cau-Bareille D, Gaudart C, Chatigny C, Santos M, Vidal-Gomel C (2012) Ergonomic analysis on work activity and training. Work 41(2):111–114. https://doi.org/10.3233/WOR-2012-1286
3. Abrahão J, Berthelette D, Desnoyers L, Ferreira L, Jobert G, Lacomblez M, Launis K, Leppanen A, Maggi B, Montreuil S, Patesson R, Paumès D, Teiger C, Vogel L, Wendelen E (1997) General introduction to the symposium: aims, context, concept, methods, practices and problems. In: Proceedings of the 13th triennial congress of the internacional ergonomics association, Finland, Tampere, vol 1
4. Daniellou F, Rabardel P (2005) Activity-oriented approaches: some traditions and communities. Theor Issues Ergon Sci 6(5):353–357
5. Jobert G (1992) Position social et travail du consultant. Educ Perm 113:157–177
6. Falzon P (2007) Natureza, objetivos e conhecimentos da ergonomia: elementos de uma análise cognitiva da prática. In: Falzon P (ed) Ergonomia. Editora Blucher, São Paulo, pp 3–19

7. Teiger C, Laville A (2013) La coopération syndicats-recherche aux sources de la dynamique «recherche-formation-action» - L'expérience du laboratoire d'ergonomie du CNAM (1965–1990). In: Teiger C, Lacomblez M (coord.) (Se) Former pour transformer le travail - Dynamiques de constructions d'une analyse critique du travail. Presses de l'Université Laval/PUL/l'European Trade Union Institute/ETUI, Québec/Bruxelles, pp 59–83
8. Re A (2013) Une nouvelle perspective pour la compétence ergonomique dans l'analyse du travail. In: Teiger C, Lacomblez M (coord.) (Se) Former pour transformer le travail - Dynamiques de constructions d'une analyse critique du travail. Presses de l'Université Laval/PUL/l'European Trade Union Institute/ETUI, Québec/Bruxelles, pp 644–647
9. Teiger C, Lacomblez M (2013) Construire des "outils" ad hoc? Mise en perspective. In: Teiger C, Lacomblez M (coord.) (Se) Former pour transformer le travail - Dynamiques de constructions d'une analyse critique du travail. Presses de l'Université Laval/PUL/l' European Trade Union Institute/ETUI, Québec/Bruxelles, pp 459–469
10. Montreuil S (2013) Des "outils" de formation: pour faire quoi? In: Teiger C, Lacomblez M (coord.) (Se) Former pour transformer le travail - Dynamiques de constructions d'une analyse critique du travail. Presses de l'Université Laval/PUL/l'European Trade Union Institute/ETUI, Québec/Bruxelles, pp 535–539

Tutor/Trainee Cooperation in Work-Based Training Situation: Which Creative Process for Mentoring Interactions in Care Situation?

An Example in Medical Radiology

Vanessa Rémery(✉)

Training Adults Section, Interaction and Training Team, Faculty of Psychology and Educational Sciences, University of Geneva, Geneva, Switzerland
Vanessa.Remery@unige.ch

Abstract. The oral presentation is integrated in the symposium «Ergonomic analysis of work activity and training: evolutions of basic paradigms and creativity in practices» (coord. by C. Delgoulet et M. Santos). It contributes to the study of mentoring interactions in care settings, specifically in the field of medical radiology. Our methodological approach is based on a video-ethnographic fieldwork to explore mentoring practices at the workplace. It is inspired by Work Analysis from a Francophone ergonomics perspective and researches in the field of Workplace Studies. We highlight varied forms of mentoring configuration that reflect a local dynamics collaboration between tutor and trainee. A trajectory of tutor's actions is outlined by two main results: reconfigurations of interactive setting and formats of participation at the workplace on the one hand, and three mentoring configurations deployed on the register of «do with», «let it do» and «do instead» . These mentoring configurations constitute possible resources for conception of training situations. They create interactive spaces in which formats of participation change over time. They introduce variation within the workplace to enhance its learning potential. Trainee's experience at the workplace is structured by different forms of participation which play a central role in learning process.

Keywords: Mentoring interactions · Medical radiology · Configuration Training · Workplace

The oral communication deals with mentoring at the workplace in Vocational Education and Training (VET), specially in Medical Radiology. This study is part of a collective research program in adult education. "Becoming X-ray Technician" is a partnership with University Hospitals of Geneva and Swiss Hight School of Health. *Interaction & Training* team's contribution focuses on guidance's interactive modalities of X-ray technicians training at the workplace.

We aim to provide elements for reflection on conditions to create learning opportunities for trainees in care situations. Our perspective is at the crossroads of several theoretical and methodological approaches: Work Analysis from a Francophone ergonomics perspective [1, 2] and research in the field of Workplace Studies focused on collaborative teamwork practices' analysis [3–7]. This oral communication

S. Bagnara et al. (Eds.): IEA 2018, AISC 821, pp. 374–380, 2019.
https://doi.org/10.1007/978-3-319-96080-7_44

emphasizes on interaction and collaborative practices between tutors and trainees to manage a patient in a conventional radiology department. An analysis of mentoring at the workplace is proposed based on video recording. We focus on tutor's actions with the trainee during a consultation with a real patient, and show different types of mediation introduced by the tutor to guide the trainee.

1 Mentoring at the Workplace: Between Adaption and Arbitration

Analysis of practices on mentoring at the workplace has been developed in adult education over the past twenty years, specially in the Francophone tradition ergonomics and Professional Didactics [8–10]. Mentoring at the workplace is devoted to the construction of a "practical" experience. Novice access to work activities under close supervision of tutor. Research underline the articulation between "production" and "training" as a major preoccupation for tutors at the workplace. In patient care situation, tutors are confronted with a hybrid activity [11] that simultaneously engages them to produce a "work" towards a patient, and teaching the job to the trainees. Temporality of production and training are source of conflict. One of the more important tutor's competence is probably his ability to transform work organisation, reorganize time, space, nature and order of tasks to create favourable conditions for learning process [12]. Tutors must adjust themselves continually to constraints of production and seize learning opportunities [13]. "Care interactions" and "mentoring interactions" overlap each other. A double frame structures the situation for the participants [14]. This framework is "polyfocused" [15] toward work and training both. Asymmetrical relationships intersect: the care relationship between technician and patient on the one hand, and the mentoring relationship between tutor and trainee on the other hand. Participants are confronted with a constantly practical issue of redefined position in interaction. That's the reason why it is interesting to observe how they arbitrate *in situ* with dilemmas. Trainee's participation at the workplace brings change in the usual teamwork routines and highlights interesting cooperation practices between tutor and trainee [16]. It is possible to observe this cooperation with a micro-analysis of participation practices [17 for example]. Particularly, in this oral communication, we analyse how mentoring is produced in a patient care situation which requires complex interactional adjustments [18].

2 Dynamic Organization of Mentoring Configurations at the Workplace

In Professional Didactics, the gap between forms of mentoring prescription that describe major functions (reception, supervision, socialization, evaluation, etc.) and the activity implemented by tutors was highlighted [19]. Far from being a spontaneous, approximate and improvised practice, mentoring is structured by a dynamic organization in the workplace. A diachronic model of mentoring organization was proposed based on different configurations (familiarization, advanced familiarization, transmission, assisted work setting, semi-assisted work setting and work setting). These configurations change over time, specially the structures of communication between tutor and trainee according to the complexity of the task and the trainee's autonomy. Mentoring at the workplace is produced through a commitment/disengagement movement that reflects "scaffolding"/"dis-scaffolding" process [20]. It is possible to trace a "situated learning trajectory" [21] when we analyse mentoring configurations from interactions between tutor and trainee to understand the dynamic process of learning over time.

Configurations of participation are a descriptor of mentoring. The concept of "configuration of participation" developed within the *Interaction & Training* team is based on Elias' theory [22] on the one hand, and Goffman and Goodwin's research about participation [23, 24] on the other hand. This concept is relevant to explore mentoring's transformations at the workplace. The "configurations perspective" put the emphasis on interdependency of relationships created between individuals. It answers the need to understand the reciprocity of actions through forms, adjusted and reorganized, that emerge from interpenetration of behaviours.

Methodologically, this perspective involves empirical work based on a video-ethnographic approach of work and training situations. This approach is developed in *Interaction & Training* team. It contributes to analyse mentoring at the workplace from video data and interviews. It is inspired by research conducted in the field of Work Analysis from a Francophone ergonomics perspective and Workplace Studies. By focusing on multimodal aspects of interaction (gestures, looks, facial expressions, intonations, movements, positioning of bodies in space, material manipulation), analysis emphases variation and distribution of actors' participation within and between short sequences of activity. Also, it highlights how tutors create participations' settings for trainees but also how trainees invest these settings reciprocally.

3 An Illustration in Conventional Radiology

The oral communication focuses on a singular situation at the hospital in department of radiology. A video-ethnographic fieldwork was authorized by the hospital's ethics commission. Consent was obtained from the participants, professionals and patients for the video recording. The situation deals with a consultation for an elderly patient in conventional radiology. It integrates in a larger data corpus in which several trainees in different departments of radiology were observed at different times. Here, the situation involves an experienced X-ray technician in a mentoring role, a first-year trainee and a patient. The collected data consist of video recording of verbal and non-verbal interactions between technicians and patients. Analysis is based on transcriptions of significant episodes on verbal and non verbal aspects of interaction (gestures, looks, facial expressions, intonations, movements, positioning of bodies in space, material manipulation).

Analysis focuses on tutor's actions to guide the trainee. The excerpts selected for the oral communication delimit several episodes that appeared significant to highlight variety of tutor's actions. Each episode describes different aspects of mentoring configuration: (a) interactive space configured by participants' position, (b) manifest event that causes tutor's action, (c) forms of guidance implemented by the tutor, (d) formats of participation and the focus of attention over time.

4 A Trajectory of the Tutor's Actions

4.1 Reconfigurations of Interactive Space and Formats of Participation

Analysis shows three forms of tutor's actions to guide the trainee in a care situation with the patient. It underlines activity's inflexions underpinned by a reconfiguration of interactive space and formats of participation. From an analysis of participant's position, three participants are involved in the situation (tutor, trainee, patient) which potentially multiplies the possibilities of interactions between them. Depending on the inflexions that affect activity, several configurations are produced: "tutor-patient", "trainee-patient", "tutor-trainee" or "tutor-patient-trainee". These configurations change according to modalities of engagement. We schematize these formats as follows (Fig. 1):

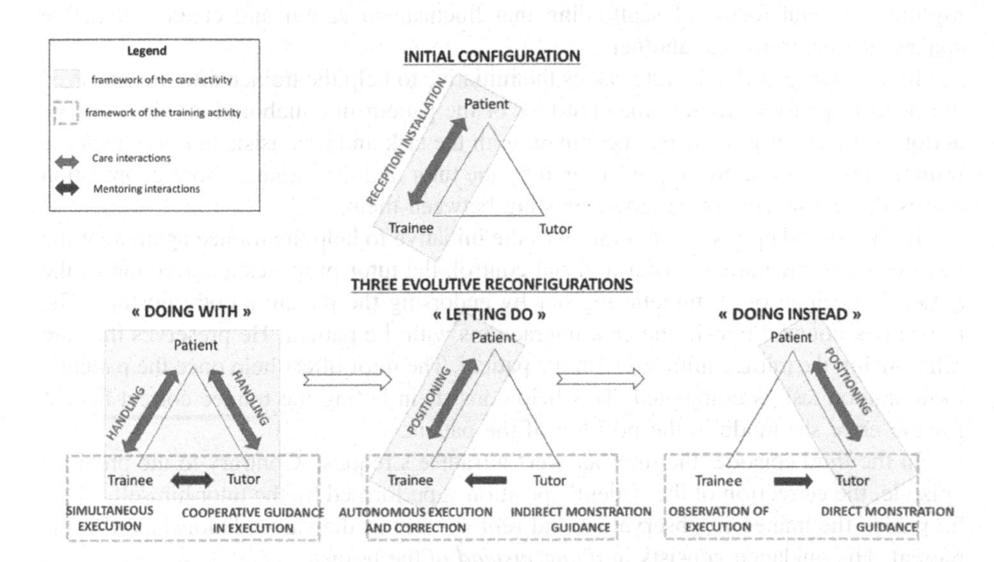

Fig. 1. The interactive space, as it is configured by participants' position before and after the tutor's intervention, shows dynamic formats of participation.

In the initial configuration, the tutor entrusts a task to the trainee which implies a privileged relationship with the patient. The trainee is in a particularly interesting position to develop autonomy at the workplace. The tutor stays away from the focus of interaction between the trainee and the patient. Nevertheless, he is physically present and extremely attentive to the trainee's actions towards the patient.

Three dynamic configurations are highlighted to reflect some of tutor's actions to guide the trainee with the patient.

In the first case, the interactive space is reconfigured around a focus of interaction including the three participants. The tutor engages in a care relationship and integrates the setting of interaction between the trainee and the patient. Both cooperate and adjust their respective actions with a common goal towards the patient, who must be carefully mobilized to avoid a fall.

In the second case, the interactive space is unchanged because the setting of trainee-patient interaction is preserved and the trainee remains in charge of the care relationship established with the patient. The tutor positions himself away from the care relationship but he proposes correction of the trainee's action, at a distance from the trainee-patient interaction setting.

In the third case, the interactive space is reversed and we see the emergence of an interaction's setting constituted by the tutor and the patient. This setting induces a momentary marginalization of the trainee. The tutor replaces the trainee in the care relationship with the patient.

4.2 Mentoring Configurations: "Do with", "Let Do", "Do Instead"

These three configurations differ from the event and the moment of tutor's action They highlight several forms of scaffolding that fluctuate in action and create interactive spaces distinct from one another.

In the first episode, the tutor takes the initiative to help the trainee after he observed the need to perform the mobilization task of the patient in collaboration with her. His action is a form of guidance concomitant with the task and it consists in *doing with* the trainee. The care relationship is ensured by the tutor and the trainee whose cooperation erases the asymmetry of expertise existing between them.

In the second episode, the tutor takes the initiative to help the trainee again after the trainee install the patient. After a visual control, the tutor proposes a correction of the patient's position on a mimetic register by endorsing the patient's body posture. The tutor does not take part in the care interactions with the patient. He preserves the care relationship the trainee initiated with the patient. The tutor offers help once the patient's mobilization task is completed. This help consists in letting the trainee correct herself for the error she made in the position of the patient.

In the third episode, the tutor answers a trainee's request. Contrary to the previous episode, the correction of the patient's position is performed by the tutor himself. Thus, he places the trainee in observation and replaces her in the care relationship with the patient. His guidance consists in *doing instead of* the trainee.

It is interesting to question the more general scope of such a model. Our previous work in the field of childhood education allows us to find some invariants that probably reflect typical elements of mentoring at the workplace [16]. First, the configurations show recurring formats that can be observed beyond the professional fields investigated, characteristic of mentoring situations involving a relationship with users, whether the user is a patient as in this article, or young children in previous studies. The existence of the user as a third party in the tutor/trainee relationship shows possible registers of participation between tutor and trainee, whose configurations are probably not infinite. Three configurations were highlighted for young children (co-facilitation;

observation; monstration) that overlap with those identified here (doing with; letting do; doing instead).

Analyse mentoring from a configuration approach shows not only the invariant but also the dynamic dimensions of this activity from a situated learning trajectory. We show the importance of tutor's resources to guide the trainee. This approach is critical towards a linear conception of tutoring that would correspond to stable formats of participation in a long period [20]. A micro-analysis is pertinent to describe local dynamic transformations. We believe that it is precisely trainee's experience of a varied configurations at a local level which could be a source of learning over a longer period.

5 Conclusion

This oral communication presents several relevant aspects. First, an ergonomics perspective on mentoring at the workplace based on a video-ethnographic fieldwork is relatively recent. Second, the results contribute to analyse mentoring in a relational job, and more specifically in patient care situations. This leads us to question tools' analysis and previous theorical models whom users were absent. Our context in medical radiology shows that interaction and participation patterns between tutor and trainee are "porous" to the presence of the patient. It impacts forms of guidance and creative learning opportunities.

We showed three modalities of mentoring process that reflect the local dynamics of tutor/trainee collaboration in the care situation. A trajectory of tutor's actions was outlined, highlighting two main results: the reconfigurations of the interactive space configured by formats of participation, and three mentoring configurations deployed on the register of "doing with", "letting do" and "doing instead". These mentoring configurations constitute possible resources for conception of training situation. They create interactive spaces in which formats of participation change over time. They introduce variation within the workplace to enhance its learning potential [12]. Trainee's experience at the workplace is structured by different forms of participation which play a central role in learning process.

References

1. Leplat J (1994) Collective activity in work: some lines of research. Le Travail Humain 57:209–226
2. Teiger C, Lacomblez M (eds) (2013) (Se) Former pour transformer le travail. Dynamiques de constructions d'une analyse critique du travail. Presses de l'Université Laval, Laval
3. Luff P et al (eds) (2000) Workplace studies: recovering work practice and informing system design. Cambridge University Press, Cambridge
4. Heath C, Luff P (eds) (2000) Technology in action. Cambridge University Press, Cambridge
5. Hindmarsh J, Pilnick A (2002) The tacit order of teamwork: collaboration and embodied conduct in anaesthesia. Sociol Q 43(2):139–164
6. Goodwin C (2013) The co-operative, transformative organization of human action and knowledge. J Pragmatics 46(1):8–23

7. Heath C, Luff P, Svennson MS (2005) Technology and medical practice. Sociol Health Illn 25(3):75–96
8. Rémery V, Chrétien F, Chatigny C (eds) (2019) Apprentissage et transmission de l'expérience en situation de travail: un dialogue entre ergonomie et formation d'adultes. Presses Universitaires de Rouen et du Havre, Rouen (à paraître)
9. Chrétien C, Métral J-F (2016) Potentiel de transmission professionnelle et configurations de tutorat en situation de travail. Recherche et formation 83:49–68
10. Thébault J (2013) La transmission professionnelle: processus d'élaboration d'interactions formatives en situation de travail. Une recherche auprès de personnels soignants dans un centre hospitalier universitaire. Cnam, Thèse de doctorat en ergonomie, Paris
11. Rémery V, Markaki V (2016) Travailler et former: l'activité hybride des tuteurs. Education Permanente 206:47–59
12. Mayen P, Gagneur C-A (2017) Le potentiel d'apprentissage des situations: une perspective pour la conception de formations en situation de travail. Recherches en éducation 28:70–83
13. Olry P (2000) La gestion des composantes temporelles pour se former au travail. Cnam, Thèse de doctorat en formation d'adultes, Paris
14. Blancard P (1996) Compétences à l'oeuvre dans l'exercice du tutorat. Recherche & Formation 22:115–126
15. Goffman E (1991) Les cadres de l'expérience. Minuit, Paris
16. Filliettaz L, Rémery V, Trébert D (2014) Relation tutorale et configurations de participation à l'interaction. Analyse de l'accompagnement des stagiaires dans le champ de la petite enfance. Activités 11(1):22–46
17. Rémery V, Filliettaz L (2017) Coordination et coopération tuteur/stagiaire dans les activités de formation en situation de travail : Un exemple en radiologie médicale. Recherche Formation 84
18. Cerf M, Falzon P (2005) Relation de service: travailler dans l'interaction. Presses Universitaires de France, Paris
19. Kunégel P (2006) Tutorat et développement de compétences en situation de travail. Essai de caractérisation des logiques tutorales. Cnam, Thèse de doctorat en formation d'adultes, Paris
20. Wood D, Bruner J, Ross G (1976) The role of tutoring in problem solving. J Child Psychol Psychiatry 17:89–100
21. De Saint-Georges, I (2008) Les trajectoires situées d'apprentissage. In: Filliettaz L, de Saint-Georges I, Duc B. (eds) Vos mains sont intelligentes: interactions en formation professionnelle initiale. Cahiers de la Section des Sciences de l'Éducation, Université de Genève
22. Elias N, Dunning E (1994) Sport et civilisation. La Violence maîtrisée. Éditions Fayard, Paris
23. Goffman E (1987) La position. In: Façons de parler. Éditions de Minuit, Paris, pp 133–166
24. Goodwin C, Goodwin MH (2004) Participation. In: Duranti A (ed) A companion to linguistic anthropology. Basil Blackwell, Oxford, pp 222–244

Virtual Simulations for Incorporating Ergonomics into Design Projects: Opportunities and Limitations of Different Media and Approaches

Esdras Paravizo(✉) and Daniel Braatz

Department of Production Engineering,
Federal University of Sao Carlos, Sao Carlos 13565-905, Brazil
esdras@ufscar.br

Abstract. Throughout the design process, project actors rely on a variety of artifacts for fostering creativity, communication and make-sensing among the project team. Among those artefacts are virtual simulations, achieved using different types of software in which different scenarios and configurations of the future situation can be discussed and tested out. In this paper, we aim to achieve a better understanding on how Ergonomics research and practice can benefit from utilizing computer-based simulation, especially using "CAD", "Game Engines", "Discrete Event Simulation" and "Digital Human Modelling and Simulation" software. Through a brief review of the relevant literature we were able to get an overall understanding of how researchers and practitioners are using these tools to incorporate ergonomics into the design process. The affordances and limitations of each tool for simulating future scenarios in an ergonomics perspective are highlighted. Additionally, a comparison framework inspired by Ziolek and Krutihof (2000) division of the human modelling environment process was developed. The four aspects of this comparison framework comprised the manikin, the environment, the analysis and the interactions. Then, each of the presented tools were compared along these dimensions. Furthermore, we discuss possible ways for articulating these simulation tools throughout the design process. The results can help practitioners and researchers when planning computer simulations targeting ergonomics aspects of future work systems.

Keywords: Computer aided design · Game engines
Discrete Event Simulation · Digital Human Modelling

1 Introduction

The design process relies on the development and use of different objects for fostering communication and joint design among project actors who usually have heterogeneous experiences, expertise and representations of the design (Eckert and Boujut 2003). These objects may take several forms and configurations and can be employed in participatory ergonomics processes (Broberg et al. 2011).

S. Bagnara et al. (Eds.): IEA 2018, AISC 821, pp. 381–390, 2019.
https://doi.org/10.1007/978-3-319-96080-7_45

When discussing objects' nature, one of the most direct consideration is whether it is a physical object, e.g. scale models, full scale mock ups, blueprints, etc. (Broberg et al. 2011), or a digital object, e.g. virtual prototypes (Aromaa and Väänänen 2016) or virtual environments (Wilson 1997). In this study, we will focus our analysis on the use of digital objects in simulation contexts for ergonomics integration into design processes.

Simulation's role for facilitating design processes from an ergonomics perspective has long been discussed (Maline 1994). Simulation can focus on observation (improving the understanding about the situation), learning (aiming to train and to develop competencies) and designing (fostering sense-making among project actors and solutions exploration and development) (Béguin and Weill-Fassina 2002).

The developments on the field of ergonomics regarding digital simulations go back to the early developments on digital human modelling (DHM) by Chaffin (1997), although several attempts have been made to model humans in computer aided design (CAD) software in the 1960s onwards (Mattila 1996). Currently, studies report the use of commercial DHM simulation software in a variety of contexts, both in the analysis of postures and workplaces and their design (Chaffin 2005; Backstrand and Hogberg 2007; Paul and Quintero-Duran 2015; Zhou et al. 2016).

The use of CAD tools in the design of facilities and products has also seen as an opportunity to foster ergonomics integration into design (Mattila 1996; Patel et al. 2016; Marconi et al. 2018). CAD tools are employed by designers to create the 2D blueprints and 3D models of the products and facilities being designed, usually requiring technical skill in the software employed and in technical drawing symbology.

Another tool usually employed in the context of facilities and operational simulation is the discrete events simulation (DES) (Banks et al. 2010). Studies have investigated the use of DES tools for modeling human performance, workload and healthcare systems design (Laughery 1999; Perez et al. 2014; Hossain et al. 2017). Recent developments on DES software in its visual aspects and representation of human manikins (e.g. Flexsim[1] software) may increasingly allow its utilization for ergonomics analysis and design.

Finally, the emergence of game engines (GEs) utilization for creating virtual environments focusing on ergonomics aspects of product development process (Aromaa and Väänänen 2016) and facilities design (Braatz et al. 2011; Zamberlan et al. 2012; Gatto et al. 2013; Paravizo and Braatz 2017). GEs are platforms employed in the development of computer games, utilizing several pre-existing modules and functionalities which make it easier to create and deploy such simulations.

In this context, we investigate how each of the four mentioned technologies (DHM, CAD, DES, GEs) can be employed for fostering ergonomics integration into design processes, highlighting affordances and shortcoming of each of them as well as pointing toward a possible articulation of those.

---

[1] https://www.flexsim.com/.

2 Methods

This qualitative, exploratory study, analyzes a number of studies employing DHM, CAD, DES and GEs through an expansion of the process for "digital ergonomics" proposed by Ziolek and Kruithof (2000). This process combines the environment (CAD data/geometry), the manikins (anthropometry) and the analysis (postures, ergonomics standards and custom analysis).

This expansion proposes four major axes for analyzing and planning of virtual simulations in ergonomics: environment (comprising both CAD data and information represented), manikins (considering anthropometry aspects and manikins modeling and animation/configuration), analysis (both quantitative focused through traditional ergonomics protocols such as RULA, NIOSH lifting equations and other biomechanical prediction models and qualitative analyzing field of view, overall postures and movements in the environment and so on) and the interactions (how users/designers can control and navigate in the environment and those that can be programmed for creating richer experiences). The expansion can be seen in Fig. 1.

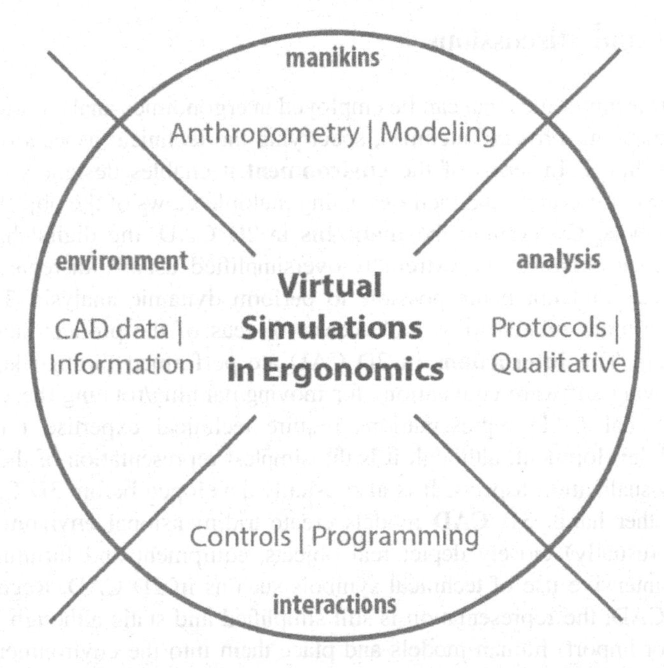

Fig. 1. Dimensions for virtual simulations in ergonomics analysis and planning

Furthermore, we subdivide each of the technologies analyzed in terms of its development complexity. Thus, eight instances of virtual simulations for ergonomics are analyzed: 2D CAD (technical drawings and blueprints in their digital format), 3D CAD (the tridimensional representation of the designed product/facility), static DHM (in which digital human models are placed and postures are configured for ergonomics analysis), dynamic DHM (in which animations, postures and process are implemented in the DHM environment), simple GE environment (with only an avatar and the designed product/facility), advanced GE environment (with complex interactions implemented, allowing for the simulation of procedures and operations of the designed product/facility), simple DES (focusing mainly on representing the facility/product and the human element) and advanced DES (incorporating operational times, manikins postures and detailed interactions in the work system).

Five studies (Duarte and Goldenstein 1996; Mattila 1996; Braatz 2015; Patel et al. 2016; Marconi et al. 2018) dealing with CAD usage for ergonomics were collected. Another 5 studies (Braatz et al. 2013; Fontes et al. 2014; Braatz 2015; Paul and Quintero-Duran 2015) regarding DHM and ergonomics were analyzed. Regarding GEs in ergonomics another 5 studies (Braatz et al. 2011; Zamberlan et al. 2012; Gatto et al. 2013; Aromaa and Väänänen 2016; Paravizo and Braatz 2017) were retrieved. Finally, on DES and ergonomics/workspace design five studies (Laughery 1999; Perez et al. 2014; Ferreira and Braatz 2015; Hettinger et al. 2015; Hossain et al. 2017) were considered.

3 Results and Discussions

The **2D CAD** technologies that can be employed in ergonomics analysis are commonly used for formalizing project outcomes, specifying the technical aspects of the design among other things. In terms of the **environment** it enables designers to create 2D representations of products and facilities, using multiple views of the object and several technical symbols. Concerning the **manikins** in 2D CAD, the digital human representations (when existing) are extremely oversimplified usually disregarding anthropometry aspects and not being possible to perform dynamic analysis. The **analysis** performed are more related to overall access to areas of the product/facility, fit and similar aspects. The **interactions** in 2D CAD are performed through keyboard and mouse, following software conventions for moving/panning/rotating the viewport.

Bidimensional CAD representations require technical expertise for its understanding and development, although it is the simplest representation of the 8 analyzed in terms of visualization aspects. It is also usually developed before 3D CAD models.

On the other hand, **3D CAD** models create tridimensional **environments** using models that (usually) closely depict real objects, equipment and furniture, thus not needing the intensive use of technical symbols such as in 2D CAD. Regarding **manikins** in 3D CAD, the representation is still simplified and static although it's possible to develop (or import) human models and place them into the environment. **Analysis** are, once again, qualitatively focused, comprising mostly reach, fit and access aspects. The **interactions** are mediated by mouse and keyboard and although users/designers now navigate around a 3D world, there's not many interactions available.

The 3D models generated in 3D CAD applications are the foundation of the other 6 virtual simulation mediums analyzed. It still requires technical expertise to develop them and operate the software, however, it's visually more engaging and self-explanatory. Nonetheless, models tend to be simplified textured only with solid colors and simple materials.

Static DHM usually employs a 3D **environment** with objects, facilities and equipment. The **manikins** are usually modelled considering biomechanical aspects and detailing joints, degrees of freedom and anthropometry aspects, being the most detailed and faithful representation of humans among the technologies analyzed. The **analysis** performed can be both quantitative, related to ergonomics protocols (RULA, NIOSH lifting equation, OWAS, etc.), biomechanical (force exerted, fatigue, etc.) or qualitative (reach envelope, field of view, etc.). The **interaction** with the system is performed through keyboard and mouse, and due to its static nature, few interactions are programmed.

Dynamic DHM in turn, present the same affordances in terms of **environment, manikins** and **analysis** that their static counterpart, however it enables designers to program dynamic **interactions** among the manikins and the environment, allowing for the simulation of operational procedures and tasks. In some DHM software, it is even possible to integrate movement capture systems for manikins control in real time (Santos et al. 2016).

DHM software are specifically designed for considering analyzing ergonomics aspects of workplaces and products and as such they present the most comprehensive collection of traditional ergonomics tools by default, thus being especially well-suited for ergonomics assessments of workplaces and postures. Nonetheless, visualization capabilities are generally simplified. We highlight that dynamic simulation in DHM software usually requires a detailed design of postures and postures interpolation, as well as setting up the interactions between the manikins and the objects/environment.

A **simple GE environment** is a highly visually developed (due to GE's high-end rendering capabilities) environment, which allows for the representation of people, equipment/objects and information in a more straightforward way. Furthermore, **manikins** are usually simplified, not considering aspects of anthropometry and biomechanical modelling as in DHM software, although it is possible to create manikins complying with those. Regarding the **analysis,** they are more related to qualitative aspects. **Interactions** may vary, most commonly employing either keyboard and mouse (using gaming conventions such as WASD/Arrow keys) or joysticks (wireless or wired) for navigation and interaction.

An **advanced GE environment** presents the same affordances its simpler counterpart does, but it expands in terms of the possible **interactions.** More advanced scenarios may also yield more realistic **environments**, due to the careful set up of advanced lighting conditions and rendering settings. GEs allow designers to program a huge number of different interactions (such as dialogues with characters, multiplayer support, environment interaction, sound and audio effects, layout changes, real time communication and exploration, etc.). Additionally, GEs can be employed for developing and deploying Virtual Reality (VR) and Augmented Reality (AR) applications, and they also can incorporate real time movement capture systems' integration.

GEs at their simplest, provide an interactive, high-end visualization, 3D environment which users and designer can explore naturally through an avatar they command through a joystick or other hardware. The multitude of possibilities for interactions programming make them a tool of almost untapped potential in the context of simulation in ergonomics, across the analysis, learning and designing goals of simulation.

Finally, **simple DES** application provides a 3D **environment,** where objects and equipment are generally more detailed in terms of their visual representation. The **manikins** are not designed to consider ergonomics aspects *a priori*. **Analysis** will focus on quantitative aspects of distances traveled, time spent and other parameters, although simple qualitative analysis may also be possible. **Interactions** once again are commanded through keyboard and mouse, although some software may incorporate joystick usage.

The more **advanced DES** expand the previously mentioned affordances on terms of the **interactions** and **analysis**, since it becomes possible to navigate the environment using head mounted displays, and also to program complex systems relationships in the model. It also is possible to create custom manikins, animations and postures fostering more complete analysis.

DES utilization may be seen as an intermediary between DHM and GEs in terms of its visual capabilities. The possibility of developing models which account for operational times, probabilistic distribution functions, failure times and frequency among other system-oriented metrics are particular to DES. Being able to incorporate ergonomics aspects in this rich scenario can be a great opportunity for process engineers and other professionals that already use this kind of software. Figure 2 synthesizes the discussions and considerations made.

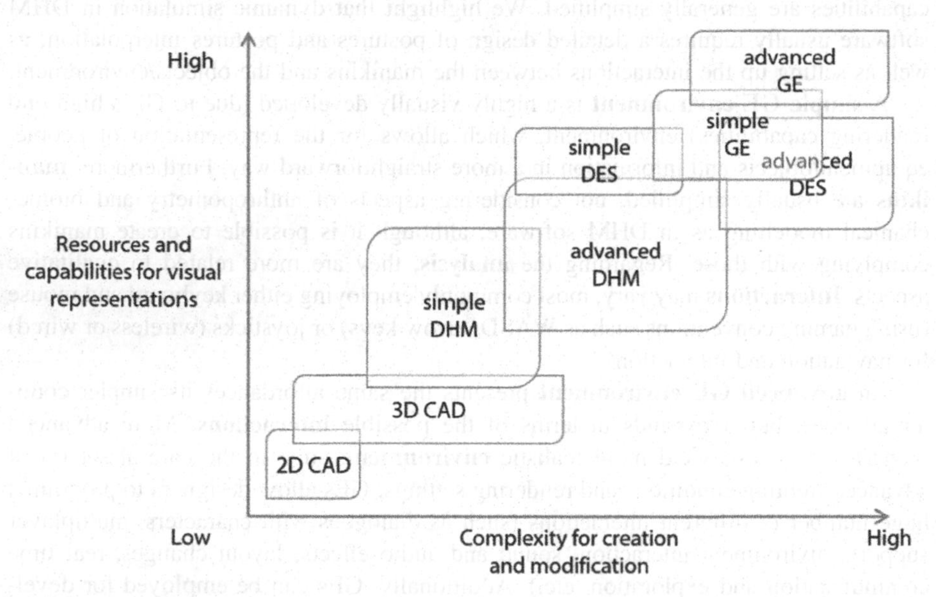

Fig. 2. Mapping of different technologies in terms of its visual representation capabilities and complexity for creation

Figure 3 presents a chart developed from the studies analyzed e from authors' experience in this area, aiming to map the main capabilities of the different computer tools in its varied application formats in the requisites for ergonomics simulation. In the figure, the closer to the center of the chart the less capabilities/affordances the analyzed technology presents for ergonomics simulation.

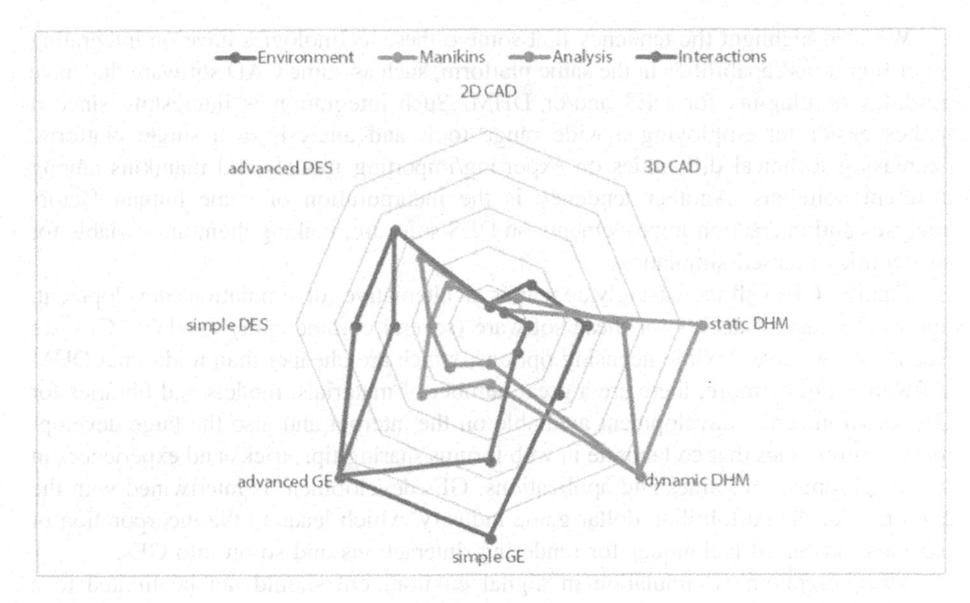

Fig. 3. Comparative analysis of the computational technologies in different requisites for ergonomics simulation.

We highlight that these analyses considered the tools and developments presented in the studies analyzed and authors' experience with software from these four technologies (for CAD Autodesk's AutoCAD was the tool mostly considered, whereas for DHM Siemens' Jack software was the one reported in the studies analyzed, for DES we based the analysis on Flexsim and the GE evaluation relied on the Unreal Engine platform). Other software may present different affordances and capabilities than those highlighted in this paper, which is a limitation of this study.

4 Conclusion

Results highlight that digital simulation in ergonomics may be achieved through a diverse set of technologies, according to the needs, interests and limitations of each specific situation.

Thus, if the simulation aims to support the discussion of a tool or layout, in a simpler way and without much time available for its development, employing CAD tools may be useful, bringing the possibility for contributions to emerge. On the other hand, if it is necessary or interesting for the project to better understand or discuss specific issues regarding the interaction between the worker and its workspace, DHM technologies are better suited. For a more comprehensive understanding of the worker and its relationship with several concurrent processes, accounting for detailed aspects of production times and even their statistical distribution and probabilities, DES are better suited. However, if the focus is on analyses and interactions with large 3D environments with extremely well-developed visuals and graphical abilities, GE appear as one the main tools.

We also highlight the tendency that some o these technologies have on integrating other functions/capabilities in the same platform, such as some CAD software that have modules or plug-ins for DES and/or DHM. Such integration is interesting since it makes easier for employing a wide range tools and analysis in a single platform, decreasing technical difficulties on exporting/importing models and manikins among different solutions. Another tendency is the incorporation of some human factors analyses and interaction improvements on DES software, making them more viable for ergonomics focused simulation.

Finally, GEs will increasingly appear as an alternative for simulations development, due to the easy of access for these software (several commercially available GEs are free to use or have flexible licensing options, which are cheaper than traditional DHM software). Furthermore, there are a great number of materials, models and libraries for GEs environments' development available on the internet and also the huge development communities that collaborate in web forums sharing tips, tricks and experiences in the development of games and applications. GEs development is intertwined with the evolution of the multibillion dollar game industry, which leads to the incorporation of the most advanced techniques for rendering, interactions and so on into GEs.

Thus, ergonomics simulation in digital environment should not be limited to a single type of technology. The greater the variety of the tools employed, the more comprehensive the (digital) simulations will be, overcoming the shortcomings of the individual technologies and combining their strengths.

References

Aromaa S, Väänänen K (2016) Suitability of virtual prototypes to support human factors/ergonomics evaluation during the design. Appl Ergon 56:11–18. https://doi.org/10.1016/j.apergo.2016.02.015

Backstrand G, Hogberg D (2007) Ergonomics analysis in a virtual environment. Int J Manuf Res 2:198–208

Banks J, Nelson BL, Carson JS, Nicol DM (2010) Discrete-event system simulation, 4th edn. Prentice Hall, Upper Saddle River

Béguin P, Weill-Fassina A (2002) Da simulação das situações de trabalho à situação da simulação. In: Duarte F (ed) Ergonomia e projeto: na indústria de processo contínuo, 1st edn. Editora Lucerna, Rio de Janeiro, pp 34–63

Braatz D (2015) Suportes De Simulação Como Objetos Intermediários Para Incorporação Da Perspectiva Da Atividade Na Concepção De Situações Produtivas. Universidade Federal de São Carlos, Tese de Doutorado

Braatz D, Lopes D, Camarotto J, Menegon NL (2011) Evaluation of different interaction methods applied in participatory workspace design. In: 21st international conference on production research

Braatz D, Tonin L, Fontes A, Menegon N (2013) Digital human modeling and simulation for ergonomics workspace design : two Brazilian cases. In: Proceedings of the 2nd international digital human modeling symposium, Ann Arbor, pp 1–10

Broberg O, Andersen V, Seim R (2011) Participatory ergonomics in design processes: the role of boundary objects. Appl Ergon 42:464–472. https://doi.org/10.1016/j.apergo.2010.09.006

Chaffin DB (2005) Improving digital human modelling for proactive ergonomics in design. Ergonomics 48:478–491. https://doi.org/10.1080/00140130400029191

Chaffin DB (1997) Development of computerized human static strength simulation model for job design. Hum Factors Ergon Manuf 7:305–322. https://doi.org/10.1002/(sici)1520-6564(199723)7:4<305::aid-hfm3>3.0.co;2-7

Duarte F, Goldenstein M (1996) O layout como imagem da organização do trabalho: a participação da ergonomia no projeto de salas de controle. In: Anais do Encontro Nacional de Engenharia de Produção. ABEPRO, pp 1–8

Eckert C, Boujut JF (2003) Introduction: the role of objects in design co-operation: communication through physical or virtual objects. Comput Support Coop Work CSCW Int J 12:145–151. https://doi.org/10.1023/A:1023954726209

da Costa Alves Ferreira R, Braatz D (2015) Articulação Conceitual Integrada para Aplicação de Ferramentas Computacionais no Projeto de Situações Produtivas. In: Anais do XXXV ENCONTRO NACIONAL DE ENGENHARIA DE PRODUÇÃO. ABEPRO, Fortaleza, pp 1–15

Fontes ARM, Parise AC, Filippi GF et al (2014) Complementarities of DHM and ergonomics analysis of work into workplace design: the task of manual packing. Int J Hum Factors Model Simul 4:266–277

Gatto LBS, Mól ACA, dos Santos IJAL et al (2013) Virtual simulation of a nuclear power plant's control room as a tool for ergonomic evaluation. Prog Nucl Energy 64:8–15. https://doi.org/10.1016/j.pnucene.2012.11.006

Hettinger LJ, Kirlik A, Goh YM, Buckle P (2015) Modelling and simulation of complex sociotechnical systems: envisioning and analysing work environments. Ergonomics 58:600–614. https://doi.org/10.1080/00140139.2015.1008586

Hossain NUI, Debusk H, Hasan M et al (2017) Reducing patient waiting time in an outpatient clinic: a discrete event simulation (DES) based approach. In: 67th annual conference Expo and Institute of Industrial Engineering

Laughery R (1999) Using discrete-event simulation to model human performance in complex systems. In: Farrington PA, Nembhard HB, Sturrock DT, Evans GW (eds) Proceedings of the 1999 winter simulation conference, pp 815–820

Maline J (1994) Simuler le Travail: une aide à la conduite de project. ANACT, Montrouge

Marconi M, Germani M, Favi C, Raffaeli R (2018) CAD feature recognition as a means to prevent ergonomics issues during manual assembly tasks. Comput Aided Des Appl 15:1–13. https://doi.org/10.1080/16864360.2018.1441240

Mattila M (1996) Computer-aided ergonomics and safety - a challenge for integrated ergonomics. Int J Ind Ergon 17:309–314. https://doi.org/10.1016/0169-8141(95)00062-3

Paravizo E, Braatz D (2017) Analysis, validation and design: using Game Engines to support ergonomics intervention and design process. In: 48th annual conference of the association of Canadian ergonomists & 12th international symposium on human factors in organizational design and management, Banff, pp 488–495

Patel T, Sanjog J, Karmakar S (2016) Ergonomics perspective in agricultural research: a user-centred approach using CAD and digital human modeling (DHM) technologies. J Inst Eng Ser A 97:333–342. https://doi.org/10.1007/s40030-016-0162-2

Paul G, Quintero-Duran M (2015) Ergonomic assessment of hospital bed moving using DHM Siemens JACK. In: Proceedings of 19th Triennial Congress of the international ergonomics association, pp 1–6

Perez J, de Looze MP, Bosch T, Neumann WP (2014) Discrete event simulation as an ergonomic tool to predict workload exposures during systems design. Int J Ind Ergon 44:298–306. https://doi.org/10.1016/j.ergon.2013.04.007

dos Santos WR, Braatz D, Tonin LA, Menegon NL (2016) Análise do uso integrado de um sistema de captura de movimentos com um software de modelagem e simulação humana para incorporação da perspectiva da atividade. Gestão Produção 23:612–624. https://doi.org/10.1590/0104-530X1758-14

Wilson JR (1997) Virtual environments and ergonomics: needs and opportunities. Ergonomics 40:1057–1077. https://doi.org/10.1080/001401397187603

Zamberlan M, Santos V, Streit P et al (2012) DHM simulation in virtual environments: a case-study on control room design. Work J Prev Assess Rehabil 41:2243–2247. https://doi.org/10.3233/WOR-2012-0446-2243

Zhou D, Chen J, Lv C, Cao Q (2016) A method for integrating ergonomics analysis into maintainability design in a virtual environment. Int J Ind Ergon 54:154–163. https://doi.org/10.1016/j.ergon.2016.06.003

Ziolek SA, Kruithof PC (2000) Human modeling & simulation: a primer for practitioners. In: Human modeling & simulation: a primer for practitioners. Human factors and ergonomics society, pp 825–827

A Framework of Participatory Ergonomics Simulation

Ole Broberg(✉) and Carolina Conceicao

Department of Management Engineering, Technical University of Denmark, 2800 Lyngby, Denmark
{obro, casou}@dtu.dk

Abstract. The aim of this paper is to develop a framework of participatory ergonomics simulation (PES). The framework is targeting researchers and practitioners and offers a six-phase model for analyzing, planning and executing PES based intervention in architectural and engineering design projects. The framework was developed based on analyzing seven PES projects that all were aimed at integrating ergonomics into design projects. The proposed framework includes six phases: contextual knowledge, simulation aim, simulation design, prototyping, simulation session, and simulation outcome. In each phase a number of elements to consider are listed. Based on the framework a number of research needs are identified.

Keywords: Participatory ergonomics · Simulation · Work systems

1 Introduction

It is acknowledged that simulation games may help researchers and practitioners in better understanding complex sociotechnical systems [1]. Many work systems are complex systems in the sense that there are multiple interactions between several elements of the system. Such work systems are difficult both to analyze and to design due to the many heterogeneous interactions and characteristics like emergence and adaption [2]. Within human factors and ergonomics simulation methods has gained momentum, especially in relation to designing work systems as part of design projects. By simulating future work systems, it is possible to analyze and identify system performance related to human wellbeing and hence provide feedback to the design process and the involved designers.

Participatory ergonomics simulation (PES) is a method in which workers and other stakeholders participate in simulations of future workplace and work practices. PES aims to (i) innovate the workplace design [3], (ii) enable evaluation of future ergonomics conditions [4], and (iii) adjust the design to improve the future working conditions and safety [5, 6].

The aim of this paper is to outline a framework of participatory ergonomics simulation in design. The framework is intended helping researchers and practitioners in analysing, planning and executing PES based intervention in architectural and engineering design projects with the aim of designing healthy and efficient work systems.

S. Bagnara et al. (Eds.): IEA 2018, AISC 821, pp. 391–395, 2019.
https://doi.org/10.1007/978-3-319-96080-7_46

An illustration of the PES project type was the design of an outpatient clinic in a new hospital building. The first author was involved in organising three workshops in which hospital staff explored the work system to be designed, including Lego figure table top simulations of patient flow scenarios, staff work tasks, and work organisation (Fig. 1).

Fig. 1. Participatory ergonomics simulation session of a new outpatient clinic at a hospital.

2 Methods

The framework was developed by an analysis of seven participatory ergonomics projects across industries. The data material was based on the authors' own engagement in four projects and additional three ones being collaboratively analysed by researchers in a workshop. All selected projects were aimed at integrating ergonomics into design projects by involving workers and designers in workshops in which the future work system was explored and simulated in difference ways. The PES projects were analysed in an inductive approach but inspired by the game design framework of Lukosch et al. [1].

3 Results

The elements identified across the seven projects were organised into a framework pointing to a phase sequence in planning and executing PES events in design projects, each including a number of elements (Table 1).

3.1 Contextual Knowledge

Before planning PES it is important to understand both the overall system design project and the real-world work system involved. The methods for gaining this knowledge include contextual interviews [7], observations, and document studies. The outcome includes identifying the main stakeholders and their role and interests in the design project, and the current challenges in the work system concerning health and safety, and overall performance.

Table 1. Participatory ergonomics simulation framework

Phase	Elements
1. Contextual knowledge	System design project: client, project management, design phases, designers Current work system: work domain, work system analysis, ergonomic work analysis
2. Define simulation aim	Work system innovation, ergonomics evaluation, redesign to improve health and performance. Simulation questions. What is being simulated
3. Simulation design	Representation of real-world work system. Simulation media: physical, digital, fidelity. Session design. Participants. Location. Define simulation scenarios
4. Prototyping	Testing the simulation with target group representatives
5. Simulation session	Session script. Session rules. Modes of simulation: scenario playing, object manipulation, bodily gestures. Facilitation of session. Debriefing
6. Simulation outcome	Recording of outcome: simulation objects, writing, photos, video. Type of outcome: design feedback, revealing user needs, system performance. Transfer of outcome to system design project: documents, drawings, diagrams, digital representations

3.2 Define Simulation Aim

The aim of PES may take different directions. Because of the complex nature of work systems, it is often necessary to focus on only parts of the system. A work system may be defined as including at least four interdependent elements: space, organisation, finance, and technology [8]. A simulation may have its main focus in one of these elements while trying to take the others into account. Simulation questions and scenario design are also directing the course of PES sessions. In a simulation of a new outpatient clinic at a hospital one of the sessions were directed by this simulation questions: What if we want the patients to complete their clinic visit in one day instead of having appointments with different medical doctors over two to three days? Based on this question different work system designs were simulated to identify the requirements.

3.3 Simulation Design

Designing the PES includes the selection of media to represent the work system. Medias might be digital or physical. Digital media includes 3 D diagram and animation software, virtual reality, and augmented reality. Physical media include scale models, full-scale mock-ups, and gameboards for table-top simulation. More research is needed on the influence of simulation media on simulation outcome. Östermann et al. [9] compared three different media for simulation of a ship bridge work system: 3 D CAD model, 1:16 paperboard model, and a 1:1 plywood mockup. All models were capable of generating useful design feedback and eliciting user needs when applied with use scenarios. In a study comparing simulation media and simulation outcome Andersen & Broberg (2015) indicated that applying full-scale physical mockup's in PES sessions enabled identification and evaluation of space and technology elements of a work system. In contrast, table-top gameboard based PES sessions enabled identification and evaluation of work organisation elements in a work system.

3.4 Prototyping

Lukosch et al. [1] recommend a testing and possibly redefining the simulation before final simulation game design and execution. None of the PES projects in this study included testing the simulation with representatives from the target group in a prototyping session. Probably it might have improved the outcome of PES sessions. This should be studied further.

3.5 Simulation Session

The simulation session may be organized by help of a session script taking into account the amount of time available, the aims, session rules and the number of participants. The issue of facilitating the session is very important. This might be the role of an ergonomist, and hence it requires skills in facilitating a group of people interacting with a simulation media in order to explore and design a future work system. Debriefing after a PES session is essential for capturing the learning outcomes among the participants, and for concluding on system (re)design proposals.

From a scientific perspective it is of importance to understand the knowledge creation process during a PES session. This is not well understood today. Andersen & Broberg [10] pointed to a combination of interacting with the simulation media, doing experiments (what if), and sharing work experiences between session participants as the sources of ergonomics knowledge creation.

3.6 Simulation Outcome

The outcome of a PES session needs to be captured and taken into account in the overall system design project. If design feedback and identified user needs are not integrated in the design process it will not impact the actual design of a new work system. The question of transferring generated PES knowledge into the design process seems to be an issue needing much more research. Andersen & Broberg (forthcoming) pointed to that participants in PES sessions acted as a ‘filter’, meaning that not all generated ergonomics knowledge was transferred into the overall design process, and hence no design action was taken.

4 Discussion

Participatory ergonomics simulation methods are becoming more widespread in both practice and research. However, there is a gap of knowledge concerning the understanding of how different simulation media support different types of outcomes, and how PES in design projects can be designed to meet the aim of contributing to human wellbeing and systems performance. In this study we propose a framework for understanding how to apply PES in the design process of complex work systems.

We point to issues that need further research:

- How does different simulation media influence the quality of simulation outcome?
- Does prototyping a PES design improve the quality of simulation outcome?
- By what mechanisms are design feedback and user needs knowledge generated during simulation sessions?
- How is simulation outcome transferred into the overall design process in order to have an impact?

This study is based on the analysis of seven PES events. Hence, we consider the proposed framework as a very initial one that needs to be revised as new evidence come to light.

5 Conclusion

In this paper we have proposed a framework aimed at helping researchers and practitioners in analyzing, designing and executing participatory ergonomics simulation as part of design of complex work systems. PES needs to be further studied as a research tool to explore and understand complex work systems.

References

1. Lukosch HK (2018) A scientific foundation of simulation games for the analysis and design of complex systems. Simul Gaming 49:279–314
2. Wilson JR (2014) Fundamentals of systems ergonomics/human factors. Appl Ergon 45:5–13
3. Broberg O, Edwards K (2012) User-driven innovation of an outpatient department. Work 41:101–106
4. Andersen SN, Broberg O (2015) Participatory ergonomics simulation of hospital work systems: the influence of simulation media on simulation outcome. Appl Ergon 51:331–342
5. Daniellou F (2007) Simulating future work activity is not only a way of improving workstation design. @ctivités 4:84-90
6. Medwid K, Smith S, Gang M (2015) Use of in-situ simulation to investigate latent safety threats prior to opening a new emergency department. Saf Sci 77:19–24
7. Holtzblatt K, Beyer H (2017) Contextual design: design for life, 2nd edn. Morgan Kaufmann, Amsterdam
8. Horgen T, Joroff ML, Porter WL, Schön DA (1999) Excellence by design: transforming workplace and work practice. Wiley, New York
9. Östermann C, Berlin C, Bligård L-O (2016) Involving users in a ship bridge re-design process using scenarios and mock-up models. Int J Ind Ergon 53:236–244
10. Andersen SN, Broberg O (2017) A framework of knowledge creation processes in participatory simulation of hospital work systems. Ergonomics 60(4):487–503

Perspectives on Autonomy – Exploring Future Applications and Implications for Safety Critical Domains

Steven C. Mallam(✉), Salman Nazir, Amit Sharma, and Sunniva Veie

Training and Assessment Research Group, Department of Maritime Operations, University of South-Eastern Norway, Borre, Norway
{steven.mallam, salman.nazir, amit.sharma, 144643}@usn.no

Abstract. As the sophistication and feasibility of implementing highly automated and autonomous technologies increases, the way in which the human element interacts and contributes to achieve a system's goals continues to transform. The role of the human, including their required training, competencies and work tasks within complex socio-technical systems is therefore constantly being redefined by technological advancement. This study explores the potential effects of autonomous technologies on future work organization and evolving roles of humans within maritime industries. Ten Subject-Matter Experts working within industry and academia were interviewed to elicit their perspectives on the current state and future implications of autonomous technologies in the maritime domain. All interviews were transcribed verbatim and assessed using Thematic Analysis. Four main themes emerged from the interviews: (i) *Trust*, (ii) *Awareness & Understanding*, (iii) *Control*, (iv) *Training and Organization of Work*. A fuzzier fifth theme also emerged from the data analysis: (v) *Practical Implementation Considerations*, which encompassed various sub-topics related to the realities of real-world implementation of autonomous ships and shipping, including regulatory, security and economic issues. The results provide valuable input and perspective of autonomous systems and the future organization of complex safety-critical systems, including the role of humans in ever evolving and increasingly technology-oriented operations.

Keywords: Maritime · Automation · Digitalization · Work organization
Macroergonomics · Transport · Safety

1 Introduction

Advancing technologies and digital transformation continue to alter how work tasks and systems are designed, organized and operated. As the sophistication and feasibility of implementing highly automated and autonomous technologies increases, the way in which the human element interacts and contributes to achieve a system's goals will continue to transform. Thus, the role of the human, including their required training, competencies and work tasks within complex socio-technical systems is continuously

S. Bagnara et al. (Eds.): IEA 2018, AISC 821, pp. 396–405, 2019.
https://doi.org/10.1007/978-3-319-96080-7_47

being redefined by technology. Maritime-related industries, such as shipping, offshore energy, commercial fishing, aquaculture and coastal tourism operate in frequently harsh, isolated and hazardous environments. As such, they have developed as engineering and technology-rich domains, typically interested in applying new solutions that can enhance both safety and efficiency of operations (Stopford 2009). As new technologies and methods develop, there is typically a delay between its initial creation, feasibility and adoption (Doraszelski 2004; Mallam et al. 2017). Additional technology can have negative impacts and unintended consequences on system operations. For example, there have been numerous maritime accidents where adoption of novel technology was found to be a contributory factor to specific incidents (Lutzhoft and Dekker 2002). Thus, as emerging autonomous technologies become increasingly viable, a better understanding of how to facilitate these applications in not only new, but existing systems must be investigated.

The purpose of this paper is to explore the perceived impact and implications of automated and autonomous technologies on the operations and human element in complex safety-critical industries, with particular focus on the maritime domain. This research is based on interview data from Subject-Matter Experts in order to elicit their perspectives on how applications of autonomous systems can, and have already, impacted their work and system processes. Two distinct cohorts were recruited: academic researchers (e.g. human factors and ergonomics researchers working within academic research) and industry practitioners (e.g. operators, engineers, computer scientists working within industry) in order to better understand emerging technologies' perceived implications, benefits and challenges. This explorative study will help address questions related to "explore-exploit" trade-offs associated with implementation of new technologies related to highly automated and autonomous systems.

2 Background

2.1 Defining Automation and Autonomy

To illustrate the contemporary state of development in autonomous systems, it is useful to depict the levels of the automation continuum and clarify associated definitions. The term "autonomy" in its crudest form refers to self-regulation or self-government. Whereas "automation" can be defined as "the execution of machine agent (usually a computer) of a function that was previously carried out by a human" (Parasuraman and Riley 1997). As a combination of these two concepts, an autonomous agent is defined in the technological context as "a system situated within and a part of an environment that senses that environment and acts on it, in pursuit of its own agenda and so as to effect what it senses in the future" (Franklin and Graesser 1997). However, these refer to the agents or system that have complete liberty in actions and their execution in pursuit of its goals. Most of the agents in modern industrial applications fall somewhere between being completely manually operated and completely autonomous. Therefore, the level of "autonomy" a system has can be viewed as a function of associated human intervention. Figure 1 illustrates a continuum of automation, ranging from an absence of automation (i.e. Level 0) through increasing levels of automation until reaching full

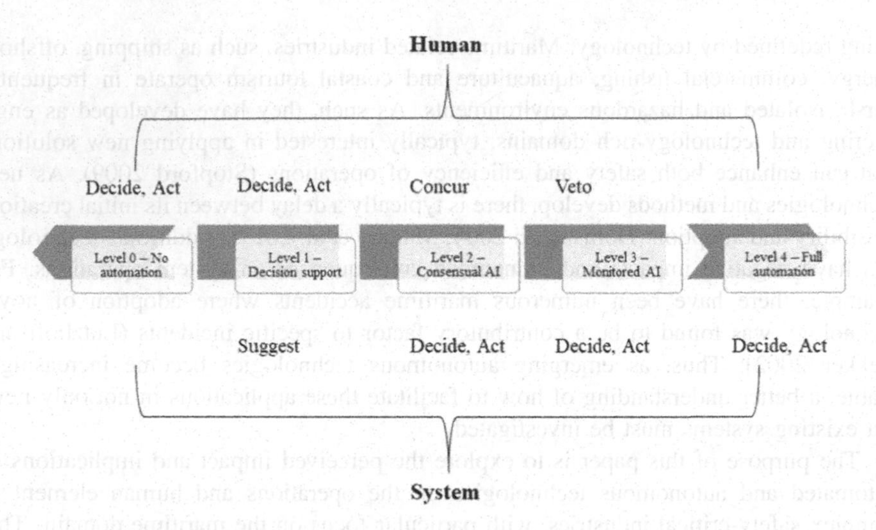

Fig. 1. Levels of automation in sociotechnical system (adapted from Endsley and Kiris 1995).

automation (i.e. Level 4), where a system has complete autonomy to pursue its goals and execute desired actions (Endsley and Kiris 1995).

2.2 Automation Lessons Learned: Proceed with Caution

The motivation, purpose and goals of implementing automation can have a range of benefits, including potentially reducing costs and increasing safety. However, new technology implementation, such as automation and autonomy, can also have drawbacks contributing to undesired outcomes. For example, numerous accidents in safety-critical domains have occurred not despite, but because of, automation and technology (Nazir et al. 2014). A prominent and reoccurring issue across domains with respect to human-technology interaction has been the loss of situational awareness and communication during operations. Issues such as loss of skills, out-of-loop syndrome, mental workload and trust in the automation system itself have been shown to contribute to accidents (Bainbridge 1983; Endsley and Kiris 1995; Hoff and Bashir 2015; Strauch 2017). In this regard, Parasuraman (1987) has stated that despite technological complexity, humans are fundamentally poor passive monitors of automated systems. Therefore, stakeholders should exercise "cautious optimism" whenever it comes to the use and adoption of escalating levels of automation and new technologies in the workplace.

2.3 Maritime Applications

Though characterized as "traditional" in comparison to other industries, the maritime domain, and specifically merchant shipping, has not remained impervious to change or advancement. The technologies and operations of contemporary merchant vessels stand in stark contrast to those of just a few decades earlier. Many operational functions, such as navigation tasks and equipment (e.g. bridge instruments such as GPS, ECDIS, DP, etc.)

are automated and operations have moved closer to automation levels 2, 3 and 4 (Endsley and Kiris 1995). Automation has allowed for the redistribution of operational duties between human and technical agents, resulting in ships operating with far less crew today than in the past (Grech, et al. 2008; NRC 1990). Within the shipping industry there has been work towards establishing unmanned vessels, with shore-side control and supervisory centres (e.g. MUNIN Project). Furthermore, shipping companies are now beginning to invest in the development of autonomous unmanned vessels, such as the Yara Birkeland (Kongsberg Maritime 2017).

While increased automation has optimized a ship's operational resources, it has also led to the coinage of terms amongst seafarers and industry stakeholders such as "Radar assisted collision" or "ECDIS assisted grounding", highlighting the threat and challenges faced in complex sociotechnical operations. Thus, it is necessary to better understand the perspectives of involved stakeholders early within the process of applying new technologies.

3 Methods

3.1 Participants

Ten Subject-Matter Experts (SME) participants were recruited for individual interview sessions. SMEs consisted of (i) academic researchers (working experience in field: AVG = 17.4 yrs; SD = 5.0 yrs; MIN = 10.0 yrs; MAX = 25.0 yrs), and (ii) industry practitioners (working experience in field: AVG = 25 yrs; SD = 10.5 yrs; MIN = 5 yrs; MAX = 35 yrs) working within various positions (e.g. researchers, training instructors, marketing, managers, analysts) where automation and autonomy are influencing their areas of interest (see Table 1). Convenience and snowball sampling were utilized to recruit participants from the researcher's network. Participants worked for various academic institutions, government and private sector companies in Northern Europe and the United States. All participants completed an informed consent form prior to participation in the interviews. This research project, and its interview questions, were registered and approved by the Norwegian Centre for Research Data (NSD) (project no. 56374).

Table 1. Participant backgrounds and expertise.

Discipline	Current position/employer	Domains of expertise
Academic researchers	• 3 Full Professors • 1 Associate Professor • 1 Research Scientist	Safety Management, Design, Maritime Technology, Human Factors, Training
Industry practitioners	• 1 Process Training Center • 1 Maritime Training Center • 1 Equipment Manufacturer • 1 National Transport Agency • 1 Classification Society	Competence Development/Assurance, Maritime Transport, Simulation/Modelling, Human Factors, Energy Sector

3.2 Procedure

The interviews were performed by two researchers following a semi-structured question script which focused on three general areas: (i) the definition and concept of autonomy and autonomous systems, (ii) perceived challenges, and (iii) the role of humans in future training and work paradigms (see Table 2). Six of the ten participant interviews were in English, while four were in Norwegian, due to both interviewer and interviewee being native speakers. Norwegian-spoken interviews were translated to English for data analysis. Each participant was interviewed individually, and interviews lasted between approximately 15-25 min. All interviews were audio recorded.

Table 2. Participant Interview Question Transcript.

1	**Introduction and Definition Questions:**
1.1	What is your definition and concept of "autonomy"?
1.2	Could you please discuss an example(s) of advancing automation and/or autonomy in your field?
2	**Perceived Challenges and Changes Questions:**
2.1	What will be the biggest challenge faced in implementing autonomous systems?
2.2	What will be the biggest challenge faced in operating autonomous systems?
2.3	How will autonomy change every day processes/operations in your field of interest?
2.4	How do you see the future impacts of autonomy in your domain?
3	**Role of Human Questions:**
3.1	What is the new role of the human/operator in autonomous systems?
3.2	How will increasing automation and autonomy affect the training of new and existing workers?
3.3	What type of skills will be required for these emerging technologies?
3.4	What solutions are needed in training?

3.3 Data Analysis

All interviews were transcribed verbatim post-hoc from the audio recordings. The interview transcripts were analyzed using Thematic Analysis (Braun and Clarke, 2006). The six step process of Thematic Analysis was used as a framework to code content and define common patterns within the data. Three members of the research team first independently reviewed and analyzed the interview transcripts. This was followed by a workshop where the researchers presented their individual findings in order to validate individual interpretations and analysis of the qualitative data to confirm specific results.

4 Results

The Thematic Analysis revealed four major themes related to the human element and autonomy in complex safety-critical systems: (i) *Trust*; (ii) *Awareness & Understanding*; (iii) *Control*, and (iv) *Training and Organization of Work.* Several other themes emerged, although with less frequency and attention across participant interviews, which can be categorized within an additional theme: (v) *Practical Implementation Considerations*. This fifth theme consists of various sub-topics which arose where participants discussed pragmatic threats and limitations, not of the technology, or technological development itself, but rather of real-world implementation issues regarding such areas as regulations, cybersecurity and economic considerations.

4.1 Trust

A predominant theme throughout participant interviews was the concept of trust. Participants stressed the need to trust the operations and decisions of autonomous systems as human operators continue to move further away from the sharp-end of processes. Trust was a pervasive topic which transcended most areas of participant interviews, and the subsequent themes presented in the results.

Trust was seen as a function of successful and proven operations, which participants discussed in terms of societal norms and acceptance in technology. The development of trust in autonomous applications in a complex safety-critical domain, such as shipping, requires the acceptance from many actors. For example, several participants discussed the concept of multi-modal shipping control configurations having to share the same physical environment (i.e. traditionally manned and operated ships, fully autonomous unmanned ships, unmanned manual shore-controlled ships, fishing vessels, pleasure crafts, etc.). The participants highlighted that the actors must be able to trust each other and each other's technological systems in order to operate confidently and successfully.

4.2 Awareness and Understanding

It was revealed that the concept of "trust" in autonomous systems was closely associated with the requirement for a certain level of awareness and understanding of a particular system, its operations and what/how decisions are made. Several participants saw the development of trust as a function of awareness and understanding of a system and its decision making. For example, in shipping, an awareness of external entities (i.e. the operations of other ships and external actors) is achieved through common rules and regulations (e.g. The International Maritime Organization's industry standard regulations all ships/operators must abide by [i.e. STCW, COLREGS, etc.]). These common regulatory frameworks standardize procedures, and thus create shared expectations and promotes trust between seafarers, ships and other stakeholders. Participants felt these qualities were necessary between humans and emerging technologies, with one participant stating:

> *"The system needs to know your intentions. If you are acting as a team member or team it is very important that you know what the other persons intention is, so you can support that and you can also predict what the other one needs...you need an idea of the intention of the other system or the person."*

Participants discussed that regardless of whether humans are in active control of a system, or in more passive monitoring and supervisory roles, they require some level of understanding of how an autonomous system works, its capabilities, strengths, weaknesses, and ultimately knowledge of how decisions are made.

4.3 Control

Participants questioned the value, relevance and purpose of humans in increasingly automated and autonomous systems of the future. Although, shipping operations in general have become increasingly automated over the past several decades, with seafarers taking on increasingly supervisory roles, participants stressed the concept of whether the relationship between humans and autonomous systems was one of collaboration or one of support. In particularly, several participants noted that systems are generally reliable and trustworthy in normal operations, however when abnormal situations occur, humans have many strengths, including the ability to adapt, be creative and think *"outside of the box"*. Furthermore, as humans take on increasingly supervisory roles one participant feared the loss of relevant skills and knowledge of operations:

> *"(we need to) generally prevent them (operators) from becoming deskilled. This is a huge problem with automation because once the system breaks down you are so untrained you have forgotten all about how to actually work with it."*

4.4 Training and Organization of Work

> *"There would be new kinds of jobs, there will be new kind of careers."*

Several of the participants (particularly those working with training and assessment research and within professional training facilities) discussed the types of skills people involved with autonomous systems will require in the future:

> *"I think so far everyone is very occupied with the technology but not with how we train people in using these systems."*

Participants asked if the traditional seafaring career, and thus, if a traditional seafaring education would be relevant in the future? They proposed differing future skills required, highlighting the likelihood that programming and serious gaming skills will be relevant, with one participant likening future seafaring to that of Air Traffic Controllers or Vessel Traffic Service operators and requiring a specialized education.

4.5 Practical Implementation Considerations

The fuzzier fifth theme encompassed several sub-topics, which related to the real-world practical considerations of implementing autonomous technologies within the maritime domain. Several topics were mentioned, including cybersecurity, economic, safety, environmental and ethical issues. However, the predominant sub-topic addressed the rules and regulations of international autonomous shipping:

> *"The technology is already there, it is just the question of implementing it. The main issue is going to be regulations, particularly in the maritime domain..."*

Participants were generally impressed with both the pace and direction of technological development, however continually brought up the conservative nature of the international regulatory body and the possibility of inhibiting real-world application within the industry. This was a general reflection of the human perspective in autonomous systems which was discussed across participants: are we moving too fast and do we know the impacts of these emerging technologies?

5 Discussion

5.1 Future Roles of the Human Element in the Maritime Domain

The results from participant interviews revealed five prevalent themes regarding the issues of implementing autonomous systems and the future role of humans in the maritime industry. With the incremental shift of operations to increasingly automated agents, future maritime operators may find themselves in progressively supervisory (and/or redundant) roles which are physically separated from sharp-end operations. Traditional seafaring skills may become unnecessary and obsolete, just as in the past differing maritime skills have altered and changed with technological advancements (Mallam and Lundh 2016).

Future maritime operators may never go to sea themselves, but instead receive training of ship operations remotely through simulator exercises or shore-side centers. As these operators likely will obtain a supervisory role, the question of required skills is highly relevant. The participants in the interviews revealed a common prediction of skills related to programming, Information Technology (IT), analytics and serious gaming being valuable assets. However, participants also described how basic knowledge and experience of seafaring and seafarer skills would be needed to comprehend the actions of autonomous systems to be able to interact with, and understand operations.

Utilizing autonomous vessels could save costs by reducing or eliminating crews and all the facilities and requirements of having humans onboard (i.e. hotel facilities – thus reducing weight, space and consumption of electricity, fresh water, supplies, etc.). Yet some would argue that as these vessels would require some level of supervision, as well as maintenance of onboard systems (if and when equipment breaks down at sea what procedures would be in place?). If the reduction of onboard crew leads to increased onshore manning, cost savings related to crew reduction would be minimal as the crew would simply be "moved" onshore. Furthermore, the increased costs of specialized equipment, sensors, IT infrastructure, transport to/from remote ships at sea for unforeseen maintenance, amongst other issues, may also negate any cost savings argued for by reducing crew numbers.

Highly automated vessels have been predicted to operate with increased safety levels in comparison to contemporary manned vessels, with the intention of removing the human, and thus "human error" (Hetherington et al. 2006). Under normal operating parameters it likely that vessel operations will be generally successful. However, limitations may be revealed if a vessel encounters a situation completely unfamiliar and unrelated to previous experiences or expectations where no official procedure is yet in place. In such abnormal situations an autonomous system may be deficient, whereas a human could employ creative and adaptive solution-oriented thinking and decision making. Humans may be relatively unpredictable factors in complex systems, but they are also able to interpret situations and find solutions where current technologies may falter. Thus, this predicted increase in safety may be highly dependent on the flexibility and adaptability of the system.

5.2 Moving Forward

The pace of the technological development of autonomous systems is rapid, resulting in current scaled pilot projects and full-scale shipping (e.g. Yara Birkeland). The obstacles of implementing such solutions are no longer related to technology, but rather related to human-technology aspects, new operating paradigms and practical considerations, such as regulatory, liability and security concerns. Despite the attention and practical demonstrations of utilizing autonomous technologies in transport domains in recent years, challenges continue to exist and be revealed. Technological development will rapidly move forward, however, it is critical to develop a better understanding of how and where humans fit into operations, work organization and ever evolving complex safety-critical systems.

6 Conclusions

Implementation of any new technology generally faces initial resistance before being accepted and integrated into daily life (e.g. mobile phones, personal computers, automobiles, etc.). Therefore, making a prediction of the integration and implementation of autonomy in maritime systems is at this stage relatively immature and explorative. However, in the further development and application of autonomous systems in maritime industries, and in particularly shipping, it is important to identify and begin developing frameworks for the organization of people and technology in future systems. Operational paradigms will continue to evolve with technological advancements. Thus, issues relating to system trust, understanding and predictability of decision making, as well as the skills required to develop, operate and maintain such technologies will become increasingly relevant to the successful and sustainable functioning of highly automated and autonomous systems.

Acknowledgements. This research has been supported by the MARKOM2020 project, funded by the Norwegian Ministry of Education and Research. The authors would like to thank all the participants in the study for their time and for sharing their valuable insights.

References

Bainbridge L (1983) Ironies of automation. Automatica 19(6):775–779

Braun V, Clarke V (2006) Using thematic analysis in psychology. Qual Res Psychol 3(2):77–101

Doraszelski U (2004) Innovations, improvements, and the optimal adoption of new technologies. J Econ Dyn Control 28(7):1461–1480

Endsley MR, Kiris EO (1995) The out-of-the-loop performance problem and level of control in automation. Hum Factors 37(2):381–394

Franklin S, Graesser A (1997) Is it an Agent, or just a Program? A taxonomy for autonomous agents. In: Müller JP, Wooldridge MJ, Jennings NR (eds) Intelligent agents III. Agent theories, architectures, and languages, ATAL 1996. Lecture Notes in Computer Science (Lecture Notes in Artificial Intelligence), vol 1193. Springer, Heidelberg

Grech M, Horberry TJ, Koester T (2008) Human factors in the maritime domain. CRC Press, Boca Raton

Hetherington C, Flin R, Mearns K (2006) Safety in shipping: the human element. J Saf Res 37(4):401–411

Hoff KA, Bashir M (2015) Trust in automation: integrating empirical evidence on factors that influence trust. Hum Factors 57(3):407–434

Kongsberg Maritime. https://www.km.kongsberg.com/ks/web/nokbg0240.nsf/AllWeb/4B8113B707A50A4FC125811D00407045. accessed 14 May 2018

Lutzhoft MH, Dekker SW (2002) On your watch: automation on the bridge. J Navig 55(1):83–96

Mallam SC, Lundh M (2016) The physical work environment & end-user requirements: investigating marine engineering officers operational demands & ship design. Work 54(4):989–1000

Mallam SC, Lundh M, MacKinnon SN (2017) Evaluating a digital ship design tool prototype: designers' perceptions of novel ergonomics software. Appl Ergon 59:19–26

National Research Council (NRC) (1990) Crew Size and Maritime Safety. The National Academies Press, Washington, DC

Nazir S, Kluge A, Manca D (2014) Automation in process industry: cure or curse? How can training improve operator's performance. Comput Aided Chem Eng 33:889–894

Parasuraman R (1987) Human-Computer Monitoring. Hum Factors 29(6):695–706

Parasuraman R, Riley V (1997) Humans and automation: use, misuse, disuse, abuse. Hum Factors 39(2):230–253

Stopford M (2009) Maritime economics, 3rd edn. Routledge, Oxon

Strauch B (2017) The automation-by-expertise-by-training interaction: Why automation-related accidents continue to occur in sociotechnical systems. Hum Factors 59(2):204–228

Enablers and Barriers for Implementing Crisis Preparedness on Local and Regional Levels in Sweden

Eva Leth(✉), Jonas Borell, and Åsa Ek

Ergonomics and Aerosol Technology, Department of Design Sciences, Faculty of Engineering, Lund University, Lund, Sweden
eva.leth@design.lth.se

Abstract. Crisis preparedness planning work can be characterized as complex and challenging being performed in large public organizations requiring communication and cooperation with many people. This paper reports on a bottom-up study investigating the pre-requisites for efficient crisis preparedness work, as well as motivational factors for implementing crisis preparedness decisions, which might be studied and understood in terms of sense-making, communication, and cooperation. As focus was on the crisis preparedness planners' need to understand their roles, assignments, and authorities in the crisis preparedness planning process, ten semi-structured interviews were conducted with local and regional crisis preparedness planners at a county council in southern Sweden. Enablers and barriers for implementing crisis preparedness were identified and concerned area such as: assuming the role of expert in crisis preparedness planning in one's organization and understanding what crisis preparedness is about in relation to the organization's core tasks; power and attaining enough mandate to be able to perform one's designated work by, e.g. seeking and finding different networks, stakeholders and partners that can legitimate the planner's work and enable trust and action; communicating demands and benefits of the planners' tasks realizing that members and working areas in the organization have different needs, cultures, and use different languages; and managing and delivering risk analysis information from the local to the regional level. For a attaining a comprehensive view of enablers and barriers in the crisis preparedness planning processes and implementation, an organizational top-down perspective will also be needed.

Keywords: Sense-making · Sense-giving · Crisis preparedness

1 Introduction

1.1 Background and Aims

The Swedish crisis management system is based on three principles: responsibility (the one responsible for an activity under normal conditions is also responsible in a crisis); parity (activities should, as far as possible, be organized and located in the same way under crisis as under normal conditions); and proximity (a crisis should be handled where it occurs and by the closest one affected and responsible for the activities). These

S. Bagnara et al. (Eds.): IEA 2018, AISC 821, pp. 406–412, 2019.
https://doi.org/10.1007/978-3-319-96080-7_48

principles mean that no actor can take the responsibility from another and that regional and local actors have prominent roles in most crises. Therefore, they need to attempt to preventing crises and prepare for those crises that nevertheless do occur.

In Sweden, public organizations such as authorities, the 21 county councils and 290 municipalities are all required by the legal framework to recurrently prepare for crises, disasters and emergencies. This includes performing risk and vulnerability analyses regarding the organizations' societal tasks and operations to prevent crises. It also includes writing and implementing plans for dealing with crises and conducting crisis preparedness exercises. The activities entail many challenges, as performed in large and complex public organizations, and requiring negotiation, coordination and cooperation of large numbers of people. Another challenge is that there is no agreement across actors on what preparedness planning is and how it should be defined, and that preparedness planners working in public administration must have an understanding of politics, economics and management as well as the issues pertaining to public health, transportation, urban development, and human resources [1].

The 21 public county councils are responsible for health and medical services, dental care, regional developments, and public transportations in their geographical areas. In the organizations, regional and local preparedness planners are responsible for the crisis preparedness process. The preparedness planners, as well as managers, professional groups, and stakeholders in the public organizations, have to understand, articulate, advocate and handle the preparedness process activities as part of their work. For success, this requires bottom-up negotiated social construction activities regarding sense-making and sense-giving between actors and levels in the organization.

In practice, challenges exist when implementing crisis preparedness decisions and plans. Successful implementation presumes that managers and co-workers actively participate in the implementation work. Values, attitudes, knowledge, and comprehension among the individuals about the crisis preparedness implementation may have consequences for the actual implementation. Additionally, the prioritization and allocation of resources for crisis preparedness can have consequences for the implementation and the capability to handle known and unknown crises.

The aims of an ongoing research work are to identify pre-requisites for efficient crisis preparedness work on regional and local levels in one county council in Sweden and to increase knowledge about managers' and preparedness planners' ways and means, incentives, and motivational factors for implementing decisions concerning crisis preparedness. Focus is on the preparedness planners' and organizational members' need to understand their roles, assignments, and authorities in the crisis planning process via processes of sense-making, sense-giving, communication, and collaboration between managers and planners on different levels in the organization.

1.2 Sense-Making, Communication, and Cooperation

The current research work seeks to explore motivational factors for implementing decisions concerning crisis preparedness, which might be studied and understood in terms of sense-making, communication, and cooperation. Traditionally, this is mainly studied top-down, focusing on managers' and middle managers' perspectives [2, 3]. Managers and middle managers are thus traditionally considered as sense-givers influencing other members and actors in an organization. However, preparedness planning is about actors and activities working in a continuous top-down and bottom-up process to prepare and enable action in crises and gaining acceptance and coordination with other actors in the organization. This puts emphasis on the preparedness planners, who need to influence both managers, colleagues and other actors to implement decisions and activities concerning preparedness planning. This additional bottom-up view is what this research study seeks to explore and understand.

Sense-making can be defined as the action or process of making sense of or giving meaning to something, especially new developments and experiences [4]. Weick [5] argue that sense-making begins with a sense maker and is grounded in both individual and social activity. Gioia and Chittipeddi [2] argue that organizational members, including the CEO, need to understand any intended changes in a way that fits into some revised interpretive system of meaning. Organizational meetings are important for sense-making among individuals as well as action, and meeting conversations could be synonymous with organizational action. Organizational sense-making is an ongoing pressure to develop generic subjectivity to premise control and interchangeability to people, and is developed through processes of arguing, expecting, committing, and manipulating [5].

As sense is generated by words, sentences, and conversations, organizational communication is a central and integrated part in crisis preparedness planning that can benefit from the use of narratives and enable action and learning [5].

Interpretations and sense-making evolve in organizations through cooperation and communication. In a communication process interpretation guide actions and actions shape interpretations [6]. Eisenberg [6] argues that effective inter-organizational cooperation should be placed on cultivating ideas and relationships between different organizations to create coordinated action in the explicit absence of agreement.

2 Methods and Material

2.1 The Interviews

This study was performed at a county council in southern Sweden. Ten semi-structured interviews were conducted with local and regional crisis preparedness planners. Interview items focused on areas such as: how planners on various organizational levels work with crisis preparedness and how individual understandings and values about the work are expressed; tactics and strategies taken to implement crisis preparedness decisions and plans despite an existing general low priority of the issue in the organization; why issues about crisis preparedness are diminished in the organization although existing requirements and issued implementations say otherwise.

The interviews were performed with one interviewee at a time and every interview lasted approximately one hour. Each interview was recorded and transcribed. The reason for using semi-structured interviews was to enable the interviewees to in their own words describe their work and its possibilities, challenges, and problems.

2.2 The Interviewees

All ten interviewees had held their position as crisis preparedness planner for five years or less, and five of them, one year or less. They had varying backgrounds and degrees in areas such as chemistry, risk and crisis management, biomedical engineering, business economics, criminology, as well as policeman and marine engineer. They worked in the council's management areas such as health care and public transport, and in several units within these areas such as headquarter, quality, real estate, safety, and disaster medicine. Thus they all belonged to different units, but the essence of their work was to be crisis preparedness planners. However, the organizational diversity was also reflected in their job titles, ranging from administrator, coordinator, strategist, to manager.

2.3 Interview Analyses

Interview data was analyzed using a SWOT-analysis focusing on strengths, weaknesses, opportunities, and threats related to the crisis preparedness planners work. The analysis was thereafter presented to the interviewees for review and additional comments.

A content analysis of interview data was also conducted, focusing on finding focus areas, enablers, and barriers when implementing crisis preparedness issues.

3 Results

3.1 Alone and Exposed or Free and Creative

The ten crisis preparedness planners' perceptions of their work differed. On the one hand, some of the interviewees expressed that no one understands their assignment, that too many meetings waste time, and that no one clearly states what work is to be done. They lack resources and mandate, and feel they are placed far down the hierarchies when their tasks ought to be managed top-down. On the other hand, some of the interviewees believe that things are better as they can choose what to do and how to do it, that they feel they can manage their own time, that meetings are essential for their tasks, that they get to choose focus areas themselves, and that they have the necessary resources for doing the job. Thus, the interviewees value their situation differently, ranging from negatively to positively.

3.2 Goal Description and Sense-Making of Own Work

Some interviewees had the possibility to choose their job title and defining their job description. Several of the interviewees stated that they set their own work goals. This indicated freedom and responsibility in fulfilling the task of being a preparedness planner, along with demands of having expertise and know-how. However, the planners did not have their own area of business, and were supposed to make risk analyses as an incorporated task in other organizational members work. For success, this requires cooperation with line management and operators that are the performers of core businesses within the organization.

3.3 Cooperation and Action

The interviewees expressed the vital need for cooperation and choosing who could assist them in the work when implementing crisis preparedness. Some of the interviewees used the line managers in order to implement the work. Some found expert groups useful and some connected with other actors in clusters, teaming up with people working with, e.g. occupational health and environment issues. At large, the preparedness planner's work was very much concerned with building trust with line operations and management and with them having ongoing sense-making processes, generated by communication, in order to drive implementation.

3.4 Communication and Motivation

The interviewees raised communication and motivation as important issues when implementing crisis preparedness. It concerned the planner having enough knowledge to be able to argue and motivate management and other functions across the organization, as well as the staff that are to perform the actual risk analyses and preparedness planning work. The preparedness planners' strategies were to convince the organization, i.e. to meet and communicate requirements of and stressing the benefits of the preparedness planning work. Descriptions of benefits for core operations were regarded as an important tactics to work with. These results is much in line with a sense-making process.

3.5 Local vs Strategic Levels

Several interviewees expressed that solving ad hoc operative problems on a local level was sometimes preferred to working more strategically on a regional level. In the strategic work the local preparedness planners should for example compile local organizational information to be delivered upwards to the regional level for input to the regional risk and vulnerability analysis. This work was often considered difficult to perform and even unnecessary, and also required the planner to engage and motivate other actors in the local organization to supplement information. As a consequence of this perceived difficulty, they prioritized local ad hoc operative tasks because they fit more into their areas of competence. Some of these tasks concerns developing local contingency plans and crisis management plans, which requires an understanding of the specific, local complexities and familiarity with core businesses. However, one should remember the relatively short time some of the informants had spent on their jobs, entailing limited understanding and trust as the organization's crisis preparedness planner.

4 Discussion

Crisis preparedness planning work can be characterized as complex and challenging being performed in large public organizations requiring communication and cooperation with large numbers of people. This paper reports on an ongoing ethnographic study on enablers and barriers for crisis preparedness at a county council using interviews as the data collection method. The findings illustrates the complexity of the implementation process that includes activities such as sense-making, sense-giving, communication, and cooperation.

Several focus areas were identified in the study. The first area concerns assuming the role of expert in crisis preparedness planning in one's organization, which in some cases include sense-making. The second concerns power and attaining enough mandate to be able to perform one's designated work, which also includes sense-giving and communication that involves managers and coworkers. The third area concerns communicating demands and benefits of the planners' task, which involves demonstrating competence, receiving trust and getting the possibility to get access to stakeholders and managers. The fourth area concerns managing and delivering information to the regional level, which involves organizational understanding, competence in crisis preparedness, sense-making and sense-giving, communication, and cooperation. Certain information shall be gathered and passed on within the organization, and eventually delivered to national authorities.

There are different barriers for the interviewees to handle. At first they have to understand what general crisis preparedness is about. Then they have to understand how to manage crisis preparedness planning in their own organization. This involves cooperation and learning activities between the regional crisis preparedness planner and the local crisis preparedness planners in the organization. They also need to learn that members and working areas in the organization have different needs, cultures, and use different languages.

To find enablers the interviewees seek and find different networks, stakeholders and partners that can legitimate their work and enable trust, action, and implementation of crisis preparedness. Another example is that they communicate and articulate their needs in a way that can be understood by the professionals in the organization, i.e. public health, transportation, urban development, and human resources. To enable this, they find their own ways to meet and learn about the organization as well as the members' and professionals' working situation.

This study focused on the crisis preparedness planners' perceptions of their work and work tasks. In a continued research, it is important to also investigate the receivers of their performed work, i.e. the members on the regional level that receives the local risk and vulnerability analyses and reports. The continued research can give more information on enablers and barriers, by investigating the thoughts and views of the members on higher organizational levels on the crisis preparedness planning processes both in practice and ideally.

References

1. McEntire DA (2007) Disaster response and recovery. Wiley, Hoboken
2. Gioia DA, Chittipeddi K (1991) Sensemaking and sensegiving in strategic change initiation. Strateg Manag J 12(6):433–448
3. Rouleau L (2005) Micro-practices of strategic sensemaking and sensegiving: how middle managers interpret and sell change every day. J Manag Stud 42:1413–1443
4. Oxford Dictionary Homepage. https://en.oxforddictionaries.com/. Accessed 28 May 2018
5. Weick KE (1995) Sensemaking in organizations. Sage, London
6. Eisenberg EM (2007) Strategic ambiguities: essays on communication, organization, and identity. Sage, London

Worker's Management and Skills Development in a Temporary Team

Fabiola Maureira[1(✉)], Tahar-Hakim Benchekroun[2], and Pierre Falzon[2]

[1] University of Concepcion, Concepcion, Chile
fmaurei@udec.cl
[2] Conservatoire nationale des arts et métiers, Paris, France

Abstract. The current communication is part of the results of the PhD thesis in Ergonomics related to a fire forest coordination and control center in Chile [1] which despite working in a complex and risky dynamic environment, is handled by a temporary team. The aim is to reveal the strategies mobilized by the manager against time to develop individual and collective operators' skills in order to become an operational team. But also, to identify the leadership style suitable for a temporary team that has to handle a dynamic complex system.

Based on the theories of situated action [2], distributed cognition [3] and distributed social cognition [4], systematic observations were carried out at the control center for six months (2010 and 2011) in order to analyse both the actions and the verbal learning exchanges among team members.

It was observed that the manager articulate situated learning activities by a flexible, dynamic, opportunistic and constructive strategy; encouraging active participation and interaction, the reflexive action on their own practices, mutual knowledge, self-training initiatives, mutual assistance, epistemo-vigilance and collaboration at learning. As a consequence, team members create opportunities, both formal and informal, by mobilizing individual or collective action modalities to learn and develop their skills.

There is scientific support in the sense that leadership style shown by the manager at combining different leadership approaches is more effective for temporary work environments.

Keywords: Leadership · Risky dynamic systems · Temporary workers

1 Introduction

Forest fires cause serious environmental damage due to the destruction of the ecosystem, in terms of flora and fauna, the deterioration of soil quality and environmental pollution, but they also have a significant socio-economic cost due to damage to the environment, property, productive processes and public health in general [5]. Chile's patrimony possesses regions with a high density of forest resources that are protected by two operating systems: the first being private, controlling the fire of the heritage of large forest companies and in their immediate environment; the second, is a private-law entity that depends on the Ministry of Agriculture, identified as the

S. Bagnara et al. (Eds.): IEA 2018, AISC 821, pp. 413–421, 2019.
https://doi.org/10.1007/978-3-319-96080-7_49

National Forest Corporation (CONAF) whose mission is the protection of the heritage of the state, small and medium-sized enterprises and national reserves [6].

The Fire Forest Department of CONAF has one operational unit identified as Fire Forest Coordination and Control Center (Control Center) which aims to prevent forest fires and reduce damage caused by forest fires but also in monitoring the rational and controlled use of fire by companies, contractors and forest owners registered in the protection system [7].

The activity of control center's operators is directly related to the control activity of dynamic environments characterized by a loop of tasks such as: information gathering, diagnosis/prognosis, action planning and possibly the means of action, decision, execution and control [8]. But also, coping with fire forest in a risk-averse dynamic system involves dealing with unexpected events, hazards and unforeseen events, under heavy temporal pressure and where the operator only has partial control of operations and may require the participation of other actors outside of CONAF.

Taking into account the complexity of work demands at the CONAF, it is paradoxical that, except for the chief coordinator, all the operators have a temporary contract in between October and April, leaving their job at the end of the summer season and they expect the return of some former workers to the following season. In that case, there is a concrete problem for field work is the loss of the human capital, the experienced knowledge and the organizational know-how. This condition is identified as a clear weakness for the system and implies an important investment every year both in training and guiding the new heterogeneous team. note that the first paragraph of a section or subsection is not indented.

2 Method

Based on the theories of situated action [2], distributed cognition [3] and distributed social cognition [4], the researcher observed the work at the control center for an extended period (six months in between 2009 and 2010) and collected data during the initial training process and throughout the peak season, in the operational setting. Data collection was supported by audio and video recordings in order to analyse both the actions and the content of the trainer's interventions and the spontaneous verbal exchanges among members of the control center (5 experienced operators, 3 trainee operators). Systematic observations were supplemented with individual interviews with each member of the team.

3 Results and Discussion

According to the situational analysis of the activity in progress, the manager demonstrate several sociotechnical strategies which conduct in coordination with the training department team with an particularly dynamic situational style during the formal training process at the beginning of the forest-fire season as well as during the season.

3.1 Sociotechnical Strategies Conducted by the Manager During the Formal Training Process

Each season starts with a formal training period, planned and organized by the manager in coordination with the team of the CONAF training and development unit. Despite having a previously defined program in terms of topics and approximate duration, the truth is that this period can never be guaranteed according to plan. In fact, during that period external constraints, such as adverse climatic conditions could cause forest fires of variable complexity affecting the training period. These environmental changing conditions well known by the training team influence the sociotechnical strategies aimed to quickly train novice operators, to update the knowledge of the experienced and to make each other operational.

During the formal training period they receive technical knowledge in order to be able to identify a number of parameters useful to the management of forest fires. They also learn the normative aspects of the task, pre-established conventions, and strategic and tactical operation procedures with other actors in joint operation plans when activating alerts in the region. But, the skills to link certain key variables that leads operators decisions related with fire fighters allocation, the treatment of calls and also the operators coordinated actions requires other learning activities such as: (a) practicing the job with the tasks of the center, where the novices are assisted by more experienced colleagues, (b) episodic events, (c) the evaluation of knowledge, (d) simulations, (e) visits to the field as well as brigades allocation decision-making exercises towards fire. Those learning activities that are chosen to develop people's skills directed towards the mobilization of intelligence in situation, towards autonomy and cooperation by the confrontation of people with real work situations will help to enrich them [9]. Both, the training team and the responsible of the control center, choose those learning activities due to their field knowledge, because it is not possible to rely on previously defined behaviors based on a prescribed tasks or rules. In fact, adaptation and success at work depends on the mobilization of people collectively committed, with initiative and responsibility, to cope with the situations they face, through the exploitation of experienced events as sources of learning for the future.

It is so that the training team based their practices with the idea that people learn through social development which evokes the Vygotsky's theory of social constructivism [10] and the situated learning theory developed by Lave and Wenger [11] that emphasized that knowledge needs to be presented in authentic contexts with the social interaction and active collaboration.

It was observed that the training team and the responsible of the control center articulate these situated learning activities by a flexible, opportunistic and constructive educational strategy in considering the dynamism of the current activity, nourished by the active participation of the members of the team of the control center and by the reflexive action on their own practices or experiences. This participation is functional (enriched exchanges resulting from the training sessions), but also meta-functional because the responsible of control center encourages experienced operators to reflect on their own practice of the profession and on experiences to help improve the training process.

This style of management feeds their mutual knowledge to develop spontaneous self-training initiatives. In fact, the manager encourages the team towards constant interaction, mutual assistance and collaboration at learning their job. But also, in considering the situation of the environment and the emerging demands made by the team members, he makes arrangements in order to gradually adapt the training modalities of future. In addition the manager transmits some key work rules for both experienced and novices team members. Some of those key work rules are related to encourage novices to ask the experienced and the latter to support the novices in their training, to share ideas in order to improve training process, and also to share what they know and what they still need to learn.

But also, the manager insists on transmitting the idea of mutual vigilance and support. This working rule contributes to the development of a collective competence, named as *"épistèmo-vigilance"* [1], a kind of mutual learning in which team members exploit both the constructive and productive dimension of the activity, in order to prevent mistakes; guiding, correcting and reinforcing their behaviors and their knowledge aimed to improve their performance as a team.

3.2 Sociotechnical Strategies Conducted by the Manager During the Season

All of those training modalities set up by the manager during the training period designed to promote a cooperative, mutual aid and assistance mode of operation and skills development, result in a certain collective behavior, a particular way of working together during the functional period. During the operational period it was observed from the beginning an active participation of team members, self-organization, and mutual support not only in order to fulfill the task, but also in order to exploit opportunities for learning from experience.

It is possible to confirm that functionally the system works, in spite of the seasonal condition of the operators. In fact, beyond the initial formal training: the team members exploit the situations to interact with the goal of building knowledge during the action. These opportunities have been identified as learning interaction situations (LISs) in which team members create opportunities, both formal and informal, by mobilizing individual or collective action modalities to learn and develop their skills.

During the systematic observations all along the season it was noticed that the novices were the team members that took the initiative, most of the time, in setting up the LIS, and the interactions were more oriented to the experienced team member than the manager (see Fig. 1). This dynamics of the exchanges contribute to the construction of self-confidence and confidence between colleagues, not only as part of the development of the action skills, but also as part of the collective functioning in the operational management of forest fires.

It is also expected to see that at the beginning of the season the LISs are more frequent because team members and particularly the novices, are getting familiar with work demands in situation. But in the middle of the season when the workload increases both in number and complexity, it might be expected that all team members should be focused on operational interactions, leaving the formative interactions only occur during off-peak periods. However, team members take advantage particularly of

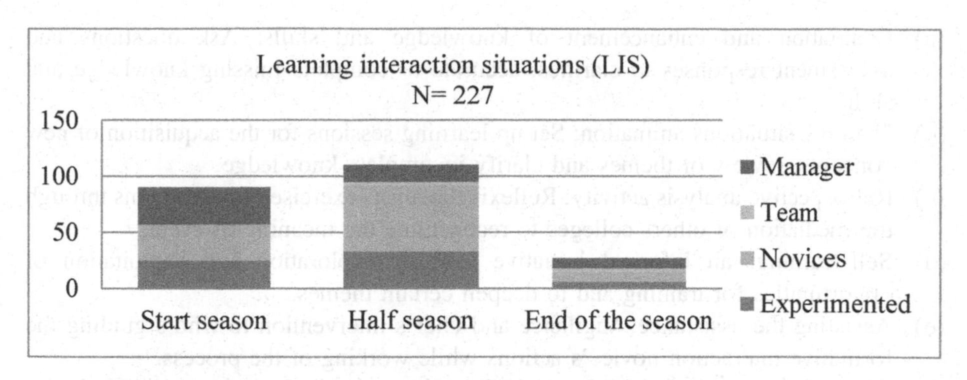

Fig. 1. Learning interactions situations activated by the team members during the fire forest season, region Biobio, Chile, 2010–2011.

the fire forest coordination and control activity as a training exercise to help themselves and to reinforce knowledge and develop their skills. The purpose of training is not moved to quiet moments or breaks; on the contrary, the group constructively exploits the current activity so that the regulation seeks both the continuation of the current action and an opportunistic exploitation to reinforce, stabilize and update the knowledge and skills of the collective action.

In contrast, towards the end of the season, there is a significant decrease in the need for learning interactions as shown in Fig. 1. So, it might be possible that the need for assistance and training are fewer as operators are already trained and have developed the necessary skills. In any case, even though the number of LISs is lower, it is not expected to have none of the LISs because they are constantly facing dynamic environment that may changes any moment so they are in a systematic learning dynamic. So they learn how to cope with different types of strategies or training modalities that are put in place during the season.

Within the LIS, different types of informal training modalities take place according to the dynamics of the exchanges and the people involved in these exchanges. These training modalities will be named as organizational accelerators of training, learning and reinforcement of knowledge (OATLR)[1]. This notion that was borrowed from management science referred to training and support programs aimed at promoting innovation and innovative business generation capable of facing competitive environments [12] identifies a set of individual and collective processes that have the property of activate and advance operators in their efforts to learn and enhance their autonomy, but also at pushing them to achieve operability and performance.

Throughout the season, seven OATLR mobilized by the team members were identified both in quiet and in more demanding working periods:

[1] The acronym OATLR it's a translation from the french notion *accélérateurs organisationnels de la formation, de l'apprentissage et du renforcement des savoirs* (AOFAR), that was developed in the PhD. Thesis [1].

(a) Evaluation and enhancement of knowledge and skills: Ask questions and assessment responses to clarifies, accurate or complete missing knowledge and skills.
(b) Training situations animation: Set up learning sessions for the acquisition of new concepts, review of themes and clarify incomplete knowledge.
(c) Retrospective analysis activity: Reflexive analysis exercise of past actions through the mediation of others colleges to reconstitute the meaning of events
(d) Self-learning: an informal initiative towards exploration and exploitation of opportunities for training and to deepen certain themes.
(e) Assisting the assistance: Vigilance and chef's intervention to who's guiding the formative interaction novice's actions while working of the process.
(f) Direct assistance focused at learning: To guide, assist and support the execution of the activity of an operator in order to train in action and through action.
(g) Cooperative learning: Meta-cooperation exercises by knowledge exchange and explanations among the team members while somebody is seeking for advice on any aspect of the job, leading to the co-construction of the response.

The formative interactions in situ, as shown in Fig. 2, identifies a significant participation of the novices, assisted rather by the experienced ones than by the manager, who take, clearly, more distance in the process of formation and development of the competences. The learning modality most used by both the novice and the experienced operators was the direct and focused learning assistance because they take advantage au maximum by asking their colleagues nearby in situ.

Nevertheless, the manager has preponderance in three accelerators: *assisting the assistance*, *training situations animation* and *retrospective activity analysis*. This finding suggests that those three OATLR are inherent to his leadership role related to the training and skills development process of its members. Thereby, setting up these training modalities, the manager supervises the assistance of the experienced to the novices (assisting the assistance), continues the training on site to deepen the gains on the initial formal training (training situations animation) and monitors the activity of operators in order to exploit opportunities to discuss their performance, in order to getting them used to reflect about their actions as a way of regulating their own performance (retrospective analysis activity).

The manager leadership style identified from the beginning when the fire forest seasons begins and throughout de operative period may be identified as constructive, dynamic, flexible, opportunistic and situated. As a part of a complex dynamic environment, with a heterogeneous team in terms of experience, budget constraints, temporal pressures and unexpected events that can become complex, the leader takes advantage of any functional activity to transform it into a constructive experience for the team members. His leadership strategies can be described as it follows:

- Team member's roles distribution with mechanisms of adjustment, reorganization, and functional restructuring of roles according to context.
- Accurate contextualized operational collective rules emphasizing self-organization, surveillance and mutual assistance, active participation, self-analyses and mutual teammate skill knowledge.

- Setting up favorable conditions for informal learning and maximum use of the time dedicated to training while working.
- Let operators an active participation in the proposal of ideas, real possibilities to contribute freely to improve their work.
- Operators recognition, as stated by Dejours [13], in terms of expressing gratitude, the symbolic retribution to the subjects, even if they are novices, by their contribution in the constructive dimension and productive activity of colleagues by highlighting the knowledge of each of them.
- Opportunistic exploitation of meta-functional dynamics [14] of reflective analysis of operators practices [15], in order to become quickly operational in the collective forest fires management.
- Organization of a proxemics work space conducive to floating listening, multi-addressing, co-presence which allows at sharing circulating information which is functionally efficient for complex dynamic systems [16, 17].

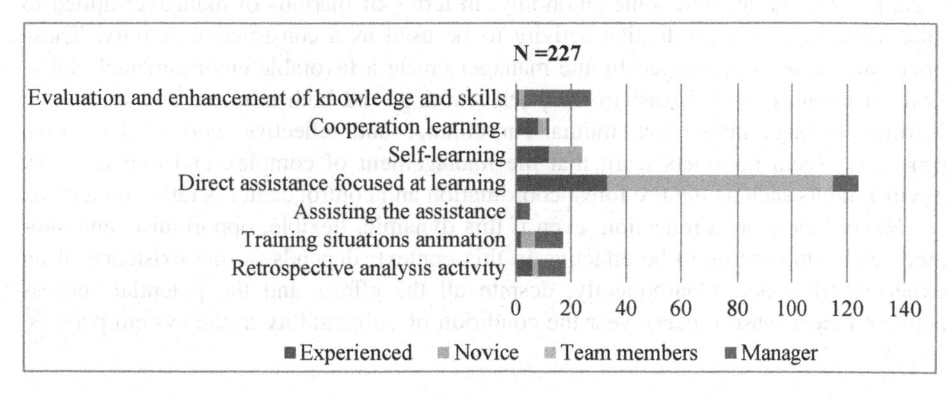

Fig. 2. Organizational accelerators of training, learning and reinforcement of knowledge (OATLR) mobilized by the team members during the fire forest season, region Biobio, Chile, 2010–2011.

There is a psychological leadership typology for executives that considers that opportunists ones are associated with a performance below average [18]. Nevertheless, it is important to distinguish that in this context the notion "opportunistic" has a positive connotation because comes from a particular type of assistance in situated action of a tutor towards a novice called as Opportunistic Learning [19] where he can accomplish the task and at the same time enrich his knowledge and know-how due to the functional and meta-functional understanding of the system.

This field research has demonstrated that the sociotechnical strategies mobilized by the manager during both, at the training process and the functional period, not only allowed the operators individual skills development but also contributed to the development of a community operator's team. The theory of distributed social cognition helps to explain the development of the collective and the emergence of a community of work and practices with shared values, knowledge and know-how, through interactions and mutual interdependence [20].

Those results aim to identify the role of management style that enhances both individual and collective learning engagement as well as the collective functional configuration, oriented to self-regulation, a key aspect mentioned by De la Garza and Weill-Fassina [21]. But also this leadership allows the construction of the working collective (team building) proposed by Caroly and Clot [22].

This leadership style confirms what Tyssen, Wald and Spieth [23] have established, in the sense that for temporary work environments, is more effective a combination of leadership approaches. The present field research also gives empirical evidence suitable for complex and dynamic risk environments, conducted by temporary team workers than need to learn under severe temporal pressure.

4 Conclusion

The leadership style showed by the manager is possible due to the existence of an organization that let him some autonomy, in terms of margins of maneuver aimed to take advantage of a productive activity to be used as a constructive activity. These operating modes encouraged by the manager create a favorable environmental condition for team members' participation, self-learning, and both informal and autonomic learning through interaction, mutual surveillance and reflective work of their own practices. Team members learn that the management of complex and dynamic risk environments such as the fire forest coordination and control center is rather collective.

Nevertheless, as a reflection, even if this dynamic, flexible, opportunists and situated leadership seems to be effective in this context; depends on the existence of an experienced leader. Consequently, despite all the efforts and the potential success achieved each season, every year the condition of vulnerability in the system persists.

References

1. Maureira F (2015) Favoriser la construction d'un collectif apprenant: les conditions organisationnelles du développement des compétences d'un collectif éphémère. In: Centre de Recherche sur le Travail et le Développement, Conservatoire nationale des arts et métiers. Paris, p 298
2. Suchman L (1997) Centers of coordination: a case and some themes. In: Resnick LB et al (eds) Discourse, tools and reasoning: essays on situated cognition. Springer, Berlin. pp 41–62
3. Decortis F, Pavard B (1994) Communication et coopération: de la théorie des actes de langage à l'approche ethnométhodologique. In: Pavard B (ed) Systèmes coopératifs: de la modelisation à la conception. Octarès, Toulouse, pp 21–50
4. Hutchins E (1995) How a cockpit remembers its speed. Cogn Sci 19:265–288
5. Urzúa N, Cáceres F (2012) Incendios forestales: principales consecuencias económicas y ambientales en Chile. Revista Interamericana de Ambiente y Turismo-RIAT 7(1):18–24
6. CONAF (2018) Incendios Forestales en Chile. http://www.conaf.cl/incendios-forestales/incendios-forestales-en-chile/
7. CONAF (2018) Despacho y coordinación de recursos. http://www.conaf.cl/incendios-forestales/combate-de-incendios-forestales/despacho-y-coordinacion-de-recursos/

8. Samurçay R, Rogalski J (1992) Formation aux activités de gestion d'environnements dynamiques: concepts et méthodes. Education Permanente 111:227–242
9. Jobert G (2011) Intelligence au travail et développement des adultes. In: Carré P, Caspar P (eds) Traité des sciences et des techniques de la formation. Dunod, Paris, pp 357–381
10. Vygotsky L (1978) Interaction between learning and development. In: Mind and Society. Harvard University Press, Cambridge, pp 79–91
11. Lave J (1991) Situating learning in communities of practice. In: Resnick LB, Levine JM, Teasley, S (eds) Perspectives on socially shared cognition. American Psychological Association, Washington, pp 63–82
12. Metz R (2012) Business Impact: El hombre que susurraba a las start-ups. MIT Technology Review
13. Dejours C (1993) Coopération et construction de l'identité en situation de travail. Futur antérieur 16(2):41–52
14. Falzon P (1994) Dialogues fonctionnels et activité collective. Le Travail Humain 57(4): 299–312
15. Falzon P, Mollo V (2009) Pour une ergonomie constructive: les conditions d'un travail capacitant. Laboreal 1:61–69
16. Benchekroun TH (1994) Modélisation et simulation des processus intentionnels d'interlocution. Conservatoire national des arts et métiers
17. Pavard B et al (2009) Conception de systèmes socio-technique robustes. In: De Terssac G, Boissières I, Gaillard I (eds) La sécurité en action. Octarès, Toulouse, pp 67–80
18. Torbert W, Rooke D (2005) Siete transformaciones del liderazgo. Harvard Bus Rev 83 (4):46–57
19. Falzon P, Pasqualetti L (2000) L'apprentissage Opportuniste. In: Benchekroun TH, Weill-Fassina, A (eds) Le travail collectif: perspectives actuelles en ergonomie. Octarès, Toulouse, pp 121–133
20. Weill-Fassina A, Benchekroun TH (2000) Diversité des approches et objets d'analyse du travail collectif en ergonomie. In: Benchekroun TH, Weill-Fassina, A (eds) Le travail collectif. Perspectives actuelles en ergonomie. Octarès, Toulouse, pp 1–15
21. De la Garza C, Weill-Fassina A (2000) Régulations horizontales et verticales du risque. In: Benchekroun TH, Weill-Fassina, A (eds) Le travail collectif. Perspectives actuelles en ergonomie. Octarès, Toulouse, pp 217–234
22. Caroly S, Clot I (2004) Du travail collectif au collectif de travail. Des conditions de développement des stratégies d'expérience. Formation et Emploi 2004(88):43–55
23. Tyssen AK, Wald A, Spieth P (2013) Leadership in temporary organizations: a review of leadership theories and a research agenda. Project Manage J 44(6):52–67

Retention Management of Nurses: A Case of University Hospital in Japan

Motoki Mizuno[1,2](✉), Yasuyuki Yamada[1,2], Takumi Iwaasa[2], Emiko Togashi[1], Michiko Suzuki[3], and Yuki Mizuno[4]

[1] Faculty of Health and Sports Science, Juntendo University, Chiba, Japan
mtmizuno@juntendo.ac.jp

[2] Juntendo University Graduate School of Health and Sports Science, Chiba, Japan

[3] Juntendo University Urayasu Hospital, Chiba, Japan

[4] Tokyo College of Transport Studies, Tokyo, Japan

Abstract. Retention management is a human resource strategy designed to improve job satisfaction and to reduce voluntary employee turnover by PDCA cycle of benefit programs and work-life balance. This study aimed to ascertain job satisfaction and intention to continue working among nurses, and to determine how their job satisfaction relate to the intention to continue working as nurses. As research methods, a paper-based survey was conducted at a university hospital located in the metropolitan area. The survey consisted of a face sheet and questions to assess the degree of respondents' job satisfaction and intention to continue working as nurse. It was distributed to 692 registered nurses employed on a permanent basis at a university hospital. As results, valid responses were obtained from 661 nurses (620 females and 41 males) with a mean age of 29.9 years (SD = 8.0) (valid response rate: 95.5%). A high rate of job satisfaction was identified with items such as communication with colleagues, a sense of pride in being a nurse, enjoyment from contact with patients, instructions from superiors, the frequency of nightshifts, working with superiors and colleagues who can be respected, one's salary and the workplace climate. However, a low rate of job satisfaction was identified with factors such as trust from doctors, their own vital role, appropriate number of nursing staff, the clinical ladder, work outside of nursing duties, break rooms and hospital facilities, break times, and the amount of overtime. Looking at the degree of the intention to continue working, the highest rate was seen in those with more than 5 years' experience, and the lowest rate was those with 1 year or more to less than 3 years' experience. As a consideration, the result of the analyses indicated that such satisfactory factors as a sense of pride in being a nurse and recognition as a nurse contributed to increase the pleasure of nursing practice and the intention of working. Therefore, it is necessary to carefully consider "recognition" as a key factor that influence retention management of nurses.

Keywords: Retention management · Human resource strategy
PDCA cycle · Job satisfaction · Teamwork

S. Bagnara et al. (Eds.): IEA 2018, AISC 821, pp. 422–428, 2019.
https://doi.org/10.1007/978-3-319-96080-7_50

1 Introduction

Along with the advancement of aging society with falling child birthrates and decline of population of people, the labor force in Japan has been shrinking, especially the chronic shortage of nurses is a crucial problem. Moreover, the long working hours, the enormous workload, and night shifts place a huge physical burden on nursing staff. These factors have contributed to high job turnover rate of around 11% for nurses in Japan. In order to retain highly skilled and talented nurses for the long-term and to enable them to demonstrate and build on their skills, there is a need to implement retention management. Retention management is a human resource strategy designed to improve job satisfaction and to reduce voluntary employee turnover by PDCA cycle of benefit programs and work-life balance (Fig. 1). This study aimed to ascertain job satisfaction and intention to continue working among nurses, and to determine how their job satisfaction relate to the intention to continue working as nurses. In particular, for the purpose of considering prospective and aggressive retention management efforts, we clarified the current situations (living environment, working background, retention will and intention, workplace environment) and factors of nursing job satisfaction and job dissatisfaction for nurses including managers belonging to the university general hospital.

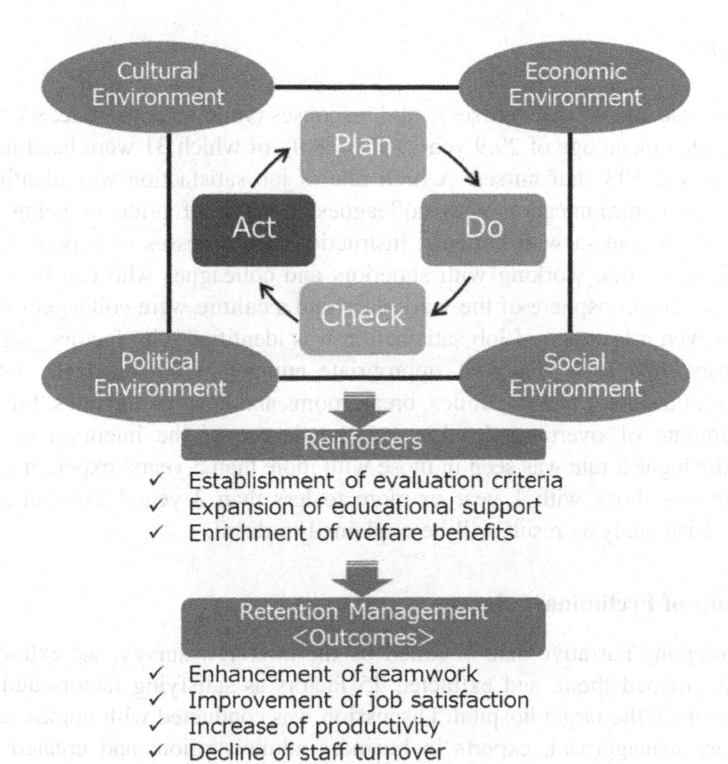

Fig. 1. PDCA cycle of retention management

2 Methods

This study was carried out in two steps of preliminary survey and main survey.

2.1 A Preliminary Survey

A semi-structured interview survey was conducted for 10 nursing staff (5 males and 5 females), the workplace environment of target hospital, workplace climate, welfare policies, employment situation of nurses, factors causing job satisfaction and dissatisfaction, on 28th of November 2017.

2.2 Main Survey

Based on the results of the preliminary survey, we conducted a questionnaire survey using the detention method. The survey consisted of a face sheet and questions to assess the degree of respondents' job satisfaction and intention to continue working as nurse. The survey was left with respondents to complete between January 13–17, 2017. Distributed to 692 registered nurses employed on a permanent basis at target hospitals excluding nursing manager.

3 Results

661 (620 female and 41 male nurses) valid responses (valid response rate: 95.5%) were obtained with a mean age of 29.9 years (SD = 8.0), of which 31 were head nurses, 35 chief nurses and 593 staff nurses. A high rate of job satisfaction was identified with items such as communication with colleagues, a sense of pride in being a nurse, enjoyment from contact with patients, instructions from bosses or superiors, the frequency of nightshifts, working with superiors and colleagues who can be respected, one's salary, the atmosphere of the workplace and a culture were colleagues help each other. However, a low rate of job satisfaction was identified with factors such as trust from doctors, their own vital role, appropriate number of nursing staff, the clinical ladder, work outside of nursing duties, break rooms and hospital facilities, break times, and the amount of overtime. Looking at the degree of the intention to continue working, the highest rate was seen in those with more than 5 years' experience, and the lowest rate was those with 1 year or more to less than 3 years' experience. In the following, each analysis result will be explained in detail.

3.1 Result of Preliminary Survey

After intercepting narrative data obtained by the interview survey, we extracted keywords and grouped them, and extracted 25 factors as satisfying factors and unsatisfactory factors in the target hospital. Discussion was conducted with nurses, experts in nursing care management, experts in business administration, and created question items of 61 items.

3.2 Results of Job Satisfaction Survey

Results are shown in Fig. 2. High satisfaction with items such as communication with colleagues, pride of being a nurse, pleasure of contacting a patient, instructions from a boss, the number of times of night shift, respectable bosses and colleagues, salaries, workplace climate were clarified. On the other hand, the result of low satisfaction with trust from the doctor, the role of indispensable self, the number of proper nurses, clinical ladder, work outside nursing, rest room and office equipment, break time, overtime hours etc. were identified.

3.3 Results of Factor Analysis on Job Satisfaction

Subsequently, factor analysis by the main factor method was carried out for 61 items. As a result, four factors were extracted by Promax rotation. We picked up only items with a factor loading amount of .40 or more and conducted factor analysis again, and as a result, 9 factor structure was confirmed (Fig. 3). The first factor is "institution and management", the second factor is "relationship with boss", the third factor is "relationship with colleagues", the fourth factor is "relationship with doctor", the fifth factor named "overtime and workload", the factor 6 is "facilities and equipment", the factor 7 is "salary", the factor 8 is "approval", the factor 9 is "pride". As shown in Fig. 2, the degree of satisfaction with each other was high, pride was high, overtime hours and workloads, facilities and equipment, and approval were low.

3.4 Results on Turnover Intention

Regarding the intention to leave off, I would like to continue working in my department in the future "," I would like to continue working in the current hospital "," I would like to continue to work as a nurse in the future "3 For one question, I asked for a reply with five methods of "very agree = 5" to "absolutely disagree = 1". The results were as follows.

"I want to keep working in my current department" average = 3.34, SD = 1.01
"I would like to continue working in the current hospital" average = 3.25, SD = 0.91
"I would like to continue working as a nurse in the future" average = 3.80, SD = 0.85

3.5 Relationship Between Job Satisfaction and Intention of Turnover

In order to investigate the factor on the turnover intention, the relationship between 9 factors obtained from the factor analysis result of job satisfaction and intention to leave off the job was statistically evaluated using logistic regression analysis.

"I would like to continue working in the current department."
→ Relationship with colleagues (OR 3.16, 95% CI 1.94 - 5.12), approval (OR 2.63, 95% CI 1.67 - 4.14)
"I would like to continue working in the current hospital."

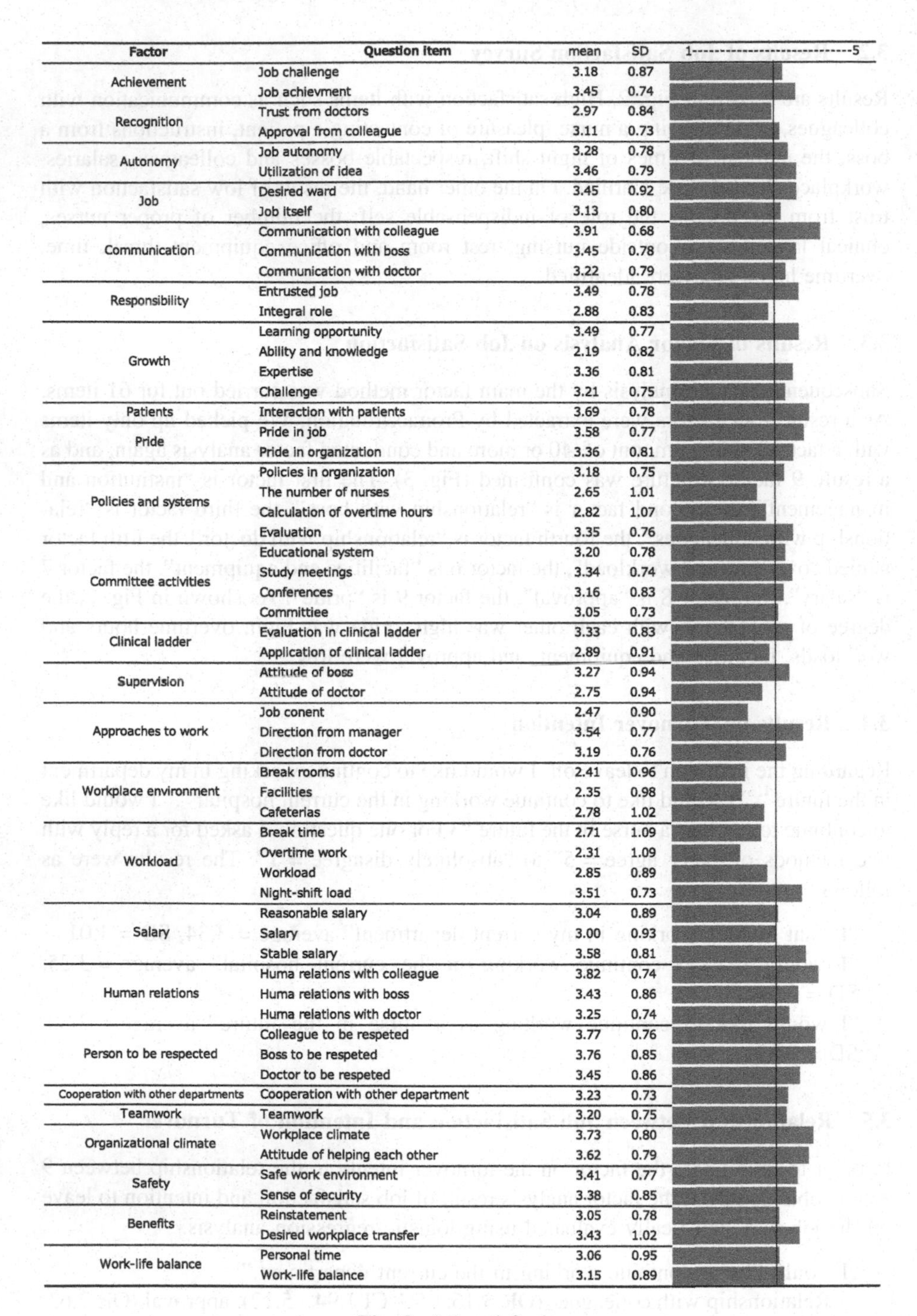

Factor	Question item	mean	SD	1--------5
Achievement	Job challenge	3.18	0.87	
	Job achievment	3.45	0.74	
Recognition	Trust from doctors	2.97	0.84	
	Approval from colleague	3.11	0.73	
Autonomy	Job autonomy	3.28	0.78	
	Utilization of idea	3.46	0.79	
Job	Desired ward	3.45	0.92	
	Job itself	3.13	0.80	
Communication	Communication with colleague	3.91	0.68	
	Communication with boss	3.45	0.78	
	Communication with doctor	3.22	0.79	
Responsibility	Entrusted job	3.49	0.78	
	Integral role	2.88	0.83	
Growth	Learning opportunity	3.49	0.77	
	Ability and knowledge	2.19	0.82	
	Expertise	3.36	0.81	
	Challenge	3.21	0.75	
Patients	Interaction with patients	3.69	0.78	
Pride	Pride in job	3.58	0.77	
	Pride in organization	3.36	0.81	
Policies and systems	Policies in organization	3.18	0.75	
	The number of nurses	2.65	1.01	
	Calculation of overtime hours	2.82	1.07	
	Evaluation	3.35	0.76	
Committee activities	Educational system	3.20	0.78	
	Study meetings	3.34	0.74	
	Conferences	3.16	0.83	
	Committiee	3.05	0.73	
Clinical ladder	Evaluation in clinical ladder	3.33	0.83	
	Application of clinical ladder	2.89	0.91	
Supervision	Attitude of boss	3.27	0.94	
	Attitude of doctor	2.75	0.94	
Approaches to work	Job conent	2.47	0.90	
	Direction from manager	3.54	0.77	
	Direction from doctor	3.19	0.76	
Workplace environment	Break rooms	2.41	0.96	
	Facilities	2.35	0.98	
	Cafeterias	2.78	1.02	
Workload	Break time	2.71	1.09	
	Overtime work	2.31	1.09	
	Workload	2.85	0.89	
	Night-shift load	3.51	0.73	
Salary	Reasonable salary	3.04	0.89	
	Salary	3.00	0.93	
	Stable salary	3.58	0.81	
Human relations	Huma relations with colleague	3.82	0.74	
	Huma relations with boss	3.43	0.86	
	Huma relations with doctor	3.25	0.74	
Person to be respected	Colleague to be respeted	3.77	0.76	
	Boss to be respeted	3.76	0.85	
	Doctor to be respeted	3.45	0.86	
Cooperation with other departments	Cooperation with other department	3.23	0.73	
Teamwork	Teamwork	3.20	0.75	
Organizational climate	Workplace climate	3.73	0.80	
	Attitude of helping each other	3.62	0.79	
Safety	Safe work environment	3.41	0.77	
	Sense of security	3.38	0.85	
Benefits	Reinstatement	3.05	0.78	
	Desired workplace transfer	3.43	1.02	
Work-life balance	Personal time	3.06	0.95	
	Work-life balance	3.15	0.89	

Fig. 2. Job satisfaction

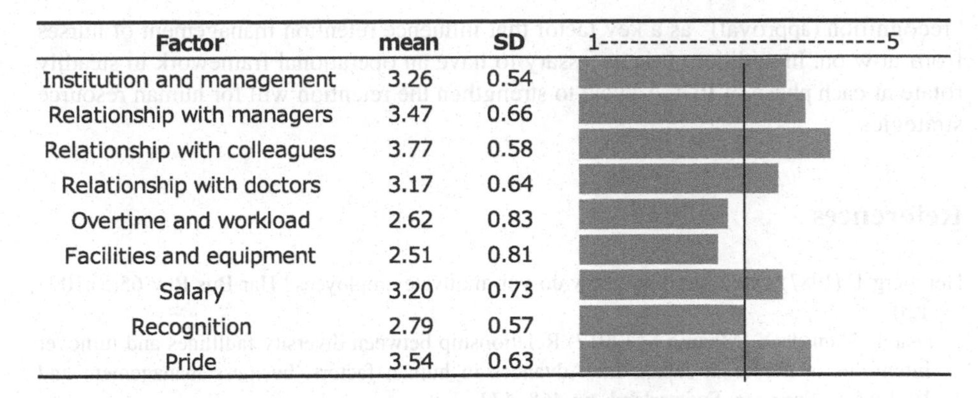

Factor	mean	SD	1 ----------------5
Institution and management	3.26	0.54	
Relationship with managers	3.47	0.66	
Relationship with colleagues	3.77	0.58	
Relationship with doctors	3.17	0.64	
Overtime and workload	2.62	0.83	
Facilities and equipment	2.51	0.81	
Salary	3.20	0.73	
Recognition	2.79	0.57	
Pride	3.54	0.63	

Fig. 3. Factor analysis on job satisfaction

→ pride (OR 2.31, 95% CI 1.50-3.56), approval (OR 2.23, 95% CI 1.45 - 3.45)
"I would like to continue to work as a nurse."
→ approval (OR 5.21, 95% CI 2.82-9.63), salary (OR 1.70, 95% CI 1.06 - 2.75)

4 Discussion

Implications of this research are as follows. The first point is that the desire to continue working (willingness to retire) is "to be a nurse (3.80) > to remain at the current department (3.34) > to stay at the current hospital (3.25)". Secondly, the key factor that affects the retention will is "approval (recognition)". Therefore, a mechanism for integrally managing retention intention and approval is required. In other words, despite the will to retire from hospital and department, the will to continue as a nurse is considered high. This means that by implementing human resource management incorporating retention and management viewpoints into the management cycle (PDCA) within the nursing organization, the intention will to be nurses at the current department or hospital is strengthened. In order to strengthen the retention will, PDCA (system to check the effectiveness) should be monitored at each phase of PDCA led by an appropriate specialized team (such as a team in charge of nursing education). For that purpose, it is necessary to have an operational framework to steadily rotate PDCA cycle. At the same time, the human resource management at each phase of PDCA is required a methodology to scientifically (quantitatively) evaluate communication activities within the nursing organization to increase satisfaction to "approval" which is the key factor to nurse's retention.

5 Concluding Remarks

The result of the analyses indicated that such satisfactory factors as a sense of pride in being a nurse and recognition as a nurse contributed to increase the pleasure of nursing practice and the intention of working. Therefore, it is necessary to carefully consider "recognition (approval)" as a key factor that influence retention management of nurses from now on. In addition, it is necessary to have an operational framework to steadily rotate at each phase of PDCA cycle to strengthen the retention will for human resource strategies.

References

Herzberg F (1987) One more time: how do you motivate employees? Har Bus Rev 65(5):109–120

Iwaasa T, Yamada Y, Mizuno M (2017) Relationship between diversity faultlines and turnover intentions of nurses in Japan. In: Advances in human factors, business management and leadership. Springer, Switzerland, pp 468–473

Mizuno M, Yamada Y, Hirosawa M (2012) Examination of hygiene factors influencing on job dissatisfaction among nurses. J Ergonomics Occup Safety Health 14(1):17–24 (in Japanese)

Mizuno M et al (2015) Practice and assessment of a team-building approach for team medical care performance from the viewpoint of positive belief in cooperation among Japanese nurses. J Health Care Nurs 11(2):8–14 (in Japanese)

Mizuno M et al (2015) Behavioral sensor-based organizational design and management in Japan: from the perspectives of communication channel in nursing organization. In: Proceedings 19th triennial congress of the international ergonomics association

Mizuno Y et al (2016) A study on communication activity and social skills of nursing organization. In: Advances in human factors, business management, training and education. Advances in Intelligent Systems and Computing, vol 498, pp 561–568

Mizuno M, Iwaasa T, Yamada Y (2017) Current tendency of retention management of nurses at university hospital in Japan. In: Final program 8th international conference on applied human factors and ergonomics

Togashi E, Iwaasa T, Mizuno M (2017) Team efficacy and self-efficacy of nursing teams at small and medium sized hospitals in Japan. In: 8th international conference on applied human factors and ergonomics, conference proceedings

Yamada Y et al (2016) Elements of workforce diversity in Japanese nursing workplace. In: Advances in social & occupational ergonomics. Advances in Intelligent Systems and Computing, vol 487, pp 167–176. Springer, Cham

Developing an International Master's Programme in Ergonomics at a Technical University in Sweden

Linda M. Rose(✉) and Cecilia Österman

Division of Ergonomics, Department of Biomedical Engineering and Health Systems, KTH School of Engineering Sciences in Chemistry, Biotechnology and Health, Hälsovägen 11C, 141 57 Huddinge, Sweden
lrose@kth.se

Abstract. The objective of this paper is to present the rationale for, and the development process of, a new international master's programme in Ergonomics, here framed as 'Technology, Work and Health', at KTH, Royal Institute of Technology in Sweden. The masters' programme is designed for two years of full-time studies (120 credits).

A challenge during this development process, has been to decide not only what to include in the programme, but also what to exclude. The systems-oriented discipline of ergonomics now covers all aspects of human work. Thus, two tracks of specialization are offered: *Work Environment Engineering* (WEE) and *Human Factors and Ergonomics* (HFE).

The programme is given for the first time in the autumn 2018. After the first year it is necessary to evaluate the programme from student and teacher perspectives, to capture effects and improvement possibilities. Naturally, this evaluation includes the usual course evaluations that are done during and after each course. But also, from a recruiting perspective, when we now how many students, from where, and of which background that eventually were awarded a MSc degree. Finally, a programme review with an emphasis on outcomes is essential to establish that the programme meets academic standards, professional discipline expectations among employers of the graduates, and student expectations and satisfaction. The results from this review will provide the incentive for necessary major and minor changes to maintain an up-to-date and high-quality curriculum in a discipline that continues to evolve.

Keywords: Learning outcomes · Teaching · Blended learning

1 Introduction

In 2009 the Swedish Government decided to form a delegation within the field of Occupational Health Services with the task to procure and coordinate longer education and one and two-year master's programme in Occupational Health Services [1]. The need for such education had been identified, to secure competence within the field after the National Institute of Working Life which had provided such education, especially for Health and Safety engineers, had been closed in 2007. The idea was that

S. Bagnara et al. (Eds.): IEA 2018, AISC 821, pp. 429–436, 2019.
https://doi.org/10.1007/978-3-319-96080-7_51

universities should provide such education. This led to that the development of the one-year master's programme, 'Technology, Health and Work Environment Development' (THAM, 60 credits) started at KTH Royal Institute of Technology in Sweden in 2009. The programme was targeted at students with varying educational and professional backgrounds, ranging from engineers and physiotherapists to people with competences in HR, and behavioral and social psychology sciences. The development process, driven by work environment/ergonomics experts and educationalists at KTH also involved discussions and co-operation with, among others, representatives from industry, other universities, and possible learners-to-be.

The ambition was to develop an education for Health and Safety engineers that was as 'holistic' as possible in a one-year master's programme, to equip the learners with a broad range of adequate skills, including applied knowledge in physical and chemical work environment, safety, change management, and the financial effects of work environments. It was also considered essential that the various parts of the education, and those responsible for the parts, should have the same value base regarding what the education should convey, pedagogical approach, the level of the courses and what was expected from the students in the different courses. To achieve this, a process with continuous meetings and workshops with the different course developers was used. In these also requirements and syllabus for the courses were developed. All course developers and examiners were strongly advised to adapt to the pedagogical approach with constructive alignment [2], based on Tyler's statement that '*Learning takes place through the active behavior of the student: it is what he does that he learns, not what the teacher does.*' [3] and emphasizing that to achieve desired outcomes in an effective way, an important task for the teachers is to form the education so that the students become engaged in learning activities [4].

The one-year THAM master's programme, was started in the autumn 2011. At approximately the same time another one-year master's programme, which had been developed and provided by the University of Linköping in Sweden since 1996, 'Ergonomics and Human – Technology - Organization' (Ergonomics HTO) was moved to KTH. This programme was based on a system perspective view and focused on the interaction between humans, technology and organizations with the aim to develop sustainable prerequisites for product development and production systems by including ergonomics aspects.

KTH is one of few universities in Europe providing higher education within the field of Ergonomics and Human Factors at advanced level. In fact, many education programs within the Ergonomics & Human Factors field have been closed during the last decade, for example in the UK and in New Zealand. This is trouble-some, especially since the increasingly complex environments and techniques lead to an increased need of changed competences [5]. This requires not less, but rather more enhanced competence in Ergonomics and Human Factors to secure adequately designed and functioning systems and to work towards the objectives of the discipline; to optimize human well-being and overall system performance [6].

For several reasons, e.g. to meet an increased demand from students for a more advanced education in the field of Ergonomics and Human Factors, adoption to the Bologna system (the European University educational system, with three year Bachelor degree programmes followed by two year Master degree programmes), economic advantages of running one two-year programme instead of two one-year programmes, to meet the CREE criteria to become certified European Ergonomist, and to support the base of eligible and competent people into research education, the development of a two-year master's programme was initiated in 2016.

The aim of the programme is to provide students with deeper understanding of the interplay between humans, technology and organizations, from sustainable work and organizational performance perspectives and provide them with knowledge and skills to contribute to sustainable systems and working life.

The aim of this paper is to present the rationale for, and the development process of, a new international master's programme in Ergonomics, here framed as 'Technology, Work and Health', at KTH.

2 Methods

In the development of the two-year masters' programme a similar methodology was used as when developing the THAM programme. This included needs identification, workshops and meetings and iterative development of course content in collaboration between the different examiners of the courses included in the programme. Examples of topics which were addressed were '*what is needed and what is possible to include, under the given prerequisites for developing the master's programme*', as well as to describe and define how well each topic needs to be understood by the students and what competence and skills they should possess when graduating. Other examples were what topics and skills to prioritize and which student groups to attract. One of the challenges was how to develop courses with appropriate breadth and depth on a topic by topic basis, but also to secure that the different courses together enable sufficient competence and form an adequate education. The programme draw on wide range of faculty expertise that were included in the development, including physical ergonomics, applied ergonomics, work organization and change processes.

In developing the programme, desired learning outcomes were specified regarding topic content and level of student skills and abilities. These were included in the programme syllabus. Thereafter course syllabi emphasizing interactive teaching and learning, and critical dialogue were developed. Here constructive alignment as defined by Biggs [2]. Continuous assessment and feedback throughout the courses plays an important part in the course design. This is regarded as important for motivation and for facilitating learning [7]. The design with blended learning [8] was chosen to integrate learning activities in classroom face-to-face with online learning activities. The purpose is to provide meaningful learning experiences for all types of learners, and further to enable students who are non-residential in Stockholm to follow the programme and allow for students to combine work and studies.

3 Results

3.1 The Programme and Its Learning Objectives

The experiences and feedback from the two previously developed and successfully run one-year programmes at the university contributed to the knowledge base when starting the development of the new programme. The needs analysis showed that the programme should provide the graduates with skills to be able to identify and assess various work environments, as well as to suggest, argue for, and implement interventions.

The programme aims to provide a deeper understanding of the interplay between humans, technology and organizations, from the perspectives of sustainable work and organizational performance. The programme was developed to support the students' development of knowledge and proficiency in proactive occupational safety and health (OSH) management, as well as in how to plan, design and analyze work environments. Further, the goals of the programme included that the graduates will possess advanced theoretical and applied knowledge in the field of physical, cognitive and organizational ergonomics, applicable for work with development of organizations and processes, with focus on improvements of work and ergonomic interventions for a sustainable working life. After graduation, the students should be able to actively influence and manage change and development projects within the field of technical work environments. The students should also have knowledge about risk assessment methods, safety strategies and project management within production and product development. Further, it was decided that that programme should be designed in way that provides an adequate foundation for research studies within technology, work and health. In addition, to structure learning outcomes, course contents, and assessments, specific learning objectives have been constructed for the following three levels: knowledge and understanding; skills and abilities; and ability to judgements and adopt a standpoint.

The masters' programme is designed for two years of full-time studies (120 European Credit Transfer and Accumulation System, ECTS, credits). Eligible students are required to have completed a BSc in technical, natural or medical sciences, including or supplemented with 15 ECTS in mathematics.

3.2 The Masters' Programme Structure, Content and Pedagogics

Two tracks are offered: Work Environment Engineering (WEE) and Human Factors and Ergonomics (HFE) (Fig. 1). WEE includes theoretical and practical knowledge regarding technical and physical work environments, as well as in organization and management of occupational safety and health. HFE is based on the International Ergonomics Association's (IEA) definition of Ergonomics, as a design-oriented discipline.

Work Environment Engineering (WEE) includes elaborations in the theory and practice of technical and physical work environments, as well as organization and management of occupational safety and health work (OSH). This specialization provides a successive deepening of knowledge in OSH management with focus on the

	YEAR 1			YEAR 2		
	WEE	WEE + HFE	HFE	WEE + HFE		
Period 1	Evaluation and Measures of the Physiological and Acoustic Work Environment (7.5)	Occupational Safety and Health Management and Change (7.5)	Ergonomics for the prevention of musculoskeletal disorders (7.5)	Ergonomics in Product Development (7.5)	Elective Course (7.5)	Ergonomics, Human Factors and Patient Safety (7.5)
Period 2	Evaluation and Measures of the Physical and Chemical Work Environment (15)		Cognitive Ergonomics (7.5) Work Environment Economics (7.5)	Planning and design of work environments (7.5)	Elective Course (7.5)	Elective Course (7.5)
Period 3	Theory and methodology of science with applications (7.5) Elective Course (7.5) or: Degree project (15)			Thesis Course (30)		
Period 4	System safety and risk management (7.5) Elective Course (7.5) or: Degree project (15)					

Fig. 1. Illustration of courses within the two available tracks within the programme

technical work environment, from the organization of work, via change management and risk assessments, to the effects of the work environment on an organizations operational and financial performance. Correspondingly, the ergonomic design of products and artefacts can be optimized from and performance influencing factors. This specialization was developed in accordance to the governmental requirements regarding advanced education and training in work environment engineering. After completion of the studies following this track, the graduate will have suitable knowledge, understanding and proficiency to work as a Work Environment Engineer.

Human Factors and Ergonomics (HFE) is based on the International Ergonomics Associations (IEA) definition of ergonomics as a scientific discipline concerned with the understanding of interactions among humans, technology and organizations of a system, and the profession that applies theory, principles, data and methods to design to optimize human well-being and overall system performance [6]. The discipline is design oriented and mainly concerned with planning and designing of workplaces, tasks, products and systems. After completion of the studies following this track, the successful graduate will have suitable knowledge, understanding and proficiency to work as a Human Factors Specialist or Ergonomist.

Through course projects and the master thesis, the graduates will train their ability to integrate the acquired knowledge and abilities from the different topics in the programme.

MSc Thesis Requirement. As a master's programme at a university, the programme includes a mandatory MSc thesis (MSc degree project). The purpose of the MSC thesis requirement is to provide the graduates with the opportunity to perform a research or development project in a particular area of interest, related to the courses within the programme. The thesis is expected to include a comprehensive and critical synthesis of the relevant literature and present either a theoretical contribution to knowledge, an empirical investigation, or both. While it is preferred that student study the entire two-year programme, including an MSc thesis of 30 credits, there is a possibility for students to complete a 15 credits thesis after the first year and receive a one-year master's degree, in Swedish called 'magister'.

To equip students with the knowledge and skills to plan, design and conduct the degree project, several courses include smaller projects that are performed alone or in groups. In addition, the students are given a 7.5 credits course in theory and methodology of science with application. In this course, students will develop core skills in qualitative and quantitative methods and techniques for collection and analysis of data.

3.3 Marketing and Recruitment

A comprehensive marketing and recruitment strategy was used, including a broad range of activities. For example, international information activities were managed via the university's professional communicators and the coordinator for international students at conferences and by on-site visits to several international universities. In addition, internet resources, using the programme and course developer's professional networks, and social media contacts were used to disseminate information about the new programme.

3.4 Application Process Results

In total up to 40 students will be accepted into the programme during its first year. The official application time for the program was between 16 October 2017 and 15 January 2018, corresponding to the deadlines that generally apply for international master's programmes. For Swedish programmes, the deadline is 16 April and for the local students that had not observed the earlier deadline and missed their opportunity to apply, it was decided to open the admission site for another round of applicants during the days immediately before April 16.

After closing the first round of application, 76 persons had applied for the programme. 10 of these did not to meet the general entry requirements for university studies at advanced level and were not further assessed. Of the 66 eligible applicants, 49 were Swedish citizens and another 3 were EU/EEA citizens, hence not required to pay a tuition fee. The second largest group of qualified applicants were from Asia (7), followed by Africa (4), Middle East (2) and USA (1). line. These applicants are required to pay a tuition fee in advance for each semester, which in practice mean that they may have to secure funding or scholarship to be able to accept the position.

If an applicant wishes to apply for more than one programme, he or she must rank the applications in order of priority. In this first round of application, 36 applicant's hade this programme as the first choice (Fig. 2).

In the days before the Swedish deadline, April 16, the admission site was opened again for local students. People who have showed interest in the programme after the official deadline were contacted by personal email and invited to apply. In this second round, another 17 students applied of which 15 were eligible and 14 had ranked this programme as their first choice. In all, 81 students have been notified that they are welcome to start the programme. Since this is the first time that the programme is given, it is not possible to compare with previous statistics for applications, and how

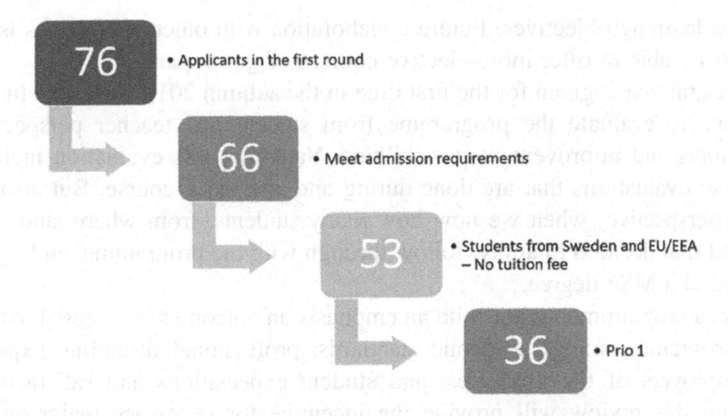

Fig. 2. Number of applicants to the Master's programme in the first round

many that eventually may accept the position, but generally the students that rank the programme first hand are more likely to accept.

4 Discussion and Concluding Remarks

In the development process it was judged important to map societal needs and educational resources, and to secure constructive alignment with suitable learning activities and outcomes to provide a high-quality education meeting the needs. Needs have been articulated by different stakeholders, for example prospective employers of the graduates, but also alumni from earlier programmes that after some years of working experience have identified strengths that should be maintained, as well as knowledge gaps in previous curricula that needs to be bridged.

A challenge during this development process, has been to decide not only what to *include* in the programme, but also what to *exclude*. The systems-oriented discipline of ergonomics now covers all aspects of human work. The successful graduate from a university master's programme in ergonomics will be expected to have a broad understanding of the full scope of the discipline, as well as a solid understanding of contexts and practices at various industries or sectors. The final content of the programme, as it is currently designed, is the product of balancing societal and industry needs and demands, with administrative requirements and availability of resources at the university and the faculty. The aim has been to ensure as good a learning environment as possible within the financial and practical constraints that follows with the need to secure teacher resources, appropriate rooms for lectures, laboratories, seminars and group work, and field study visits.

To enable students to shoulder responsibility for their own learning and allow for a certain flexibility for specialization, there are room for several elective courses. These can be chosen among the courses that are included in the other track (WEE or HFE), among other relevant courses that are given at the university, or at another university, after making sue that the elective course will contribute towards fulfilling the programme learning objectives. Future collaboration with other universities is seen as essential to be able to offer more elective courses of good quality.

The programme is given for the first time in the autumn 2018. After the first year it is necessary to evaluate the programme from student and teacher perspectives, to capture effects and improvement possibilities. Naturally, this evaluation includes the usual course evaluations that are done during and after each course. But also, from a recruiting perspective, when we now how many students, from where, and of which background that decided to apply, follow through with the programme, and eventually were awarded a MSc degree.

Finally, a programme review with an emphasis on outcomes is essential to establish that the programme meets academic standards, professional discipline expectations among employers of the graduates, and student expectations and satisfaction. The results from this review will provide the incentive for necessary major and minor changes to maintain an up-to-date and high-quality curriculum in a discipline that continues to evolve.

References

1. Regeringskansliet (2009) Delegationen för kunskapsområdet företagshälsovård. Kommittédirektiv 2009:18, Stockholm. (in Swedish)
2. Biggs JB (2011) Teaching for quality learning at university: what the student does. McGraw-Hill Education (UK)
3. Ralph TW (1949) Basic principles of curriculum and instruction. Syllabus for education, Chicago
4. Shuell TJ (1986) Cognitive conceptions of learning. Rev Educ Res 56(4):411–436
5. Dul J et al (2012) A strategy for human factors/ergonomics: developing the discipline and profession. Ergonomics 55(4):377–395
6. IEA (2018) What is ergonomics? http://www.iea.cc/what_is_ergonomist.html. Accessed 11 May 2018
7. Lee M, Kirschner PA, Kester L (2016) Learning analytics in massively multi-user virtual environments and courses. J Comput Assist Learn 32(3):187–189
8. Garrison DR, Kanuka H (2004) Blended learning: uncovering its transformative potential in higher education. Internet High Educ 7(2):95–105

Designing an Organizational Readiness Survey for Total Worker Health® Workplace Initiatives

Michelle M. Robertson[1,2(✉)], Diana Tubbs[2(✉)], Robert A. Henning[2(✉)], Suzanne Nobrega[3(✉)], Alec Calvo[2(✉)], and Lauren Murphy[1(✉)]

[1] Northeastern University, Boston, USA
{m.robertson,l.murphy}@northeastern.edu
[2] University of Connecticut, Storrs, CT, USA
{diana.tubbs,robert.henning,alec.calvo}@uconn.edu
[3] University of Massachusetts, Lowell, USA
suzanne_nobrega@uml.edu

Abstract. The aim of the present study was to develop a means to assess organizational readiness for a Total Worker Health® initiative, a comprehensive approach for improving employee safety health and wellbeing. A systematic literature review was conducted to integrate past multidisciplinary peer-reviewed theoretical and empirical work. Although the initial search of the organizational change literature revealed nearly 300,000 related titles, there was considerable conceptual ambiguity and only a limited number were focused on major occupational health and safety initiatives. A revised set of inclusion criteria identified 30 relevant publications that yielded the following eight key organizational characteristics and predictors: (1) culture, (2) communication, (3) leadership, (4) change history, (5) job design, (6) teams and relationships, (7) flexible organizational practices and policies, and (8) positive organizational climate. The findings from this review of the literature and our subsequent conceptual model provide the foundation for developing an organizational readiness assessment tool that researchers and practitioners will be able to use prior to implementing comprehensive workplace safety, health and wellbeing initiatives.

Keywords: Organizational readiness · Total Worker Health
Organizational culture

1 Total Worker Health (TWH)

Total Worker Health® (TWH) is an approach focused on integrating work-related safety and health protection with health promotion [1]. This is still a relatively new concept, although similar approaches are seen in Europe, and research findings on the best programmatic ways of achieving TWH in an efficient and cost-effective manner remain rather limited. Total Worker Health programs require implementation of new policies and practices by the organization, however, few methods have been published

S. Bagnara et al. (Eds.): IEA 2018, AISC 821, pp. 437–445, 2019.
https://doi.org/10.1007/978-3-319-96080-7_52

to date for specifically assessing an organization's readiness to implement these policies and practices.

2 Organizational Readiness

Organizational readiness for change becomes a concern whenever a new workplace initiative or program is at risk of not being fully successful. The peer-reviewed literature on readiness for change offered a number of different ways to measure readiness in a context-specific manner. However, there was a fair amount of disagreement across the various conceptualizations of organizational readiness for change in this literature, as well as how this could be defined and then operationalized in some way to benefit the actual organizational change efforts. In their present form, these divergent perspectives create a tenuous guide for researchers who want to study organizational readiness for change, and also for practitioners who simply want to assess it prior to an implementation effort.

Our literature review of the state-of-the-art in this area was expected to help identify the evidence-based factors associated with organizational readiness for change prior to implementing a workplace Total Worker Health program. The results of this novel literature review lay the foundation for designing an organizational readiness survey and assessment guide to indicate how ready an organization is to improve its safety and health practices in accord with Total Worker Health principles.

3 Method

A systematic literature review was conducted to find articles related to organizational readiness for change. A keyword search of four electronic bibliographic databases was conducted to identify candidate articles. The following databases were searched: PsycINFO, PubMed, ABI Inform Global, and Google Scholar. Those databases were chosen because they included multidisciplinary peer-reviewed articles. Google Scholar also includes sources that are not peer-reviewed, but it is nonetheless a good search engine as well as a useful citation index. The search strategy combined four groups of keywords using "AND" to reflect the particular elements of interest. Also, the key terms were searched in a tiered process. The first set of terms were "organization" and "workplace," to define that organizational readiness takes place in the workplace. The second set of terms were "readiness," "change," and "pre-implementation," to indicate the state of affairs prior to change that can be identified and described. The third set of terms were "intentions," "assessment," "antecedents," and "determinants," which define the degree to which an organization is ready and areas for improvement. The fourth and last set of terms were "safety," "ergonomics," "wellness," and "well-being" to show the change efforts are related to our particular domain of interest. There were a number of criteria used to include or exclude articles in this review. The inclusion criteria were as follows: (1) review articles from 2010 to 2015, (2) readiness for organization as a whole, (3) readiness in general and for safety and wellness interventions, (4) empirical and case studies (was the study clearly described?),

(5) conceptual/theoretical papers (was the conceptual basis of the need for change clearly explained?), (6) scale development articles, and (7) peer-reviewed and scholarly books (particularly highly cited works). The exclusion criteria were as follows: (1) review articles before 2010, (2) readiness related to a specific subset of employees (e.g., management, sales teams), (3) readiness related to patients and their treatment (e.g., hospitals, clinics) or students or communities to makes changes, and (4) readiness and its association with workplace changes in developing countries (e.g., e-government in the Republic of Yemen).

4 Results

Cumulatively, the combined searches yielded approximately 300,000 titles: PsycINFO 868 titles, PubMed 2,825 titles, ABI Inform Global 34,558 titles, and Google Scholar 259,000 titles. We selected articles based on the number of times they were cited relative to the year they were published (capturing breadth with review articles), and empirical work that has measured and described pre-implementation readiness particular to safety/wellness initiatives (capturing depth). Given the broadness of our criteria, and the large number of titles found, we selected publications by going through web pages of search results in order, as well as retrieving additional important sources from the list of references from the articles that had already been reviewed. "Readiness for safety initiatives" was not a strong theme that emerged from our search. While there was evidence of that topic as well as progress being made with it, the literature associated with it was sparse and outdated. Surprisingly, the readiness indicators that emerged from safety-specific works were similar to other organizational readiness indicators in topically unrelated papers.

Overall, there were a limited number of publications in the area of organizational readiness measures that related specifically to workplace safety/ergonomics and human factors/wellness change initiatives. Readiness for change does of course depend, in part, on the specific change involved. Still, everyday functioning in the workplace can have much impact on change efforts even though such functioning may seem at first to be unrelated. In order to reduce the number of articles to only those that were most relevant to this study, we also identified classic works by prominent authors, clarified key theories and definitions, delineated points of debate and disagreement, extracted broad themes and the way knowledge is structured, and retrieved additional sources that were cited in review articles and recent works. That led to the identification of 30 key articles used in the present study.

A conceptual model of organizational readiness, depicted in Fig. 1, was created using the 30 identified sources. The first layer of organizational readiness is context, which is the organization's current functioning and how it operates day to day. The second layer is content, which is the particular change effort choice. Included in the content layer is visionary communication that clearly explains the change, why it is needed, and what can be expected. The quality of the communication is very important as is its consistency and the actions aligned with the message. The third layer of organizational readiness is related to the individuals within an organization. Individuals need to be motivated to support the change, to accept the change when it occurs, to believe they can change their behaviors to accommodate the change, and to already be working together in ways that would support a change effort if it occurs. It should be noted that the different levels of an organization, including the organization itself, the group (teams), and individuals, are implied within and across the three layers.

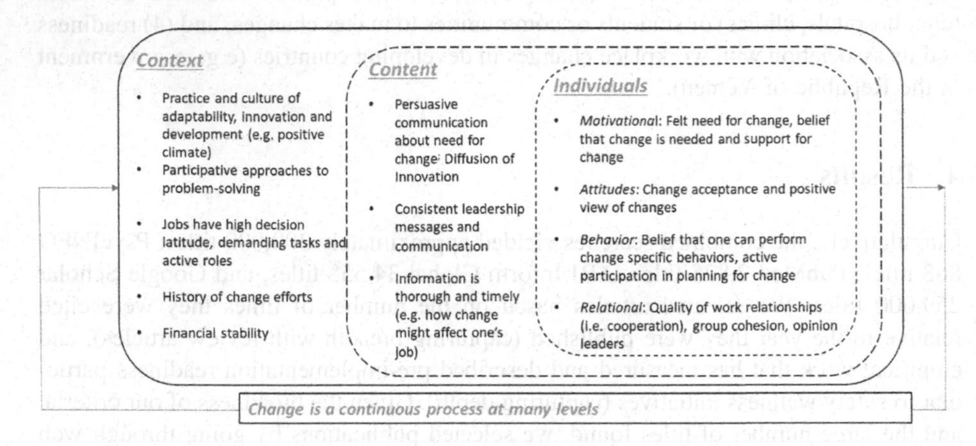

Fig. 1. Conceptual model of organizational readiness.

Dominant in the literature is the idea that organizational readiness for change is necessarily content specific. Yet this generally implies that a specific change effort has been described, designed, and planned for, and that this is the only way to assess readiness. By this logic, readiness greatly depends on what is specifically on the horizon. A less explored approach is to consider the state of the organization and its level of readiness for change in general; that is, without only one pre-determined change initiative in mind. This latter approach is our own, and we think this introduces a continuum of readiness to consider that is more reflective of the readiness needed to support a set of change initiatives that are programmatic in nature. In contrast, the content-focused view of organizational readiness is more of a capacity for a highly specified change effort than a readiness for change generally.

Organizations with the same contextual attributes can differ in their effectiveness to implement change. The design of the change effort becomes just as important as the capacity to change. Some structural features that shape change perceptions are the organization's financial, human, material, and information resources. Also, non-structural factors that are likely to generate a sense of readiness are: (1) consistent leadership messages and actions, (2) information sharing through social interaction, and (3) shared experiences with past change efforts. Overall, the key organizational characteristics and predictors of organizational readiness are: (1) culture, (2) communication, (3) leadership, (4) change history, (5) job design, (6) teams and relationships, (7) flexible organizational practices and policies, and (8) positive organizational climate. Following are discussions of a few of these contextual factors.

4.1 Organizational Culture

Organizational culture is frequently identified as a contextual determinant of organizational readiness for change. A culture that embraces innovation, risk-taking and learning supports readiness better than organizations that value stability and control or efficiency and productivity [2]. The reason for this is that having a practice of changing a task or process helps employees keep positive attitudes toward change and prevents the type of inertia often encountered when meaningful change is needed and sought, and employees' reactions are stagnant. Similar types of organizational culture features that have empirical support as readiness antecedents are cultures that value cohesion and morale through training and development, open communication, and participative decision-making [3]. In their temporal study design, Jones and colleagues [3] found evidence that readiness for change mediated the relationship between employees' perceptions of a human relations culture and the organization's capacity to change (a composite of three dimensions: engagement/participation, development/resource creation, and performance management/proactive management of factors that drive performance). A case study that investigated the preconditions of readiness in a multisite organization found that the strongest predictor of readiness was having a positive culture through employees who are motived to see a change [4]. A positive culture creates and instills an contextual atmosphere where employees are enthusiastic about change and don't need a lot of convincing to take on change. Organizational culture reflects the shared assumptions about what important, how things are done, and how people should behave in organizations [5].

There are four distinct ways culture affects employees: (1) artifacts, or visual symbols of culture such as clothing, and physical arrangements; (2) norms, or unwritten rules of behavior that guide employees on how to behave in certain situations; (3) values that tell members what is important in the organization and what deserves their attention; and (4) deep assumptions, the taken for granted assumptions about how organizational problems should be solved [5]. Given that workplace culture can be a good indicator of readiness or even change outcomes, it can also be the subject of change. Cummings and Worley [5] advise that culture change should be considered only after other less difficult and costly solutions have been applied or ruled-out. This advice is based on the view that culture change is a radical change that goes far beyond making the existing organization better. In this way, an organization that is more likely to be ready for change is one in which culture has been carefully cultivated.

To illustrate these perspectives, we propose an adapted model depicting the continuous cycle of an emerging organizational culture serving as both a determinant and outcome of organizational change as well as its interactions with the external environment and organizational outcomes [5–7]. These systems components are elaborated further below in the open systems perspective of change (see Fig. 2).

Culture is also related to another commonly mentioned readiness factor that creates a receptive environment for change, that is, an organization's *financial resources* [2]. Burnett and colleagues conducted qualitative interviews [4] that elucidated the link between readiness and financial stability for leaders. Hospital employees in leadership positions felt that since their organization met its targets for 6 years and had been in "financial balance," they could start on a journey to make workplace improvements [4].

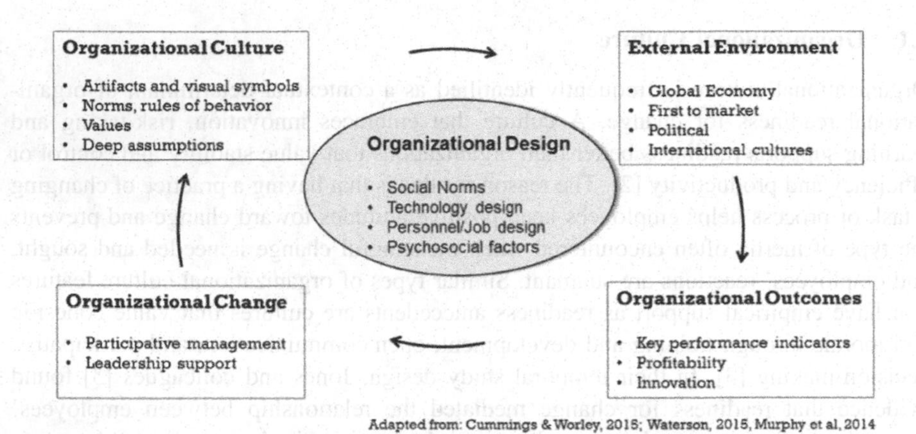

Fig. 2. Open systems perspective of change

In relation to readiness, financial resources, and culture, another study found that aspects of organizational culture, like openness in communication, openness to change, and clarity of mission and goals, were likely to occur in an organization that has consistent budgets and a stable environment, and where institutional resources are more likely to be predictable. In comparison, when the budget was decreasing and when the environment was unstable, the culture (or climate) appears to shift into survival mode rather than adopt a change-and-adaptation mode. Interestingly, the same study found that a more certain organizational environment resulted in significantly lower pressures for change [8]. This paradox is in line with the idea that in reality, there is a lack of harmony between readiness for change and a felt need for change within a given organization [9].

Regarding *perceived readiness for change*, organizations that might be assessed as "ready" for a change could be organizations that have less of a need for change because they already engage in workplace improvements or modifications on an ongoing basis. Conversely, organizations that are assessed with a poor level of readiness probably have a much stronger need for change. Zhang et al. [9] therefore investigated how to select organizations based on the feasibility and need for change, and found that responses to readiness items rarely predicted change-agent behaviors for managers. For example, when asked about how supportive they felt for a specific participatory intervention, managers expressed great support but failed to follow through with the support promised. Zhang et al. [9] recommend that assessment methods include brief hypothetical case studies to highlight the need for concrete support, rather than just verbal. A few concrete examples of support are; making time to attend meetings, and providing resources and accountability for funds. Barrett and colleagues [10] conducted a case study that elucidates the discrepancy Zhang et al. [9] described between readiness and felt need for change.

By using a classic *stage-of-change framework* by Prochaska and DiClemente [11], Barrett et al. [10] found that this framework was applicable to understanding readiness attitudes and beliefs within and between levels in a manufacturing company that was attempting to make health and safety improvements. The six stages of change are: (1) Pre-contemplation; in which no change is considered and there is no awareness of problems. In order to progress past this stage, individuals must be persuaded that there is an issue to be addressed. (2) Contemplation; change is being considered in the long-term future. In order to progress past this stage, individuals need educational and practical information through training so that motivation to change is reinforced. (3) Preparation; definite plans to change in the short-term future are under way. In this stage strategies to raise awareness of what is needed to implement change, and removal of barriers to change (i.e. workplace performance) are critical. (4) Action; engagement in change behaviors. Full support to achieve and maintain new behaviors is needed for this stage to be fulfilled. (5) Maintenance; proactive behaviors to prevent relapse. At this stage, the organization must be actively monitored for relapse. Lastly (6) Relapse; failure to continue with recent modifications or new behaviors. This stage can occur at any point, so progression back through the cycle towards action and maintenance must also be supported. Barrett and colleagues [10] found that attitudes and beliefs related to change can vary considerably both horizontally amongst individuals in the same level (e.g., production workers) and vertically between job roles (e.g., middle management and senior management). This variation in the respective stages people are in regarding change has important implications for targeting workplace interventions associated with individual's particular needs [10] so that those who are in the pre-contemplative stage, for example, are adequately informed and prepared to join their other coworkers at both the contemplative and actions stages. Bringing alignment to readiness by acknowledging that some employees may be on a different stage than others is a way of both preparing adequately to initiate change and successfully managing the change process itself.

4.2 Sociotechnical Systems

Sociotechnical Systems (STS) also offers a unique approach to seeking alignment at the individual and organizational level. STS theory is based on two assumptions. The first assumption is that effective performance (i.e., productivity, quality, employee well-being, job satisfaction) is a function of the extent to which the people (social part) and tools, technologies and techniques (technical part) are jointly optimized [6]. Joint optimization is the deliberate design of the relationship between the social and technical components so that they both work well together and produce positive outcomes. The second assumption is that the sociotechnical system is open to its environment so that it interacts with external variables by receiving inputs of energy, raw materials, and information, with the STS providing the environment with products or services. Here too are design implications so that the interface between the STS and its environment is effective and not constricting or limiting to the work system [5].

An open systems perspective is a common and ideal way to diagnose an organization, namely, a process that allows understanding as to how an organization is currently functioning, and which also provides the information necessary to design change interventions. Since organizational culture is cited often as a proxy for understanding the context for potential interventions, an open systems approach offers a systematic way for assessing and exploring organization culture. To clarify, diagnosis in this sense does not assume there is a problem with an organization (like a medical diagnosis) and can address areas of potential improvement and development that are collaboratively addressed [5]. In broad terms, a typical diagnosis model considers the interactions between four major facets at every level of an organization: inputs (information and energy), transformations (social and technical components), outputs (finished goods, ideas), and external environment.

The theoretical underpinnings of sociotechnical systems form the basis of an emergent sub-discipline of human factors: macroergonomics and systems ergonomics. Macroergonomics is defined as "a top-down sociotechnical system approach to the design of work systems and application of the overall-work-system design of the human-job, human-machine and human-software interfaces" [12]. Therefore, macroergonomics offers a sociotechnical framework for studying both the macro and micro issues associated with large-scale organizational change. Because macroergonomics is concerned with the optimization of work systems through consideration of relevant social, technical and environmental variables and their interaction, culture change is often the outcome of a macroergonomic intervention. Some examples of macroergonomic methods for changing culture are: (1) mandating behavior change through major policy changes; (2) changing behaviors of organizational leaders, whereby leading by example; (3) hiring new employees whose value systems are aligned with the idealized culture; (4) training existing employees to align attitudes and behaviors; and (5) changing culture through means of a comprehensive work system redesign [12]. In this sense, a macroergonomic approach to culture change identifies the different levels of an organization (e.g., individual, group/managerial, and organizational) as interacting within and between levels of the organization, which is the most realistic perspective.

5 Concluding Remarks

A comprehensive literature search, analysis of findings, and the creation of a conceptual model was the first step in the development of guidelines for assessing organizational readiness. These results have led to the development of an integrated approach to understanding the level of organizational readiness in relation to the employees' capacity to change and providing a roadmap for the development of an appropriate assessment tool. The long-term goal is to provide an organizational diagnostic tool that can be used to assess an organization's level of readiness to initiate, manage, and sustain integrated well-being and safety initiatives of a programmatic nature. A tool like this can be designed to help organizations with low scores in some dimensions improve their readiness for change by identifying specific resource and training needs necessary to support these new integrated workplace initiatives.

References

1. NIOSH (2016) National occupational research agenda (NORA)/national total worker health® agenda (2016–2026): A national agenda to advance total worker health® research, practice, policy, and capacity (DHHS (NIOSH) Publication 2016–114 edn.). Department of Health and Human Services, Centers for Disease Control and Prevention, National Institute for Occupational Safety and Health, Cincinnati. https://www.cdc.gov/niosh/docs/2016-114/pdfs/nationaltwhagenda2016-1144-14-16.pdf
2. Weiner BJ (2009) A theory of organizational readiness for change. Implement Sci 4:67–75
3. Jones RA, Jimmieson NL, Griffiths A (2005) The impact of organizational culture and reshaping capabilities on change implementation success: the mediating role of readiness for change. J Manage Stud 42:361–386
4. Burnett S, Benn J, Pinto A, Parand A, Iskander S, Vincent C (2010) Organisational readiness: exploring the preconditions for success in organisation-wide patient safety improvement programmes. Qual Safety Health Care 19:313–317
5. Cummings T, Worley C (2015) Organization development and change, 10th edn. Cengage Publishing, Mason
6. Waterson P (2015) Sociotechnical design of work systems. In: Wilson JR, Sharples S (eds.) Evaluation of human work. CRC Press
7. Murphy LA, Robertson MM, Carayon P (2014) The next generation of macroergonomics: Integrating safety climate. Accid Anal Prev 68:16–24
8. Lehman WE, Greener JM, Simpson DD (2002) Assessing organizational readiness for change. J Subst Abuse Treat 22:197–209
9. Zhang Y, Flum M, West C, Punnett L (2015) Assessing organizational readiness for a participatory occupational Health/Health promotion intervention in skilled nursing facilities. Health Promot Pract 16(5):724–732
10. Barrett JH, Haslam RA, Lee KG, Ellis MJ (2005) Assessing attitudes and beliefs using the stage of change paradigm—case study of health and safety appraisal within a manufacturing company. Int J Ind Ergon 35:871–887
11. Prochaska JO, DiClemente CC (1982) Transtheoretical therapy: Toward a more integrative model of change. Psychother Theory Res Pract 19(3):276
12. Hendrick HW, Kleiner B (2002) Macroergonomics: Theory, methods, and applications. Lawrence Erlbaum Associates, Mahwah

Similarity of Risk Factors for Musculoskeletal Disorders and Poor Product Quality in Manufacturing

W. Patrick Neumann[1,4(✉)], Ahmet Kolus[2,4], and Richard Wells[3,4]

[1] Ryerson University, Toronto, ON M5B 2K3, Canada
pneumann@ryerson.ca
[2] King Fahd University of Petroleum & Minerals, Dhahran 31261, Saudi Arabia
akolus@kfupm.edu.sa
[3] University of Waterloo, Waterloo, Canada
wells@uwaterloo.ca
[4] Centre of Research Expertise for the Prevention of Musculoskeletal Disorders (CRE-MSD), University of Waterloo, Waterloo, Canada

Abstract. The key Musculoskeletal Disorder Risk Factors (MSDRF) are high forces, awkward postures, vibration and time (expressed as repetition, duty cycle or duration). A recent systematic review of the literature identified 207 human factor variables that were linked to poor product quality in manufacturing operations; what were called Quality Risk Factors (QRF). QRF were identified at the Product design, Process design, Workstation design and Individual levels. The purpose of this paper is to compare the QRF documented to MSDRF. The descriptions of the quality risk factors were extracted verbatim from the papers included in the original review. A rubric was developed to rate the strength of the QRF to MSDRF linkage. The three authors each rated the relationship independently. We then met to discuss the ratings and resolved any differences in rating. The findings demonstrated that product design and workspace design aspects had the strongest relationships between QRF and MSD risk factors. This implies that design that is sensitive to operators' capacities should lead to both improved operational system performance as well as reduced risk of developing MSD. This knowledge may open new avenues for engaging in MSD prevention through design.

Keywords: Musculoskeletal Disorders · Product quality · Manufacturing

1 Introduction

The key Musculoskeletal Disorder Risk Factors (MSDRF) are high forces, awkward postures, vibration, time (expressed as repetition, duty cycle or duration) with secondary risk factors of contact stress, vibration, and cold (NRC 2001). Kolus et al. (2018) recently performed a systematic review of the literature that examined factors associated with poor product quality in manufacturing operations; they termed these Quality Risk Factors (QRF), i.e., factors that increase (or decrease according to how they are expressed) product quality. There were 207 QRF identified which included 54

S. Bagnara et al. (Eds.): IEA 2018, AISC 821, pp. 446–451, 2019.
https://doi.org/10.1007/978-3-319-96080-7_53

Product design, 77 Process design, and 50 Workstation design related variables. The factors associated with product and process design greatly outnumbered individual risk factors for poor product quality. Examples from within the product design category included load, task, visibility and complexity. Examples from within the workspace category included tools and equipment, space and reach environment. Examples from within the process category included instructions and procedures, work organization, time/pace and production system.

The risk factors for poor product quality appeared to overlap those for MSD considerably. A formal process was undertaken to determine if any relationships existed between the quality and MSD domains.

The purpose of this paper is to answer the research question: Are factors related to poor product quality in manufacturing also risk factors for the development of MSD?

2 Methods

A ratings approach, based on the judgment of the authors, was applied. In order to answer the research question we developed a rubric that could be used to systematize the rating process. Four questions were considered in determining the linkage between a quality "risk factor" and an MSD risk factor:

- Biological feasibility: Is there a biologically feasible pathway or theory linking the QRF and MSDRF?
- Complexity of linkage: How many intervening steps are there on the pathway?
- Assumptions: Are many assumptions are necessary to construct the pathway?
- Empirical relationships: Are there empirical studies that directly show relationships between the quality "risk factor" and a specific MSDRF?

Table 1. Rubric for rating the relationships between the quality risk factor and a specific MSDRF

Rating	
8–10	Overall, biologically feasible and likely • Short, biologically feasible pathway from a quality "risk factor" to MSD risk factor(s) • Few assumptions • Empirical studies of the relationship available
5–7	Overall, seem reasonable, biologically feasible and possible • Biologically feasible steps in the pathway from a quality "risk factor" to an MSD risk factor exist • Some assumptions required • Few empirical studies that directly show relationships between the quality "risk factor" and a specific MSD risk factor
2–4	Overall, seems possible • Multi-step, speculative but biologically feasible pathway from a quality "risk factor" to an MSD risk factor can be constructed • Many assumptions required • Little empirical study or theory that show direct relationships between the quality "risk factor" and a specific MSD risk factor

(*continued*)

Table 1. (*continued*)

Rating	
0–1	Overall, no relationship • No known pathway from a quality "risk factor" to an MSD risk factor • No empirical studies or theory that show direct relationships between the quality "risk factor" and a specific MSD risk factor
NA	Not rated • The descriptions in the original papers did not supply enough information to rate the factors

Based upon the four considerations described, the rubric shown in Table 1 was developed for rating the linkage between quality risk factors and MSDRF. The descriptions of the quality risk factors were extracted verbatim from the papers included in the original review by Kolus et al. (2018). The rating was determined by a dialogue approach using Table 1. The three authors each rated the existence of a linkage independently. We then met and discussed any items with differences greater than 2 points in rating, making adjustments to our scores accordingly. In some cases we referred back to original articles to clarify the particular situation. The median score then calculated for each item. Within each category of Product design factors, process design factors, and workspace design factors items were clustered according to the type of design decision under consideration – for example "tools and equipment" related variables – and a median score was calculated for this group.

3 Results

The ratings of the relationships between the quality risk factors and specific MSDRF for the three major classes of QRF, product design, workspace and process are shown in Tables 2, 3, and 4 respectively. There were a total 16 QRF in the Product design category, 19 in the Workspace category and 25 in the process category. Median ratings were generally higher in the product design categories (8/10), intermediate in the workspace (7/10) and lowest in the process categories (5/10).

Table 2. Ratings of the relationships between product design related QRF and specific MSDRF

Product design related quality risk factors	Examples of quality risk factors	Number of quality risk factors	Median rating (0–10)
Load	Force to insert components, tool or material weight	6	10
Task	High precision required	1	7
Visibility	Viewing distance, insufficient contrast	3	8
Complexity	Memory intensive, many choice options	3	5.5

(*continued*)

Table 2. (*continued*)

Product design related quality risk factors	Examples of quality risk factors	Number of quality risk factors	Median rating (0–10)
Other	Meticulous and rigorous work, physically and cognitively difficult tasks	3	6

Table 3. Ratings of the relationships between workspace related QRF and specific MSDRF

Workspace related quality risk factors	Examples of quality risk factors	Number of quality risk factors	Median rating (0–10)
Tools and equipment	Working height, adjustable tables, fixtures, finger work to press components, high tool weight	10	8
Space and reach	Sufficient space, poor material movement, greater accessibility	5	7
Environment	Poor illumination, high noise level, appearance	4	2.5

Table 4. Ratings of the relationships between process related QRF and specific MSDRF

Process -related quality risk factors	Examples quality risk factors	Number of quality risk factors	Median rating (0–10)
Instructions and procedures	Better, clearer work procedures	4	3.5
Training	Technique training	1	5
Work organization	Monotony, no rotation, repetitiveness, lack of movement	3	10
Management	Wage policy, supervision, policy	8	4.25
Time/pace	More rest time, faster work pace	6	6.5
Production system	Lean, batch	2	6.75

4 Discussion

The research question was answered in the affirmative. The findings showed that product design factors had the strongest link to MSDRF with process factors having moderate links and workspace factors being intermediate. The medians do obscure some exceptions. Low ratings were seen for "Environment" which was rated as a 2.5 in the workspace category, much lower than others in the category and the effect of work organization was rated as 10 in the process category, much higher than other process factors.

Strong relationships were identified when characteristics of the work system actually required high forces, awkward postures, vibration, adverse timing (expressed as repetition, duty cycle or duration) or contact stress in body tissues. The relationships could usually be supported by evidence as causal. For example, for the QRF of high hand force for part insertion, there is a clear biologically feasible pathway with few intervening factors between part hand insertion force and MSD of the hand/arm. High hand force has also been directly linked as a strong MSD risk factor for carpal tunnel syndrome (e.g., Harris-Adamson et al. 2014).

Moderate strength relationships were seen when characteristics of the work system created conditions that likely could *in turn* require high forces, awkward postures etc. For the QRF of poor working height, a multi-step, biologically feasible pathway could be constructed with some assumptions: Poor working height (low or high) could possibly lead to awkward postures and extended reaches which could possibly lead to long term static load on neck/shoulders which again could lead to neck myalgia.

Low strength relationships were seen when characteristics of the work system were not so clearly linked to MSD risk factors. For the QRF of clearer work procedures, a multi-step, speculative, but biologically feasible pathway could be constructed using many assumptions.

Strengths of the study were the creation of the rubric and independent ratings of the relationships, basing the quality risk factors on a recent systematic review and the theory driven approach in this and the original study. Limitations included the restriction of quality risk factors to those actually studied and captured in the original systematic review, and the low amount of detail in the original journal articles. This analysis has implications for the potential of ergonomics to be applied in early design stages, including product design, in order to simultaneously improve both quality performance and employee well-being. Such an approach may help ergonomists overcome the perception of ergonomics being exclusively health focused that is limiting its application in companies (Theberge and Neumann 2013).

5 Conclusions

Overall, the study showed many linkages between QRF and MSDRF. The strongest linkages were seen between QRF and MSDRF in product design. Ergonomics can be sold as an important contributor to the design process. This demonstrates that product design and assembly that is sensitive to operators' capacities should lead to both improved operational system performance as well as reduced risk of developing MSD.

Acknowledgements. This project was funded by the AUTO21, Canada's automotive research program. AUTO21 is supported by the Government of Canada through a Networks of Centres of Excellence program.

References

Harris-Adamson C, Eisen EA, Kapellusch J, Garg A, Hegmann KT, Thiese MS, Silverstein B, et al (2014) Biomechanical risk factors for carpal tunnel syndrome: a pooled study of 2474 workers. Occup Environ Med, oemed-2014

Kolus A, Wells R, Neumann P (2018) Production quality and human factors engineering: a systematic review and theoretical framework. Appl Ergon (in press)

National Research Council and Institute of Medicine (2001) Musculoskeletal disorders and the workplace: low back and upper extremities. National Academy Press, Washington, D.C.

Theberge N, Neumann WP (2013) The relative role of safety and productivity in Canadian ergonomists' professional practices. Relations industrielles/Ind Relat 68(3):387–408

A Developmental Framework to Analyze Productive and Constructive Dimensions of Collaborative Activity in Simulation Workshops

Flore Barcellini(✉)

Le Cnam, CRTD, 41 rue Gay-Lussac, 75005 Paris, France
flore.barcellini@lecnam.net

Abstract. This communication proposes a methodological framework to produce knowledge about collaborative activity performed during simulation workshop, on their productive and constructive dimensions. Producing knowledge about these situations may have two type of outcomes: an epistemic one dealing with a better understanding of (1) the productive and constructive potentiality of simulation, in line with the Constructive Ergonomics Approach (Falzon 2014; Barcellini et al. 2014; Béguin 2014; Béguin 2003) and (2) the role of ergonomist in designing and managing simulation workhops. The proposed methodological framework is anchored in more than thirty years of research about collaborative design situations (e.g. Darses et al. 2001). Grounded in this initial framework we propose additional methods to address the constructive dimensions of collaborative design activity (Barcellini et al. 2015): (1) the analysis of roles of participants – in relation to the functions of their verbal interactions - and their various forms all along design interactions, as a revelator of the constructive potentiality of interactions; (2) the analysis of the development of the project in which participants are engaged, thanks to the analysis of intermediary objects produces.

Keywords: Simulation · Collaboration · Constructive activity
Methodology

1 Why Revealing Productive and Constructive Dimensions of Activities Performed During Simulation Workshop?

Research in ergonomics about design articulates two complementary approaches: understanding of collective design work in various technological or organizational situations (e.g. Détienne 2006; Visser 2009; Barcellini et al. 2013); and proposal of ergonomics's model of action, such as project management regarding design of work situation (e.g. Barcellini et al. 2014) or methods (e.g. simulation, Beguin 2003; Daniellou 2007).

This communication is grounded in these two approaches as it proposes to mobilize methodology used to analyze collective design work at stake in simulations workshops set up during ergonomics action.

S. Bagnara et al. (Eds.): IEA 2018, AISC 821, pp. 452–456, 2019.
https://doi.org/10.1007/978-3-319-96080-7_54

Producing knowledge about these situations deals with a better understanding of the productive and constructive potentiality of simulation, in line with the Constructive Ergonomics Approach (Falzon 2014; Barcellini et al. 2014; Béguin 2014; Béguin 2003). This approach advocates for a renewal of ergonomics scientific discipline goals, which may target explicitly design of work resources supporting development of individuals, collective and organization. In this sense, Ergonomics' goals is not anymore "to fit job to the workers" but also the to support a sustainable development of work activity through actions of the design of organizations and resources proposed to workers to act. This proposal is directly in line with Activity-Centered Ergonomics (e.g. Daniellou and Rabardel 2005) which outline that any work activities incorporate a productive – in relation with goals pursued by workers – and a constructive dimensions – acting at work generates individual and collective learning and development for workers (e.g. Samurcay and Rabardel 2004). Thus, in Constructive Ergonomics approach, one finality of ergonomists' action is to foster this developmental process, in particular through specific design actions (e.g. Barcellini et al. 2014). Here, development is taken with a large scope as it concerns individuals, collective and organizations (Falzon 2014).

In this frame, our goal is to propose a methodology to produce knowledge about development potentialities of actions set up by ergonomists, in particular simulation workshops. To do so, we propose to better understand actual collaboration at stake in design situations and its role in learning process. In the following, we briefly review previous research of collaborative design work in order to ground and then present the approach we propose.

2 How to Analyze Developmental Potentialities of Simulation Workshops?

2.1 A Socio-Cognitive Perspective of Collaborative Design Work: Analyzing the Productive Dimension of Collaborative Design Work

Here, we assume that simulation workshops are seen as a collaborative design situations (e.g. Detienne 2006), that are designed and managed by ergonomists. By *collaborative design situations*, we mean situations in which participants are joint to fullfil an common goal, in this case design of a future work situations. They co-elaborated as they interact to build a common representation of the task they have to perform and knowledge regarding their work. These interactions mainly take place during meetings considered as the driving force of design in which participants negotiate among different viewpoints at stake in design process (e.g. Bucciarelli 1988). Lots of research in Ergonomics and Design studies (e.g. Cahour 2002; Détienne 2006; Olson et al. 1992; Luck 2012) focused on these meeting in various contexts (software design, aeronautic, engineering, robotics; architecture).

The proposed methodological framework is anchored in these more than thirty years of research about collaborative design situations and the development of specific methods to analyze it (e.g. the COMET Method Darses et al. 2001). These methods consider mainly verbal interactions as the mean of co-elaboration of knowledge in design and functions of these interactions regarding the design task (e.g. generation and evaluation of design solution, clarification, synchronization…). This framework was mainly used to understand the performance of the design, i.e. its productive dimension, with regards to the design of an artefact. It helps in revealing three main subsets of collaborative activities:

- *Generation-evaluation activities* related to the process of solving and evaluating various aspects of design problems. These activities are based on argumentative process by which participants negotiate among various constraints and viewpoints inherent to collaborative design situation (Wolff et al. 2005; Détienne et al. 2005).
- *Clarification activities*, concerning the construction of common references, or common ground, within a group of participants. Here too, argumentative process is of prime importance, as it supports the elaboration of a common ground between participants (Clark and Brennan 1991; Baker 1996).
- *Group management activities* related project management activities that concern the coordination of people and resources - e.g., the allocation and planning of tasks – are of this kind, or meeting management activities – e.g., the ordering and post-poning of topics of discussion – are another example of this kind of activity.

2.2 Contradictions and Argumentation as a Driving Force for Learning and Development in Collaborative Design Situations

Some research specifically targeting understanding of learning process in design situations (e.g. Beguin 2003) reveals that learning among designers may originate from contradictions experienced by participants, due in particular to the various viewpoints present in design process. An issue to analyse the learning process at stake is thus to better understand how participants cope with these contradictions, in particular thanks to argumentive process. Actual collaboration – co-elaboration and thus learning – is intimately linked to this argumentative process. The field of cooperative learning (e.g. Dillenbourg et al. 1995) reveals in particular that quality of interactions – in particular argumentation and management of disagreements - among participants is the driven force of actual collaboration and learning, as it helps in supporting understanding of participants. Moreover, these research stress the importance of distribution of activities among participants: actual collaboration is viewed as an interactive situations in which alternance of activities is symmetric and smooth, i.e. roles are distributed among participants.

2.3 The Actual Role Analysis in Design as a Proposal to Understand Actual Collaboration and Learning Among Participants

Following proposals of the cooperative learning field, analyzing learning in collaborative design situations implies to characterize quality of interactions among participants through the distribution of roles among them. To do so, we develop a method called "Actual Role Analysis in Design" (ARAD) approach proposed by Baker et al. (2009); Détienne et al. (2012b); Barcellini et al. (2013) to study collaborative design. The ARAD approach has been designed to capture actual collaborative design activities, i.e. that are no predefined but effectively performed by participants, and that emerge from actual interactions between participants. It identifies roles that correspond to distinctive and regular individual behaviors emerging in the interaction. Four types of role are considered to embrace different facets of participation. They are characterized on the basis of the structure of the interactions during design meetings (interactive role), and according to the orientation of the interactions among the participants engaged in discussions in relation to collaborative design activities (Barcellini et al. 2008, 2013; Détienne et al. 2016).

3 Conclusion

The approach has been used in various design contexts: distributed and asynchronous design situations such as in Open Source Software Design (Barcellini 2008) or participatory design in agro-ecological context (Barcellini et al. 2016).

In this communication, we will propose an instantiation of this framework to analyse collaborative interactions between ergonomists and others participants in simulation workshops, in relation with the developmental potentiality of simulation we discussed in Barcellini et al. (2014) or Béguin (2003). We will mainly propose to understand this potentiality not only with regards to the development with one or another artifacts under design (space, organization, process) but with regards to the development of a future activity performed thanks to these artifacts. We will be mainly interested in the type of knowledge that co-elaborated, and the role of ergonomist and the setting she/he propose to suppport this co-elaboration. This epistemic outcome may have a pragmatic one dealing with teaching of ergonomics (e.g. Barcellini and Van Belleghem 2014), in particular the role of ergonomist in designing and managing simulation workshops.

References

Baker M (1996) Argumentation et co-construction de connaissances. Interaction et cognitions 1 (2–3):157–191

Barcellini F, Détienne F, Burkardt JM (2013) A situated approach of roles and participation in Open Source Software Communities. Hum Comput Interact. https://doi.org/10.1080/07370024.2013.812409

Barcellini F, Van Belleghem L, Daniellou F (2014) Design projects as opportunities for the development of activities. In: Falzon P (ed) Constructive ergonomics. Taylor and Francis, New York, pp 150–163

Barcellini F, Van Belleghem L (2014) Organizational simulation: issues for ergonomics and for teaching of ergonomics action. In: Broberg O, Fallentin, N, Hasle P, Jensen PL, Kabel A, Larsen ME, Weller T (eds) Proceedings of human factors in organizational design and management–XI and nordic ergonomics society annual conference–46

Barcellini F, Prost L, Cerf M (2015) Designing a tool to assess agricultural sustainability: designing the concept of sustainability? Appl Ergon 50:31–40

Béguin P (2003) Design as a mutual learning process between user and designers. Interact Comput 15(5):709–730

Béguin P (2014) Learning during design through simulation. In: Broberg O, Fallentin N, Hasle P, Jensen PL, Kabel A, Larsen ME, Weller T (eds) Proceedings of human factors in organizational design and management–XI and nordic ergonomics society annual conference–46, pp 867–872

Bucciarelli L (1988) An ethnographic perspective on engineering design. Des Stud 9:159–168

Cahour B (2002) Décalages socio-cognitifs en réunions de conception participative. Le Travail Humain 65(4):315–337

Clark HH, Brennan SE (1991) Grounding in communication. In: Resnick L, Levine JM, Teasley SD (eds) Perspectives on socially shared cognition. APA, Washington, pp 127–149

Daniellou F, Rabardel P (2005) Activity-oriented approaches to ergonomics: some traditions and communities. Theor Issues Ergon Sci 6(5):353–357

Daniellou F (2007) Simulating future work activity is not only a way of improving workstation design. @ctivités 4(2):84–90. http://www.activites.org/v4n2/v4n2.pdf

Darses F, Détienne F, Falzon P, Visser W (2001) COMET: A method for analysing collective design processes. Rapport de recherche INRIA, projet Eiffel, September

Détienne F (2006) Collaborative design: managing task interdependencies and multiple perspectives. Interact Comput 18(1):1–20

Détienne F, Martin G, Lavigne E (2005) View points in co-design: a field study in concurrent engineering. Des Stud 26:215–241

Détienne F, Baker M, Fréard D, Barcellini F, Denis A, Quignard M (2016) The Descent of Pluto: interactive dynamics, specialisation and reciprocity of roles in a Wikipedia debate. Int J Hum Comput Stud

Dillenbourg P, Baker M, Blaye A, O'Malley C (1995) The evolution of research on collaborative learning. In: Spada E, Reiman P (eds) Learning in humans and machine: Towards an interdisciplinary learning science. Elsevier, Oxford, pp 189–211

Falzon P (2014) Constructive ergonomics. Taylor & Francis, Boca Raton

Luck R (2012) 'Doing designing': on the practical analysis of design in practice. Des Stud 33:521–529

Olson GM, Olson JS, Carter MR, Storrosten M (1992) Small group design meetings: an analysis of collaboration. Hum Comput Interact 7:347–374

Samurcay R, Rabardel P (2004) Modèles pour l'analyse de l'activité et des compétences, propositions. In: Samurcay R, Pastré P (Coords.) Recherches en didactique professionnelle. Octarès, Toulouse

Visser W (2009) Design: one, but in different forms. Des Stud 30(3):187–223

Wolff M, Burkhardt JM, De la Garza C (2005) Analyse exploratoire de points de vue: une contribution pour outiller les processus de conception. Le travail humain 68(3):253–286

Simulation in Diagnosing and Redesigning Knowledge Transfer Systems in the Offshore Oil Industry

Carolina Conceição[✉] and Ole Broberg

DTU Management Engineering, Technical University of Denmark, Produktionstorvet 424, 2800 Kongens Lyngby, Denmark
{casou, obro}@dtu.dk

Abstract. How can simulations help diagnosing and redesigning knowledge transfer (KT) systems? This paper addresses this research question and discusses how simulations carried out over a set of workshops helped diagnosing and redesigning KT from operations to engineering design in a study involving an offshore oil company. During three simulation activities, end-users participated in a multi-voiced diagnosis of KT challenges and in a cross-organizational development of new solutions for KT systems. The participatory and interactive approach on the workshops brought participants of different professional worlds together to discuss and negotiate the possible solutions for improving the existing KT from offshore rigs' operation to onshore engineering design of new rigs. Participatory simulations helped (1) validating a conceptual model for KT from operations to engineering design and (2) developing a set of requirements for redesigning these KT systems. The outcome was a KT model based on the main challenges and presenting practical solutions to face them.

Keywords: Participatory simulation · Knowledge transfer · Ergonomics

1 Introduction

This study falls within the field of participatory ergonomics, where end-users are involved in planning and designing their own work (on the physical and organizational levels). Simulation, or participatory simulation, allows access to different stakeholders' knowledge in order to identify and evaluate key challenges for future work systems. Different studies have analyzed and discussed participatory simulation events with the design (or redesign) of work systems [1–5], mainly on the physical and organizational levels.

In this study, we look into participatory simulation on the process of redesigning knowledge transfer (KT) systems, but the basic components of the approach are the same as described by Daniellou et al. [4] and Daniellou [5]: the choice of participants and the participatory approach (different representatives), and the choice of simulation media and scenarios. We use participatory events to gather and benefit from the different stakeholders' knowledge.

S. Bagnara et al. (Eds.): IEA 2018, AISC 821, pp. 457–465, 2019.
https://doi.org/10.1007/978-3-319-96080-7_55

Ergonomists focus on the need of involving potential end-users of the work system being designed [6] and the participatory approach takes on the involvement of representatives of different levels in the company. The goal is to set the stage for end-users to foreseen themselves in a situation that does not yet exist [7]. Simulation media, such as models, diagrams, mock-ups and other boundary objects, supports this process [8], and scenarios help guiding the participants to revive their current work activities in the new work system.

In this paper we present a study involving the KT from rigs' operations to engineering design of new rigs. We first present the different simulation events and the media used to facilitate the participation of end-users of KT system, both from the offshore rig side that would be registering knowledge in the system, as well as from the onshore engineering design side that would be retrieving this knowledge during the design process. After we present the outcome from these events and how it helped us mapping KT transfer challenges and improving a KT model with requirements for new KT systems.

2 Methods

The methodological approach was based on a case study [9]. The goal was to seek characteristics that may be generalizable to scientific knowledge production, rather than to individualize the case or to generate statistical data. The case study was carried out over a two-year period. We investigated the KT between operations on offshore oil rigs and onshore engineering design as our unit of analysis in the case company.

For this study, three researchers worked in collaboration with a large company operating in the drilling sector. We focused our study in two specific departments of the company: the technical organization and the operational organization. Both departments are mainly situated in the company's headquarters. However, the operational organization included the rigs, and as such its crews working offshore. The technical organization included the engineering teams working with the design and optimization of rigs.

2.1 Data Collection

We carried out two workshops of two hours each where we used three simulation activities to help identifying the main KT challenges and developing requirements for new KT systems. The workshops happened three months apart from each other, towards the end of the project. At that point of time, a number of interviews, a survey and observations onboard a rig had already been conducted to gather knowledge of the existing KT systems at the case company. Both workshops were video recorded and had one researcher as the main facilitator and the other two supporting the workshop facilitation.

The first workshop had five participants working within onshore engineering design and the second had seven participants from both offshore rig operations and onshore engineering design. Three participants were in both workshops and are marked in italic in Table 1, which shows the full list of participants. The participants from the commercial organization, which was not the focus of this study, were part of the workshop as they were our facilitators in the case company and directly involved with the possibility of improving the knowledge transfer from rig operations to engineering design.

Table 1. List of participants from Workshops 1 and 2

Workshop	Participants	Organizational unit
1	*Project manager*	Technical
	Project manager	
	Regulations compliance responsible	
	Offshore working participating in projects	Operational
	Innovation manager	Commercial
2	*Project manager*	Technical
	Project engineer	
	Project engineer	
	Offshore working participating in projects	Operational
	Offshore section leader	
	Offshore section leader	
	Innovation manager	Commercial
	Innovation manager	

The first workshop had two activities using game boards and game pieces provided by the researchers: the first one to draw a knowledge landscape and the second to draw a design process grid. Both activities started by an individual task where participants had to write down their own ideas before a more interactive part of sharing these ideas. After, there was a brief simulation giving the chance of discussing and building up on the initial ideas presented. To set the stage for the participatory simulation activities, the researchers asked the participants to consider what the main challenge was in transferring knowledge from operating rigs to the design of new rigs.

The knowledge landscape activity aimed at mapping KT channels in the company. The simulation activity was planned based on the existing channels at the company, but participants were asked to build the landscape based on their ideas to improve KT. Participants were asked to think individually on what their suggestions were to improve the KT from operating rigs to engineering design of new rigs. After they shared their ideas one by one, while using the game pieces to build the knowledge landscape based on how to implement their suggestions. To finalize the activity, they were asked to simulate the flow of knowledge coming from the rigs in order to validate the knowledge landscape and as such built on each-others ideas to finish building the landscape. If issues were identified, they were asked to focus on what could be added to support the knowledge transfer management. Figure 1 represents the overall idea behind this activity: the questions on the top were meant to trigger the discussions and based the preparation of the landscape activity, and the list on the bottom represents the game pieces that were made available for the participants.

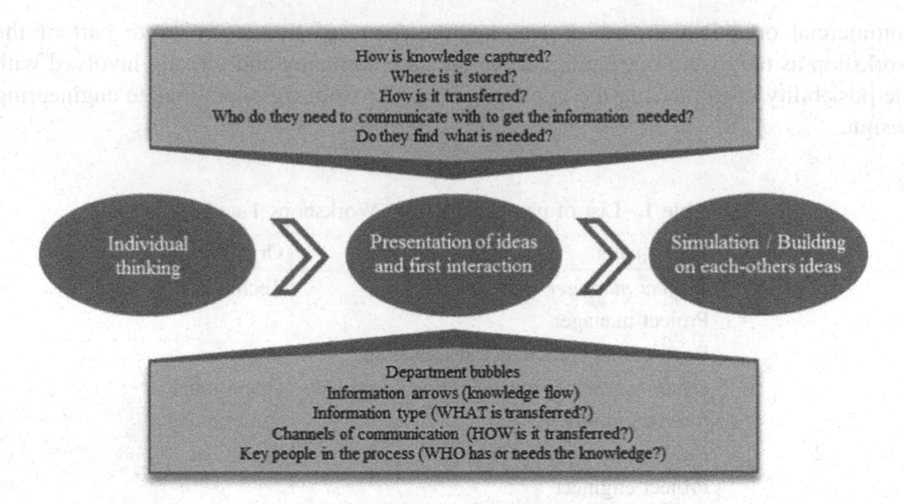

Fig. 1. First activity workshop 1

The design process grid activity aimed at mapping the different knowledge needed over the design process. Participants were asked to think individually on (1) which were the design activities needed and when; (2) who was part of which design activity and when; (3) who needed which kind of knowledge in design activities and when. After they presented their ideas using the game pieces to build the design process grid in light of the knowledge from operations needed to complete design and the knowledge landscape built in the previous activity. To finalize, they were asked to consider an ongoing project they were all familiar with to validate the grid and be able to built on each-others ideas to finish building the grid and structuring the knowledge landscape on it. Figure 2 represents the idea behind this activity, similar to Fig. 1.

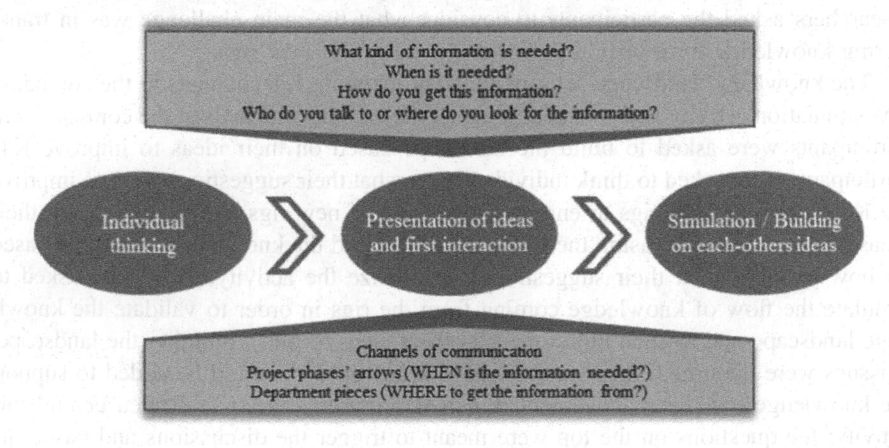

Fig. 2. Second activity workshop 1

The second workshop had one activity that consisted in testing a suggested KT system and unfolding the requirements for this KT system by simulating the flow of knowledge with real-case scenarios. We used a game board with a conceptual model representing the KT system and game cards the participants used to point out challenges and suggestions for improvements. The cards included the actors involved in the KT, the objects required in an information and communications technology (ICT) system, the functionalities desired in the KT system and blank post-its for comments on the system. We initially filled out part of the game board with our initial ideas of how to structure the KT system. After, the first part of the activity consisted of asking the participants to think individually and after present two issues on the conceptual model considering their own role in the organization (and on the knowledge flow). A joint discussion followed based on the possible changes to be made to the system.

The case company had already in use on the rigs an artifact that consisted of a paper leaflet where workers could register health and safety issues observed onboard. It was not, however, used with the purpose of informing engineering design of those issues. We had tried onboard a new version of this artifact, capture cards, where the main purpose was to transfer knowledge to engineering design and we took the examples collected onboard to the simulation activity. The simulation itself started by taking a capture card filled out by workers onboard a rig and participants had to make it registered and go through the overall system. New issues were identified and participants suggested new changes. After, there was a second round of simulation through the "changed" system with other capture cards, where participants could identify final issues that could be resolved. Figure 3 shows the overall structure of the activity in workshop 2.

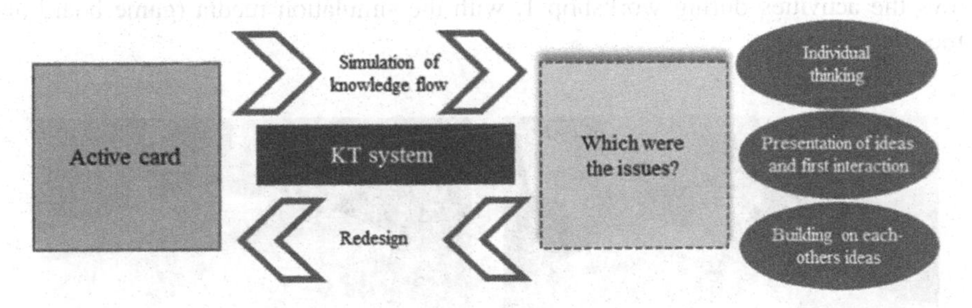

Fig. 3. Activity workshop 2

2.2 Data Analysis

After each workshop, the video recordings were analyzed by two researchers and the selected and systematized transcripts were read by the third researcher. For the first workshop, the recordings were seen through a first time to take overall notes of the main points discussed and to complement the information registered in the game boards after the two activities. This allowed us to complement and better understand the knowledge landscape and how it fitted into the design process timing. After, the recordings were analyzed a second time in order to transcribe most of the talking in order to systematize the transcriptions into challenges for KT, suggestions for improvement and needs within KT mentioned by the participants.

For the second workshop, the recordings were seen through a first time again to take notes on the main comments complementing the game board and the changes and improvements suggested to the KT system proposed. On a second analysis of the recordings, the two researchers doing the analysis identified quotes from the different participants that corroborate the whole purpose of the KT system from operations to engineering design and the suggestions for this new system.

3 Results

The first workshop was targeted in (1) getting a better understanding of the challenges involved in the KT process and how these challenges were timed with the design process of new rigs and (2) validating our initial assumptions to build a conceptual KT model. In the first activity, we had a collaborative mapping of the knowledge landscape in order to identify the key challenges involved. In the second activity, we had a joint discussion on how the knowledge landscape and the KT needs fit into the design process.

Overall, in this first workshop, we had a multi-voiced identification of knowledge needs when the goal is to design new rigs optimizing rig crew work practices, safety and well-being onboard, productivity and customer value. The participants found the first activity more abstract and initially more difficult to relate to and simulate the knowledge flow. The second activity, on the other hand, the participants found it was easier and quicker to relate to and to see themselves in the suggested scenario. Figure 4 shows the activities during workshop 1, with the simulation media (game board and game pieces) used.

Fig. 4. Activities from workshop 1

From the analysis and results of the first workshop, we were able to also validate the initial thoughts on a conceptual KT model that would be the basis for the improved KT system proposed. The knowledge landscape mapping helped validating the main steps involved in the KT process and the design process grid helped validating the different phases of the design project where different knowledge was seek by the engineering design teams. The systematized quotes showed us that the challenges could be classified into four KT steps, part of the conceptual model.

In our pragmatic model, KT implies the knowledge to be (1) *captured* on the operating units, (2) *transformed* into an engineering design context, (3) *transferred* to the appropriate project team members, and finally (4) *applied* throughout the design process of new installations. It is a four-step process involving challenges going from not having specific performance indicators encouraging rig workers to focus on capturing knowledge targeted to design to not having this knowledge available to be applied at the right time in the projects, making it at times impossible to implement in terms of design specifications. Challenges also pass through the large amount of knowledge registered in the systems without standards to categorize and store this knowledge, to the difficult access and retrieval of knowledge in the ICT systems.

This was the basis for developing further the model and the system suggestions for workshop 2, which was targeted in (1) unfolding the four-step model towards the redesign of KT systems and (2) creating new KT solutions. We had an interactive and collective development of means and structures that could enhance the KT from operations to engineering design. The participants liked the participative setting where they could build on each-others ideas and get closer to professional worlds other than theirs. Figure 5 shows the outcome of workshop 2, with the conceptual model filled out with systems features and requirements.

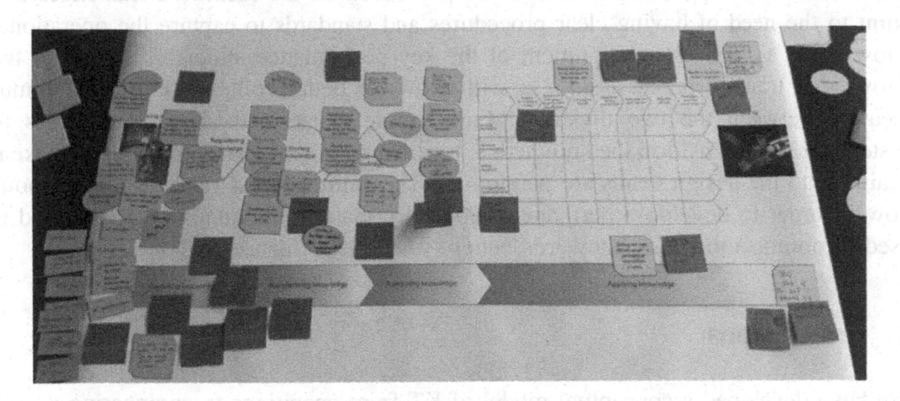

Fig. 5. Outcome of workshop 2

The main result was a supplemented and validated set of requirements for improving and redesigning KT. This set of requirements was structured together with the conceptual model targeting a more generalizable model to be applied not only by the case company, but for other companies facing similar challenges, such as companies with geographically dispersed organizations. Participatory events with simulations were paramount to the development of the final KT model. Simulating the capture cards knowledge flow through the system revealed weak points and made emerge new possible solutions. Figure 6 shows the final model developed.

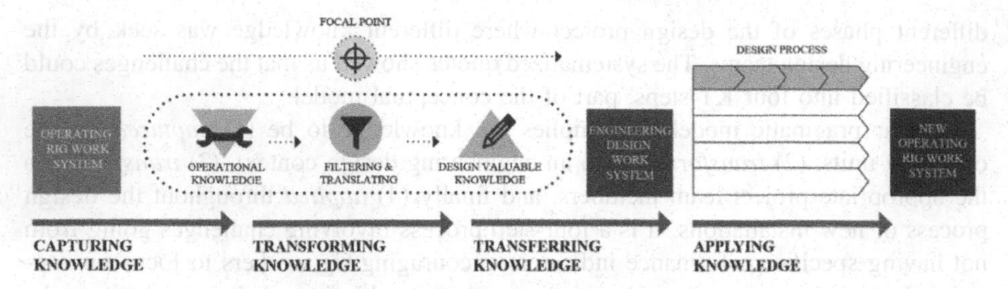

Fig. 6. KT model developed

The final KT model follows the four steps described previously – capturing, transforming, transferring, and applying – starting from operating rigs, passing through the engineering design team and the design process, to arrive in new rigs. One of the key elements, discussed repeatedly during the workshops is: the knowledge as it is registered in ICT systems from the rigs is not immediately formatted and suitable for use by the engineering design teams. There needs to be a filtering extracting from the large amount of knowledge and "translating" this knowledge into a more applicable knowledge in design. During the simulation, it was clear that this could not be done in a fully automated way, emerging the need of a focal point that would be responsible for this transition.

Overall, the requirements were developed based on the identified challenges and point to the need of having clear procedures and standards to capture the operational knowledge, as well as an alignment of the key performance indicators related to the knowledge transfer process, since it will allow for better collaboration and communication between the two divisions. Furthermore, clear methods and resources to systematize and transform the knowledge, together with appropriate methods to make it available to the design teams are paramount. The entire process requires a continuous flow in order to develop a permanent repository that is continuously updated and is used to optimize the design towards better system performance.

4 Conclusions

We have developed a conceptual model of KT from operations to engineering design and we used simulation activities to validate and unfold this model into diagnosing and redesigning KT systems. The workshops supplemented the empirical data obtained throughout the project and allowed us to test and validate our initial insights and ideas. As a tangible result, a conceptual model and a set of requirements for KT systems firstly targeted to the case company and, secondly, generalized in order to be available for other companies facing KT challenges. The KT model was developed pragmatically based on the literature and tested with the help of simulations using a single case company in the offshore oil sector; more studies are needed to consolidate it. Such model, however, is of importance for companies with complex design processes and dispersed design teams, like in the offshore oil sector.

References

1. Van Belleghem L (2017) What are the design requirements for an organisational simulation support? In: Proceedings of the 48th annual conference of the association of Canadian ergonomists & 12th international symposium on human factors in organisational design and management, pp 400–406, Banff
2. Andersen S, Broberg O (2016) Participatory ergonomics simulation of hospital work systems: the influence of simulation media on simulation outcome. Appl Ergon 51:331–342
3. Barcellini F, Van Belleghem L, Daniellou F (2014) Design projects as opportunities for the development of activities. In: Falzon P (ed) Constructive Ergonomics. Taylor & Francis, Boca Raton, pp 187–205
4. Daniellou F, Le Gal S, Promé M (2014) Organisational simulation: anticipating the ability of an organization to cope with daily operations and incidents. In: Broberg O, Fallentin N, Hasle P, Jensen P, Kabel A, Larsen M, Weller T (eds) 11th International Symposium on Human Factors in Organisational Design and Management & 46th Annual Nordic Ergonomics Society Conference. IEA, Copenhagen, pp 781–785
5. Daniellou F (2007) Simulating future work activity is not only a way of improving workstation design. @ctivités 4(2):84–90 (2007)
6. Barcellini F, Prost L, Cerf M (2015) Designers' and users' roles in participatory design: What is actually co-designed by participants? Appl Ergon 50:31–40
7. Garrigou A, Daniellou F, Carballeda G, Ruaud S (1995) Activity analysis in participatory design and analysis of participatory design activity. Int J Ind Ergon 15(5):311–327
8. Broberg O, Andersen V, Seim R (2011) Participatory ergonomics in design processes: the role of boundary objects. Appl Ergon 42:464–472
9. Yin R (2009) Case study research: design and methods, 4th edn. SAGE Publications, London

Exploring the Relationships Among Safety Climate, Job Satisfaction, Organizational Commitment and Healthcare Performance

Sabina Nuti(✉), Milena Vainieri, Giorgio Giacomelli, and Nicola Bellè

Health and Management Laboratory, Sant'Anna School of Advanced Study, Piazza Martiri della Liberta, 33 Pisa, 56127 Pisa, Italy
sabina.nuti@santannapisa.it

Abstract. Using data from large-scale survey of some 50,000 healthcare professionals in Italy, we explore the relationships among safety perceptions, self-reported levels of job satisfaction and organizational commitment and objective performance measures. This work aims at contributing a nascent stream of research that investigates how safety climate predicts variables beyond safety outcomes [1, 2], such as self-reported levels of job satisfaction and organizational commitment. Whereas most previous studies in this area exclusively rely on personnel data, we also analyse how safety perceptions, organizational commitment and job satisfaction jointly predict healthcare performance as measured through objective multidimensional indicators [3].

In order to test our hypotheses, we used data collected from an organizational climate survey of the employees of 68 public health authorities from eight Regional Health Systems (RHSs). This survey is an individual-based questionnaire, which contains measures regarding safety climate, job satisfaction, organizational commitment as well as other self-reported attitudes and behaviours. The survey is administered via computer assisted web interviewing (CAWI) on a census basis [4]. The survey has been conducted in a network of RHSs that adopt and fund, on a voluntary basis, a common performance management system (Inter-Regional Performance Evaluation System, IRPES), aimed at collecting the performance data of health authorities for benchmarking [5]. Our findings show that safety perceptions significantly predict job satisfaction, organizational commitment as well as objective measures of healthcare performance. We use structural equation modelling to illuminate multivariate associations among these constructs.

Keywords: Performance · Safety climate · Organizational commitment Job satisfaction

1 Introduction

Safety climate, i.e. the degree to which employees perceive that safety is prioritized in their organization [6], is considered a key variable in promoting safe work environments.

S. Bagnara et al. (Eds.): IEA 2018, AISC 821, pp. 466–472, 2019.
https://doi.org/10.1007/978-3-319-96080-7_56

Whereas the research exploring the effects of safety climate factors on health- and injury-related outcomes has boomed during the last decades, relatively little information exists on how these may affect employee outcomes beyond physical and psychological safety. However, some scholars have recently run counter these trend [1, 7] and initiated a research branch devoted to investigate the role of these factors in predicting employee engagement and organizational commitment. This paper tries to advance this stream of research by exploring the role of safety climate factors in predicting both individual and organizational outcomes. Namely, this study makes the following contributions. Firstly, it explores how safety climate factors might predict employee job satisfaction. Secondly, it explores how safety climate factors might predict employee organizational commitment. Lastly, it assesses how safety climate factors might contribute to predict organizational performance in healthcare, thus expanding the analysis upon safety climate to the domain of objective outcomes. The paper is structured as follows. Section 2 gives some preliminary considerations inferred from the literature and presents the research hypotheses. Section 3 reports the methodological issues. The results of the analysis are summarized in Sect. 4, while Sect. 5 discusses the results.

2 Theoretical Background

Organizational climate is generally defined as the set of practices and policies upon which the members of an organization agree, as well as the shared values characterizing the organization. In other words, it pertains to the 'shared perceptions of the way things are around here' [8]. In line with a quest for climate measures to be specific to the predicted outcome [9], organizational climate scholars have developed constructs aimed at measuring specific dimensions, such as safety climate [6]. Safety climate refers to a climate for physical health and safety, particularly employees' perceptions of management's commitment with regards to safety policy, procedures and practice [10].

Right until very recent times, scholars have bound the exploration of safety climate to its capacity to predict safety performance [11–13], such as safety compliance and safety participation [14] by means of safety knowledge and motivation [15].

More recently, some scholars [7, 16] have defined the concept of psychosocial safety climate to describe the set of policies, practices and procedures for the protection of workers' psychological health and safety, thus focusing on the organizational factors aimed at protecting psychological health safety. Hence, the discourse on safety climate has been extended beyond safety performance: the introduction of psychosocial measures as an organizational resource influenced by senior management [7], indeed, has pushed scholars to investigate the role of management in enhancing individual safety perceptions and, by this mean, in fostering outcomes such as work engagement. This trend has been reinforced by Huang et al. (2016), who have shown that workplace safety have an impact not only on traditional safety outcomes, but also on outcomes such as job satisfaction, engagement and turnover. The present study aims at contributing to this research strand by analysing the relationship between safety climate and individual outcomes such as job satisfaction and organizational commitment in a public service environment where professionals cope with intense job demands set by end users and face high risk of burnout [17, 18], such as healthcare.

H1. Safety climate (SC) perceptions are positively associated with job satisfaction perceptions.

H2. Safety climate (SC) perceptions are positively associated with organizational commitment perceptions.

Furthermore, when considering dimensions beyond safety outcomes, scholars tend to concentrate either on variables measured via self-reported data (e.g. job satisfaction) or on objective measures of individual outcomes (e.g. turnover). To the authors' knowledge, no previous studies have explored the relationship between SC and organizational outcomes measured via objective data. The present contribution aims to fill this gap, by analysing the direct and indirect effect of SC on organizational performance.

H3. Safety climate (SC) perceptions are positively associated with organizational performance.

3 Methods

To test the hypotheses, we use a data set from 68 health authorities from 8 Regional Health Systems (RHS) in Italy. Safety climate (SC) is measured using data from the routine organizational climate survey administered to employees working in all the health authorities via computer-assisted web interviewing on a census base [4], and represent the average assessment that employees give to the items of the questionnaire on a 5-point Likert scale. When measuring SC, we consider items measuring both physical and psychological safety climate. The individual responses are then transformed into a 100-point scale by following the methodology already applied in other studies [19], and average responses grouped by health authority are used in the analysis. We replicate the same process to measure job satisfaction and organizational commitment, which are both gathered from the same organizational climate survey as single-item variable.

The variable measuring organizational performance is gathered from the IRPES that publicly discloses multidimensional healthcare performance indicators [5]. The IRPES has been active since 2008, and it includes about 60 composite indices and about 200 simple indicators, which measure the performance of each healthcare authority considering multiple dimensions (population health, regional strategy compliance, quality of care, patient satisfaction, staff satisfaction, financial efficiency). About 100 out of 200 indicators are assessed in benchmarking considering international or national/local standards by using five coloured evaluation bands (where dark green is the best performance and red is the worst). As a result, for each evaluation measure, five different performance levels define the performance of each health organization in each category, from worst to best on a scale from 0 to 5. Table 1 reports the composition of the variables.

Table 1. Variables composition

Variable	Items
Safety climate (SC)	*'The technical facilities of my unit are adequate.'*
	'The work environment of my unit is adequate with regard to cleanliness, spaces, vent, maintenance, furniture.'
	'My work environment is safe (e.g. electrical system, fire-resistant system).'
	'I have received information and training about work-related risks and their prevention.'
	'I know about professional harassment at my workplace, such as professional downgrading, reduction of decisional autonomy, excessive control.' (reverse coded)
	'I know about harassment at my workplace, such as physical or verbal abuse, that create a negative working climate.' (reverse coded)
	'I feel distress or discomfort in relation to my daily work activity.' (reverse coded)
Job satisfaction	*'I feel satisfied with my job.'*
Organizational commitment	*'I feel proud being a member of this organization.'*
Performance index	*Average value of 60 composite indexes evaluation level (0–5)*

To test our hypotheses, we fitted the structure equation model (SEM) illustrated in Fig. 1. The model includes four items (i.e. Phys. SC_1 through Phys. SC_4) aimed at eliciting respondents' perception with regards to physical features of the work environment that have an impact on SC and three items (i.e. Psyc. SC_1 through Psyc. SC_3) related to psychological dimensions. The seven SC items were used to extract a SC index. The model then estimates the associations between the SC variable and three items measuring job satisfaction, organizational commitment and performance respectively.

4 Results

Model estimates reported in Table 1 seem to suggest that, holding the other variables constant, the SC variable is positively associated with the three other variables included in the SEM. More precisely, when the SC variable increases by one standard deviation, job satisfaction goes up by 0.27 standard deviations ($p = 0.035$), organizational commitment by 0.54 standard deviations ($p < 0.0005$) and the performance index by 0.41 standard deviations ($p = 0.002$).

However, the results of goodness of fit analysis turn out unsatisfactory, indicating a poor fit of our initial theoretical model with the data. Therefore, we supplemented our SEM estimates with OLS analysis using an average of SC items instead of the variable calculated by the measurement component of the SEM. Table 2 reports standardized coefficients from the OLS regression (Table 3).

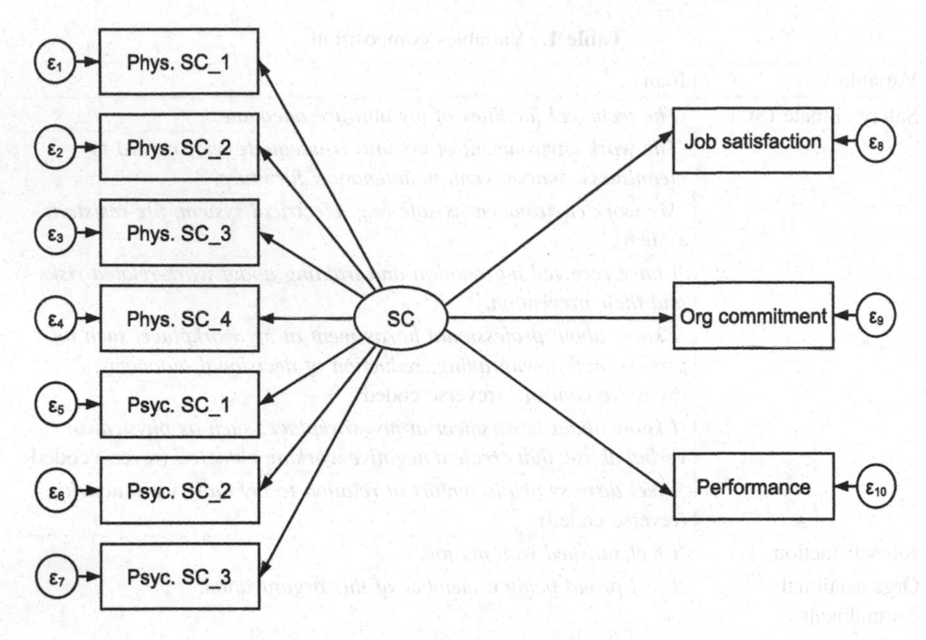

Fig. 1. The structure equation model (SEM).

Table 2. Estimated standardized coefficients from the SEM with maximum likelihood

	Beta std	SE	z	p	[CI 95%]	
Safety climate → Job satisfaction	0.27	0.13	2.11	0.035	0.02	0.52
Constant	13.38	1.21	11.09	0.000	11.02	15.75
Safety climate → Organizational commitment	0.54	0.10	5.47	0.000		
Constant	9.20	0.74	12.43	0.000	0.35	0.74
Safety climate → Performance index	0.41	0.13	3.06	0.002	7.75	10.65
Constant	5.63	0.51	11.10	0.000	0.15	0.66

Table 3. Estimated standardized coefficients from the OLS regression

	Beta std	SE	z	P	[CI 95%]	
Safety climate → Job satisfaction	0.32	0.10	3.21	0.001	0.12	0.51
Constant	10.13	1.79	5.66	0.000	6.62	13.64
Safety climate → Organizational commitment	0.23	0.09	2.62	0.009	0.06	0.41
Constant	6.80	1.25	5.44	0.000	4.35	9.25
Safety climate → Performance index	0.25	0.11	2.18	0.029	0.03	0.47
Constant	3.07	1.42	2.15	0.031	0.28	5.86

Despite differences in size and p values of this model with the respect of the previous one, the pattern of results remains the same and provides support to all three hypotheses.

5 Discussion

The study provides with novel empirical evidence on the associations between SC perceptions and job satisfaction, organizational commitment and organizational performance in the context of healthcare organizations. A first contribution lies in bringing together physical and psychological features of the work environment that may have an impact on safety climate perceptions. Investigations on those two constructs tend to be separated in previous scholarship about safety climate. Moreover, the results of the study show an association of safety climate with objective measures of organizational performance: this contribution is in line with recent work that aims at stretching beyond exclusive reliance on either self-reported attitudes or measures of safety outcomes. The results of this study should be interpreted in light of several limitations that point towards directions for future improvements. For instance, further work is needed that to improve our estimation strategies, as demonstrated by the poor fit of the structural equation model illustrated in Fig. 1. Moreover, a refinement of our estimation models will require taking into account the hierarchical nature of the data with observation nested within Regions.

References

1. Huang YH, Lee J, McFadden AC, Murphy LA, Robertson MM, Cheung JH, Zohar D (2016) Beyond safety outcomes: an investigation of the impact of safety climate on job satisfaction, employee engagement and turnover using social exchange theory as the theoretical framework. Appl Ergon 55:248–257
2. Smith TD (2017) An assessment of safety climate, job satisfaction and turnover intention relationships using a national sample of workers from the USA. Int J Occup Saf Ergon 3548:1–8
3. Nuti S, Vainieri M. Strategies and tools to manage variation in regional governance systems. In: Sobolev B (ed) Handbook of health services research. Springer
4. Pizzini M, Furlan S (2012) L'esercizio delle competenze manageriali e il clima interno. Il Caso del Servizio Sanitario della Toscana. Psicol Soc 18(3):429–446
5. Nuti S, Vola F, Bonini A, Vainieri M (2015) Making governance work in the health care sector: evidence from a 'natural experiment' in Italy. Heal Econ Policy Law 11(1):17–38
6. Zohar D (2010) Thirty years of safety climate research: reflections and future directions. Accid Anal Prev 42(5):1517–1522
7. Dollard MF, Bakker AB (2010) Psychosocial safety climate as a precursor to conducive work environments, psychological health problems, and employee engagement. J Occup Organ Psychol 83(3):579–599
8. Reichers AE, Schneider B (1990) Climate and culture: an evolution of constructs. In: Schneider B (ed) Organizational climate and culture, vol 1. Jossey-Bass, San Francisco, pp 5–39

9. Schneider B (2000) The psychological life of organizations. In: Wilderom CP, Ashkanasy NM, Peterson MA (eds) Handbook of organizational culture and climate, pp 17–21
10. Rasmussen K, Glasscock DJ, Hansen ON, Carstensen O, Jepsen JF, Nielsen KJ (2006) Worker participation in change processes in a Danish industrial setting. Am J Ind Med 49(9):767–779
11. Nahrgang FPM, Jennifer D, Hofmann DA (2007) Predicting safety performance: a meta-analysis of safety and organizational constructs. In: 22nd annual conference of the society for industrial and organizational psychology, pp 1–21
12. Brondino M, Silva SA, Pasini M (2012) Multilevel approach to organizational and group safety climate and safety performance: co-workers as the missing link. Saf Sci 50(9):1847–1856
13. Barbaranelli C, Petitta L, Probst TM (2015) Does safety climate predict safety performance in Italy and the USA? Cross-cultural validation of a theoretical model of safety climate. Accid Anal Prev 77:35–44
14. Griffin MA, Neal A (2000) Perceptions of safety at work: a framework for linking safety climate to safety performance, knowledge, and motivation. J Occup Health Psychol 5(3):347–358
15. Neal A, Griffin MA, Hart PM (2000) The impact of organizational climate on safety climate and individual behavior. Saf Sci 1(34):99–109
16. Zadow A, Dollard MF (2015) Psychosocial safety climate. In: The Wiley Blackwell handbook of the psychology of occupational safety and workplace health. Wiley, pp 414–436
17. Demerouti E, Bakker AB, Nachreiner F, Schaufeli WB (2001) The job demands-resources model of burnout. J Appl Psychol 86(3):499
18. Zellars KL, Perrewe PL, Hochwarter WA (2000) Burnout in health care: the role of the five factors of personality. J Appl Soc Psychol 30(8):1570–1598
19. Vainieri M, Ferrè F, Giacomelli G, Nuti S (2017) Explaining performance in health care: how and when top management competencies make the difference. Health Care Manag Rev

From Diagnosis and Recommendation to a Formative Intervention: Contributions of the Change Laboratory

Rodolfo Andrade de Gouveia Vilela[1(✉)], Susana Vicentina Costa[1(✉)], Amanda Aparecida Silva Macaia[1(✉)], Marco Antonio Pereira Querol[2(✉)], Sayuri Tanaka Maeda[3(✉)], and Laura Elina Seppanen[4(✉)]

[1] School of Public Health, University of São Paulo, São Paulo, Brazil
ravilela@usp.br, susipsi@hotmail.com, as.amanda@gmail.com

[2] Federal University of Sergipe, Aracaju, Brazil
mpquero@gmail.com

[3] School of Nurse, São Paulo University, São Paulo, Brazil
sayuri@usp.br

[4] Finland Institute of Occupational Health – FIOH, Helsinki, Finland
laura.seppanen@ttl.fi

Abstract. Ergonomic Work Analysis is important methodology for understanding the present situations and support recommendations for transforming work. However, more emphasis to the historical analysis of the activity system is needed for understanding the origin of problems so that expansive solutions can be designed. This article shows how a historical analysis can be used to redesign the relationship between two organizations, by seeking to solve problems of collaboration between them. This paper used the data and results obtained from a Change Laboratory (CL) intervention, which took place in the Health Center & School, which is part of a Faculty of Public Health (São Paulo University) in the State of São Paulo, Brazil, in 2015. The double stimulation strategy and tools used during the CL sessions, which served as first and second stimuli and produced concrete results and learning were revisited. A timeline was collectively constructed, helping to visualize the historical events involved and how they contributed to the emergence of contradictions, which led to the crisis in the relationship between the two institutions in their actual activity. The contradictions were related to changes in curricula, isolation of the service from the Unified Health System, an emphasis on research, and financial difficulties at university. Both the CL participants and researchers expanded the way in which they understood the origin of the conflicts faced in the activity and its possible solutions. They were able to visualize alternatives for the future. The historical analysis is a powerful tool for redesigning the future and to ensure interventions.

Keywords: Change laboratory · Historical analysis · Contradictions
Work ergonomic analysis

S. Bagnara et al. (Eds.): IEA 2018, AISC 821, pp. 473–481, 2019.
https://doi.org/10.1007/978-3-319-96080-7_57

1 Introduction

Brazilian higher education is based on the indissolubility of teaching, research and extension tripod. The activity of community assistance is one of the extension modalities practiced in the University of Sao Paulo (USP). Changes in the country's health systems and in the higher education activity, lead to an emphasis in teaching and research but not in extension. This context led to a crisis in the collaboration relationship between academic activities between USP's School of Public Health (SPH) and its School Health Centre (SHC). After several failures to resolve the crisis, the Change Laboratory (CL) was chosen as a method for understanding the crisis and building solutions. In this paper we present how the historical analysis taken during the CL helped the practitioners to understand the historical changes and contradictions that led to the problems identified among institutions.

Ergonomic Work Analysis makes an important contribution for understanding daily situations and serves as a basis for ergonomists' recommendations for the transformations of work situation. In general, this kind of approach has as its presupposition that the understanding of the present is sufficient to reflect on and foresee the future. However, understanding the origin of actual problems requires a historical perspective. In order to overcome this limitation, we propose the CL method, as a theoretical-methodological approach that can be used to formulate hypothesis of historical contradictions in and between activity systems.

From theoretical point of view, contradictions can be understood as opposing forces that develop historically in and between activities and can culminate in "crises", but are also driving forces for the transformations and development of institutions [1]. Grasping these historical contradictions may help participants to expand their understanding of the origin of current problem and to become protagonists that can transform their activities.

In 2015 a CL was demanded by SPH. Such demand was related to the growing distance between the faculty and the Centre within its academic activities. The origin of the demand was the health of the workers, manifested as conflicts between them and users in the reception of the unit. The researchers wanted to carry out a pilot intervention with the methodology in the Brazilian context. This joint interest between researchers and practitioners was a fertile ground for the application of the method [2].

We start by presenting CHAT theoretical approach; second we present the CL Methodology and the historical context where the intervention took place. Finally, we will present the results, the discussion and the final considerations.

2 The Theoretical Approach

An Activity System as Unit of Analysis. In CHAT approach an activity system (AS) is the basic unit of analysis. Such unit helps the interventionist to have a clearer view of the object of the intervention and the structure of the activity being transformed. It is an important analytical tool that helps to identify the material and cultural mediations occurring on the system. An activity may be understood as a set of actions

directed towards the transformation of an object. An object is a material and ideal thing that is being collectively transformed from a raw material to a thing with the capability to satisfy a human need [3].
The process in which the subject produce the object is mediated by cultural artifacts (mediators) (Fig. 1) The subject makes use also of societal rules (external, internal, explicit or not, deadlines, etc.), a division of labour (a division of tasks, hierarchy) and community involved with the object as social partners, clientele and etc.).

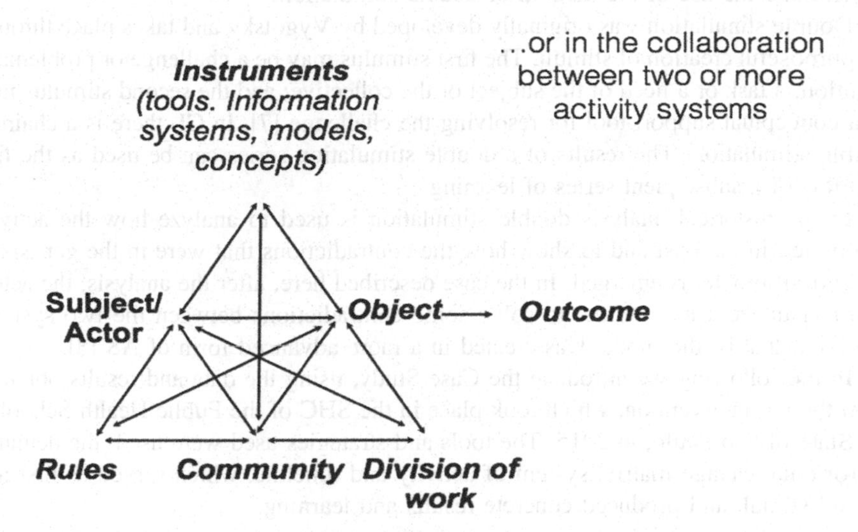

Fig. 1. Triangular model of the human activity system [3] (Engeström 1987, p. 78)

Historicity and Contradictions in the Activity System (AS). In CHAT the driving force of change in an AS is the contradictions within and between activity systems. Production in the capitalist system is faced by a primary contradiction between use value and exchange value [4] (Marx 1983). This contradiction is related to the fact that in this production system the produced object is a commodity, which has a use and exchange value, which is intrinsic and contradictory.

Changes in the elements of the AS generate incompatibility between them, which is called a secondary contradiction. These contradictions arise in the system from historical events that introduce changes in one or more of its elements. Such contradictions may aggravate leading the system to the state of 'crisis'. A historical analysis may help to create a hypothesis of the contradictions, which may be leading to disturbances, conflicts and dilemmas [3].

Once a new model of the activity is designed and implemented tertiary contradictions between the applications of a new principle and the old principle that tends to resist for innovations may emerge. A new principle is as a new concept in activity or within some element, or a new way of organizing or developing actions and the activity. Finally, quaternary contradictions occur between the new and the neighboring activity systems [3].

Expansive Learning and Double Stimulation. The CL method involves questioning the current practices, analysis of the present and the future, creation, testing and implementation of a new model of the activity, with building Agency. The method may be seeing a tool for promoting expansive learning. Expansive learning occurs precisely when the subjects of the activity system, collectively, creatively overcome contradictions, by expanding the object of their activity [5, 6]. This process takes place through expansive learning actions, such as the historical and empirical analysis of the activity system, and the use of the method of double stimulation.

Double stimulation was originally developed by Vygotsky and takes place through the purposeful creation of stimuli. The first stimulus may be a challenge or problematic situation, a task or a need of the subject or the collective; and the second stimulus may be a conceptual support tool for resolving the challenge [7]. In CL there is a chain of double stimulation. The results of a double stimulation series can be used as the first stimulus of a subsequent series of learning.

In the historical analysis double stimulation is used to analyze how the activity functioned in the past and to show how the contradictions that were in the genesis of the current problems emerged. In the case described here, after the analysis, the actors could visualize a new model to solve some contradictions between the two systems involved, that is, the model was created in a more advanced form of AS [8].

In the following we introduce the Case Study, using the data and results obtained from the CL intervention, which took place in the SHC of the Public Health School in the State of São Paulo, in 2015. The tools and strategies used were used: the demand, mirror data, change matrix, system of activity and timeline, which served as first and second stimuli and produced concrete results and learning.

3 The Empirical Environment and the Methodological Process

Representatives from Health Centre & Scholl (HCS) and the School of Public Health (SPH) participated in the intervention. The Centre was originally created to be a field of practice for the academic activities of the university. SHP offers undergraduate courses in Public Health and Nutrition, as well as postgraduate courses. The SHC was responsible for a territory with 110 thousand inhabitants, whose main activity is the provision of primary and secondary healthcare. In the twentieth century the SHC was a health unit school considered a model, but in recent years it faces difficulties due to the lack of human and financial resources, as it happens with the University of Sao Paulo.

After negotiations, the CL started in 2013 with an ethnographic study covering the functioning of the two institutions. This study guided researchers for the elaboration of hypotheses about the development of the system and its historical and current contradictions. Besides documentary analysis, interviews and participatory observations of the work in the institutions were carried out. Between September and December 2015, 11 CL sessions of 2 h/week were held in a space provided by Centre. SPH teachers, employees, users of the Centre, and graduate students participated in the intervention. The sessions were recorded in audio and video. This material was analyzed and some sessions were transcribed. Five of these eleven sessions were devoted to the historical analysis, in between 7 to 18 people participated.

The historical analysis started in the third session of the CL by questioning participants about the activities of collaboration between the institutions at the beginning of the relationship in 1925. In the following sessions, the participants chose to construct a historical line, a tool which enables participants to identify the occurrence of events in AS and their respective dates [6]. The analysis of the historical line was stimulated through the identification of historical events that produced gradual, notable and qualitative changes in the structure of the activity [9]. Participants were asked to identify the change periods and to name them, discussing changes in institutions and between them from one period to the next.

Both the identification of contradictions in the two AS and the indication of possibilities for the solution were the results of the analysis of the records of discussions and interpretations of the participants during the sessions. These records and interpretations were performed during the sessions and are empirical evidence of CL results.

4 Results

4.1 Learning Outcomes

The series of double stimulation in the phase of historical analysis produced concrete results and learning. Participants analyzed the collaboration between the two institutions since the origin of the Faculty in 1925 until 2015. In this process, both researchers and participants understood the reasons that constituted the crisis over the period. In the Fig. 2 we show the results of the double stimulation series organized according to the first and second stimuli.

The historical line was enriched by the participants themselves, who as they were stimulated in and by the process itself, collected other data to be analyzed. They understood that SHC was created by SPH in 1925 with the objective of training public health professionals to control infectious diseases through health education and health care of the population of the territory. At the same time, SPH expected to educate professionals to control these same diseases. SPH teachers taught the practice while attending the population at the health Centre. Thus, in 1925 there was a collaborative relationship between the Faculty and the Centre. Each AS dealt with a different dimension of the their objects, i.e., the student and the user. This means that their objects were shared between the two systems.

In 1970 the creation of the post-graduation in SPH impelled its professors to emphasize in the activities of research. This led to the beginning of the rupture between the two institutions. In 1982 the SHC left the space of the faculty and changed to a nearby facility, what characterized the realization of the break. The end of the 1980s and the beginning of the following decade occurred three significant events: (a) the incorporation of the employees that provided services in the SHC to the staff of USP, but without the respective creation of these positions in their human resources framework; (b) the choice of the health center staff in not joining the Unified Health System; (c) the reform in the USP's statute, which has put more emphasis on

Session	Objective	1 Stimulus	2 Stimulus	Concrete results	Learning outcomes
3	Analyze the collaboration at the beginning of the relationship between SPH and SHC as activity systems.	Question: What was the activ. the SHC and the FSP in the beginning of their relations?	System of Activities and Matrix of Changes	Change Matrix - 1925	Strong collaborative relationship
4	Analyze collaboration after 1925.	Matrix filled for the year 1925.	Question: In what periods will the Activity systems be analyzed?	Timeline	Complexity of collaboration between systems
5	Analyze collaboration as research teaching and extension.	Timeline with SHC and SPH managers	Question: What were the points of activity changes in the Timeline?	Restructuring the Timeline	Managers influenced the relationship of distancing or approximation
6	Define the periods of change in the restructured timeline.	Restructuring the Timeline	Points of change hypotheses in the collaboration between SHC and SPH	Timeline divided by phases and named	Understanding of systemic process of distancing
7	Passage from Historical Analysis to and Analysis of Current and Future Activity	Timeline divided by named phases and employees' speech about future possibilities	Question: What concrete ways of collaboration in the future?	Framework of future scenarios	Changes in concepts related to the work process and collaboration

Fig. 2. Double stimulation with concrete results and learning

international scientific research with evaluation of teachers and postgraduate programs based on publication goals than to extension activities that did not produce publications.

The 1990s and 2000 culminated in the period of consolidation of the separation due to a series of events such as: municipalization of health services; changing the epidemiological profile of the population in the territory; changes in the work process and in the training of health professionals; establishment of a new contract for offering services in specialties with the municipality, dispute over SPH's funds to maintain the primary care service and assist in the payment of hired employees to keep the unit in operation. The period between 2000 and 2011 was called questioning about the crisis. At that moment, the withdrawal was identified as a crisis between institutions. Although the crisis has been questioned and has mobilized actions of approximation between the two institutions, they were considered by the participants as incipient actions, isolated and non-institutional.

The period between 2011 and 2015 was recognized as a period of Crisis marked by the difficulty in renewing the service contract with the municipality. Such difficulty caused loss of financial resources and employees hired with this resource. At that time, there were also eight employees joining the University's voluntary dismissal program, with consequent increase in the workload of the remaining workers and longer waiting time for assistance to users. The SHC Users Association also started to claim improvements in the assistance to the population, opening of pharmacy and maintenance of the linkage of SHC with SPH. Participants analyzed that at this time the ASs were separated and the relation survived by a tenuous line of collaborative actions represented by the students' Nutrition course.

In 2012, the creation of the undergraduate course in Public Health was another event that reinforced the fragility of the relationship between AS. The incompatibility of the academic activities between the Public Health Undergraduation Program and the work developed at the center became evident, according to the words of a teacher of the course:

> *"So for my students [...] there is nothing interesting about this Centre. [...] the way it is, it is unfeasible [...]. The SHC, in fact, is no longer part of the design, in practice, of the SPH professor's. One day it did! The University as a research institute does not recognize SHC as a space [like a laboratory]".*

One of the main annoyances of the teachers was that the center did not participate fully in the Unified Health System.

4.2 Contradictions in the Two Systems of Activity

From the historical analysis, we have identified primary and secondary contradictions within and between the activities of the health care activity conducted by Centre and the research and teaching activity conducted by the Faculty.

Throughout the development of AS of the Faculty, the user ceased to be a direct motivation for their activities as it was in 1925 at the beginning of the collaboration between the two institutions. With the creation of post-graduation, the main goal of the University was linked to the production and scientific dissemination of research. Teachers expressed that conducting outreach activities such as assistance reduces time to research and teaching and is well evaluated. The actions of teaching, research and extension are practically exclusive and therefore constitute a primary contradiction inside the object.

Another primary contradiction was hypothesized in the HCS object due to the absence of the student. SHC should be organized to act both for the student and for the user, as objects of a Health Center that is both a Care Centre and a School Centre, as in 1925. The SPH understood that SHC does not have many of the tools needed to provide comprehensive care and are not organized to achieve the desired outcomes of student learning.

Several secondary contradictions were identified in HCS. An example is a contradiction between the need for full care of the user of the territory (object) and the type of service rendering services of specialties (rules). Another example is a contradiction that directly involves the relationship of collaboration between institutions. It concerns the contradiction between the little integration between the conceptual tools developed in SPH, focused on UHS, and the practice of actions in SHC. Between the 1980s and 1990s, forces from different directions pressed the health Centre for a change. The primary contradictions in the objects suggest that the AS's are in different stages of development that culminated in contradictions between them, expressing itself in a sense of impasse by the professionals of the Centre, as reported below:

[...] when the entire teachers' movement (SPH) went to university reform, health reform and the implementation of Unified Health System in Brazil, SHC was isolated from this process with the State Health Department, which was also isolated of the world (Teacher, session CL, 2015).

5 Discussion and Concluding Remarks

In the study, the historical analysis supported practitioners in understanding the origin of the crisis in terms of two sets of contradictions. The first is the result of historical changes in the rules, such as a greater academic investment for research actions and the evaluation of teachers, preferably for the production of scientific papers. These new rules consequently generated change in the objects of SPH and SHC activities (health care, teaching, research). They have influenced the manner, intensity and frequency of use of instruments such as teaching, research and extension. In general, the rule where research was priorized, led to the removal of this triad and collaborated with, among other factors, for the gradual abandonment of SHC as a field of practice for SPH students.

The second group of contradictions is internal to the activity of health care and is due to direct changes in its object. The users (object) increased in quantity and diversity of demands, and epidemiological data evidenced progressive ageing of the local population, demanding coherent needs to that phase of life. However, the whole structure of the activity remained little changed. Such a change has led to the emergence of a series of incompatibilities that are interpreted as contradictions between the elements of SHC's Activity System.

The uncovering of the historical roots that gave rise to the crisis occurred in a collaborative way with the participants enriching the analysis and building agency of practitioners. The CL helps to reinforce the role of the participants to start driving the transformation in their activities. This aspect is related to the fact that CL, as a formative intervention methodology, integrates diagnosis, protagonism and solution creation by the participants as part of the same process. Our findings suggest that this kind of interventionist methodology help to understand the motives and to find new solutions.

References

1. Engeström Y (2001) Expansive learning at work: toward an activity theoretical reconceptualization. J Educ Work 14:133–156
2. Pereira-Querol M, Jackson Filho JM, Cassandre MP (2011) Change laboratory: uma proposta metodológica para pesquisa e desenvolvimento da aprendizagem organizacional. RAEP 12(4):609–640
3. Engeström Y (1987) Learning by expanding: an activity-theoretical approach to developmental research. Orienta-Konsultit, Helsinki
4. Marx KO (1983) Capital: Crítica da Economia Política. vol. 1, T 1, Abril Cultural, São Paulo
5. Engeström Y, Sannino A (2010) Studies of expansive learning: foundations, findings and future challenges. Educ Res Rev 5(1):1–24

6. Virkkunen J, Newnham DS (2013) The change laboratory: a tool for collaborative development of work and education. Sense Publishers, Rotterdam
7. Engeström Y (2007) Putting Vygotsky to work: the change laboratory as an application of double stimulation. In: Daniels H, Cole M, Wertsc JV (eds) The Cambrigge campanion to Vygotsky. Cambridge University Press, Cambridge
8. Miettinen R (1999) The riddle of things: activity theory and actor network theory as approaches to studying innovations. Mind Cult Act 6(3):170–195
9. Sewell JR, William H (1996) Historical events as transformation of structures: inventing revolution at the Bastille. Theor Soc 25:841–881

Application of the HTO Concept for a Powered Pallet Truck

Jörgen Eklund[1,2](✉)

[1] Division of Ergonomics, KTH Royal Institute of Technology, 141 57 Huddinge, Sweden
jorekl@kth.se

[2] Helix Competence Centre, Linköping University, 581 83 Linköping, Sweden

Abstract. Truck drivers suffer work injuries to a higher extent than most other occupations. The HTO concept and the interaction framework were applied in a pre-study leading to a redesign of a new powered pallet truck. The old truck was evaluated using a literature study, interviews, observations, injury statistics and benchmarking, as part of the HTO analysis. The analyses showed that the driver often stood on the rear part of the platform with the heels outside the platform, making them vulnerable to injury. The injury statistics also showed that drivers of powered pallet trucks had more heel injuries than drivers of other truck types. There were two reasons for this. The steering arm was slightly too long, and the vibration damping was better the further back the drivers stood on the platform. This study led to redesign of the steering arm and platform suspension in the new truck generation. The combination of the HTO concept and the interaction framework supported the analysis in identifying relationships that otherwise would not have been obvious.

Keywords: Pallet truck · Folding platform · Product design

1 Introduction

1.1 The Project

A forklift truck manufacturer planned to develop a new model of its powered pallet truck with a folding platform. The manufacturer intended to develop their products in order to improve effectiveness of the fork lift – driver system, better profitability for the user organizations and better ergonomic solutions for the drivers. The company contacted the researchers in the field of ergonomics in order to get support for the development of ergonomics aspects of the truck design. There was a high level of awareness in this project that the driver-truck system created the basic conditions for ergonomics, safety and efficiency. The focus of the project was on improving the driver – truck interface. A project group was formed including designers, engineers, marketing personnel and ergonomics researchers. Also master students were involved.

The HTO concept was found suitable and consistent with the system perspective. It was used as the basis for the project plan (Eklund 2003) and as a basis for the interactions between the company and the researchers. Further, the interaction

S. Bagnara et al. (Eds.): IEA 2018, AISC 821, pp. 482–485, 2019.
https://doi.org/10.1007/978-3-319-96080-7_58

framework (Nolimo Solman 2002) was also used as a basis for the project. This project was part of a larger collaboration project, which included several sub-projects, more staff and several truck types. In this paper, however, only the sub-project dealing with a new powered pallet truck is reported. Also, several sub-studies were performed based on the powered pallet truck.

1.2 Truck Driver Work

According to the statistics, lift truck drivers experience more occupational diseases including musculoskeletal disorders and more accidents than the average swedish worker. The risk is approximately 7 times higher for a lift truck driver compared to the average for all jobs (Arbetsmiljöverket 2017). The body parts afflicted from musculoskeletal work injuries are in general mostly the back, shoulders and neck (Nolimo and Eklund 1999). Despite improvement of the ergonomics design of lift trucks, the risks shown from injury statistics seem to prevail.

1.3 Aim

The aim of this paper is to describe how the HTO concept can be applied in the analysis and design of system interactions, illustrated in a case dealing with development of a new powered pallet truck with a folding platform.

2 Methods

The first phase in the project was to carry out a literature review regarding ergonomics in lift trucks. Further, competitor trucks were benchmarked regarding the ergonomics solutions. Parts of this benchmarking study were performed using the internet. Also, national and international accident statistics were analyzed, regarding different types of lift trucks, body parts injured and work activities involved in the injury. Lift truck drivers were interviewed in order to identify tasks and situations that could cause risks, problems or disturbances. Further, observations were performed of the truck driver work. In particular, problematic work activities were identified by focusing on how the drivers interacted with the technology (the truck), the organization (work patterns) and the environment (the warehouse) according to the HTO model (Eklund 2003). These interaction analyses of work activities and the interviews focused on identifying difficult work activities and situations that developed to disturbances, e.g. those demanding extra time to perform, causing errors, demanding recurrent corrections, giving rise to near accidents, or those that were exerting, tiring or uncomfortable. The framework proposed by Eklund and Nolimo (1998) and Nolimo Solman (2002) was the basis for the detailed analysis and interviews. The following factors were used as indicators of deficient interactions.

- Productivity (time needed for task, time needed for learning, variability, disturbances)
- Quality (errors, deficiencies, adjustments, rework, precision, variability)

• Safety (accidents, near misses, injuries, unplanned events)
• Physiological and mental effects (pain, disease, discomfort, effort, wellbeing)
• Experiences (comfort, feeling, tiredness, usability)

In the HTO analysis, the analysis was based on how technology supported or hindered the drivers in their work, and if other technology could support them better. In the next part of the analysis, the same types of questions were repeated but with a focus on the organization, and in the third part, the focus was on environmental factors.

3 Results and Analyses

The injury analysis showed that injuries on the heels were more common for the drivers of powered pallet trucks, compared to drivers of other truck types. The results from the interviews showed that the drivers pointed to discomfort from the back, shoulder and neck. In particular, they mentioned loading and unloading lorries as an important source of back discomfort. This was because they in those tasks drove on uneven ground which gave rise to vibrations and impact forces that triggered back pain. It was also pointed out that too narrow aisles are strainful for the drivers and reduce productivity.

The observations showed that the pallet truck drivers often stood with their heels outside the platform, which was assumed to be the cause of the increased rate of heel injuries. It could also be seen in the observations that the steering arm seemed to be a little too long, thus preventing the drivers to stand on the center of the platform. Also, the platform design had an influence, since the platform was hinged in its front, which meant that the vibration damping was better the further back the driver stood on the platform.

The results converged to a diagnosis of the main problem with the design of the old pallet trucks. The driver often stood on the rear part of the platform with the heels outside the platform, making them vulnerable to injury. There were two reasons for this. The steering arm was slightly too long, and the vibration damping was better the further back the drivers stood on the platform.

4 Discussion

This study illustrates how the different methods used within the HTO concept contributed to one another in order to identify problems and causes of those problems. Also, the systems approach which is prevalent in the HTO concept supported the overarching analyses.

The HTO concept enabled identification of work tasks with deficient interactions between the driver and other components of the system. It also identified causes of the deficient interactions, which gave a direct input to the planning of preventive actions. As a result of the HTO analyses, new design proposals were put forward, including those of two thesis workers. This meant that the steering arm was shortened and vibration damping was improved for the whole platform in the new design.

5 Conclusions

The HTO concept was found suitable in order to identify problems in the interaction analysis between the driver and the lift truck in identifying relationships that otherwise would not have been obvious. The systems view and the multi-method usage supported the analysis and identification of causes for the interaction problems. The old powered pallet truck investigated was found to force the driver to stand on the rear part of the platform, due to a slightly too long steering arm and too little vibration damping in the front part of the platform. This caused an increased risk of heel injuries. The results were so clear that they led to a redesign of the new truck generation.

References

Eklund J (2003) An extended framework for humans, technology and organization in interaction. In: Luczak H, Zink KJ (eds) Human factors in organizational design and management VII. Re-designing work and macroergonomics - future perspectives and challenges. IEA Press, Santa Monica, pp 47–54

Nolimo Solman K (2002) Analysis of interaction quality in human machine systems: applications for forklifts. Appl Ergon 33(2):155–166

Occupational accidents and work-related diseases, 2016 (2017) Arbetsmiljöstatistik Report 2017:1. Arbetsmiljöverket

Nolimo K, Eklund J (1999) User quality in a human truck system. In: Proceedings of the international conference on TQM and human factors, Linköping, vol 1, pp 373–378

Eklund J, Nolimo K (1998) Interaktion mellan trucken och dess förare - effektivitet och ergonomi. In: Proceedings från Nordiska Ergonomisällskapets årskonferens NES 1998, Lund, 16–18 September 1998, pp 66–69

New Public Management, Performance Measurement, and Reconfiguration of Work in the Public Sector

Bruno César Kawasaki[1(✉)], Ruri Giannini[1], Selma Lancman[2], and Laerte Idal Sznelwar[1]

[1] Polytechnic School of the University of São Paulo, Av. Prof. Luciano Gualberto 380, 05508-010 São Paulo, Brazil
bruno.kawasaki@usp.br

[2] Medical School of the University of São Paulo, R. Cipotânea 51, 05360-160 São Paulo, Brazil

Abstract. In this paper, we discuss performance measurement systems in public organizations and their effects on both public sector workers and quality of the services provided to the population. Although performance measurement is often regarded as prerequisite for an efficient provision of public services in the context of New Public Management, a review of academic studies actually shows that its consequences on work and organizational performance are inconsistent and potentially deleterious. Hence our core questions are: how performance measurement became widespread in public sector? Which are the critical aspects in the design and use of performance measurement systems, regarding their relation to workers' activities? How can they positively or negatively impact behaviours, tasks and activities, and relationships among workers? We discuss these questions based on the literature of public sector and ergonomics. Then, we present a case study that illustrates how performance measurement impacted the work of labour judges. Despite difficulties in accurately assessing or predicting the effects of performance measurement, managers can be benefited by considering their benefits, risks, and limitations.

Keywords: New Public Management · Performance measurement
Work evaluation

1 Introduction

In this paper, we discuss performance evaluation systems in public organizations and their effects on both public sector workers and quality of the services provided to the population. We are particularly interest in evaluation systems that heavily rely on performance measurement or performance metrics. Although performance measurement is often regarded as prerequisite for an efficient provision of public services and may be promoted by political and electoral agendas, a review of academic studies actually shows that their consequences for organizational performance are inconsistent and potentially deleterious. We revisit these studies and analyse them from the point of view of work activities, based on concepts of ergonomics.

S. Bagnara et al. (Eds.): IEA 2018, AISC 821, pp. 486–493, 2019.
https://doi.org/10.1007/978-3-319-96080-7_59

Which are the critical aspects in the design and use of performance evaluation systems, regarding their relation to workers' activities? How can they positively or negatively impact tasks and activities, and relationships among workers? These are the core questions of our investigation.

We assume that organizational performance and work organization should be analysed vis-à-vis workers' activities. By stating that performance evaluation systems may generate opposite effects from those desired, we are not necessarily defending their extinction. We are fundamentally interested in understanding their limitations as a prerequisite for a proper implementation or development, in case organizations want to use them.

In Sect. 2, we begin by explaining how performance measurement, among other management principles and techniques, spread from private to public organizations in the context of the New Public Management.

In Sect. 3, we start by presenting a literature review on work and performance evaluation in the context of public sector. Then, we contribute to the discussions by introducing concepts of ergonomics.

Section 4 is dedicated to a case study in which we discuss how performance evaluation can affect the work of labour judges.

In the last section, we present the main conclusions, limitations, and recommendations for future researches.

2 The Emergence of New Public Management

The diffusion of performance evaluation systems in public organizations can be understood in the context of New Public Management (NPM), a form of managing the public sector which emerged from the 1980s and diverges from the welfare state.

The NPM is manifest in different rhythms, intensities, and forms depending on the country and sector analyzed, but some recurrent traits can be observed: attention to outputs and performance rather than inputs; provision of services to the population through a network of public, private, and third sector organizations; promotion of competition between service providers, as a means of rising governmental efficiency; a stronger role of non-elected state agencies in charge of monitoring public expenditures and regulating public service provision (Clarke et al. 2000, pp. 2–5). These traits are associated with the introduction of managerial principles and practices such as performance evaluation systems, which quantify the performance of individuals and organizations and turn them into object of managerial action.

NPM may have heterogeneous, ambiguous, and contradictory effects. In addition, it is not necessarily the only way to conceive coordination and provision of public services. In spite of these characteristics, a process of managerialization can often be observed, meaning the subjection of public administration to principles and practices of management. This process has been supported by the belief that turning governmental structures into 'business-like' organizations would naturally generate more transparency, efficiency, effectiveness, and economy to the public system (Clarke et al. 2000, pp. 5–7). However, particularly regarding performance evaluation practices that have disseminated in the private sector, academic literature on human resources management is not consensual in confirming neither its theoretical fundamentals nor its practical effectiveness.

According to Fleetwood and Hesketh (2011, p. 8), nothing actually demonstrates how or why human resource metrics can enhance organizational performance, although it has propagated and acquired an aura of scientific method.

Lacombe and Albuquerque (2008, pp. 5–8) state that several authors rely on insufficiently explained and tested theoretical basis in order to propose methodologies of human resource measurement. Besides, theoretical and methodological problems have been found in a growing number of studies aimed at assessing and measuring the effect of human resource management on organizational outcomes. The inherent intangibility of such outcomes is a partial explanation for the difficulties faced.

3 Performance Evaluation Systems and Work in Public Sector

In this section, we present a literature review on work and performance evaluation in the context of public sector. We were ultimately interested in identifying critical aspects in the design of performance evaluation systems regarding their potential impacts on workers' activities. Papers were selected in May 2018 using *Web of Science* database and the following search term: "public sector AND (work OR performance OR evaluation)".

A great number of quantitative studies investigate relationships between performance (or productivity), performance measurement, and work motivation. They present conflicting evidences on the correlation between work motivation and work performance, and between performance measurement and work performance. Conflicting evidences are also noticed in studies which compare private sector work and public sector work, hence it has not been possible to sustain that they are different based on quantitative methodologies. These results can be explained by the heterogeneity of organizations and the lack of consensus on how to define and measure concepts like work motivation, work performance, and organizational performance (Hood 2012; Wright 2001). Considering these limitations, we decided not to focus on quantitative studies based on one or few organizations. Instead, we focused on authors who developed discussions from a qualitative point of view and/or whose opinions draw on a review of quantitative studies.

Based on personal experience as a former employee of the Organisation for Economic Co-operation and Development (OECD) and two national competition agencies, Davies (2018) notices that competition agencies' outcome assessment is heavily based on activity measures, such as number of cases, fines, or decisions adopted. It may be useful for accountability purposes; however, they require significant amount of time and effort, and say nothing about the quality of decisions. The study of single concluded cases is much more effective for learning purposes and to appreciate the quality of decisions taken, but would not be adequate for accountability purposes.

Davies (2018) even disagrees that outcome assessment, since it has been based on simplistic activity measures, is in fact a measure of outcomes. He understands that sophisticating activity measures will not essentially solve the limitations and recommends more modesty about what outcome assessment can achieve.

Curristine et al. (2007, pp. 2–3, 12–17, 32) discuss potential benefits and dangers from the increasing use of performance information in budget processes by OECD countries. Although it is appealing to reward good performance, the authors recommend avoiding government systems that automatically link performance results to resource allocation. This mechanical approach ignores government priorities, budgetary constraints, and underlying causes of poor performance. Furthermore, it can generate perverse incentives and encourage agencies to manipulate data, as they can be used to cut back their programs. Curristine et al. (2007) state that there is not an universal model of performance budgeting; it needs to be adapted to the political and institutional context.

Wright (2001) reviews work motivation theories and discusses how goals affect motivation. The processes involved are of two types: goal content and goal commitment. Goal content refers to goals' characteristics and how they can influence performance, whereas goal commitment refers to the conditions under which individuals are determined to achieve the goals even if confronted with setbacks or obstacles. Concerning how goals should be structured, literature provides support for the following hypothesis:

- Specific goals improve performance. They reduce ambiguities, set priorities, and facilitates accurate performance evaluation as basis for feedbacks and rewards.
- Goal difficulty has a curvilinear effect on performance. Goals should be difficult but achievable. If goals are too difficult, which may be a consequence of multiple, conflicting goals, individuals may give up trying to reach them;
- Goals viewed as important promote goal engagement. When job goals are aligned to organizational goals, or when the individual understands the social relevance of his work, individuals are encouraged to accept job goals as personal ones.

Hood (2012) argues that costs and benefits of performance measures are rarely taken into account to assess whether benefits actually exceed costs, i.e., a positive balance which would justify their existence. He observes that numbers are mainly used for three management purposes: target, ranking, and 'intelligence'. The same numbers can be used for different and possibly coexistent purposes. Targets are used to set and monitor minimum thresholds of performance or activity. They can encourage performance improvement, but may unintentionally harm performance by fostering three types of undesired behaviours:

- Threshold effect: lack of incentives to perform above the target threshold;
- Ratchet effect: workers and organizations are induced to perform below maximum level if they expect targets will be raised;
- Output distortion: workers and organizations tend to focus on the targets and neglect other values.

These behaviours illustrate how some sort of 'gaming' behaviour may arise if individuals take the system of metrics and rewards as the primary guide for action.

Ranking involve the use of numbers to compare the performance of different units (individuals, groups, organizations, cities, countries etc.) in order to encourage performance improvement relative to other parts or rivals. Hood (2012) argues that rankings are not exempt from fostering the three types of undesired behaviours, although threshold and ratchet effects are more likely to be observed in targets rather than in rankings.

Finally, ‘intelligence’ means ‘using numbers for the purpose of background information for policy development, management intervention, or user choice’ (Hood 2012, p. 4). This use is well-established on both public and private management systems and is not susceptible to produce threshold effect, ratchet effect, nor output distortion. However, as ‘intelligence’ depends on the ability of managers, choosers, or system operators, it becomes critical in situations of uncertainty in which it is hard to choose priorities.

Hood (2012) defends that whether management by numbers will actually enhance performance or not largely depends on the culture in which it will be applied. He claims that targets, ranking, and ‘intelligence’ are respectively most likely to be provided by some form of hierarchist, individualist, and egalitarian culture. In hierarchist cultures, individuals expect a clear source of authority, a capacity for blame absorption over political risks, and a shared sense of purpose. In individualist cultures, competition is expected and seen as the natural answer to performance problems. Egalitarian cultures foster cooperation, tries to avoid inhibitions generated by ranks or status, and are prone to contest any data which may be damaging to any one of the players. They are capable of avoiding threshold and ratchet effects and output distortions. Traits from different types of cultures may be noticed in the same group or organization.

Hood’s concepts of ‘intelligence’ and egalitarian culture resonate with Deming’s (1993) proposal for management of people. This author defends that instead of judging, ranking, or classifying people, managers should: (1) help them to optimize the system so that everyone will be benefited in an environment of cooperation; (2) explain them functioning and meaning of a system; (3) foster interest, challenge, and joy in work; (4) try to discover if anyone is outside the system, in need of special help; (5) create confidence and not expect perfection.

3.1 Concepts of Ergonomics

In this subsection, we introduce concepts of ergonomics to discuss performance evaluation systems. Here, we make no distinctions between public and private sector.

Ergonomics is traditionally oriented to analyze and act upon work situations, seeking to address both health issues and operational issues like productivity, efficiency, and quality. One of its core concepts is the distinction between task and activity; both should be taken into account to understand work. Tasks refer to organization’s demands, rules, goals, and work prescriptions in general, whereas activity refers to the real work, or everything that has to be actually done in attempt to accomplish previously assigned tasks. Activity includes efforts, thinking, interpreting, developing strategies, acting, using knowledge and abilities, working individually and collectively, etc. (Abrahão et al. 2009; Guérin et al. 1997).

It is up to the worker to deal with situations of conflicting tasks, or with unforeseen situations in which previously assigned tasks are of little or no help. It may be necessary to disobey some prescriptions in benefit of the whole situation. Elements as fatigue, justice, ethical concerns, and professional values, also influence workers' decisions although they are often not considered in organizations' formal prescriptions. Hence, Schwartz (2000, p. 41–2) argues that working cannot be reduced to a mere execution of tasks or orders. In other words, previously set tasks and prescriptions are necessary, but do not suffice to deal with the complexity of real work situations. By complexity, we mean the coexistence of incommensurable issues that have to be considered in a decision, in the same sense as Morin's (1990).

Management practices, such as measurement of work performance, can positively contribute to make goals clearer and can be understood as a part of the tasks. Performance evaluation systems may even be demanded by workers. However, an excessive focus on performance metrics may inflate the importance of some issues in detriment of others, which may lead to decisions that are deleterious in the long term or to the organization as a whole. Priorities may become clearer, but it does not necessarily lead into a better appreciation of work's complexity. This can only be done by appealing to personal ability, intuition, experience, subjectivity, and judgement. Unfortunately, there is no previous one best answer for each situation.

Ergonomics also emphasizes the importance of cooperation for the accomplishment of tasks. Procedures such as performance measurement, ranking, and linking performance evaluation to a system of rewards tend to reinforce a meritocratic rationale in which success is primarily explained by individual efforts, neglecting other factors such as cooperation, luck, and individuals' different conditions. Thus, performance measurement may foster a kind of competition that reduces possibilities of cooperation between members of the same organization, although competition and cooperation may coexist.

4 The Case of Judges in Brazil

In this section, we discuss a research with Brazilian labour judges conducted by the authors of this paper. We first describe their work and then we discuss how it is affected by the performance evaluation system adopted.

In a short description, the work of a labour judge consists in solving conflicts presented by two parts (usually the institution which hires a person and the person who was hired) with agreements or sentences. The judge primarily performs his/her work during a court hearing, when he/she listens to the parts, witnesses, and lawyers and collects all necessary information to analyze the case in a fair way.

Collecting evidence, reading the petition, listening to parts, witnesses and lawyers, analysing, and deciding can be understood as the prescribed work of a labour judge, or what he/she is expected to do. On the other hand, what is the nature of the real work, or what sorts of difficulties have to be overcome so that tasks are accomplished?

First of all, testimonies may contain lies, and dealing with their (possible) existence demands judges' sensibility and practical knowledge. The understanding of what is true and fair in each case is unique. There is no equal case when it comes to judgement of labour conflicts, although there may be similar cases and jurisprudence.

Second, judges usually deal with the difficulty or impossibility of collecting enough evidence to support a thesis. They may feel obliged to take a decision that contradicts their feelings and personal beliefs. There is the risk of judging according to their convictions and having the sentence reformulated latter. Additionally, labour judges have to choose among different lines of thought and approaches in each case. Different justifiable choices can lead to different results, which generates anxiety.

The performance evaluation system used to monitor judges' work is based on indicators of time consumed to perform some tasks: time to conduct a court hearing and time to judge a case and pass a sentence. Time targets make no distinction between daily, simple cases and complex ones. The court publishes monthly reports on the performance of all judges, highlighting the ones who are late to complete hearings or sentences.

This performance evaluation system does have positive aspects, like enabling some assessment on how large volumes of cases are being handled by the court, thus promoting transparency and accountability. However, it does not consider the quality of work nor the various levels of effort and time each case requires.

Consequently, labour judges usually have to decide between either prioritizing time target in detriment of the quality of work, or prioritizing quality in detriment of time target. Complex cases are more challenging and interesting, but some judges avoid them or do not spend the necessary amount of time to solve them (output distortion). Judges who do not face difficulties in hitting targets have no incentives to improve their measured performance (threshold effect). Some judges may choose to perform near target levels for they fear a future target raise (ratchet effect).

Judges compensate the limitations of the institutional evaluation system by using other evaluation systems. They create their own parameters for self-evaluation, such as feedbacks from parts involved in a case and comparisons with judgements made by senior judges for a same case. The creation of self-evaluation mechanisms can be explained by judge's desire to have an evaluation system and, at the same time, an enormous discrepancy between a simplistic performance measurement and everything that takes to accomplish the real work.

5 Conclusions

At least in the current moment, it is impossible to predict the impacts of performance measurement on work. The best we can do is identifying its potential benefits and, especially, its dangers, considering the widespread misbelief that it is necessarily beneficial to work and organizational performance. The actual impacts of performance measurement depend on how performance is defined and its interaction with individuals, i.e., how metrics are used and how people react to it. Parameters such as metrics purposes, specificity of goals, oversimplification of work evaluation, 'gaming' behaviours, and inclination to cooperate can be used to appreciate the effects of performance evaluation, although an unique answer cannot be provided.

For organizations that wish to implement or develop performance measurement, we recommend extreme caution and highlight the importance of incentivizing workers to give their sincere opinion about it, considering that imposed decisions are likely to be rejected in the long run. Organizations in which competition among workers is already strong may have extra difficulties in openly discussing performance metrics, since individuals are likely not to reveal their difficulties until they feel a trusting environment.

Concerning the limitations of this study, we focused our literature review on public sector. However, those studying public organizations can be benefited by studying private organizations as well, though some sort of adaptation in the theory may be needed. We have also not explored other disciplines, such as work psychodynamics and psychology of work, which would help in analyzing how performance measurement can affect work from the point of view of subjectivity. Finally, as we used *Web of Science* to conduct a great amount of the literature review, we may have ignored important books on public administration and performance measurement, since this database is primarily constituted by journal articles.

References

Abrahão JI, Sznelwar LI, Silvino A, Sarmet M, Pinho D (2009) Introdução à Ergonomia: da prática à teoria. Blucher, São Paulo

Clarke J, Gewirtz S, McLaughlin E (2000) Reinventing the welfare state. In: Clarke J, Gewirtz S, McLaughlin E (eds) New managerialism, new welfare? SAGE Publications Ltd, pp 1–26

Curristine T, Lonti Z, Joumard I (2007) Improving public sector efficiency: challenges and opportunities. OECD J Budgeting 7(1)

Davies J (2018) "Outcome" assessment: what exactly are we measuring? a personal reflection on measuring the outcomes from competition agencies' interventions. De Economist 166(1):7–22. https://doi.org/10.1007/s10645-017-9307-6

Deming WE (1993) The new economics: for industry, government, education. MIT-CAES, Cambridge

Fleetwood S, Hesketh A (2011) Explaining the performance of human resource management. Cambridge University Press, Cambridge

Guérin F, Laville A, Daniellou F, Duraffourg J, Kerguelen A (1997) Comprendre le travail pour le transformer: la pratique de l'ergonomie. ANACT, Montrouge, Hauts-de-Seine

Hood C (2012) Public management by numbers as a performance-enhancing drug: two hypotheses. Public Adm Rev 72(s1)

Lacombe BMB, Albuquerque LG (2008) Evaluation and measurement in human resource management: a study in Brazil. Revista de Administração 43(1):5–16

Morin E (1990) Science avec conscience. Points sciences

Wright BE (2001) Public-sector work motivation: a review of the current literature and a revised conceptual model. J Public Adm Res Theory 11(4):559–586

Simulating Work Systems: Anticipation or Development of Experiences. An Activity Approach

Pascal Béguin[1(✉)], Francisco Duarte[2], Joao Bittencourt[2], and Valérie Pueyo[1]

[1] University of Lyon, UMR 3500 EVS, Lyon, France
pascal.daniel.beguin@gmail.com
[2] University Federal of Rio de Janeiro, COPPE, Rio de Janeiro, Brazil

Abstract. Simulation is ubiquitous during design. However, due to its concern (the living, cultural and social work activity), ergonomics is confronted with ontological and epistemological questions, distinct from those of the engineers. From our perspective, designing entails coping with unpredicted future work, experiencing novelty due to design but also contradictions and debates about values between heterogeneous actors. Based on this view, the aim of this communication is to make a contribution for developing simulation as a resource for an ergonomics participatory design process.

Keywords: Simulation · Activity · Work · Participatory design Ergonomics

1 Introduction

Historically, activity-centered ergonomics approach has developed a method named "Ergonomics Work Analysis" (EWA) with the aim to understand the activity constructed by a given operator, in order to "*define the problems*" encountered by the workers [1, 2]. The main contribution of such a method is to offer those involved in the company a new way of understanding the difficulties encountered by the workers, based on a fine analysis of their activity. Very often, the whole areas of real activity in a company is ignored, particularly the importance of the variability of production operations, the strategies adopted by the operators to realize a task, or to deal with and to avoid incidents. The analysis of activity produces a new "point of view", which compared to the previous descriptions of work, make it possible to propose some changes -named "*recommendations*"- better suited to the needs and constraints of the workers [3].

However, the active participation of ergonomists in the design of large-scale industrial or informatics projects during the 1980s was to lead to question EWA [4]. Although well equipped to analyze activity in existing work setting, ergonomics in the French-speaking world was at pains when participating in design processes. Workers in a given setting -i.e. with given organizations, tools and tasks- perform activities, but designing new tasks, tools or organizations will change activities. Therefore, what can

S. Bagnara et al. (Eds.): IEA 2018, AISC 821, pp. 494–502, 2019.
https://doi.org/10.1007/978-3-319-96080-7_60

the ergonomist analyze? This "*paradox of design ergonomics*" [5] leads to question EWA. And due to the need to produce knowledge about working activities that do not yet exist, simulation will quickly appears as an unavoidable method [6, 7].

Nowadays, and beyond the French-speaking world, the terms "*ergonomic simulation*" or "*humans simulation for ergonomics*" are commonplace (a search on the web yields hundreds of commercial proposals). But this generalization overshadows profound differences between many uses and backgrounds of ergonomic simulations in design processes. The aim of this communication is to contribute to document epistemological and ontological questions about simulation, based on an activity-centered approach and on our own experiences as ergonomist in France and Brazil.

2 To Anticipate and Predict the Future or to Develop Knowledge and Experience?

When one defining simulation, a tenacious idea is to assert that the purpose is to anticipate and predict some of the features of a future workstation or a work system. And sure, it was initially the position developed in French-speaking ergonomics.

The «*paradox of design ergonomics*» evoked previously has initially led to the notion of a «*future activity*» [8]. In this view, the ergonomist try to predict some characteristics of future activities, through a model or a reconstitution of a future situation. However, the notion of «*future activity*» has been quickly questioned and replaced by that of a «*possible future activity*'». To take a simple example, if an architect designs a staircase to enter a building, he/she is ruling out the possibility for a disabled person to enter the building. However, if the architect designs a wheelchair ramp, he/she is opening up the possibility of an activity for a disabled person to enter the building on his/her own. The idea of "possible future activity" asserts that, given the choices made by the designers, a range of possible or impossible activities can appears. And this leads to an important idea for design: the organizational and technical environment must be designed in such a way as to open a wide range of possibilities, so as to enable the operator to face the diversity of future contexts of action. But at the same time, this idea assumes that the activity itself cannot be predicted none completely anticipated.

Because activity is constructed by a given operator as a response to a given context, the activity that such and such an operator will develop on a piece of equipment or organization cannot be fully anticipated or predicted. Indeed, the key question is ontological: what is the very nature of activity? Due to its concern (the living, cultural and social work activity), ergonomics is confronted with epistemological questions different from those of the engineers. Activity is not comparable to a mechanical system, in particular because inventiveness and creativity are ontological characteristics of activity in situations. Ombredane and Faverge [1], two authors often considered as the founders of French language ergonomics, argued that: "*while some significant aspects of the task are foreseen, there is an indefinite number of others which are unforeseen and liable to be discovered by the worker*" (p. 22). Since then, worker inventiveness in situation has been constantly highlighted, to the extent that one of the features of this current of ergonomics is to consider operator inventiveness as a central component of work activity [9]. Weill-Fassina et al. [10] synthesized this position by writing that "*actions cannot be reduced to the effecting of responses to stimuli received more or less passively, to motor actions, and to execution procedures. They display the processes through which operators explore, interpret, use, and transform their technical, social and cultural environment*" (p. 22). Anticipating and predicting creativity and inventiveness is certainly an excitant scientific and philosophical question, but actually totally out of the scope of our possibility.

However, focusing exclusively on concern such as anticipation and prediction is grounded on a partial vision of the design process, those of decision-making and uncertainty reduction. Midler [11] shows that design can be grasp as the convergence and articulation between two dimensions: on the first hand the possibilities for action, which correspond to uncertainty reduction, and on the other hand a developmental curve of knowledge (see Fig. 1). It is the articulation between these two dimensions (of knowledge and the possibilities for action) that is characteristic of one design process. As Midler notes, "*at the start of the project, we can do anything but we don't know anything*", whereas at the end "*we know everything but we can't do anything anymore.*"

Our main point is that simulation could gain to be understood as a method for developing knowledge [12, 13] and experiencing novelty coming from versions of something to design [14], rather than a method for anticipating and predicting a future (which do not exist yet). Through an analysis of simulation methods in a large range of scientific domain, Morgan and Morrison [15] found that simulation brings on learning in two ways: (i) through the building of a model, and (ii) in using the model. Even if simulation is not explicitly serving a learning goal, the use of a model appears undoubtedly associated to any questioning on learning and experiencing through a model.

It is impossible to fully anticipate and predict the future. But what is possible is to stage some idea or design hypothesis through a model, a mock-up or a prototype, in order to experience and learn from them in order to identify what could cause problems or troubles and to suggest possible resolution.

3 Simulation as a Copy of a Work Situation or as a Resource for Designing?

A tacit position (often associated with anticipation) is to consider that simulation methods must be figurative: they must copy and even substituted a model to a given situation. This is reinforced by the way in which the simulation is most often thought out and built up: "*a technique that substitutes a synthetic environment for a real environment, in such a manner to enable work to take place under laboratory controlled conditions*" [16], or "*a strategy where a researcher substitutes the user with his/her model*" [17]. Indeed, this is clearly linked to the notion of "*fidelity*" or "*ecological validity*", and more extensively to the generalization of results: in what manner is the data resulting from such simulations representative of real work situations? In such an approach, simulation is explicitly apprehended and thought out as a copy of a

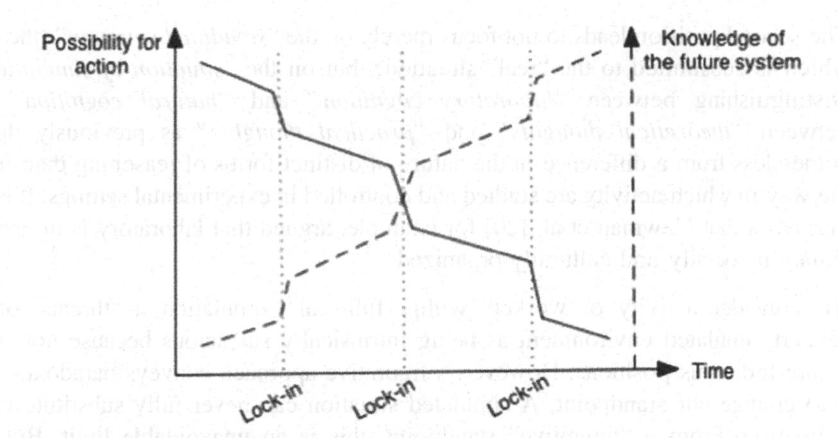

Fig. 1. The temporal articulation between possibilities for action and developmental curve of knowledge during design (according to Midler).

"*reference situation*" (whether it is an existing situation –as in training- or a future situation –as in design-). We named this position *"figurative"*.

Full-scale simulations are particularly representative of this figurative standpoint. Although these have existed for a longtime taking on different forms[1], the arrival of current technical possibilities to combine image and sound thereby creating virtual environments leads to considerable effort centered on simulator, questioning the many factors to take into account for a simulated situation: man-man interactions, organizational factors (e.g. time pressure), nature of resources and prescriptions, etc. And numerous efforts are made for defining valid and accurate substitutions of a real situation. But the workers acting in such full-scale simulated environments may appear disturbing, because their activities threaten fidelity. Villemeur [18] underlines this in a review centered on simulation in the nuclear field: *"even if the functioning of a nuclear plant is very well simulated, the situation is not perfectly representative, because the team is expecting to be confronted with a type of accident that they must manage according to defined procedures"*. It is clearly paradoxical to consider activity, performed individually or collectively, as non-pertinent or disturbing.

Faced with this paradox, there are two possible positions:

– The first one is a radical *"ecological"* approach, which definitively deny all interest to simulation methods. Such a position is for example those of Charlesworth [19] who considered that obtaining valid measurements, as well as identifying *"real cognitive processes"*, are only possible in natural environments. That line of reasoning leads to denying interest to simulation method, because simulation cannot be considered as *"natural situations"*. Consequently the activity observed in such setting would be regarded as intrinsically suspicious.

---

[1] We could talk about mobile mock ups used in the middle ages, the goal being to allow cavalry to develop motor-sensorial abilities when mounted on their horses.

- The second position leads to not focus merely on the "*simulated situation*" (the one which is substituted to the "real" situation), but on the "*situation of simulation*". Distinguishing between *"laboratory cognition"* and *"natural cognition"*, or between *"theoretical thoughts"* and *"practical thoughts"* as previously done, comes less from a difference in the nature of distinct forms of reasoning than from the way in which activity are studied and controlled in experimental settings. It is on that basis that Newman et al. [20] for example, argued that laboratory is in itself a situation socially and culturally organized.

To consider activity of workers within full-scale simulation as threats, or to apprehend simulated environment as being intrinsically suspicious because not "natural", are ludicrous positions. However, a figurative approach conveys paradoxes that need to change our standpoint. A simulated situation can never fully substitute for a real situation. From a "figurative" standpoint, this is an unavoidable limit. But for developing simulation methods, it is insufficient to question and thought out simulation solely from the angle of its relations (either fidelity or validity) to another situation, located in the real world, but which doesn't exist yet!

The figurative approach needs to be completed with an "*operative*" standpoint. Operative approach consists to question and to apprehend the use of simulation as resources for workers acting and interacting during design settings; those settings being to grasp in their technical, social and cultural specificities. From an operative standpoint, it is not a matter if what is simulated does not contain all the properties of the real world. On the contrary, it can be useful that a model forgets some dimensions, and pop out others one, allowing them to be conceptualized, manipulated or discussed during design. What is important is to support activity of the actors involved in design -this is the meaning of the term "*operative*"-. In this approach, "*ecological validity*" does not disappear, but must be situated in the context of the design process. For example, methods developed with theoretical modesty (e.g. mock-up associates to use) may appear preferable to more scientifically or technically founded methods (e.g. cognitive or full-scale simulation), because they are better suited to the temporal dynamics of the design process.

To simulate is before all defining operative resources, which must be apprehended in the specificity of the cognitive, cultural and social dimensions of their use. The two points below may help to question simulation from an operative standpoint:

- Models of a situation must not only be analyzed as a thing, none as a technical or a cognitive device to interact with. They are instruments, mediators for finalized actions. A key point of activity theory is to argue that, when using an instrument, a user is consciously involved in finalized activities during which the instrument allows the transformation of an object in order to reach a goal [21]. The way an instrument mediates action, the goal of the finalized action with that instrument and the object of action need to be questioned.
- Simulated situations are collective and multi-voice settings. To simulate could gain from being apprehended as a sort of bridge laid down between heterogeneous actors, as a common ground and carrier of exchanges by means and through simulated situations [7]. The aim, nature and dynamics of these collective dimensions must be analyzed and question in their cognitive and social dimensions.

4 Simulating for or with the Users?

Daniellou [22] has proposed to classify simulation of work systems in considering the status of the users or the workers. He distinguish (i) methods where the workers are not present but modeled (by means of a manikin or a computer program, etc.), (ii) methods where workers, individuals or collectively, are requested to take part in a controlled experiment, and their behavior analyzed, and (iii) methods where the users are participants in a participatory ergonomic process, simulation being one of the components of a more global involvement of different actors in the design process.

In many ways, this classification encounter the points discussed above. For example, the second methods, where workers are requested to take part in a controlled simulated situation, are largely grounded in what we named above a "*figurative approach*"[2]. However, such a classification goes a step further. Daniellou considers that methods where the workers are not present but modeled presuppose a triple modeling: a model of the Human, a model of the work system, and a model of the activity to perform. This means that such an approach requires a very great modeling capacity for those who implements it. Is it currently the case? The second method solves part of that difficulty: because a human is enrolling, such a method seems less demanding in modeling capacity (there is no need for a Human model, and to a lesser extent for a model of activity, because activity is expected to be performed). But the workers are placed in a strictly experimental context: their behaviors are analyzed, but the workers are not supposed to influence in any way the experimental settings, neither the design of their future work situation. We share the idea that, in discussing simulation, the questions are not only ontological or epistemological, but also axiological: are the workers the right to be human actor (and not only human factor) during the design of their future work situations?

Our main point is that understanding simulation as an individual and collective resource between heterogeneous actors engaged to develop their knowledge and to make experience of something to design, offers opportunity to develop simulation as a method for an ergonomics participatory process. Simulation is a method that can be use in order to contribute to a "*dialogical learning process*" between users and designers [23]. Basic components for such an approach are:

- As is it argued by the famous metaphor of a "*conversation with the situation*" proposed by Schön [25], each designer carries out, over the course of one's activity, some learning. The designer, aiming for a goal, projects ideas and knowledge. But the situation "responds" and is a source of "surprise", because it presents some unexpected forms of resistance that are a source of novelty, leading to learning. However, since design is a collective process, the other stakeholders of the process also "respond to" and "surprise" the designer, contributing to the learning process.
- In this context, the result of a designer's work is at best *an hypothesis*, steering the learning of other actors of the design process. But this learning may be possible or

[2] Whatever the designer's–experimenter's efforts to achieve a simulated situation that respects all the figurative elements of the situation, the activity that is set up within the controlled situation will be simulacrum for the worker.

impossible, leading – depending on the case – to validate, but also sometimes to refute, or more simply to put in motion the initial hypothesis.

- This approach leads to an interesting model, if one extends it to interactions between designers and workers. A novelty designed by the designers may lead to learning on the part of workers during simulation. But the worker can validate or invalidate the hypotheses produce by the designer. There is, therefore, a mutual learning process that should be supported and supervised by the ergonomist, due to the fact that design is a social process [25].
- Following such an approach, what is to be designed evolve over the course of dialogs between workers and designers. This concept of "dialog" should not be understood in the narrow sense of "verbal communication". It is a process with a dialogical structure: the result of the work of one person is put back to work within the activity of another, leading to a creative response source of novelty.
- The vectors of such a dialog are, for example, models of a setting, blueprint, mock-up, prototypes. We join here the issue of intermediary object [14], which posits that the objects produced during design are media, used for representation and communication between the stakeholders of design. In our own words, the intermediary objects constitute a vector of the dialogues between workers and designers.

5 Conclusion

From our perspective, designing entails coping with unpredicted variability, mobilizing personal and collective resources, experiencing contradictions and debates about values between human actors. The ergonomist has a role in the understanding of this density of the work activity and in the new avenues that may be opened to its development. Based on this background, our understanding of simulation can be resume with the points below:

- Simulation is ubiquitous during design. However, due to its concern (the living, cultural and social work activity), ergonomics is confronted with ontological and epistemological questions, which lead to relativizing and even abandoning any predictive ambition.
- Understanding simulation as a method explicitly apprehended and thought in relation to a future situation is not sufficient. To design is to define a specific and finalized work setting, which must be apprehended in the specificity of its cognitive, cultural and social dimensions. In those design settings, simulation must be grasp as instruments.
- Even if simulation is not explicitly serving as a learning goal, the use of these instruments appears undoubtedly associated to any questioning on learning, and development of knowledge and experience. Simulation brings on learning as a mediator in two ways: (i) through the building of a model, and (ii) in using the model.
- During design, simulation is a resource and a media for a dialogical and mutual learning process, where the result of the activity of one actor will be validate, refute, or set in motion, based on activity performed by an other actor involved in the process.

– For ergonomists, one challenge is to contribute, through and by simulation methods to a mutual learning process between workers-users and designers. From that point of view, simulation may support and contribute to a participative method.

References

1. Ombredane A, Faverge JM (1955) L'analyse du Travail. PUF, Paris
2. Wisner A (1995) Understanding problem building: ergonomics work analysis. Ergonomics 38:596–606
3. Guerin F, Laville A, Daniellou F, Duraffourg J, Kerguelen A (1991) Comprendre le travail pour le transformer. ANACT, Montrouge
4. Daniellou F (2007) The French-speaking ergonomists' approach to work activity: cross-influences of field intervention and conceptual models. Theor Issues Ergon Sci 6, 2005(5):409–427. https://doi.org/10.1080/14639220500078252
5. Pinsky L, Theureau J (1984) Paradoxe de l'ergonomie de conception et conception informatique. La Revue des Conditions de Travail n° 9, pp 25–31
6. Maline J (1994) Simuler le travail. Une aide à la conduite de projet. ANACT, Montrouge
7. Béguin P, Weill-Fassina A (1997) De la simulation des situations de travail à la situation de simulation. In: Béguin P, Weill-Fassina A (eds) La simulation en ergonomie, connaître, agir, et interagir. Octares Editions, Toulouse, pp 5–28
8. Daniellou F (1992) Le statut de la pratique et des connaissances dans l'intervention ergonomique de conception. Thèse d'habilitation à diriger des recherches, Université de Toulouse-Le Mirail, Toulouse, France
9. de Montmollin M (1986) L'intelligence de la tâche, éléments d'ergonomie cognitive. Peter Lang, Berne
10. Weill-Fassina A, Rabardel P, Dubois D (1993) Représentations pour l'action. Octarès, Toulouse
11. Midler C (1993) Situation de conception et apprentissage collectif. Réponse à Schön et Llerena. Les limites de la rationalité, Tome 2. Les Figures du collectif. La Découverte, Paris, pp 169–180
12. Béguin P (2005) La simulation entre experts. Double jeu dans la zone de proche développement. Dans Apprendre par la simulation. De l'analyse du travail aux apprentissages professionnels. P. Pastré, Toulouse, Octarès. P. Pastré (coord.), pp 55–77 Octarès, Toulouse
13. Béguin P (2014) Learning during design through simulation. In: Broberg O et al (eds) Ergonomics challenges in the new economy. Proceeding of the 11th international symposium on human factors in organisational design and management and 46th annual nordic ergonomics society conference, 17–20 Août 2014, pp 867–872
14. Bittencourt J, Duarte F, Béguin P (2017) From the past to the future. Integrating work experience within the design process. Work J Prev Assess Rehabil 57/3, 379–387. https://doi.org/10.3233/wor-172567
15. Morgan M, Morrison MS (1997) Models as a mediating instrument. In: Morgan M, Morrison MS (eds). Models as mediators. Perspectives on natural and social sciences. Ideas in context. Cambrige University Press, Cambridge
16. Sanders AF (1991) Simulation as a tool in the measurement of human performance. Ergonomics 34(8), 995–1025

17. Bronckart JP (1987) Les conduites simulées. In: Piaget J, Mounoud P, Bronckart JP (ed.) Introduction Psychologie. Gallimard, Encyclopédie de La Pléiade, pp. 1653–1662
18. Villemeur J (1988) Sureté de fonctionnement des systèmes industriels. Fiabilité-Facteurs humains - Informatisation. Eyrolles. Collection de la Direction des Etudes et recherches d'Electricité de France, Paris, France, 370 pages
19. Charlesworth WR (1976) Human Intelligence as adaptation: an ethnological approach. In: Resnick LB (ed) The nature of intelligence. Erlbaum, Hillsdale
20. Newman D, Griffin P, Cole M (1989) The construction zone: working for cognitive change in school. Cambridge University Press, Cambridge
21. Rabardel P, Béguin P (2005) Instrument mediated activity: from subject development to anthropocentric design. Theor Issues Ergon Sci 6(5):429–461. https://doi.org/10.1080/14639220500078179
22. Daniellou F (2007) Simulating future work activity is not only a way of improving workstation design. Activités 4-2. https://journals.openedition.org/activites/1704. https://doi.org/10.4000/activites.1704
23. Béguin P (2003) Design as a mutual learning process between users and designers. Interact Comput 15(5):709–730. https://doi.org/10.1016/S0953-5438(03)00060-2
24. Schön D (1988) Educating the reflexive practitionner. Jossey-Bass, San Francisco
25. Bucciarelli L (1994) Designing engineers. MIT Press, Cambridge

Implementing Sustainable Work Discussion Spaces (WDSs), a Challenge for Managers and Organizations

Lauriane Domette[1,2(✉)] and Pierre Falzon[1]

[1] Centre de Recherche sur le Travail et le Développement (CRTD), Conservatoire National des Arts et Métiers (CNAM), 41 rue Gay Lussac, 75005 Paris, France
laurianedomette@yahoo.fr
[2] Plein Sens, 5 rue Jules Vallès, 75011 Paris, France

Abstract. Facing a socio-economic context which tends to increase conflicting objectives while keeping managers away from the reality of their teams' work, some authors promote a management type based on a regular discussion on work issues. This implies to equip managers and organizations accordingly. In this perspective, about thirty managers from three establishments of the French Post Office were trained to conduct "Work Discussion Spaces" (WDSs) – i.e. meetings devoted to discussing work issues, regularly run by managers with their teams to concretely improve their work situations. A follow-up study was then conducted for two years, in order to apprehend the impact of the WDSs on work situations, management and organizations. This paper aims to show how these WDSs were actually implemented by the managers, how they transformed managerial practices, and what conditions are needed to implement sustainable WDSs. If WDSs have contributed to concrete improvements in work situations and to making some managers move toward more participatory practices, their effects remain local. These results call for a reflection within organizations, regarding the expected role of managers, the tools at their disposal and the leeway they have to better articulate strategic and operational logics.

Keywords: Work discussion spaces · Management · Organization

1 Problem Statement

The transformation of organizations and the implementation of new models of production have generated an intensification of work that reduces opportunities to discuss compromises made between increased and conflicting objectives (Detchessahar 2011; Daniellou 2015).

Additionally, the expansion of a "disembodied" (Dujarier 2015) and "impeded" (Detchessahar 2011) management, based on indicators, tends to keep managers away from the reality of work by involving them in numerous meetings or reporting tasks.

This is why some authors advocate alternative models of management, organization and even governance that enable employees to voice work concerns, such as a management "of labour" (Conjard and Journoud 2013), "by labour" (Bonnin 2013) or

S. Bagnara et al. (Eds.): IEA 2018, AISC 821, pp. 503–510, 2019.
https://doi.org/10.1007/978-3-319-96080-7_61

"by discussion" (Detchessahar 2013). They aim at better taking into account the "actual work" in the decision making and organizational process.

According to them, implementing a "Work Discussion Spaces" methodology, requires to design both:

- **An "engineering of discussion":** the discussion has to be regularly led by managers (trained in activity analysis) with their teams, about work situations faced by these teams that need to be improved or developed with concrete actions collectively built.
- **An engineering of the "organizational space"** in which this discussion takes place: allowing teams to gather regularly means to recognize this exchange time as a working time, by planning it in the organization of the production; it also implies that all the hierarchical levels are involved and have enough autonomy, leeway and power to support the implementation of the proposed action (Fig. 1).

Fig. 1. Synthesis of the criteria needed to design the discussion (Detchessahar 2011, 2013; Van Belleghem and Forcioli Conti 2015; Rocha 2014; Conjard and Journoud 2013)

Such a methodology also implies to develop managers and teams' empowerment, by implementing a "subsidiarity logic" – i.e. decision-making at the lowest relevant level of the organization.

2 Research Objective and Questions

In this perspective, we assume that implementing "Work Discussion Spaces" (WDSs) may contribute to concretely improve work situations, support managers in their role of regulating work by better articulating strategic and operational issues, and make organizations evolve toward more participation by promoting subsidiarity logics.

3 Methodology

To test these hypotheses, Work Discussion Spaces were experimented in three sites of the French Post Office: a mail sorting platform, a post offices network and a financial center. A field study was carried out in five steps: a diagnosis (analyzing work, its organization, existing discussion opportunities); a design step (establishing the WDSs' principles and adapting them to the situations); a training step (29 managers were instructed on work analysis and animation methods); manager's support/tutoring and experimentation's assessment steps (Fig. 2).

Fig. 2. The fieldwork methodology carried out

As a result of the diagnosis and design step, and in line with the literature recommendations, the WDSs were designed in a way that they could be adapted to the specificities of each context. The following WDSs' principles were suggested to managers in the training step: a one-hour Work Discussion Space per month, co-facilitated by them with their teams (8 to 10 participants), complementing the existing exchange opportunities (work situations could be reported by employees before the WDS in a meeting called "Communication Space-Time" – i.e. 1/2 h of top down-information followed by 1/4 h of reported concerns – and feedbacks on the decisions taken or actions implemented could be provided in the usual service meetings). Furthermore, to focus the discussion on "actual work" and lead it to concrete actions, managers were equipped with a five-step methodology (Cf. Fig. 3): introduction with "game rules" (expressing freely, listening without judgment, accepting divergence of opinion as an opportunity to open up possibilities and errors as an opportunity to learn…); elicitation interview and collective analysis of the situation before developing proposals and committing to implement them.

Then, a follow-up study of the WDSs' effects and evolution after two years of autonomous implementation has been organized. Eight semi-structured interviews were conducted: three with post office managers, one with the project manager of the Industrial Mail Platform and four with managers from the financial center.

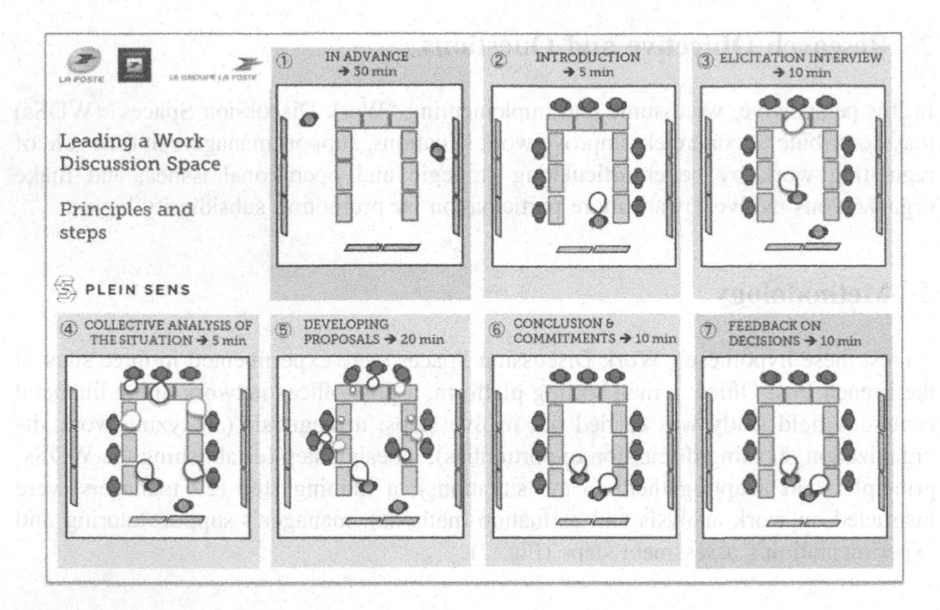

Fig. 3. Sequence of the discussion steps

4 Results

In this section, we will focus on three main results.

First, we will wonder how managers used the Work Discussion Spaces: how many WDSs did they conduct? What topics were discussed? What effects were generated? Were the WDSs adapted to each specific context, to be integrated into the productive activity, and how?

Then, we will analyze how Work Discussion Spaces transformed managerial practices.

Finally, we will consider how the implementation of sustainable WDSs challenges not only managers, but also the entire organization.

4.1 How Did Managers Use the Work Discussion Spaces?

In two years, as far as we know, 37 Work Discussion Spaces were conducted. They dealt with various subjects, such as *quality of service* (e.g. avoid jams in mail sorting machines or misdirected packages), *resources development* (e.g. change a script unsuitable for the situations encountered), *organizational and collective performance* (e.g. better define the role and task distribution between two teams; attend decision committees to better process a file; collectively catch up with a delay).

Most of the situations could be concretely improved in different ways. To give only three examples:

– *Quality of service*. In the Industrial Mail Platform, implementing a control sheet to avoid reversals of sorting plans has significantly reduced the number of errors.

"The numbers speak for themselves, we even have a very significant curve and history" [Production Manager, Industrial Mail Platform, our translation]

- *Resources development.* In the Financial Center, changing a script unsuitable for the situations encountered helped the agents to better interact with customers.
  "About the customer relationship, it helped them: they are better able to answer a customer who wants to know too much in advance, when we are not able to give reliable information. Having the script with sentences reassured them." [Manager, Financial Center, our translation]
- *Collective development.* In a service of the financial center, geographically isolated from the main building and where the agents' activities are relatively independent of each other, working on "how to collectively catch up with a delay" started to create team cohesion.
  "What stands out in my mind, in one word, is 'exchange' (...) an exchange that does not exist at present." [Manager, Financial Center, our translation]

To implement these WDSs, managers needed to adapt them to the specificities of their contexts, in terms of duration, frequency, integration in existing exchange opportunities or modalities of animation.

In one team where the Work Discussion Spaces are still going on, some modalities of the "prescribed WDS-instrument" fit well with their local specificities. For instance, the team had about ten agents, which corresponds to the number of participants recommended to allow both the expression of each and some group dynamic. But *regarding the duration and frequency*, they adapted the methodology to their specificities, so that WDSs could be integrated in the organization of work. Indeed, they used to conduct a weekly meeting called "Communication Space-Time", consisting of thirty minutes of top-down information and fifteen minutes of free discussion about work issues; they decided to replace one Communication Space-Time with one WDS each month, so that they could identify subjects to be discussed during the fifteen minutes of free discussion and analyze properly the topic chosen by the team during a WDS. Concerning the duration, they used to organize these Communication Space-Time every Thursday at lunch time, closing the post office to the public for forty-five minutes; this is how WDSs have been part of the organization, forty-five minutes a month instead of a Communication Space-Time.

In other teams, some adaptations are still being researched, as for example in a service composed of three teams comprising fifty people each. One of these three teams solved this problem by organizing WDSs spread over five days: the entire service was gathered to collect the topics that were then simultaneously discussed in small groups of five to six participants, to allow each person to express their views if they wished to. Furthermore, this work led the managers of all the teams in this service to meet, in order to share what was said in the different WDSs, to capitalize on these experiences, to arbitrate the proposals, and to give feedback to the entire service about the situations handled, the actions proposed and the decisions taken *in fine*.

As another example of adaptation regarding the *modalities of animation*, most of the managers decided to conduct the WDSs in pairs. Thus, those who were less comfortable in animation could develop skills with the support of their colleague or manager. In one situation, two managers supervising two different teams belonging to the same department decided to lead their WDSs together, one month with one of the teams, and the other month with the other team. In this way, they could regulate their workload but also share the product of the WDSs between the teams.

4.2 How Did Work Discussion Spaces Transform Managerial Practices?

As a second result, we can notice two ways in which WDSs transformed managerial practices.

An evolution in the positioning of managers. Some managers mentioned that conducting Work Discussion Spaces represented a significant change in their positioning, toward more participative management. Indeed, WDSs required them to help the team in defining not only the situations to be dealt with, but also the proposals for actions. However, the management is historically and culturally oriented toward a more "top-down" way: managers are usually required to set issues and provide solutions themselves. These different positionings are likely to create a "paradoxical injunction" if they are not articulated.

Managers' strategies to make the discussion efficient. Managers also encouraged their teams to choose "simple" topics to be discussed and find solutions that they could actually implement by themselves, so as to guarantee the effectiveness of the discussion. These motivations have the virtue of overcoming the skepticism of the teams, unsatisfied with the lack of effects following previous experiences of expression. But on the other hand, they could lead to managerial practices, paradoxically less favorable to the expression of the participants, such as the choice of the situations to be discussed by the managers only, or their reframing during debates when more strategic dimensions of the activity were addressed.

4.3 Work Discussion Spaces Sustainability: A Managerial and Organizational Challenge

The last result reveals that the Work Discussion Spaces did not last in time. These findings challenge not only managers, but also the entire organization.

A competition with other managerial tools. One of the main reasons for not carrying on with the WDSs in the Financial Center and the Industrial Mail Platform, is that managers are requested to implement many other participatory tools: lean management including daily briefs, one hour per month team meetings, working groups with variable durations and frequencies depending on the projects to be completed, "Communication Space-Time", co-development… Consequently, managers no longer have time to implement all these tools and do not identify the differences and complementarities between them.

A stable approach… in an ever-changing organization. Moreover, we can notice that, out of the twenty-nine managers trained in the implementation of the WDSs, seven (i.e. 24%) are no longer there, due to internal mobility during the two years of the experiment. This may have contributed to the process becoming unsustainable, especially when the successor was not trained initially. Similarly, on one of the pilot sites, a change in the top management (less favorable to the continuation of the WDS), limited the possibilities for the project leader to impel them.

Opportunities of discussion on work issues already existing. Among the reasons mentioned by the teams that did not implement WDSs, one manager said that she already benefitted from an organizational and social situation that allowed her to regularly discuss work issues in a less formal way, "as topics arise".

In a similar way, another manager, after conducting one WDS, said that the discussion on work issues continued less formally, during service meetings for example.

An intervention which supports managers and impels the WDSs' implementation. Finally, we can also assume that our solicitations had an impact on the impelling of the implementation of the WDSs. Indeed, our research methodology led us to conduct interviews before and after the WDSs, in order to support managers and collect data about the topics discussed, the output of the discussions, and the way the WDSs were implemented. We can notice that the moment when most of the WDSs stopped corresponds to the period when we stopped soliciting managers about WDSs. These results could argue in favor of supporting management in the sustainable implementation of the Work Discussion Spaces.

5 Discussion-Conclusion

This paper aimed to apprehend the impact of Work Discussion Spaces' implementation on management.

If this methodology has made it possible to concretely improve work situations, its impact remains local. Indeed, managers tend to favor "simple" subjects to be discussed, and even to reframe the debate when it moves toward more strategic questions, to be able to concretely implement their teams' proposals and ensure the effectiveness of the discussion. But by doing so, the development of subsidiarity logic (i.e. decision-making at the lowest relevant level of the organization) seems limited. These results question the expected role of managers: are they in a position to transmit information and support the transformation in a top-down way? To what extent can the feedbacks from the teams on work issues have an impact on the prescription system or on how the transformations are conducted?

Furthermore, WDSs may challenge the managerial culture. Indeed, they require a "participatory attitude" while another positioning is required the rest of the time. This can be experienced as contradictory injunctions.

Finally, most of the WDSs did not seem to last in time for various reasons. It raises research questions in terms of organization, management and methodology: how to design a stable device in an ever-changing environment? How to make managerial expectations more coherent regarding the current overabundance of managerial tools? How to adapt better the WDSs' methodology to each specific context?

References

Bonnin D (2013) Les modèles d'organisation et de management en question: pour un modèle de management par le travail engendrant la confiance. In: Karsenty L (ed) La confiance au travail. Octarès, Toulouse, pp 167–185

Conjard P, Journoud S (2013) Ouvrir des espaces de discussion pour manager le travail. Manag Avenir 5(63):81–97

Daniellou F (2015) Agir sur l'intensification du travail. In: Thébaut-Mony A, Davezies P, Vogel L, Volkoff S (eds) Les risques du travail. Pour ne pas perdre sa vie à la gagner. La Découverte, Paris, pp 246–255

Detchessahar M (2013) Faire face aux risques psychosociaux: quelques éléments d'un management par la discussion. Négociations 1(19):57–80

Detchessahar M (2011) Santé au travail. Quand le management n'est pas le problème, mais la solution…. Revue française de gestion 5(214):89–105

Dujarier A-M (2015) Le management désincarné. Enquête sur les nouveaux cadres du travail. La découverte, Paris

Rocha R (2014)Du silence organisationnel au développement du débat structuré sur le travail: les effets sur la sécurité et sur l'organisation. Thèse de doctorat en ergonomie. Université de Bordeaux, Bordeaux

Van Belleghem L, Forcioli Conti E (2015) Une ingénierie de la discussion? Chiche! Actes du 50ème congrès international de la Société d'Ergonomie de Langue Française, Paris, France

When Employee Driven Innovation Becomes an Organizational Recipe – Implications for What It Means to Be an Innovative Employee

Kristiane M. F. Lindland(✉)

University of Stavanger, Stavanger, Norway
kristiane.m.lindland@uis.no

Abstract. Despite the fact that creativity and solution-seeking activities to all times have been crucial for workplace progress, output and employee wellbeing, it is only in recent years that employee driven innovation has emerged as research theme in innovation literature. How to foster employee-driven innovation has been a theme of interest, both for researchers and for practitioners. In the Nordic countries, public sector is seeing EDI as a way of renewing its services, and additionally enhancing employee wellbeing. Based on findings in a longitudinal study of the implementation of EDI in Norwegian municipalities, there is reason to question whether *the way EDI is implemented*, actually can lead to less employee wellbeing through alienation.

Keywords: Employee-driven innovation · Public sector · Job crafting
Employee wellbeing · Meaning

1 Introduction

The creativity and solution-seeking activities of employees have to all times been crucial for progress and output in any work place and a vital source for employee wellbeing. However, it is mainly within the last two decades that research on employee driven innovation (EDI) as a phenomenon has emerged in the innovation literature. Much of the research on EDI has revolved around defining the phenomenon and identifying what kind of work-conditions that foster and constrain EDI (Kesting and Parm Ulhøi 2010). In the Scandinavian countries, EDI is increasingly recognized as a valuable tool for renewing the public sector. In addition to making better use of the competence, creativity and experiences of the employees, EDI, in itself is seen as contributing to enhanced employee wellbeing (Hansen et al. 2017). Consequently, the strategic efforts to implement EDI as a form of organizational "recipe" as implementation guide is spreading throughout the public sector. The aim of this paper is to explore how EDI as an organizational recipe for public sector relates to EDI as practice of employees. Further, how potential mis-matches in interpretations of EDI potentially can lead to employee wellbeing as well as discouragement and alienation in relation to employee driven innovation.

S. Bagnara et al. (Eds.): IEA 2018, AISC 821, pp. 511–520, 2019.
https://doi.org/10.1007/978-3-319-96080-7_62

1.1 Employee Driven Innovation and Employee Wellbeing

Employee driven innovation can be defined as « the generation and implementation of ideas, products and processes created by an employee or through shared effort of one or more employees" (Smith 2017, p. 1). In the Nordic countries, researchers have often connected employee driven innovation to the Nordic model, where management, employees and public authorities collaborate closely in the development of work organizations. This collaboration, which is also formalized through various rules and regulations, has led to a work culture characterized by shared responsibility for reaching organizational goals, employee influence and involvement in development and change processes, and short power distance between owners, management and employees. These characteristics of the Nordic work environment contributes to a work culture where employees to a larger extent than in more hierarchical societies, can take own initiatives for improving both work processes, work conditions and work output, without being sanctioned for doing so.

Despite the fact that EDI probably have been part of work life as long as there has been employees, it is only in recent years that EDI has been addressed in the innovation literature. The motivation for fostering employee driven innovation in organizations has typically emphasized the possibilities for making better use of the competences, creativity and experiences of employees in developing better solutions. Through EDI organizations can potentially improve their efficiency, enhance quality and user satisfaction. It is also seen as positive for employees to have more autonomy and influence over own work situation.

Given the prospective positive effects of EDI, researchers have by far focused on identifying the factors that apparently foster EDI, and thus how leaders can enable employees to initiate and implement innovative ideas. Kesting and Parm Ulhøi (2010) propose the following factors as central for enhancing EDI: work autonomy; insight in goals and strategies of the organization; high degree of leadership support; acceptance for failure; innovation as an expressed goal; slack in time and resources; and short power distances between leaders and employees. Smith (2017) found leadership support, autonomy, cooperation and innovation climate as the main enablers. For leaders, this probably implies that they must conduct their leadership role differently than the traditional top-down approach to innovation (Aasen et al. 2012). Hansen et al. (2017) found in their research that leaders had to foster employee autonomy, encourage the ideas of employees, create an informal work environment, reduce power distances and support the employees.

Employee wellbeing is understood as having positive effect on organizations through job retention and enhanced performance (Harter et al. 2002) which again appears to be fostered by autonomy support (Baard et al. 2004). Slemp et al. (2015, p. 3), building on Baard et al. (2004) and Moreau and Mageau (2012), define autonomy support as "an interpersonal orientation of one´s manager or work supervisor that involves acknowledging and understanding employee perspectives, providing employees with opportunities for volition over what they do and how they go about it, encouraging employee initiative, and being open to new initiatives". Slemp et al. (2015) propose that autonomy support predicts job crafting, which again predicts workplace wellbeing. Job crafting is about what the employee does to make a better fit between herself and the demands of the job. "By engaging in job crafting, employees can essentially reshape their job such as it becomes more closely aligned with their motivations for work, as well as their individual skills and preferences. This process affects the nature of the job itself, including the demands experienced on the job as well as a personal sense of efficacy for meeting those demands" (Slemp et al. 2015, p. 3). In other words, job crafting is something the employee does to fit the job to herself, her motivations and capabilities, and thus make it easier to handle the work tasks. Slemp et al. (2015) propose that autonomy support influences on job crafting, which again influences on employee wellbeing.

The purpose of fostering EDI in organizations is on the other hand not to improve the fit between employees and their tasks, neither to improve employee wellbeing, but rather about improving organizational output. However, employee wellbeing might be a consequence of employee driven innovation. Hence, this implies that leaders by performing a leadership style that supports employee autonomy can contribute to foster both employee wellbeing and a culture for EDI. The connections between job crafting and EDI, on the other hand, appears to be lesser addressed in the innovation literature.

According to the research on EDI and on employee wellbeing, there is reason to assume that an organizational recipe for fostering EDI, also will lead to employee wellbeing. Consequently, I explore and discuss whether the practical implementation of EDI as an organizational recipe, aligns well with autonomy support and job crafting.

2 Study Design

I have based this paper on a longitudinal study of the implementation of EDI as a strategic way of enhancing service-innovation in four Norwegian municipalities. Although the main trends in the findings I point to in this paper, goes for all four municipalities, the empirical situations I refer to, are taken from two of the municipalities.

In total, we conducted around 60 interviews in the two municipalities, where a few of them were group interviews. In addition, we did observations on various events in the two municipalities referred to in this paper. A survey repeated twice to all municipal leaders and employees were also conducted. However, the findings from the survey have been less relevant for this paper.

The main goal of the project was to explore how the four municipalities implemented EDI as a tool for innovation in the municipal services, and furthermore; how this could develop into a form of work practice. Based on existing research on EDI, we had some implicit assumptions about what factors that would enable, and what factors that would constrain the implementation. However, underway in the project, we saw that some of our previous assumptions were challenged. The theme for this paper has emerged through an abductive process, where the empirical material indicated needs for other analytical categories than first planned. Based on the findings, I have developed some plausible understandings of how EDI leadership acts in municipalities are interpreted by employees, and possible influences this might have on employee wellbeing. Hence, these interpretations are potential understandings of prospective consequences, rather than unambiguous, realized outcomes.

RFF Nord and RFF West in Norway financed the research project, and the project ran from the beginning of 2015 throughout 2017. The International Research Institute of Stavanger (IRIS) had the project management, and conducted the study in cooperation with NORUT.

3 Results

Norwegian municipalities are – together with many municipalities in Europe - facing an increasingly aging population, more demand for public services and reduction in both people to deliver the services and in tax income. The four municipalities we followed in our study had all decided to make use of EDI as a tool for improving public services. In the beginning of our project, we saw that the focus for the municipal management was on how to get municipal leaders at different levels and employees onboard and committed to the EDI work. The strategy for doing so was mainly embedded in developing a shared understanding of the need to find better future solutions, and why EDI would be a good way of addressing these future demands. Another central part of the implementation and spreading of the EDI work, was training leaders and employees in a shared method for how to develop the future solutions. The choice of methods were not taken out of the blue, just as the definition of EDI and the reasoning for it also was embedded in a concept as a form of an organizational recipe for how to do this. All four municipalities used external consultants to facilitate the implementation of the organizational "recipes".

The inspiration and "recipes" were sourced from other municipalities. Especially Aarhus in Denmark and Vestre Toten in Norway were central role models for the EDI-strategy, and most of the political and administrative leaders in the municipality of Alfa and the municipality of Beta had visited them and often referred to them as examples of how far you could go with this. One municipal manager in Alfa commented, "Vestre Toten has implemented the use of Lean very stricktly. We have chosen to be more flexible in our methods approach". The two municipalities in this case chose other methods that Lean, to start with. Nevertheless, the loyalty to using the chosen methods were acknowledged as central for how to build EDI competence.

The EDI work was "sold in" to leaders and employees in the municipal organizations through innovation days where successful EDI projects from other municipalities were presenting success-stories from their work. EDI was underlined as a smarter way to work. Through making better use of the competence, creativity and experience of employees, better solutions to municipal demands could be met. EDI would also very possibly lead to more influence on own job situation and more interesting work environment. We do not know whether the "selling-in" was tweaked regarding whether it was the municipal politicians or the municipal employees who were the target group. However, the politicians expressed expectations of EDI leading to more efficiency and better use of municipal budgets. Some employees, on the other hand, saw possibilities for realizing their own ideas for better services, without necessarily saving money. Other employees expressed this as just another trend recipe that would soon fade, and yet others did not get the message in the first round.

Part of the implementation of both a culture for EDI and for the methods, was to encourage employees to send in ideas for EDI projects the employees wanted to realize in their work situations. The administrative management expected these projects to work as vehicles for also spreading interest for EDI as well as spreading the solutions that were developed through the projects. That happened only partly. First, to hear about a project did not necessarily lead to others wanting to make their own projects. It was also unclear who should be responsible for spreading the developed solutions, as the involved employees not necessarily had time for doing this as part of their ordinary job.

Due to the slowness in implementing the methods and the awareness of EDI as a tool for improving the municipal services, Alfa municipality and Beta municipality increased the pressure for every municipal unit to be involved in the work. In Alfa, they organized an internal resource group that would provide training in Lean methods for the municipal units that wanted help. Every sector had to allocate a 20% position to the resource group, and could freely use the group in their units. The municipal management expressed an explicit expectation that over a given period, all units would have made use of the resource group. Implicitly, it also became important to register which units that had used the group. To conform to the EDI project managers indicated that leaders and employees out in the units might need a bit of pressure. "Most people are reluctant to change", was often said. The pressure in Alfa was also increased by starting leadership training also for the middle managers, as they were recognized as important for the realization of EDI out in the units.

In Beta, they increased the pressure for taking part in the EDI project by making all municipal units develop an obligatory innovation plan where they described what and how they would work with innovation in their own unit. This initiative did not turn out well. Many of the innovation plans were variations of "window dressing", where the plans were poorly connected to everyday work tasks, and appeared to be made for the sake of obliging to the system. Nevertheless, the plans were seen as visual symbols of the innovation work. When I asked one of the sector managers whether she felt they had succeeded the EDI work in her sector, she replied, "Well, no. We did not manage to develop the innovation plans very well." Despite the fact that she and her sector was one of the areas where they really had started developing a culture for EDI in their daily work, her answer related to their ability to oblige to the formal reporting structures. This division between what they did of EDI in their daily work and what they did to fulfill the demands of the formalized structures of the EDI project, appeared to be two separate things, where the top management owned and directed the formal project, while the more informal EDI work was embedded in the local work activities.

During the EDI project, some employees also started recognizing how other initiatives they took informally, actually could classify as EDI. In both Alfa and Beta, employees had asked the managers about whether also these self-initiated "projects" should be registered as EDI activities. The answer was that the registration was only for the projects that were part of the official project. This can be seen as another indication of that the fully employee-initiated projects were treated as something different from what the top management saw as EDI work.

Although the top management expressed the wish for employees to take initiative for finding better solutions in their work-environment through bottom-up-processes, the tools they used for increasing the pressure on spreading the EDI-work in the units were dominantly top-down oriented tools. Nevertheless, that is not an indication of employees refraining from using the EDI projects as possibilities to address needs and challenges they experienced in their work. Regarding the projects we followed over the 3-year project period, they were all embedded in employees experienced needs for better solutions. One example of such a project was from the unit for employment and social benefits, called NAV. Many of the employees in this unit are social workers with ethical ideals connected to their professions. Some of the social workers in Beta wanted to develop a service for young people who were outside both education and jobs. The social workers saw this user group as being of special importance, as the consequences of not getting into work or training at young age could lead to lifelong un-employment. They sent in a project idea to the EDI-project and got approval for starting up a "youth-guide". In contrast to other service workers at the unit, they would only have a few clients, which they could follow up very closely. The youth guides had mobile phones their clients could reach them on 24/7. They also had frequent social activities with the clients with the purpose of preparing and training their clients for situations they needed to overcome in order to get into work life. This service might appear as less innovative to us outsiders. Nevertheless, in the NAV system, this way of working is quite radical. Due to many clients per service worker, they have very little time to follow up the specific client. Client assistance is also increasingly digitalized, in order to reduce the work pressure on the service workers, and clients never get the direct number to their service worker. However, the experiences from giving the clients in the youth-guide project direct access to their guide, was only positive. "Our experiences are that our young clients call us less that ordinary clients do to their service worker. It seems that when they know they can reach us anytime, which in itself reduces the need to call us", says one of the youth guides. The youth guide project also gave the youth guides more freedom to collaborate employees in other relevant public units, such as youth club leaders and the police. This is an example of how employees used the EDI projects as tools for crafting their work situations in accordance to their professional ethics, motivation and capability in a system that often challenged these employee needs.

Also in Alfa, employees from early on used the innovation projects as opportunities for addressing needs they saw in their work situation. They used the projects as arenas to work on challenges that needed to be solved to make their work easier, but which there was little room or legitimacy for going deeper into in their everyday work. Through the early projects that we followed, the most important result of them was that the participants developed new ways of working. Consequently, they could also tackle new challenges through the work-processes and work-relations they had developed. As the pressure for more units having EDI-projects increased, the problems that were to be addressed sometimes appeared to be less embedded in experienced needs. This will probably also have influenced on their impact on learning and development of better solutions.

Summed up, the municipalities we followed implemented EDI as forms of organizational recipes, adapted from other municipalities, but later adjusted to the experienced needs underway in the project. The bottom-up processes that the municipal managers wanted to foster were challenged by the top-down structures developed underway, to control and enhance the spreading of the EDI-work.

4 Discussion

What we found in the project, leads to questions about whether the way EDI was implemented in the municipalities, both as part of the planned implementation strategy and the more iterative initiatives they took underway for fostering EDI, actually would lead to employee wellbeing. I will discuss this in relation to two aspects. First, how the implementation of EDI possibly contradicts both the factors that foster EDI and the factors that characterizes employee wellbeing. Second, how EDI potentially can develop into two different forms, existing in parallel.

EDI as a way of working resonates well with employee wellbeing, as autonomy support is underlined as an important condition for EDI. However, initiating and implementing EDI as an organizational recipe it is typically done through a top-down approach, and where the strategies and structures for doing so, can potentially lead to more restrictions than functioning as enablers. The way the municipalities implemented EDI as a concept and strategy for innovation in the municipalities, delivered two contradictory messages to the employees. On the one hand, EDI was presented as a way of working that made use of the creativity, ideas and experiences of employees, and employees were encouraged to take initiatives for defining needs, suggest solutions and test them out. On the other hand, the bottom-up approach that the leaders expressed, was simultaneously contradicted through their increasing demand for all employees to be involved in the EDI work through proposing "needs" for new solutions through the structured EDI work. To go further into discussing the consequences of this way of implementing EDI, I will discuss how the two central aspects Slemp et al. (2015) underline as predictive of employee wellbeing, are fostered through the implementation process.

In the specific EDI-projects we followed, we saw many informal examples of autonomy support from middle managers and the sector managers. In the most successful projects, the leaders already appeared to have a supportive approach to their employees. However, the EDI-strategy might have contributed to legitimize tighter collaboration between leaders and employees, and strengthen the efficacy of employees in pursuing their own ideas. The projects were also forms of learning labs for how far the employee autonomy could stretch out and thus how far the autonomy support could reach out. We saw that the relations between managers and employees and between employees themselves were in dynamic development. While the development of autonomy support is broadly about enabling employees in initiating and driving forward EDI-processes through bottom-up processes, the organizational recipes for *how* the EDI-work should be conducted, constrained the employees. Although the employees developed the ideas for what to suggest, the management, and later on dedicated persons would judge what ideas to take further. In itself, this selection method can be understood as a lack of autonomy. When the push comes for every unit to make use of the resource group in Alfa, and for making an innovation plan for every unit in Beta, these management acts might contribute to confirm for the employees the top-down approach in deciding that there are problems to address. Slemp et al. (2015) underlines how control and supervision actually undermines autonomy support. Hence, through what the managers did to boost the EDI work, they un-intentionally also weakened the autonomy support, which is the foregoing factor for both job crafting and then again employee wellbeing, as well as for EDI.

The most problematic aspect of the top-down approach as a weakening of autonomy support is that it weakens trustful relationships. The relation between employees performing autonomy and leaders performing autonomy support relies on a trustful relationship developed between leaders and employees. If that autonomy is reduced in practice, while the idea of autonomy is upheld, this might express both an undermining of employee judgement and an upholding of double messages. These factors have negative impact on employee autonomy by weakening employee confidence.

Autonomy support is suggested as a predictor of job crafting (Slemp 2015). In relation to the EDI project, we can see at least two forms of job crafting. On one hand, EDI strategies can be legitimized possibilities for employees to develop and organize their work tasks in ways that make them more meaningful and motivational for the specific employee. We could see this as a positive form of job crafting.

Job crafting can also be a way of narrowing the gap between external expectations of job fulfillment and the capabilities of the employee to fulfill these expectations. If the EDI work is experienced as a mandatory part of the work, then finding practical ways of fulfilling specific demands in the EDI project can thus be a form of job crafting. It can also contribute to employee wellbeing, as it can reduce frustrations. However, job crafting needs not necessarily lead to good EDI.

On the other hand, job crafting *can* be a source for EDI, as employees in their job crafting use their creativity, insight and efficacy to shape their work in better ways. We can also imagine that employees who have developed good capabilities for job crafting, will find it easier to adopt the idea of EDI as a form of both job crafting and service crafting towards users of public services.

What we have seen here is that EDI projects in the municipalities have two more or less different meanings. First, we have the structured and formalized top-down approach; next, we have the informal bottom-up approach. Although this is a simplified way of formulating it, it helps us identify possible conflicting demands. For the formal EDI approach, obedience and compliance with the directives given by the management, characterizes a good EDI employee. For the informal, bottom-up EDI work, employees will work out from what they see as reasonable, meaningful and efficient, independent of the formal structures and directives. What characterizes a good EDI employee in such a regime, is an autonomous employee with insight and interest, and who takes responsibility beyond her own, specific work tasks. It will also often imply being in conflict with existing ways of doing things, and thus possibly upsetting both managers and colleagues. In order to realize EDI as an organizational recipe, employees, middle managers and managers will thus need to handle these two approaches and their contradictions continuously. Alternatively, managers need to concentrate on providing autonomy support to employees, while also providing them with the insight and legitimacy they need for making informed choices. The question is whether the general public would accept such an approach for developing the municipal sector. Municipal managers are also dependent on having legitimacy for their work, and this legitimacy is mainly embedded in procedures, directives and control routines.

5 Conclusions and Implications

The aim of this paper has been to explore how EDI as an organizational recipe for public sector relates to EDI as practice of employees. Further, how potential mis-matches in interpretations of EDI potentially can lead to employee discouragement and alienation in relation to employee driven innovation. I discussed potential for employee wellbeing in relation to autonomy support and job creation. Findings indicated that despite managers underlining that EDI work happens through bottom-up processes, EDI was in fact implemented as an organizational recipe on the initiative of the managers. The EDI work that employees had initiated and organized more informally, were not taken into account when they documented the EDI activities. Several employees understood this as there being a division between EDI as practice and EDI as a manager-driven project. The increasing urge to direct, control and document the project activities contributed to possible ambiguities about what autonomy the employees actually had in the EDI work. However, the reason for this increased control might just be that the managers fall back on the tools usual for conducting management in the public sector, namely through procedures, control and documentation. Despite the unintended consequences of these additional demands, control is contradictive to autonomy support, which again influences on potential for job crafting. Consequently, if EDI is implemented through top-down processes where the leeway for autonomy related to both tasks and methods are restricted, there is reason to believe that employee wellbeing will be reduced, and be replaced by alienation to the project and its activities.

These findings and prospective outcomes of how the EDI project was managed, underscores how leadership acts get their meaning through their contextual interpre-tation, not through the intentions of the managers. Learning EDI managers how to manage for fostering EDI demands more than adopted methods and strategies, it demands close attention to how employees already are performing autonomy, and thus fostering this autonomy further. This will again imply coping with the risk of employees acting in other ways that managers intended. Autonomy support will imply to avoid controlling activities and expressed lack of trust.

It is easy to say that EDI can foster employee wellbeing as well as improving public services, and further, that this will demand other ways of working, for both managers and employees. The challenges are often that the municipal context and work culture in itself, can restrict to what degree the necessary changes can be realized. However, the expectations to how the municipal sector can be arenas for innovation can change through new experiences. Hence, EDI projects in municipalities have a function in itself as a step in developing possible futures.

In addition to enhancing employee wellbeing, employee autonomy and job crafting can both be factors that fosters EDI, and also function as sources of EDI initiatives in themselves. More research on how autonomy support and job crafting influences on EDI as well as employee wellbeing, is needed.

References

Aasen TM, Amundsen O, Gressgård LJ, Hansen K (2012) Employee-driven innovation in practice – promoting learning and collaborative innovation by tapping into diverse knowledge sources. In: Lifelong learning in Europe vol 4

Baard PP, Deci EL, Ryan RM (2004) Intrinsic need satisfaction: a motivational basis of performance and well-being in two work settings. J Appl Soc Psychol 34(10):2045–2068

Hansen K, Amundsen O, Aasen TMB, Gressgård LJ (2017) Management practices for promoting employee-driven innovation. In: Workplace Innovation, pp 321–338. Springer, Cham

Harter JK, Schmidt FL, Hayes TL (2002) Business-unit-level relationship between employee satisfaction, employee engagement, and business outcomes: a meta-analysis. J Appl Psychol 87(2):268–279

Kesting P, Parm Ulhøi J (2010) Employee-driven innovation: extending the license to foster innovation. Manag Decis 48(1):65–84

Moreau E, Mageau GA (2012) The importance of perceived autonomy support for the psychological health and work satisfaction of health professionals: Not only supervisors count, colleagues too! Motiv Emot 36(3):268–286

Slemp GR, Kern ML, Vella-Brodrick DA (2015) Workplace well-being: the role of job crafting and autonomy support. Psychol Well-Being 5(7):1–17

Smith R (2017) Work(er)-driven innovation. J Workplace Learn 29(2):110–123

Work Macroergonomics Analysis (AMT Method): Identification of Ergonomic Demands in Sewing Laboratory

Luiza Debastiani e Silva(✉), Bruna Marina Bischof, Raquel Pizzolato Cunha de Oliveira, Ricardo Schwinn Rodrigues, and Elton Moura Nickel

Santa Catarina State University, Florianópolis, Brazil
luizadebastianis@gmail.com, brunamar@gmail.com, raquel.pizzolato@gmail.com, schwinn.ricardo@gmail.com, eltonnickel@gmail.com

Abstract. The sewing laboratory is a space used by students almost every semester throughout the fashion course degree. It is common for students to staying in the laboratory for several hours nonstop, which justifies the importance of carrying out a study on this environment in order to identify issues. Therefore, the objective of this paper is to perform a macroergonomic analysis in the sewing laboratory of the bachelor fashion design program at the UDESC - Santa Catarina State University (Florianópolis, Brazil), in order to identify the Ergonomic Demand Items through the Work Macroergonomics Analysis (AMT) taking into consideration the Human Factors involved in the activities practiced in this workspace. The methodological procedures involved the Work Macroergonomics Analysis method, proposed by Fogliatto and Guimarães (1998). As a result, the macroergonomic approach was considered efficient in the identification of ergonomic demands, which help in the evaluation and propositions of suggestions for improvement the sewing laboratory, increasing the efficiency and effectiveness of the practices performed there, as well as health and wellbeing of the students, technicians and teachers who use this workspace.

Keywords: Macroergonomic analysis · Sewing laboratory
Ergonomic demands

1 Introduction

The bachelor's degree in Fashion design seeks the training of students regarding the uptake of the technological and social reality of the textile and garment production chains, according to the UDESC Fashion Department's online page [1]. In addition, according to the information available on the website, fashion designer profession require the theoretical-practical domain in order to execute innovative designs that meet the needs of a contemporary fashion market. Therefore, it is understood the importance of practical classes, allied to the theoretical knowledge, on the formation of well-skilled professionals.

S. Bagnara et al. (Eds.): IEA 2018, AISC 821, pp. 521–529, 2019.
https://doi.org/10.1007/978-3-319-96080-7_63

Among practical classes, disciplines involving the use of the Sewing Laboratory are offered in virtually every semester throughout the UDESC fashion design bachelor's degree. Thus, it is crucial for the UDESC Sewing Laboratory to be similar to a professional atelier, approximating it to clothing industry standards and providing a more realistic practical experience for students.

In this sense, it's interesting to note the indications of the *Serviço Brasileiro de Apoio às Micro e Pequenas Empresas* (SEBRAE) - Brazilian Service to Support Micro and Small Enterprises- [2] regarding the gear that a professional atelier studio, in Brazil, should dispose. Those being: cabinets with shelves, trims organizer, cutting and finishing table, cutting machine, industrial straight sewing machine, overlock, galley, interlock sewing machine, sewing machine, knitting machine, chairs and air conditioner, macaws, table, chairs, armchairs and air conditioner, plus a specific places for reception, and an office that are not necessary for the practical sewing classes.

The sewing classes length, can reach up to four hours in a row, which are often added with the time provided for extra class activities, where students can use the workspace accompanied by a supervisor, which usually happens in the hours before or after class. Thus, it is common for students to staying in the laboratory for several hours nonstop, which justifies the importance of carrying out a study on this environment in order to identify issues, as better use of space, efficiency of activities plus circulation, and the students health. There are four levels of organizational analysis [4]: technological needs investigation, design of organizational structure linked with appropriate intervention, implementation of the process, and measurement and evaluation of organizational effectiveness. For this, there are several macroergonomic tools and methods widely studied and recognized by ergonomics literature.

In order to perform this study, it was sought to use the macroergonomic approach, which proposes an restructuration of a whole system through participation and optimization with the focus on increasing the efficiency and safety of the system as a whole. This approach, differs from the more micro-oriented approaches such as Ergonomic Work Analysis.

Macroergonomic approach is preferred in this particular research, rather than microergonomics, even if the focus of the analysis is only one part of the organization: the sewing laboratory - by reason of the systemic view allows one to observe factors that beyond this reduced, departmental, scope. A global understanding may provide greater clarity about the perception of punctual errors.

Therefore, the objective of this paper is to perform a macroergonomic analysis in the sewing laboratory of the bachelor fashion design program at the Santa Catarina State University (Florianópolis, Brazil), in order to identify the Ergonomic Demand Items through the Work Macroergonomics Analysis (AMT) taking into consideration the Human Factors involved in the activities practiced in this workspace. The methodological procedures involved the Work Macroergonomics Analysis method, proposed by Fogliatto and Guimarães (1999) [3].

2 ATM e DM Methods

Participatory and holistic, the, Work Macroergonomics Analysis (AMT) comprises workplace, organizational and environmental factors covering safety and life quality, aiming the improvement of the system as a whole [3]. The stages for applying this method are similar to ergonomics interventions, comprising: project launch; ergonomic appraisal - exploratory phase, which ergonomic demands are detected through observation and contact with users of the system under consideration; ergonomic diagnosis - detailed ergonomic evaluation and data analysis obtained on the previous stage; ergonomic design; evaluation or ergonomic validation - construction of models or prototypes for ergonomic testing through the collaboration of workers next to specialists; ergonomic detailing and optimization [3, 5].

The macroergonomic approach developed by Guimarães [3], used in this work, is aided by the Macroergonomic Design (MD) tool, which aims to design a product, workstation or system according to the ergonomic demands pointed out by users [6]. According to its creators Fogliatto and Guimarães [7], this tool contemplates seven stages:

1. User identification and organized data collection regarding the ergonomic demand;
2. Prioritization of Ergonomic Demand Items (EDI) identified by the user;
3. Incorporation of expert opinion, incorporating new items that were not pointed out by users, correcting discrepancies;
4. List of design items (DI) to be considered in the workstation ergonomic design;
5. Relate EDI and DI and identify design items to prioritize
6. Ergonomic treatment of design items; and
7. New project implementation and follow-up.

Both the Work Macroergonomics Analysis and its tool Macroergonomic design, provide stages, allow a complete macroergonomic and participatory analysis. However, due to resources such as time and availability of changes, the ergonomic assessment and ergonomic diagnosis steps based on the AMT method [3] were used to reach the proposed goal of identifying the Items of Ergonomic Demand through the macroergonomic approach. This research was carried out in 6 phases with the aid of a questionnaire, interviews, writing instruments and audio recording through a mobile device.

Permission to carry out the analysis proposed here occurred through contacting one of the teachers and a technician responsible for the classes at the State University of Santa Catarina Sewing Laboratory. Once the authorization was granted, the researchers considered it coherent to observe the students' experience. By observing at different course levels, in order to obtain an overview of the ergonomic conditions evaluated and to verify if the EDIs remained or changed according to the experience in the sewing laboratory. The UDESC fashion course lasts eight semesters and is an annual entry, which means that only odd or even levels are offered semiannually. At the moment of this analysis, the levels offered were the odd ones, and as in the first level the students still do not have sewing classes, were searched the classes of the third, fifth and seventh levels.

For this research, the strategy "b" of the Macroergonomic Design tool [7] was used to identify the Ergonomic Demand Items (EDI). Initially, informal interviews were carried out with students from the different levels, along with direct observations on work and environment at different Sewing Laboratory classes moments. Were possible to identify some ergonomic issues related to the posts, students sources of constraint and analysis of the environment as a whole.

Regarding the observations and interviews carried out, the researchers performed the ergonomic evaluation pointing EDIs cited by students - users of the system - or by freely observed and intuited. Those were then grouped by assumptions, as classroom, environment, workplace and health. After that it were list in a questionnaire form, considering the degree of importance.

Ergonomic diagnosis was then performed, detailing the ergonomic appreciation through the listing of EDIs in order of affinity, and then measuring their degree of importance. For this measurement, a continuous scale of fifteen centimeters was used for each EDI, with two anchors at the ends marking the states of "unsatisfied" to the left, in the section of 0 cm to 5 cm; "Very satisfied" on the right, in the section of 10 cm to 15 cm; as well as another central anchor marking the "satisfied" level in the range of 5 cm to 10 cm, as shown in Fig. 1. Six students from each phase were invited to mark their opinions about EDIs through these scales: each student should point a position on the scale line, within the fifteen centimeters arranged for that purpose.

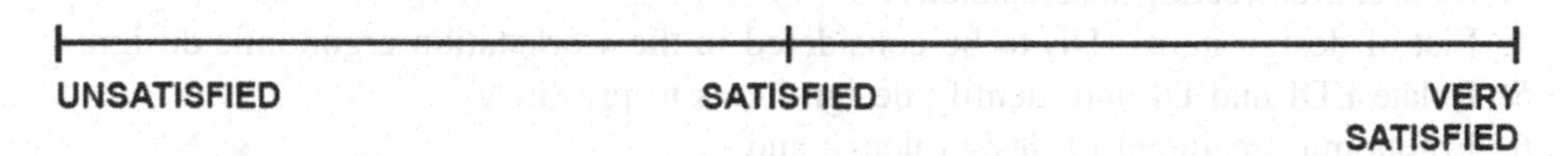

Fig. 1. Anchors used in the questionnaires. (Source: Developed by the authors from Fogliatto and Guimarães [7])

In order to carry out the last step, corresponding to the analysis and the diagnosis of the Sewing Laboratory, each student's EDIs answers were measured with a decimal precision ruler, from left to right, and tabulated in order to enable the overview of the data obtained. Following are the results and the discussions that could be carried out from the application of the method described here.

3 Results and Discussions

The Ergonomic Appraisal in the Sewing Laboratory was carried out through direct observation, it was possible to identify ergonomic issues regarding the work stations, sources of student constraints, and moreover, an analysis of the work environment as a whole. Firstly, will be presented the general analysis of the environment and the Ergonomic Demand Items, spotted out by the group of students, through an interview on the Ergonomics Assessment stage; Then, will be presented the analysis of EDIs with lower satisfaction indexes, for understanding if there are relations between EDIs, whether they are causal and, finally, suggestions for improvement.

With respect to the work environment, it was identified that the Sewing Laboratory of the Santa Catarina State University comprises of a single and wide room which accommodates industrial irons, cutting table, sewing machines, as well as small offices reserving space for the fitting room, the laboratory technician room and the teachers room. In addition, the place uses of artificial lighting over natural. Six white fluorescent lamps that illuminate the whole room. For air circulation, the laboratory has small swing-type windows located on the same wall as the entrance door, on opposite wall there are large windows, covered with curtains. This configuration contributes for an environment virtually absent of natural lighting. In addition, the large windows on the laboratory are permanently closed, therefore the environment relies exclusively on air conditioners - facing the sewing machines and cutting table - for its ventilation.

The layout of the laboratory environment is divided into two areas delimited by the furniture and appliances arrangement. On the first half of the room the students are faced with a large table of 1,80 m × 6,24 m, with a height of 0.92 m - surrounded by chairs. After crossing the table space, they face the sewing machines, arranged in five rows. On further wall - below the large windows - there is a single row of sewing machines and on the opposite wall - where is the door - are five industrial iron appliances. Therefore, the students flow within the laboratory is constant, between the sewing machines, the large table and the irons areas.

After the environment analysis, individual interviews were conducted with six students from each analyzed curricular phase - third, fifth and seventh - totaling eighteen interviewees. In conformance with the orientation of Fogliatto and Guimarães [7], firstly, was requested for the student to "talk about the laboratory (atelier)", and were given the option to discourse freely about the atelier, its functions delegated by the teacher and their perceptions about the environment. Immediately after the speeches were finished, the interview was continued, in an induced manner, addressing the following topics: illumination, temperature, air circulation, noise, occurrence of physical and psychological constraints, flow needed to perform activities, environment cleaning, safety, conservation, opinion on the current aspect and quality of furniture, machines and their adequation to the chairs and table dimensions. The interviews pointed out that, in general:

- Students are uncomfortable with the strong artificial light that illuminates the studio uninterruptedly - some believe that, although uncomfortable, it is necessary for the execution of the tasks. In addition, they associate the strong light with the pains in the eyes and headache;
- Students do not feel constrained in any way with the individual orientation made before everyone, even if their mistakes are exposed in the making of the clothes pieces, yet they complain about the lack of monitoring in the classroom, because they believe the teacher can not attend to all. In addition, they have shown dissatisfaction with the way in which individual guidance is organized, since some stay longer than others being oriented;
- Opinions about the temperature, vary, but the majority considers the environment very cold; furthermore, they point out that, although the air circulation is satisfactory, it would be interesting if the ventilation could be more natural. There was a

case of complaints concerning dry eyes and throat due to air conditioning, the student in question suggested the use of an air humidifier during classes;

- Room size (whole laboratory), quality of machinery, furnishings and cleanliness had all positive ratings; only one student reported problems due to residues of fabric and dust. In addition, all students feel safe in using the machinery, even if, at times, the iron releases hot steam randomly;
- The disposition the machinery and furniture in the environment was considered good and well organized, except for the fact that a student, left-handed, complained that all the irons are positioned on the right side ironing table. As for the flow, the movement required to carry out the activities in the laboratory was considered good, due to the fact that the students sit for a long time, and therefore they see as a positive thing the fact to have to cross the different areas of the laboratory;
- Not all students complained about the height of the cutting and sewing tables, however, many confessed to have back and neck pain. In addition, few students pointed out difficulties in regulating the chair seats.

After the Ergonomic Appreciation stage, the Ergonomic Diagnosis were helded, in it, the ranking of ergonomic demands occurs by measuring the degree of importance of each demanded item [7]. Thus, in this stage the observation and the interviews of the previous stage, served as input to format the questionnaire, the EDIs identified were grouped into four categories: classroom, environment, workplace and health. Being subdivided into:

- Class: satisfaction, orientation, organization of orientation, orientation time, relationship with colleagues, flow of external people in the studio;
- Environment: Cutting table (height and space available), cutting table chairs (height, quality and positioning), furniture (quality and positioning), layout of workflow, noise (conversations and machinery), noise level, music during class, cleaning, artificial lighting, natural lighting, lighting machinery, temperature and air quality, air circulation, positioning of the air conditioning, air conditioning noise;
- Workstation: leg room in the machine, machine chair height, machine table height, machine quality, table top height, industrial iron (weight);
- Health: pain (neck, arms, hands, head, legs, spine, feet and eyes), fatigue, psychological stress, safety iron and machinery).

The eighteen students - previously interviewed - answered the questionnaire, marking the scale (see Fig. 1) on how they felt about the covered topics. The use of this tool is interesting in macroergonomic appreciation, since anonymity, confidentiality and freedom of expression of thought [8] makes the respondent with a sense of protection and free to express their real opinion.

From the questionnaire responses, a table[1] was created containing the average of each phase and the general average of the phases referring to the answers of the topics covered, where it was possible to observe that the topics marked as unsatisfied (values smaller than 5) were, respectively:

---

[1] The table can be accessed from the link: https://docs.google.com/spreadsheets/d/1Zpg9wenVnTcE3shfbIHuqGx311ervsjn-RSK-ZAI7T4/edit?usp=sharing.

- Third stage: headache (4.52), neck pain (4.03), eye pain (3.80), psychological stress (3.68) and spinal pain (2.93);
- Fifth stage: psychological stress (3.75);
- Seventh stage: head pain (3.97), neck pain (3.47), psychological stress (3.20), spinal pain (3,12), natural light (3,10) cutting table - space available (2.98).
- General: neck pain (4.92), spinal pain (3.73) and psychological stress (3.54).

It is interesting to note that there is a hegemony regarding the dissatisfaction with psychological stress, which is punctuated in all levels analyzed. It is understood that there may be correlation with the deadlines offered for each activity, together with the relative lack of order in the orientation, pointed out by the students during the interviews. It is also noted that neck pain and pain in the spine are among the general dissatisfactions, in accordance with the observations made, which indicated that the positions with the head low and the spine are curved in all the workstations - in the iron, at the cutting table and especially on the sewing tables - this problem could be minimized with the proper adjustment of the chairs, but most of them are broke, and the students confessed not having the habit of adjusting them.

In order to expand the analysis, a chart (Fig. 2) was elaborated with topics that obtained a general average below 8.5 cm - using as criterion satisfaction (7.5 cm) plus 1 cm. In this way, it is possible to observe that the rated EDIs show less satisfaction regarding the psychological stress, pain, illumination, circulation and air temperature, noise and height of the furniture. In addition, the graph makes it possible to make some interesting correlations, such as headache being related to noise level and lack of natural light - something that was observed, talked about in the interviews and confirmed in the questionnaire. It is interesting to note that although not figured far below the anchor "satisfied" (7.5 cm), an item that was not very talked about in interviews and appears on the graph is the "flow of external people in the studio", proving the importance of using the observations and interviews allied to the questionnaire for a more realistic perception of the analysis.

Psychological stress, which has the lowest overall average, may be connected to fatigue, to the organization of monitoring activities, and to monitoring activities time, which may result in a productivity loss by the students. To prevent this from happening, it would be interesting to take introductory classes every time a new item was to be made by the students, where the teacher explained the step-by-step and indicated the main mistakes made. This would only be possible for the third and fifth levels. And then, for all levels, is suggest to prepare an ordered list of students to be attended with predefined time.

The available space at the cutting table is an EDIs that is on the verge of dissatisfaction, which, although not mentioned in the interviews, was observed by the researchers. When seated on the table, students' feet stand on the "shelf" under the table - where unused bags and materials should be kept, however they stand on the tables taking up space. this ergonomic demand could easily be avoided if students used the appropriate space to store their personal belongings.

Lastly, the space for legs in the sewing machine; the height of the table and the chair of the machine; and, the regulation of the cutting table chair can be correlated to the EDIs of greatest dissatisfaction corresponding to the pains of the head areas, spine

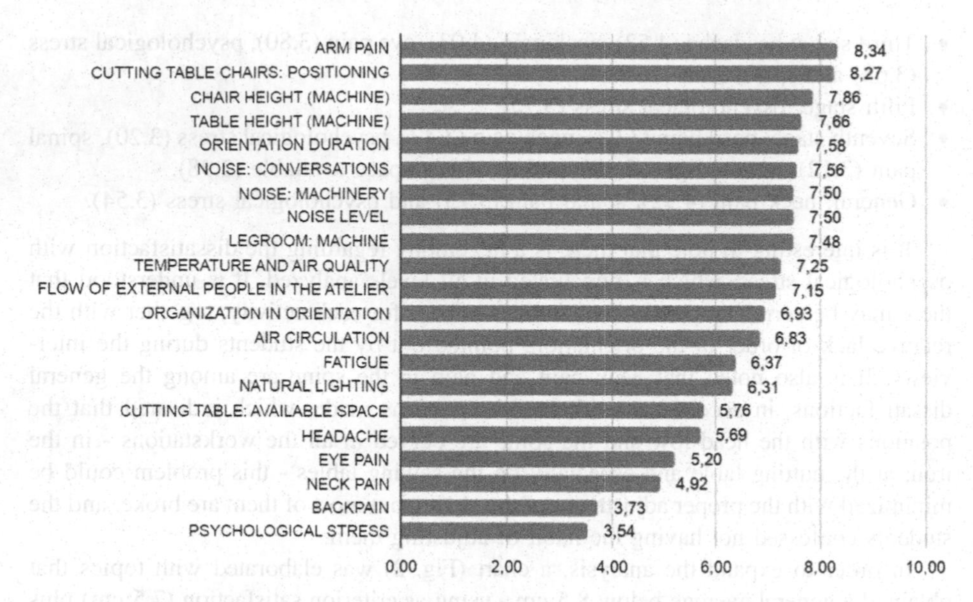

Fig. 2. General averages below 8,5 (Source: Developed by the authors)

and neck. Although they are more difficult to solve because it is a furniture exchange, it is advised that the students are submitted to pauses from time to time ordered by the teacher to relax the muscles and stretch, as well as make postural recommendations or labor gymnastics.

4 Considerations

As a result, the macroergonomic approach was considered efficient in the identification of ergonomic demands, which help in the evaluation and propositions of suggestions for improvement the sewing laboratory, increasing the efficiency and effectiveness of the practices performed there, as well as health and wellbeing of the students, technicians and teachers who use this workspace.

Carrying out this analysis, it was noticed that one of the advantages of the employed method is the possibility to evaluate the IDEs through more than one platform, as observation by the evaluators, verbal interview with the students involved, and the anchors questionnaire. Thus, correlations and causalities can be better perceived and examined, as well as suggestions for more efficient improvements can be developed.

Many of the issues raised by this macroergonomic analysis could be resolved if the furniture and equipment of the UDESC sewing laboratory went through maintenance more frequently. By this maintenance, it is understood not only the repair of its final functions, but of all its structure and good functioning. An example is the cutting table or sewing machine chairs whose height adjustment mechanism does not work.

Problems reported by students such as neck and back pain could easily be solved if the adjustment mechanisms were running and available for use.

Other questions raised by the students could also be solved in a simple way, such as the use of light and natural ventilation i would be enough just open the curtains and the windows. The flow of external people into the laboratory, at least during classes, could be resolved with some indication on the door placed in these periods, with the words: “we are in class, please do not enter”. However, there are more complex points to improve: those related to the exchange or reform of university goods, such as the height of cutting tables and machines and space to place and support the feet, others related to changes in organizational culture, as changes in waiting time for assistance or aspects of pedagogical and didactic evaluation, such as delivery times considered short and that could be related to psychological stress.

Finally, it extends to the State University of Santa Catarina and the Postgraduate Program in Design and Graduate degree Program in Fashion, the most sincere thanks on the part of the authors of this text, who could benefit from financial, material and intellectual support made available by the university. Likewise, thanks to the students, who kindly agreed to participate and collaborate with the present research, as well as the teacher and the technician of the sewing studio for their proactivity in accepting that this analysis be carried out in their place of work.

References

1. UDESC Fashion Department’s online page. http://www.udesc.br/ceart/moda. Accessed 26 Apr 2018
2. SEBRAE, Como montar um Atelie de Costura. http://www.sebrae.com.br/sites/PortalSebrae/ideias/como-montar-um-atelie-de-costura,fe787a51b9105410VgnVCM1000003b74010aRCRD. Accessed 26 Apr 2018
3. Guimarães LB de M (1999) Abordagem ergonômica: o método macro. In: Guimarães. Ergonomia de Processo. UFRGS/PPGEP, Porto Alegre, cap 1. v 1
4. Guimarães, LB de M (2004) Ergonomia de Processo, V2, 4th edn. FEENG, Porto Alegre
5. Moraes A, Mont’Alvão C (1998) Ergonomia: Conceitos e Aplicações. 2AB, Rio de Janeiro
6. Krug SR (2000) Aplicação do método de Design Macroergonômico no projeto de postos de trabalho: estudo de caso de posto de pré-calibração de medidores de energia monofásicos. Porto Alegre
7. Fogliatto F, Guimarães L (1999) Design Macroergonômico de Postos de Trabalho, Produto & Produção. Porto Alegre, vol 3, no 3, pp 1–15
8. Hendrick HW, Kleiner BM (2009) Macroergonomics: theory, methods, and applications. CRC Press, New York

On Human Terms – A First Evaluation of a Massive Open Online Course (MOOC) in Ergonomics

Martina Berglund[1(✉)] and Anna-Lisa Osvalder[2,3]

[1] HELIX Competence Centre and Division of Logistics and Quality Management, Linköping University, 581 83 Linköping, Sweden
martina.berglund@liu.se
[2] Design & Human Factors, Chalmers University of Technology, 412 96 Gothenburg, Sweden
[3] Ergonomics and Aerosol Technology, 221 00 Lund, Sweden

Abstract. The Massive Open Online Course (MOOC) 'Work and Technology on Human Terms' (www.onhumanterms.org) was launched in July 2017 with the aim to contribute to safer and healthier workplaces by increasing the knowledge about how products, systems, and work organizations can be designed on human terms. The purpose of this paper is to present the results of a first evaluation of the MOOC. The online course was used in four different university courses in Ergonomics in Sweden, two given at Chalmers University of Technology in Gothenburg and two given at Linköping University. The MOOC material was used in different ways in the courses: (1) suggested voluntary, alternative material for the students' self-studies, (2) scheduled activity for self-studies with appointed chapters, and (3) mandatory, selected course material being discussed in follow-up seminars. Data for the evaluation was collected through questionnaires and semi-structured interviews with students and teachers. The results showed that the MOOC served as a repetition of lectured material and gave increased understanding of the theories. The recorded interviews with practitioners and researchers in the MOOC highlighted the importance of the subject in real working life. The knowledge tests were appreciated as rehearsal of understanding. However, the MOOC in parallel with the other course material was also considered to be too much work by some students. A recommendation is to carefully consider how to use and integrate the MOOC as a meaningful, individual, theoretical learning activity for the students. Thereby the lectures in classroom could focus more on discussions and problem-solving regarding the topics and less on basic theory.

Keywords: Higher education · Blended learning · Human factors

1 Introduction

In July 2017, the Massive Open Online Course (MOOC) 'Work and Technology on Human Terms' (www.onhumanterms.org), was launched. It had been developed by the publishing company Prevent and five Swedish Universities of Technology. The aim of

S. Bagnara et al. (Eds.): IEA 2018, AISC 821, pp. 530–538, 2019.
https://doi.org/10.1007/978-3-319-96080-7_64

the course is to contribute to safer and healthier workplaces by increasing the knowledge about how products, systems, and work organizations can be designed on human terms. The potential participants of the MOOC are university students in e.g. engineering, economics, human relations as well as professionals holding equivalent university degrees.

Ergonomics is a wide field that encompasses multiple disciplines. Corlett and Clark (1995) defined Ergonomics as "the study of human abilities and characteristics which affect the design of equipment, systems and jobs. It is an interdisciplinary activity based on engineering, psychology, anatomy, physiology and organizational studies." From a teaching perspective, it can be a challenge to cover the width of the field. This applies for course design, teacher competences and finding suitable study material. In addition to that, there is a need to open up for different learning styles among the students. Integration of a MOOC may open up the possibility for flexible learning based on the student's individual conditions. Furthermore, it can be used as a complement to traditionally organized courses. The purpose of this paper is to present the results of a first evaluation of the MOOC 'Work and Technology on Human Terms'. The online course was used in four different university courses in Ergonomics in Sweden, two given at Chalmers University of Technology in Gothenburg and two given at Linköping University.

2 Theoretical Background

2.1 Study Approaches and Learning

There are different study approaches, which influence the outcomes on students' learning (Ramsden 2003). In a surface approach, the students memorize facts and details without much reflection about the implications of their meaning (Marton and Säljö 1976). In a deep approach to studying, on the other hand, the students reflect on the meaning of theories and how they can be applied to real-life situations. Research shows that deep learning approaches are related to higher quality outcomes as well as higher students' satisfaction (Ramsden (2003). A deeper understanding of the subject is crucial to be able to apply principles and concepts in a new situation (Silén, 1998), such as applying Ergonomics in design and product development.

A deep approach may be encouraged through four principal factors (Biggs 1989):

- An appropriate motivational context
- A high degree of learner activity
- Interaction with others, both peers and teachers
- A well-structured knowledge base.

2.2 MOOCs

Since the first Massive Open Online Course (MOOC) was developed and offered in the late 2000s by Canadian professors (Blackmon and Major 2017) there has been an increasing interest and number of MOOCs. There are different types of MOOCs, but common features are (ibid, referring to Major and Blackmon 2016):

- **Massive:** There are no limits of participants and the courses could potentially have large numbers attending.
- **Open:** The courses are often free and accessible to anyone.
- **Online:** The courses are offered online.
- **Courses:** No credits are offered. However, the courses have a specific content and often a syllabus or some other structure for elaborating the course material. There are often assignments and self-assessments.

Earlier studies have reported experiences from MOOCs, e.g. from integrating an entire MOOC in a campus course (Bruff et al 2013). It offered a possibility for blended learning with some positive elements, such as self-paced learning for students. However, there were also challenges in matching the online material with activities in class (ibid). Other studies similarly reported both opportunities and challenges when integrating parts of MOOCs in university courses. Opportunities included that the students are offered different types of course contents resulting in enriched discussions among the students, and the teachers have the possibility to redesign courses without developing online material themselves (Israel 2015). Challenges included the difficulty to embed MOOC material in existing courses.

3 Method

3.1 Courses in Which the Use of MOOC Was Evaluated

The MOOC material was used and evaluated in four university courses in Sweden; two at Linköping University (LiU) and two at Chalmers University of Technology. At Linköping University, two 6 credit courses in Product Ergonomics were evaluated. One was for third year students in furniture design, furniture carpentry, and upholstery, educational programs to a great deal also based on arts and crafts. The course involved 15 students. The other course in Product Ergonomics was taught to second year engineering students in design and product development. This course involved 68 students. The DPD program was developed to meet demands on future engineers to be able to solve technical and functional problems as well as usability issues. The program includes classical engineering subjects such as Mathematics and Engineering as well as e.g. Ergonomics, Industrial Design and Interaction Design. The goal is to educate product development engineers with a human centered approach (ISO-standard 2010) and an understanding of aesthetic values when developing new products and services.

At Chalmers University of Technology, two 7,5 credit university courses were evaluated. One was a course in Human-Machine Systems (30 students) for engineering students within mechanical engineering in the third year. This course deals mostly with working environments in general and specifically on work place design for operators in complex technical environments, such as control rooms in process industry, production systems, vehicle cabins, intensive care, aviation, shipping etc. Here, cognitive ergonomics and risk and safety are deeply studied. The second course was in Ergonomics/ Human Factors (50 students) for industrial design engineering students in the second year. This course deals much with ergonomics design of products, where physical and cognitive design aspects are in focus as well as environmental aspects such as noise, lighting and climate aspects.

3.2 Data Collection for the Evaluation of the MOOC Material

Data for the evaluation were collected during the academic year of 2017/18, and through a mix of questionnaires and interviews, which varied based on possibility in the different courses, see Table 1 below. To learn about the students' views, questionnaires with open and closed end answers were used. The questionnaires with closed end answers dealt with what parts of the MOOC material that had been used by the students and to what extent, ratings of the content of the MOOC material, and its importance for learning. The open end questionnaires dealt with in what way the MOOC material had been used (or why it had not been used, the students' views on the material, advantages vs. drawbacks, suggestions for further development including the teachers' roles to facilitate the use of the MOOC material. In two of the courses, the students' views were also collected in semi-structured interviews with the same topic as the open questionnaire.

Regarding the teachers' views and experiences, the paper is based on the authors' own experiences from teaching in all the analyzed courses. In two cases, the course in Product Ergonomics for engineering students in Design and Product Development and the course in Ergonomics for engineering students in Technical Design and Industrial Design Engineering, a questionnaire with open answers was also sent to the teacher team. This questionnaire dealt with how the MOOC material had been used by the teachers, its advantages and drawbacks, ideas for further use of the MOOC material, etc.

An overview of the collected data is found in Table 1.

Table 1. Collected data for the evaluation.

Course	Type of data collection	Type and no. of respondents
Ergonomics, Chalmers (autumn 17)	Open end questionnaire	20 students
	Closed end questionnaire	20 students
	Open end questionnaire	3 teachers
Product Ergonomics, LiU (autumn 17)	Closed end questionnaire	7 students
	Semi-structured interview	4 students
Product Ergonomics, LiU (spring 18)	Open end questionnaire	16 students
	Closed end questionnaire	11 students
	Open end questionnaire	6 teachers
Human Machine Systems, Chalmers (spring 18)	Closed end questionnaire	10 students
	Structured interview	10 students

4 Results

4.1 The MOOC 'Work and Technology on Human Terms'

The MOOC-course 'Work and Technology on Human Terms' (www.onhumanterms.org) was introduced in August 2017. The aim of the course is to be a modern, highly accessible study material in Ergonomics, which can contribute to safer and healthier workplaces as well as improved organizational performance. The course was developed by the publishing company Prevent in collaboration with five Swedish Universities of Technology. It can be regarded as a complement to the earlier published book (Bohgard et al. 2011).

The MOOC is in English and free of charge. It comprises totally 20 h of study, and the student is free to start and finish as he/she pleases. The course also includes knowledge tests and the student receives a certificate after being approved. The course covers the subjects Human-Technology-System, Psychosocial and organizational environment, Physical environment, Information and interaction in technical systems, Methods and design processes, and Economic and legal conditions in Sweden. Theory and models are presented with animations supported by interviews with experts and researchers active in the various disciplines. The content of the MOOC is directly related to different types of businesses, which is reflected in interviews with product developers, managers and safety representatives. Practical application of the theory is presented in four workplace cases: the IT-company, the Food Company, the Construction Company and Home Care. For a further description of the course, see Lagerström et al. (2017).

4.2 The Uses of the MOOC in the Evaluated Courses

The MOOC material was used in different ways in the courses: (1) suggested voluntary, alternative material for the students' self-studies, (2) scheduled activity for self-studies with appointed chapters, and (3) mandatory, selected course material being discussed in follow-up seminars. In the first case (1), the students were introduced to the MOOC during lectures, in which the teachers described the structure of the MOOC, and how to register. The students were then advised to use the MOOC as they wished. It was clear that it was a voluntary, complementary material. In the second case (2), the MOOC material was also used for self-study, but in this case the self-studies were scheduled throughout the course and the students were instructed what parts to study at certain scheduled times to match the other course activities. In the third case (3), parts of the MOOC-material was mandatory. The students had assignments in which the MOOC-material had to be elaborated by the students on their own, and the material was later followed up and discussed in seminars. For a further description of the uses of the MOOC course, see Osvalder and Berglund (2018).

4.3 The Students' Views on the MOOC

The students perceived that the MOOC material was relevant for the evaluated courses. It covered all areas and was considered as a good introduction to the field. However, it was considered shallow. Depending on what was focused in the evaluated courses, different parts of the MOOC was elaborated and considered more relevant. In general, the introduction with a focus on humans and technology at work was highly interesting along with psychosocial work environment.

Since the students at Chalmers were focusing on Technical Design and Product Development, cognitive and physical ergonomics, the parts about human-technology, risk and safety were mostly in focus. Here, the MOOC provided overall basic knowledge and the examples were relevant regarding simple technical products. There were, however, a lack of examples related to more complex sociotechnical systems. In one course at Linköping, many students similarly mentioned physical ergonomics, implying physical load but also physical factors as important parts on which they focused. Since the course had a project work at a building site, many students also looked at the case 'The construction company' and found it useful.

The questionnaire results showed that the students at Chalmers University of Technology rated the overall MOOC-course to 3.8 on a 1–5 scale (5 highest score), although too shallow and extensive for a course in a specific area of Ergonomics. The course was also considered trustworthy (4.2 on the 1–5 scale). The ratings of specific parts and features are shown in Table 2.

Table 2. Evaluation of the MOOC-course by students at Chalmers University of Technology.

Evaluated item	Mean rating on a 1–5 scale (5 highest rating)
Animations	4.4
Interviews with practitioners and researchers	3.0
Text material	4.0
Speaker presenting text material	3.6
Study questions	3.3
Depth of content	3.3
Knowledge questions	3.5
Case studies	3.1
Modern material	4.4
Navigation	2.2
Pedagogical	4.1
Useful in working life	3.4
Enough depth of theory	3.1
Easy to understand the content	4.4
Resulting in increased understanding for the subject	3.8
Useful in university education	3.8

At Linköping University, the ratings varied on some aspects regarding the course for students in furniture design, furniture carpentry, and upholstery and the engineering students, why their ratings are presented separately, see Table 3.

Table 3. Evaluation of the MOOC-course by students at Linköping University.

Evaluated item	Mean rating for students in furniture design, furniture carpentry, and upholstery	Mean rating for engineering students
Animations	3.8	4.3
Interviews with practitioners and researchers	4.6	3.2
Text material	3.0	3.6
Speaker presenting text material	2.4	1.4
Knowledge questions	-	4.0
Case studies	3.0	4.5
Modern material	3.8	3.4
Navigation	2.8	3.4
Pedagogical	3.0	-
Useful in working life	2.5	4.0
Enough depth of theory	2.5	4.0
Easy to understand the content	3.2	3.0
Resulting in increased understanding for the subject	4.7	4.3
Useful in university education	3.8	4.0

Since the courses had both lectures and offered web course, the students often chose to either follow lectures or the web course. It was considered too much to follow both. In case they missed any lecture, they used the web based instead. In general, many students just looked at obligatory sections, or central sections of the course, they did not study other sections of general interest. The main criterion for selection was what was focused in the university course.

The course book related to the MOOC-course was considered as good among those who read it. It was the basis for profound knowledge. Some students focused on reading the book and considered the MOOC as a waste of time, while some students considered the MOOC material as sufficient to pass the examination. One student even pointed out that the MOOC material had been crucial for passing. Major advantages with the MOOC was that it could be used at one's own pace and whenever suitable.

4.4 The Teachers' Experiences and Views on the MOOC

As described above, the teachers used the MOOC in the evaluated courses in different ways. They also used it for their own purpose. All teachers looked through the MOOC regarding their own topic (in which they were teaching) to see how the topic was presented and what was highlighted, type of self-study questions, etc. For some teachers, the MOOC material was also used as part of their lectures (e.g. by showing an interview or an animated part) or served as inspiration for examination questions.

All teachers described positive aspects in the MOOC, although there were varying views. Many pointed out that the MOOC served as a good introduction to the different topics. However, there were also a couple of teachers who found parts of the material too shallow and wanted deeper theorizing. The animated parts were mainly considered positive, being short, clear and easy to understand. So were also the interviews with practitioners, which could help the students relate theory to practice.

Other views were that the MOOC material could present the different topics within Ergonomics in alternative ways, as complementary material, and the student could elaborate the material at their own pace.

Areas of improvement included more depth in the MOOC material, suggestions on how the MOOC could be used for specific user groups, e.g. engineering students with a focus on Design and Product Development, and development of a teacher site related to the MOOC.

5 Discussion and Conclusion

The results showed that the students especially liked the animated parts that served as a repetition of lectured material and gave increased understanding of the theories. The recorded interviews with practitioners and researchers in the MOOC highlighted the importance of the subject in real working life. The knowledge tests with yes/no questions were appreciated as rehearsal of understanding. However, having the MOOC as a scheduled activity in parallel with the other course material, lectures and course book, was considered to be too much work for the students within the given time limit for the university course. Among the students who had the MOOC as voluntary complementary, there were also few students who used the material. Sometimes this was due to difficulty or hesitation to register, and for others lack of time.

The MOOC material served as a good introduction to the different fields within Ergonomics. Some students asked for deeper theory. However, it is a challenge to find the right mix for introducing the wide field of Ergonomics for a large group of potential students and practitioners, while at the same time providing the depth for students diving deeper into specific areas. Working with the MOOC material does require the students' own initiative in selecting what to work with, doing knowledge tests etc. and hopefully that may be one of facilitating deep learning.

Worth mentioning is also that Ergonomics is a subject taught to many different types of students. Not all of these learn best mainly by reading literature. Here, MOOC material in combination with questions to prepare regarding specific topics and follow-up discussions could be a meaningful way of learning.

One conclusion of this evaluation is that if a course offers both lectures, web based course and literature, the students will choose the sources of information that suit them best for their studies. They will probably not use all sources. As the learning aims for the course often differ from that of the web course, it will be difficult for the students to match what is written in the course PM.

Another conclusion is that a MOOC cannot be run parallel to regular teaching without a proper course design so that the MOOC is integrated in a suitable way. It is advisable to avoid lectures and MOOC material abut the same topic. The students will not use all sources in parallel.

The recommendation to teachers is to carefully consider how to use and integrate the MOOC as a meaningful theoretical learning activity for the students by themselves. Thereby the lectures in classroom could focus more on discussions and problem-solving regarding the topics and less on basic theory.

References

Biggs J (1989) Approaches to the enhancement of tertiary teaching. High Educ Res Dev 8(1):7–25

Blackmon SJ, Major CH (2017) Wherefore art thou MOOC? Defining Massive open online courses. Online Learn J 21(4):195–221

Bohgard M, Karlsson S, Lovén E, Mikaelsson L-Å, Mårtensson L, Osvalder A-L, Rose L, Ulfvengren P (2011) Work and technology on human terms. Prevent, Stockholm

Bruff DO, Fisher DH, McEwen KE, Smith BE (2013) Wrapping a MOOC: student perceptions of an experiment in blended learning. J Online Learn Teach 9(2):187–199

Corlett EN, Clark C (1995) The ergonomics of workspaces and machines: a design manual, 2nd edn. Taylor and Francis, London

ISO-standard 9241-210:2010 (E) (2010) Ergonomics of human-system interaction – Part 210: human-centred design for interactive systems

Israel MJ (2015) Effectiveness of integrating MOOCs in traditional classrooms for undergraduate students. Int Rev Res Open Distrib Learn 16(5):102–118

Lagerström G et al (2017) Development of the online course "Work and Technology on Human Terms". In: Proceedings for NES 2017 joy at work, Lund, Sweden

Marton F, Säljö R (1976) On qualitative differences in learning: I - outcome and process. Br J Educ Psychol 46(1):4–11

Osvalder A-L, Berglund M (2018) On human terms – integration of a massive open online course (MOOC) in ergonomics in university courses. In: Proceedings of the 20th Triennial Congress of the IEA, Florence, August 26–30

Ramsden P (2003) Learning to teach in higher education, 2nd edn. Routledge Falmer, London/New York

HTO at Vattenfall from External Requirements to Strategic Advantage

Fredrik Barchéus(✉)

Vattenfall, Evenemangsgatan 13, 169 79 Solna, Sweden
fredrik.barcheus@vattenfall.com

Abstract. The Swedish energy company Vattenfall has employed the concept Human-Technology-Organization since the late 80's. As mandated by regulations, all nuclear companies retain competence in the area, and Vattenfall also have a specialized HTO-unit that functions as a central competence resource for Vattenfall and its subsidiary companies.

Although the main work is done in the areas of Human Factors Engineering and event investigation, the trend is that efforts are increasingly being called for regarding organizational matters. One reason for this is the recent strategy to decommission several nuclear plants which has triggered a need to manage a new kind of venture as well as retaining competence by stimulating workforce motivation.

The HTO-unit has been commissioned to specify a strategy for work in the HTO area in order to maximize the benefit for the company. One of several initiatives from this strategic assignment is a new support function for selected groups in order to disseminate and promote a systemic perspective in the organization. The support function is carried out through participation at selected meetings and by challenging the participants at the meeting by questioning based on an established set of principles of systemic thinking. The expectation from this support function is to make decisions more well founded and thus of higher quality.

The paper describes a selection of cases in which consequences from the wider HTO perspective has been evident, comprising benefits as well as some identified pitfalls that can serve as relevant experiences for readers interested in pursuing similar endeavors.

Keywords: Human-Technology-Organization · Energy sector
Nuclear · Strategy

1 Introduction

Vattenfall is the largest energy company in Sweden with an annual electricity production of over 110 TWh to more than six million European customers. The company is based in Sweden but has business in seven European countries. In 2015 Vattenfall reoriented its strategy towards becoming fossil free in one generation. Although Vattenfall invests heavily in wind power and also provides customized sun power solutions for residential homes, the main electricity production comes from nuclear and hydro power, which is organized within Business Area Generation. This department harbors

S. Bagnara et al. (Eds.): IEA 2018, AISC 821, pp. 539–548, 2019.
https://doi.org/10.1007/978-3-319-96080-7_65

two of Sweden's nuclear power companies, as well as the company tasked with handling of spent nuclear fuel, alongside the Vattenfall-internal business units Hydro and Projects & Services. In accordance with regulations the nuclear companies have dedicated competence in the area Human-Technology-Organization, HTO. The supporting unit Projects & Services retains a group of HTO-specialists that functions as a central competence center for Vattenfall and its subsidiary companies.

The main work is done in the areas of Human Factors Engineering and event investigation, but the trend is that efforts are increasingly being called for regarding organizational matters. One reason for this is the recent strategy to decommission several nuclear plants which has triggered a need to manage a new kind of venture as well as retaining competence by stimulating workforce motivation. Another reason for increased organizational focus is the acknowledgement of Human Factors as an important discipline in such matters.

The HTO-unit also provide training in safety management, Human Factors Engineering and event investigation, for appointed personnel groups with the ambition to reach all areas in the company.

2 Historical Background of the Area Human – Technology – Organization at Vattenfall

As owner of several nuclear plants Vattenfall has followed the development of the nuclear industry regarding Human-Technology-Organization. Thus, the main applications are Human Factors Engineering, event investigations and safety culture. Following a heightened focus on the human element after the Three Mile Island accident, the Swedish nuclear authority, developed and expanded its competence in Human Factors in the 1980's.

Vattenfall explicitly incorporated the HTO-perspective in the late 1980's when HTO-competence was employed at managerial level. At that time there was a surge in application of HTO-perspective at Vattenfall including its nuclear subsidiaries and key personnel were enforcing new procedures in Human Factors Engineering, safety culture and event investigation.

After a reorganization, the HTO-unit was defined as one of several specialist areas in a separate technical consultancy company. The intention was that the company should support refurnishing and other projects just as any other technical consultant. The organizational form required an ability to respond to tenders according to customers' requirements. The outspoken strategy was then also to focus on the nuclear sector, with its emphasis on safety.

The pendulum swung back in 2014 when the consultancy company was reintegrated into the main company Vattenfall as a part of the ordinary operations. The intention was to increase the capability by treating the prior consultants as a centralized pool of support personnel and mandating subsidiaries to give priority to these internal resources instead of procuring engineering consultancies outside of the organization. As an effect of this, the HTO-unit was given the opportunity of a more strategic potential not only limited to explicit customer requirements but with a certain capability to formulate a strategic vision of HTO as a way to gain a more foundational traction within the organization.

With the introduction in the 80's there was a great focus on HTO, also on the managerial level, but it seems as though with further institutionalization application has been more restricted to lower operational levels. While e.g. project managers have become better at including HTO-competence, there is less ownership at executive level.

3 Three Applications of HTO

According to Swedish nuclear regulations each nuclear operator is required to consider humans in design of nuclear systems (including refurbishment), explicit management of safety culture, and connections between humans – technology – organization when performing event investigations (Table 1).

Table 1. Extract from regulations on holders of permits for nuclear operations [1]:

Regulation	General advice on the application
Chapter 2. Basic safety provisions Section 9. The licensee shall ensure that: […] 6. the personnel working in the nuclear activity are provided with the necessary conditions to carry out work in a safe manner,	[…] analyses and evaluations of the man-technology-organisation interaction should be conducted and recurrent evaluations performed
Safety programme Section 10. After it has been taken into operation, […] the safety of a facility shall be continuously analysed and assessed in a systematic manner. […] An established safety programme shall be in place for the safety improvement measures, i.e. technical as well as organisational measures arising as a result of this continuous analysis and assessment	[…] analysis and evaluation of facility safety should particularly take into account technical and organisational experience […] Organisational experience for example refers to results from analyses of man-technology-organisation interaction, evaluations of the organisation and the personnel's working conditions as well as self-assessments of the safety climate and safety culture
Chapter 3. Facility design Section 3. The design shall be adapted to the personnel's ability to, in a safe manner, monitor and manage the facility and the abnormal operation and accident conditions which can occur	[…] experts on the man-technology-organisation interaction should be engaged to take part in the design, analysis and evaluation of the solutions
Chapter 4. Assessment and reporting of the safety of facilities Safety review Section 3. A safety review […] shall be performed in order to verify that applicable safety aspects have been taken into account and that applicable safety requirements with respect to the design, performance, organisation and activities of the facility are met	The safety review should comprise a review of technical factors as well as are view of the man-technology-organisation interaction. Thus, personnel with adequate technical competence within the areas in question as well as personnel with competence in behavioural sciences should participate in the review work. […]

(*continued*)

Table 1. (*continued*)

Regulation	General advice on the application
Chapter 5. Operation of the facility Investigation of events and conditions Section 4. […] investigation […] shall be conducted systematically. As far as possible and reasonable, the investigation shall determine the sequence and causes of an event or the causes of another demonstrated safety deficiency as well as establish the measures needed to restore the facility's safety margins and to prevent the recurrence of safety deficiencies. […]	[…] The investigation methodology should be characterised by all relevant aspects and circumstances having been taken into account, including technical factors as well as those relating to the man-technology-organisation interaction

As nuclear companies organize in order to meet these requirements the organizations become, in some sense, replications of the outline of the regulations. One company may apply an interpretation that results in an organization where safety culture is managed within the Human Resource department, Human Factors Engineering is retained within the construction department and event investigation is performed inside the quality department. This may not necessarily be the optimal way of organizing, as they are have their idiosyncrasies that must be acknowledged. Benchmarking attempts between companies have resulted in failure when another company tried to allocate responsibility for safety culture to their Human Resource department. Thus, each nuclear operator has adapted its organizational structure to external requirements as well as the internal logic of the organization, both mechanisms inducing an apparent separation between Human Factors Engineering, event investigation and safety culture.

Conversely, specialists in the HTO-area see these three areas as applications of HTO rather than as disciplines in their own right, whereas the way of organizing may make the connections hard for people in the organization that are outside the HTO-area. The result may be that if one raises issues from a Human Factors Engineering point of view, i.e. in the development phase, a response may be that it is strange that the HTO-specialist raises the issue since "HTO is something you do when investigating incidents". Partially this may be a pedagogical issue, but it may also lie deeper.

4 HTO – An Integrational Perspective

As a discipline, HTO supports the indivisible connection between human centeredness and systems thinking. In design of technical systems one of the main tasks becomes to identify the users, their tasks and conditions in terms of organizational procedures and roles, identifying the interactions among them. In researching an incident the perspective may be expanded from a human actor (typically an operator performing an "error") and identifying the technical and organizational conditions.

For specialists in HTO the interrelation between these perspectives both become a value adding uniqueness as well as a source of confusion. An organization that has sorted competences according to a certain logic, e.g. the outline of regulations, may find it discouraging that separate paragraphs in regulations should be coupled together. In the case of experience reporting systems there are things that an organization has to do as (Table 2).

Table 2. Requirement on experience reporting in the nuclear sector [1]

Regulation	General advice on the application
Chapter 2. Basic safety provisions Section 9. The licensee shall ensure that: […] 7. experience of importance for safety from the facility's own nuclear activity and from similar activities is continuously utilised and communicated to the personnel concerned, […]	Efficient procedures should be in place for continuous experience feedback within the nuclear activity. In the light of experience gained, it should be continuously investigated whether the facility and its activities comply with the applicable conditions and regulations

From a HTO perspective this only covers part of the intention since there are other requirements about "good design" of both the organization and the reporting tool. For the purpose of example we may call the two stances integrative and separatist. The separatist interpretation is principally directed toward satisfying the stated requirement. According to this position actions on the Human, Technology, and Organizational sides may be taken independently.

The integrative interpretation is directed toward satisfying the underlying rationale of the stated requirement while appreciating related requirements, such as the ones concerning provision of necessary conditions for human work, appropriate management systems etc. According to this position relations between Humans, Technology, and Organizations are intertwined and actions should be considered related to these aspects in conjunction (Table 3).

Table 3. Example of separatist and integrative perspective

	Separatist	Integrative
Human	Individuals have an obligation to report any deviances that may be of importance.	In order to learn from past experience it is of essence to have mechanisms for sharing insights. Such mechanisms include means to express, analyse and disseminate knowledge as well as creating opportunities to change conditions in case of need. All personnel, organisational support and technical tools should endorse candid reporting as well as facilitating understanding in order to promote change
Technology	There should be a system in which to report, incorporating means of classifying and finding reports.	
Organisation	There should be procedures for reporting, incorporating clear responsibilities.	

Consider the following example: An organization experiences underreporting and as a remedial action the management instigates a grand "marketing" program to increase reporting. (i.e. enforcing personal accountability) A deeper investigation may discover inappropriate conditions or defect assumptions such as the following (Fig. 1);

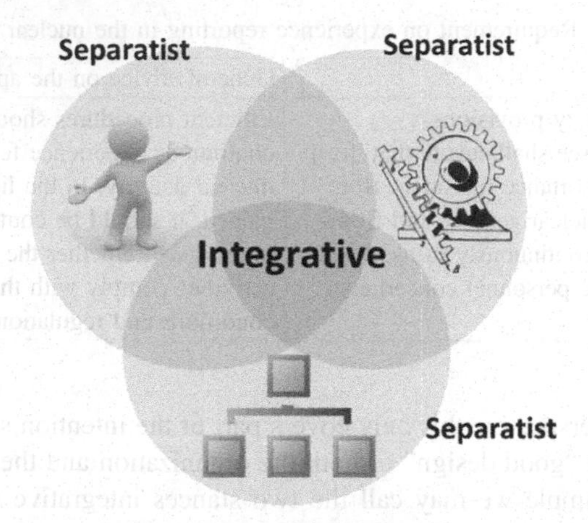

Fig. 1. Illustration of integrative and separatist stance

- the department receiving the reports is understaffed so little feedback is given to reporters, resulting in a feeling of futility,
- personnel are overworked so little attention is allowed for "administrative" tasks including reporting,
- the reporting tool is experienced cumbersome so prospective reporters refrain from reporting and rather use their energy on workarounds and local solutions,
- many actually do report albeit directly to their immediate managers, out-side the reporting system,. The managers become overloaded with "minutiae" and only administer a minority of reports resulting in them being seen as gate keepers causing diminished morale.

There are distinct couplings between usability of the tool, required effort of the user, allotted time for reporting and feedback, and procedures for reporting and feedback which cannot be addressed other than in conjunction.

One project in which the author participated aimed at designing a web based re-porting tool and was evaluated using moderated, scenario-based usability testing. The results did incorporate many classic usability issues, such as finding information or entering specific information in the right fields, but also issues concerning organizational structure which was not in the scope of the design project. The intention from the designers was that the reporter should assign the report to the appropriate handler. This required all potential users to have a fairly well developed image of the entire organization, which they didn't. Charts of the organization were provided to the re-porters by the tool but these were depicted in different ways depending on the current task – process oriented or organization structure oriented (e.g. "who is responsible for handling the issue?" vs. "where did the problem occur?"). This resulted in confusion and many test subjects indicated that this would make them skip the reporting altogether. In this case it was suggested that either making the organization more obvious or not requiring the reporter to assign the report would be more effective than to mitigate the difficulty through the visual design of the computerized tool.

5 New Directions

A few years ago there was an analysis to see the benefits of Human Factors research that Vattenfall sponsored. The result from the analysis was that there were large differences in the way that the area of HTO was interpreted and implemented throughout the larger organization. Based on this information, management ceased all related research funding in order to address first things first. A three year strategic development project was set up under the umbrella of the HTO-unit with the purpose of identifying the best way to perform work activities in order to maximize value for Vattenfall.

The first part of the project aimed at investigating the current implementation of HTO in Vattenfall and its subsidiaries. This was done through a series of interviews with key personnel as well as some benchmark with companies similar to Vattenfall. The findings from the investigation were divided into three areas in accordance with a view on HTO that has been prevalent at Vattenfall since the 90's; (1) Competence and organization of HTO, (2) Methods and applications, (3) HTO as a perspective.

5.1 Competence and Organization of HTO

Based on the interviews a picture emerged showing that work that falls under the area of HTO is very compartmentalized and generally an effect of "the usual" organizational silos. Thus, if Human Performance is allocated to a particular organizational unit, that unit may or may not maintain communication to another unit tasked with competence in another application of HTO, say event investigation. This is in line with the description given in previous sections in this paper (Fig. 2).

It is signifying of regulations and guidelines on HTO that they favor or even mandate a holistic or systemic view, so the separation between different applications appears somewhat counterintuitive. At Vattenfall the situation seems to be (1) that business units that "do HTO" do it in a separated and even compartmentalized fashion, and (2) that business units that have no established culture of HTO don't do it at all, or apply it in more general terms with little concretion. To state it bluntly; some acknowledge the benefits but have no money, others have the money but don't appreciate the benefits.

It is apparent from the benchmarking exercise that this kind of division is widespread, but also that many organizations employ a central unit mandated with the overall coordination of HTO-related areas. This unit may serve the purpose of a strategic control function e.g. for directing research needs or mandating application and coordination of the specialized areas as well as advocating and enforcing HTO throughout the organization.

Since coordination and collaboration are generally regarded as ideal organizational behavior, there are already many different groups and forms of meetings within Vattenfall within the areas of HTO. However, they are mostly organized around the logic

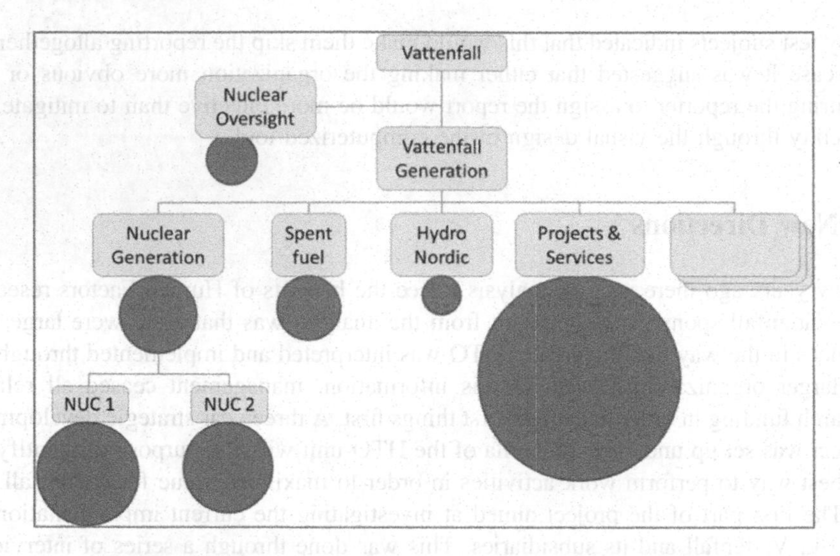

Fig. 2. Allocation of specific HTO-resources in Vattenfall

of the nuclear sector and other members may need to attend to many different meetings in order to get a coherent picture that is consistent with the own organizational logic. Moreover, there is very little collaboration between business units where HTO is established and those where it is not. Many of the different coordination activities also overlap so even though there are explicit initiatives to increase information exchange and collaboration, the initiatives themselves are not as well coordinated.

5.2 Methods and Applications

Many of the typical activities and methods instantiating Human Factors have been installed over time since the introduction several decades ago and are by now established in the nuclear sector as well as within the hydro sector. These include safety management, which has been institutionalized since a decade through a seminar course for management. Besides this training effort, there is a general risk awareness in the organization but a systematic application of a HTO perspective is lacking.

Human Factors Engineering is mandated for nuclear activities and although all subsidiaries have processes in place the maintenance area does have potential for further application as do IT-development as the financial aspect of cumbersome administrative systems can be significant.

Equally well established is the area of experience feedback and event investigations, at least in the nuclear field. Other areas commonly do employ some kinds of feedback systems, but not always as rigorously. There are also opportunities form further development as investigation from a HTO-perspective is regarded quite hefty and many smaller investigations only cover technical aspects.

Safety culture and Human Performance are likewise well established in the nuclear sector, but maturity is less developed in other parts of the organization although there are attempts to raise awareness of these concepts there as well.

One area which has been increasing in focus the last years is organizational issues, partly connected to the upcoming decommissioning processes but there is also a general increase in the acknowledgement of the potential contribution from HTO to organizational change, not least to gain further benefits from greater acceptance.

5.3 HTO as a Perspective

As stated previously, the HTO-perspective entails a thoroughly systemic view founded in individuals' situational context. In this manner it is very close to concepts such as organizational learning (e.g. [2]). Taking this as a backdrop, the results from the interviews show a multitude of instances where decision making groups incorporating HTO-competence have been positively affected in either the decision making or in the communication of the decisions. Currently there is great variety in the application which is largely haphazard and dependent on retaining the right individual in the appropriate setting.

6 Intent and Next Steps

Following from the results of the initial investigation a number of initiatives have been planned. Based on the perceived fragmentation of HTO-related applications, further efforts should be made to increase communication between different organizations within the HTO-area. This is being achieved through an internal conference that has been established to provide a recurring forum for further dissemination of information as well as serving as a community building event. An index of interrelated fora has also been established and already existing fora have been extended across organizations to provide regular opportunities for benchmarking and discussions.

Acknowledging that the biggest impact would result from wide application of a systemic perspective at management level, the most beneficial activity that was suggested was to provide "HTO-mentors" for groups to strengthen this ability. This function is currently implemented in two specific recurring management meetings, and is performed by the HTO-mentor through role-modelling a questioning behavior based on a set of principles on systems thinking. The method is supported by a manual incorporating the principles, which have been adapted from the Waters Foundation [3]. Examples of such principles include the following.

- Seek to understand the big picture,
- Consider long-term, short term and unintended effects,
- Consider an issue fully by resisting the urge to come to a quick conclusion,
- Change perspective to increase understanding.

Though the support function has been endorsed by senior management there has been a bit of uphill struggle to gain application. In many cases the groups are not convinced of the applicability, and HTO-mentors also report difficulties since the discussions in the groups may be specialized and the groups may be fairly cohesive.

From these experiences a general conclusion is that further work should be directed towards immediate benefits for the stakeholder community and to strengthen the relationships between short term and strategic impact of HTO for the company as a whole.

References

1. SSMFS 2008:1 The Swedish radiation safety authority's regulations concerning safety in nuclear facilities. https://www.stralsakerhetsmyndigheten.se/en/publications/regulations/ssmfs-english/ssmfs-20081/
2. Senge P (1990) The fifth discipline: the art and practice of the learning organization. Doubleday/Currency, New York
3. Waters Foundation. Habits of a Systems thinker. http://www.watersfoundation.org

A Rose by Any Other Name

Connotations of Human – Technology – Organization

Fredrik Barchéus(✉)

UXbridge, Eddagränd 11, 18775 Täby, Sweden
fredrik.barcheus@uxbridge.se

Abstract. The paper discusses overlaps between Human Factors/Ergonomics, HF/E, and related areas that may affect the effectiveness and efficiency of its application. On one hand concepts such as Human Systems Integration indicate that humans shall be considered throughout the system lifecycle. On the other hand such integration is dependent not only on the HF/E professional but highly dependent on other activities to be integrated with. The concept Human-Technology-Organization, HTO, is introduced to broaden the systemic perspective to contain also "hindsight rationale" based on established accident models. The paper conveys examples of situations where disciplinary overlaps have been identified and managed through trade-offs between retaining HTO tasks and pragmatically supporting the overlapping disciplines.

Keywords: Human-Technology-Organization · Disciplinary overlap Coordination

1 Introduction

This paper will discuss preconceptions and interpretations of the concept Human – Technology – Organization, HTO, by sharing some reflections on similarities and common traits of disciplines related to HTO. It will be pursued by exemplifying instances of overlaps between disciplines and by proposing ways of dealing with them and ask to what degree we as a community are prepared and willing to let go of some tasks in the name of increased overall effectiveness and efficiency.

The outset of this paper is the basic tenet that the concept Human – Technology – Organization, HTO, is an integration of Human Factors/Ergonomics as a design activity with the purpose of optimizing overall system performance and human welfare [1], and a distinct underlying acknowledgement of systems thinking [2].

These two aspects of HTO can be seen in principle as a cohort of two separate streams – design and analysis. The separation between these two streams may be clarified by the following example from the Swedish nuclear regulation [3] (Table 1).

The former wording is directed towards design and captured in principles, methods, standards, and guidelines, whereas the latter concerns investigation of events and is represented in a distinctly separate set of methods and principles.

While affirming that Human Factors and Ergonomics, HF/E, has been advanced as a discipline in its own right (e.g. [4]), I would like to call to front some common

S. Bagnara et al. (Eds.): IEA 2018, AISC 821, pp. 549–557, 2019.
https://doi.org/10.1007/978-3-319-96080-7_66

Table 1. Examples of requirements on HTO in the Swedish nuclear regulation [3]

	Regulatory requirement	General advice on the application
Chapter 3, Sect. 3	The design shall be adapted to the personnel's ability to, in a safe manner, monitor and manage the facility and the abnormal operation and accident conditions which can occur	To ensure a knowledgeable evaluation of design solutions where the capability of personnel is an important prerequisite, experts on the man-technology-organisation interaction should be engaged to take part in the design, analysis and evaluation of the solutions
Chapter 5, Sect. 4	[…] investigation […] shall be conducted systematically	The investigation methodology should be characterized by all relevant aspects and circumstances having been taken into account, including technical factors as well as those relating to the man-technology-organisation interaction

denominators with other areas that are either explicitly stated or tacitly implied. In the HF/E community it is generally agreed that integration of HF/E into other activities is intrinsic and an ideal goal, but coordination may also generate an overhead cost, and be managed by other disciplines. (Notably, some standards even prescribe non-duplication [5].)

2 Related Activities in Design

Systems design is a cross disciplinary activity and as such riddled with obstacles in the form of translation issues, both linguistically and culturally. This is even more evident in the case of HF/E.

Several standards try to clarify inclusion of human activity into the applied setting, but there is a difficulty to define concepts such as Human Systems Integration, HSI, by listing a set of distinct areas on the one hand and emphasizing its integrating intention on the other (e.g. [5–9]). It is somewhat reassuring to see that actual implementations work to overcome separation; the Constellation project [10] defined HSI as a cross-cutting coordination activity in its own right, rather than as simply a cohort of application areas.

2.1 Operator Interface Design

Perhaps the most apparent connection between overlapping HF/E-activities in a development project is between HF/E and design of physical things like machinery, or more specifically their human interfaces.

The entire rationale of HF/E is to actively promote a focus on the actual use of a system continuously through the development phase. One core principle in Human Centered Development is participation in design activities by representatives from the user population. If no such representation exists it is the responsibility of the HF/E-competence to advocate the user's position. However, in many cases there is some kind of user representation already present, but it may be insufficiently coordinated or documented, or there may be other reasons for employing a HF/E-competence e.g. stated by standards or regulations (Table 2).

Table 2. Requirements on the employment of HF/E practitioners

MIL-STD-46855A [5] 4.1.2	"The human engineering program shall be executed by a qualified human engineering practitioner(s)…"
SSMFS 2008:1 [3] 3.3	"…experts on the man-technology-organisation interaction should be engaged to take part…
NUREG-0711 [11] Appendix	"…the HFE design team's personnel should satisfy the minimum qualification specified below for each area of expertise", "Human Factors Engineering • Minimum qualifications: - Bachelor's degree in Human Factors Engineering, Engineering Psychology, or related science…"

In such cases there is a probability that HF/E-activities are seen as redundant and/or superfluous. Quite obviously, the organizational control of a development project is under the mandate of the project manager and any such situation should doubtlessly be addressed to the appropriate authority. In reality, the management of such coordination may differ depending on the development activity, the personal acumen of the project manager, as well as on the available processes at hand and the organizational structure.

In one project an engineer responsible for a refurbished human interface continuously asked operators about their opinion, generating some nice-to-have interfaces. Although well intended, this by-passed the HF/E-process, which relied to a greater extent on the use of scenarios. Later, the project was halted partially because the interface was overly ambitious and costly.

2.2 Design for Maintenance

While the operator is typically addressed as a HF/E issue in development maintenance seems to be downgraded, but might also be more tacitly integrated into the design process.

In one development project I emphasized user inclusion whereby one engineer complained, "but we do that". As the appointed HF/E-resource I attended continuously recurring design meetings and soon discovered that present at the meetings were also appointed operators and maintenance personnel. The engineers duly addressed appropriate questions to the operating personnel and shortly after, I stopped going to the meetings. Instead, I tried to find documented evidence of analyses and trade-offs, but had difficulties in locating any. When inquiring the maintenance representative I found out that he had some informal notes that upon review were deemed satisfactorily, so I simply asked him to store them in the appropriate form. In this case I refrained from insisting on the appointed work and attempted to increase traceability of already ongoing work by an administrative action.

2.3 Competence and Manpower

Standards on HF/E not only address design but also organizational entities such as procedures, training, competence requirements, manning, manuals and instructions. This may be seen as a typically human centered viewpoint; to address the direct and indirect context in which actual human activities take place. However, the tasks of performing training, or to conduct competence and manning analyses, or writing instructions may be assigned to other functions than HF/E. The question is; do they anticipate performed analyses in these areas from HF/E? Some standards are explicit regarding coordination of disciplines [4], whereas others are less prescriptive and only states that it should be done [11]. Any coordination is fully up to the applicant.

Standards may be used as reference in implementation into organizational processes. In such translation there is ample opportunity for interpretation, and owing to organizational logic disciplines like design, competence analysis, training, and instructions may be more than reasonably subject to separation into silos with their own intra-disciplinary processes, methods and traditions.

It is not uncommon that HF/E is used as a "quality stamp" for development projects. If a formal process states that integration of the human shall be done and that the HF/E competence shall be responsible for the integration, but the organization does not support such integration; what would be an appropriate action from the HF/E competence?

A very pleasant experience occurred in one project when the HF/E competence made the HR-department aware that the HF/E process prescribed competence analysis in conjunction to the design process. (The method that the HR-department used was basically interviewing post-holders about their tasks and responsibilities, and this was not possible for the systems that were under development, since there was no one in the organization one could readily interview.) I informed HR that the HF/E design process performed competence analyses for new systems, and that these analyses could be used as inputs in the implementation stage when HR took over responsibility for personnel issues. This was highly endorsed by the HR-department.

2.4 Training

A traditional approach to training is to consider it as an activity related to deployment of a system or identified lack of competence given an existing system. From an integrated HF/E perspective it is self-evident that training should be considered in system conception for trade-offs with other areas, perhaps most notably interface design. If the operator interface is designed so that it is easy to accomplish the goals then training isn't needed. There are also cases where interface design is more expensive than the total training cost.

It is not uncommon that system manufacturers provide training for their limited system only from a functional point of view, and not operational, and that system integrators expect piece-wise training to be sufficient. In some projects the person responsible for training isn't employed until the deployment phase, yet another evidence of training being seen as merely the actual training sessions.

A generic requirement on training is common as a kind of catch-all clause. The danger of this is that training continues to be seen as an appropriate way of considering the human. What I personally lack are specific trade-offs directed to training in the design stage; e.g. "Activity A is critical and needs be trained annually, Activity B follows industry practice and is coupled to ongoing training, Activity C may be redefined using interface X which is cheaper than training."

2.5 Manuals and Instructions

Like training, manuals and instructions are often coupled to specific systems and could benefit from a top-down approach in order to retain system-wide consistency. Even when operators are intimately involved in writing their own instructions, the logic is to base them on existing systems. Hence, consideration of the instructions, and the tasks they prescribe, are not discussed until late in a development project. Perhaps the most daring tip ever given was "first write the instruction, then design the system".

In practice instructions may also be (erroneously) used in conjunction with reference to organizational issues such as safety culture e.g. "Technology cannot ascertain absolute safety. Humans are also influential, and must adhere to instructions."

3 Health and Safety

A particular example of overlapping boundaries is HTO in relation to work environment or health & safety. Sweden has a history of a strong workers' movement and is highly regulated regarding the control of work environment. All companies in Sweden are required to have processes in place for managing their work environment, mandated by the Swedish Work Environment Authority. In construction work, not only does a company need to assure health & safety for the construction work as such, but the mandate goes even further and requires a person responsible for the inclusion of the work environment for all human actors during the use/operation of the construction. The latter part is very much in line with what the job of HTO would be in the design phase. Before the regulation was extended to the operational phase of the construction, work environment tended to include considerations during the actual design and construction phase. "The CAD-personnel gets headaches because of sun glare in their computer screens."

The issue at stake here is not that work environment as a discipline is in conflict with HTO but that interpretation, and thereby implementation, of activities may be incongruent. For example, a strong indicator of similarities is that quite commonly basic courses in HTO in fact concern design for a good work environment. Courses in work environment, on the other hand, may be solely directed towards legal requirements, with very little design content.

One example is a current large construction project that has set much focus on work environment (health & safety). Much coordination is done by HTO with the work environment engineer in order to maintain a common understanding of the overlapping areas between work environment and HTO. To give an example; one question was the viability of a requirement of low placement of equipment to prevent the need of potentially having to use ladders, which should be restricted. I saw it as a design task, thereby typically Human Factors Engineering, whereas the engineer saw it as a work environment issue. Both were indeed right. The question now became how to coordinate this in terms of e.g. ownership of the risk. The tools may differ. Work environment typically applies industry specific checklists and experience to find potential flaws, whereas HTO applies design principles and task analyses.

4 HTO in Analysis

The analysis stream has a long history of modelling emerging from linear models, through epidemiological models, to systemic models [12].

Linear models tend to be coupled to the technical systems which they seek to investigate. "If component X fails then event Y can happen, so let's add component Z as a back-up". Although not much in fashion in the HTO community they are still abundantly used, presumably for their ease of use and applicability to technical architecture. Epidemiological models [13] extended causes to negative events beyond the immediate maneuvers by operators at the sharp end and explained that such maneuvers were only a last in a series of conditions for the event, and that most such conditions were latent in the system. As a response to the still lingering linearity several outspokenly systemic models and methods have been produced [14, 15]. These models aim to reduce focus on any individual component in a system in favor of being used as a map of potential interactions, that in turn may contribute to detrimental effects. In many cases contributing factors to incidents are administratively inclined and some 60% may be related to organization and information [16].

Preferably analysis should be done before an event rather than after. Unfortunately one then don't have the benefit of hindsight. Such analysis may not only be performed in the name of discrete risk analyses, but indeed continuously.

One example as a project where work had been divided into several packages: instructions, organization, processes, safety criteria, definition of a certain set of "parallel activities", definition of an experience feedback process, as well as some other distinct areas. HTO had its own package. My judgement was that HTO was in fact present in many of the other packages, which was also a second opinion from colleagues. Rather than trying to redefine each of the already defined areas into a separate HTO deliverable, the solution was to work with the other packages and (trying to) appropriately address would-be HTO issues in their deliverables. The totality was then described in a separate HTO report, frequently citing the other reports.

5 Quality

Quality as a discipline has been abundant in industry for the past half-century, more or less. Although the immediate goal of Quality aims at controlling a process in order to obtain a product, many of the basic tenets of Quality are intimately coupled to HTO.

- Humans are fallible and should not be made responsible for errors that are beyond their control.
- Most of quality problems reside in "the system" (i.e. conditions for humans) rather than in humans.
- In order to manage problems, we should look for underlying contributors (root causes, e.g. 5-Why?)
- Products should be designed based on real human needs. (Deming [17] saw this as the task of Marketing as opposed to Advertising. Today, Marketing may have been superseded by HTO.)
- Central to well-functioning operations is a general understanding of the expected outcome and appreciation of the entire process.
- Focus should be on the place where things happen (gemba). (This could be compared to context of use.)

Following the release of ISO 9001 in the 90's, and its accompanying offspring lean, it seems that western implementations of Quality Management Systems often rely overly on formal "boxes-and-arrows" processes manifesting organizational silos. This is unfortunate and was probably not the intention of the originators, who assumingly were more oriented towards the fundamental tenets. Personal communication with a Japanese committee, tasked specifically with investigating application of HTO in their operations, revealed that they saw HTO as covered by their Quality processes. It might surely bee, given that the Japanese version of Quality is a much more integrated construct than in the Western hemisphere. From what I personally see in quality plans, controls, as well as audits this is not the case. All too often they lack effectiveness being formalistic rather than human centered and focused on overall system performance.

6 HTO as a Self-referential Meta-activity

The history of the incident analysis branch has, I conclude, made professionals in HTO sensitive to certain deviations in current ongoing activities in which they are part. Knowing from literature and experience that latent conditions often reside in organizations [8], that such conditions may interact and exacerbate each other, that singular conditions may be subliminal [14], that they are often administratively mundane [16], and that they are in fact actionable rather than intrinsic and intractable, the HTO professional may be uniquely positioned to continuously identify typical detrimental organizational behaviors.

So, if I raise a question regarding physical space in a maintenance task from an anthropometric view, and an engineer states that it has already been taken care of, referring to a standard which I (professionally) deem as invalid, several factors arise to me.

First, I apply the view of a Human Factors Engineer in identifying the issue.

Second, another (technically inclined) engineer has also identified the issue. Although from a formal viewpoint this was handled in the wrong process, it must be regarded a good thing that the issue was identified. Four eyes are always better than two.

Third, apparently the technical standards used haven't been correlated with HF, indicating a weakness in the requirements management process.

Fourth, I did not know that the issue had been identified. This is not good from the standpoint that there are, evidently, no institutionalized procedures. I could define appropriate processes, guerilla style, but ideally they should be defined by management and coordinated throughout the project.

Fifth, the fact that communication was in fact absent until my action could serve as an indication that communication in general is insufficient, and in retrospect could be identified as a contributory condition to an accident. This diagnosis comes from the extension of my knowledge of accident models to the current case.

The lacking communication could of course be caused or influenced by many things; there may be insufficient processes, there may be a lagging safety culture, I could be misinterpreting the organizational communication procedures and be in need of training, etc.

The immediate task I have is the design task, so my task is not to control the communication or training. On the other hand, if I were to analyze the situation after a factual even and discover that standards were deviant and processes and training insufficient, I would indeed note this. So, given that I don't control the project as such but may find (potential) issues, I want that the one responsible to act on them.

In a current project I have argued for Configuration Management to be installed in order to manage task analyses, since the project have several different versions of their processes on which to base the task analyses. The consequence could be that the equipment, which should be based on task analyses, becomes cumbersome in operation because the processes actually put in place were not the ones used as inputs to the task analyses. In this case I argue that it is more effective to aid the project with its Configuration Management (which is arguably a management activity) than to actually perform the task analyses that were specifically defined as a HTO-activity.

In relation to the design activities listed in Sect. 2 in this paper, the significant trait of HTO is that the "hindsight rationale" for the activity is constantly at the forefront of the analyst.

The conundrum is even more apparent in situations where a HTO-competence is employed as support for organizational development. In such cases the typical HTO design activities aren't specified, so the central task is to focus on analyzing structural issues in the organization coupled to potential problems. In any case there is arguably a fundamental difference between a subjective opinion, and an informed diagnosis of potential detrimental consequences based on HTO-disciplinary fundamentals.

7 Conclusion

In this paper I have tried to convey some recurring distinctions between disciplines relating to Human Factors/Ergonomics in general and the concept Humans-Technology-Organization in particular. The main observation is that the HF/E discipline has a number of potential overlaps with other disciplines that must be acknowledged and managed in real time. Explicit overlaps do exist in development but I argue that HF/E is particularly sensitive to these.

A logical corollary of increased intra-disciplinary knowledge within professions and organisational complexity is that work become more specialised over time. This is true up to a certain point, when the cost of coordination between different disciplines becomes higher than the gain of increased specialisation. If the increased specialisation is manifested as entirely new disciplines, they merely need to interface to the existing ones. But in other cases the specialisation comes as a redefinition of already existing knowledge and structures. Is this the case of current HTO?

A general recommendation for beginning HF/E practitioners is to quickly establish strategic relations with stakeholders in order to collaboratively define the overlaps and document any concerns for further discussion and escalation. I also call for further cross-disciplinary knowledge transfer in order to enhance understanding and communication across disciplines, thereby facilitating the necessary coordination.

Referencens

1. International Ergonomics Association. https://iea.cc/whats/index.html
2. Wilson JR (2014) Fundamentals of systems ergonomics/human factors. Appl Ergon 45:5–13
3. Swedish radiation safety authority regulatory code SSMFS 2008:1 The Swedish radiation safety authority's regulations and general advice concerning safety in nuclear facilities. ISSN 2000-0987
4. Wilson (2000) Fundamentals of ergonomics in theory and practice. Appl Ergon 31:557–567
5. MIL-STD-46855A, Department of Defense Standard Practice Human Engineering Requirements for Military Systems, Equipment, and Facilities
6. ISO/IEC/IEEE 15288:2015. Systems and software engineering - System life cycle processes
7. International Council on Systems Engineering (2015) Systems engineering handbook – a guide for systems life cycle processes and activities, 4th edn. INCOSE-TP-2003-002-04
8. Booher HR (2003) Handbook of human systems integration. Wiley, Hoboken
9. Boy GA (2013) Orchestrating human-centered design. Springer, London
10. Zumbado JR (2013) Human systems integration in practice: constellation lessons learned. BiblioGov. ISBN 978-1289067144
11. O'Hara JM, Higgins JC, Fleger SA, Pieringer PA (2012) Human factors engineering program review model. NUREG-0711, Rev. 3 United States Nuclear Regulatory Commission
12. Hollnagel E (2004) Barriers and accident prevention. Ashgate, Aldershot
13. Reason (1990) Human error. Cambridge University Press, Cambridge
14. Hollnagel E (2012) FRAM – the functional resonance analysis method. Ashgate, Farnham
15. Leveson NG (2011) Engineering a safer world: systems thinking applied to safety. MIT Press, Cambridge
16. Rollenhagen C, Alm H, Karlsson K-H (2017) Experience feedback from in-depth event investigations: how to find and implement efficient remedial actions. Saf Sci 99:71–79
17. Deming WE (1986) Out of the crisis. MIT, Cambridge

From Safety I to Safety II: Applying an HTO Perspective on Supervisory Work Within Aviation

Martina Berglund[1,2(✉)] and Oscar Arman[2]

[1] HELIX Competence Centre and Division of Logistics and Quality Management, Linköping University, 581 83 Linköping, Sweden
martina.berglund@liu.se

[2] Division of Ergonomics, KTH Royal Institute of Technology, Hälsovägen 11C, 141 57 Huddinge, Sweden

Abstract. In aviation, there is a strong focus on safety to prevent accidents. This paper deals with how supervisory authorities in aviation can apply a Safety II perspective. In particular, the aim is to analyze how the concept of HTO (Humans, Technology, Organization) is related to a possible shift from Safety I to Safety II within supervisory work within aviation. Data for this case study research was collected through semi-structured interviews with inspectors at the civil aviation authority in Sweden. The study showed that the important building stone of proactivity in Safety II could be promoted by the Safety Management System (SMS), the Safety Performance Indicator, and systems for reporting incidents and near-accidents. These systems constituted examples of Technology. Similarly, the Humans consisted of the inspectors, and the Organization included international and national regulations that the inspectors needed to follow during inspections. In the analysis, it was clear that an internal HTO-perspective could be taken. The study indicated that the shift towards Safety II should first be done within the supervisory authority by applying an internal HTO-perspective. This could later be developed to an external HTO-perspective also including the operator organizations.

Keywords: Humans · Technology · Organization

1 Introduction

Aviation is a safety-critical sector with advanced technological and automated systems, including interactions with humans in organizational contexts. In aviation, there is strong focus on safety to prevent accidents. One way to look at safety is by focusing on mitigating and learning from incidents, accidents and any erroneous actions. This is named Safety I. Another perspective is Safety II, which instead highlights what goes right. This also includes an organization's proactivity and ability to succeed under varying conditions. This paper deals with how supervisory authorities in aviation can apply a Safety II perspective. In particular, the aim is to analyze how the concept of HTO (Humans, Technology, Organization) is related to a possible shift from Safety I to Safety II within supervisory work within aviation. The paper is structured as follows.

S. Bagnara et al. (Eds.): IEA 2018, AISC 821, pp. 558–565, 2019.
https://doi.org/10.1007/978-3-319-96080-7_67

In the theoretical background, Safety I and Safety II are presented and a short overview of the HTO-concept. In the following chapter the methodology is described. The result chapter is divided into a first descriptive part of how supervisory work within aviation in Europe can be conducted in practice. This should be considered as support for further reading, before the presentation of the interview results. The paper then continues with a discussion before the conclusions are presented.

2 Theoretical Background

2.1 Safety I and Safety II

Traditionally, the concept of safety can be associated with 'a lack of things that go wrong'. An inevitable consequence of associating safety with things that go wrong generates a lack of understanding regarding the things that go right and why they work properly (Hollnagel 2012).

A consequence of regular patterns and actions that generate repeatedly expected results when things 'simply' work leads to reduced attention and complacency which according to Hollnagel (2012) can be described by two effects:

1. There is no difference between the expected and what has actually happened. Hence, there is nothing that captures attention or creates a reaction.
2. There is no motivation to understand why things went well. "This obviously worked out fine because the system – Human and Technology – worked as it should, seeing nothing that went wrong".

Hollnagel (2012) adds that the first claim, that is, an absence of notable differences between outcomes, is acceptable. The other claim, on the other hand, could describe the motives of an inadequate safety effort by just paying attention to what goes wrong. The precondition for Safety I is defined as a state where the number of negative outcomes would be as low as possible, and achieving a low number of negative outcomes would be the purpose of safety organizations. This approach to safety work is seemingly reactive as it starts as a response only when something has gone wrong or has been identified as a risk (Hollnagel 2014).

Safety I: Avoiding that Things Go Wrong. If for a given type of failure there is a statistical probability of 1 to 10,000, Safety I choses to focus all resources on that particular occasion. But for each time an error is expected, in this example there would be 9,999 occasions where it is expected that the outcome would be the wanted one, as in positive and safe.

Safety I could be summarized through Hollnagel's (2014) manifestations, mechanisms and foundations behind Safety I as a safety paradigm:

- The manifest of Safety I is to see what could or goes wrong.
- The mechanisms behind Safety-I relate to causality and an assumption of specific root causes and possibilities to look back in time by isolating causes.

- One assumption is that systems can be broken down into smaller components and that this process is fully reversible where humans could understand the behaviour of a system through only its meaningful elements.
- The prerequisite for functionality and performance is binary: either it works or not.

Hollnagel (2014) believes that Safety I as a concept is no longer sustainable, or at least not generally applicable as today's sociotechnical systems are not as easily deconstructable or predictable as older two-dimensional and linear systems (Hollnagel 2014).

Safety I and the Regulator Paradox. A last consequence of quantifying safety by measuring what goes wrong is that its positive development leads to an inevitably paradoxical situation. Hollnagel (2014) means that the paradox lies within the fact that the safer an activity or a system becomes, the less there is left to measure. In the end, when safety systems are perfectly designed and completely safe – provided this is possible and achievable – there is nothing left to measure. This is known as 'the regulator paradox' (Hollnagel 2014). Hence, an advancement towards discovering and investigating new dimensions of safety should be made.

Safety II: Making Sure that Things Go Right. Again, if the statistical probability of a failure would be in 1 of 10,000 occasions, Safety II choses to focus on the 9,999 occasions where an outcome did not lead to a failure.

Safety II can be defined as a state where as much as possible goes right and hopefully, everything goes right. In line with Resilience theory (Hollnagel et al. 2008) Safety II also describes itself as the ability to succeed under expected and unexpected conditions. Consequently, the number of acceptable outcomes can be as many as possible (Hollnagel 2014). Resilience acknowledges the fact that unacceptable and acceptable outcomes often have the same foundation where the foundation is everyday work and performance variability.

Hollnagel (2014) summarizes the foundation of Safety II in a number of points:

- In complex and sociotechnical contexts, systems can not be meaningfully broken down into individual components.
- The functions of the systems are not binary or bimodal. There is variety and flexibility in daily performances – not only good or bad outcomes.
- Events and outcomes are based on human variation in actions and behaviours, which should constitute the starting point for successes and failures.

The Relation Between Safety I and Safety II. To be clear; both Safety I and Safety II lead to a reduction of unwanted consequences, but they are fundamentally different in their approaches. As people's interactions with technical systems in daily life are becoming more and more complex, the Safety I approach could soon be insufficient as a tool to improve safety – which it may be already today.

According to Hollnagel (2014), the recommended solution is not one where Safety II would replace Safety I but a solution where Safety II would complement the current safety paradigm. It is more about a combination of the two ways and approaches to safety.

2.2 Humans, Technology, Organization – A Systems Perspective

The HTO concept is a systems perspective with a focus on the interaction between Humans (H), Technology (T) and Organization (O). It was first developed to improve safety within the nuclear power industry (Rollenhagen 1997), and has therefore spread to other domains and wider use. It can be used as a conceptual tool, an analysis framework, a meta methodology, a pedagogical tool, and a design tool (Karltun et al. 2017). One important aspect is that it is not enough merely to understand H, T and O, but also how these interact with each other (Fig. 1).

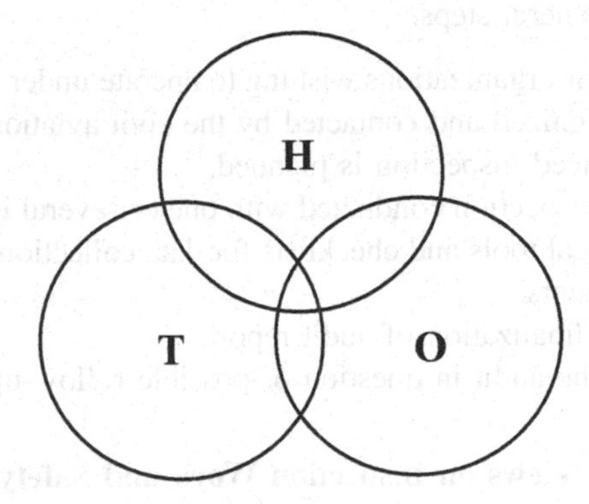

Fig. 1. The HTO concept based on a systems view on the interaction between Humans, Technology and Organization.

Depending on the problem or issue at stake, different views can be taken on H, T and O, which in return will generate different foci on the interactions. For example, the human at work (H) can be viewed as a biological energy processing system, an information processing system, a subject with unique history, and a member of social groups (Daniellou 2001), or as an exposed being, a cooperative, a learning being, or an actor (Westlander 1999). Technology (T) can include tools and machinery to perform work the physical work system and setting software systems (Wäfler 2001), including technical knowhow, procedures and methods. The organization (O), finally, consists of: formal aspects, such as goals, hierarchy, written instructions and follow-up systems as well as informal aspects, such as organizational culture, individual attributes, management style, and informal work practice.

3 Method

Data for this case study research was collected through semi-structured interviews with inspectors at the civil aviation authority in Sweden. The inspectors were selected based on their expertise within different areas related to supervisory work, such as regulations, airports, security, airline companies, etc. Six interviews were conducted, a first one by telephone and the following five in person. Each interview lasted 90 min.

Interview topics included inspection and safety in aviation, accidents and incidents, and different safety perspectives. The interviews were recorded and their contents analyzed and categorized.

4 Results

4.1 Inspection Work in Practice

The civil aviation authority of Sweden illustrates how their supervisory work can be characterized in four general steps.

1. Relevant operators or organizations wishing to operate under the relevant regulatory framework are recognized and contacted by the civil aviation authority of Sweden unless an unannounced inspection is planned.
2. Physical audit and inspection conducted with one or several inspectors. The audit is supported by technical tools and checklists for data collection to verify findings and regulatory requirements.
3. Analysis work and finalization of audit report.
4. Formal closure of the audit in question or possible follow-up procedures.

4.2 The Inspectors' Views on Inspection Work and Safety

All of the interviewed inspectors responded that the purpose of supervisory work is about regulatory fulfillment, which in the long run should contribute to increased safety.

"Basically, we work towards fulfilling EU and EASA regulations - no more, no less. We are not consultants for improvement. The operators and license holders have to find the solutions themselves"

However, the inspectors also expressed that increased supervision might not be synonymous with higher levels of safety. Supervisory and regulatory work must be about quality rather than quantity. One supervisor also expressed that supervision in some form is black and white, approval or disapproval, perhaps lacking several areas in between.

Four out of five respondents meant that accidents do not occur due to an isolated factor, but is often contributed by a series of events leading to unwanted outcomes in complex system with many different components interacting depending on each other.

4.3 Attitudes and Knowledge About Safety I and Safety II

All interviewed inspectors had to at least some extent knowledge about the concepts Safety I and Safety II, although in-depth knowledge were not established. Safety I/II as concepts was more or less new to the majority of the respondents a part from basic introductory courses in Safety I/II given internally by the agency.

The respondents who had more in-depth knowledge of Safety I/II concluded that the current paradigm the agency were working in could be considered as a Safety I perspective. The present safety work conducted was described as a paradigm of trying to prevent future accidents by reactively looking for shortcomings in the activities being inspected. However, a predominant positive attitude towards Safety II was observed among the inspectors, who suggested that a properly and comprehensively implemented Safety II could improve air safety.

4.4 Challenges and Opportunities to Work with Safety II

Challenges for a regulatory authority in working with Safety II lied mainly in regulatory frameworks, resource requirements and concretizing and defining Safety II as a definition.

The inspectors stated that as long as there is no support for Safety II in the regulatory frameworks for supervision, Safety II can not be exercised as a tool for supervision. It also follows that in order to develop and implement a new security perspective, resources will be required. It may be in the form of training for the inspectors, costs related for technical system development, adaptation to new types approaches to safety, all requiring additional resources for the Transport Agency. Interviews also revealed challenges with Safety II in the definition and practical implementation of Safety II within the line of regulatory work.

Opportunities in working with Safety II was described as a shift towards more dialogue and openness with operators and the development of tools for proactive safety work. The relationship and transparency between licensees, operators and the Agency was highlighted among the inspectors as very important. One inspector expressed that one opportunity with the implementation of Safety II could be an improvement in the way positive feedback towards operators are made, where Safety II could promote a more continuous and open feedback-loop to develop and improve an operators safety work.

Finally, technical tools and platforms for proactivity was mentioned as elements already in place today supporting a more proactive way of safety work. For example Safety Management Systems (SMS), Safety Performance Indicators (SPI) and incident reporting systems all support a more proactive way of dealing with safety. These technical tools all align with Safety II in their nature.

SMS works as an operator's quality system. The Transport Agency supervises and oversees an operator's SMS, where it is inspected, e.g. how well an operator identifies their own risks. SMS can act as preventive support in addition to safety inspections. Risks are identified internally within the operator and contributes to a proactivity in allowing the licensee to identify risks themselves.

To make safety concerns more quantifiable Safety Performance Indicators together with incident reporting systems are able to assess an operators safety levels and their ability to reach safety targets. These indicators are able to monitor and evaluate both past and presents safety events and could also be seen as tools aligned with Safety II according to the inspectors.

5 Discussion

The study showed that the supervisor authority faced some challenges in working with Safety II. These were related to the regulatory frameworks, resource requirements, organizational inertia, just culture aspects, gaps between planned and practiced work, etc. However, the study also showed that the important building stone of proactivity in Safety II could be promoted by the Safety Management System (SMS), the Safety Performance Indicator, and systems for reporting incidents and near-accidents.

Working with safety requires a systems perspective. The aspects highlighted regarding how to achieve Safety II could be placed in relation to the described HTO concept, see Fig. 2. In this case the above mentioned systems constituted examples of Technology. Similarly, the Humans consisted of the inspectors and the Certificate Holders. The Organization, finally included international and national regulations that the inspectors needed to follow during inspections, but also the organizational culture in the inspected organizations.

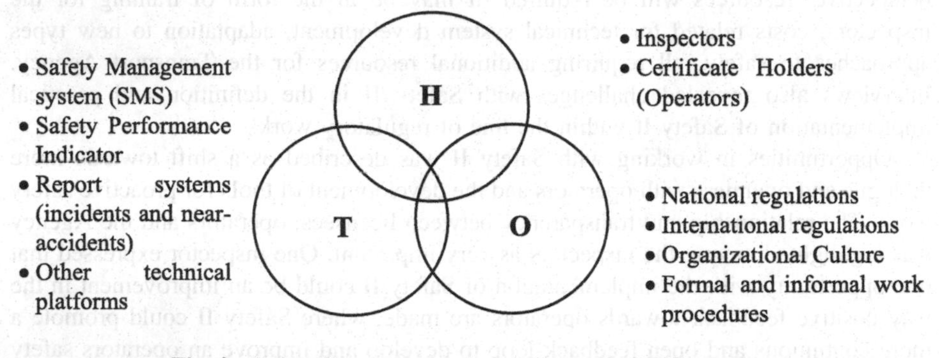

Fig. 2. Safety II components in relation to the HTO concept.

Applying the HTO concept could be one way to conceptualize the Safety II components, creating a model to facilitate analysis and reflection. With an HTO perspective, a number of potential interactions between different components are highlighted. Here it is possible to reflect on the current state regarding these components and their interactions and what measures are required to move towards Safety II. In the analysis of this specific case, it was clear that an internal HTO-perspective could be taken. A step towards proactivity to develop Safety II could be to develop the SMS and educate the inspectors, which together with organizational rules, could provide improved decision support to the inspectors. The study also indicated that the shift towards Safety II should first be done within the supervisory authority by applying an internal HTO perspective. This could later be developed to an external HTO perspective also including the operator organizations.

6 Conclusion

The study showed that the important building stone of proactivity in Safety II could be promoted by the Safety Management System (SMS), the Safety Performance Indicator, and systems for reporting incidents and near-accidents. These systems constituted examples of Technology. Similarly, the Humans consisted of the inspectors and the operators, and the Organization included e.g. international and national regulations that the inspectors needed to follow during inspections. Furthermore, the study indicated that the shift towards Safety II should first be done within the supervisory authority by applying an internal HTO perspective. This could later be developed to an external HTO perspective also including the operator organizations.

Acknowledgement. The authors would like to acknowledge Klara Hall who participated in the study of the inspectors at the civil aviation authority in Sweden within her master's thesis work.

References

Daniellou F (2001) Epistemological issues about ergonomics and human factors. In: Karwowski W (ed) International encyclopaedia of ergonomics and human factors, part 1. Taylor & Francis, London, pp 43–46

Hollnagel E (2014) Safety–I and Safety-II: the past and future of safety management. Ashgate Publishing Limited, Farnham

Hollnagel E (2012) A tale of two safeties. Nuclear Saf Simul 4(1):1–9

Hollnagel E, Nemeth CP, Dekker S (2008) Resilience engineering perspectives: remaining sensitive to the possibility of failure. Ashgate Publishing Limited, Aldershot

Karltun A, Karltun J, Berglund M, Eklund J (2017) HTO – a complementary ergonomics approach. Appl Ergon 59:182–190

Rollenhagen C (1997) The relationships between humans, technology, and organization: an introduction. Studentlitteratur, Lund

Waefler T (2001) Planning and scheduling in secondary work systems. In: MacCarthy BL, Wilson JR (eds) Human performance in planning and scheduling. Taylor & Francis, London, pp 411–447

Westlander G (1999) Focus on the human in research on operations development. In: Ahlin J (ed) Forskningsperspektiven. NUTEK, Stockholm, pp 20–33 (1999). (in Swedish)

Co-constructing Organizational Autopoiesis: The Developmental Laboratory as a Model and Means of Enabling Interventions

Gianna Carta(✉) and Pierre Falzon

Research Centre on Work and Development, Conservatoire National des Arts et Métiers, 41 rue Guy Lussac, 75005 Paris, France
gnn.carta@gmail.com, pierre.falzon@lecnam.net

Abstract. This paper presents a model for an ergonomic intervention with developmental and autopoietic aims. It is based on the organizational change of an entity in charge of the signalling devices maintenance for the Parisian subway. Its developmental goal is to empower actors to redesign their work processes in an enabling, autonomous and sustainable way based on the emerging needs of a cross-functional activity that is to be imagined. The intervention was conceived and equipped as a formative process. The methodology put in place is called the "Developmental Laboratory" (DL). The DL impels two levels of cross-functional inquiry. The DL1 concerns the production work or even the redesign of maintenance processes (functional dimension). The users have jointly constructed new more able organizational solutions (processes, tools, methodologies, forms of coordination and management, etc.). The DL2 deals with the practices underlying the previously organizing process (metaflective dimension). In other words, the results and processes in DL1 were investigated in DL2. The users were therefore mobilized in co-design and formalization of new organizing standards and routines. The place of the real work analysis and the roles of Enabling Ergonomist are investigated. These appear to be key elements to translate the developmental potential of the actors (Ergonomist included) and their practices into actual productions.

Keywords: Organizational autopoiesis · Developmental Laboratory Processes participatory design · Enabling interventions

1 Introduction

This intervention-research aimed at designing and implementing an enabling methodology for organizational development. This methodology had a double goal, developmental and sustainable. First, an intervention environment was set in order to allow the actors to design organizational solutions meeting the current needs of cross-functional activity. Second, it attempted to make actors able to redesign their processes in a continuous, enabling and autonomous way, beyond the end of the ergonomics intervention. The design of the intervention apparatus builds upon the autopoiesis model [1]: a support was provided for the design of organizational components and

S. Bagnara et al. (Eds.): IEA 2018, AISC 821, pp. 566–576, 2019.
https://doi.org/10.1007/978-3-319-96080-7_68

coordination network; particular attention was paid to the development of self-reference skills, aiming at achieving a continuous organizational development.

This intervention-research is grounded in assumptions regarding the nature of organisations and the means and purpose of participatory interventions on organizations.

We approach organizations according to a process perspective [2, 3]. Organisations are not seen as stable structures, imposing rules, norms and procedures, but rather as self-designing systems [4], as permanent processes of articulation between various actors defending various interests. Organizational processes result from a collective sensemaking process [5]: agents produce "organizational work". In this view, cross-functional practices are central to organizational design, and actors play a central role as creators of their work practices.

In this perspective, ergonomic interventions aim at facilitating "design-in-use" processes, so as to empower actors [6]. The intervention was designed according to the Constructive Ergonomics perspective [7]: the objective of ergonomic actions is both to foster processes of development throughout the ergonomic intervention and to design work systems that promote development. Enabling interventions integrate development as a means and as a purpose. Intervening means setting up, during the intervention itself, a dynamic apparatus that encourages experimentations [8] and debates about the multiple representations of work [9, 10]. Thus, intervention is seen as a formative process [11] that allows actors to learn and to develop themselves and theirs processes. Development is a means and a goal of the intervention.

1.1 The Context: The Insourcing of the Signalling Devices Maintenance Activity

This organizational development project was conducted for the Railway Transport Company of Paris (RATP). The Subway Signalling Maintenance Unit performed a partial insourcing process of the maintenance function in order to achieve strategic and economic goals. This process required to develop operators' competence and bring together operational and supporting functions, i.e. Maintenance Engineering, Service Quality, Logistic, Railway Work Management. This reorganisation led to several difficulties shortly after being implemented, impacting availability, reliability and resilience of the maintenance system. The ergonomics intervention started at this stage. A work analysis of the maintenance system was performed. The results highlighted a techno-centred approach to the organisation. It led to management practices characterized by reactive routines, centred on technical and mitigation solutions, a top-down change management, a lack of cross-functional analysis of organizational dysfunctions, distance between HR policy and operational needs, project management practices disconnecting departments, a supremacy of the construction service, a very limited risk assessment. This led to a breakdown of cross-functional actions, and impeded knowledge management and management development. To cope with this issue, a participatory project was proposed in order to design enabling solutions regarding processes, IT tools, standards, methods, collaboration rules, etc., and to develop an enabling and autopoietic organizational functioning.

2 Methodology

The organizational development was conducted through an ergonomics approach to project management, with various meetings (kick-off, steering committee and follow-up meeting) and workshops for designing cross-functional processes [12].

The Enabling Intervention focused on the global life cycle of signalling devices. The transverse activity of the involved professionals, i.e. Designers, Constructors, Commissioners and Maintainers, was analyzed. Concerning the Maintenance Department, the production units (maintenance centres) and the support functions both at local level (Maintenance Engineering, Quality Service Assurance, Logistic, Railway Site Work) and central level (HR, Safety and Quality) were enrolled in the project. Globally, 44 managers of the Department were involved. They belonged to five hierarchical levels, from operational to top managers. At the production level, 78 maintenance agents were involved as well. The project lasted from April 2011 to August 2015 and included three phases (see Fig. 1):

- a « *preimplantation* » phase lasted almost two years. A preliminary diagnosis of interactions and transactions between professions over the life cycle of signalling devices provided information on the requirements for design of the developmental laboratory (DL; see below).
- a « *cross-functional organizing* » phase, that lasted eight months, aimed at supporting the actors in the participatory design of the new and enabling maintenance processes.
- an « *autopoietic mechanism anchoring* » phase, that lasted fourteen months, aimed at developing a self-reference practice, which is supposed to enable a future and durable organizational redesign.

We propose to see organizational interventions as the implementation of a "developmental laboratory" (DL). When a DL is implemented with an enduring, sustainable ambition, it is composed of the embedded DLs.

The DL has been designed as a "model environment" aimed at giving the framework for an enabling sensemaking and the model for the cross-functional design practices. The DL looks like and functions as a FabLab allowing the actors to draft, prototype, test and formalize new maintenance processes and management practices. The constructive dimension of processes design was used as a lever. Through a scaffolding process [13], we helped actors to reach pedagogical and functional achievements. Actors were requested to apply a cross-functional inquiry method [14] mediated by boundary objects[1] [15]. Particularly, two embedded levels of practices inquiry have been set up (see Fig. 2).

The first one (DL1) focused on the maintenance processes (operational tasks) and aimed at developing functional (operational) knowledge. Enabling methods and tools drove the actors from the analysis of the current situation to the design of new organizational solutions (i.e. processes, tools, methods, cooperation and management

[1] A boundary object is an object shared between actors of a collective task. For instance, for a team of architects, the provisional plan of a building being designed.

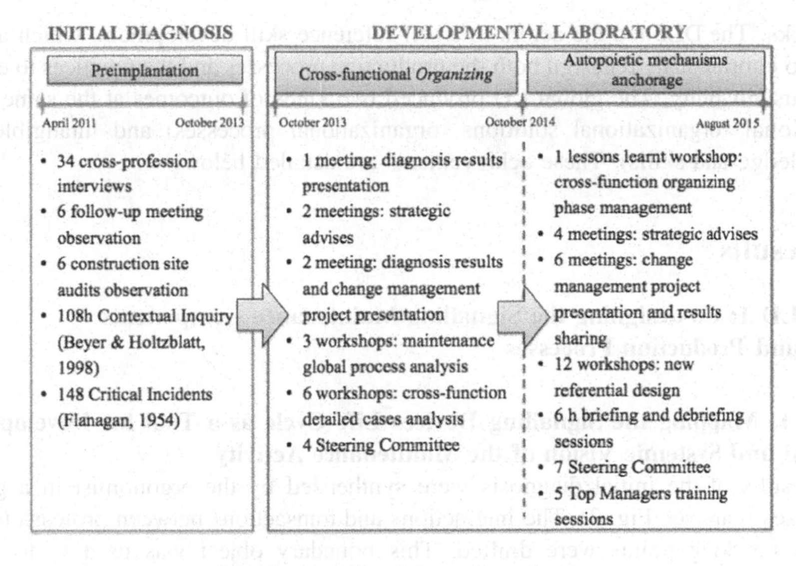

Fig. 1. Enabling Intervention phases

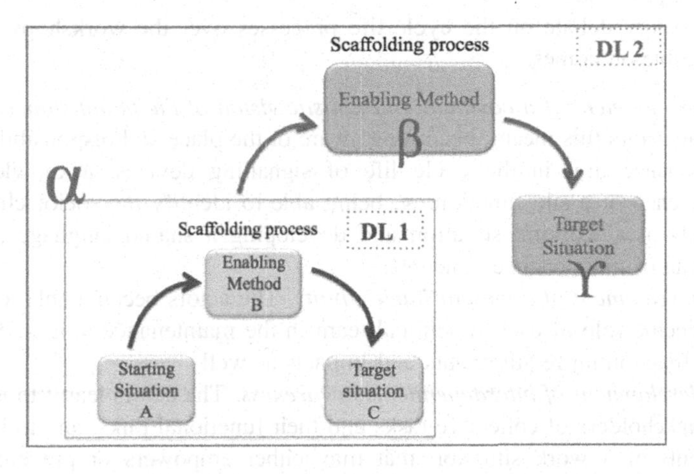

Fig. 2. Development Laboratory model

practices, etc.). During these achievements, the actors developed a global and shared vision of the maintenance processes and a new way to act together as well.

The second one (DL2) focused on the practices that allowed actors to redesign maintenance processes. In other words, the first level (DL1) was itself analyzed in the next level (DL2). DL2 aimed at developing metareflexive knowledge. Enabling methods and tools allowed the actors to carry out an inquiry on the way they managed the redesign of maintenance processes. The result expected was the definition of success criteria for enabling processes of redesign.

Accordingly to autopoiesis theory, DL1 aimed at supporting the actors in shaping the Maintenance Unit components (functional teams) and underlying coordination networks. The DL2 led the actors to a self-reference skill development, which allows them to continuously redesign both the production processes and the practices to ensure their sustainability. The global LD produced two kinds of outcomes at the same time: operational (organizational solutions, organizational processes) and intangible one (knowledge and skills). These achievements are detailed below.

3 Results

3.1 LD 1: Co-designing the Signalling Maintenance Components and Production Processes

Stage 1: Mapping the Signalling Devices Life Cycle as a Tool for Developing a Shared and Systemic Vision of the Maintenance Activity
The results of the initial diagnosis were synthetized by the ergonomist in a global processes map (see Fig. 3). The interactions and transactions between professions and related blocking points were drafted. This boundary object was used to trigger a sensemaking process through a collective analysis, debate and reshaping of the global maintenance processes.
The participatory debate on the cycle life processes over the workshops resulted in three intangible outcomes:

(1) *The development of a common and holistic vision of the maintenance processes.* For the actors this meant: becoming aware of the place and responsibilities of the maintenance unit in the cycle life of signalling devices, acknowledging the inefficiency of a silo functioning, being able to identify the sociotechnical components of a working situation and developing a shared language and understanding of maintenance concepts.
(2) *The development of a mutual intelligibility.* The actors became able to recognize the specific role of each functional team in the maintenance unit, and the reciprocal functioning requirements and impacts as well.
(3) *The development of interdependence awareness.* The actors learnt to identify the key stakeholders of collective tasks and their functional links, and to identify the elements of a work situation that may either empowers or prevents a cross-functional activity.

From a functional point of view, this participatory processes analysis led to the identification of the main dysfunctional areas of the maintenance process. For each area, in a participatory way, the actors identified 3 critical cases, which were then collectively analyzed in stage 2.

Stage 2: Mapping Incident Stories as a Tool for Co-designing Maintenance Unit Components and Production Processes
Starting from the dysfunctional areas, the actors were asked to map the processes and events underlying the cases previously selected. The story of the incident was progressively shaped interactively (see Fig. 4). This allowed actors to produce a shared

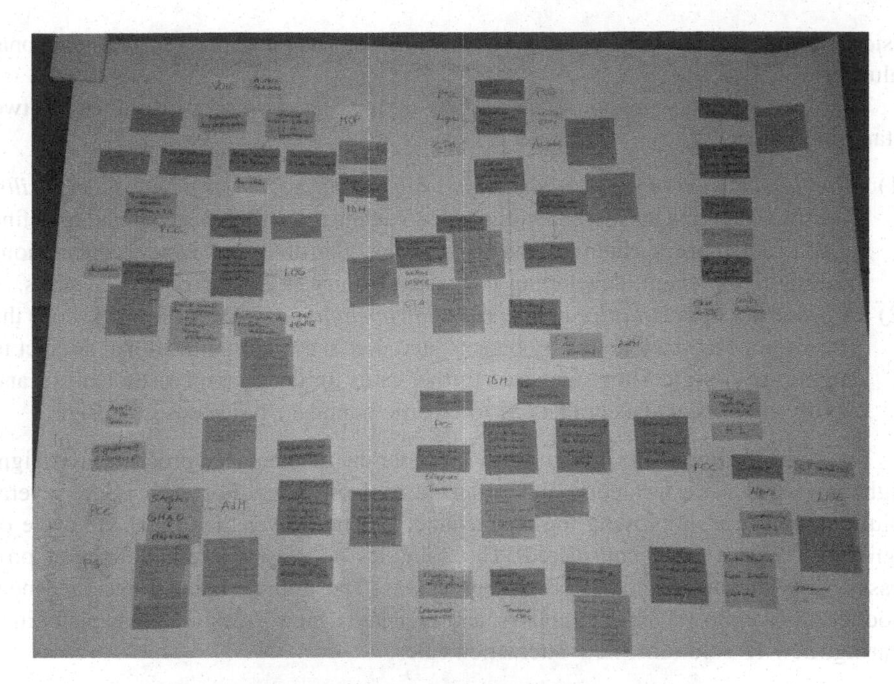

Fig. 3. Signalling devices cycle life processes mapping

Fig. 4. Incident story mapping

vision of the situation and of related blocking points and to design organizational solutions.

The incident story mapping performed during the workshops resulted in two intangible outcomes:

(1) *The development of a cross-functional problem setting and problem solving skills*. For the actors this meant: becoming able to define the problem statement, to define a shared temporal chain of events and the plausible stakeholder interactions leading to the event, developing reciprocal inquiry skills and co-design skills.
(2) *A gradual rise in initiative in the cross-functional inquiry*. Over the workshops the actors proposed new boundary objects, and used current organizational artefact to support co-design. They suggested further cases for deeper processes analysis and involved external stakeholders as to enlarge boundary processes co-design.

The functional outcome was an action plan for the maintenance processes redesign. Eighteen issues were tackled in the organizational solutions. For each issue, several solutions aiming at improving the maintenance practices over the global life cycle of signalling devices were considered. The solutions addressed internal function processes, cross-functional and cross-unit processes. The improvements concerned new production workflows, new templates and standards, new roles and responsibilities sharing, new IT solutions, new interaction rules.

3.2 LD 2: Anchoring the Autopoietic Functioning in the Maintenance Division Culture

Stage 3: Analysis of Lessons Learnt as a Means to Encourage Self-reference
A continuous work analysis was conducted both on the outcomes (functional and intangible) and on the way to reach these outcomes, allowing the ergonomist to set the requirements for the design of this stage of the DL2. Starting from the elements which led to poor performance during the previously DL1, the ergonomist translated them into a "lessons learnt" grid. The grid was supposed to help actors to conduct a retrospective analysis of the cross-functional design management. This grid used five key performance indicators in order to analyze the participatory processes design dimensions:

(1) Relevance: pointing out the elements leading to a poor or successful performance during problem setting tasks;
(2) *Effectiveness*: analyzing the quality of participatory action;
(3) *Efficiency*: analyzing planning efforts;
(4) Impact: pointing out the improvements (achieved or missed) of organizational solutions;
(5) Durability: discussing the future involvement of top management to ensure a continuous improvement of the maintenance processes.

The lessons learnt performed on processes design management resulted in three intangible outcomes:

(1) *The development of a cross-functional metareflective skill.* The actors learned to take themselves as objects of analysis. Starting from their choices and behaviours leading to good or poor performance, they pointed out the factors to be implemented in future enabling design situations. Particularly, the facilitation of the workshops, the availability of the resources for active involvement of the stakeholders, the visibility of the current situation and of the design progress, a continuous communication flows on results were selected;
(2) *The development of the awareness of being change makers.* The actors recognized their responsibility in the contribution to the success of the management of processes change. They have shifted to a "waiting for expert solution" mind-set to an "active change-maker" one;
(3) *The ability to transfer the concepts to other situations.* The actors have generalized the golden rules for participatory processes design to the maintenance unit management practices. Particularly, a participatory management of the unit, a work analysis as basis for change management, a cross-functional inquiry routine to manage organizational issues have been identified as success factors to be sustained over time.

Stage 4: Transferring the Steering of Processes Design as a Means to Empower Autopoietic Functioning
The next step towards the autopoietic functioning target was the transfer of the steering of the continuous processes improvement to the actors. The intangible and functional outcomes produced throughout the ergonomics methodology were operationalized into two levers of improvement:

(1) the success criteria and the best practices previously identified were shaped in new functional standards and methods aiming at driving the cross-functional processes inquiry and the knowledge management regarding the experience;
(2) the actors have performed a detailed cross-functional processes mapping in order to point out new processes to be designed and to act on current processes to be improved.

Regarding these two levers of organizational development, the actors took the lead and managed the design sessions, i.e. the workshop facilitation, the problem statement definition, the root-cause analysis, the prototype drafting, the process deployment and test stage, the formalization of the standard and the communication plan.

During this stage the ergonomist gradually shifted towards an observer status, using several strategies:

(1) *Coaching the "champions" and the "sponsors" to become leaders of continuous improvement actions.* Champions were those who played a key role in sustaining and promoting collective involvement in autopoietic functionings. Sponsors were the function managers and the unit manager. The content of the coaching sessions

were set up to meet the operational specific requirements gathered by the ergonomist during work analysis. These sessions focused on techniques and strategies to lead and facilitate collective sessions of processes design, and generally collective inquiry tasks.

(2) *Briefing and debriefing sessions.* These sessions aimed at supporting the leaders of workshops for processes design. Briefing sessions focused on the way to set up the workshop facilitation (agenda, methods for supporting collective design, choice of boundary objects). Debriefing sessions concerned a metareflective action on the cause-effect relation between the expected targets, the leading and facilitation strategies, and the achieved results. Both kinds of sessions aimed at developing the scaffolding competences (steering and collective work animation and development) of the concerned leaders.

(3) *Renewing stakeholder involvement.* The methodology and the achieved outcomes were analyzed by the ergonomist in order to set up several events aiming at fostering the sensemaking of the project. During these events, doubts, lacks of information, loss of motivation, fears, passive or negative attitudes were collectively discussed, analyzed and fought by action plans definition.

3.3 The Enabling Ergonomist's Roles

The roles played by the ergonomist in such enabling interventions went beyond the traditional one, which tends to focus on health or physical design issues. The enabling ergonomist is a promoter and guarantor of the organizational autopoiesis. To fulfil this goal, enabling ergonomists should spread their action over three levers of organizational development:

(1) *Operational lever.* The enabling ergonomist played three roles. 1/Project director, by defining planning, resources and methods and by ensuring the compliance with project milestones. 2/Supporter of the collaborative design, by fostering and tooling a cross-functional analysis and production. 3/Work specialist, by making sure that real work remains at the centre of process design.

(2) *Political lever.* Through the initial and the on-going work analysis, the enabling ergonomist gathered the operational data needed for stimulating organizational development and for negotiating margins of action. The aim was to develop an autopoietic management and to integrate an ergonomics approach into the organizational strategy and policies.

(3) *Pedagogical lever.* The enabling ergonomist engineered learning spaces. In order to develop a collective autopoietic competence, on the basis of situated operational requirements, learning mediators tools were designed, collaborative learning was supported and promoted, a reflexion and debate of real work were triggered and sustained, actors were taught of techniques and methods for an enduring organizational development.

4 Discussion and Conclusion

This intervention-research presented a model for organizational development in a sustainable aim. From this contribution, three conclusions could be drawn for ergonomics intervention on organisations.

First, acting for a sustainable achievement entails that the ergonomist is to perform at three parallel levels.

- At a political level, the need for and the legitimacy of ergonomics intervention in organizational development should be justified and renegotiated with decision-makers. The ergonomist has to perform a continuous translation of ergonomics messages in a more "acceptable industrial language" and has to develop a marketing activity so that ergonomics contribution to organizational strategy can be acknowledged.
- At an operational level, the ergonomist has to bring together human requirements in work situation design and the global coherence of the methodology for organizational development.
- At a pedagogical level, the ergonomics contribution is not limited to training at enabling tools nor to the design of "learning by doing" situations. Enabling interventions should imply methods for triggering metareflective competences, for enhancing implicit knowledge externalisation and for making it sustainable.

Second, bringing together multiple levels of intervention means that the ergonomist needs to develop and master various competencies, which enrich traditional ergonomics framework (health, human errors and physical issues for design). The organisation should be analyzed and understood under a systemic and global view. The ergonomist should thus master organizational issues (strategy, management models, performance measurement) and pedagogical issues (knowledge models, learning methods and techniques).

Third, real work analysis (ergonomics diagnosis) is not limited to the first stage of an ergonomics intervention. It is rather an on-going process, allowing the ergonomist to develop a fine understanding of work situations. In fact, the analysis of real work points out the main dysfunctions and opportunities for improvement both in organizational processes and in intervention methods and practices. Additionally, real work analysis highlights the main competencies that the actors (ergonomist included) should develop both for performing the task expected and for making it sustainable. Thus, on-going real work analysis can be seen as a developmental tool contributing to enabling processes, developing intervention methods and empowering actors.

References

1. Varela F, Maturana HR, Uribe R (1974) Autopoiesis: the organization of living systems, its characterization and a model. North Holland Publishing Company, Amsterdam, pp 187–196
2. Hernes T (2008) Understanding organization as process. Theory for a tangled world. Routledge, New York

3. Hernes T, Maitlis S (2010) Process, sensemaking and organizing: an introduction. In: Hernes T, Maitlis S (eds) Process, sensemaking and organizing. Oxford University Press, Oxford, pp 27–37
4. Weick KE (1977) Organisation design: organisations as self-designing systems. Org Dyn 6(2):30–46
5. Weick KE, Sutcliffe KM, Obstfeld D (2005) Organizing and process of sensemaking. Organ Sci 16(4):409–421
6. Seim R, Broberg O, Andersen V (2014) Ergonomics in design processes: the journey from ergonomist toward workspace designer. Hum Factors Ergon Manuf 24:650–670
7. Falzon P (ed) (2015) Constructive ergonomics. CRC Press, Taylor & Francis, London
8. Petit J (2006) Experimentation as means of intervention for the ergonomist in the management of an organizational change? In: Proceedings of the 16th world congress on IEA, Maastricht
9. Béguin P (2003) Design as a mutual learning process between users and designers. Interact Comput 15:709–730
10. Rocha R, Mollo V, Daniellou F (2015) Work debate spaces: a tool for developing a participatory safety. Appl Ergon 46:107–114
11. Engeström Y, Sannino A, Virkkunen J (2014) On the methodological demands of formative interventions. Mind Cultur Act 21(2):118–128
12. Barcellini F, Van Belleghem L, Daniellou F (2015) Design projects as opportunities for the development of activities. In: Falzon P (ed) Constructive ergonomics. CRC Press, Taylor & Francis, London, pp 187–204
13. Lorino P, Mourey D (2013) The experience of time in the inter-organizing inquiry: a present thickened by dialog and situations. Scand J Manag 29:48–62
14. Bruner JS (1983) Child's talk: learning to use language. Norton & Company Inc., New York
15. Broberg O, Andersen V, Seim R (2011) Participatory ergonomics in design processes: the role of boundary objects. Appl Ergon 42:464–472

Accounting Standard for Ergonomics: Relation of Ergonomics and Accounting

Rafael Bezerra Vieira(✉), Mario Cesar Vidal, José Roberto Mafra, Rodrigo Arcuri, and Luiz Ricardo Moreira

Universidade Federal do Rio de Janeiro, Rio de Janeiro, Brazil
rbvieira84@gmail.com, mcrvidal@gmail.com, arcuri.rodrigo@gmail.com, mafra@facc.ufrj.br, luizricardo@ergonomia.ufrj.br

Abstract. The presentation of costs in ergonomics has been studied for some years, and has been increasingly used by professionals in the field. These costs, whether regarding the lack of ergonomics or regarding the ergonomic intervention can generate equity variations in the companies, which must be demonstrated in a reliable way to the stakeholders. This paper aims to demonstrate that the ergonomic intervention causes changes in the equity of the companies and, therefore, should present a unique accounting standardization. This paper presents a case in the food industry, more specifically a pasta production plant that outputs several pasta products. After the ergonomic intervention, the company presented accounting values that did not represent the reality of the company, indicating a variation of 5% when they should present a variation of less than 1%. The value of the company's assets would be lower, so that the expenses would be overestimated in the first year, presenting a lower profit than the real, presenting a profit lower than the real in the first period, and above the real in later periods. Incorrect disclosure of these variations may cause problems for investors who will not have access to the information needed to make the best decision.

Keywords: Ergonomics · Cost · Accounting

1 Introduction

The disciplines of accounting sciences and ergonomics have the potential to go hand in hand, but few joint researches are found. When performing an ergonomics intervention, the ergonomist is directly affecting the object of study of the accounting sciences, the equity of the company. These values have already proved relevant in several studies and in the experience of the professional ergonomist. Nevertheless, the general public does not have access to this information.

Accounting has as one of its objectives to reduce the information asymmetry, by showing to external agents all relevant information related to the company's equity. Thus, accounting enables external agents to have an appropriate amount of information for decision making about the company.

S. Bagnara et al. (Eds.): IEA 2018, AISC 821, pp. 577–582, 2019.
https://doi.org/10.1007/978-3-319-96080-7_69

When performing an ergonomic intervention, often its result directly affects the equity of the company, either with the direct financial results of the intervention, with the amounts invested in the project, or even with amounts that the company stopped spending or earning.

More and more countries around the world have adopted the same accounting standards to meet the goal of reducing information asymmetry, enabling companies and users of financial statements (such as investors) from all over the world to understand each other.

Currently, the most accepted accounting standards are the international accounting standards (IAS and IFRS), which are issued by the International Accounting Standards Board (IASB) and followed (completely or partially) by 113 countries as of this time. It is worth to note that, in this scenario, there are no statements or references to the possibilities of equity change that ergonomic intervention can bring to the company.

In this context, the objectives of the present work is to present the need to produce such information and the need to develop of possible applicable accounting standards to be used by companies in ergonomic interventions, aiming to provide users of financial statements the equity variation caused by the intervention.

2 Theoretical Foundation

Costs application in ergonomic interventions has been discussed in the last two decades (Oxenburgh 1997; Hendrick 1996; Hendrick 2003; Mafra 2006; Lahiri et al. 2005; Rickards and Putnam 2012). These works described methodologies for the identification and presentation of these costs at the initial moment of ergonomic intervention, focusing on the justification of hiring specialists on ergonomics to company managers. This direction was necessary as a way of demonstrating the importance of the integration between the ergonomist and the high management of the company.

These works (as well as many others) also presented data confirming that ergonomic interventions indeed create value, generating changes in the equity of companies. These changes will be transmitted to various stakeholders through the accounting information. Yamamoto and Salotti (2006) define the accounting information as one that changes the state of the art of the knowledge of its user in relation to the company (…) thus, the accounting information has the consequence of ratification or alteration of the opinion of its users about the activities of the company.

By "users", one should understand: (a) the company's internal agents, which have privileged access to the information at the time of its elaboration, as employees and the senior management, whose interests are the maintenance of the proper functioning of the company, as well as its growth; and (b) those external to the company, which depend on the public disclosure of information, represented mainly by investors and creditors. For the proper functioning of a capital market, external users should have access to reliable information that has the potential to contribute to decision making.

In order for the accounting information to reach this objective, the accounting-financial reports should represent faithfully the economic phenomena that it proposes to represent, even though it is not possible to represent it perfectly, but then as close as possible (IFRS 2010).

For the accounting information to be reliable, there is a need for it to be generated by the specialists in the area. Thus, these costs identified in the ergonomic intervention can be used as a basis for the measurement of the actual values that impact the company's equity. This information must be transmitted to the interested parties in a trustworthy and reliable manner.

In this context, it should be emphasized that the accounting-financial information on changes in equity caused by (a) the absence of ergonomics and (b) by the ergonomic interventions are not standardized, and may generate different equity variations in similar situations. Therefore, there is a need for accounting standardization and awareness of the parties on the equity variations caused by these two aspects.

3 Case Presentation and Discussion

As stated earlier, this work aims to demonstrate that there is a need to establish guidelines and accounting standards to be used by companies in cases of ergonomic interventions, aiming to highlight the equity variation caused by the intervention. It is up to the ergonomics team to prepare the technical report that measures the property effects related to Ergonomics.

The result of the ergonomic work involving the costs of pre-intervention, intervention and expectation of return should be used as a basis, before the end of the work, to determine the values to be evidenced to the users of the financial statements. To do this, one of the last steps should be an analysis with the responsible accountant, in order to determine the values accurately, so it can then be evidenced in the financial statements.

A case was used to demonstrate the need for such standardization. The example is a case in which the team of the Complex Systems Ergonomics research Unit (GENTE) laboratory of COPPE/UFRJ performed an ergonomic intervention, which beared significant values. The results are representative in accounting terms, however they have not been demonstrated by the company in the year in which the intervention occurred, as there is no such standardization. It provides a basis for actual comparison of the effects of the absence of such statement.

The case concerns the food industry, more specifically a producing plant that outputs several pasta products. It is a company of good reputation, market acceptance and good standard of quality. With low absenteeism and only small environmental problems, the team encountered a major problem in the production of packages. That sector had a reject rate of about 7 to 9%, but that loss cost only one cent (R$ 0.01) per package and the managers of the company discarded the packaging as a priority for intervention, thus not investing to solve the problem. Daily production was about 3200 units in each of the nine production lines, generating 874,800 units per month (9*3,200*27 days = 874,800), resulting in loss of R$ 612.13 (0.07*874,800.00*0.01 = R$ 612.13). After the ergonomic intervention, it was identified that, with the estimated investment of R$ 174,500, the equipment could be repaired, and the 61,236 packages that would be discarded, could be in the market for sale. It was also identified that in one of its lines there was market space to absorb more production. Since this production could be absorbed by the market, company managers were failing to realize that the revenue generated by mitigating the loss paid the investment in approximately two and a half months, and the increase would be constant, since the 61,236 packages would be sold with a profit margin of R$ 1.20 per unit, generating a profit of R$ 73,483.20 in the period (about 12% of the monthly profit).

In accounting terms, immediately, the intervention generated the investment of R$ 174,500.00 and the mitigation of the rejected packages while at the same time increasing the expense of raw material in the confection of the product. The investment consisted of: R$ 12,000 maintenance, R$ 31,000 to relocate the plant and R$ 131,500 to new equipment. In the presented case, the amount spent on maintenance and relocation, due to the rapid return and low value, should be considered an expense of the period, while the investment in equipment would be recorded as an asset (equipment) and depreciated according to its use. In this way, users would have the notion of a return on investment over time, in an appropriate way, and the profit earned in the period would be greater and more coherent.

Hypothetically, if in this case the return on investment was greater than a year, all this value could be recognized as an asset of the company, being separated between fixed assets and intangible assets. The firs, depreciated according to the use of the machine, and the second depreciated according to its expectation of return. The creation of the intangible asset would be justified by the ergonomist's technical opinion on the expected return on that investment. Thus, users could identify the variation and perceive the allocation of cost appropriately, so the profit earned in the period would be even greater. For example, we assume that the amortization time would be the same as the depreciation, five years.

Account	Registered values		1st possibility		1nd possibility	
	1st year	2nd year	1st year	2nd year	1st year	2nd year
Expenses of the Period	R$43.000.00	R$0.00	R$12.000.00	R$0.00	R$0.00	R$0.00
Permanent assets	R$131,500.00	–	R$162,500.00	–	R$162,500.00	–
Intangible Assets	R$0.00	–	R$0.00	–	R$12,000.00	–
Increase in Sales Revenue	R$881,798.40	R$881,798.40	R$881,798.40	R$881,798.40	R$881,798.40	R$881,798.40
Loss reduction	R$43,000.00	R$0.00	R$12,000.00	-	R$0.00	-
Depreciation and amortization	R$2,191.66	R$2,191.66	R$2,708.33	R$2,708.33	R$2,908.33	R$2,908.33
Profit/Loss for the period	R$836,606.74	R$879,606.74	R$867,090.07	R$879,090.07	R$878,890.07	R$878,890.07
Impact	5.12%	0.25%	1.67%	0.31%	0.33%	0.33%

In both cases, the explanatory notes should show detailed account of the ergonomist's work, highlighting the following points: first year - amounts invested, estimated usage time (demonstrating calculation of depreciation and amortization), income expectation or expense reduction; second year onwards - same information, however, comparing the expectation with the performance of the period, adjusting the values according to reality, keeping the explanatory note until the assets are written down or their relevance diminished.

In the situation where the company decides not to carry out the investment suggested by the ergonomic intervention, it is still up to the company to show details of the intervention in the explanatory notes, generating information about the possible consequences of such a decision, in addition to the points presented. In the case presented, it would be for the company to estimate the value of the investment and the increase in sales revenue, and to disclose in the explanatory notes such information and reasons that led the company not to invest in the proposal. Mainly the value that the company would be failing to receive (absence of ergonomics), this information is of extreme relevance for users in general, since the company would be failing to earn a relevant revenue for its business.

4 Conclusion

This work aimed to present the need to create an accounting standard for possible patrimonial variations caused by the observation of Ergonomics, either by its absence and the consequences of interventions. The process needed to achieve this goal is slow and will require considerable efforts from all parties, but it begins with the awareness of ergonomists, senior management, accountants and others.

The development and application of such a standard, with the capacity to show the financial consequences of the ergonomic intervention and its good practices, would result in the presentation of the direct values of the intervention, the values that the company would be losing from the intervention, the potential future consequences of the ergonomic proposal, the direct values invested with the execution of the proposal, among others. As well as several indirect results such as the presentation of ergonomics and its consequences to all users of the financial-financial statements, among other results.

In the case presented, the company showed a variation that would not represent its reality, since the value of the asset would be lower, so the expenses would be overestimated in the first year, thus displaying a profit lower than the real one. Such a situation could be misleading the investors.

References

Oxenburgh MS (1997) Cost-benefit analysis of ergonomics programs. Am Ind Hyg Assoc J 58(2):150–156

Hendrick H (1996) Good ergonomics is good economics. In: Proceedings of the human factors and ergonomics society 40th annual meeting. Santa Monica, CA

Hendrick Hal (2003) Determining the cost-benefits of ergonomics projects and factors that lead to their success. Appl Ergon 34(5):419–427. https://doi.org/10.1016/S0003-6870(03)00062-0

Mafra JRD (2006) Metodologia de custeio para a ergonomia. Rev Contabilidade Finanças 17 (42):77–91

Lahiri S, Gold J, Levenstein C (2005) Estimation of net-costs for prevention of occupational low back pain: three case studies from the US. Am J Ind Med 48(6):530–541. https://doi.org/10.1002/ajim.20184
Rickards J, Putnam C (2012) A pre- intervention benefit- cost methodology to justify investments in workplace health. Int J Workplace Health Manag 5(3):210–219. https://doi.org/10.1108/17538351211268863
Yamamoto M, Salotti B (2006) Informação Contábil: Estudos sobre a sua divulgação no mercado de capitais, 1st edn. Atlas S.A, São Paulo
IFRS (2010) Conceptual framework for financial reporting. Statement. International Accounting Standards Board, England

Transformation of Work Systems - Towards Remotely Supervised Controlled Work

Pernilla Ulfvengren(✉)

INDEK, KTH Royal Institute of Technology, Lindstedtsvägen 30, 10044 Stockholm, Sweden
pernilla.ulfvengren@indek.kth.se

Abstract. The world of operational and work systems is changing not only because of increased commercial pressures to cut costs but also because the deployment of new technologies. Digitalization creates opportunities for new ways of organizing and managing existing work systems. Being able to achieve change and to design future systems are the core capabilities to meet future digitalization without compromising sustainable work systems. It is argued that the increase of automation and robots in theses work systems calls for a new approach. Paradoxically with increased automation and more advanced technology it is now more than ever essential to view these operational systems as people systems.

Keywords: Sociotechnical system · Remote control · System change

1 Introduction

1.1 Examples

The world of work systems and operational functions is facing a radical transformation phase where new technologies, en masse, needs to be internalized into already dynamic sociotechnical systems. The reasons for his are many new technologies fives hope to cut costs in an increased commercial competition but also because the deployment of new technologies allow new way to perform existing functions, even those already highly automated and remote. Technologies related to digitalization, automation and robotics and their connectivity at large, creates opportunities for new ways of organizing and managing existing work systems. Being able to achieve change and to design future systems are the core capabilities to meet future operational challenges without compromising the sustainability of work systems.

Already in the early work by Emery and Trist (Trist 1981) principles and approaches for managing change of work systems and the socio-technical concepts were developed. At the time, the fact that work was based on a few, but longitudinal studies, was mentioned as a limitation to generalizability. Today, history of ergonomics and human factors have much evidence to show that those principles and approaches are general. Their identified need to study to study work as systems as well as at different work levels such as the human-technology interface, the organizational level and the macrosociety level are as relevant and needed today. Rasmussen (1986)

S. Bagnara et al. (Eds.): IEA 2018, AISC 821, pp. 583–594, 2019.
https://doi.org/10.1007/978-3-319-96080-7_70

developed this further with the socio-technical system level model. In his model the dynamics of the system is emphasized. The system is constantly changing. This is due to learning at all levels of the system as well as variations in external influences in terms of competence, technology development and market trends and societal changes.

What we see now is a drastic game change for work systems due to a combination of several external influences. The maturity of technology allows the realization of new levels of automation, robotics and connectivity in operations. This has given organizations the opportunity to develop new concepts that are more cost effective and delivers operational performance that is more efficient. Increased competition at a global scale has led to increased outsourcing of manufacturing which meets the needs of industry. At a societal level globalization may be both a need and desire as well as a threat for job opportunities and national growth. This has led to nations wide initiatives for Industry 4.0 and cyberphysical systems. These are examples of on-going forces on and reactions by a sociotechnical system in production systems today. It could be argued that if there is any change in the way organisations and work systems as well as technology is developing today compared to the early organizational design studies, it is the pace and complexity in the dynamics of the system.

So, more than ever ergonomics may gain to look at its origin theoretical base, when other industrial revolutions occurred. System is the fundamental construct of the field of Ergonomics (Meister 1997). From the beginning the sociotechnical (S-T) concept has developed in terms of systems. As the "open-system theory" (Von Bertalanffy's paper 1950 Open systems in Psychology and Biology) due to the interdependencies with external factors to almost any system boundary. It influenced both theory building and field projects on individual as well as higher system levels such as self-regulation and environment relations. Also Cypernetics (Weiner 1950) with partially closed loops within subsystems or within particular system levels, were used as analogy of for example small group self-organisation. "The system is critical to HFE theorizing because it describes the substance of the human-technology relation. Anyone observing the stew of humans in real-life interactions with technology and wishing to organize it conceptually classifying it into described units must inevitably see this activity in terms of system" (Meister, ibid).

The growing complexity of work system design and development may require a recurrent application of engineering practices specifically apt at managing complexity in systems engineering. However, systems engineering standards and practices (INCOSE, IEEE…ISO…) lack integration of human knowledge with purpose of applying "system requirements" for humans at work for engineering work systems, i.e. dynamic socio-technical systems.

Core for any system analysis process is the knowledge and ability to analyze the problem situation and extract variables that affects the problem and how they do so, i.e. to understand the system functionality or how the system works. Early work analysis is listed 1–9, for general purposes put in system terms by Trist (1971):

1. An initial scanning is made of all the main aspects – technical and social – of the selected target system – that is, department or plant to be studied.

2. The unit operations – that is, the transformations (changes of state) of the material or product that take place in the target system- are then identified, whether carried out by men or machines.
3. An attempt is made to discover the key variances and their interrelations. A variance is key if it significantly affects (1) either the quantitity or quality of production, and (2) either the operating or social costs of production.
4. A table of variance control is then drawn up to ascertain how far the key variances are controlled by the social system – the workers, supervisors, and managers concerned. Investigation is made of what variances are imported or exported across the social-system boundary.
5. A separate inquiry is made into social-system members' perception of their roles and of role possibilities as well as constraining factors.
6. Attention the shifts to neighbouring systems, beginning with the support or maintenance system.
7. Attention continuous to the boundary-crossing systems on the input and output side – that is, supplier and user systems.
8. Target system and its immediate neighbors are then considered in the context of the general management system of the organization as regards the effects of policies or development plans of either a technical or a social nature.
9. Recycling occurs at any stage, eventually culminating in design proposals for the target and/or neighboring systems.

In recent work, a similar approach to research as the classic work by Emery and Trist has been taken to further develop theory on socio-technical system functionality and in particular the human role in this functionality. The author participated in three of the contributing projects and many case studies developing the current framework (McDonald and Corrigan 2015).

The novel human factors model and framework called SCOPE, System Change and OPerations Evaluation (McDonald 2015; Ulfvengren and Corrigan 2015).

"Today both manufacturing and service providers are looking for opportunities with new technologies such as digitalization, robotics and increased automation. Paradoxically, with increased automation and advanced technologies, it is now more than ever essential to view these operational systems as people systems. No matter what the degree of automation, people are involved, at some level of the system, in a co-ordination role. But many processes involving advanced technologies, as well as those with low technology involvement, are also essentially "people processes" with the functional links that produce the product or service being mediated or co-ordinated partially or entirely by people… Understanding the functionality of how such a system works is the key to managing the system more effectively, and to comprehending how it is possible to change the system to achieve better outcomes, or how to design a future system to operate in a way that transcends current practice" (McDonald 2015*).*

The research logic assumes at its most general, the capability of a resilient organisation has to encompass the following three characteristics (McDonald 2015):

1. To be able to mobilise its resources (especially its knowledge and the information that supports this)

2. To anticipate future challenges, and to respond and adapt to such challenges (whether they are fully foreseen or not). In order to do this, it has to understand itself in two ways:
   - how the system functions (how the interaction of human, social and technical aspects makes it possible to deliver value in the short, medium and long term); and
   - what it is doing (how data from this activity are converted to knowledge about how the system is performing).
3. To have sufficient consensus to participate in and support the effective leadership of change.

SCOPE will be applied to identify different elements and relations in the system in a systemic and systematic way, called a Structured Enquiry (ibid.). The framework comprises three interlocking, interpenetrated sub-systems—the functional process, social relations (teams and trust) and the role of knowledge and information logic mediating between these. This provides a strong basis for planning, evaluating and providing support to change initiatives and implementation.

In this study various levels of socio-technical models will be applied to make an initial screening of current trends of transformation of work systems into remotely supervised controlled work. This study describes various types of systems transforming due to new technology development entering existing work system and transforming them into new systems, but with same functionalities as before. The work systems described and analysed are piloting UAVs (unmanned aerial vehicle), remote air traffic control and remote workstations of technical experts in manufacturing. Each are different with respect to automation and human presence at the remote location or object and hence challenges differ for design and implementation for these systems. Various competences needs to be changed in relation to the new context of remote operations, in addition to expertise of for example flying, air traffic control and technical fitting skills in factory. Process and system analysis methods are in part applied and discussed to cover relevant socio-technical system aspects prior to design and implementation. This is relevant because there is much general S-T knowledge from early adopters of automation and remote work that will become highly relevant in areas not so commonly operated remotely.

1.2 Objective

The objective of this study is to identify examples of challenges and affecting factors specific for transformation of current work systems into remote supervised controlled work.

2 Remotely Supervised Control Work Concepts

2.1 Remote Manufacturing

A remote concept in manufacturing addresses an upcoming need due to lack of local expertise in expert functions in a factory. Here it is a combined need of a rare competence with few experts and no real growth of this group in sight and the resources required for and from these experts. They many times need to travel from factory to factory to perform various try-out tasks on remote sites. In addition, this job, when performed manually on site is very physically demanding with arms in overhead position. With the remote concept this competence would be made available to more production plants and the heavy work performed in factory would be replaced by a robot. It is anticipated to require fewer resources from both companies and individual workers. It would mean reduced travel costs, and less environmental impact. Travel adds to persons' long-term stress and fatigue as well as social aspects such as family logistics, which are even more reasons to keep travels down.

The worker's skill is identified as the valuable driver of the development. There is a need to make this expertise available to a greater extent. Current level of automation and robotics is far away from automating this work without keeping the expert try-out operator in the loop for each try-out. The production technologies are not yet developed that would manage the function without the need for advanced hands-on try-out. But the remote concept will make hands-on try-out expertise available remotely.

The remote concept will enable expertise to be available remotely. Realizing this innovation requires a transformation of a whole work system.

2.2 Remote Flight Operations

A future remote flight operational concept is a radically new way to manage the operation of remotely operated vehicles. In early aviation, a single pilot operated an aircraft. This one-to-one relationship evolved drastically as aircrafts become more and more complex, getting into a climax in which pilot, co-pilot, flight engineer and radio operator were required to operate large and complex aircrafts. Then, technology and automation came into play. This initiated a continuous reduction in the number of persons to two-pilot crews, reaching the point in which single pilot operations are viable in a very near future.

The remote concept redefines the current paradigm by switching the relationship from one crew to one aircraft, to a team of pilots to a fleet of aircrafts. The introduction of remotely piloted aircraft systems (RPAS) has induced a new line of development. However, in most cases, RPAS concepts have simply reproduced classical pilot schemes with a single pilot flying a small RPAS, and with one pilot with a co-pilot (sometimes called payload operator) in more complex operations. Nevertheless, RPAS operations have paved the way to new elements that differentiate manned aircraft from remotely piloted aircraft. For example will a crew on ground allow for surveillance over long time since crew may make handovers on ground, shift work.

2.3 Remote Air Traffic Control Towers

The conventional configuration of ATS is to have air traffic controllers physically present at towers for ATS, overlooking the airport and its surrounding airspace. Challenging this conventional setting is a solution that has been brought forth in recent years – the Remote air traffic service (RATS). The basic concept of RATS is to supply ATS to one or multiple airports by providing services like communication, navigation, and surveillance from a remote location. This is anticipated to provide improved service capabilities, reduced costs, and improved safety at airports compared to conventional solutions (Adrem and Klintberg 2018a; Van Beek 2017). The ability to provide ATS to multiple airports from one air-traffic control centre will facilitate in particular airport services in more rural areas. This transformation is different since it is already a remote supervisory controlled work, since the controllers remotely control the air space through the means of pilots flying aircrafts. The obvious difference is that now not only the distant traffic will be monitored on a display, but also the near aircrafts will be on a screen and the controllers will not be physically close to the approaching and landing aircraft and crew. The aircrafts in the tower area will be visual through real time cameras and not symbols as the distant traffic.

3 Theoretical Framework

3.1 Sociotechnical System Models

In Rasmussen's work (Rasmussen 1986) the various levels of a socio-technical system are described as in the model in Fig. 1. Relevant for the industrial revolution are the external factors influencing the system and then how the system reacts to the internalization of these external or environmental stressors. The fast pace of technological change and development, the need for changing competency due to technological change entering the system, or a decline in new generation with skills and expertise still needed but no longer being available in the work force. The parallel demands of changing markets and financial pressure that makes companies and societies to develop strategies with the help of anticipated advantages with new technology. On top of this we have global challenges and needs to accommodate changing demographics of our work forces as well as demands for sustainable development for the sake of environment and work. On the other side of Fig. 1 the various "silos" of research and knowledge is listed.

The logic of the model is the dynamics of external influences from both knowledge and the ones presented above. There are also dynamics between each level in the system. It is a peoples' system where individuals and groups learn and shift their understanding of the system they are functioning in. The work system and process analysis needs to integrate various levels of people in the system, and the functional role of people in a system in which their activity make sense as well as the function between people and technology, in an integrated functional socio-technical system.

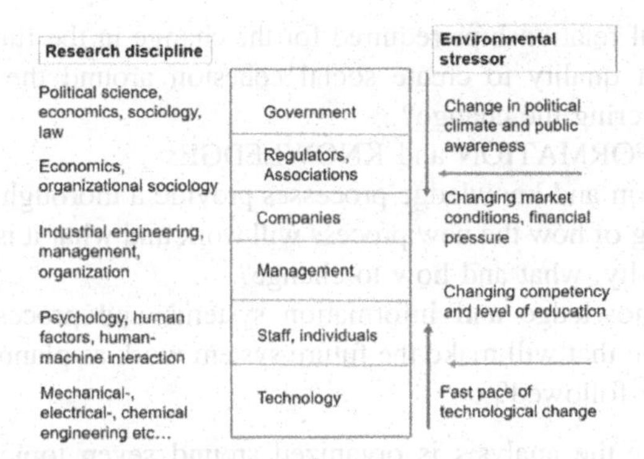

Fig. 1. The socio-technical system in a dynamic society, from Rasmussen (1997a, b).

3.2 SCOPE and Structured Enquiry

The novel human factors model and framework called SCOPE, System Change and OPerations Evaluation (McDonald 2015; Ulfvengren and Corrigan 2015). The framework comprises three interlocking, interpenetrated sub-systems—the functional process, social relations (teams and trust) and the role of knowledge and information logic mediating between these (Fig. 2).

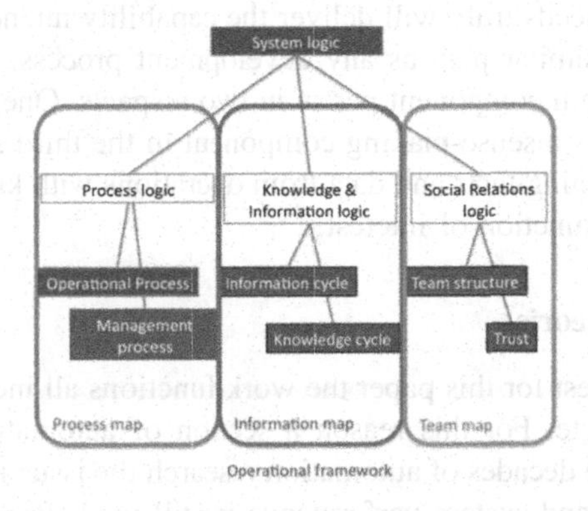

Fig. 2. Schematic architecture of overall system analysis, from McDonald (2015).

A structured enquiry is a set of questions related to each of the systems in Fig. 2.

- System I - PROCESS FUNCTIONALITY:
  - Will the capability of the new process deliver strategic performance targets?
  - Will the process that is under development and implementation satisfy the external demands on the process and be aligned with the company strategy?
- System II - SOCIAL RELATION and COHESION:
  - Do the social structure and trust sustain stable process performance?

  - Are the social relationships required for the change in the functional process of good enough quality to create social cohesion around the new process and process delivering the change?
- System III - INFORMATION and KNOWLEDGE:
  - Do information and knowledge processes provide a thorough, valid and current understanding of how the new process will work and what it is doing and hereby understand why, what and how to change?
  - Are there knowledge and information systems and processes in place that engage people that will make the future system work as planned and the change process to be followed?

Each element of the analysis is organized around seven topics: goals, previous state, description, evaluation, risk, impact, requirements/recommendations. Each topic is addressed to each system respectively. This provides a strong basis for planning, evaluating and providing support to change initiatives and implementation.

The application of SCOPE and the structured enquiry relates to a socio-technical system in a generic way. It may be focus on operational processes in the lower level of the system-level model in Fig. 1 or to organizational processes and management system in a higher level interacting with lower levels. At the same time the seven topics include similar to systems engineering, a strong focus on identifying objectives and goals as well as driving forces and needs for change. Then the analysis moves to identifying current (as-is) processes delivering the function and asks if the suggested solution to identified needs truly will deliver the capability intended. Then the process of enquiry follows a similar path as any development process. This framework does include the social system component per se in two respects. One is the trust and teams aspects and the other is a sense-making component in the third system of information and knowledge, combining facts and data from operations with know-how of those that is part of the process function of interest.

3.3 Automation Theoris

In the systems of interest for this paper the work functions all include high automation levels to operate remote. For this reason a section of automation basic theories are included. Despite three decades of automation research the issue in relation to operators and work satisfaction and system performance is still vital (Parasuraman et al. 2008). Studies conducted in the eighties when problems from automation emerged are still valid, like "ironies of automation" (Bainbridge 1983; Baxter et al. 2012). Bainbridge warned for unanticipated effects like tasks becoming less sustainable or more difficult to perform given the lack of transparency with automation. However, there exists several central concepts, models and frameworks that may support a successful transformation from manual work to remotely supervised controlled work.

In the context of automation "function allocation" means that the agent, human or machine, that is; best suited should perform the function. Task allocation principles between humans and technology used in automation may be applied for both for robots and remote work. One is the left-over principle and the other is the compensatory principle. In the left-over principle, you automate what you can and what is economic, the rest is left-over to the human operator. In the compensatory principle you refer to the classical MABA-MABA (Men are better at - Machines are better at) according to Fitt's list which divides tasks between what fits the technology or humans best given their capabilities.

The model for Level Of Automation, (LOA) (Sheridan 2002; 1990; 1987) describes ten levels of automation, from least; the human does everything, to most; the computer does everything. This has been adapted to manufacturing and assembly (Dencker et al. 2009) work with seven levels of automation: (1) Totally manually, (2) Static hand tool, (3) Flexible hand tool, (4) Automatic hand tool, (5) Static work station, (6) Flexible work station and (7) Totally automatic. Later, models that consider the human-technology as a joint system in which you like to maximize the overall system performance outcome have been developed.

The Cognitive Function Analysis is a methodology to model cognitive tasks and activities in order to assign them in a holistic manner, rather than according to Fitt's list which is static (Boy 1998; 2009). Machine performance of human functions induces new cognitive functions according to Cognitive Function Paradigm (Boy 1998). By automating a function, the number of decisions needed in situations can increase because the operator not only needs to diagnose the state of the situation but also needs to diagnose the reliability of the automation at different levels (Wickens and Hollands 1999).

The SRK-model (Rasmussen 1983) allows work to be analysed in terms of cognitive levels like basic skills for routine work, rule-based work which require training and knowledge-based work which is applied in novel situations and requires expertise and experience. For example in relation to automation it is the skill-based tasks that are more often suitable for automation. Supervisory control and work content models have been particularly adapted to manufacturing (Fasth et al. 2010) applying and combining Sheridan's and Rasmussen's models which define operator tasks at various levels in terms of Planning, Teaching, Performing, Intervening and Learning. Mårtensson (1996) focused especially on operators' quality of work and performance and lists six requirements and provide a set of criteria to meet each of these in her work; Versatile job content; Responsibility and participation; Information processing; Influence on the physical work performance; Contact and Co-operation and Competence development.

4 Results and Analysis

4.1 Human-Robot Collaboration and Interface Design of Work Station

Everytime automation is used to transfer tasks from humans to technology like robots or tools it is essential to carefully address trade-offs like in all design and consider unanticipated effects and consequences. Many times the task allocation may not be designed optimally due to technology feasibility or costs. In this remote concept a thorough analysis of both the current work and the work remaining work station will need to be conducted to strive for a sustainable work system.

The interface technology and tools that the operator will use will need to be flexible and designed with user-centered design principles to ensure high usability and flexibility for user. This is to ensure the remote manufacturing concept will be sustainable from a human and work perspective and at the same time economically beneficial. This means a wide range of knowledge will have to be utilized. Cost analysis of production and work is tangible but requires a good inventory and definition of system boundaries. For social aspect of the work system methods are not so straight forward. From a shop floor work perspective ergonomics principles apply and for expert work, human factors with cognitive focus apply. There will be a combination of automation aspects with allocating tasks both to robots, local technicians and remote operators. For sustainable work systems relevant models and aspects from the field of ergonomics and work science are models and methods for work and task analysis of work content, work satisfaction, workload. From a human-centered automation approach, aspects of task allocation, level of automation and de-skilling, when changing or reducing work content, are essential for anticipating consequence of automation. The concept suggests a rare challenge of combination of automation of skill-based tasks transferred to robots and other technology as well as development of remote knowledge-based supervisory controlled work.

For the remote flight operations similar work stations to current ATC could be used in combination and complementary interface for flight operations control.

In the remote towers the interface will not change to current design.

4.2 Shift Work and Handover

The first differentiating element with remote flight operations is the handover process that may exist when the RPAS crew in charge of departure/arrival operations transfer the aircraft control to the crew that will perform the required surveillance or combat operations (as this scheme is mostly used in military operations).

Managing a flight operating concept (FOC) for RPAS will also require the design of a new concept on how the RPAS responsibility could be transferred from one pilot to another as the RPAS operation evolves. That nominal transfer of responsibility should also cover specially situations, like emergencies or high workload situations.

Based on the state of the art regarding the way in which RPAS are operated, the remote concept would need to explore a more general RPAS-pilot relationship paradigm, in which the individual allocation of each pilot to a certain RPAS is extremely relaxed. Once such detachment is accepted, the currently assumed N pilots to 1 single aircraft relationship vanishes into a much more flexible and potentially efficient N pilots to M aircraft relationship. N-to-M will require the design of new remote control stations, especially suited to manage multiple vehicles, potentially operating in different areas and flexible enough to provide support to manage different phases of flight. Moreover, in addition to RPAS flight, supervision roles may be necessary in order to guarantee the adequate operation as remote FOC.

Compare the 1 to 1 concept with future NtoM with respect to human performance challenges from the pilot's perspective. Identify main differences and implications for New pilot role and work organization will include aspects in terms of Human factors and supervisory control, i.e. automation, situation awareness, decision making, technologies supporting collaboration (Collaborate HCI), team work (task allocation, shared situation awareness, collaborative decision making, hand-over etc.), fatigue models (vigilance, stress), training (competence requirements for various tasks).

4.3 Change of Roles and Cyber Security

As the ultimate operators of the systems, air traffic controllers arguably provide an important perspective on (mainly) potential operational-related obstacles. It has been highlighted reduced sensory information, lack of awareness of local airspace conditions, physiological impacts, workload changes, specific training and certification requirements etc. Svensk Pilotförening, the Swedish Air Line Pilots Association, raise, in a press release from 2017, concerns about the readiness among the users of RATS to the major changes that its implementation entails (Svensk Pilotförening 2017).

Concerns about cyber security relates to anything that is remote. EASA (2017 p. 59) argues that "[RATS] relies on IT infrastructure for data exchange to support, amongst others: visual presentation, communications (in particular aeronautical mobile service and surface movement control service) and management of aerodrome equipment/systems/assets. This makes it vulnerable to potential security threats to computer systems or the data exchanged", and primarily highlights the RATS vulnerability to loss of visual presentation data and potential interruption of communication with the remote airdrome.

5 Discussion and Conclusion

Technology an work system change, human factors remains…

References

Rasmussen J (1997a) Safety Science; Risk Management in a Dynamic Society. 27

McDonald N, Cogn Tech Work (2015) 17:193. https://doi-org.focus.lib.kth.se/10.1007/s10111-014-0296-9

Bertallanffy L (1969) General system theory: foundations, development, applications

Bainbridge L (1983) Ironies of automation. Automatica 19:775–780

Baxter G, Rooksby J, Wang Y, Khajeh-Hosseini A (2012) The ironies of automation … still going strong at 30? In: Proceedings of ECCE 2012 Conference, 29th–31st August, Edinburgh, North Britais

Boy GA (2009) The orchestra: a conceptual model for function allocation and scenario-based engineering in multi-agent safety-critical systems. Paper presented at the European conference on cognitive ergonomics, Helsinki, Finland

Boy GA (1998) Cognitive function analysis for human-centered automation of safety-critical systems. Paper presented at the CHI 1998, Los Angeles, CA

Dencker K, Stahre J, Mårtensson L, Fasth Å, Akillioglu H (2009) Proactive assembly systems – realizing the potential of human collaboration with automation. Special Issue of the annual reviews in control. Elsevier http://dx.doi.org/10.1016/j.arcincontrol.2009.05.004

Fasth Å, Bruch J, Dencker K, Stahre J, Mårtensson L, Lundholm T (2010) Designing proactive assembly systems (ProAct) – criteria and interaction between automation, information, and competence. Asian Int. J. Sci. Technol. Prod. Manufact. Eng. (AJISTPME) 2(4)
Mårtensson L (1996) The operator's requirements on work in automated systems. In: The international journal of human factors in manufacturing, vol. 6 (1) 1996. Wiley
Parasuraman R, Wickens CD (2008) Humans: still vital after all these years of automation. Hum. Factors, 5(3): 511–520. https://doi.org/10.1518/001872008x312198. Copyright © 2008, Human Factors and Ergonomics Society
Rasmussen J (1983) Skills, rules, knowledge; signals, signs, and symbols, and other distinctions in human performance models. IEEE Trans Syst Man Cybern 13:257–266
Rasmussen J (1997b) Risk management in a dynamic society: a modelling problem. Saf Sci 27 (2–3):183–213
Sheridan TB (1987) Supervisory control. In: Salvendy G (ed) Handbook of human factors. Wiley, New York, pp 1243–1263
Sheridan TB (1990) Task allocation and supervisory control. In: Helander M (ed) Handbook of human-computer interaction. North-Holland, Amsterdam, pp 159–173
Sheridan TB (2002). Humans and automation: system design and research issues. Wiley in cooperation with the Human Factors and Ergonomics Society (HFES), Santa Monica, CA
Wickens CD, Hollands JG (1999) Engineering psychology and Human performance. Prentice Hall, Upper Saddle River. p. 07458, ISBN 0-321-04711-7

On Human Terms – Integration of a Massive Open Online Course (MOOC) in Ergonomics in University Courses

Anna-Lisa Osvalder[1,2(✉)] and Martina Berglund[3]

[1] Design and Human Factors, Chalmers University of Technology, 412 96 Gothenburg, Sweden
alos@chalmers.se
[2] Ergonomics and Aerosol Technology, Lund University, 221 00 Lund, Sweden
[3] HELIX Competence Centre and Division of Logistics and Quality Management, Linköping University, 581 83 Linköping, Sweden

Abstract. The aim of the Massive Open Online Course (MOOC) 'Work and Technology on Human Terms' was to be a modern, highly accessible study material in Ergonomics/Work science to contribute to safer and healthier workplaces. The course targets engineering students, but also students in economics, human relations, physiotherapy, occupational health, as well as professionals. The purpose of this paper is to describe the course and how it has been integrated in six Swedish university courses dealing with various ergonomic issues, and reviewing how the lecturers and students experienced the outcome. The MOOC covers the subjects: Human-Technology-System, Psychosocial and organisational environment, Physical environment, Information and interaction in technical systems, Methods/design processes, and Economic/legal conditions. Theory and models are presented in in photos, illustrations, animations, documentary clips, expert interviews etc. It contains four workplace cases, knowledge tests and a textbook. The course material was integrated in three ways; as voluntary, alternative material for the students' self-studies, as a scheduled activity for self-studies with appointed chapters, and as compulsory, selected course material discussed in seminars. The type of integration that is most beneficial depends on the content and organization of the course and type of students. It opens up the possibility for flexible and blended learning based on the student's individual conditions. It also saves lecturer time by excluding some traditional theoretical lectures in favor for supervised discussions, problem-solving in workshops, and practical projects.

Keywords: Work science · Blended learning · On-line course

1 Introduction

The Massive Open Online Course (MOOC) 'Work and Technology on Human Terms' (www.onhumanterms.com) was developed by the publishing company Prevent in collaboration with lecturers and researchers at five Swedish Universities of Technology; Chalmers, KTH, Lund, Luleå and Mid Sweden (Lagerström et al. 2017). The course is

S. Bagnara et al. (Eds.): IEA 2018, AISC 821, pp. 595–601, 2019.
https://doi.org/10.1007/978-3-319-96080-7_71

available globally, in English and free of charge. A main goal is to increase the knowledge about how products, systems, and work organizations can be designed on human terms. The ambition of the course is to be a modern, highly accessible study material in ergonomics/work science to contribute to safer and healthier workplaces, as well as to maintain or improve organizational performance, so that people are able not only to uphold good health during their working lives, but to retire in good health as well. The MOOC can be regarded as a complement to the earlier published book with the same title (Bohgard et al. 2015).

Ergonomics/work science is an interdisciplinary activity based on engineering, psychology, anatomy, physiology and organizational studies. The MOOC targets students at universities who are pursuing degrees in engineering, economics or human relations. It is also aimed at students in the medical field such as physiotherapists and occupational health professionals. In addition to these primary target groups, professionals holding equivalent university degrees can benefit from the course. The parties of the Swedish labor market emphasize that the course is a suitable tool to strengthen professionals' skills in work science and other related topics.

In August 2017, the course was launched. Universities should be able to use the entire training course or selected portions of it as a teaching component in their respective courses, supplemented with e.g. laboratory exercises and case studies. Since the autumn semester 2017 it has been used in selected university courses dealing with ergonomics/work science at the Swedish Universities of Technology involved in the development of the MOOC.

The purpose of this paper is two-fold. First the content of the MOOC 'Work and Technology on Human Terms' is presented as well as how it has been integrated in some Swedish university courses dealing with ergonomics, work environments and health, human-machine systems and product design. Then results are presented regarding how the involved university lecturers, course developers and students experienced the outcome of this integration.

2 Description of the MOOC

The MOOC 'Work and Technology on Human Terms' comprises of 14 chapters dealing with (1) The Human-Technology-System (2) Psychosocial work environment, (3) The organization of production and work, (4) Physical loads (physical ergonomics), (5) Physical environmental factors, (6) Chemical health risks in working life, (7) Human-machine systems (cognitive ergonomics), (8) Safety and risk, (9) Methods, (10) Development process, (11) Occupational injuries, (12) Work environment and economics, (13) Work environment legislation, and (14) Standardization. It also contains four workplace cases for practical application of the theory; the IT-company, the Food Company, the Construction Company and Home Care. The course also includes self-correcting knowledge tests in most chapters, where theoretical statements should be assessed as false or true.

In the chapters theory and models are presented using a mixture of photos, illustrations, animations, documentary clips supported by interviews with experts and researchers active in the various disciplines in order to maintain a high degree of credibility. There is also an English speaker, and all Swedish interviews are subtitled in English. The content is closely linked to the situation on the ground in different types of businesses, as reflected by company reports and interviews with product developers, managers and safety representatives, in order to highlight the relevance of the content and to increase user comprehension of when and how the knowledge can be applied.

The course comprises totally 20 h of study on-line. The student is free to start and finish parts of the course whenever and independently of other users. The course also includes knowledge tests and the student can receive a certificate after being approved. It is primarily designed for use on a computer, but it can also be used on tablets and mobile phones in a responsive interface. The course has no chat feature or other social media component. The chapters from the English textbook "Work and Technology on Human Terms" (Boghard et al. 2015) are also included as additional study material for each chapter.

3 Integration of the MOOC in University Courses

This study deals with the first attempt to use and and integrate the MOOC material in six university courses during the academic year 2017/18. The courses were given to engineering and design students at three Universities of Technology in Sweden; two courses at Chalmers, two in Lund and two in Linköping. The courses at Chalmers and Linköping dealt mainly with product development, i.e. ergonomic design and human-machine systems, where physical loads, human cognition, interface design, safety and risk, methods and development processes were essential. One of the two courses at Lund dealt with work science in general, i.e. all aspects that can affect the working environment and health, while the other course focused on management and organization. All six courses provided 6 or 7.5 university credit points, and were studied during a period of about 8 weeks. Data regarding the courses are presented in Table 1.

Table 2 shows how the MOOC was integrated in the different university courses presented in Table 1. Some courses used nearly all parts of the MOOC while others included only few selected important chapters, this in spite of that all courses dealt with ergonomics/work science and was give to engineering students. In most courses the basic theory regarding human-technology systems, psychosocial work environment, physical and cognitive ergonomics were mainly focused, and self studies of these chapters were suggested as complementary or rehearsal material to given lectures. Sometimes scheduled time was specified in the course schedule for working with the MOOC, but often it was said to the students to study the MOOC whenever they could during the period the university course was given. Sometimes compulsory tasks were given related to the MOOC. The case studies were seldom used. The self-evaluating knowledge tests had to be taken by the students if they should complete the actual chapter they studied, i.e. they were compulsory to do to be able to proceed.

To summarize, the MOOC material was used in the following ways in the six evaluated courses: (1) suggested voluntary, alternative or complementary material for the students' self-studies in parallel with the course book, given lectures and lecture slides, (2) scheduled activity for self-studies with appointed chapters, where the students were instructed with what parts to study at certain scheduled times to match other course activities, and (3) compulsory, selected course material being discussed in follow-up seminars or used for compulsory hand in tasks, where the MOOC material had to be elaborated by the students on their own.

Table 1. University courses

Course name	Programme and year	University	Students
Human machine systems	Automation and Mechatronics, Mechanical Engineering, 3d year	Chalmers	40
Ergonomics	Industrial Design Engineering, 2d year	Chalmers	60
Product ergonomics	Furniture design, art and craft, 3d year	Linköping	15
Product ergonomics	Design & Product Development, 3d year	Linköping	68
Working environment, health and safety	All programmes, 4th year	Lund	30
Management, work organisation and project management	Industrial Engineering and Management	Lund	30

Table 2. Integration of the MOOC in University courses

Course name University	Usage of the MOOC chapters and	Case studies & knowledge tests used
Human machine systems (Chalmers)	Chapters 1–8 were proposed as theoretical course material in parallel with lectures regarding these topics and the on-line course book. Chapter 2 was compulsory including hand in tasks (no lecture). Chapters 9–10 &13–14 were said to be useful also in the practical workplace project. Lecture time was set in the course schedule for studying chapter 2. For the other MOOC chapters, self-studies were proposed in the schedule jointly with the given lecture of that topic	Knowledge tests had to be taken to complete chapters

(*continued*)

Table 2. (*continued*)

Course name University	Usage of the MOOC chapters and	Case studies & knowledge tests used
Ergonomics (Chalmers)	Chapters 1–4 & 7–8 were proposed as theoretical course material in parallel with lectures regarding these topics and the on-line course book. Chapters 9 and 10 were said to be useful also in the practical product evaluation and redesign project. Self-studies of the different MOOC chapters were proposed in the schedule jointly with the given lecture of that topic. Chapter 1 was compulsory to study during the introduction lecture of the course	Construction Company used in examination Knowledge tests used in lecturers
Product ergonomics (Linköping)	The MOOC was said to be a voluntarily, alternative study material to use for the furniture design students. The MOOC material was introduced during a lecture and an interview with a practitioner was shown	Knowledge tests had to be taken to complete chapters
Product ergonomics (Linköping)	All lecturers studied the MOOC to be aware of how their theme was presented. They referred to the MOOC during lectures and showed films, animations etc. The students were encouraged to use the MOOC as a complementary study material. Chapter 1 was used as compulsory material followed up in hand in tasks an in a seminar	Knowledge tests had to be taken to complete chapters
Working environment, health and safety (Lund)	Prior parts of the textbook was used, now the students have to study the complete MOOC, and then choose and motivate which parts they will focus on in their project work, which the final examination is based on	Knowledge tests had to be taken to complete chapters
Management, work organisation and project management (Lund)	Only chapter 1 and 2 of the MOOC was used as compulsory theoretical material for a reflective hand-in assignment before a follow-up seminar was held including additional theory and discussions regarding psychosocial working environment issues	

4 Review of the Integration of the MOOC

4.1 Method

As a base for the review of how the integration of the MOOC worked in the six university courses (Tables 1 and 2), information was combined through a mix of results from discussions with lectures and responsible course developers, as well as student opinions of the MOOC regarding content, design and usefulness collected by questionnaires and in interviews. The specific result from the students' evaluation of the courses given at Chalmers and Linköping is presented in Berglund and Osvalder (2018). Regarding the lecturers' experiences, the review is also based on the authors' own experiences as course developers and/or lecturers in the four courses given at Chalmers and in Linköping.

4.2 The Lecturers' and Students' Experiences of Integration of the MOOC

This was the first time all lectures and course developers introduced an on-line course, or MOOC, as a part if their on-going university course in ergonomics/work science. Therefore most lecturers thought that the MOOC could serve as an additional, or alternative or complementary course material for the students, which was voluntarily to use in parallel with given lectures, lecture slides and course book. However this did not work, since the students thought is was to much to use three sources of information for the same topic – lecture slides, MOOC and course book.

In the beginning of the course the students were a bit curious and studied the first chapter(s) in the MOOC, but when the course proceeded with a lot of theory to study, hand in assignments and project work etc., they stuck mostly to one source of information to learn the subject. They found it time consuming to use several sources. However, they used the source they thought to be the best for their own learning and skills. Students who would like to achieve the highest grades used the more complete and heavy material in the course book. Students who usually did not reach the highest grades in courses were more interested in the MOOC, since it was a more easy way to learn the subject with various types of presentation forms and the material was shallower. Many students however went to regular lectures and simultaneously used the lecture slides for own notes. They thought that the information given during lectures by the examiner or other teachers were most useful for the coming examination tasks. However in case they missed a lecture, they used the MOOC course instead.

To conclude, using MOOC as voluntary, alternative material for the students' self-studies is not the best way to integrate a MOOC in an existing course, where other sources of information also are available. All students will probably not use it, but weak students will certainly benefit form this type of more 'easy to follow' educational material.

When scheduled activities, or compulsory activities for self-studies of the MOOC are integrated in the course, which also matches other course activities such as seminars, workshops or project work, the benefits of using the MOOC material instead of giving traditional lectures increase. By studying the basic theory using the MOOC the students can be prepared and motivated to elaborate ideas together with other students during discussions and workshops lead by the lecturers or by themselves. The lecturers then have more time to supervise, give spontaneous thoughts and go deeper into some areas and leave the basic knowledge to be learnt from the MOOC. The introduction with a focus on humans and technology at work along with chapter 2, psychosocial work environment are proposed to be compulsory chapters to study in the MOOC followed by workshops.

5 Conclusion

A few courses included in this study used nearly all parts of the MOOC, while others included only few selected important chapters. All courses were dealing with the area of ergonomics/work science and were offered to engineering students at technical universities. The usage of the different parts of the MOOC was directed by the aim of the university course and the type of engineering programme it was given to. The conclusion is that the aims of the MOOC are seldom the same as the aims of the university course, therefore the complete MOOC is seldom possible to integrate.

Having the MOOC as a scheduled activity in parallel with other sources of course material, such as lectures, lecture slides and course book, was considered to be too much work for the students within the given time limit for the university course. The conclusion is to carefully integrate parts of the MOOC to be compulsory and then skip lectures on these topics. This will allocate time for lecturers to supervise more and arrange workshops and seminars where additional and deeper material can be highlighted. Thereby the lectures in classroom could focus more on discussions and problem-solving regarding the topics and less on basic theory.

The concluding recommendation to course developers and lecturers is to carefully consider how to use and integrate the MOOC as a meaningful theoretical learning activity for the students.

References

Berglund M, Osvalder A-L (2018) Proceedings of the 20th Triennial Congress of the IEA, Florence, 26–30 August 2018

Bohgard M, Karlsson S, Lovén E, Mikaelsson L-Å, Mårtensson L, Osvalder A-L, Rose L, Ulfvengren P (2015) Work and Technology on Human Terms. Prevent, Stockholm

Safety on Working Place? The Ergonomic Approach in Pulverit

S. Bernardini[(✉)] and R. Trigona

24 U Member – Umania Srl, via Stezzano 87, 24100 Bergamo, Italy
{silvia.bernardini, raffaella.trigona}@umania.it

Abstract. Working on ergonomics, make the difference in a lots of industrial approaches, especially for people who works in training field. What's important in training methods are not the results, but the people to whom the training is focused on. The paper we will present a big and complex training project, created and organized into a field of safety on working place in Pulverit Spa, a chemical firm, which produces powder paints. The training path indicates clearly the necessity to focus on people especially for the tangle of changes and innovations we're living (not only technological as Industry 4.0) and we have to face.

Keywords: Self observation – observation · Responsible action – care
Anti fragility – fragility

1 Introduction

This paper s the synthesis of a wide training project, created, developed and realized between 2015–2017 in the safety context in Pulverit, a chemical manufacturer of powder paints.

Starting from the real problems of safeties in using machinery and instruments, the training project faces the organization system with an ergonomic approach to verify the most adaptable condition for workers and productivity. First of all, the company is described under different points of view in most part of the work: how it appears from the outside (customer point of view), in terms of safety, control and quality; how it appears from inside (operators and trainers point of view).

The training activity the trainers will describe here, is based on precise methodological choices which contribute, the trainers believe, on one hand to activate processes of personal evolution to acquire an always wider professional consciousness about safety on working place, on the other hand the building of "micro-communities of practice" able to incentive a renewed company design.

These activities connected with practice, comparison, dialogues and interaction among people revealed strategic to let the workers grow as collaborative co-operator, orienting them towards an effective working community.

When and how can you talk about "safety" on working place in a chemical farm with a high accident risk level?

This was the first question that help Pulverit as "company" to move on a ground that goes over the law and can touch deeply the consciousness the workers.

S. Bagnara et al. (Eds.): IEA 2018, AISC 821, pp. 602–608, 2019.
https://doi.org/10.1007/978-3-319-96080-7_72

The Italian Law Code is very clear about this: "A company is safe when the working place is full of all those hints, instruments and activities of prevention that warrant a reasonable level of protection against the possibility of unexpected events, which can be dangerous for the health of the operators" (D.Lgs. 81/2008)

So, all about "safety of the workers" has the goal to improve working condition, avoid possibilities of hurts and accidents and, moreover damage to the health of all the people (workers or guest) present also occasionally on the working place.

The intention of the judge is to underline how the "danger" is the source of a damage for people health, that means a damaging quality interior to the material object or of a situation. It's not the danger itself which creates a damage to the workers, but the exposure to the danger.

It's called "risk" and it can have an objective nature (a probability: connected to the context or activities), or it can have a subjective nature, that depends on the reaction of the operator to the event. In this second case, the risk is out from the probabilistic calculation and depends on a more complex context of personal perceptions: uncertainty, safety, benefits, exposure, information and training. Lots of anthropological studies underline that the concept of "risk" in different actors, is strongly influenced from the cultural, environmental and social context, that constantly compare information in a specific working community.

So, if the goal of the company, in the respect of the law, is that of avoid dangers, the aim of the trainer is that of getting lower in a responsible way the risks of the workers, considering them co-operator, and really receiver of the risk communication as active actors of the process.

To the risk is associated the concept of Opportunity, because an uncertain event can be both source of negative effects (Downside risk) but also positive effect (Upside risk).

The activities of training or self training proposed had the aim to activate perception processes of reflection on the risk, and relative actions that can be putting on going. The methodologies used introduce the maieutic value of the training, based on a deep dialogue that privileges the comparison and context instead of a hard-predefined schedule. We like the comparison with Socrate's midwife: with the vision that the trainer should give birth to working communities instead of singular personalities, facilitating a cooperative and conscious dimension.

So, the training experience shows that in Pulverit the right way to face and avoid risks is an appropriate way of dialogue and communication, that allow people to develop and grow (if they choose to do it) in the "new Pulverit company community"

The steps of the training porposed have been:

- **Creation** of three groups of different levels, distinguished for roles and jobs description (production area, commercial area, research and development area), with the commitment to meet once in a month for 8 months from November 2015 to June 2016. The first step builds micro-community undefined but characterized from the ansversely of the roles, in which everybody has been invited to begin a self-training path and show in the group the personal swot analysis. The nice result of this activity was not the building of the community but the attention to the purpose, and the favourite perception of people that decide to "cultivate" it as essential instrument to transfer knowledge produced in that specific context (Wenger et al.

2002) with the aim to adopt in a conscious and correct way the right measure to keep safe and healthy the working place, and to face directly problems, unexpected events and risks.
- In the second phase the groups activate themselves in **different processes**: an individual path on consciousness of its proper way to work in the company, with special focus on safety; a collective path to acquire information useful to the good functioning of its working area, to analyse and find solution to the single safety problems
- In the third phase, the trainers adopted **new strategies** of good practice to improve the work organization, throughout the analysis of the two previous paths and reflecting under different point of view (internal and external) on the afforded problems.

On the base of the growing consciousness (individual and collective of the community) you show a possible evolution of the company on long term (vision) that warrant the workers a more involving family feeling and that limited the perception of a uncertain social and economical context the society is actually living.

2 Different Moment of a New Ergonomic Meaning in the Working Practice

In the last years took place a studying area called Practice-based studies (Bruni and Gherardi 2007; Gherardi 2008), this area is modifying the approach to work. You rediscover the work as "situated action" and what is relevant is what people in concrete do when they work. The working practice is seen as "… action modalities and knowledges emerging on site thanks to interaction dynamic":

- They occurred in specific context
- They involved all the dimension of the human being: body, thought and feeling
- They involved a Human-Machine interaction (from traditional instrument to new technologies)
- They develop thanks to a collection of rules (mainly implied)
- They Incorporate different acknowledge (scientifically, technological, aesthetical and ethics)
- They can be realized also throughout soft skills and transversal abilities connected with communication and relationship competence

In the perspective of the situated activity, the work is settled as a process of knowledge incorporated into practice, that find their root in the anthropological and ethnographic tradition.

Under this point of view you can define better the practice:

- It's not routine (a collection of orders) and from the sum of chronological activities
- The aim is practical, and you realize produce and perform something (object, service, handworks), overwhelming problems and managing different resources
- It's a way to build interconnections, a network of relationships (between and among people and machines)

– At the base of a practice there's a "practice community) (Wenger)

The contribution of Etienne Wenger is really important as he get evidence to how the learning of an expert competence is a social process of participation to a community.

And the communities, at the same time, "are groups of people who share an interest, a collection of problems, a passion towards a special theme and who want to go deeper their knowledge and experience in this area through continuous interaction".

A real turning, epistemological and paradigmatic, in which the practice connected with reflection on it is basic (Argyris and Schoen 1996): this theory has been adopted by trainers in Pulverit.

This has meant:

– **Active participation** to session of analysis of working practice. The trainers drive the operator of Pulverit in the description and analysis of expert working practice, to indicate speech, dynamics, behaviour and processes of action that produce new competence and roles. In this case, the example of the work on technical operator show how the full work involved not only the functioning and maintenance of the device but also the social relationship with other people. This reflection actuated a new ergonomic approach based on both men and device, to analyse problems that needs new solution.
– **Reflection and storytelling**. During the path, the trainers compared the workers with the narration of previous working places, developing a problem solving on the road: what emerged was similarity and difference between practice, opportunities and different solution, but mainly reflection on what can be shared with positive results, in unformal situation without the necessity of a strong and rigid rule.
– The **care** of the practice community. You activate a shared learning process, with important reflection in the building practice of a real Pulverit Community

3 From "S" to "H"

The different phases of the path, driven by this epistemological premises can be focalized with some key-words

Self Observation – Observation

The ability to reflect on itself, to observe itself and the others, to put in relationship and listen to. In Pulverit the learning room became the relation room, where who participate to the training decide to involve himself creating a new relationship between trainers and trainee. The knowledge became an experiential value, which applied learning tools inside the context, building a conscious and strong relationship. The learning became an active conscious process which underlined strengths and weakness "outside", performing self-esteem and critical abilities of each participant (as experience is important but only if it's not reiteration of an old habit)

Responsible Action and Care

The "care" is always fatigue, among attention, learning and self-involving. It's fundamental for all our action: for what you do and for "how" you do it, the base which can never be shared by the responsibility towards itself and others.

Be responsible in Pulverit has meant "care" and translate the consciousness of personal value into something to be transferred to each other in the right and best way. This had an impact on ergonomic process inside the company, improving the working place and empowering the interaction Human-Machine.

Antifragility and Fragility

Been able to react to something unforeseen in un uncertain context means to be "Antifragile" (Taleb).

We're living in a real uncertain society, full of risks under the economic and financial point of view, so even if the working context has a certain stability, the decision making has no experience of the unexpected and this cause discomforts difficult to manage.

The antifragile can be used to sum up information under stress to adapt to uncertain situation, and the modality of adaptation are strongly solicited when there's the exposure to untidy situation.

Resilience and strength allow to each one to tolerate the shock and preserve towards the goal. The Antifragility allow each one to improve while adapting. What the trainers observe in Pulverit are at least 4 components that conditioned the antifragile process:

1. Technical and tactical component: acquire and take effective consciousness of its proper technical competences
2. Reflexive and self reflexive component: be conscious of what you really do with responsibility
3. Social-cultural component: understand deeply the meaning of the word community, and be responsible in shared doing
4. Antifragile component: develop new transversal competence in context or strong uncertainty.

The different working group went through different discontinuous phases and should compare with a good number of unpredictable cases.

What make the difference among the three resilient groups has been the ability to optimize their performance during a favourable working time. The antifragile instead are those who have the ability to adapt in a cooperative way and be transilient during an unfavourable working time in un uncertain context, for contingent reason: economic crises of the special chemical market, technological innovation, but also damage or substitution of parties or instruments, etc.

But how you decline "transilience" and in which way it can be an "antifragile" factor?

Talking about "care", you can say that is fundamental to learn the possibilities of "resilience", that means facing positively moments of hard difficulties and re-organize positively its personal life and its personal professional activity. Nowadays is not sufficient to be resilient, but you have to develop another meta-competence called "transilence". This term, taken from science fiction indicates that characteristic which flows competences and energy from a side to the other, creating communicating vessels, so that you don't remain isolated in rigid schedules (Vitullo, Zezza), developing "antifragility". Under this point of view, Transilience is a fundamental competence of antifragility.

This concept drives to the letter H as Human Being. So, the methodology the trainers called **FROM S(afety) TO H** is inspired the methodology **"theory U"** by Otto Scharmer referring to creativity and innovation.

The term "Human" really appears in the company to indicate the Human resources, but the new challenge and complexity of the context proceed towards new steps and towards a new focus on the adjective "Human" instead of "Resources" that are really low in such a critical social and economic situation.

This is the new starting point of different typologies of training, more "human" and "experiential", which stimulate the outgoing from a comfort zone, to activate new processes of exploration, metamorphosis and learning, both personal and professional.

4 Conclusions

To conclude, the case history shows some important points, that reveal how ergonomics should drive also innovative training:

- Define the rhythm of the meetings among different roles and competences
- Drive the individual and personal self-reflection and the comparison which stimulate "antifragility"
- Predispose specific metaphoric storytelling, exercise and experiential activities individuating emerging goals and company goals to be reached
- Allow an ergonomic approach, to activate better conscious decision making, and improving transilience and antifragility in an uncertain context.

References

AAVV (2005) Qualità della vita e sicurezza nei luoghi di lavoro. Strategie, ruoli, professionalità e interventi. Franco Angeli, Milano

Agosti A (2006) (a cura di): La formazione. Interpretazioni pedagogiche e indicazioni operative. Franco Angeli, Milano

Argyris C, Schön D (1996) Organizational Learning II: Theory, Method and Practice. Addison Wesley, Reading

Bauman Z (2000) Liquid Modernity. Polity Press, Cambridge

Beck U Risk Society Towards a New Modernity. Sage Publication, London

Bruni A, Gherardi S (2007) Studiare le pratiche lavorative. Il Mulino, Bologna

Cappelletto F (2009) Vivere l'etnografia. SEID, Firenze

Gherardi S (2008) Dalla comunità di pratica alle pratiche della comunità. Breve storia di un concetto in viaggio. In: Studi organizzativi 1

Gherardi S, Nicolini D (2004) Conoscenza e apprendimento nelle organizzazioni. Carocci, Roma

Mortari L (2003) Apprendere dall'esperienza. Il pensare riflessivo nella formazione. Carocci, Roma
Mortari L (2006) La pratica dell'aver cura. Bruno Mondadori, Milano
Scharmer O (2009) Theory U. Leading from the Future as it Emerges. Berrett-Koehler, Oakland
Tacconi G (2015) La didattica al lavoro. Franco Angeli, Milano
Taleb N (2012) Antifragile. Random House, New York
Traversi Guerra L, Trigona R (2015) Sulla via creativa. Este Milano
Vitullo A, Zezza R (2014) Maam. La maternità è un master. Rcs, Milano
Wenger E, McDermott R, Snyder WM (2002) Cultivating Communities of Practice. A Guide to Managing Knowledge. Harvard Business School Press, Boston

The Culture of Innovation: Needs and Opportunities

A. Augusto(✉)

24 U Member – Umania srl, via Stezzano, 87, 24100 Bergamo, Italy
info@umania.it

Abstract. At such a peculiar historical time as the one we are currently living in, companies strongly ask for "innovation": innovation of product/service, innovation of processes, innovation of the business as a whole. Such need might arise from far-sightedness, curiosity or from concern, always and in any case, from a discomfort of inadequacy of what already exists.

Keywords: Innovation · Culture

1 Introduction

This short article deals with the culture change occurred during a business innovation process and, specifically, during a consulting and coaching session carried out in a Law Firm, this being an example of a business strongly linked to "tradition" that needs to set off a journey to change its processes, client concept and products.

This type of action is aimed at empowering the organization performance, always keeping in mind the "wellbeing" of people belonging to it, in order to improve the quality of work.

Culture *is a powerful, latent, and often unconscious set of forces that determine both our individual and collective behaviour, ways of perceiving, thought patterns, and values* (Schein); *it encompasses all patterns of thinking, feeling and acting, up to the ordinary and menial things of life, it is not innate, it is learned from one's environment* (Hofstede, Minkov).

In a nutshell, what we do is influenced by our way to interpret events and by past behaviours that proved to be effective in responding to those events.

Therefore, our past strongly affects the choices we make and, whenever no similarity with past occurrences can be identified, we very often attempt to distort reality so that we are enabled to find a reassuring and already tested solution: Otto Scharmer defines this as *downloading the past* (Senge, Scharmer, Jaworski, Flowers).

2 Need for Organizational Change in a Law Firm

In September 2017, I was contacted by an important Italian Law Firm that was going through a slow and yet unavoidable phase of decline.

S. Bagnara et al. (Eds.): IEA 2018, AISC 821, pp. 609–616, 2019.
https://doi.org/10.1007/978-3-319-96080-7_73

The Law Firm had nearly 100 years of history, one of its founders was still working there, but it was destined to kick off a generational shift. Its past success was strongly related to the prestigious figures of the founders - famous jurists, important academics and brilliant settlers of litigations - yet those achievements were faced with the change in market trends, new emerging professional figures and widespread, fierce competition in the sector.

The current market of legal advice and legal counsel has stemmed from clients' changed requirements; clients are much more evolved than in the past and more demanding as for time and modes to tackle legal issues, in particular when these issues relate to business organizations.

Failure to adapt to those changes was having as first, evident consequence: the perspective drop of turnover (decrease in number of legal case files). This had consequently caused confusion about how to approach and communicate to new and potential clients, to identify the necessary resources (in particular human resources) able to support and relaunch the business and to adapt processes to new technologies.

The main hurdle the Law Firm needed to overcome had surfaced since the very first time we met. It had to become able to innovate its business, create and implement new services that could regain the "pioneering" and state-of-the-art image it used to have, rather than being stuck on its current one, characterized by homologation and customization (no perception of any distinctive trait to stand out from competitors).

It is fundamental that the consultant understands, at the very beginning, who his or her client is, his or her identity, motivations and values. As soon as the consultant gains this knowledge, the client becomes aware of the same things through the consultant's "eyes", as if looking at someone looking back at you would enable you to truly see yourself for the first time, free of assumptions and dogmas that otherwise are seen as indisputable.

This process was implemented by using a well-known and simple methodology by Alexander Osterwalder, called "*Business Model Canvas*". The main elements of a business model can be identified by means of a table (canvas):

- Key partners
- Key activities
- Key resources
- Value propositions
- Customer relationship
- Channels to be used to promote the company
- Customer segments
- Cost structure
- Revenue streams

This initial *identikit*, which may seem unnecessary, proved to be very useful since it was outlined together with one of the charter members. For the first time, he had stopped and thought about all the aspects of his job, with no interruption from the continuous and confusing flow of the "things to do"; just like a driver who focuses on the numerous actions he performs as he is driving his car as well as on their goals, free of repetitive, automatic gestures and habits.

This preliminary analysis produced a paradoxical effect: on the one side, the top managers realized what limitations prevented the organization from changing; on the other one, they upheld those organization aspects and "defended" their effectiveness, as if it were their clients' fault because they could not see how valuable those features were (a typical effect of the constraints that the "software of the mind" imposes on changing cultural assumption).

The underlying concept of human organizations, including companies and law firms, is increasingly borrowed from the "Systems Theory", which was originally developed in biology during the 1920s, and it was later on applied to any structured organization.

The founder of the General Systems Theory, the Austrian biologist Ludwig Von Bertalanffy, divided systems into two groups:

- Open System: it has external interactions, exchanging matter and information (input flows – transformation –output flows), as well as energy, with the outside; it is possible that disorder decreases and remains constant, due to the taking of matter and information from the outside;
- Closed System: it exchanges neither matter nor information with the external environment, but only energy. Closed physical systems tend to pass from order to increasing disorder, aiming to a maximum level that, at the end of the process, identifies the final balance (entropy, as gauge of disorder). In closed systems, the level of entropy tends to grow constantly and, therefore, increasing disorder will lead to the system dissolution.

Living organisms, because of their open nature, are able to keep vital and steady, also when balance is completely missing, through some particular processes enabling each organism to create its own internal environment capable of responding adequately to stimuli from the external environment. This property of open systems is called "homeostasis". Homeostasis is based on the self-adjustment mechanism.

A company (and organizations in general) takes on its typical systemic configuration as soon as its components (people and equipment) start interacting to respond to stimuli that generally come from the external environment. As open systems, they interact among them and with the external environment, thus implementing the several activities and necessary processes to achieve the system goals. The goal of the system is survival, based on the capability to generate value by creating competitive edges, which have to be defendable and last as long as possible. Losing the systemic properties leads to the system dissolution.

After carrying out the "canvas"-based analysis, which makes it possible to observe all the organization elements and the relationships existing among them, the Law Firm clearly appeared more and more similar to a "closed" system and it was losing its vitality.

It had become aware of this and it was trying to change but, as previously stated, it was necessary to interrupt the "downloading of the past" and change those behaviours that no longer suited the new situation.

The first result to achieve was to drive the Law Firm towards an "open" system and trigger the homeostasis in it, that is to say building the capability to respond adequately to the external environment.

However, such process cannot take place if the top management does not give the impulse and set it off. Change will never come from co-operators, since they adapt to their leaders' culture otherwise, if they do not, the system perceives them as foreign bodies and expels them.

Schein maintains that culture stems from founders, who set the mission, values and assumptions of their organization based on their vision.

The founders' vision being still valuable, the organization mission, values and assumptions have to change, and only the organization leaders can carry this out. The organization as a whole is devoted to pursuing the mission and favour the integration and efficiency of each component as much as possible, so that the system can be as efficient as possible. The "product" of the organization is what enables it to survive, by means of the clients' satisfaction, the latter being co-authors of the success (or failure) of this complex, social system called company (or professional firm).

As I had the chance to ascertain, that vision was still solid, oriented to excellence and willing to give its contribution to the growth of legal doctrines and, as a consequence, of justice as cornerstone of respect of the social pact among men.

The mission, however, was no longer up-to-date, because of its conception of a no longer existing customer (co-builder) with purchasing behaviours of services that had deeply changed and of very differentiated requested benefits.

Such assumptions had to be "turned over" through the acquisition of a new awareness of the market.

Chip and Dan Heath highlighted how, at the root of any process of change, it is necessary to motivate "the Elephant", that means our emotional and irrational side, adopting the example by psychologist Jonathan Haidt, who used to compare the rational side of our brain to a Guide (Rider) sitting on a huge Elephant. The Guide's control is precarious, because he is much smaller than the Elephant: any time they disagree about which direction to go, the Guide is going to lose.

The law firm's founders' Elephant was "heavy" and reluctant to start this journey, they would have appreciated a simpler and more immediate solution, a stroke of genius that could solve any problem involving few effort and no big changes. Despite being aware of the decline that was being experienced, the Guide was not able to drive the heavy Elephant.

It was possible to motivate the Elephant, only rekindling the leaders' Vision: the Law Firm's survival was going to have to maintain the possibility of growth for the legal doctrine, the sense of Justice, the preservation of responsibility towards the Civil Society improvement. After the perception of this "sense of urgency", that Kotter (Cohen) considers being the first stage of any change, the first step was taken, the Guide was given free rein to plan a new mission.

The second stage was necessarily a training stage, as the leaders needed some ideas about how to observe and understand the customers' needs.

As we have already said before, understanding the market was the weak link of the chain towards innovating the business, that still consisted in trustingly waiting for an assignment (sooner or later customers will need us).

As for the *canvas* generation, the leaders needed to find a "protected" space, where they could work out these new ideas, otherwise the reluctant "Elephant" would have replied: "I have so many things to do, I cannot take up time from work; if we do not take our work forward, we won't earn". However, the motivation allowed the elephant to acknowledge how important it was "to whet the axe before cutting the trees down".

Hence, a certain temporal space was set, in which it was possible to work on the innovation process, where leaders had no chance of being distracted, and a new positive habit was created.

Every Friday afternoon a new strategic activity of business development took place; at the beginning, training courses were held about methods of "value proposition". The training sessions were an opportunity and an incentive to reflect, from different perspectives, about customers' needs, about which benefits they are looking for, about the tasks they have to carry out and the difficulties they have to face.

An exercise I consider useful and that I think had worked also in this intervention is to make an analysis of all the existing clients, trying to find out which "benefits" they are looking for, while turning to the Law Firm for their services, what kind of help do they need, while carrying out specific "tasks" or what "discomfort" they would like to get rid of.

In this way, the leaders became more aware of what they were really being asked by clients and of the things customers retained of higher value.

It was of no minor importance to understand, who were those clients who did not require legal services or did not retain the Law Firm an interlocutor able to offer what they needed. In order to carry out this analysis, it was necessary to "go out into the world", that means to organize interviews with potential customers, in order to acquire information under this new perspective. For the first time, the leaders were not approaching customers with the aim of offering services, but with the purpose to understand their needs; a change in the paradigm that required a lot of practice. After every programmed meeting, that had its specifically dedicated space on Friday afternoon, I was reported with the outcomes of the interviews, so that I could provide an independent feedback with the purpose of improving the learning process.

The "system" was becoming open, that means it was able to fully implement external inputs, and to conceive and conceptualize the necessary outputs to realize the new mission.

At the end of this "opening-up" process, the leaders were more aware of the requirements of the new and potential customers and, not of secondary importance, of the existing customers' needs that had never been met.

As a natural consequence, the organizational structure was redesigned: are staff, technology, working methods suitable to meet such needs?

The leaders had approached the new assumptions and, therefore, it was time to get the co-operators, the most important resource for activities of this type, involved.

In this case too, it was felt necessary to identify a precise moment of the week when the new mission was communicated, and the necessary feedbacks implemented.

To the leaders' great surprise, the organization was willing to accept the new way, even if they had so far done nothing but adapting themselves to the implicit and explicit culture consolidated inside the Firm.

They only needed to be able to take part in that awareness that the leader themselves had developed over months of work. The co-operators' Elephant too needed to be motivated through their involvement in the Firm's development and reorganizational processes: the most powerful energy derived from their feeling that they were partners of a hope.

Co-operators entered a training programme for the acquisition of the same concepts integrated by the leaders, and, at the same time they took part in parallel and equally important training sessions dealing with being more efficient in working methods.

3 A New "Space"

At the end of this journey, it was necessary to create a new "space", separated from everyday activities, so that innovation could arise without being influenced by the logics and schemes of consolidated activities. A new space (physical and of time) needs to be created to create.

In that occasion, I coordinated the creation of a new, innovative service in line with the new acquired awareness on the client and the resources of the Law Firm. The leaders and a team of co-operators, that had previously been identified as the most suitable, attended those sessions.

Training the team is an important activity. Its components have to be chosen according to a cross-cutting criterion, that means they need to be as rich as possible in terms of different skills, varied experiences (also personal ones) and above all, they have to mirror a range of age that is as large as possible with different and enriching points of view.

The innovation process is very complicated, above all if the context is experiencing it for the first time: the risk of trying to make comparisons with what had already been done is high and the risk of not proposing any meaningful idea for the activity progress is even higher. The only way is never look back to the "past" as a starting point, to be oriented, instead, from the very beginning, to the future (that, by definition, is not yet here), towards something that wants to arise, but we are not yet aware of.

The desired future was the result of that current strong Vision, of that identity that was aimed at and, therefore, the steps to be taken towards that direction became simple preliminary requirements.

During a few meetings, a path was defined, to be undertaken in order to launch a new idea of business, a new service that had never been provided before and deeply consistent with the previously carried out analysis.

Would the market be interested in what the Firm considered a "product" of great value? *Design thinking* was the reference theory to draw from, while considering this hypothesis.

"*Design thinking*" is a methodological approach that allows all the staff of a company to take part and give their contribution to problem solving. The process implies, through a (as broad as possible) team work, to bring out internal and external needs of a company, so that it becomes possible to conceive and prototype new products/services and, afterwards, to test them directly with the customer.

After some research and brief implementations, the creation of a "prototype service" to offer to clients was decided. When you offer a prototype, you have the opportunity to have immediate feedback with no big investment, and, through the clients' feedbacks, it is possible to quickly and more consciously make the necessary changes, so that the prototype becomes a requested service of value for the customer.

The team programmed and organized short meetings with potential customers, where the prototype was explained and all the major useful observations in response to the presentation were taken in. It is extremely important to be able to grab these information, to "measure" them, in order not to uselessly waste energies on looking for something that is of no interests for anybody.

As for the leaders and co-operators, a consequent effect of such interviews was their listening approach towards people, being interested in their points of view and so, being able to better understand how they were perceived by "non-customer" of the Firm.

Few meetings were enough to refine a new legal service, that, as it meets a new customer's need, developed and boosted a new line of business.

Above mentioned Kotler thinks that obtaining "small successes" over short periods of time is extremely important in processes of change, as it allows to lessen cynicism, pessimism and scepticism, the strongest enemies of innovation.

The launch of the new service allowed the Firm to create a habit to meet on Friday afternoons, a space of observation and of thinking to design the future, so that creation and innovation won't be exceptional occurrences anymore but established habits instead.

4 Conclusions

From the described experience we can come to the conclusion that culture influences the processes of innovation and change: if there are no working assumptions, values and methods consistent with the necessity of change, it is rather difficult that this culture will find the necessary lifeblood for its growth.

Culture is the result of a powerful, emotional and stimulating Vision, that cannot come but from the leaders. Without a generative and endorsing drive of the entire process by the leadership, no innovative process could ever succeed.

References

1. Colombo C, Donadio A, Galardi A, Marini V, Solari L (2015) The human side of digital. Guerini Next, Milano
2. Brugnoli C (2002) Pensiero sistemico e decisioni strategiche. Egea, Milano
3. Clark T, Osterwalder A, Pigneur Y (2014) Business model you. Hoepli, Milano
4. Golinelli G (2010) L'Approccio Sistemico vitale (ASV) al governo dell'impresa, 2nd edn. Cedam, Padova
5. Heath C, Heath D (2010) Switch. How to change things when change is hard. Kindle
6. Hofstede G, Hofstede GJ, Minkov M (2010) Cultures and organizations: software of the mind, 3rd edn. McGraw-Hill Education, Milano
7. Kotter JP, Cohen DS (2012) The hearth of change: real life stories of how people change their organizations. Harvard Business Review Press, Watertown
8. Osterwalder A, Pigneur Y, Bernarda G, Smith A (2015) Value proposition design. Edizioni LSWR, Milano. 2nd edn. Berret-Koehler Publishers (2016)

9. Scharmer O, Kaufer K (2016) Leading from the emerging future. From ego-system to eco system economics, 2nd edn. Berret-Koehler Publishers, Oakland
10. Schein EH (1999) The corporate culture survival guide. Wiley, Singapore
11. Senge P, Scharmer O, Jaworski J, Flowers BS (2005) Presence. Exploring profound change in people, organizations and society. Nicholas Brealey Publishing, London

The Effect of Support from Superiors and Colleagues Between Occupational Stress and Mental Health Among Japanese Sport Facilities Workers

Myunghee Park[1(✉)], Hirokazu Otake[2], Iwaasa Takumi[3], and Motoki Mizuno[4]

[1] Jungwon University, 85 Munmu-ro, Goesan-eup, Goesan-gun, Chungbuk 28024, South Korea
himee01@gmail.com
[2] Kanagawa University, Rokkakubashi, 3 Chome−3−27-1, Yokohama-shi, Kanagawa-ku, Kanagawa-ken 221-8686, Japan
[3] Graduate School of Health and Sports Science, Juntendo University, Chiba, Japan
[4] School of Health and Sports Science, Juntendo University, Chiba, Japan

Abstract. This study examined the occupational stress experienced by office worker and exercise instructor working in the Japanese fitness club and the association of these occupational stressors on mental health and the influence of support of superiors and colleagues on worker's mental health. The survey was conducted on 488 workers in the capital area sports facilities in Japan, among which 426 valid respondents (219 office staff and 207 instructor staff) were grouped and data were analyzed. The survey included (1) Occupational stress measured by the New Brief Job Stress Questionnaire, (2) Mental health measured by General Health Questionnaire. To examine associations between job stress and mental health, *t*-tests and linear regression analyses models were conducted. As a result, the two groups showed different occupational stress factors. For office staff group, the quantitative demand, qualitative demand, and meaning of job had a stronger relationship with mental health. For instructor staff group, the qualitative demand, interpersonal relationships, and job aptitude had a stronger relationship with mental health. Superior support showed a buffering effect on interpersonal relationships and mental health for office worker group, and colleagues support showed a buffering effect on job insecurity, quantitative demand, and work circumstances for instructor staff group.

Keywords: Job demand · GHQ · Buffering effect · Support at work

1 Introduction

Recently, changes in the working environment have increased the mental and physical burden on workers and intensified occupational stress [1, 2]. According to the "annual Health, Labor and Welfare Report" by the Japanese Ministry of Health, Labor and Welfare, approximately 60.9% of workers are feeling strong anxiety, concern and stress

S. Bagnara et al. (Eds.): IEA 2018, AISC 821, pp. 617–626, 2019.
https://doi.org/10.1007/978-3-319-96080-7_74

in relation to their current occupations of work life. The biggest cause of this is 'human relations at work' [3]. Furthermore, the advancement of the information-oriented society has led to less face-to face communication and lack of proper social support at work, and the poor relations with superior or colleagues result in isolation in interpersonal relations [4]. Empirical studies thus far have implied social isolation as a major risk factor for mental health [5]. Many studies have been conducted on the role of social support as well as its effectiveness [6, 7]. The NIOSH (National Institute for Occupational Safety and Health) Model of Job Stress introduces the process in which job stress factors such as human relations problems cause stress reactions like depression, which may lead to health problems or diseases. It also introduces social support from superior or colleagues as a crucial factor that alleviates the stress reactions due to job stress factors [8]. Recently the level of consciousness about health is increasing, and accordingly more and more users are using sports facilities. There is an increase of paid training program members expecting more definite fitness effects in addition to just working out. Big fitness clubs that also run relaxing services like massage or beauty care are also increasing every year along with sales. Accordingly, sports facility workers have the pressure of having to provide high-quality human services with advanced and specialized skills. According to the Fitness Industry Association of Japan [9], facility management and customer management are the factors that have the greatest impact on general satisfaction of members for fitness clubs. Moreover, a case analysis by [10] points out exercise effect and training program as the factors. In other words, sports facility workers require qualitative improvement as a type of content of the facility and have become relevant to the sales performance of that club as well.

The labor force of sports fitness clubs consists of 78,000 workers as of March 2017, among which 8,000 are permanent workers and 35,000 are temporary instructors. The staff is divided into office and instructor staff, and there are many temporary workers [11]. Moreover, temporary workers are put in charge of paid training programs such as studio programs, and thus they are also demanded of high expertise and commitment. Meanwhile, fitness center workers also face the risk of health or job such as job insecurity, which raises the need for more active measures for mental health [12, 13].

Thus, the objective of this study is to clearly determine the effects of job stress faced by sports facility workers in Japan and support from superior or colleagues on mental health. More specifically, this study set the following hypotheses to examine job stress of sports fitness club workers and support from superior or colleagues, compare the differences between office staff and instructor staff, and verify the buffer effect of support on mental health.

Hypothesis 1. Job stress has less effect on mental health for the instructor staff than office staff.

Hypothesis 2. Support from superior (H2a) or colleagues (H2b) has a buffer effect between stress and mental health.

2 Method

2.1 Subjects and Method

A survey was conducted on 488 workers in sports facilities run by 11 sports facility companies (5 private facilities and 6 public facilities) in the capital area, and 426 subjects (198 female, 228 male) excluding 62 with incomplete responses were finally selected as the subjects of analysis (effective response rate 87.2%). The survey was conducted from May 8 to June 7, 2017 after sending emails for cooperation to the relevant staff in each institute and obtaining approval. The questionnaire was distributed considering the ratio of permanent and temporary workers, and office staff (office, sales, counter, facility managers, etc. excluding department managers) and instructor staff (instructors, trainers). The subjects were to respond after agreeing to the intention of the survey and cautions written on the cover of the questionnaire.

2.2 Survey Items and Contents

All items in structured questionnaire consisted of attributes relevant to job stress, job insecurity stress, job satisfaction and life satisfaction as well as demographics and higher score indicated higher level of positive state, of respondents.

Occupational Stress Evaluation Items. This study used the simple job stress questionnaire developed based on the US NIOSH model of job stress [14]. There are total 17 items for stress factors: 3 for demand of quantity, 3 for demand of quality, 1 for demand of physical, 3 for job control, 1 for application of skill, 3 for interpersonal conflict, 1 for circumstance of job, 1 for aptitude for job, and 1 for meaning of job, all rated on a 4-point scale (Not at all – Very much so). This scale did not include items about job insecurity that is recently a social issue, and thus used 4 items about job insecurity from the Korean version of job stress scale (KOSS-33) after back translation [15].

Mental Health Evaluation Items (GHQ). The mental health scale [16] is used for items measuring the mental health of the subjects. This scale was developed based on the General Health Questionnaire (GHQ) by [17]. It consists of two factors, anxiety/depression and obstacles to activity, and total 12 items (6 items for each factor) rated on a 4-point scale ('No more than usual', 'Not at all', 'Rather more than usual' and 'Much more than usual'). Higher score indicates better mental health.

Facet. As studies on job stress and health are reporting the effects of interaction among individual, job and family factors [18], the subjects were to fill out facet sheets on their individual or job attributes that may affect mental health. The items used were gender, age, years of education, marital status, form of employment, income (monthly), working hours, type of occupation, and facility type of company.

2.3 Data Analysis

Sports facility workers are divided into office staff and instructor staff for analysis. To determine the disposition of the two groups and test the hypotheses, the χ^2 test is conducted on gender, age, years of education, marital status, form of employment, length of service, income (monthly), working hours, type of occupation, and facility type, after which Pearson's product moment correlation was estimated. To examine the stress factors related to mental health in the two groups, a multiple regression analysis was conducted with the aptitude as the control factor, stress factors as independent variables, and mental health as the outcome variable. Lastly, a moderated regression analysis was conducted for each group to verify the buffer effect of support from superior and colleagues about mental health on job stress. In step 1 we entered demographics, in step 2 we entered the stressor (e.g., demand of quantity) and the proposed moderator (e.g., superior), followed by the interaction of the stressor and support at work in step 3. A changed in R^2 in step3 and a significant beta weight for the interaction indicate the presence of a moderator effect.

In all regression models, the residuals were distributed normally around the regression line, Adjusted R^2s are reported.

3 Results

3.1 Individual Attributes of the Sample

There was no statistically significant difference in gender. There were fewer office workers in their 20 s and more in their 50 s than instructors, and the office staff had the average age of 35.76 (SD = 12.07) and the instructor staff had the average age of 30.91 (SD = 9.96). There was no statistically significant difference in years of education ($p > .05$), and most of the subjects were graduates of university or graduate school in both groups. As for marital status, many office workers were married, whereas many instructors were single. There was no significant difference between the two groups in the form of employment ($p > .05$). Many office workers earned at least 300 thousand yen for monthly income. There was no significant difference in the length of service and working hours. 34% of both groups worked at least 40 h a week, and the average working hours was 30.10 h for the office staff (SD = 15.30) and 28.24 h for the instructor staff (SD = 16.21). There was no significant difference in facility type ($p > .05$). Relationships among occupational stress, support from superior and colleagues, and mental health.

A regression analysis was conducted to determine the relationship among job stress, support from superior and colleagues, and mental health (Table 1). The result showed that among the job stress factors of the office staff, job insecurity ($\beta = .225$ $p = .001$), demand of quantity ($\beta = .228 p = .003$), and meaning of work ($\beta = .156 p = .046$) were factors related to mental health with the significance of 0.05, and the explanatory power of the regression analysis was 41.7%. For the instructor staff, factors related to mental health were demand of quantity ($\beta = -.037 p = .032$), demand of physical ($\beta = .090 p = .016$), interpersonal conflict ($\beta = .289 p = .000$), aptitude for job ($\beta = .011, p = .005$), and support from colleagues $\beta = 0.210 p = .016$), and the explanatory power of the regression analysis was 34.9%, which is lower than the office staff. Therefore, Hypothesis H1b was supported.

Table 1. Regressions of demographic and job relate variables on mental health by job type.

	Variables	Office staff		Instructor staff	
		β	p-value	β	p-value
Demographics					
Age		0.010	0.892	0.086	0.264
Gender	(Standard: woman) Man	0.161	0.010	0.134	0.048
Education	(Standard: high school) University and Graduate	–0.107	0.082	0.040	0.527
Marital status	(Standard: Married) Single	–0.101	0.170	0.087	0.282
Contract	(Standard: Regular) Dispatch	–0.037	0.551	–0.120	0.088
Facility type	(Standard: Public) Private	–0.053	0.368	–0.013	0.861
Income per month	(Standard: under 200,000) 200,000–under 300,000	–0.014	0.823	–0.090	0.167
Work hour per week		–0.053	0.382	–0.003	0.969
Job - related variables					
Demand of quantity		0.228	0.003	–0.037	0.032
Demand of quality		0.150	0.052	0.175	0.635
Demand of physical		–0.060	0.359	0.090	0.016
Job control		0.113	0.087	–0.010	0.881
Application of skill		0.024	0.713	0.051	0.480
Interpersonal conflict		–0.052	0.485	0.289	0.000
Aptitude for job		0.035	0.591	0.011	0.005
Circumstance of job		0.128	0.087	0.236	0.877
Meaning of job		0.156	0.046	–0.048	0.599
Job insecurity		0.225	0.001	0.159	0.861
Support from superior		0.116	0.207	–0.015	0.870
Support from colleagues		0.136	0.129	0.210	0.018
	F-value	7.903		6.092	
	Adjusted R^2	0.417		0.349	
	P-value	0.000		0.000	

*P < .05, **P < .01, ***P < .001

3.2 Buffer Effect of Support from Superior and Colleagues

Buffer Effect of Support from Colleagues in the Instructor Staff. The moderated regression analysis approved the significant buffer effect of support from colleagues between job insecurity and mental health. As shown in Figure b, low support from colleagues shows higher relevance between job insecurity and mental health. On the contrary, high support from colleagues shows lower relevance in job security than mental health (insecurity × colleagues support on mental health $\Delta R^2 = \Delta R^2 = .242$, $\beta = -.133$, $p < .05$). Likewise, support from colleagues had a buffer effect on the relevance between mental health and demand of quantity. As shown in Figure c, high support from colleagues lowers the relevance between demand of quantity and mental health (demand of quantity × colleague support on mental health $\Delta R^2 = .194$, $\beta = -.145$, $p < .05$). Likewise, support from colleagues had a buffer effect on the relevance between mental health and work environment (circumstances × colleague support on mental health $\Delta R^2 = .193$, $\beta = -.155$, $p < .05$). Therefore, Hypothesis H2b was supported in the instructor staff and H2a was not (Fig. 1).

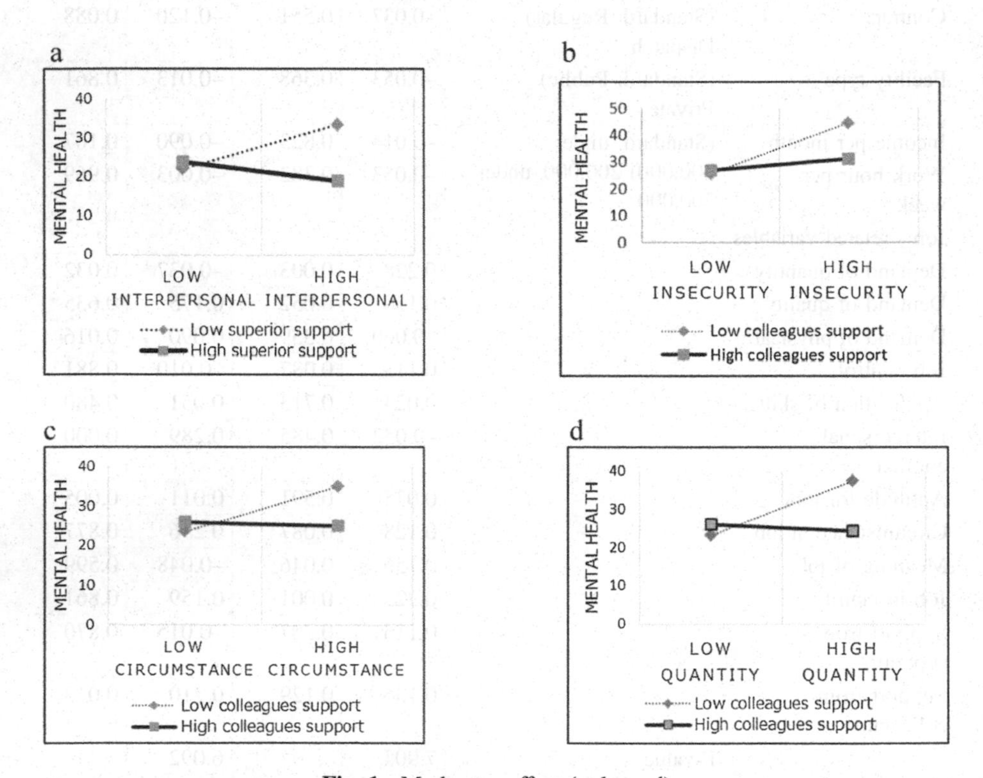

Fig. 1. Moderator effect (a, b, c, d)

4 Discussion

4.1 Relevance Between Job Stress and Mental Health

Demand of quantity has a significant relevance to mental health in the two groups, Fitness club workers, even temporary workers are required to have high expertise and commitment as they are put in charge of studio programs or paid training programs [19]. Furthermore, with the recent diversity of the job market, a high level of skills and responsibility are demanded from permanent workers, increasing pressure of work related to responsibility and role [20].

Next, the instructor staff is showing a more favorable state in job insecurity. According to the Labor Force Survey conducted in September 2017, there is a high rate of temporary workers in service, and the rate of temporary work is lower in specialized jobs [21]. In other words, this shows the reality in which specialized jobs are more likely to have permanent posts. The instructor staff has relatively low job insecurity as they are specialized jobs with exercise capacity and skills. Furthermore, the job openings for personal trainers are increasing nationwide, and there is active head-hunting for popular instructors in aerobics as well. For example, there are numerous recruitments and career move websites specialized in the fitness industry, which also shows why they have lower job insecurity than the office staff [22]. On the other hand, the office staff has insecure job status and has a significant relevance with mental health.

The instructor staff has a significant relevance between interpersonal relations and support from colleagues, which represents the importance of human relations at work. This result also supports the result of the Workers' Health Environment Survey that presents human relations problems at work (41.3%) as a factor of anxiety, concern and stress in current work life [3]. Moreover, aptitude for the job is included as a job stress factor affecting mental health of the instructor staff. The fact that it has a significant relevance with mental health even though the score of aptitude is higher than the office staff indicates that, as a job that requires expertise and demand of quality, not meeting the aptitude may lead to a negative impact on mental health.

4.2 Buffer Effect of Support from Superior and Colleagues in Occupational Stress and Mental Health

Buffer Effect of Support from Superior. [23] reported in a study of workers in Dresden, Germany that there is a buffer effect of support from superior on stress and depressive symptoms. Depressive symptoms increased when there was not enough social support, and high social support reduced depressive symptoms caused by stress. [24] studied various organizations and industrial workers in England and China and reported that the rate of absenteeism is low among workers with high support from senior. As such, the buffer effect on stress from interpersonal relations and mental health for the office staff supports the results of previous studies.

Buffer Effect of Support from Colleagues. The regression analysis approved that support from colleagues had a buffer effect related to mental health and job stress factors such as job insecurity, demand of quantity, and work environment for the instructor staff. The Resilience Improvement Support Guidelines for fitness club workers developed by [25] emphasize the importance of friendly relations with colleagues. Previous studies on job insecurity implied that support from superior had a significant relevance with job security perception of workers [26]. Uncertain social status tends to make workers be under the control of the manager and superior [27]. The result shows a buffer effect of not superior support but colleague support on the relationship between job insecurity and mental health, where high support from colleagues weakens the effect of job insecurity on mental health. Previous studies overseas rarely reported the buffer effect of support from colleagues on job insecurity and mental health, and similar reports were also not found in Japan.

The instructor staff works face-to-face with customers and thus the work is directly linked to customer satisfaction. In other words, relationship with colleagues includes not only cooperative relations as coworkers but also rivalry. Work evaluation and criticism from customers are likely to lead to job insecurity with rivalry as the medium. Colleagues may build trust while working or having conversations within the same space. As reported by [28], there is effectiveness as a buffer between role burden and conflict at work. Moreover, according to [29], support from colleagues has a buffer effect on physical safety environment and mental health, thereby having the effect to weaken the stress factors in the work environment.The result of this study supports the buffer effect of support from colleagues on work environment and mental health in previous studies.

Next, support from colleagues is reported to alleviate mental stress of workers feeling pressured by time [30]. If there are limited working hours and time constraints, cooperative relationship with colleagues promote a high level of support from colleagues also known as teamwork, and as a result the work proceeds efficiently and smoothly. With a high level of support from colleagues, there will be a buffer effect that reduces the impact of demand of quantity on mental health.

5 Conclusion

This study verified the buffer effect of support from superior or colleagues between occupational stress and mental health of sports facility workers. Support from superior is proved to have a buffer effect between interpersonal relations and mental health of the office staff. Moreover, colleagues support showed a buffering effect on job insecurity, quantitative demand, and work circumstances for instructor staff group

References

1. Benach J, Muntaner C, Santana V (2007) Employment conditions and health inequalities: final report to the WHO Commission on Social Determinants of Health (CSDH). World Health Organization (WHO), Geneva
2. Kawakami N (2011) International trends and priorities in Japan for the promotion of workplace mental health. Job Stress Res 18:233–240
3. LNCS Home page. http://www.mhlw.go.jp/toukei/list/h24-46-50.html. Accessed 10 Sept 2017
4. Takano T (2014) Practice of workers' mental health and industrial mental health. Occup Health Rev 27(1):39–67
5. House JS, Landis KR, Umberson D (1988) Social relationships and health. Science 241:540–544
6. House JS (1981) Work stress and social support. Addison-Wesley, Reading

7. Cobb S (1976) Social support as a moderator of life stress. Psychometric Med 38(5):300–314
8. Hurrell Jr JJ, Mclaney MA (1988) Exposure to job stress - a new psychometric instrument. Scand J Work Environ Health 14(Suppl 1):27–28
9. Fitness Industry association of Japan (1997): Heisei FY 1996 foundation vision realization project. CS (customer satisfaction) fitness club research implementation. diagnostic manual preparation project
10. Nakazi K (2006) Longitudinal case analysis on customer satisfaction and membership continuity of members at fitness club. Jpn J Manag Phys Educ Sports 20(1):1–15
11. LNCS Homepage. http://www.meti.go.jp/statistics/tyo/tokusabido/result-2.html. Accessed 11 Feb 2018
12. Vocational Ability Development Association (2011): Comprehensive vocational ability evaluation system maintenance committee Fitness industry activity report. Central Industry Capacity Development Association
13. Mizuno M (2007) Jitsen sangilyou & sosiki sinnrigaku. Souseisya, Tokyo
14. LNCS Homepage. http://www.mhlw.go.jp/bunya/roudoukijun/anzeneisei12/. Assessed 15 June 2015
15. Chang SJ, Koh SB, Kang D, Kim SA, Kang G, Lee CG, Chung J, Cho JJ, Son M, Chae CH, Kim JW, Kim JI, Kim HS, Roh SC, Park JB, Woo JM, Kim SY, Kim HS, Roh SC, Park JB, Woo JM, Kim SY, Kim JY, H M, Park J, Rhee KY, Kim HR, Kong JO, Kim IA, Kim JS, Park JH, Hyun SJ, Son DK (2005) Developing an occupational stress scale for Korean employees. Korean J Occup Environ Med 17(4):297–317
16. Shinno M, Mori N (2001) Reliability and validity of the Japanese version GHQ mental health questionnaire 12-item version (GHQ-12) based on survey of corporate workers. Clin Psychiatry 43(4):431–436
17. Goldberg DP (1979) The detection of psychiatric illness by questionnaire. Mauds monograph. Oxford University Press, London
18. Mizuno M, Yamada Y, Hirosawa M (2012) Examination of hygiene factors influencing on job dissatisfaction among nurses. J Ergon Occup Safety Health 14(1):17–24
19. Mizuno M (2007) Jitsen sangyou & sosiki sinrigaku. Souseisya, Tokyo
20. Alcalay R, Pasick RJ (1983) Psychosocial factors and the technologies of social. Sci Med 17 (16):1075–1084
21. LNCS Homepage. http://www.stat.go.jp/data/roudou/sokuhou/tsuki/index.htm. Accessed 11 Oct 2017
22. LNCS Homepage. http://www.fitnessjob.jp/index.html. Accessed 7 Sept 2017
23. Dormann C, Zadf D (1999) Social support, social stressors at work, and depressive symptoms: testing for main and moderating effects with structural equations in a three-wave longitudinal study. J Appl Psychol 84(6):874–884
24. Luo L, Cooper CL, Hui YL (2013) A cross-cultural examination of presentism and supervisory support. Career Dev Int 18(5):440–456
25. Shoji N, Moriguti H, Honda R, Kitamura M (2015) Development of the guideline to support for enhancing resilience in Japanese fitness club. J Ergon Occup Safety Health 16(1):19–25
26. Russing A (2005) Can control at work and social support moderate psychological consequences of job security? results from a Quasie experimental study in the steel industry. Euro J Work Organ Psychol 8:219–242
27. de Castro AB, Fujishiro K, Schweitzer E, Oliva J (2006) How immigrant workers experience workplace problems: a qualitative study and effect. Int Arch Occup Environ Health 61:249–258

28. Jimmieson NL, Mckimmie BM, Hannam RL, Gallagher J (2005) An investigation of the stress-buffering effects of social support in the occupational stress process as a function of team identification. Group Dyn 14(4):350–367
29. Yagil D, Luria G (2016) Friends in need: The protective effect of social relationships under low-safety climate. Group Org Manage 35(6):727–750
30. Hagihara A, Tarumi K, Miller AS (1998) Social support at work as a buffer of work stress-strain relationship: a signal detection approach. Stress Med 14(2):75–81

Benefits of the Human-Technology-Organization Concept in Teaching Ergonomics – Students Perspective

Anette Karltun(✉) and Johan Karltun

Industrial Engineering and Management,
Jönköping University, Jonkoping, Sweden
anette.karltun@ju.se

Abstract. The human-technology-organization (HTO) concept has been used for creating systems understanding of ergonomics in three engineering educations at the School of Engineering in Jönköping. Students from courses given in two undergraduate and one graduate program (n = 122) participated in the study, which involved a course evaluation questionnaire to assess the understanding of ergonomics as discipline and HTO as a means for creating systems understanding. The questionnaire included both ranking and personal comments to the questions. The results show that the students in general considered knowledge of ergonomics and HTO as beneficial for their future work and that the HTO concept did contribute to their understanding of workplace ergonomics. However, there was a significant difference between undergraduate and graduate students in all these aspects where undergraduates ranked all these aspects lower than graduates. This was also reflected in personal comments on the questions. Conclusions that can be drawn are that understanding systems is generally difficult and the HTO concept can assist in helping students to overcome these difficulties. However, the differences between the student groups must be explicitly considered as well as increasing students' awareness of the relevance of ergonomics for engineers.

Keywords: HTO concept · Engineering education · Systems understanding

1 Background

In engineering education, providing courses in ergonomics is one way of trying to meet the demands for understanding and improving future workplaces. The increasing technological complexity also make it more obvious that to understand, manage and develop technical systems, understanding the workplace as a system becomes more and more important. The systems character of ergonomics/human factors was for example described by Wilson (2014) where he concluded that the view should be holistic and complemented by concerns for context, interactions, complexity, emergence and embedding of professional effort.

One of the systems' approaches within ergonomics is HTO, where this acronym stands for considering a work activity to be an interaction between three separate subsystems, the human (H), the technology (T) and the organization (O) (Karltun et al. 2017).

S. Bagnara et al. (Eds.): IEA 2018, AISC 821, pp. 627–636, 2019.
https://doi.org/10.1007/978-3-319-96080-7_75

By focusing the interaction between these three major interdependent sub-systems, rather than the sub-systems themselves, the reasons for system performance becomes more obvious. In addition, the interaction indicates a dynamic system with ongoing activities, which implies that the understanding of the relationship between activities and organizational processes and system performance as a whole is facilitated (Karltun et al. 2017).

The authors have worked for a long period of time (>10 years) with developing and applying the HTO concept in research and education, inspired by developments in safety critical businesses like nuclear power plants and hospitals (Karltun et al. 2017; Karltun 2014). We have found that this concept brings advantages also outside such specific businesses. Main advantages are a more explicit emphasis on systems understanding, on interactions between identified sub-systems and stressing the human as a sub-system of equal importance as technology and organization.

Ergonomics as subject is very wide and encompasses parts from for example psychology, medicine, engineering and sociology (Daniellou 2001), which is needed to reflect the complexity in working activities as well as in businesses where these activities take place. It thus puts large demands on the teaching of the subject. There are also challenges involved in teaching a systems perspective for workplace understanding and design. To emphasize such a perspective during the ergonomics courses, students are encouraged to reflect on the facets of each HTO sub-system, their interactions and their relevance to different cases.

Students may also find it difficult to realize why they should engage in systems' thinking as it is demanding, and ergonomics is often not a focus in engineering students' mental models of their future work. Their future employers (The Confederation of Swedish Enterprise) however, consider that knowledge in ergonomics and work environment development is what engineering students in Sweden often lack (Jeppsson 2018).

There are as far as we know, few studies presented about how we reach the students with teaching ergonomics as a systems discipline, except for the ones we previously published (Berglund and Karltun 2015; Karltun et al. 2017).

The aim with this paper is to present the students' experiences from participating in courses where the HTO concept was used to explain and develop the systems' character of ergonomics to undergraduate as well as to graduate students.

2 Method

The study was based on a course evaluation questionnaire specifically devoted to assessing the perceived relevance of ergonomics as subject and the impact of using the HTO concept as a method for creating understanding of the systems' perspective inherited in workplace ergonomics.

The students participating in the investigation came from educations in logistics and management (undergraduate), sustainable supply chain management (undergraduate), about 10% exchange students not following a program (undergraduate) and production development and management (graduate) respectively, in total 122 students distributed as 87 undergraduates and 35 graduate students.

The questionnaire had some introducing questions related the students background regarding which program the students were engaged in, if they were Swedish students, non-paying exchange students or tuition fee paying students from outside Europe.

The three main questions used to assess students' experienced impact of using the HTO concept as a method for creating understanding of the systems perspective inherited in workplace ergonomics were:

A. Do you think you will benefit from your knowledge of ergonomics in your future work?
B. How much do you think the HTO concept contributed to your understanding of workplace ergonomics?
C. Do you think you will benefit from your knowledge of the HTO concept in your future work?

These questions were designed with a five alternative response scale rated from 'to a low extent' (1) to 'to a high extent' (5). Each of the questions were followed by a space were the students were asked to give a text comment to motivate their scaled answer. A majority of the students also gave such answers.

The five scale questions were analyzed quantitatively using descriptive and inferential statistics. The comments were then analyzed question by question and the rationale for the examples chosen in the paper was to show representative answers and give a picture of the span in the answers as well.

3 Results

The results of the three questions used are presented below. Each question is presented sequentially, first the quantitative results on the five-scaled questions and then the motivations for the quantitative answer as a representative sample.

3.1 Question A: Do You Think You Will Benefit from Your Knowledge of Ergonomics in Your Future Work?

Both student groups regarded the benefits of ergonomics knowledge in future work important but to somewhat different degrees as shown in Table 1.

The answers in Tables 2 and 3 below gives a more substantial understanding of the benefits perceived in relation to question A (Table 4).

Table 1. Students' perceived benefits from ergonomics in their future work. Answers scale 1–5 (to a low extent - to a high extent) n = 122.

Question A: do you think you will benefit from your knowledge of ergonomics in your future work?			
	Undergraduate students (87)	Graduate students (35)	Students T-test on difference
Average	3.65	4.71	$p < .001$
Median	4	5	

Table 2. A representative sample of undergraduate students' comments regarding perceived benefits from ergonomics in their future work related to Question A. The programs were logistics and management (L&M) and sustainable supply chain management (SSCM).

Undergraduate students' comments	Program	Previous work experience time; type of work; country
In the past, I personally didn't understand the reasons and problems concerning ergonomics but after taking this informative course I have learned a lot	SSCM	>3 years; sales, administration, customer service industry; India, Sweden, UK
It will be good to have in the back of my head regarding potential issues but no more	SSCM	>3 years; industrial, carpenting, logistics; Sweden
I think we will benefit from it because much of the future work will probably be about improvements of different kinds	L&M	1–3 years; production work; Sweden
Partly, I will probably have theories in mind, but much focus and time I think will be given priority on value-adding, etc.	L&M	1–3 years; grocery store mostly; Sweden
The importance of how you sit and stand and so on	SSCM	1–3 years; car rental, office; Sweden
I can use the HTO modules to increase efficiency in work	*	<6 months; amusement park; USA

*Exchange student from Singapore

Table 3. A representative sample of graduate students' comments regarding perceived benefits from ergonomics in their future work related to Question A. The program was production development and management.

Graduate students' comments	Previous work experience time; type of work; country
It is an inevitable topic that we need to include in the production system. This can make drastically changes and can make profit in any organization	1–3 years; production engineer; India
This course helped me understand how to develop a systems perspective and broadened my perspective of the ergonomics field. I earlier used to think ergonomics is only related to design and mental process at the most. Now I have a better view about ergonomics and how Human, Technology and Organisational aspects interact	1–3 years; mechanical industry; India
I think this course provided a good overview on areas, which I have not (unfortunately) reckoned in engineering education. Also, the way of the exam enabled us to store the information deeper and it has become more natural for me to think about and consider ergonomics	6 months - 3 years; production engineer; England
I think ergonomics could be applied to a lot extent in a production, its performance and the relationship to HTO	<6 months; café, healthcare; Sweden

3.2 Question B: How Much Do You Think the HTO Concept Contributed to Your Understanding of Workplace Ergonomics?

The answers on this question show that the students think the HTO concept contributed to their understanding of workplace ergonomics but to a different degree in the undergraduate and graduate group related to the quantitative responses shown in Table 5.

Table 4. Student's experience of benefits of the HTO concept in workplace ergonomics. Answers scale 1–5 (to a low extent - to a high extent) n = 122.

Question B: How much do you think the HTO concept contributed to your understanding of workplace ergonomics?			
	Undergraduate students (87)	Graduate students (35)	Students T-test on difference
Average	3.16	4.49	$p < .001$
Median	3	5	

The comments show however, that many of the undergraduate students actually think that the HTO concept has contributed to understanding workplace ergonomics but not as deeply as the master students' comments show (Tables 5 and 6).

Table 5. A representative sample of undergraduate students' comments regarding the HTO concepts contribution to their understanding of workplace ergonomics (Question B). The programs were logistics and management (L&M) and sustainable supply chain management (SSCM).

Undergraduate students' comments	Program	Previous work experience time; type of work; country
The HTO concept contributed enormously as one must think of all parts of HTO to evaluate a work place	SSCM	>3 years; sales, administration, customer service industry; India, Sweden, UK
The HTO concept contributed a lot, if something is wrong with the ergonomics, the HTO simplifies and gives us a broad understanding of the reasons	SSCM	<6 months; production line; Sweden
The need for a holistic view is immanent and HTO allows you to easily understand workplace ergonomics	L&M	>3 years; construction industry, hospital, manufacturing; Australia, Germany
Good way to understand the parts and how these interact with each other's and also easy to educate others to understand it	L&M	>3 years; logistics; Sweden
The HTO concept is very important to understand in our job	L&M	<6 months; warehouse; Sweden

Table 6. A representative sample of graduate students' comments regarding the HTO concepts contribution to their understanding of workplace ergonomics (Question B). The program was production development and management.

Graduate students' comments	Previous work experience time; type of work; country
The HTO concept helped me to understand how workplace ergonomics can be viewed in various ways	1–3 years; mechanical industry; India.
The HTO concept gave a whole new perspective to the ergonomics and how it actually affects all parts	<6 months; café, healthcare; Sweden
The HTO concept gives me a good mental model of where to look and how it can help	6 months - 1 year; McDonalds, transport planning; Sweden
The HTO concept gave me a wider perspective on how important workplace ergonomics is and how the HTO-factors affect each other	1–3 years; process consultant/engineer; Germany
Understanding of each of the HTO sub-systems is very important in order to design a suitable workplace for all	<6 months

3.3 Question C: Do You Think You Will Benefit from Your Knowledge of the HTO Concept in Your Future Work?

Table 7 shows the quantitative results of question C, illustrating that the students think that they will benefit from their knowledge of the HTO concept in their future work. As in the answers on question A and B we can also here see a difference between undergraduate and graduate students as groups considered.

Table 7. Student's experience of benefits from their knowledge of HTO in their future work. Answers scale 1–5 (to a low extent - to a high extent) n = 122.

Question A: do you think you will benefit from your knowledge of ergonomics in your future work?			
	Undergraduate students (87)	Graduate students (35)	Students T-test on difference
Average	3.27	4.43	$p < .001$
Median	3	5	

Regarding the comments below we found that the undergraduate students were more superficial in their comments to question C and there was a larger span in opinions compared to the graduate students. However, both groups show benefits from the HTO concept in many comments regarding their future work (Tables 8 and 9).

Table 8. A representative sample of undergraduate students' comments regarding perceived benefits from their knowledge of HTO in their future work (Question C). The programs were logistics and management (L&M) and sustainable supply chain management (SSCM).

Comment	Program	Previous work experience time; type of work; country
Of course, I will benefit from the HTO concept in my future work and it will help me a lot	SSCM	>3 years; sales, administration, customer service; India, Sweden, UK
Yeah, you can not only blame the human (H), you have to consider the other aspects T and O	L&M	6 months - 1 year; carpententing, logistics; Sweden
I can't say where I will work but definitely not with ergonomics and HTO	SSCM	>3 years; warehouse, assembly line, customer support; Sweden
Yes, because it is tool of analysis that can be adaptable to any situations	*	<6 months
As studying industrial engineering with focus on production it will be definitely beneficial to have it in mind!	#	>3 years; construction industry, hospital, manufacturing; Australia, Germany

*Exchange student from Italy
Exchange student from Germany

Table 9. A representative sample of graduate students' comments regarding perceived benefits from their knowledge of HTO in their future work (Question C). The program was production development and management or follow the course as exchange students.

Comment	Previous work experience time; type of work; country
Yes, in designing, work load distribution, work place, error analysis and better decision making	–
Yes, for better workplace design and help industries work differently	1–3 years; mechanical industry; India
The HTO perspective is very helpful when understanding a system holistically + OD (an organization development model we also used in the course) as support in planning organizational change (this is what I want to do – help improve companies)	1–3 years; process consultant/engineer; Germany
I will definitely see that there are interactions and that they can be used in analysis, development, design etc. I am very grateful for this knowledge!	<6 months; café, healthcare; Sweden

4 Discussion

The students' experiences from participating in courses where the HTO concept was used to explain and develop the systems' character of ergonomics show that undergraduate students in logistics and management and sustainable supply chain management as well as graduate students in production development and management perceived knowledge about ergonomics and the HTO concept beneficial.

Both undergraduate and graduate students regarded the benefits of ergonomics knowledge (Question A) in future work important but to different degrees. These differences were statistically significant according to the Students T-test. In the motivations, we could recognize a difference in the characteristics and spread of the answers between the groups. There was a greater span in the answers given by undergraduate students regarding their perceived benefits of ergonomics in future work compared with graduate students. The depth of understanding of the application and benefits of workplace ergonomics were also identified in the comments. These illustrated that some undergraduates had less interest in ergonomics as well as a different view of it.

How much the students thought the HTO concept contributed to their understanding of workplace ergonomics according to Question B differed somewhat more between the groups in the scaled answers, also statistically significant differences according to the Students T-test. The comments show however, that many of the undergraduate students actually thought that the HTO concept had contributed to understanding workplace ergonomics, but they don't seem to have grasped it as well as the master students.

Undergraduate and graduate students thought they would benefit from their knowledge of the HTO concept in their future work (Question C). Similar to the answers in question A and B, there was a statistically significant difference according to the Students T- test. Moreover, the undergraduate students were more superficial in their comments to this question and there was a larger span in opinions compared to the graduate students' comments.

The result gap between the two groups indicates that the undergraduate students did not grasp the subject as deeply as the graduate students. This is reasonable as they had less study maturity and generally had less work place experience. Another reason for the differences could be due to program focus and what they perceive to be their responsibility in their future work.

A further reason may be the bigger group and spread of students in the undergraduate group which inevitably changed the pedagogic conditions. Another pedagogic aspect referred to by a graduate student in Table 3, was the benefits of the examination method used in the graduate course. The experience of this oral exam method was proven successful in contribution to a more active learning (Karltun and Karltun 2014).

Ergonomics is a discipline that is very broad in theory and application. The HTO concept has provided a useful tool to help the students to relate to overall systems performance associated with interactions between the sub-systems Human, Technology and Organization. It has also shown to be useful as an entry to system's thinking in education on undergraduate as well as graduate level (Karltun et al. 2017).

However, systems knowledge includes the understanding of dynamic changes and is challenging to achieve. It involves a 'development capacity' that requires a deeper kind of learning as double loop learning and second order learning (Kjellström and Andersson 2017). Double-loop learning requires a shift in understanding, from simple and static to broader and more dynamic, such as taking into account the changes in the surroundings and the need for expressing changes in mental models (Argyris 2002). First-order learning refers to people learning something within their current understanding. But change at the system level often requires transformational or second-order learning, which means that persons need to change their way of understanding (Mezirow 1991). This takes time and it is also related to workplace experiences to be applied within workplace ergonomics. We consider the differences in the perceived benefits between undergraduate and graduate students also relate to these aspects.

5 Conclusions

One conclusion that can be drawn from this study is that to improve the teaching of ergonomics as a systems topic, using the HTO concept is a way of doing so. The difficulty in moving the understanding of work systems, from simple linear models of cause and effect to system models illustrating the complexity of work and organizations' operations, must be acknowledged. By further analysis and development of appropriate actions there are possibilities of moving a larger portion of the students towards more elaborate systems understanding of ergonomics. Moreover, differences between student groups must be explicitly considered and the relevance of ergonomics within the engineering educations should be further motivated for the students.

References

Argyris C (1999) On organizational learning. Blackwell, Oxford

Berglund M, Karltun A (2015) Emphasizing the interactive systems view in a master's programme in Ergonomics and HTO. In: Proceedings 19th Triennial Congress of the IEA, Melbourne, 9–14 August

Daniellou F (2001) Epistemological issues about ergonomics and human factors. In: Karwowski W (ed) International encyclopedia of ergonomics and human factors, part 1. Taylor & Francis, London, pp 43–46

Jeppsson P (2018) Speech (in Swedish) presented at AFoU dialogue seminar 180130. https://www.youtube.com/watch?v=og1HOFrS4eQ. Deputy CEO The Confederation of Swedish Enterprise, Stockholm

Karltun A (2014) A novel approach to understand nested layers in quality improvement. In: Proceedings of human factors in organizational design and management – xi nordic ergonomics society annual conference – 46, Copenhagen, 17–20 August 2014, pp. 343–360

Karltun A, Karltun J (2014) Interactive oral assessment supporting active learning. In: Proceedings of the 10th international CDIO conference, Universitat Politècnica de Catalunya, Barcelona, Spain, 16–19 June

Karltun A, Karltun J, Eklund J, Berglund M (2017) HTO - a complementary ergonomics approach. Appl Ergon 59:182–190
Kjellström S, Andersson A-C (2017) Applying adult development theories to improvement science. Int J Health Care Qual Assur 30(7):617–627. https://doi.org/10.1108/IJHCQA-09-2016-0124
Mezirow J (1991) Transformative dimensions of adult learning. Jossey-Bass, San Francisco
Wilson JR (2014) Fundamentals of systems ergonomics/human factors. Appl Ergon 45(1):5–13

Contributions of Simulation for Developing New Activities: Simultaneous Approach of Activity and Space in the Case of a Fire Department Situation Room Design Process

Special Session: Ergonomics Simulation Tools in Design and Organizational Change Projects

Nadia Heddad(✉)

Université Paris 1 Panthéon Sorbonne, FCPS, 21 rue Broca, 75005 Paris, France
nadia.heddad@icloud.com

Abstract. This text discusses the contribution of simulation in designing space and organization particularly for an activity which does not exist yet but that need to be anticipated. The design concerns the spatial organization for a new building that could shelter a crisis activity in an operational coordination center for firefighters in the city of Paris. The project was done before the recent attacks in Paris. The difficulty was specially focused on designing a situation room for managing crisis with firefighters with low experience of crisis situation. The objective was to assist the firefighters to imagine the space for an unpredictable activity. By exploiting tools and media used daily in their activity in addition to simulation on cardboard mock-ups, it became possible to build benchmarks to think the criteria for the space of their possible future activity. By anchoring their gestures around magnets during simulations, they were able to project themselves in different crisis situations. By projecting different organizations with magnets on whiteboard combined to drawings and mock-ups, they finally succeed to question their way of working in a reflexive approach. Their contribution to the design process was then possible. The architect involved in the participatory design process could finally get access and take into account the different requirement of work in an operational coordination center designed for firefighters having to prepared to deal with possible and unimaginable crisis.

Keywords: Simulation · Organization · Space

1 The Project and Its Issues

This project was done before the recent attacks in Paris. The project was to design the spatial organization that could shelter a crisis activity in the new operational coordination center for firefighters in Paris. The propose here is to discuss the contribution of simulation to design spatial organization for an activity which does not exist yet but that need to be anticipated.

S. Bagnara et al. (Eds.): IEA 2018, AISC 821, pp. 637–641, 2019.
https://doi.org/10.1007/978-3-319-96080-7_76

The center is singular. It ensures the coordination of the rescue operations on a territory with 10 million people in a dense urban area. The size, the urbanism and the density of the covered area give to the project a particularly strategic character.

The difficulty was specially focused on designing the situation room. Firefighters are used to organize and regulate from the operational center, actions and rescues on different operational areas in the city. But they are not used to contribute to design adapted work spaces to crisis management in their own space in the operational center.

An architect was involved in the design process, but it was not easy for the firefighters to contribute to the definition of their needs in such a situation as a crisis.

Trying to imagine their activity on the drawings proposed by the architect in charge of the building design was not easy. Having access to architectural drawings was not necessary helping them to imagine their activity in unexpected situation. Their representation of future activity was partial and sometimes even non-existent. The anchoring with their experience of crisis was hard to take place. Crisis is hopefully not happening so often and firefighters whom really bean confronted to crises in the fire center were rare. In fact, firefighters in the center are young et their professional career is usually short. The probability of facing a crisis is in reality quite low in a firefighter's working life. Their experience was limited to crisis simulation exercises whom were taking room in the center without really an appropriate space. The simulation exercises where designed to help the firefighters to act in interaction with the different actors located outside the center, in a direct related to the simulated situation and not in a reflexive approach on the spatial organization of the crisis room itself.

So, the question is, how to design a situation room for people with no experience of real and various crises situation? What kind of spatial organization is to promote to assist and help crisis management in case of a terrorist attack, a plane crash, or any major accident in the city?

Dealing with a natural disaster, an attack or a major technological accident (plane crash, toxic cloud, etc.) are in fact among the possible events but impossible to predict or anticipate. However, firefighters must be able to be prepared to face these various situations and the building must be designed in the perspective of these requirements.

2 Imagining and Simulating an Unpredictable Activity

A Working Group bringing together firefighters was build and the discussion was opened with questioning the actual and future organization of the different spaces in the operational center.

The choice of using known media and tools during crises exercises by the firefighters in a simulation perspective became a way to help them to anchor their thinking on the organization et it's spatial translation.

The use of magnets system and a simple white magnetic board representing different machines and functions usually used in the situation room or during exercises of crisis simulation was a way to hang up the discussion on the spatial organization and link it to their experience.

Instead of representing a situation outside the center, these tools were hijacked and combined with mock simulation. The use of simulation combined with tools used in the daily life of the organization allowed to represent and discuss the operational organization to be set up in case of crisis. These different supports were used, not to represent situation during a fire, but to invite them to introduce a reflection on their organization and its possible spatial translation in a project configuration.

By using these same tolls and media in another way simulating not the outside event, but their own organization in the situation room during crisis became a way for them to represent their own organization and the needs and requirements for the working space. They became able to explore, represent and arbitrate on different possibilities to organize their space in the situation room (Fig. 1).

Fig. 1. Pictures illustrating the principle of magnets affixed on a whiteboard used in the relief coordination work in the field from the operational center.

3 Designing Space by Simulating Possible Future Activity on Mock-Ups

The next step was to simulate on a cardboard model the unfolding of different categories of crisis situation management. Firefighters played on mock-ups different scenarios of possible or impossible crisis situation trying to imagining as much as possible unpredictable events (Fig. 2).

The Working Group focused on integrating in the design process, the simultaneous design of the activities and their possible spatial configurations (Fig. 3).

The use of role-play on mock-up and the discussion of the organization of the interactions between different stakeholders allowed the development of a common operative reference [1]. It was the way of questioning how to place the multi-skilled actors in the space (firefighting, medical, media, information and data recording means…).

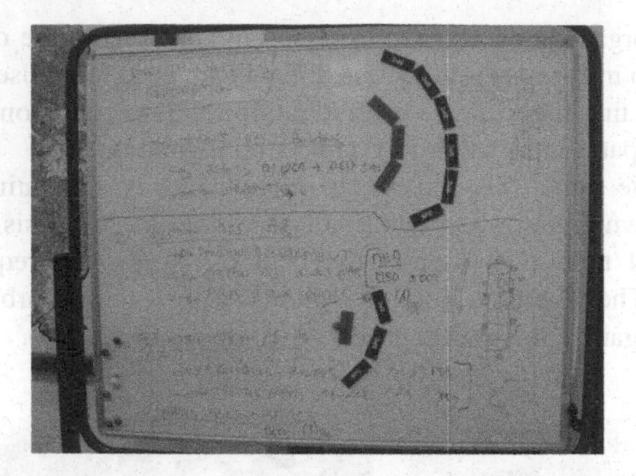

Fig. 2. Picture illustrating the production of a proposal to organize space by using and combining magnets and text.

Fig. 3. Picture illustrating the mock-ups used in the working group.

By questioning their way of working, they developed a critical analysis of their organization and learned by using simulations [2] to build a reflexive approach in the way of using their space organization to support their activity [3].

The combination of different media (mockups, plans, diagrams, videos and magnets on a whiteboard) helped them to anchor a critical analysis on existing and future organization in a collective action for designing the organization in the new center [4]. A collective discussion of the modalities and forms of communication in the activity of management unpredictable events was then possible to take place. Helped by different forms of simulation, the firefighters were able to think simultaneously their activity, the organizational and the new spaces.

4 Conclusion

To conclude here, the simulation approach allowed a dynamic approach of the spatial organization. It allowed the exploration of the possible forms of the activity at the same time of their translation into the physical space. This project was carried out before the recent terrorist attacks in Paris. It has allowed firefighters to have the means to deal with the crisis.

References

1. De Terssac G, Chabaud C (1990) Référentiel opératif commun et fiabilité. In: Leplat J, De Terssac G (eds) Les facteurs humains de la fiabilité dans les systèmes complexes. Octarès, Toulouse, pp 110–139
2. Béguin P (2014) Learning during design through simulation. In: Broberg O, Fallentin N, Hasle P, Jensen PL, Kabel A, Larsen ME, Weller T (eds) Proceedings of human factors in organizational design and management – XI and nordic ergonomics society annual conference – 46, pp 867–872
3. Heddad N (2017) Space of activity: a simultaneous construction. Le travail humain 80(2):207–233 DOI 10%.391/th802.0208
4. Heddad N (2017) Thinking and designing activity space: towards a methodological framework. Activités 14(1). http://journals.openedition.org/activites/2958, https://doi.org/10.4000/activites.2958

Effectiveness of Onsite Occupational Health Clinics in Management of Work Related Musculoskeletal Disorders in 12000 Information Technology Professionals

Deepak Sharan[1(✉)], Joshua Samuel Rajkumar[2], and Jerrish A. Jose[2]

[1] Department of Orthopaedics and Rehabilitation, RECOUP Neuromusculoskeletal Rehabilitation Centre, 312, 10th Block, Anjanapura, Bangalore 560108, KA, India
deepak.sharan@recoup.in

[2] Department of Physiotherapy, RECOUP Neuromusculoskeletal Rehabilitation Centre, 312, 10th Block, Anjanapura, Bangalore 560108, KA, India
{joshua.samuel,jerrish}@recoup.in

Abstract. Onsite clinics play a crucial role in the provision of occupational health services. A retrospective analysis was conducted from 2006 to 2017, covering 12000 employees of a single multinational IT company with software development centres in an Industrially Developing Country. The employees (8574 males and 3426 females, between the ages 20 to 60 years), were diagnosed by an experienced occupational health physician (OHP) to have a WRMSD in specific regions following extensive usage of desktop and/or laptop computer. All the employees then underwent an ergonomic workplace analysis and protocol-based rehabilitation for the WRMSD by specially trained occupational physiotherapists. A total of 58% of the employees worked for at least 5–9 h per day and 42% for 10–14 h per day. Most of the male workers complained of low back and radiating pain in upper or lower limbs, compared to female workers who complained predominantly of neck and shoulder pain. 94% of workers reported complete resolution of symptoms and 6% reported partial resolution of symptoms but could work without restriction. The onsite occupational health clinics were effective in the management of WRMSD in the IT companies.

Keywords: Onsite clinics · Work related musculoskeletal disorder
Information technology

1 Introduction

The prevalence of work related musculoskeletal disorders (WRMSD) in Information Technology (IT) companies has been reported to vary from 75% to 78%. Onsite clinics play a crucial role in the provision of occupational health services. However, the effectiveness of onsite clinics in the management of WRMSDs is an under studied area. The aim of this study was to find out the outcome of management of WRMSDs in on-site occupational health clinics in an IT company over a 11-year period.

S. Bagnara et al. (Eds.): IEA 2018, AISC 821, pp. 642–646, 2019.
https://doi.org/10.1007/978-3-319-96080-7_77

2 Methodology

A retrospective analysis was conducted from 2006 to 2017, covering 12000 employees of a single multinational IT company with software development centres in an Industrially Developing Country. The employees (8574 males and 3426 females, between the ages 20 to 60 years), were diagnosed by an experienced occupational health physician (OHP) to have a WRMSD in specific regions following extensive usage of desktop and/or laptop computer. All the employees then underwent an ergonomic workplace analysis and protocol-based rehabilitation for the WRMSD by specially trained occupational physiotherapists. The protocol based rehabilitation is called Skilled Hands-on Approach for the Release of myofascia, Articular, Neural and Soft tissue mobilization (SHARANS) protocol, which included soft tissue mobilization techniques like Trigger Point release technique, myofascial release technique, articular mobilization of the relevant joints, stretching exercises, postural training and home care advice, core stabilization exercises, EMG Biofeedback, McKenzie's MDT exercises, progressive resisted exercises, aerobic training, job retraining, and ergonomic advice. Total duration of treatment varied from 1 week to 8 weeks based on the level of discomfort. The rehabilitation was done in the onsite clinic located in the office premises, wherein the employees booked an appointment for consultation with the physician using an online tool. After the consultation with the OHP, ergonomic workplace analysis and rehabilitation was initiated. The employees were reviewed by the OHP monthly and at the completion of rehabilitation.

3 Results

Most the employees fell under the category of software and application engineers (54%), managers (33%) and technical support staff (7%). 63% of the employees used laptops, 30% used desktops and 7% used both. A total of 58% of the employees worked for at least 5–9 h per day and 42% for 10–14 h per day. Most of the male workers complained of low back and radiating pain in upper or lower limbs, compared to female workers who complained predominantly of neck and shoulder pain. Both the population had eye strain and increased fatigue in common. 80% had overall body pain, 67% neck pain, 58% lower back pain, 46% shoulder pain and others with upper arm, thigh, knee and foot pain. 75% were diagnosed to have Myofascial Pain Syndrome, followed by Thoracic Outlet Syndrome (43%), Fibromyalgia (34%), Tendinopathies (15%) and Type 1 Complex Regional Pain Syndrome (8%). After the rehabilitation, the VAS scale showed significant reduction in pain levels ($p < 0.01$). 74% had reported reduced productivity due to the WRMSD, which improved markedly after the rehabilitation. 94% of workers reported complete resolution of symptoms and 6% reported partial resolution of symptoms but could work without restriction. No employee had to take leave for more than 7 days or leave the job due to WRMSD.

4 Discussion

WRMSD is one of the commonest occupational diseases in India. Since WRMSD is not a reportable condition in India, its frequency in various industries is not known. However, several studies are available regarding WRMSD in the IT sector in India.

The prevalence of WRMSD in Indian IT professionals has been reported to be 76% (Sharan et al. 2011).

The largest published study on WRMSD in India covered 3053 IT Professionals who were treated in on-site occupational health clinics in various companies across Bangalore, Delhi and Hyderabad, India (Sharan et al. 2012). The mean age of the participants was 29 ± 4.5 years. The distribution of laptop and desktop users were almost equal (OR: 1.21, CI: from 0.70 to 2.1). 69% of the populations were working for 5–9 h and 30% of the populations were working 10–14 h. The major risk factor was poor office ergonomics (54%), including lack of keyboard tray (25%), lack of mouse tray (35%), lack of foot rest (60%), Improper monitor height (80%). The other risk factors were lack of breaks (64%) and stress (37%). Myofascial Pain Syndrome (49.20%) was the commonest condition diagnosed, followed by Thoracic Outlet Syndrome (25.02%), Fibromyalgia (8.5%), Cubital Tunnel Syndrome (0.16%), Complex Regional Pain Syndrome (0.20%), and Wrist Tendonitis (3.70%). The body regions affected were neck (64.9%), lower back (56.5%), shoulder (42.1%) and thigh (34.2%).

In another large study on 4,500 IT professionals, 22% were reported to have a high risk of an adverse workstyle and 63% reported pain symptoms. Social reactivity, lack of breaks, and deadlines/pressure subscales of workstyle questionnaire were significantly correlated with pain and loss of productivity. Workstyle factors and duration of computer use per day were significant predictors of pain. Based on the study findings, it was recommended that intervention efforts directed towards prevention of WRMSD should focus on psychosocial work factors such as adverse workstyle in addition to biomechanical risk factors (Sharan et al. 2011).

Strong associations between boredom, workload, and social support to musculoskeletal discomfort have been reported in computer professionals from India (Bhanderi et al. 2007).

On-site occupational health clinics are recommended for the primary prevention and treatment of WRMSD. The advantage of on-site clinics (Sharan 2012) includes:

1. Convenience of employees, saving time and greater compliance,
2. Promoting earlier reporting of symptoms and hence a quicker and easier recovery,
3. Better follow up regarding recovery, work modifications and return to original work,
4. Allows on-site workstation assessment and modifications,
5. Maintenance of recovery by monitoring posture, breaks and exercises,
6. More effective coordination with members of Human Resources, Facilities, Health and Safety team, etc.,
7. Improved awareness levels regarding Ergonomics among management and employees.

Comprehensive ergonomic programmes that involve primary prevention of WRMSDs through ergonomic changes in jobs, early detection of WRMSDs through surveillance, and prompt treatment of WRMSDs with an emphasis on early return to modified work have been recommended according to several publications.

The on-site occupational health clinics also offer an excellent return on investment (ROI) (Sharan 2012). In a study, the total cost of intervention in the company includes nominal wages (INR 15000 per month) to the Therapists/Ergonomists for performing the assessment and treatment plus in some cases addition of accessories (keyboard tray, mouse tray). In 95% of these cases success was achieved by just adjusting the existing components of the workstation. The benefits include reduced complaints of pain and discomfort leading to increase in productivity and decrease in absenteeism in the employees. After completing the on-site program, 95% of employees reported no pain and that their productivity increased. Therefore, for 150 employees (approx. wages INR 180 per hour) with an investment of one therapist employed, the productivity increased by a conservative estimate of 20% based on employee feedback. With this data, an approximate amount of INR 25000 per worker per annum was saved by the company because of the intervention.

The recommended primary prevention measures in the IT sector include:

1. Work systems organisation and design: encouraging stretch breaks during the working hours, on site fitness programmes.
2. Workplace organisation and design: individual ergonomic workplace analysis of each worker covering work posture, work station arrangement, equipment design and positioning, environment conditions (e.g., lighting, noise, temperature), provision of ergonomic accessories (e.g., tray for keyboard and mouse), and adjustment of chairs according to the body dimensions of the workers.
3. Training and education: healthy computing practices, role of breaks, nutrition, stress prevention, personal fitness, etc.

Although the focus of prevention efforts against WRMSD should be on primary prevention, the reduction or elimination of workplace risk factors, it is also important to ensure that workers have access to appropriate and timely on-site medical care if they do become injured.

The goals of on-site occupational health clinics are to:

- Reduce or eliminate symptoms;
- Prevent progression of WRMSDs;
- Reduce the duration and severity of functional impairment; and
- Prevent or reduce the severity of disability.

Crucial elements of this program include:

- Surveillance;
- Timely access to appropriate ergonomics, medical and rehabilitation professionals;
- Ergonomic workplace analysis of injured workers;
- Availability of appropriate job modification; and
- Follow-up of treated workers and coordination with primary prevention efforts.

5 Conclusions

The onsite occupational health clinics were effective in the management of WRMSD in the IT companies. A comprehensive ergonomic programme that involves primary prevention of WRMSD through ergonomic changes in jobs, early detection of WRMSDs through surveillance, and early treatment of WRMSDs with an emphasis on early return to modified work is recommended.

References

Sharan D, Parijat P, Sasidharan AP, Ranganathan R, Mohandoss M, Jose J (2011) Workstyle risk factors for work related musculoskeletal symptoms among computer professionals in India. J Occup Rehabil 21(4):520–525

Sharan D, Ajeesh PS, Rameshkumar R, Jose J (2012) Risk factors, clinical features and outcome of treatment of work related musculoskeletal disorders in on-site clinics among IT companies in India. Work 41(Suppl 1):5702–5704

Sharan D (2012) Return on investment of an on-site employee health clinic in multinational information technology company in India. Work 41(Suppl 1):5921–5923

Bhanderi D, Choudhary S, Parmar L, Doshi V (2007) Influence of psychosocial workplace factors on occurrence of musculoskeletal discomfort in computer operators. Indian J Community Med 32(3):225–226

The BRICS$_{plus}$ Network: A Historical Overview and Future Perspectives of the Network's Role in Human Factors and Ergonomics

Jonathan Davy[1(✉)], Jose Orlando Gomes[2], Aleksandr Volosiuk[3], Arnab Jana[4], Gaur Ray[5], Anindya Kumar Ganguli[6], Wei Zhang[7], and Andrew Todd[1]

[1] Department of Human Kinetics and Ergonomics, Rhodes University, Grahamstown, South Africa
j.davy@ru.ac.za

[2] School of Engineering, Federal University of Rio de Janeiro, Rio de Janeiro, Brazil

[3] Department of Information Systems, St. Petersburg Electrotechnical University, Saint Petersburg, Russia

[4] Centre for Urban Science and Engineering, Indian Institute of Technology Bombay, Mumbai, India

[5] Department of Biosciences and Bioengineering, Indian Institute of Technology Bombay, Mumbai, India

[6] Faculty in Ergonomics, RamMohan College, University of Calcutta, Kolkata, India

[7] Department of Industrial Engineering, Tsinghua University, Beijing, China

Abstract. In December 2016, at Tsinghua University in Beijing China, the first meeting of the newly constituted BRICS$_{plus}$ Network of Brazilian, Russian, Indian, Chinese and South African Ergonomics societies was held. The Network was created to promote and develop ergonomics in BRICS countries, collaboratively represent the BRICS countries at the level of the IEA and ensure that professional standards of ergonomics, in teaching, research and practice, are maintained. This paper offers a short overview of the history of each BRICS country's ergonomics society, in which we highlight the impact of key individuals, events and milestones on the development of ergonomics in the respective countires. We then provide an overview of the current challenges faced by BRICS countries, with respect to the dissemination of ergonomics, which is proceeded by a snapshot of how the intended activities of the network will address these. We conclude by emphasizing how the BRICS$_{plus}$ network has the potential to create globalized ergonomics graduates, educators and researchers, who will continue to foster the advancement of ergonomics in their countries and more globally.

Keywords: BRICS · History · Education · Research

S. Bagnara et al. (Eds.): IEA 2018, AISC 821, pp. 647–656, 2019.
https://doi.org/10.1007/978-3-319-96080-7_78

1 Background

On 15th of December, at Tshingua University in Beijing, China, the network of ergonomics societies from Brazil, Russia, India, China and South Africa (BRICS) held its first official meeting. The Network was created to promote and develop Human factors and Ergonomics (HFE) in BRICS countries, collaboratively represent the BRICS countries at a global scale with the ultimate goat at the level of the International Ergonomics Association (IEA). Secondly, the network aims to ensure that professional standards of Human Factors and Ergonomics (HFE) are maintained across these countries. More specifically, the network aims to foster collaborative efforts in research and education programs across BRICS countries and countries that are similar in socioeconomic status to BRICS nations. With respect to research specifically, it was envisaged that this could include joint research programs through funding avenues such as the BRICS Multilateral Research programs across all countries or bilateral agreements between BRICS countries. From an education perspective, the network has the potential to also provide opportunities for student and research exchange, facilitated by Memorandums of Understandings (MOUs) between different universities and the establishments of funding avenues to support this exchange. Lastly, the network has the potential to initiate and roll out the first Joint Education Program (JEP) across all the BRICS countries, with the express goal of ensuring that future researchers and practitioners are adequately prepared to take HFE forward. Ultimately, when realized, these initiatives can ensure high-level ergonomics research and practice across BRICS countries and globally. To date, the network has developed a set of by-laws, and has had a second meeting, hosted by The Brazilian Ergonomics Society (ABERGO) in Porto Alegre, Brazil in September 2017 and plans to have a 3rd meeting in July of 2018 in Saint Petersburg, Russia before making application to the IEA to be officially recognized as an affiliated network.

While it is important to understand the short history of the network, its activities to date and the envisaged goals behind its creation, it is equally as important to understand the history of each of the BRICS countries, with particular reference to the development of each country's ergonomics society and ergonomics in general. In doing so, one can understand better the similarities (and differences) between BRICS countries, which in turn, provides some justification for why such a network came into being. Therefore, the aim of this short paper is to provide a brief overview of each BRICS country's ergonomics society, particularly with respect to role they have played in establishing ergonomics in their respective countries. Additionally, we touch on the dominant areas of ergonomics research in each country. The paper concludes with a synthesis of challenges faced by BRICS countries with respect to the enhancement of the discipline of ergonomics. We offer some further perspectives on how the network may be important in growth of HFE in BRICS countries and globally.

1.1 Brazilian Ergonomics

Moraes and Soares [1] and later in Soares [2] provide a comprehensive overview of the history of ergonomics in Brazil. More specifically, they identified six keys are that contributed to the development of the discipline of ergonomics. The first source of ergonomics development began with Sérgio Kehl at the École Polytechnique of University of Sao Paulo at the beginning of 1960. He introduced the subject "Product and Man" in the course "Product Design" for production engineers training. The model of the Ecole Polytechnique was applied to the minimum content of the production engineering courses of the other engineering schools. Sérgio Kehl also founded the GAPP (Associate Research and Planning) group, which began offering ergonomics consulting services and disseminated this knowledge to several Brazilian mixed economy companies, such as the Compagnie Sérérurgique Nationale (CSN) and the São Paulo Metro [1].

The second source was the work of Professor Itiro Iida at COPPE/UFRJ. In 1978, Iida visited and taught at Rio de Janeiro using the book "Ergonomics: course notes". There he taught as part of a specialized higher education program in Production Engineering/Production Management Sector and Product Engineering. His course has been the foundation for spreading the knowledge of ergonomics and has helped produce many Masters Graduates. In addition to this, Iida, with the assistance of the Industrial Technology Secretariat of the Ministry of Industry and Trade in 1976, published "Ergonomic Aspects of the Urban Bus", which has influenced the improvement of public transport. Finally, the influential work entitled 'Ergonomics - Project and Production', published in 1990 is still of great importance in the field of ergonomics in the Portuguese-speaking academies.

The third source was initiated by ESDI. In this school, Prof. Karl Heinz Bergmiller initiated ergonomics education for product design and project management, according to Tomás Maldonado's model from the Ulm School in Germany. From this, ergonomics became part of all Industrial Design courses and has been so since 1979. The inclusion of ergonomics in Industrial Design course content accounts for the large number of associated designers affiliated to ABERGO currently.

The fourth source of ergonomics development originated from Europe, courtesy of academics such as Rozestraten and Stephaneck who settled at University of Sao Paulo Ribeirão Preto. There, they created a research program in ergonomic psychology, with an emphasis on the visual perception applied to road traffic studies.

The fifth source involves the actions of Prof. Franco Seminério in the FGV ISOP in Rio de Janeiro. In 1974, this institution organized the first Brazilian Ergonomics seminar, a fundamental event in the history of ergonomics in Brazil. In 1975, Seminério created the country's first specialized course in ergonomics. This course was offered to the ergonomists (involved in establishing ergonomics groups in companies) in the country until 1990.

Finally, the initiative of Prof. Franco Seminério, who invited Professor Alain Wisner, from CNAM to come to Brazil, was important for Brazilian ergonomics. He later became a great proponent of Brazilian ergonomics by guiding the first ergonomic work of the Getulio Vargas Foundation/Rio de Janeiro on the sugarcane plantations in the city of Campos, State of Rio from Janeiro. Since then, many Brazilian students have visited CNAM to obtain training in ergonomics. Currently, ergonomists trained at CNAM are distributed across several regions of the Brazil. Many are also involved in teaching and research centers, as well as the consulting sector. In short, the influence of Prof Wisner on Brazilian students being trained at CNAM has and continues to have the largest impact on the development of Brazilian ergonomics, and in particular the application of analysis techniques unique to Francophone countries [1, 2].

ABERGO, the Brazilian Ergonomics Society was founded in 1983, became affiliated to the IEA in 1991 and has grown substantially since then [2]. In 2002, ABERGO initiated the process of certifying Brazilian ergonomists. In an analysis of the chronology of ergonomics in Brazil, it is important to underline twelve major events: the Brazilian Ergonomics Congress which took place in Rio de Janeiro (1974, 1984, 1989, 1995, 2000, 2010), in São Paulo (1987 and 1991), in Florianópolis (1993 and 1997), in Salvador-Bahia (1999), in Gramado-RS (2001), Recife-PE (2002, 2012 IEA2012), Ouro Preto (2004), Curitiba (2006), Porto Seguro (2008), Sao Carlos (2014) and Belo Horizonte (2016). It is also necessary to note that the 2010 ABERGO Congress was aligned with the congress of the ULAERGO in Rio de Janeiro. Furthermore, ABERGO initiated (in 2011) and continues to host regular conferences with industry every second year. In 2017, this conference coincided with first meeting of the BRICS_{plus} Network Executive in 2017 in Porto Alegre. Such events have and continue to channel the diffusion of the technical and scientific knowledge of ergonomics in Brazil. Currently, the ABERGO Ergonomics Congress has emerged as a popular event, with the active participation of national and international researchers.

1.2 Russian Ergonomics

The Russian society of ergonomists, known as the Inter-Regional Ergonomics Association (IREA) originated from the Soviet Ergonomics Association (SEA) established in 1986. The founder and first president of SEA was Professor A.I. Gubinsky. SEA became a federated society registered with the IEA around this time. SEA organized the first international conference 'Ergonomics in Russia, Commonwealth of Independent States (CIS) which was a once off activity in 1993. IREA was established on August 1995 as an assignee of the SEA, with the first president being Professor P.Ya. Shlaen.

Ergonomics in Russia has its origins in the 1920s, with the main areas of interest being industrial design, physiology, biomechanics, and 'psycho-technology'. Russian avant-garde artists were the source of some of the original ideas behind industrial design, to the point where they proposed the idea of 'industrial art' (Vlasov, 1995; in [3]). They were inspired by the idea of building a new society where humans deal with modernity under comfortable work conditions. They asserted that artists must be involved in the design of industrial machines, workplaces and interiors.

The Institute for the Study of Brain and Psychical Activity was established under the leadership of the famous Russian physiologist V.M. Bekhterev in 1918, with the Department of Occupational Psychology forming part of this [3]. In 1920 V.M. Bekhterev and V.N. Myasischev established 'ergology' or ergonology' a new science aimed at studying labour. Later, A.A. Ukhtomsky began research in the area of occupational physiology, establishing a laboratory for studying work in industrial plants in Petrograd. The Central Institute for Labour was created at much the same time. During this time, N.A. Bernstein made a significant contribution to ergonomics by founding the basics of biomechanics [3].

In the thirties, 'psycho-technology' became popular in the USSR and this was followed by rigorous research in the areas of aviation ergonomics. Many large-scale plants then set up psycho-physiological laboratories aimed at studying and improving human work within machine and production lines. Two journals, The 'Psycho-technology Society' and 'Soviet Psycho-technology' were created for the dissemination of work in this area. However, in the middle of 1930s all activity in the field of psycho-technology was stopped as the result of political instability in the USSR.

From the 1950s, the area of occupational psychology in the USSR was being re-established. As a result, engineering psychology and ergonomics were introduced. Around this time, as was the case in other developed countries, defense was the main area of focus and included specific research and development in the fields of aviation, space and defense technology. Research laboratories in engineering psychology were established in Leningrad State University in 1959 (under leadership of B.F. Lomov), and in Institute of Automatic Equipment in 1961 (under leadership of V.P. Zinchenko). Noteworthy were the contributions by Russian psychologists such as B.G. Ananyev, A. N. Leontyev, P.Ya. Galperin and others [3].

At the beginning of the sixties equipment and environmental design became areas of significant and intense focus. In Moscow the Research Institute of 'Technical Aesthetics' (industrial design) (VNIITE) comprising of the Department of Ergonomics was established. Over time, VNIITE founded 10 regional branches in the biggest cities of the USSR, with projects in the civil sector promoting and popularizing ergonomics and industrial design. It was during this time that Professor V.M. Munipov and colleagues published the first ergonomics textbooks and other materials.

During the seventies and eighties tremendous progress was made in the areas of defense, which saw the development of standards for defense equipment design. Furthermore, Professors A.I. Gubinsky and V.G. Evgrafov, contributed significantly to the development of the theory of human-machine systems, with A.I. Gubinsky being the first to coin the ergonomics terms such as operator, error, reliability, efficiency. He also proposed a theory for the integrated analysis of human-machine systems (HMS), including analysis and assessment of HMS and operator reliability and efficiency.

Economic recession during the 80s and 90s, negatively affected funding availability for research in the areas of ergonomics. In the 2000s, the economic situation in Russia improved. As a result activity in different fields of ergonomics was strengthened. Gradually many industrial plants replaced production equipment and technology, implementing new human-oriented approaches to the design and manufacture of goods.

The slogans of "Good ergonomics - good economics", "Diversity in ergonomics" and "Ergonomic is a style of life" all emphasize that a human-oriented approach. Professional ergonomists in modern Russia feel a responsibility for all of these issues [3].

1.3 Indian Ergonomics

Indian ergonomics was born in the Department of Physiology of the Presidency College, Kolkata in the 1950s where extensive work was conducted on the energy metabolism of rickshaw pullers and the body surface area of Indians. In the early sixties this work was extended to the Industrial Physiology Division of the Central Labour Institute, Mumbai as well as the work physiology and ergonomics division of the Central Mining Research Institute, Dhanbad. It then spread into various sectors including academics, defense, agriculture, design and industry [4].

The first teaching in ergonomics took place at the Department of Physiology, University of Calcutta around the year 1971, where a post-graduate science course in Work Physiology and Ergonomics was offered. Ergonomics was introduced as part of a design curriculum at the Industrial Design Centre of the Indian Institute of Technology Bombay (IITB), in the year 1979 as a post-graduate course for engineers, and at the National Institute of Design, Ahmedabad at the graduate level [4]. All these centers contributed to the intermingling of ergonomics with product and interaction/interface design in their respective design curricula in association with their doctoral programs.

In 1979 the importance of the role that ergonomics could have in national development was emphasized for the first time when Dr. Sen (President of the section of Physiology at the 69th session of the Indian Science congress) spoke about "Ergonomics, Science and Technology of man at work: its role in our national development". Further developments occurred in 1983 when consensus was reached for formation of the Indian Ergonomics Society, which was followed by a proposal in 1985 and the official formation of the society in the January 1987. Under the auspices of this society, national and international conferences (HWWE) are organized at different locations across the country every year.

Ergonomics research in India in the past five decades has focused on the following main areas; Physical work capacity and work stress of different occupations; anthropometry of the people of India, and the development of anthropometric databases for the Indian population; load carriage – both in defense and in the informal sector; improvement of working conditions in adverse environments, including hot and humid environments; certain aspects of agriculture (upon which most rural people still depend) including high profile areas like tea cultivation and tool design; low-cost improvements for some traditional and informal sectors; industrial ergonomics - including material handling, assembly and welding operations, shop-floor problems; product design; women at work; the electronics and information technology (IT) sector. Ergonomics research, teaching and practice were introduced in these areas at different times through different institutions [4].

1.4 Chinese Ergonomics

The Chinese Ergonomics Society (CES) was founded in 1989 and became a federated society of IEA in the early 90s [5]. The CES has seen a rapid growth in the number of active members and organizations due to the economic development. Currently CES has eight branches with a total of 985 active members from over 200 organizations, compared to only seven branches with a total of only 272 members from 89 organization six years ago. In 1989, CES was founded in Tongji University, Shanghai, with several major contributing organizations, including Tongji University, the Institute of Psychology, Chinese Academy of Sciences, Hangzhou University (currently part of Zhejiang University), Beijing Medical College (currently as part of Peking University), Southwest Jiaotong University. The founding president was Professor Rongfang Shen from Tongji University, with Professor Run-Bai Wei acting as Secretary General [6]. Accordingly the CES secretary office was located in Tongji.

In the middle 90s, after election, Professor Kan Zhang from the Institute of Psychology, Chinese Academy of Sciences, Beijing, was elected as new president. The secretary office then moved from Tongji University, Shanghai, to IP-CAS, Beijing. In 2004, Professor Sheng Wang, from Peking University, was elected as the new president. The secretary office moved from IP-CAS to Peking University. In his presidency, Prof. Wang successfully organized the 2009 IEA congress in Beijing, with over 1,000 participants from all over the world.

In November of 2012, Professor Wei Zhang, from Tsinghua University, was elected and is still the president. The secretary office accordingly moved from Peking University to Tsinghua University. In 2015, nominated by the CES and some IEA council members, Professor Sheng Wang, the CES' immediate past president, was granted the title of IEA fellow. Currently, the CES has eight branches, headed by eight organizations. They are (1) human-machine interaction technical committee, headed by Beijing Institute of Technology; (2) cognitive technical committee, headed by Zhejiang University; (3) biomechanics technical committee, headed by Aerospace Medical Research Institute; (4) macro-ergonomics technical committee, headed by Northeast University; (5) safety and environment technical committee, headed by Wuhan Safety and Environment Research Institute; (6) standardization technical committee, headed by China National Institute of Standardization; (7) transportation ergonomics technical committee, headed by Anhui Sanlian Research Institute; and (8) occupational ergonomics technical committee, headed by Peking University. Most these committees were formed when CES was constituted [6].

Annual conferences are held by some of the technical committees, such as the macro-ergonomics technical committee. As a result of the conferences and economical/technical development in China, CES has witnessed a significant increase in active membership, and more active research and consulting services. The CES also organize frequent exchanges or jointly organized conferences with other ergonomics societies, such as the Ergonomics Society of Taiwan (EST) and the Hong Kong Ergonomics Society (HKES). CES has also been willing to collaborate with other ergonomics societies to promote ergonomics research and services to the industrial and service problems which are often popular in developing countries, although they might be less common in developed countries.

1.5 South African Ergonomics

The Ergonomics Society of South Africa (ESSA) was formed and officially constituted in February 1985 at the first conference, entitled "Ergonomics '85", with the committee largely drawn from academics and practitioners who volunteered to run the society. In this case, the charter members included Tony Golding, Jan van Tonder and Brian Hill [7]. It was only in 1994 that ESSA was recognized by the IEA as a Federated Society, a sign of the end of the Apartheid regime in South Africa. The role of ESSA was (and still is) divided into four core areas which include; building awareness and understanding of what ergonomics is and its associated benefits, the certification of professional ergonomists (which only came into being in 2014), the provision of resources for the advancement of ergonomics in South Africa and facilitation of strategic partnerships between the society, educational institutions and cognate disciplines. Since its inception, the society has continued to remain active, hosting, as of 2017, thirteen conferences, which have offered an opportunity for practitioners, academics and students to share their work. Despite this, the society's membership has hovered between 50 and 70 members, something that the society wishes to address.

Ergonomics was and continues to be largely restricted to the academy (originally at the University of Cape Town under Mike Cooke, George Jaros and Bob Bridger and then later at Rhodes University under Pat Scott and Jack Charteris) [7]. Most recently, ergonomics is now offered the University of Witwatersrand, under Andrew Thatcher. The practice of ergonomics within South Africa, although still limited, is set to steadily grow in the upcoming years, mainly driven by the increasing awareness and the potential for ergonomics regulations within the country. This is supported by the increasing number of registered professionals and general interest in ergonomics in key industries (aviation, rail, military, built environments and mining). Furthermore, there is an increased provision of postgraduate courses offered in Ergonomics across South African universities, including Rhodes University and the University of Witwatersrand. Rhodes University is also the only institution that offers some training in ergonomics in undergraduate courses as well.

With respect to ergonomics research, [7] provide some examples of foundational applied ergonomics research in South Africa. This includes research on heat stress in mining in the 1960s, alongside research in the areas of the military, forestry, construction and the automotive industry [7]. More recently, Professor Andrew Thatcher, the immediate ex-President of ESSA, has contributed very meaningfully to the development of regional and global understanding of green ergonomics [8] and the role of HFE in issues of sustainability [9] This has been complimented by the international work performed by firstly Professor Pat Scott and more recently Andrew Todd, who have sat on the IEA's executive committee as the Chair for the international development standing committee. These activities have brought an international perspective to the activities of ESSA and its members. In addition to this, there has been an expansion in applied research and design, most notably in the rail industry, office ergonomics space and in the South African defense forces. There is also a growing interest and practice within the aviation industry, sparked by ESSA's continued partnership with Air Traffic Navigation Services (ATNS) in hosting combined conferences aimed at exploring Human Factors and Aviation safety on the African continent.

While the outlook for ergonomics, as a discipline, is positive for South Africa, there are still some challenges that it and ESSA need to navigate. They relate to a continued poor public understanding of what ergonomics is in South Africa, something that ESSA is attempting to address by having more opportunities for practitioners to attend workshops aimed at building awareness, accompanied by the potential for ergonomics regulations to be promulgated in the coming years. Alongside this, is the need to professionalize of the society further, to ensure that it can endorse ergonomics short courses and training, while being in a position to engage with other tertiary institutions to introduce additional ergonomics education. As noted in James and Scott's [7] history of the society, there is still the challenge of those not qualified to practice ergonomics, doing so. This has reduced due to the increased interactions between the society and other cognate disciplines and societies, but it still needs to be addressed.

2 Current Perspectives and Future Activities

It is evident that each BRICS country has had its own unique history with respect the development of ergonomics, with the extent of ergonomics education, training, research and practice very evidently further developed in Russia, Brazil, China and India when compared to South Africa. What is apparent is that the focus areas of the different countries has been diverse and responsive to the needs of the country at that particular time, and there is a very definite need for the BRICS countries to share and learn from this information to further enhance ergonomics teaching, research and practice. However, a challenge that has arisen out of this is that, particularly in Russia and Brazil, a large portion of educational and research materials are published in either Portuguese or Russian respectively. There is need to gain insights into the different methodologies and approaches adopted in these countries and in the others, to ensure that ergonomics education, training and research in these countries is globalized and cutting edge. In the context of a rapidly changing world of work, due to mega trends such as globalization, technological developments, climate change and demography, it now important that HFE educators, researchers and practitioners have this globalized perspective. This is critical to ensure that ergonomists in BRICS countries are well placed to cope the complexity and uncertainty that these changes may bring to work and life in their countries and other low to middle-income countries. These are some of the reasons for the establishment of the BRICS$_{plus}$ Network of ergonomics societies.

In line with this, and as part of the BRICS collaboration, there is a strong focus by the governments on the education sector. In July of this year (2018) the annual BRICS network university conference will be hosted in Stellenbosch, South Africa with the theme "Unlocking BRICS universities partnerships: Postgraduate opportunities and challenges". From a Human Factors and Ergonomics perspective the BRICS$_{plus}$ network has made excellent progress in understanding the opportunities and challenges for the universities. An outcome of these activities is the BRICS$_{plus}$ HFE University network and a proposed joint educational program that will combine and share the best human factors and ergonomics practices in each country. Furthermore, the joint efforts will also provide a community of support for joint methodology for researchers and practitioners, allowing for the development of scientists at the highest level. In order to achieve these proposed outcomes there are several sub-projects of this particular network:

- Evaluating the actual education programs in ergonomics, determining the needs and demands in each country.
- The development of a BRICS summer school, starting from 2019 in Brazil and Russia
- The development of joint research projects with sharing across laboratories
- The creation of a joint textbook on activity analysis for postgraduate students in BRICS countries. The book will be prepared by a team consisting of academics from all five BRICS countries and published simultaneously in English, Portuguese, Mandarin and Russian.

- The development of a basic module "*The basic principles of Human Factors and Ergonomics*" for use in all five countries, to be given in English by academics from all countries.

3 Concluding Remarks and Outlook

It is evident from the aforementioned activities that the BRICS$_{plus}$ network has managed to achieve a great deal in a very short amount of time. It is particularly encouraging that the network will have its third face-to-face meeting in July 2018 at the conference of the IREA in St. Petersburg, Russia. The future of ergonomics white paper [10] emphasized the importance of creating a demand for high quality human factors and ergonomics research and practice and we believe that the BRICS$_{plus}$ network is creating a solid foundation to achieve this within BRICS countries. This is due to the network taking the mantra of social innovation research seriously; Learn globally, connect regionally, act locally [11].

References

1. Moraes AD, Soares MM (1989) Ergonomia no Brasil e no Mundo: um quadro, uma fotografia. Univerta. ABERGO/ESDI-UERJ, Rio de Janeiro
2. Soares M (2009) Overview of ergonomics in Latin America. In: Scott PA (ed). Ergonomics in developing regions: needs and applications. CRC Press
3. Anokhin AN (2009) Ergonomics in Russia. In: Scott PA (ed). Ergonomics in developing regions: needs and applications. CRC Press
4. Ganguli AK (2009) Growth of Ergonomics in India. In: Scott PA (ed). Ergonomics in developing regions: needs and applications. CRC Press
5. Page K, Ankrum D, Enos L, Fraser M, Jamal S, Meier C, Noy I (2000) Ergonomics in China: perspectives from the 1998 people to people international ergonomics delegation, 17–31 October. In: Proceedings of the human factors and ergonomics society annual meeting, vol 44, no 33, pp 6–89. SAGE Publications, Sage CA, Los Angeles, CA
6. Shen RF, Wei RB, Wang ZM (1990) The first national congress of the chinese ergonomics society. Ergonomics 33(7):979
7. James JP, Scott PA (2009) Ergonomics in South Africa, and beyond the borders. In: Scott PA (ed). Ergonomics in developing regions: needs and applications. CRC Press
8. Thatcher A (2013) Green ergonomics: definition and scope. Ergonomics 56(3):389–398
9. Thatcher A, Yeow PH (2016) A sustainable system of systems approach: a new HFE paradigm. Ergonomics 59(2):167–178
10. Dul J, Bruder R, Buckle P, Carayon P, Falzon P, Marras WS, Wilson JR, van der Doelen B (2012) A strategy for human factors/ergonomics: developing the discipline and profession. Ergonomics 55(4):377–395
11. Greenhalgh T, Robert G, Macfarlane F, Bate P, Kyriakidou O (2004) Diffusion of innovations in service organizations: systematic review and recommendations. Milbank Q 82 (4):581–629

Professional Affairs

Gaining Recognition as an Official Profession at the National Level: The CIEHF Experience

David O'Neill(✉)

Dave O'Neill Associates, Clophill, Bedford MK45 4AB, UK
doneillassoc@yahoo.co.uk

Abstract. Professional bodies should have an outward-looking role to establish the credentials of the profession they represent as well as their more introspective role of developing their discipline and monitoring the competence of their practitioners. The process followed by the CIEHF to gain official Government recognition of ergonomics is described. The focus in this paper is on the 'preliminary procedure' requested by the Privy Council to assess eligibility for petitioning for a Royal Charter. This has been selected as it is regarded as the most informative and instructive part of the whole process for the interest of other professional bodies. The nine questions posed by the Privy Council concerning the uniqueness and value of the scientific discipline and the body's development and performance are listed and the Institute's responses are presented. The other parts of the process – the Petition and governance documents are also briefly discussed. The paper concludes with the key issues that may be useful to other bodies.

Keywords: Accreditation · Legal incorporation · Royal Charter Privy Council

1 Background

Learned societies, professional bodies and certification organisations all have a primary role of being the guardians of their discipline or profession and, as major part of that, ensuring that their practitioners are competent and abide by a code of professional conduct. However, they have a further function, which some might say is equally important, which is to establish (or maintain) their discipline and the professional activities arising therefrom, as a valuable contributor to society and national wellbeing in general. To achieve this more outward-looking function there has to be an authority that makes objective evaluations of such bodies and organisations in order to recognize and approve the contributions made. Such authorities are normally a part of the Government and so have appropriate status and wide jurisdiction.

In this paper I will describe the process that the Chartered Institute of Ergonomics and Human Factors (CIEHF) went through (which I led as the Institute's Chief Executive at the time) to petition for a Royal Charter, the highest level of recognition in the UK. Because of space limitations, I will focus on the nine criteria which had to be satisfied to demonstrate eligibility for making the Petition. This 'preliminary procedure' is probably the most helpful and instructive for transfer to other Societies.

S. Bagnara et al. (Eds.): IEA 2018, AISC 821, pp. 659–666, 2019.
https://doi.org/10.1007/978-3-319-96080-7_79

2 UK Government Authority

The authority empowered to recognize professional organisations in the UK is called the Privy Council and operates at the very heart of Government. Historically, and still in the UK, it is a body of close advisors which gives considered and confidential advice to the Head of State (in UK, the Monarch) on State affairs. In the UK the Privy Council (see www.privycouncil.independent.gov.uk) has existed for more than six centuries and comprises about 700 members, mainly senior politicians from both Houses of Parliament. Many other countries have similar Government or quasi-Government authorities.

Official recognition for a body of people such as a Society or an Institution is the granting of a Royal Charter, which, following a successful Petition by the body, is conferred by the Privy Council on behalf of the Monarch. According to the Privy Council Office (PC Office), Charters are rarely granted these days but, nevertheless, the Office provides guidance for those considering making a Petition. Although each Petition is dealt with on its own merits there are some general criteria that have to be met. These are summarized in the next section.

2.1 Expectations for Chartership

Any professional body or institution, including those representing a scientific (or otherwise) discipline, should ensure that, as a minimum, the four criteria given below are satisfied. However, even if these criteria appear to have been met, it does not mean that a Petition will be successful or that a Charter will automatically be granted.

(a) The institution concerned should comprise members of a unique profession, and should have as members most of the eligible field for membership, without significant overlap with other bodies.
(b) Corporate members of the institution should be qualified to at least first degree level in a relevant discipline.
(c) The institution should be financially sound and able to demonstrate a track record of achievement over a number of years.
(d) Incorporation by Charter is a form of Government regulation as future amendments to the Charter and Byelaws of the body require Privy Council (i.e. Government) approval. There therefore needs to be a convincing case that it would be in the public interest to regulate the body in this way.

Demonstrating that these criteria have been met is clearly quite a demanding exercise so the PC Office advises on a procedure involving some preliminary steps to help decide on the likelihood of success. Following these steps (see next section) also gives an insight into how the Petition may be structured. If this procedure is followed it assists the PC Office in assessing the likelihood of success and, if appropriate, should result in the advice to prepare a Petition.

2.2 Preliminary Procedure

This is a relatively informal approach and, as mentioned above, helps determine whether there is a sufficiently strong case for the (more formal) Petition. To make this early judgment on the chances of Petition being successful the PC Office requests a document which covers the nine factors listed below.

(a) The history of the body concerned.
(b) The body's role.
(c) Details of number of members, grades, management organisation and finance.
(d) The academic and other qualifications required for membership of the various grades.
(e) The body's achievements.
(f) The body's educational role both within its membership and more widely.
(g) An indication of the body's dealings with Government (including details of the Government Department(s) with the main policy interest, or which sponsor(s) the body, together with contact details of officials who deal with the body), and any wider international links.
(h) Evidence of the extent to which the body is pre-eminent in its field and in what respects.
(i) Why it is considered that the body should be accorded Chartered status, the reasons why the grant of a Charter would be regarded as in the public interest and, in particular, what is the case for bringing the body under Government control as described above.

Although this is described as a preliminary, it may well be that this whole procedure takes more time than the drafting and submission of the Petition. The information provided through this informal approach allows the PC Office to make its enquiries to validate the statements and claims provided.

In parallel with the preparation of this 'informal approach', the professional body's rules or constitution must be brought into line with how a Chartered body will operate. This will usually require some amendments to the existing statutes and probably some re-structuring to conform with the particular requirements of this form of Government regulation. Most important of these are the draft Charter and Byelaws, which should be submitted with the 'informal approach'. The PC Office can then raise queries and provide feedback on both of these submissions. Generally speaking, more work is required to be done on them.

When the PC Office is satisfied that there is a strong case for the granting of a Charter and that all the governance documents (i.e. Charter, Byelaws etc.) the professional body is advised to move on to drafting the formal Petition. The time taken for the PC Office's considerations to be concluded can vary from a few weeks to many months as it will usually depend on receiving responses from Government Departments on information given in the 'informal approach'.

Because the process of Petitioning for a Charter is in the public domain, it is open to public scrutiny. Therefore, whilst writing the 'informal approach' and drafting the governance documents, professional bodies are encouraged to make their intention to make a Petition known to other bodies and individuals who may have an interest, in order to minimise the risk of an objection or counter-petition. Any Petition which appears to be controversial is unlikely to succeed.

3 The 'Informal Approach' Submitted by the CIEHF

This section describes the information presented to the PC Office to show how the nine factors in Sect. 2.2 justify making a Petition for a Royal Charter.

3.1 The History of the Body Concerned

The emergence of ergonomics as a scientific discipline in the 1940s, mainly due to military issues of using more complex equipment was described. This led to research by senior psychologists and physiologists and the growing appreciation of the importance of the interactions between people, their equipment and their environments. Then, in 1949, at a meeting of distinguished physiologists and psychologists at The Admiralty, the term ergonomics was coined and, subsequently in that year, this same body of scientists, together with some like-minded colleagues formed the Ergonomics Research Society (ERS). This was the first such professional body in the world. Since then, the ERS has evolved through the Ergonomics Society (ES) to the Institute of Ergonomics and Human Factors (at the time of the Petition). In 1985 the ES became a Company limited by guarantee and also a Registered Charity. Some of the possible confusion or misconception over the terms 'ergonomics' and 'human factors' was also explained. Although ergonomics has its roots in the underlying scientific disciplines of physiology, anatomy, psychology and engineering, the discipline that is now established is characterised by the integration of all these areas of knowledge. The main subdivisions of the discipline are now generally regarded as Physical Ergonomics, Cognitive Ergonomics and Organisational Ergonomics.

3.2 The Body's Role

The principal role is to foster the evolution of ergonomics and in so doing to uphold standards of competence within the discipline and its practitioners. The Charitable Objects from the current Memorandum of Association (legal document stating the purpose of the Institute's existence) and to be transferred across to the (draft) Charter (an equivalent legal document).

The Institute upholds the principles that design should serve to complement the strengths and abilities of people and minimise the effects of their limitations. In so doing it is necessary to understand and design for the variability represented in the population, spanning such attributes as age, size, strength, cognitive ability, prior experience, cultural expectations and goals. Qualified ergonomists are the only recognised professionals to be competent in optimising **performance, safety and comfort**. The Institute is the only body in the UK managing and representing these combined factors.

3.3 Members, Grades, Management Organisation and Finance

The breakdown of the 1600 members, across the various grades – Student, Graduate, Associate, Technical, Registered Member, Fellow and Honorary Fellow (plus those with Retired status) -, was given. Only those in the grades of Registered Member and Fellow have been accredited by the Institute as professional ergonomists or human factors practitioners with a full understanding of all the underlying disciplines and their interactions. In this respect, most of the competent professionals in the UK are Registered Members or Fellows.

The organisational structure is specified in the Memorandum and Articles of Association (now Charter and Byelaws), which state that the Institute is managed by a Council elected by the membership. The Council members are also Trustees of the Charity. Day-to-day administration is carried out by employed staff who are managed by the Chief Executive on behalf of the Council. Professional conduct and accreditation matters are delegated to the Professional Affairs Board (PAB).

The Institute is of sound financial standing as annual audits and returns to Companies House and the Charity Commission show. The Institute holds reserves (cash and equities) in the order of one year's operating costs.

3.4 Academic and Other Requirements for Membership of the Various Grades

For all grades except Associate member, education equivalent to university first degree level is the usual minimum requirement to ensure academic competence. Ergonomics education in the UK is mostly gained from a Master's course following on from a first degree in a cognate subject. For election to Registered Member or Fellow, individuals have to provide confirmation of sufficient knowledge across all the underlying disciplines combined with evidence that they have **integrated** and successfully applied their ergonomics knowledge in a professional capacity. The evidence is usually submitted to the Institute via a log book, mentor's report and supporting referee reports which are then assessed by peer-review under the auspices of the PAB. For Fellows, the knowledge requirements are the same as for Registered Member, but evidence has to be provided some senior professional responsibility.

Progression from any of the lower grades to Registered Member is possible on presentation of satisfactory evidence.

Members all grades of membership except Associate (and those with Retired status) must commit to Continuing Professional Development (CPD).

3.5 The Body's Achievements

Of the Institute's many achievements over its 60+ year history the following five examples were selected for inclusion in the 'informal approach'.

(i) The Professional Registers to endorse individuals and businesses as being adequately qualified and experienced to practise within their stated areas of competence.
(ii) The learned journals founded by the Institute (see Table 1)

Table 1. Journals founded by the institute.

Journal	Publisher	First published
Ergonomics	Taylor & Francis	1957
Applied ergonomics	Elsevier	1970
Behaviour and information technology	Taylor & Francis	1982
Work and stress	Taylor & Francis	1987
Theoretical issues in ergonomics science	Taylor & Francis	2000

(iii) The establishment of annual scientific meetings. The first scientific meeting of the founders was on 21 April 1950 and the meetings have continued on an annual basis. Currently, the Annual Conference spans three or four days with up to four parallel sessions of scientific presentations to accommodate the number of submissions considered to be of appropriate scientific content and standard. We also organise regular conferences on specific topics within ergonomics, often in conjunction with our Special Interest Groups (SIGs). These have undoubtedly helped to promote knowledge and awareness of the discipline in the key industrial sectors. All scientific meetings have comprehensive publications which provide both valuable educational material and a record of the development of the discipline.

(iv) The Institute has staged events to engage members of the public and introduce them to ergonomics/human factors. The fiftieth anniversary of the Institute (1999) was celebrated by an exhibition at the Science Museum in London entitled 'The Human Factor'. This was an opportunity to make the essentials and aims of ergonomics and human factors more accessible to the public via attractive interactive displays. This exhibition proved to be so popular that it has subsequently been hosted by other museums in the UK (and around the world). For the Institute's sixtieth anniversary (2009) the Design Museum (London) hosted a further exhibition "Ergonomics – Real Design' which was also a very popular attraction with over 10,000 visitors a month. (Images from both these events can be found by searching on 'Google'.)

(v) The Institute supports and is represented on BSI and ISO Standards Committees etc. that have been set up to deal with ergonomics/human factors issues.

3.6 Educational Roles

Educational activities play a prominent part in the Institute's activities, ranging from education publications to collaboration with Universities to promote and disseminate ergonomics internally and with schools. The Institute is the national body for accrediting UK ergonomics courses (undergraduate and postgraduate degree courses, certificated courses, degree modules and short courses). The Scientific Meetings (3.5 iii) also perform an educational role, particularly for members' CPD. Major involvement with Centre for the Registration of European Ergonomists (CREE – see 3.7) also furthers the educational function.

3.7 The Body's Dealings with Government and Any Wider International Links

The Institute responds to Government consultations and reviews and has provided critical input from Registered Members on Government committees and select panels.

The diversity of ergonomics/human factors and its very wide field of application make it relevant to most Government Departments. Four Departments were selected for reference in the 'preliminary procedure', as summarised in Table 2.

Table 2. Ergonomics activities in government departments.

Department	Sector	Involvement
Department for Work and Pensions (DWP)	Health and Safety, HSE, HSL High hazard industries Risk assessment Manual handling, DSE Work capacity/incapacity	"Working for a Healthier Tomorrow", "Common Sense – Common Safety"/ OSHCR, Unique optimization of productivity and wellbeing.
Department for Transport (DfT)	Aviation Land transport Maritime transport	Air traffic control, cockpit design, Vehicle design, railway and road safety, Complex transport systems.
Ministry of Defence (MoD)	Human factors integration (HFI) Human capability Security	Cognitively complex and physically demanding military systems, Training and simulation.
Department of Health (DoH)	Patient safety Healthcare systems Clinical human factors Occupational health	Errors, equipment usability, Organisational factors, system failure Patient handling Staff wellbeing, workload

The Institute has made major contributions to the establishment and then running of international bodies (IEA, FEES, CREE) and, because of its sound reputation has attracted a substantial number of overseas ergonomists into its membership (c. 10%).

3.8 The Extent to Which the Body is Pre-eminent in Its Field

When the ERS was founded in 1949, we were the first professional body anywhere in the world committed to the development of this new scientific discipline. Our founding members were the first to recognise and define the discipline and subsequently the profession of ergonomics. Since then, many international (see above) and other national bodies have started up across the world.

The Institute is the custodian of ergonomics and human factors in the UK and speaks on behalf of the discipline and its professionals to cognate organisations. As the only professional organisation for the furtherance of ergonomics in the UK, the

Institute is responsible for the national accreditation of ergonomics education and training courses as well as the practitioners.

3.9 Why It Is Considered that the Body Should Be Accorded Chartered Status

The discipline and profession of ergonomics/human factors has developed from being a mutual interdisciplinary interest in human performance of its eminent and predominantly academic founders to a learned society for advancing the knowledge and application of ergonomics.

Ergonomics is now recognised as the only subject with an academic base that focuses jointly on system performance and human wellbeing. Good system performance, except in the rare situations where there are no human components in the system, requires good ergonomics inputs at all stages from conception and design to commissioning and operation. Ergonomics is being increasingly recognised as central to the successful accommodation of changes in the way we live and work, whether imposed by new technology, demographic changes or other factors.

Ergonomics is not a regulated profession so there is no restriction on those who may profess to offer or sell ergonomics services. Chartership of a body and its professionally competent members is well recognised in the UK (and many other countries) as the gold standard for professional practice. With Chartership the Institute would be better able to effectively police the profession and thereby protect the wider public from incompetent practitioners or charlatans.

4 Transferability to Other Societies or Countries

The key issues and points for you to 'take home' are given below.

- Identify your national body responsible for recognition/accreditation.
- Read and fully understand their criteria for recognition/accreditation.
- Initiate some personal interactions with the body to ensure your understanding is correct before you start the application process.
- Produce brief, carefully worded documents to show how you meet the criteria.
- Ask for feedback at appropriate stages during the application process.

Acknowledgements. Although I led the process of Petitioning for the Royal Charter I would like to acknowledge the outstanding support and assistance I received from some senior ergonomists and Institute Trustees. Their knowledge, wisdom and patience were key factors in our success. I am also very grateful for the help and advice given by staff at the PC Office.

The IEA Certification and Endorsement System

Maggie Graf(✉)

Chair, IEA PS&E Certification Sub-Committee, Meierskappel, Switzerland
dir@iea.cc

Abstract. The IEA wishes to actively encourage the certification of professional ergonomists and human factors (HFE) specialists, as it enhances the visibility and status of the profession, encourages a high standard of education, harmonization of educational programs and continual learning by certified people. Additionally it serves as a quality guarantee for employers and a protection from poorly qualified people. However, comparatively few of the member societies of the IEA have certification systems and, even where such systems exist, many qualified HFE specialists choose not to be certified. This paper describes the IEA recommendations for certification of HFE specialists and the IEA requirements for the endorsement of certification systems. Further information can be found on the IEA website [1].

Keywords: Certification · Endorsement · Ergonomics education

1 Definition of Terms

1.1 Certification

The term certification refers to the accreditation of a specific person following a peer review procedure where their qualifications and experience are assessed according to agreed criteria. Some people use the term accreditation. For the IEA it means the same as certification.

Generally, certification criteria are set by a national or regional ergonomics society and the task of assessing candidates is delegated to a group of well-regarded and experienced HFE experts from the society. Some federated societies of the IEA have admission criteria for membership of the society, whereas others operate as an interest group open to anyone who works in HFE or is interested in the field. Where there are admission criteria, different levels of membership may be offered.

The IEA does not prescribe any particular system, however, it is advised that where admission criteria are used, at least one of the membership levels should be a "professional" level, which can only be granted to people with at least three years of university level education, including substantial HFE content, and at least two years of professional experience.

S. Bagnara et al. (Eds.): IEA 2018, AISC 821, pp. 667–671, 2019.
https://doi.org/10.1007/978-3-319-96080-7_80

1.2 Endorsement

The term endorsement refers to the official recognition of a system that oversees the certification of people. The IEA does not endorse people, but rather it endorses the national or regional boards that are responsible for certifying HFE specialists.

2 IEA Recommendations for Certification of HFE Specialists

2.1 History and Aims

The first recommendations for the certification of HFE specialists where developed around the turn of this century. The main aim was to try to protect the profession from less qualified people who could damage the reputation of the profession. Additionally, it was hoped that the recommendations would encourage the harmonization of training programs and promote professional awareness.

Experience in the countries and areas with long-established certification systems has shown that certification has other advantages. It increases the visibility and acceptance of the profession by others, particularly government agencies and other professions. It is welcomed by large companies, as it makes their recruitment easier and more transparent. It encourages practitioners to keep up to date with scientific and technical developments, as most certification systems require some form of continuous professional development.

2.2 Process Requirements

The recommendations of the IEA are based on the International Standard "Conformity assessment – General requirements for bodies operating certification of persons (ISO/IEC 17024) [2]".

The expected standards for each level of HFE practice, for which a professional designation will be granted, must be clearly defined. Any candidate who is able to show evidence that they have met the standard must be accepted at the appropriate level. For the evaluation, measures may include education and training documentation, referee/mentor reports, examples of work (publications or project descriptions) or an examination. In order to assess competencies, it is strongly recommended that a range of several measures is used.

Evaluation of each candidate must be by at least two unbiased, highly qualified and respected assessors. The IEA recommends that when a certification system is being established, the society or societies contact other certification boards to ask for examples of assessment documents. If there are society members who are already certified by an IEA endorsed system, these people would be very valuable as the first assessors. If not, the IEA can assist with the assessment of the first assessors by requesting an IEA endorsed system to assess the first candidates. If the competence of these HFE experts is assessed as equivalent to the professional ergonomists that they generally certify, this will give your new assessment body credibility and it will assist with later IEA endorsement, if this is required.

The IEA does not make any recommendations in terms of the cost of certification but it does recommend that the assessment board is independent of any specific interest group, particularly training course providers and the society itself: Society membership should not be a requirement for certification.

2.3 Minimum Requirements for Certification

The IEA recommends that all certification systems should have a "professional" level that requires at least three years of university level education in HFE or a related discipline (e.g. engineering, psychology, biomechanics, etc.). For this level, specific education in ergonomics relevant content should total at least one full year of academic education. Consider making a number of routes possible, such that people without a formal education in ergonomics are not excluded. The candidates should also be able to demonstrate that they have had at least two years of unsupervised professional practice. A period of supervised training of one year is considered ideal, but the IEA recognizes that this is not always possible in practice. The outcome of the education and experience should be that the candidate can demonstrate a high level of competence in at least one of the main areas of HFE – physical, cognitive or organizational – and shows knowledge of the others, even if his or her work does not generally involve them.

It is possible to offer other levels of certification, either lower "technical" levels, or higher levels. The IEA strongly advises against establishing a system that only consists of lower level certification, as it may prove to be extremely difficult to introduce higher levels at a later stage: Society members will very likely resent being "downgraded" by the addition of "better" levels. Even if there is nobody suitably qualified for the professional level at the time of establishment, it gives a goal for your members to aim for.

2.4 Code of Conduct and Handling of Disputes

There should be some review committee and process established, such that disputes are handled adequately. The people on the review committee do not need to be HFE experts but they should be well-respected by all parties.

All candidates should be required to sign a Code of Conduct based on the IEA model (available on the IEA website) as part of their application for certification. There should also be a procedure in place for handling any reported cases of professional miss-conduct. This should include the possibility of losing certification in extreme cases.

2.5 Recertification and Ongoing Professional Development

The IEA strongly recommends limiting certification to a specific period. Most IEA endorsed societies limit certification to five years. Recertification should be dependent on evidence of ongoing professional development. The field of HFE is growing all the time and practitioners should be aware of new findings. Additionally, the exchange of experiences and knowledge provided by participation at congresses and society meetings is very valuable to maintaining the standard of the profession and should be encouraged in this way.

3 IEA Requirements for Endorsement of a Certifying System

3.1 History and Aims

The IEA first commenced endorsing certification systems in 2001 with newly developed criteria and procedures. The aim for the IEA was to try to harmonize the requirements for certification across the globe and encourage the mutual recognition between these systems so that qualified professionals could move between countries. The Board of Certification of Professional Ergonomists (BCPE) in the USA was the first, closely followed by the Japan Ergonomics Society and the New Zealand Ergonomics Society.

3.2 Criteria for IEA Endorsement

General Features. A certification body should be national or international in scope and be endorsed by a federated member or group of members. Only one certification body will be recognized by the IEA to represent a federated member, so it is important that the certifying body has the mandate of the local federated member society to perform the certification. The certification body must be independent of any educational body and not be operated for commercial profit, however, it must be financially viable. The rules of the organization must be written and must contain information about the legal status of the body and describe the appointment of certifying personnel, their responsibilities and reporting structure. All procedures and criteria should be documented and open to anyone wishing to become certified. Information about candidates and, if appropriate, examination results, must be kept confidential and a procedure for complaints or settling disputes must be defined.

Certification Personnel. The society or group of societies must have a pool of potential assessors for running the certification board. The people on the certification board should be knowledgeable about the area or areas of ergonomics that they are to assess and they should be certified themselves. The IEA strongly recommends that the original members are assessed by an independent IEA endorsed system and will assist with this, if necessary.

Rules About Applicants. Certification must not depend on whether the person is a member of any society and must be non-discriminatory in terms of gender, age, ethnicity, religion or physical status. The certification criteria must be based upon the IEA Core Competencies and must require the signing of a Code of Conduct. Certification must be for a limited period of time and re-certification must include an assessment of continuing professional development to ensure that the candidate is keeping up to date.

Records and Accounts. The certifying body must maintain an up to date register of all ergonomists who have been certified and the register of currently certified people must be publicly accessible, except where an ergonomist has explicitly requested non-disclosure. Records should be kept of the outcome of all applications.

4 Further Information

This article does not include all of the requirements for certification or IEA endorsement of a certification system. People who are interested in obtaining information that is more detailed should consult the documents on these topics produced by the IEA and available on the IEA website. They are to be found under the section "Resources", tab "Projects". There are twelve documents on the following questions:

- What are the knowledge requirements of a professional ergonomist (Core Competencies)
- What does a certification system need? (IEA recommendations)
- How do I get a certification system endorsed by the IEA? (Criteria and application forms)
- What does the IEA do with endorsement applications? (The procedure that the IEA uses to assess compatibility with the IEA criteria)

Additionally, the members of the IEA Board, particularly the Professional Standards and Education Chair and the Sub-committee chair for Certification are all very willing to offer assistance. Your problems are probably not unique and we can explain how other countries and regions have overcome most issues that get in the way of professional certification. The other presentations in this session have addressed some of these.

References

1. IEA Website, Resources, Projects. http://www.iea.cc. Accessed 27 Apr 2018
2. ISO/IEC 17024, Conformity assessment – General requirements for bodies operating certification of persons (2012)

Revising the IEA Core Competencies for Professional Ergonomists

Frederick Tey[1] and Maggie Graf[2(✉)]

[1] Chair, IEA Professional Standards and Education Standing Committee, Singapore, Singapore
[2] Chair, IEA Certification Sub-Committee, Geneva, Switzerland
dir@iea.cc

Abstract. This symposium presents the work that has been done by the members of the current IEA executive and particularly the members of the Professional Standards and Education Standing Committee to update the IEA Core Competencies in Ergonomics that were developed in 2001. The original documents can be found on the IEA website under Resources [1]. There is an introductory document, a full version which includes units, elements and performance criteria, and a summary version. In the symposium the proposed revision will be described and feedback from interested educators is desired. The revision aims to simplify the document, include newer developments in the profession according to the recommendations of the IEA strategy paper on the future of ergonomics and remove outdated elements.

Keywords: Education · Competencies · Professional standards

1 Background

1.1 Reasons for the Revision

Around 17 years ago the IEA produced a set of documents to describe the competencies that a professional ergonomists should have. These competencies result from education and experience and were grouped into nine areas (called "units"). The units were further described in sub-categories called "elements" and each element had performance criteria. The core competencies are intended for use by educational institutes to design courses for ergonomists. They are also used by certification bodies to define criteria for the certification of individuals. They assist in promoting and describing the profession to others and support the continuous development of people working in the profession. The IEA core competencies are therefore a central and important aspect of the IEA work.

Since their first development, the world of ergonomics and human factors has changed and people are more specialised within the profession. This has led to challenges and ultimately to changes in the certification practices of the IEA endorsed certification bodies, such that they all no longer follow the core competence recommendations but have each made their own adaptations, mainly to allow for more specialisation in practice. The requirements are also viewed as too restrictive by some societies that wish to set up certification systems. These societies often see the same or

S. Bagnara et al. (Eds.): IEA 2018, AISC 821, pp. 672–677, 2019.
https://doi.org/10.1007/978-3-319-96080-7_81

similar challenges as the older societies. Additionally the IEA groups that have worked on recommendations for the future of ergonomics infer that ergonomics and human factors specialists have some competencies that are not well described in the older core competency documents, particularly in relation to management competencies, and some of the content seems out of date. For these reasons, the members of the PS&E recommended to the IEA in September 2017 that the Core Competencies get reviewed and the proposal was accepted. A working group was suggested.

1.2 The Revision Process

During 2017 and the beginning of 2018 an extensive consultation process was undertaken, primarily of the various certification systems known to the IEA. A preliminary draft revision was presented to the IEA Board at the Board meeting in March 2018 and then sent to the members of the Certification Sub-Committee and a selection of leading educators from different parts of the world for comment. Comments from this consultation will be incorporated into the document before the IEA Conference in 2018. This symposium gives the opportunity for more public discussion. It is planned that this last input will permit the finalisation of the revision such that the incoming IEA Board can start with up-to-date core competency criteria.

2 Summary of Proposed Changes

One aim of the project leader was to reduce the complexity of the documents, both in terms of length and language, as non-native English speakers have some difficulty with the terminology and a quick overview should be made possible. A further aim was to remove redundancies and improve the internal coherence. The aim was not to delete content, but rather to consolidate the contents and make the documents more readable. Less may be more in this case. The "Elements" were rearranged into eight instead of nine Units. Elements from the former Unit 1 were rearranged and partially moved to other units (see Table 1). Some new elements were added to introduce recommendations from the Future of Ergonomic paper and other sources. The language was also reviewed such that it does not refer to workplaces alone, as HFE has applications in other life spheres.

The proposed new Unit 1 describes "Foundation knowledge". This term is used to describe knowledge, not specific to HFE, that should be acquired, either at the beginning of the HFE training or before the training begins. This knowledge is assumed before the fundamentals of HFE practice are developed in the remaining Units. The Foundation knowledge can be adapted according to the domain of specialisation that is being offered by course leaders (physical, cognitive or organisational ergonomics). For example, if the course aims to produced specialists in physical ergonomics, then the foundation disciplines of anatomy and physiology, biomechanics and anthropometry (among others) will be emphasised. However, a basic knowledge of all the elements described should be assured, such that all professional ergonomists are at least able to recognise issues in other domains and develop a holistic approach to solutions.

Table 1. Comparison of current and proposed competency units.

Current core competency units (order slightly changed)	Proposed core competency units	Short name
1. Investigates and analyses the demands for ergonomic design to ensure appropriate interaction between work, product and environment, and human capabilities and limitations	1. Has a basic understanding of the sciences necessary for an ergonomic assessment. This knowledge must be sufficiently broad to conduct assessments that include physical, cognitive and organisational aspects	Foundation knowledge and skills
2. Analyses and interprets findings of ergonomic investigations	2. Conducts ergonomic assessments using appropriate methods for measurement and analysis	Measurement skills
3. Determines the compatibility of human capacity and planned or existing demands	3. Determines the compatibility of human capabilities to planned or existing demands using a systems approach	Evaluation skills
4. Makes appropriate recommendations for ergonomic design or intervention	4. Makes appropriate recommendations for ergonomic (re-)design or intervention	Recommendation skills
5. Develops a plan for ergonomic design or intervention	5. Develops a plan for ergonomic (re-)design or intervention in collaboration with the appropriate people, using a systems approach where relevant	Collaboration and planning skills
6. Implements recommendations to optimise human performance	6. Implements recommendations to optimise human well-being and performance	Implementation skills
7. Evaluates outcome of implementing ergonomic recommendations	7. Evaluates outcome of implementing ergonomic interventions appropriately	Scientific skills
8. Documents ergonomic findings appropriately	8. Documents ergonomic findings, records interventions appropriately and demonstrates professional behaviour, in accordance with ethical principles	Professional behaviour
9. Demonstrates professional behaviour		

The first competence unit of the original Core Competencies was very generally formulated and corresponds closely to the definition of an ergonomist, according to the IEA. In the proposed version, this Unit is limited to skills necessary to recognise the important ergonomic and human factors of a situation and understand the risks, without including any detailed analysis or intervention aspects.

Several Units allow adaptation to local needs and some degree of specialisation. For example, specialists in cognitive factors would have expertise in different measurement and analysis tools than specialists in organisational factors. The depth of knowledge in Unit 7 – Scientific skills - can also be adapted to local needs, depending on whether the specialists being trained will work more in science or as consultants in practice.

The original core competencies refer to ergonomists and ergonomics, without mention of human factors. Although there is no consensus within the IEA concerning these terms, in the proposed documents the term HFE expert will be used to express a holistic view of the profession. It is clear that this was the intention of the original core competency authors.

3 Proposed Order of Elements

The elements within the Units have been rearranged and duplications removed. The language has been changed in some elements to make the meaning clearer or to modernise the element. The elements in Table 2 correspond to the proposal following review by the members of the IEA executive. This proposal may be altered according to feedback from the consultations, which take place before the IEA Congress. One new element has been added (4.3 *in italics*).

Table 2. Proposed competency units with elements.

Unit 1. Foundation knowledge and skills
1.1 Understands theoretical concepts and principles of physical and biological sciences relevant to ergonomics and human factors (HFE)
1.2 Understands theoretical concepts and principles of social, behavioural and emotional science relevant to HFE
1.3 Understands basic engineering concepts, with a focus on systems theory and design
1.4 Understands and can apply the basics of experimental design and statistics
1.5 Understands basic management concepts relevant to HFE
1.6 Understands the pathology relating to environmentally or occupationally generated disorders
1.7 Understands the requirements for safety, principles of industrial safety, the concepts of risk, risk assessment and risk management
Unit 2. Measurement skills
2.1 Appropriately identifies factors influencing health and human performance in a variety of contexts
2.2 Demonstrates an understanding of methods of measurement relevant to ergonomic appraisal
2.3 Analyses current guidelines, standards and legislation, regarding aspects influencing the activity
2.4 Applies a systems approach to the analysis
Unit 3. Evaluation skills
3.1 Evaluates products or work situations in relation to user requirements and expectations
3.2 Appreciates the extent of human variability influencing design
3.3 Identifies potential or existing high risk areas and high risk tasks
3.4 Makes justifiable decisions regarding relevant criteria that would influence a new design or a solution to a specified problem

(*continued*)

Table 2. (*continued*)

Unit 4. Recommendation skills
4.1 Understands the hierarchies of control systems and design methodology for systems development
4.2 Outlines appropriate recommendations for ergonomic design of organisational management, human environment, tools or products
4.3 Understands the importance of a participatory approach to designing solutions
4.3 Understands the role of education and training in relation to ergonomic principles and can develop appropriate programs
4.4 Understands the role of appropriate personnel selection in HFE and can make appropriate recommendations
Unit 5. Collaboration and planning skills
5.1 Adopts a holistic view of HFE in developing solutions
5.2 Considers alternatives for optimisation of the ergonomic quality
5.3 Develops a balanced plan for risk control
5.4 Develops strategies to introduce a new design to achieve a healthy and safe human environment
5.5 Communicates effectively with the client and professional colleagues
5.6 Incorporates approaches which would improve quality of life where appropriate
Unit 6. Implementation skills
6.1 Manages change efficiently and effectively
6.2 Relates effectively to clients at all levels of personnel
Unit 7. Scientific skills
7.1 Effectively evaluates the results of ergonomic design or intervention
7.2 Can carry out evaluative research relevant to HFE
7.3 Remains prepared to modify solutions in accordance with results of evaluation, where appropriate
Unit 8. Professional behaviour
8.1 Provides understandable feedback to the client on findings and recommendations appropriate to the project or problem
8.2 Shows a commitment to ethical practice and high standards of performance and acts in accordance with legal requirements
8.3 Recognises personal and professional strengths and limitations and acknowledges the abilities of others
8.4 Demonstrates life-long learning, to ensure that HFE knowledge and skills are up-to-date
8.5 Has a clear concept of professional identity and recognises the impact of HFE on peoples' lives

4 Performance Criteria

Changes that are more substantial have been proposed for the performance criteria. These will be presented in detail during the symposium at the IEA Conference. The main changes relate to the introduction of criteria related to systems theory and the system approach, activity analysis, more management skills, modelling (to replace computer skills), understanding emergence, greater emphasis on participation and more details on professional behaviour skills.

5 Level of Competence Required

The IEA does not wish to be too prescriptive about the details of educational programs or the evaluation of specific competency elements. Competencies are the result of both education and experience, however, it is the view of leading educators that the educational components necessary to develop the core competencies described in Units 2-8 could be adequately covered in one year of full-time post-graduate training, or the equivalent in part-time or modular training programs.

Figure 1 shows how the level of competency within each unit can vary according to the specialty domain. An awareness is necessary for all elements within a unit.

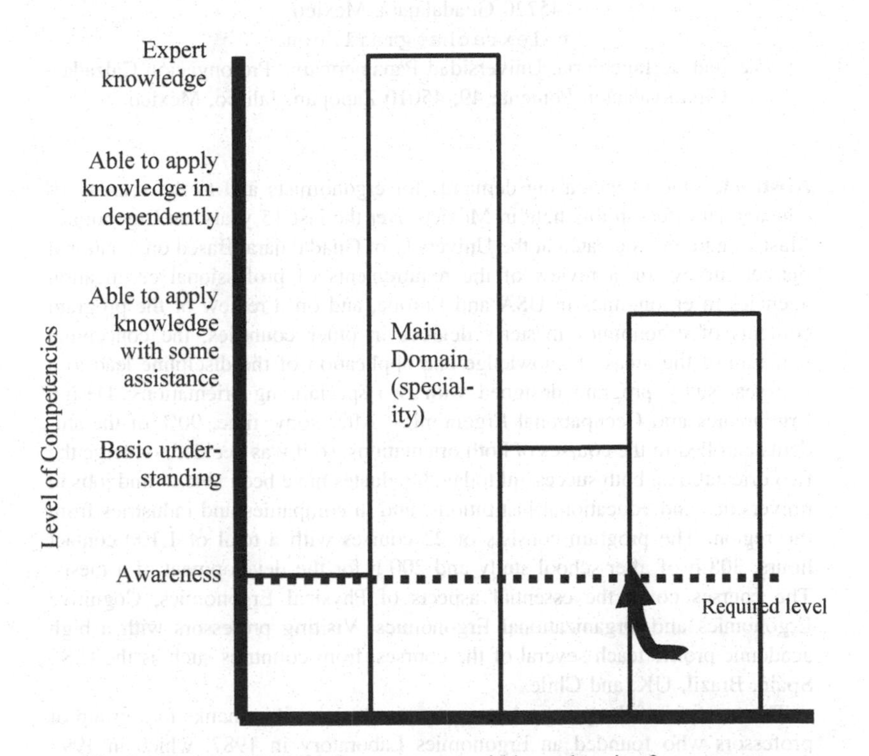

Fig. 1. Required levels of competence within each unit of competence.

Reference

1. IEA Website Resources. https://iea.cc/project/index.html. Accessed 01 May 2018

Experiences in the Development and Implementation of an Academic Master Degree in Ergonomics in Mexico

Rosalío Avila-Chaurand[1(✉)], Lilia Prado-León[1], Elvia González-Muñoz[1], Irma Cecilia Landa-Avila[2], and Sergio Valenzuela-Gómez[1]

[1] Maestria en Ergonomia, Universidad de Guadalajara, 45220 Guadalajara, Mexico
rolexracha@gmail.com

[2] Facultad de Ingeniería, Universidad Panamericana, Prolongación Calzada Circunvalación Poniente 49, 45010 Zapopan, Jalisco, Mexico

Abstract. Due to increasing demands for ergonomists and to the scarcity of educational offers in this field in Mexico over the last 15 years, an Ergonomics Master degree was created at the University of Guadalajara. Based on a national market survey, on a review of the requirements of professional certification agencies in ergonomics in USA and Europe, and on a review of the program contents of ergonomics master's degrees in other countries, the conceptual structure of the areas of knowledge and application of the discipline lead to a two-year study program designed with two specializing orientations: Design Ergonomics and Occupational Ergonomics. After some time, 90% of the students enrolled in the courses of both orientations, so it was decided to merge the two orientations, both successful today. Graduates have been able to find jobs in universities and educational institutions, and in companies and industries from the region. The program consists of 22 courses with a total of 1,100 contact hours, 308 h of after-school study and 200 h for the development of a thesis. The courses cover the essential aspects of Physical Ergonomics, Cognitive Ergonomics and Organizational Ergonomics. Visiting professors with a high academic profile teach several of the courses, from countries such as the U.S., Spain, Brazil, UK, and Chile.

The creation of the postgraduate program was possible thanks to a group of professors who founded an Ergonomics Laboratory in 1987, which in 1997 became a Research Center. They began a self-training process through courses taught by foreign professors. They started to do research and also teaching processes at the undergraduate and postgraduate levels. Thus, they acquired the necessary knowledge and skills design and implement a postgraduate program. The attendance of these professors in national and international ergonomics congresses enabled the contact with other institutions and professors from foreign universities that show a very high availability to participate and actively collaborate in the process.

Even with ups and downs the training process was followed: one of the professors could undertake his Doctorate in ergonomics at the University of Loughborough and one of our first graduates is also starting there her PhD.

S. Bagnara et al. (Eds.): IEA 2018, AISC 821, pp. 678–692, 2019.
https://doi.org/10.1007/978-3-319-96080-7_82

We are aware of IEA's efforts in favor of ergonomists training, but we would have appreciated greater support to incorporate first level visiting professors in diverse areas. The postgraduate program has been maintained for 4 generations. We hope to start a doctorate in collaboration with other universities in the world, and that the political changes coming in our country will favor the arrival of honest officials who understand the importance of supporting training processes in all scientific areas.

Keywords: Ergonomics · Professional education · Postgraduate degree Mexico

1 Program Context

In the past 15 years in our country, the need for knowledge, application, transmission and dissemination of ergonomics in their different fields and areas of study, has spread from transnational and national corporations' needs to implement programs and ergonomic solutions in processes and products, to undergraduate degrees in Industrial Design, Graphic Design, Interior Design, Architecture, Industrial Engineering, Mechanical Engineering, Mechatronics, Psychology, and Medicine, as well as various graduate degrees such as the Masters in Occupational Health, Masters in Engineering and Masters in Design and Product Development; where both companies and public and private institutions have suffered from the lack of professionals in the field of ergonomics, to be supplied by professionals who have only very basic training in self-supported short courses, whose breadth and strength is far from real training. Based on this situation and in order to meet the demand of ergonomics professionals in our country, we propose the implementation of a graduate program with two specialization options.

1.1 Objectives

To design and implement a Graduate Program in Ergonomics at Master of Science-level degree, to contribute to the training of ergonomists with the knowledge, skills, values and attitudes that enable them to solve problems with a high level of efficiency and effectiveness in the fields of Occupational and Design Ergonomics in Mexico.

The design, content and implementation of the program will address the basic needs of vocational training in ergonomics to meet the demands of the industrial, business and educational institutions in our country, preparing students to solve specific problems and develop an internationally high standard of performance.

1.2 Methodology

Curriculum Design methodology included the following steps:

1. Building a theoretical curriculum model based on reviewing curriculum design theories from the most important authors in the field.
2. Developing a Basic Ergonomics Conceptual Framework.
3. A review of ergonomics accreditation standards in various parts of the world.

4. A study of the labor market nationally.
5. The review of graduate curricula in Ergonomics from around the world.

Building a Theoretical Curriculum Model

The first stage took into account such recognized theoretical guidelines for curriculum design as Taba (1993), Tyler (1973) Johnson (1967), Sarramona (1987), Lafourcade (1969), Posner (1979), Furlan (1979), Coll (1991), Arnaz (1995) and Bruner (1971), to establish a Curricular Model that would direct further actions. The model arose from an analysis of social needs as they determine demand for professionals in the field of ergonomics (Fig. 1).

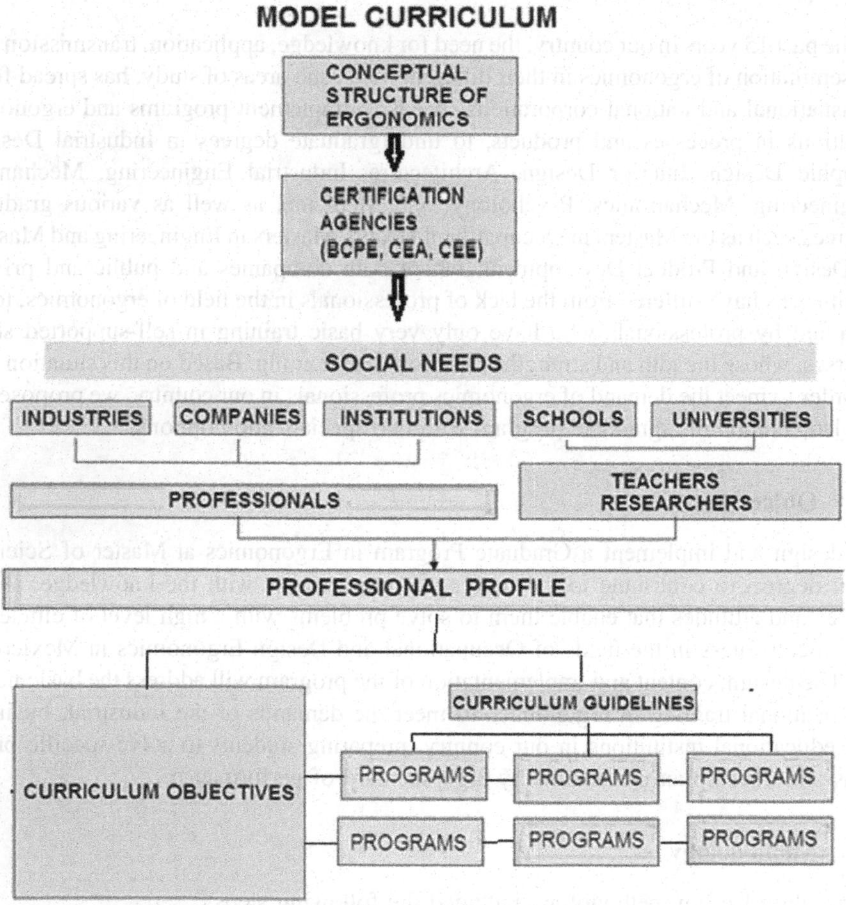

Fig. 1. Curriculum model

Developing a Conceptual Framework - Basic Ergonomics

Professional training processes encounter many of their methodological bases in the body of knowledge from various disciplines they draw upon to solve the problems they address. Integrated knowledge of their Conceptual Framework is thus essential. According to specialists in curricular design, a conceptual framework is formed by the series of principles, theories, hypotheses, laws and concepts at various levels that actual practitioners of a science produce in the course of their professional activity. This aggregate knowledge gives way to fields, areas and specialized groups within which professional work takes place, making way for specialized methods and techniques. Professional training processes encounter many of their methodological bases in the body of knowledge from various disciplines they draw upon to solve the problems they address. Integrated knowledge of their Conceptual Framework is thus essential. According to specialists in curricular design, a conceptual framework is formed by the series of principles, theories, hypotheses, laws and concepts at various levels that actual practitioners of a science produce during the course of their professional activity. This aggregate knowledge gives way to fields, areas and specialized groups within which professional work takes place, making way for specialized methods and techniques.

Thus, beginning with an extensive review of the principal books, manuals, journals and products as presented at international conferences, as well as guidelines by the main professional accreditation agencies for Ergonomics, the BCPE (2009), CEA, and CREE (2007); the first stage meant establishing a Conceptual Framework for Ergonomics that would establish the main areas of knowledge in this interdisciplinary field. Given space limitations, plus the wide range and multiplicity of concepts involved, this work will present only those that are most essential. See Figs. 2, 3 and 4.

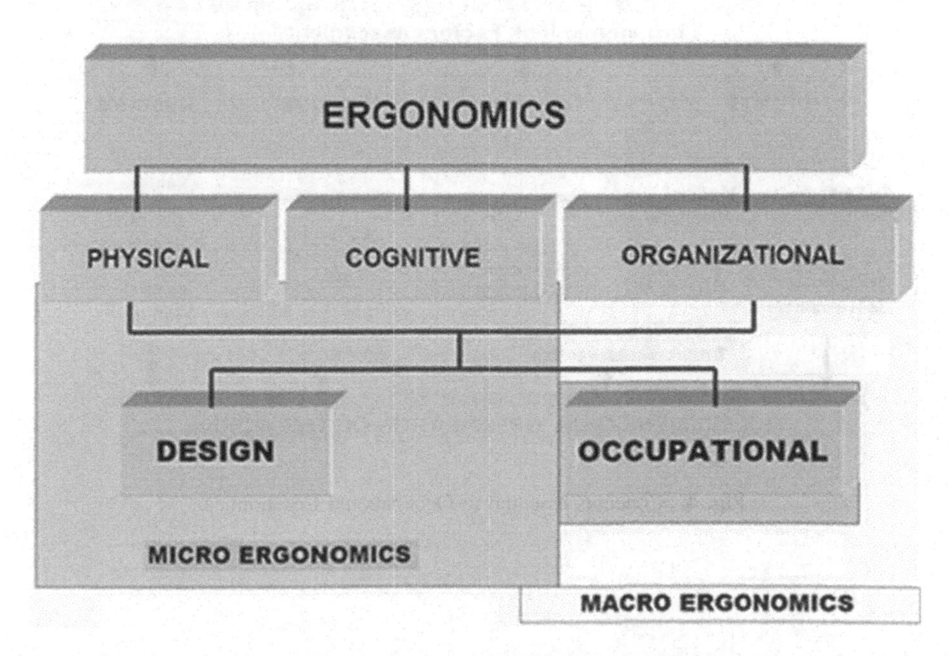

Fig. 2. Fields of ergonomic knowledge and application.

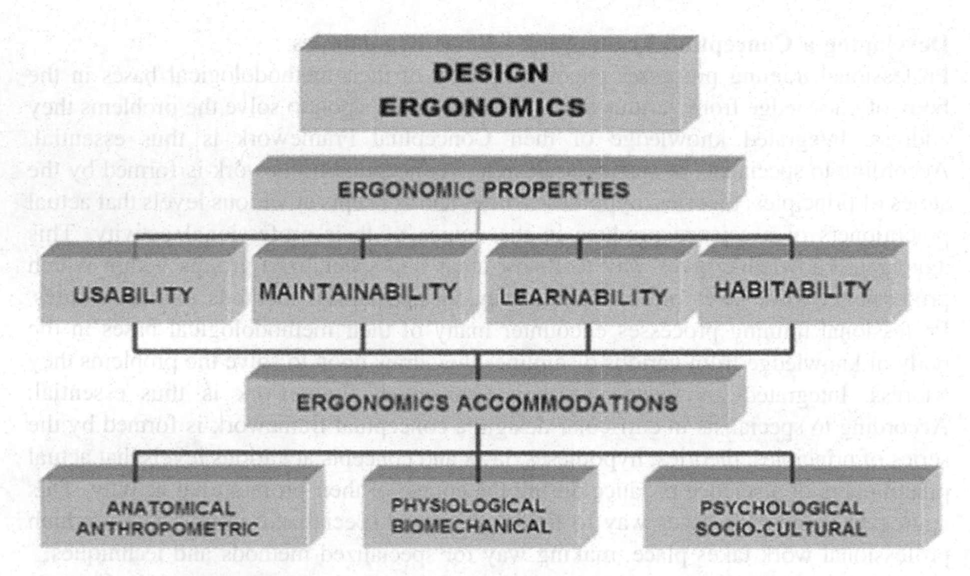

Fig. 3. Concepts essential to Design Ergonomics

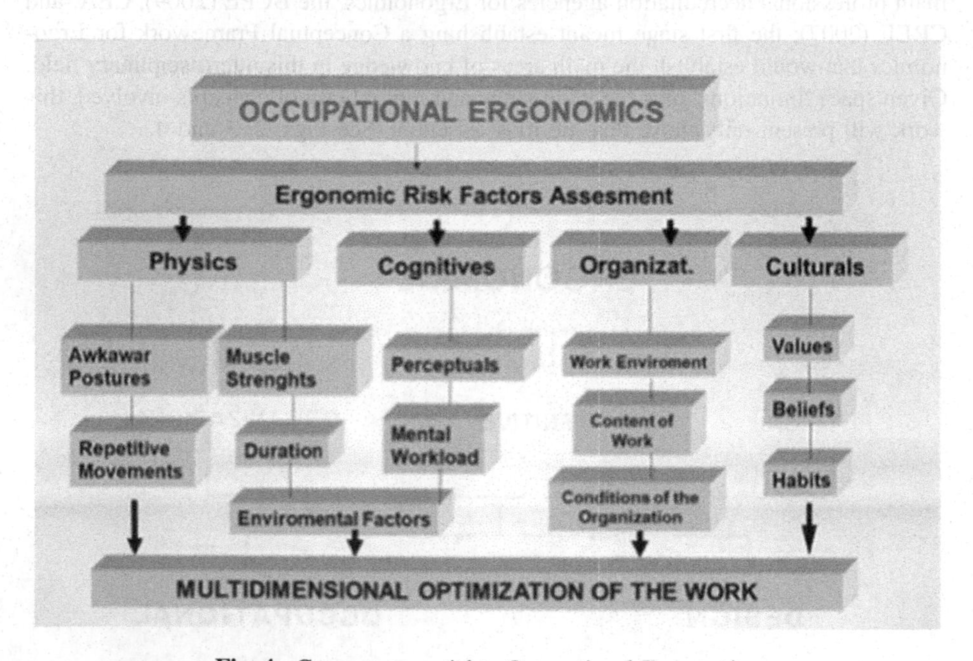

Fig. 4. Concepts essential to Occupational Ergonomics.

A Review of Ergonomics Accreditation Standards in Various Parts of the World

In reviewing accreditation agencies, five (5) areas of learning were identified:

1. General Ergonomic Principles: general study of ergonomics, its historic evolution, objectives, principles, concepts, methodologies, techniques, research and application fields. Current paradigms and schools of thought.
2. Knowledge of Human Functioning: studies into the characteristics of human beings as related to man-object-setting systems. Capacities and Limitations: physical, physiological, biomechanical, perceptual, cognitive and sociocultural aspects.
3. Work Analysis: methodological and technical tools for researching, teaching and applying ergonomics. General analytical methods, techniques, procedures, verification lists. Instruments, measuring equipment.
4. Population and Technology: Analysis and solutions for ergonomic design and occupational problems within a historic context and specific setting, taking into account economic, social and technological differences, as well as sociocultural idiosyncrasies.
5. The Application stage: Strategies for applying concepts, principles, methods and techniques to resolve particular case studies.

A Study of the Labor Market Nationally

A study was conducted into the potential labor market for ergonomics in Mexico, with the University of Guadalajara Center of Opinion Studies (CEO, for its acronym in Spanish) using a national sample of 26 of the most important universities, and 50 industrial plants and other companies of various sizes, to arrive at the following results:

Universities

(a) 77% of them offered regular courses in Ergonomics,

(b) Importance given to ergonomics within design and engineering departments: for 85% of them 'important' to 'very important', for 15% 'unimportant' to 'not very important'.

(c) 77% would hire a professional specializing in ergonomics to teach classes to design majors. Of those, 67% considered ergonomics' main field of application to be Design Ergonomics, 17% Occupational Ergonomics and 16% Environmental Ergonomics.

(d) 77% knew of no other institution offering courses, diploma or graduate programs in ergonomics, while 23% were only aware that the UNAM did.

(e) As for inaugurating a Masters in Ergonomics program in our country, 55% considered it a good idea and 45% a very good one; it is important to note that there was not one negative response.

(f) Prospective benefits of having specialists in Ergonomics:

69%: Better product design.
23%: Higher teaching quality.
8%: Better specialized applications.

(g) Interest by professors in undertaking a Masters in Ergonomics: 81% thought they would be interested, and 19% had no interest.

Industrial Plants and Other Companies

Over the past several years, transnational corporations like Kodak, Hewlett Packard, Siemens, BorgWarner, Ford, General Motors, Honda, Jabil Circuits, Fresenius Medical Care, ABB, and others have begun to implement complete ergonomic programs for their subsidiaries in Mexico and other countries; with all facing the problematic lack of specialists in this field.

Given the demand for applied ergonomics in assembly plants along the border with the U.S., the Association of Labor Medicine, National Federation of Associations and Societies for Health in the Workplace and Mexican Society of Ergonomists, have for several years offered short courses in Occupational Ergonomics to companies in the region, though with all the limitations that the brevity of these programs implies.

Based upon the above and following the proposed Curricular Model, a nationwide sample of 50 firms (30 medium and 20 large) was surveyed, with the following results:

(a) 60% have some knowledge of ergonomics and apply it in their work processes.
(b) 62% believe ergonomics to be 'important' or 'very important' to their firms.
(c) 48% of the companies would hire ergonomic specialists.
(d) 80% consider the opening of a Masters in Ergonomics in our country to be from 'good' to 'very good' news.
(e) Prospective benefits thought to result from having ergonomic specialists:

24% Greater efficiency in the workplace
20% Better working conditions
14% Reduction in workplace injuries and illnesses
12% Reduction in accidents
10% Greater job satisfaction
20% Other (reduction in insurance premiums, better workstations, etc.)

Review of Graduate Ergonomics Curricula from Around the World. Using the conceptual framework, and having defined theoretical, methodological and technical training guidelines, twelve (12) Masters in Ergonomics programs were analyzed, at various U.S. and European universities and one Brazilian institution; to identify the most commonly and frequently offered courses and program content coinciding with our Conceptual Framework. From these courses, we selected those that assured acquisition of knowledge essential to the discipline, promoting understanding and skills necessary for solving ergonomic problems in our social context and meeting the requirements of accrediting organizations. Basic required courses were complemented with courses in scientific, statistical and pedagogic research methodologies, to provide the minimum elements for developing skills in research and instruction; with care taken to integrate common knowledge from the areas of Occupational and Design Ergonomics, given the known demand for these two specializations. Finally, courses were grouped and arranged according to standards and prerequisites set by the University of Guadalajara.

2 Program Descriptions

2.1 First Stage (2011–2015)

The result was a Master of Science graduate degree to prepare Ergonomics professionals: with its first stage of curricular organization providing a basic common trunk branching first into the general knowledge, and basic skills of Ergonomics and, secondly into methodologies, techniques and procedures essential for basic and applied scientific research. In its second stage, students will have the opportunity to choose between two specialties, one in Occupational Ergonomics, including the basic knowledge and skills for comprehensive analysis of jobs, workstations and systems, and the other in Design Ergonomics, with professional training for generating and transmitting knowledge regarding anatomical adaptations, anthropometric, physiological, psychological and sociocultural professional approaches involved in the design of consumer products, machine tools, graphic communications products, and living spaces. See Fig. 5.

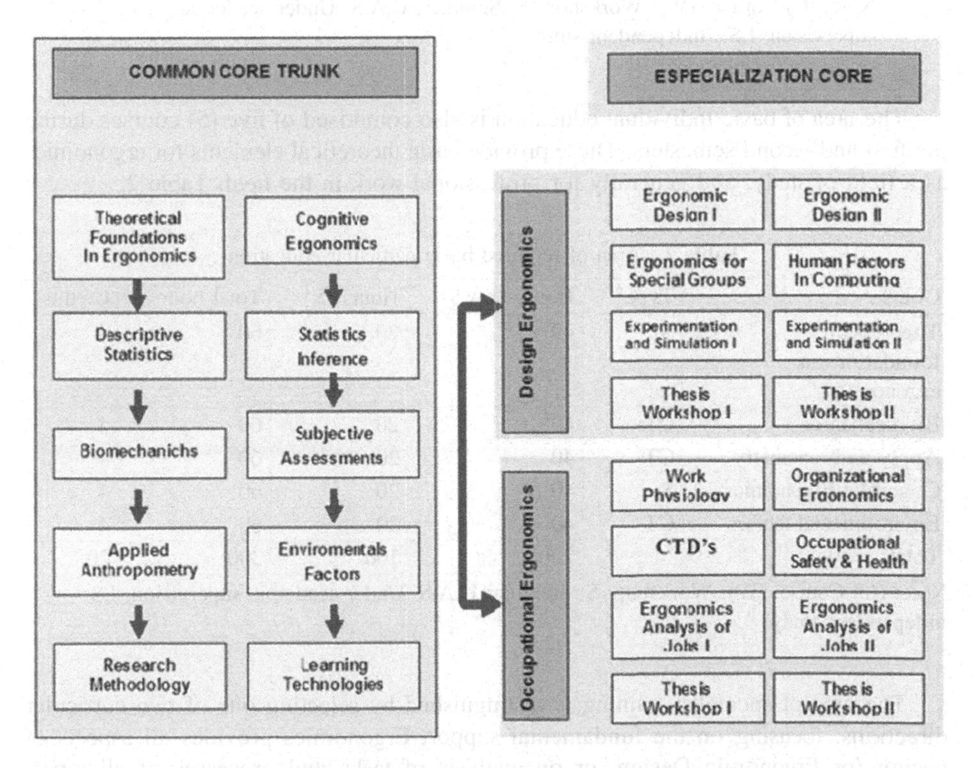

Fig. 5. First curricular map of the Masters in Ergonomics.

2.2 Academic Organization

For academic administration of the Masters, courses were grouped and arranged according to requirements set by the University of Guadalajara for educational stages, organizing them into the following didactic areas: Basic communal education, comprised of five (5) courses during the first and second semesters, offers methodological tools for research and technological approaches to learning. Table 1.

Table 1. Area of required basic communal education.

Course	Type[3]	Hours U.A.S[1]	Hours IS[2]	Total hours	Credits
Research methodology	CW	50	14	64	4
Descriptive statistics	CW	50	14	64	4
Inferential statistics	CW	50	14	64	4
Subjective assessments	CW	50	14	64	4
Learning technologies	CW	50	14	64	4
Total		250	70	320	20

Note: (C) Course; (W): Workshop; S: Seminar; U.A.S. Under academic supervision, I.S.: Independent study

The area of basic individual education is also comprised of five (5) courses during the first and second semesters. These provide basic theoretical elements for ergonomics as a field of study, and generally for professional work in the field. Table 2.

Table 2. Area of required basic particular education

Course	Type[3]	Hours U.A.S[1]	Hours IS[2]	Total hours	Credits
Theoretical foundations in ergonomics	S	40	20	60	4
Biomechanics	CT	40	20	60	4
Apply anthropometry	CT	40	20	60	4
Cognitive Ergonomics	S	40	20	60	4
Environmental factors	CT	40	20	60	4
Total		200	100	300	20

Note: (C) Course; (W): Workshop; S: Seminar; U.A.S. Under academic supervision, I.S.: Independent study

The area of specialist training is distinguished by selecting one of two curricular directions: focusing on the fundamental support Ergonomics provides all aspects of design for Ergonomic Design, or on analysis of tasks and processes of all types, principally in industry, for the Occupational Ergonomics specialization. Both areas are made up of six (6) course-workshops and two Thesis workshops starting in the third semester. Thesis workshops refine the research projects which lead to granting of the degree, Tables 3 and 4.

Table 3. Courses in Occupational Ergonomics.

Course	Type[3]	Hours U.A.S[1]	Hours IS[2]	Total hours	Credits
Work physiology	CW	40	20	60	4
Cumulative traumatic disorders	CW	60		60	4
Analysis of Jobs I	CW	40	20	60	4
Analysis of Jobs II	CW	40	20	60	4
Organizational Ergonomics	CW	60	–	60	4
Occupational health and safety I	CW	40	20	60	4
Thesis Workshop I	CW	40	20	60	4
Thesis Workshop II	CW	40	20	60	4
Thesis defence	W	60	60	120	15
Total		420	180	600	47

In this way the school program consists of 4 semesters, with a total of 1220 h of face-to-face classes and independent study, and 200 h of research. At the end of the 4 semesters the students have to finish and present their thesis work in front of a court of 5 professors, which complete the requirements to obtain the university degree of Master in Science in Ergonomics, Orientation in Ergonomic Design or Orientation in Occupational Ergonomic.

In this first stage of the program graduated two generations with a total of 17 students. During the second generation it was observed that all the students enrolled in both the subjects of Ergonomics of Design and Occupational Ergonomics, and that the graduates of the first generation had received job offers in both fields of application, so it was decided to integrate the contents of both orientations in one single, eliminating some non-essential courses and starting the thesis work from the first semester, as shown below Table 5:

Table 4. Courses in Ergonomic Design

Course	Type[3]	Hours U.A.S[1]	Hours IS[2]	Total hours	Credits
Simulation and experimentation I	CW	40	20	60	4
Simulation and experimentation II	CW	40	20	60	4
Ergonomics for special groups	CW	40	20	60	4
Ergonomic Design I	CW	40	20	60	4
Ergonomic Design II	CW	40	20	60	4
Human factors in computing	CW	40	20	60	4
Thesis Workshop I	CW	40	20	60	4
Thesis Workshop II	CW	40	20	60	4
Thesis defence	W	60	60	120	15
Total		380	220	600	47

2.3 Second Stage (2015 to Present)

Thus, the Curriculum was integrated as it is shown in Table 6.

Table 5. Integrate course list.

Course	Type[3]	Hours U.A.S[1]	Hours IS[2]	Total hours	Crédits
Work physiology	S	50	14	64	4
Analysis of Jobs I	W	50	14	64	4
Analysis of Jobs II	W	50	14	64	4
Simulation and experimentation I	CW	50	14	64	4
Ergonomics for special groups	S	50	14	64	4
Ergonomics Design I	CW	50	14	64	4
Ergonomics Design II	CW	50	14	64	4
Human factors in computing	S	50	14	64	4
Thesis Workshop I	W	50	14	64	4
Thesis Workshop II	W	50	14	64	4
Thesis Workshop III	W	50	14	64	4
Thesis Workshop IV	W	50	14	64	4
Thesis defense	W	60	60	120	15
Total		660	228	888	63

Table 6. Programming by semesters

Semester	Course	Type[3]	Hours U.A.S[1]	Hours IS[2]	Total hours	Crédits
1	Theoretical foundations in ergonomics	S	40	20	60	4
	Biomechanics	CW	40	20	60	4
	Apply anthropometry	CW	40	20	60	4
	Cognitive Ergonomics	S	40	20	60	4
	Descriptive statistics	CW	50	14	64	4
	Thesis Workshop I	W	50	14	64	4
	Total		260	104	364	24

(continued)

Table 6. (*continued*)

Semester	Course	Type[3]	Hours U.A.S[1]	Hours IS[2]	Total hours	Crédits
2	Research methodology	CW	50	14	64	4
	Environmental factors	CT	40	20	60	4
	Inferential statistics	CW	50	14	64	4
	Subjective assessments	CW	50	14	64	4
	Learning technologies	CW	50	14	64	4
	Thesis Workshop II	T	50	14	64	4
	Total		290	90	380	24
3	Work physiology	CW	50	14	64	4
	Analysis of Jobs I	W	50	14	64	4
	Ergonomics Design I	W	50	14	64	4
	Simulation and experimentation I	CW	50	14	64	4
	Ergonomics for special groups	S	50	14	64	4
	Thesis Workshop III	W	50	14	64	4
	Total		300	84	384	24
4	Analysis of Jobs II	W	50	14	64	4
	Ergonomics Design II	W	50	14	64	4
	Human factors in computing	CW	50	14	64	4
	Thesis Workshop IV	W	50	14	64	4
	Total		200	56	256	16
	Thesis defense	W	80	40	120	15
Total hours of study in the master	Face-to-face learning				1128	88
	Research activities				376	15

2.4 Feedback from Employers

When the second-generation students graduated, a small market study was carried out for the employers of the first and second-generation graduates, finding the following results: 80% of the graduates had found work within 6 months of graduation.

We contacted a random sample of 5 employers who were interviewed by telephone, asking them to rate on a scale of 1 to 5 where 1 means nothing developed and 5 very well developed, a series of characteristics about their professional formation and their performance at work, finding the following Table 7:

Table 7. Feed back evaluation from employers

Specific knowledge	Average evaluation
Methods and techniques for the design and evaluation of visual interfaces	4.4
Anthropometric measurement techniques	4.4
Definitions, types of fields of research and application in ergonomics	4.4
Anatomy, physiology and biomechanics of the musculoskeletal system	4.2
Methods and techniques of basic and applied research	4.2
Methods and techniques for the design and evaluation of living spaces	4.0
Methods and techniques for the design and evaluation of industrial design products	4.0

60.0% of the interviewees indicated that the preparation of graduates of this postgraduate program in function of the attention to the needs of their company/institution in that field or area of knowledge, is adequate.

2.5 Feedback from Graduates

In the same way, 10 graduates were interviewed by telephone who were asked to rate on a scale of 1 to 5, if they thought they had the following professional skills (Table 8):

Table 8. Feedback graduates self-evaluations.

Specific skills and competences	Average evaluation
Advising and consulting for the ergonomic design of consumer products	4.6
Advising and consulting for the ergonomic design of graphic interfaces	4.6
Advising and consulting for the ergonomic design of Interior living spaces	4.4
Advising and consulting for the ergonomic design of WEB interfaces	4.4
Conducting usability tests	4.2
Design, development and management of basic and applied research projects	4.2
The graduates were reasonably confident, as well as satisfied with their work	4.0

2.6 Current Status of the Program

Currently the program is in a quite positive phase, since 2012 it was registered in the National Postgraduate Program of Quality (PNPC) and corroborated in 2015.

Through this recognition (PNPC) some funds were obtained with which the furniture, library and the computer equipment and laboratory equipment were improved.

Contributed to these achievements, the positive evaluation of the National Council of Science and Technology (CONACYT), as well the participation of teachers and advisors of high-level academic professors from universities such as the University of Pernambuco, Brazil, the Complutense University of Madrid, Spain, the University of Texas, USA, the University of Concepción, Chile, and lately the University of Loughborough, England.

In this moments the program is in a stage of maturity, so the possibilities of growth and development are good, among which is the possibility in the near future to implement a program at the doctorate level, in collaboration with the universities mentioned.

3 Expectations of Help from the IEA

The possibilities for growth and development may increase if international organizations such as the IEA establish a cooperation program motivating its members to participate actively in the teaching, research and dissemination activities of our program, as well as helping to obtain international funds for teachers of other universities can travel and perform academic work stays in our program.

4 Conclusions

After several years of study, reflection, and review by specialists in ergonomics and Higher Education, a graduate program has been developed that meets the essential requirements of ergonomics training suited to our national needs and lays the groundwork for higher-level development in the future. The program has a core group of educators with more than 15 years of experience in research, teaching, application and dissemination of ergonomics in our country, supported by a select group of visiting professors from countries such as the U.S., England, France, Brazil, Chile and Spain. It also has a comprehensive library and equipment, basic instruments and software for various specialized uses; as well as an established network of industrial plants, other firms, public and private institutions that will provide opportunities for professional practice that address issues within our own social reality.

We are aware that the program is perfectible and hope to be able to improve it over time with help from experienced specialists in the field of ergonomics on both sides of the Atlantic; but we feel satisfied at having been able to initiate this process that we believe will substantially impact application of Ergonomics in our country.

References

Arnaz J (1995) La planeación curricular. Edit. Trillas, México
Bruner J (1971) La importancia de la Educación. Ed. Paidós, Barcelona
Coll C (1991) Psicologia y curriculum. Edit. Paidós, México
Furlan A (1979) Aportaciones a la didactica de la educacion superior. UNAM, México
HETPEP – CREE Official document rev2 – June 2007
Johnson M Jr (1967) La teoría del currículum (definiciones y modelos). Educ Theory 17(2)
Lafourcade PD (1969) Evaluacion de los aprendizajez. Ed. Kapeluz, Buenos Airex
Posner GJ (1979) Instrumentos para la investigación y desarrollo del currículo: aportaciones potenciales de la ciencia cognoscitiva. Perfiles Educativos. México, octubre-noviembre-diciembre, Ergonomía Ocupacional. Investigaciones y Soluciones. vol 4, no 6
Taba H (1993) Elaboración del Curriculum. Ed. Troquel, Buenos Aires
Sarramona JL (1987) Curriculum y educación CEAC. ISBN 84-329-9225-9
Tyler R (1973) Principios básicos del currículo. Buenos Aires, Troquel. (And more of 250 Ergonomics books in English, Spanish and French)

Issues of Certification Program of Professional Ergonomist in Japan and a Challenge of Collaboration Among Asian Countries

Kazuo Aoki(✉)

College of Science and Technology,
Nihon University, Chiyoda-ku, Tokyo 101-8308, Japan
aoki.kazuo@nihon-u.ac.jp

Abstract. The certification program for professional ergonomists of Japan Ergonomics Society (CPE-J) started in 2003, and 126 certified professional ergonomists (CPEs) were certified in the first year. We have now 204 CPEs, 95 certified associate ergonomics professionals (CAEPs) and 12 certified ergonomics assistants (CEAs). In 2007, the program of CPE-J received IEA accreditation, and with this endorsement the program has become an international program. Applying for IEA endorsement was a very complicated process, requiring submission of a lot of English documents.

There are three certification program of different level, CPE, CAEP and CEA in Japan. The CAEP program is not satisfied the criteria of IEA endorsement but many young ergonomists have got CAEP and will try to get CPE in the near future. CEA program is for the ergonomist who did not have a Bachelor degree and could not try to get CPE in the future. I hope there will be a new rout for CEA to get CPE.

The examination for certification can be conducted not only in the Japanese language but also in the English language. So, Non-Japanese ergonomists can apply for the certification examination of CPE-J in English but there has been no candidate from Asian countries. It is supposed that the problem of the language as the reason of it. The collaboration of Ergonomics Societies in Asia, the first Asian Conference of Ergonomics Design (ACED) has held in 2014 in Korea with participants from 10 Asian countries. The second conference was held in 2017 in Japan and a discussion was started on the collaboration of CPE program among Asian countries.

Keywords: Certified ergonomics professional · Asian countries
Collaboration

1 Introduction

In this paper, a history and an overview of the certification system of Japan is shown first. Then a preliminary plan of collaboration among Asian countries related the certification system of professional ergonomist will be discussed.

S. Bagnara et al. (Eds.): IEA 2018, AISC 821, pp. 693–698, 2019.
https://doi.org/10.1007/978-3-319-96080-7_83

2 History of CPE-J

The preparation of certification program for professional ergonomists of Japan Ergonomics Society (CPE-J) started in 1994 and the first Professional Ergonomist was certificated in 2003. It takes about ten years from the start of preparation to realization of the certification. At that time, Board of Certification in Professional Ergonomics (BCPE) and Centre for Registration of European Ergonomics (CREE) already existed. BCPE was incorporated in 1990 and CREE was established in 1994. In 2007, the program of CPE-J received IEA accreditation, and with this endorsement the program has become an international one.

3 Number of Certificated Ergonomist of Japan

There are three kind of certified ergonomist in Japan at 2017. One is CPE which number increases gradually, and other one is CAEP which number is growing fast from 2007. The last one is CEA which number is very small (Fig. 1).

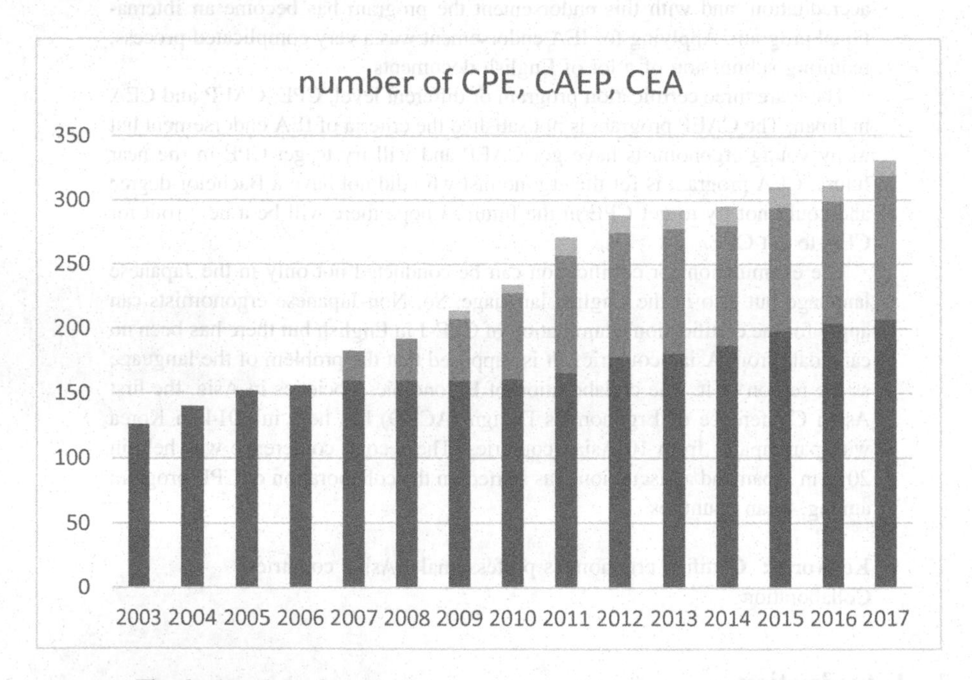

Fig. 1. Number of certificated ergonomists of Japan from 2003 to 2017.

4 Types of Certification of CPE-J

There are three certification program of different level, CPE, CAEP and CEA in Japan (Table 1). The CAEP program is not satisfied the criteria of IEA endorsement but many young ergonomists have got CAEP and will try to get CPE in the near future. CAEP

Table 1. Types of certification of CPE-J [1].

Type of certification	The Certified Professional Ergonomist (CPE)		The Certified Associate Ergonomics Professional (CAEP)	The Certified Ergonomics Assistant (CEA)
	Option A	Option B		
Evaluation	written exam, essay, interview	portfolio examination (in principle)	portfolio examination	portfolio examination
Time of examination	Once a year (September, in Tokyo)	Once a year	4 times a year	4 times a year
Application deadline	Around the end of July	Around the end of May	At any time	At any time
Requirements	(i) college/university degree + ergonomics education (at least 3 years) + ergonomic job practices (at least 2 years) (ii) college/university degree + ergonomic job eperience (at least 7 years)	college/university degree + full-time ergonomics job experience (at least 10 years) + at least 3 cases of demonstrable full-fledged ergonomics projects	(i) college/university degree + at least 3 years ergonomics education (with at least 12 credits) (ii) college/university + practical ergonomic job experience (at least 5 years)	at least 6 credits in ergonomics at a junior college, technical college or equivalent institute, or equivalent education in a corporate training program.
Fee	30,000 JPY	30,000 JPY	10, 000 JPY	10,000 JPY
Others	Please choose either Option-A or B to apply		Available also for students	Available also for students

can be got after graduation of university or college with three years learning of ergonomics program. Then they can try to get CPE after 5 years ergonomics job or experience. CEA program is for the ergonomist who did not have a Bachelor degree and could not try to get CPE in the future. I hope there will be a new rout for CEA to get CPE.

5 Collaboration Among Asian Countries

We are now preparing a collaboration system among Asian countries on the certification of Professional Ergonomists. CREE, Center for Registration of European Ergonomist, is the organization to register professional ergonomist in the area of the Council of Europe as European Ergonomist. There are 501 registered European Ergonomists in 21 European countries and two other countries at 2018.

CPE-J certification program prepared an examination system by English for the candidates from non-Japanese ergonomists, but there was no candidate from Asian countries. The reason is considered to be language problem and the program which can take in their own language is needed.

As a start of collaboration among Asian countries, the 1st Asian Conference on Ergonomics and Design (ACED) was held in 2014 at Jeju, South Korea. There were 272 presentation from 14 countries in Asia and Pacific region. In 2017, the 2nd Conference was held in Japan as ACED2017. There were 130 participants from 15 countries and about 400 participant from Japan. Figure 2 shows the sponsoring organizations of ACED2017.

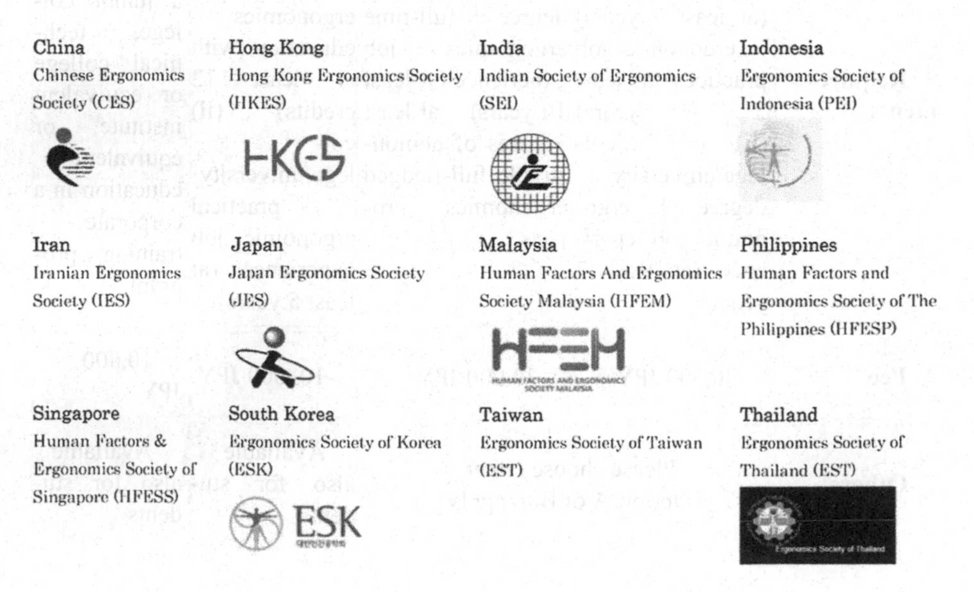

Fig. 2. The sponsoring organizations of ACED2017 [2]

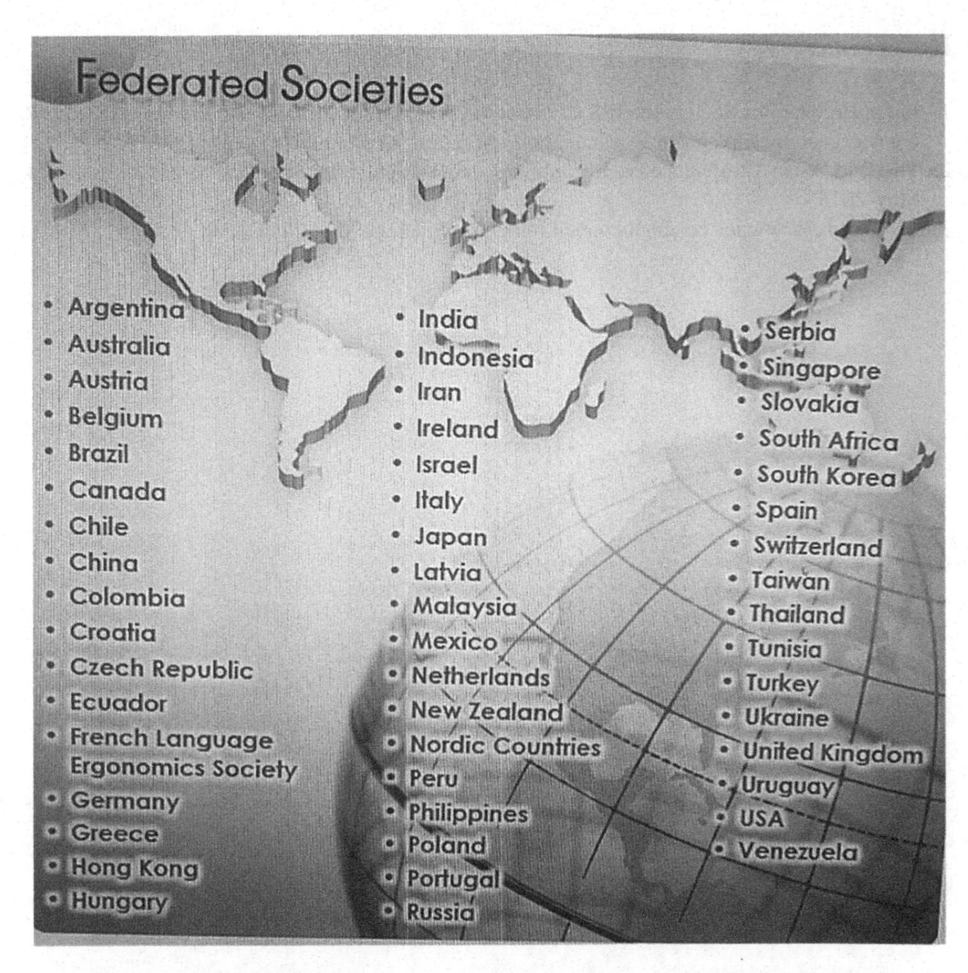

Fig. 3. The federated societies of international ergonomics association (IEA) [3]

After the 2nd conference in Japan, the discussion about the collaboration on the certification of Professional Ergonomist started.

There are a lot of Asian countries in the federated societies of IEA as shown in Fig. 3 and the need of certification of Professional Ergonomist is considered to be very high in Asian countries. The reason is that many Asian countries will develop economically in the future and the needs of Professional Ergonomist will be increase more and more.

References

1. Japan Ergonomics Society the JES Certification Program for Professional Ergonomists. http://www.ergonomics.jp/cpe/en/e-examination. Accessed 24 May 201
2. The 2nd Asian Conference on Ergonomics and Design. http://aced2017.org/. Accessed 24 May 2018
3. IEA. https://www.iea.cc/about/council.html. Accessed 24 May 2018

Education of Ergonomists in France: From the CE2 Actions to the Master Program in Ergonomics at Aix-Marseille University (Aix-en-Provence)

Nathalie Bonnardel(✉)

Aix-Marseille University, PSYCLE (Centre of Research in the Psychology of Cognition, Language and Emotion) EA 3273, Aix-en-Provence, France
nathalie.bonnardel@univ-amu.fr

Abstract. To contribute to a reflection on Education and Training of Ergonomists, the actions performed in France by the CE2 (Collège des Enseignants-Chercheurs en Ergonomie) are first presented. Then a program of Master's degree in Ergonomics is described: the professional and research Master specialized in Ergonomics: human factors and information system engineering, which is proposed at Aix-Marseille University (in Aix-en-Provence, France). This program is conceived in order to fit French and European recommendations about education and training of ergonomists. It includes classes in cognitive, organizational and physical ergonomics as well as specific trainings on risks prevention, work conditions, information system engineering, user interface and UX design, etc. Therefore, graduates from this Master's degree can access to a large variety of positions in Ergonomics and a part of graduates pursue their studies with a Ph.D thesis.

Keywords: Education · Training · Ergonomists

1 Introduction

The trainings in Ergonomics are the object of reflections and evolutions in France. To favor this evolution, the content of the trainings are discussed in different French associations, such as the SELF (Société d'Ergonomie de Langue Française), ARPEGE (Association pour la Recherche en Psychologie Ergonomique et Ergonomie) and the CE2 (Collège des Enseignants-Chercheurs en Ergonomie). These associations play a crucial role in Ergonomics and their actions are complemented by other associations and, especially, ARTEE (Association pour la Reconnaissance du Titre d'Ergonome Européen en Exercice), or by the collective ORME (Organisation Représentant les Métiers de l' Ergonomie), which contribute to the practice of Ergonomics. Due to the topic of this symposium, the actions performed by the CE2 will be more particularly presented. Then, I will describe the Master's degree specialized in Ergonomics: human factors and information system engineering, which is proposed at Aix-Marseille University, in Aix-en-Provence. More precisely, I will evoke the history of this training and present the content of this program in Ergonomics.

S. Bagnara et al. (Eds.): IEA 2018, AISC 821, pp. 699–705, 2019.
https://doi.org/10.1007/978-3-319-96080-7_84

2 Role of the CE2 Towards Trainings in Ergonomics in France

The CE2 (College of professors and assistant professors in Ergonomics) gathers full professors, associate and assistant professors of French high education institutions (Universities, engineering schools, CNAM) who are trained in ergonomics and who are training students in ergonomics. Members of this association are committed to promote and defend training and research in ergonomics.

The main actions in the recent years are in favor of the recognition of ergonomics training, associated to recommendations for the content of masters' degree in Ergonomics, work on the 'ergonomics professions sheet' and its variations as well as reflections on ergonomics certifications. In addition, it proposed a picture of training programs in Ergonomics to whom the members of the CE2 contribute (see Fig. 1).

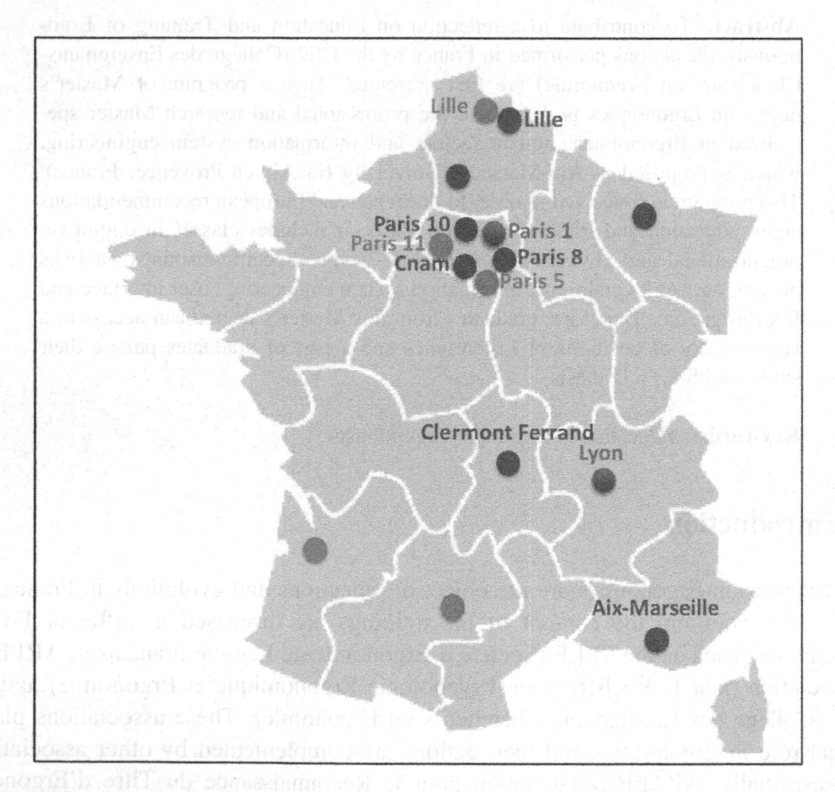

Fig. 1. Overview of training programs in Ergonomics in France, to whom the members of the CE2 contribute. (Color figure online)

Depending on the universities or institutions, these trainings are associated to Human and social sciences (in purple in the Fig. 1), to Sciences, Technologies, Health (in orange) or to Law, Economics, Management (in blue). In addition, some of the trainings in Ergonomics are directly dependent on a Mention in Ergonomics whereas other trainings are defined as 'parcours-type' in Ergonomics and associated to another Mention, such as Psychology for instance.

The actions of the CE2 also lead to the production of recommendations for Masters' degree trainings in Ergonomics in relationship with other reference texts, such as IEA recommendations, CREE criteria and ergonomics professions sheet.

Finally, the objectives of CE2 are to address more global issues, such as the following:

- how to maintain a common guideline in an evolving and ongoing accreditation process?
- where and how to intervene at a more global level (national, European, etc.)?
- what are the ergonomics practitioners' difficulties and what are their needs (in partnership with different ergonomics associations)?

All these actions are thus fully useful for professors and assistant professors in Ergonomics.

3 Master Degree in Ergonomics: Human Factors and Information System Engineering (Aix-Marseille University, Aix-en-Provence)

After this global presentation of actions related to training in Ergonomics in France, we are going to focus on the professional and research Master specialized in Ergonomics: human factors and information system engineering, which is proposed at Aix-Marseille University (in Aix-en-Provence, France) and directed by Prof. Nathalie Bonnardel.

3.1 History of the Program

In 1983, the Université de Provence (Aix-Marseille I, Aix-en-Provence) offered its first graduate program in ergonomics in the form of a specialized diploma, first called a DESS (*Diplôme d'études supérieures spécialisées*) in ergonomic psychology, and it was then renamed DESS in cognitive ergonomics.

This training was created by Pr. Claude Bastien and it was mainly based on cognitive psychology and ergonomics, with a specific focus on expertise, and studies and interventions in the workplace. Another specificity was progressively developed in this program, which consisted in providing students with a training in computer science and programming, in order to allow them to better interact with other stakeholders in their future profession. In 1994, I joined the pedagogical team of the DESS, which comprised mainly Claude Bastien (full professor), Jean-Paul Caverni (full professor), Jean-Luc Péris (associate professor) and Annie Piolat (full professor) as well as numerous professionals. I progressively became co-responsible and responsible of the DESS in Cognitive ergonomics. Thus, I introduced new orientations in the content of the program, in order to prioritize the processes of work activity analysis, to train students to use a variety of methods and techniques in ergonomics, to allow them to access to information elements related to the history of ergonomics, and to lead them to develop competencies in the design and evaluation of user interfaces as well as assistance to human activities.

These orientations were based on my own background, comprising:

(1) a Ph.D. thesis in Cognitive ergonomics performed both at the University of Provence (Aix-Marseille I) and at the 'Institut National de Recherche en Informatique et en Automatique' (INRIA, Rocquencourt, France), in the laboratory (or 'project') in 'Ergonomics psychology', which was created and first directed by André Bisseret. Then, at this period of time, this laboratory or project became co-directed by Pierre Falzon and Dominique Scapin.
(2) classes in Ergonomics at Conservatoire des Arts et Métiers (CNAM, Paris, France).
(3) professional practice as ergonomist in a SSII.
(4) a two years of post-doc at the University of Colorado at Boulder and Institute of Cognitive Science (USA).

In addition, it was very important for me to develop closer relationships with ergonomics associations and societies, such as the SELF (Société d'Ergonomie de Langue Française), EACE (European Association for Cognitive Ergonomics, in which I became a member of the executive commitee), CE2 (Collège des Enseignants-Chercheurs en Ergonomie, ARPEGE (Association pour la Recherche en Psychologie Ergonomique et en Ergonomie, in which I am currently a member of the Administration committee), and IEA (International Ergonomics Association).

Due to national changes in France and the introduction of the LMD (Licence, Master, Doctorat) system, in 2004–2005, the DESS in Cognitive ergonomics became a professional Master specialized in 'Ergonomics: Human factors and information system engineering' (in French: 'Ergonomie: facteurs humains et ingénierie des systèmes d'information'), and it was associated to a general mention of Master in Psychology. In 2008–2009, another change was introduced since this professional Master became both a professional and research Master, which allowed graduates to either work into companies or to pursue their studies with a Ph.D. This program being associated to a Master's mention in Psychology, the graduates stemming from this training can exert as Ergonomists and/or as Psychologists-Ergonomists, depending on their previous background.

This program includes traditional classes in cognitive, organizational and physical ergonomics as well specific trainings on risks prevention, work conditions, information system engineering, user interface and UX design, etc. Therefore, graduates from this training can access to a large variety of positions in Ergonomics. Most of the graduates directly work into companies just after obtaining their diploma but about one-third of them pursue their studies with a Ph.D. thesis in ergonomics psychology or in ergonomics.

Finally, a fusion among the three universities in our area (Aix-Marseille I, Aix-Marseille II, and Aix-Marseille III) intervened and this Master's program in 'Ergonomics: Human factors and information system engineering' is now proposed at Aix-Marseille University. It is also to note that, in France, a new process of accreditation of training programs has been applied (instead of habilitation, which was previously settled). Thus, although the pedagogical objectives and the content of this program globally remain the same, some changes recently occurred in the content and structuration of this Master's program, which will be effective from 2018–2019.

3.2 Description of the Training and Program in Ergonomics

The content of the education and training in the professional and research Master specialized in 'Ergonomics: Human factors and information system engineering' aims to allow students to develop the competencies targeted by associations evoked above, such as the CE2, and in accordance with the standards that are required by the ARTEE (Association pour la Reconnaissance du Titre d'Ergonome Européen en Exercice) to obtain, after a professional experience, the title of 'European ergonomist'. In addition, in accordance with this objective, some members of the pedagogical team have this title. Therefore, this program was developed in line with the European and international requirements for professional ergonomists.

This Master's program allows to train professionals who are capable of carrying out ergonomic interventions, aiming at optimizing workers' safety and well-being as well as the efficiency of production systems, through the combination of cognitive ergonomics, organizational ergonomics and physical ergonomics. It also allows to train ergonomists who have competencies in work activity analysis, in the design and evaluation of workstations and user interfaces (comprising programming and design of mock-ups). In addition, new contents were recently introduced in order to train students to UX design and design thinking as well as to develop ergonomics approaches related to new technologies of information and communication (e.g., virtual reality, with the use of Oculus-rifts, and robotics).

The Master program extends along two years, each comprising 60 credits (ECTS) that are associated to courses delivered by professors in ergonomics, in cognitive and social psychology, in computer science as well as by a large number of professionals in ergonomics.

During the **1st year** (about 400 h of class), the students follow courses that give them both theoretical and practical knowledge related to human work and activities, comprising classes in cognitive and organizational ergonomics, in cognitive and social psychology, in programming and user interface design and evaluation, and they have to realize both a 1st internship in a company and a research work on a topic related to Ergonomics.

During the **2nd year** (about 400 h of class), the students follow courses that give them more theoretical and practical knowledge in physical ergonomics, cognitive ergonomics, and organizational ergonomics. They acquire a variety of methods and techniques related to work analysis/activity analysis, work environment analysis, adaptation to workers/operators, safety and risks management, user interface design and evaluation, UX design, prospective ergonomics and design thinking. They are also trained to use a variety of methods and techniques, including the classical ergonomics methods as well as specific techniques, for instance based on the use of eye-tracking.

The students in this Master program have also to follow classes on English in ergonomics as well as classes on labor law and rights (e.g., related to Internet and digital data). They are also trained to develop ergonomics approaches related to new technologies of information and communication, such as virtual reality. This last objective is favoured by the fact that students can access to the research laboratory PSYCLE (director: N. Bonnardel) and to the UserLab of the technical platform H2C2 proposed at the 'Maison de la Recherche" (Aix-en-Provence). The Userlab comprises numerous equipments of high technology that can be used for analyzing human activity as well as human-computer interactions.

In addition, the students have to participate in tutored projects and to carry out an ergonomic intervention in response to a real or simulated request from a company or based on establishments defined by the professors, which lead the students to progressively adopt a 'professional posture'. Finally, during this second year of Master, the students also have to realize another research work in Ergonomics (usually in line with the one developed in the 1st year of Master) and to perform a long internship (about 6 months) in companies. These works are supported by groups of regulation to engage the students in reflective practice, and they result in written reports and oral presentations, with a committee composed of both professors and professionals.

Therefore, this 2 years' Master program specialized in 'Ergonomics: Human factors and information system engineering' allows students to understand and perform different kinds of interventions, to analyze work situations and activities in both professional and extra-professional contexts, to acquire a variety of methods and techniques, to develop competencies in complementary fields, and to promote new projects in ergonomics.

More information about this program and the content of the training in the 1st and the 2nd year of Master in Ergonomics: humans factors and information system engineering can be found on the following site: https://centrepsycle-amu.fr/master-ergonomie/.

3.3 Review of Difficulties and Positive Aspects of the Development of the Program

In order to introduce a reflection on the development of this Master's program, difficulties encountered when creating this program are evoked as well as its positive aspects.

Difficulties Encountered When Creating the Program

It has been difficult to obtain the department's recognition of the importance of having professors who are specialized in ergonomics psychology and ergonomics as well as the formal recognition of Ergonomics as a disciplinary.

We observe a major work overload for the main professors who give classes in this program: Nathalie Bonnardel (full professor), Brice Isableu (full professor), Ladislav Motàk (associate professor), Patrice Petitjean (PAST – associate professor who also exerts as ergonomist in his own consulting company) and Jean-Luc Péris (associate professor). They also ensure regular supervision of the students for practical works, internships in companies and research works in ergonomics. We also note difficulties in obtaining the support needed to create new positions for professors specialized in ergonomics.

Positive Aspects of the Development of the Program

Since its creation, this program has been improved to correspond to European and international standards for training programs in Ergonomics.

A large number of professionals, specialized in complementary methods and fields (e.g., work and activity analysis, risks prevention, work conditions, nuclear plants, medicine, information system architecture, user interface design and evaluation, UX design) are fully involved in the training of the students in ergonomics. They regularly intervene in this program and also contribute to tutored and applied projects in Ergonomics.

Professors and researchers specialized in cognitive, organizational and physical ergonomics as well as in cognitive and social psychology are fully involved in the training of the students. Their classes are complemented by other interventions proposed by colleagues from other departments and laboratories of Aix-Marseille University, such as Computer Science and Management Science.

This Master's program is associated to the laboratory PSYCLE (research center in the Psychology of Cognition, Language and Emotion, EA 3273) and it benefits from the use of numerous equipments that are available in the UserLab (laboratory of analysis of usability and usages) of the technological platform H2C2, such as eye-tracking (Tobii), systems of movement and facial/emotion analyses (FaceReader), and data capture and analyses (The Observer, etc.), systems of virtual reality (oculus-rifts, etc.).

Most of the students who get their diploma are directly hired by companies at the end of their internships but about one-third of them pursue their studies with a Ph.D. thesis in ergonomics psychology or in ergonomics.

The Master degree in Ergonomics: humans factors and information system engineering (or previously DESS in cognitive ergonomics) has, for years, successfully trained a large number of ergonomists and psychologists-ergonomists capable of carrying out interventions in workplaces, who exert as ergonomists in a large variety of professional areas and are fully recognized by companies.

To conclude, this Master's degree specialized in Ergonomics provides students with rich and stimulating contents in a friendly pedagogical context, which lead them to become Ergonomists who are able of working in a large variety of professional fields.

Global Ergonomics Month: Experiences, Stories and Initiatives Around the World

Michelle M. Robertson[1(✉)], Nancy Black[2], Karen Jacobs[3], Claire Dickinson[4], Frederick Tey[5,6], and Gyula Szabó[7]

[1] Communications and Public Relations, International Ergonomics Association (IEA), Milano, Italy
cpr@iea.cc

[2] Association of Canadian Ergonomists (ACE), Renfrew, Canada
pastpresident@ace-ergocanada.ca

[3] Human Factors and Ergonomics Society (HFES), Outreach Division, Santa Monica, USA
kjacobs@bu.edu

[4] Chartered Institute of Ergonomics and Human Factors, (CIEHF), Loughborough, UK
cdickinson@kub-uk.net

[5] Human Factors and Ergonomics Society Singapore (HFESS), Singapore, Singapore
psechair@iea.cc

[6] Professional Certification, Standards & Education Standing Committee (IEA), Milano, Italy

[7] Communication and Promotion Committee, Federated European Ergonomics Society (FEES), Budapest, Hungary
szabo.gyula@bgk.uni-obuda.hu

Abstract. Invited International Ergonomics Association (IEA) Federated Society and Network Presidents will participate in this symposium and present their respective Global Ergonomics Month (GEM) activities, initiatives, and awareness efforts among local communities, industries, government agencies, and schools. IEA acknowledges these outstanding efforts and strongly supports these local and national initiatives. Together, these world-wide initiatives help to promote the science, application and the profession of Ergonomics and Human Factors. Audience engagement to discuss what GEM activities could be further promoted, along with considering possible joint worldwide activities and initiatives in celebration of the upcoming IEA 60th anniversary.

Keywords: Global ergonomics month · International ergonomics association Case studies

1 Global Ergonomics Month (GEM)

October is the designated Worldwide Month of Ergonomics and Human Factors. Several International Ergonomics (IEA) Federated Societies and Networks were actively involved in promoting Ergonomics and Human Factors awareness among the

S. Bagnara et al. (Eds.): IEA 2018, AISC 821, pp. 706–710, 2019.
https://doi.org/10.1007/978-3-319-96080-7_85

local community, industries, government agencies, and schools. Typically, these planned events and activities are grassroots, community-based activities that target key stakeholders at various levels, such as industries, schools, government, and society.

These events and activities are designed to encourage the exchange of ideas regarding Human Factors and Ergonomics practices through various formats by:

- providing information and materials
- distributing case studies
- highlighting methods and discussions of experiences
- providing community services

One of the IEA Communication and Public Relations Committee's initiatives is to actively support and encourage IEA Federated Societies and Networks and other relevant global organizations involvement in promoting the science, application and profession of Ergonomics and Human Factors and to share these event experiences during the month of October. IEA acknowledges these admirable efforts and strongly supports these local and national initiatives.

For example, exchanging and distributing informative and educational HFE materials and other related planned activities (e.g., design competition) used for these GEM type efforts should be highlighted on the IEA website (http://www.iea.cc). This provides further opportunities for other IEA Federated Societies, Networks and Affiliated professional/scientific organizations to coordinate and utilize these available materials in planning their own grassroots, GEM activities.

Invited International Ergonomics Association (IEA) Federated Society and Network Presidents will participate in this symposium and present their respective Global Ergonomics Month (GEM) activities, initiatives, and awareness efforts among local communities, industries, government agencies, and schools. Audience engagement to discuss what GEM activities could be further promoted, along with considering possible joint worldwide activities and initiatives in celebration of the upcoming IEA 60th anniversary.

2 Global Ergonomics Month: Events and Activities

2.1 Association of Canadian Ergonomists (ACE)

The Association of Canadian Ergonomists (ACE) and related organizations (Work Safe British Columbia) hosted a variety of events and webinars. Their focus was across Canada and it involved sharing experiences and training in the several areas of ergonomics. Some of these included: Quebec City: several gatherings are occurring via the CNESST (ergo-relevant MSD sessions) organized by ACE members as part of other

related associations. CNESST = commission des normes de l'équité de la santé et de la sécruité au travail http://www.cnesst.gouv.qc.ca/salle-de-presse/Pages/Salle-de-presse.aspx.
Ontario: Occupational Health Clinics for Ontario Workers Inc. & Ontario Ministry of Labour. Special page, listing events: http://www.ohcow.on.ca/global-ergonomics-month.html; https://www.labour.gov.on.ca/english/hs/topics/pains.php. New Brunswick: Worksafe NB 2017 HEALTH AND SAFETY CONFERENCE: http://www.worksafenb.ca/health-and-safety-conference. British Columbia: https://www.worksafebc.com/en/about-us/news-events/calendar/2017/October/worksafebc-ergonomics-forum. Dr. Nancy Black, Past-President of the Association of Canadian Ergonomists, will present ACE and other related organizations activities and initiatives for GEM.

2.2 Human Factors and Ergonomics Society (HFES)

Human Factors and Ergonomics Society, in the United States, launched their 14th National Ergonomics Month (NEM) at their annual international conference in October 2017. Highlighting the theme of Human Factors and Ergonomics (HF/E) professionals as they work towards bettering the lives of people around the world. Examples of recent, significant HF/E contributions to the design, test, operation, maintenance, and sustainability of products, tools, systems, and environments for human use are: (1) Patient safety & design of medical devices/facilities, (2) Robotics and artificial intelligence, (3) Security & cybersecurity, and (4) Surface and air transportation–semiautonomous vehicles.

NEM has inspired a variety of activities by students and professionals in local communities. Groups participating in NEM Action Plan contests have implemented their plans through a variety of creative outreach endeavors that emphasize the contributions of HF/E science and practice through teaching, learning, networking, service, and fun! Elizabeth Phillips (elizabeth_phillips1@brown.edu) and Joseph Keebler, (joekeebler@gmail.com), National Ergonomics Month Committee chairs; and Dr. Karen Jacobs, Chair, HFES Outreach Division. See www.hfes.org.

2.3 Chartered Institute of Ergonomics and Human Factors in the United Kingdom

The Chartered Institute of Ergonomics & Human Factors (CIEHF) in the United Kingdom has completed a comprehensive set of clear and resonant stories that illustrate the impact of ergonomics and human factors, entitled “The Human Connection.” This publication is intended to be of value to a wide range of audiences, including: government, policy makers, industry, third sector groups, educators, research funders, regulatory bodies and collaborators. These case studies can serve as a valuable resource to increase understanding of the complexity, range and value of the discipline of Ergonomics and Human Factors and could be used to heighten the awareness of Ergonomics/Human Factors during Global Ergonomics Month. http://www.ergonomics.org.uk/the-human-connection-document. Dr. Claire Dickinson, Past-President of the Chartered Institute of Ergonomics and Human Factors will present an overview of these case studies.

2.4 Human Factors and Ergonomics Society Singapore (HFESS)

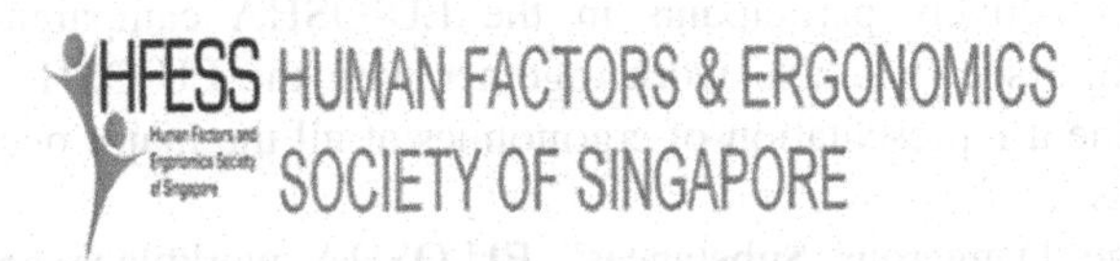

The Human Factors and Ergonomics Society Singapore (HFESS) promoted the Month of October for Human Factors and Ergonomics with a “Critic and Design Competition.” This activity goal was to highlight the awareness among the public community and related Human Factors and Ergonomics (HF/E) design associations. They invited the industry, local community, schools, and the HF/E community to participate in the contest. Their motto was “integrating HFE with sustainable design to improve human performance and work effectiveness. The contest consisted of answering five case scenario questions by identifying the HFE issues and then providing proposed design recommendations. The question topics ranged from interface and workplace issues, organizational factors of HFE, and human error.

For GEM 2016, HFESS conducted an ERGO-Trophy Challenge where people where encourage to design a trophy that could be used in one’s daily life. HFESS also provided several HFE talks. For more information and related materials, conduct nicolette@hfess.org. Dr. Frederick Tey, Past-President of HFESS will present and discuss their Federated Society activities and initiatives.

2.5 Federation of European Ergonomics Societies (FEES) - IEA Networks: European Month of Ergonomics (EME)

The Federation of European Ergonomics Societies launched the first European Month of Ergonomics (EME) in 2009 with the "Know your ergonomist" campaign. Later, the Communication and Promotion Committee of FEES decided to dedicate EME to support the actual Healthy Workplaces campaign of the European Agency for Safety and Health at Work (EU-OSHA). Today, EME has two objectives; (1) communicate the ergonomic content of the Healthy Workplaces Campaign, and (2) promote the ergonomic profession and approach. With the merging of the FEES and EU-OSHA messages, EME illustrates the capability of ergonomics professionals to improve the health and safety working conditions at work in accordance with the context of the current EU-OSHA campaign. For more information, see www.healthy-workplaces.eu.

FEES promotional activity is not limited to October; however, it is recognized as the official month of ergonomics. As an official Campaign partner of EU-OSHA at the EU level, FEES actively participants in the EU-OSHA campaign activities. The national ergonomic societies also work together with the EU-OSHA focal points to provide a permanent representation of ergonomics at all the major occupational safety and health events.

The "Manage Dangerous Substances" EU-OSHA workplace health and safety campaign was launched on April 2018. FEES become an official campaign partner again, and the current "ERGONOMIST TO MANAGE DANGEROUS SUBSTANCES BETTER" EME campaign aims to identify, elaborate and demonstrate how the ergonomic profession can contribute to the healthy and safe handling of hazardous substances.

The Communication and Promotion Committee of FEES solicits educational institutions to initiate research and development programs, in particular student projects, master thesis and PhD works, to produce solutions of high ergonomics quality in regard to the management of hazardous substances, and to submit these results for consideration of "The Healthy Workplaces Good Practice Awards."

Dr. Gyula Szabó, Eur Erg, Chair of Communication and Promotion Committee of the Federated European Ergonomics Society (FEES) will present and discuss these various initiatives. (See IEA Congress 2018 proceeding paper: What can an Ergonomist do to manage dangerous substances better?).

Professional Ergonomics Education in Argentina

Gabriela Cuenca[1(✉)] and Michelle Aslanides[2,3(✉)]

[1] Departamento de Ingeniería Industrial, Universidad Tecnológica Nacional FRBA, Buenos Aires, Argentina
gcuenca@frba.utn.edu.ar
[2] Universidad Favaloro, Buenos Aires, Argentina
miaslanides@gmail.com
[3] Universidad Austral, Pilar, Argentina

Abstract. This paper will explain the case of the Professional Ergonomic Education Program from Argentina, answering the questions that the organizing committee of the symposium "Professional Ergonomics Education" has asked us as parts of the panel. Following these questions as guidelines to our paper, we will start from the description of our program, then follow by an analysis of it's development in our real context, and finish with what we expect from IEA in order to help us develop professional education in the near future.

Keywords: Professional education · Ergonomists · Argentina IEA · Standards

1 Introduction

This is our contribution to the symposium "Professional Ergonomics Education" that was organized for the first time in an IEA congress in 2018 in Florence. We are very pleased to be part of this effort to bring together all the knowledge from different parts of the world in order to develop our discipline and the conditions under which our professionals are educated.

This paper will explain our case, the one from Argentina, and we will also answer the questions that the organizing committee has asked us as parts of the panel. The questions were the following:

1. Do you know an education program in your country that teaches ergonomics? Please indicate the complete name of the program and its references.
2. Since when does the program exist?
3. Can you give us the name of the person who created or who is in charge of the program?
4. Which are the contents of the program?
5. Des the program fit IEA criteria to educate professional ergonomist?
6. Which are the difficulties you have encountered creating the program and developing it? Which are the contributing factors of these difficulties?
7. The positive aspects of the development of the program and its contributing factors.

S. Bagnara et al. (Eds.): IEA 2018, AISC 821, pp. 711–717, 2019.
https://doi.org/10.1007/978-3-319-96080-7_86

8. If you have experienced health issues related to the this activity (stress, etc.).
9. What do you think IEA could do to help the development of the professional education process in your country?

Following these questions as guidelines to our paper, we will start from the description of our program, then follow by an analysis of it's development in our real context, and finish with what we expect from IEA in order to help us develop professional education in the near futur.

2 The Professional Ergonomics Education Program of the UTN FRBA

2.1 History of the Program

Argentina's first postgraduate degree in Ergonomics "Especialización en Ergonomía" (Res 1105 6/8/2006) started in 2006 at the Universidad Tecnológica Nacional, Facultad Regional Buenos Aires (UTN FRBA). UTN has indeed many different regional faculties, but Ergonomics was designed and directed by Gabriela Cuenca for the first time in the one based in Buenos Aires: FRBA. Gabriela Cuenca is one the argentine ergonomists educated in France who went back to the country some years after democracy started again in1983. She created first an undergraduate Ergonomics course in the Engineering Program of UTN FRBA. After some years, and taking into account a new labor regulation about MSD prevention and assessment mentioned "Ergonomics" (res. 295/2003), she could finally create the Master Degree in Ergonomics. Therefore, many students other tan engineers could follow the program. From 2011 to 2014 Michelle Aslanides relayed her and added a supervised intervention methodology training. Since 2015 it is coordinated by Lucie Nouviale and Martín Rodriguez.

2.2 Contents of the Program

The content design of the program took from the beginning into account IEA's core competencies and CREE`s guidelines. Gabriela Cuenca based the contents on the example of the CNAM in Paris, where she studied with Alain Wisner. After Gabriela, Michelle Aslanides tried to continue developing little by little the contents concerning intervention methodology in practice, the practical guidance of professional ergonomists mentoring the students, supervised trainings in companies. This is still something that needs to be developed. The information about the program in the following link http://posgrado.frba.utn.edu.ar/carreras/esp_ergono.php. where you will find the following contents corresponding to 420 h of courses:

Basic principles of ergonomics	30 h
Physiology and neurophysiology	40 h
Biomechanics	30 h
Psychodynamic and psychopathology of work	30 h
Cognitive ergonomics	30 h
Work organization	40 h

Accidentology	30 h
Product and workplace ergonomics	30 h
Environmental ergonomics	40 h
Workshop of environmental measurements	30 h
Assessment tools workshop	30 h
Applied statistics	30 h
Practical interventions during the training	30 h
Total	**420 h**

2.3 The Program's Stakes, 12 Years After

Until today UTN FRBA has educated almost 100 professional ergonomists and its master degree is the only one operating in the country. The fact there is only one program in the country is the consequence of an internal policy of the UTN FRBA that has established not to open new master degrees in other parts of the country, as it could fragilize the one in Buenos Aires. This turns the number of educated professionals very low, which makes our profession weaker and less recognized than others like occupational safety and health professionals. This is one of the reasons res. 886/2015 refers not explicitly to Ergonomists but to people with "education in ergonomics", as we will see afterwards.

3 Professional Ergonomics Education Development

3.1 Opportunities that We Have Experienced

A New Potential Profession is Born

All these years of efforts educating professionals give us the opportunity to say that we "exist" as professionals in our country. Even though as we mentioned before, the path is difficult, a first step is University recognizes the need of this kind of profession, and that the master program exists against all obstacles. Our effort will allow us progressively to create different kind of associations that will enlarge the professional ergonomists community: students and professionals associations, professional councils, teachers and researcher's associations, etc. This is our first step, and it's well done thanks to the open and wide vision of some engineers in the UTN FRBA university that helped a lot this project supporting all the efforts done by Gabriela Cuenca in the first lonely steps. In this sense, we would like to thank three UTN teachers for their support: Dr. Ing. Arturo Rodriguez Ponti, former ADEA vice president and founder member, Dr. Fernando Napoli (Academic and Posgraduate Education Director), and Ing. Raúl Sack (Vice-dean and Industrial Engineering Department Director).

New Professionals Can Develop New Regulations from New Practices

The regulations we mentioned before, especially the ones that opened the path to the Master Degree creation, despite all their defects, opened new opportunities for the development of professional ergonomists. These requests, based on a legal a technically reductive logic, can be redefined at each intervention by the professional ergonomists to open the scope of their intervention and capacity. It is not always simple, but still possible, and a challenge. For instance, some of us extend the content of our reports to other dimensions of work analysis that are not in the S&H regulations. Others even don't really sell themselves as ergonomists, for not to be associated only with MSD prevention and with the "control" or "police" logic that is perceived by workers and CEOs, usually associated with the S&H approach. These practices, if they become more and more developed, could lead to a new way of regulating the activities of MSD prevention and other diseases and health issues. Our stake is to become more and more visible also as professionals in charge of the improvement of reliability in organizations. But that is the second following step.

Students Develop New Views About Work Analysis in their Own Profession
The positive aspect of the UTN ergonomics education program, among bringing more ergonomists to the field to improve health and reliability in working situations, is that students point of view about working activities change and influence the way they practice their own professions when they don't become ergonomists after the end of the program. Today most of them are very good ergonomists but they are as well better physicians, psychologists, engineers, physiotherapist than they were before studying ergonomics.

3.2 Obstacles to the Development of Ergonomists Professional Education

Obstacles from the Regulatory Perspective
There is a growing but not explicit need for educating ergonomists in Argentina that comes from the labor regulation. Indeed, since the Labor Ministry of the Nation has developed some rules in 2003 that establish that a workplace assessment in terms of MSD and psychosocial risks is mandatory in the local industry. Even though these rules open a good opportunity for ergonomists to develop and put into practice their skills, there is a non-explicit need since these professionals are not mentioned in the rules. The first rule didn't mention at all ergonomists, but it didn't mention any other professional either. We were even. The last regulation is Res. 886/2015, approved by ADEA council, and it mentions the need to perform the workplace assessment by medical team and Safety & Health staff in a first stage, then only "people with ergonomics education", in a second stage. This regulation, is a good opportunity for ergonomists but the fact it only mentions the need of professionals "with an education in ergonomics" opens the possibility to work in these projects to all kind of professionals who will probably not always be well trained in ergonomics terms. It also opens the education "market" to all kinds of courses that are "enough" to fulfill the regulatory needs.

Ergonomists in Minority Lose Opportunities
ADEA is an organization founded by ergonomists and by other professionals et the beginning, that always remained open to different professions. In a context where regulations would enable working opportunities for all of them, lobbies started to operate when these rules were born, especially in the last case. Professional

Ergonomists, as we said at the beginning, are a minority in the giant world of safety and health, medical and other occupational health professionals. We believe medical and S&H professionals have won that "battle" and determined with a quantitative argument "ergonomists were not enough to mention them as professionals in charge of the assessments". At least these were some of the arguments that were given to us as professional ergonomists during all the debates that took place when the regulation was designed and afterwards. Up till now UTN FRBA is the only University to offer an Ergonomics Educational Program, as we said before, and the fact no other courses exist gives us ergonomists less strength to be recognized by the present regulations: we are only 120 professionals in all the country. Our discipline is recognized, but this lack of professionals opens the opportunity to other professionals to do our practical job. We are trying to improve our visibility but this work is hard when we are so few to try to change the way things happen.

The Market of Courses "Out of Professional Control" is Growing

The national regulation is opening the opportunity to a market of short education programs that are selling ergonomic knowledge without any academic criteria, and with a commercial trend leading to a low level of practitioners that are competing against professional ergonomists.

This trend, in a context where ergonomics is recognized by some institutions as a scientific discipline in Argentina, as the example of the master degree creation shows, but where professionals aren't enough recognized yet is a stake for the program: how can the program still be competitive when the context asks for less academic level in the occupational health professionals? And what's more, how can the program face the fact other professionals without any ergonomic education are fulfilling the tasks that ergonomist should be developing. The situation is nowadays very confusing: "every-one is ergonomist" without any kind of professional control. Ergonomists are united in ADEA, but ADEA is not a professional association, it is only an "Ergonomics" society, not an "ergonomists" one. No one at this time can easily hold the flag of ergonomics profession, from an institutional perspective.

The Stake of Keeping Ergonomists Inside the Loop of Regulation Opportunities

This confusing situation is due to the fact regulators have decided to use the term "ergonomics" in our Safety and Health legislation which rules the industrial world. The two regulations we mentioned before, res. 295/03 and 886/15 are naturally inviting professional ergonomists to be "part of the party", but unfortunately these professionals did not validate in their majority such regulation because they were not really partic-ipating of ADEA validating process. They would have stopped and corrected the regulation proposal since they quickly saw the regulation would establish a legal boundary to our professional scope and, what's worse, it would leave us out of the market. Indeed, only safety and health or medical domain specialists are allowed to assess workplaces in the first stage of the regulated methodology. This means this legislation gives all the power to act "in the name of ergonomics" to safety and health and medicine specialists and avoids ergonomists to do their natural work. One critical example of this situation is from now on, all the first steps of the interventions in the organizations will be "filtered" by medical and S&H professionals, by their mental models and by the regulated assessment methodology. One of the main aspects of the ergonomists intervention will be outsourced to other professionals who don't really work the same way and who won't never think about redesigning the request, as we do in our ergonomics practice to help the company to think differently the causes of occupational health diseases and accidents.

Opening the Scope of Ergonomics to the Other Concepts and Methods
Our regulation was designed based on intervention methods that consider only simple physical work load assessment checklists that only apply to certain working situations and doesn't help understanding other kind of physical or mental work load.

Indeed, today our regulation (res. 886/15) mentions the term "ergonomics" and assesses musculoskeletal disorders (MSD) without using ergonomic intervention methods at all. The analyst should fill in a check-list to identify and assess the risks. This standardized method generalizes its application to all kind of situations. It doesn't take into account many situations that don't fit the theoretical scenario, inducing the regulation's user to potential errors. This MSD assessment practice reduces and trivializes ergonomic practice considering it as a non-reflexive application of rigid and error prone assessment tools. We don't agree with this view of our profession, and one of the stakes of the program should be showing the way ergonomics profession can solve these problems from a work situation modeling and intervention perspective.

We have presented a proposal of improvement to the Occupational Risks Agency of the Ministry of Labor and have insisted on the fact professional ergonomists should analyze work situations. This would be a good point to convince the medical and safety & health professionals to follow the ergonomics program since, otherwise, why what would they want the be trained as ergonomists if ergonomists would never find a job?

This limitation is not only a problem since it broadens the scope of the health problems we can explicit and solve, but also because it restricts the limits of ergonomics scope in terms of improving companies' reliability. The safety and health model does not integrate reliability as a goal unless accidents arise because of the lack of reliable prone working conditions. Ergonomics does, and it is one of the main differences a professional ergonomics program should offer compared to the short courses that grow in the market. We are sure this vision will succeed because more and more students are looking for interesting contents that are going to enable them to face real working situations analysis and change, and not only for a certificate that is only an administrative requirement.

4 IEA Expected Contribution in Professional Ergonomics Education

In this context, IEA referentials as core competencies and certification standards are key to help ergonomics professionals that are trying to modify national legislation or creating new educational programs in this hard context to be able to establish the minimal requirements for a professional education and practice. The support from IEA when a master degree is opened could help as a way to reach international and profesional recognition when other professions are trying to use the term "ergonomics" to enlarge their own field of practice, as we saw it happens in our case, excluding profesional ergonomists from the scope of the regulations and of the assessment situations in the field. IEA should also support Professional Ergonomists Societies to establish the Professional Certification Processes that could help justify the required high level of Professional Ergonomics Education Programs.

References

Aslanides M, Cuenca G, Del Rosso R (2011) Modelo de certificación nacional del profesional ergónomo. Documento interno de la Comisión de certificación de ADEA

Aslanides M, Cuenca G, Del Rosso R (2010) Documento de Reconocimiento del ergónomo profesional en la Argentina. Comisión de certificación ADEA

Aslanides M, Cuenca G, Del Rosso R (2010) La Ergonomía en Argentina. Revista de la Cámara Argentina de Seguridad pp 3–4

Dul J et al (2012) A strategy for human factors/ergonomics: developing the discipline and profession. Ergonomics 55(4):377–395

Poy M (2006) Aspectos funcionales de los riesgos y desvíos de las normas de seguridad en el trabajo: un aporte a la comprensión de las relaciones entre actividad humana y seguridad. Tesis doctoral en psicología Universidad de Palermo. Argentina

Poy M, Gomes JO, Soares M (2006) L'analyse de l'activité : expériences sud-américaines. In: Vallèry G et Amalberti R, éditeurs. Octarès Éditions Les cas de l´Argentine et du Brésil In L'analyse du travail en perspectives: Influences et évolutions. Toulouse, pp 79–96

Soares M (2009) Ergonomics in Developing Regions. Needs and Applications. Scott CRC Press, Boca Raton. https://doi.org/10.1201/9781420079128.ch23 Print ISBN: 978-1-4200-7911-1 eBook ISBN: 978-1-4200-7912-8

Soares M (2006) Ergonomics in Latin America: background, trends and challenges. Appl Ergon 37(4):555–561

Professional Ergonomists Education: Lessons Learned from Worldwide Existing Programs

Michelle Aslanides[1,2](✉), Nelcy Arévalo[3](✉), Raouf Ghram[4](✉), Bouhafs Mebarki[5], and Frederick Tey[6,7]

[1] Universidad Favaloro, Buenos Aires, Argentina
miaslanides@gmail.com
[2] Universidad Austral, Pilar, Argentina
[3] Fundación para la investigación y el desarrollo de la Ergonomía en América Latina- Ergoideal, Bogotá, Colombia
nelcyarevalo@gmail.com
[4] Higher Institute of Human Sciences of Tunis, Tunis, Tunisia
raouf_ghram@yahoo.fr
[5] The University of Oran 2, Oran, Algeria
mebarkibouhafs@gmail.com
[6] DSO National Laboratories, Singapore, Singapore
fredericktey@gmail.com
[7] IEA Professional Standards and Education Committee Chair, Singapore, Singapore

Abstract. This research is the result of a collaboration established between ergonomists from all over the world that are concerned with education of professional ergonomists. We have started a collective state of the art of ergonomics training programs for professionals, according to IEA standards. The team that organizes this symposium has already started this research and published it at 2017 SELF Congress, and continued working in new results from other corners of the world. We have started the project by sending a mail to all the federated societies as they appear in the IEA website, to some of the IEA executive committee members, and finally contacting the ergonomists we know are or were directors of master degrees in ergonomics all over the world. We asked them to give us their feedback through some questions concerning the education program in which the consulted ergonomist has been involved. The stake is to deeply understand the training case through the answers to the questions sent. We got answers from more than 14 countries concerning more than 20 education programs. The panel of the symposium will present the case of 8 programs from different continents. We hope this research will add our "grain of sand" in the necessary collective work of IEA's "Professional Standards and Education Standing Committee".

Keywords: Education · Professional Ergonomics · Worldwide Program · Core competencies · IEA

S. Bagnara et al. (Eds.): IEA 2018, AISC 821, pp. 718–726, 2019.
https://doi.org/10.1007/978-3-319-96080-7_87

1 Introduction

Our community is, since at least two decades, working to define the boundaries of our discipline and profession through defining three aspects: (1) the term[1] ergonomics itself, (2) the core competencies[2] required for ergonomists education, and (3) the certification[2] process that should help define professional ergonomists through the contents of their education practice and/or the contents of worldwide ergonomics education programs. These efforts were and still are made by people who work in the different IEA Standing Committees concerned by these issues mainly through certification efforts (Dul et al. 2012) (IEA 2001a, 2001b) and by Professional Ergonomists that are developing the education programs in different corners of the world (Aslanides et al. 2010a, 2010b, 2011; Arévalo et al. 2017; Soares 2006, 2009). Thus, with some colleagues, we believe there is a "need for quality assurance regarding the training and credentials of professional ergonomists. Responsibility for such quality assurance often is assumed by professional ergonomist certifying bodies" (Smith 2012). Therefore, we have organized this first symposium to start a collective work concerning the state of worldwide Professional Ergonomists Education Programs and their desirable evolution. This is a first step of this challenging project, co-organized by some ergonomists in charge of education programs in four different countries: Argelia, Argentina, Colombia and Tunisia and by the IEA "Professional Standards and Education Standing Committee" chair. This paper presents our global methodology and some of the results of our work, including mainly the answers to our questionnaire that won't be presented in the symposium panel. The rest of the results will be presented in the Symposium by each author and will described in eight separate conference papers that will be published in the IEA 2018 Congress proceedings.

2 Methodology

2.1 Goals

We wanted to deeply understand the training program situation, its history, it's actual state, the positive and the negative aspects that are related to its existence and the factors that contribute to those results. The idea was to obtain more qualitative than quantitative data, but also try to establish some trends and categories of answers to get a global idea of Professional Ergonomics Education Programs all over the world including some quantitative data in our analysis.

From the qualitative perspective, one of the goals was to convince some of the master program directors to join us in Florence to be able to discuss about Professional Ergonomists Training issues, and to motivate them to bring their experience and share it during the symposium "Professional Ergonomics Education". The other was to motivate them to think about what IEA can do to help the development of Professional Ergonomics Education Programs. From the quantitative perspective, we wanted to

[1] https://www.iea.cc/whats/.

[2] https://iea.cc/project/index.html.

reach as much as possible programs around the world to get some information we could analyze to be able to establish the main trends in terms of success of the programs, their positive aspects, and their threats.

2.2 Methods

We have started the project by addressing a mail to all the federated societies as they appear in the IEA website, to some of the IEA executive committee members, and finally contacting the ergonomists we know are or were directors of master degrees in ergonomics all over the world. We asked them to give us their feedback through some questions concerning the education program in which the consulted ergonomist has been involved. The questions were the following:

1. Do you know an education program in your country that teaches ergonomics? Please indicate the complete name of the program and its references.
2. Since when does the program exist?
3. Can you give us the name of the person who created or who is in charge of the program?
4. Which are the contents of the program?
5. Des the program fit IEA criteria to educate professional ergonomist?
6. Which are the difficulties you have encountered creating the program and developping it? Which are the contributing factors of these difficulties?
7. The positive aspects of the development of the program and its contributing factors.
8. If you have experienced health issues related to the this activity (stress, etc.).
9. What do you think IEA could do to help the development of the profesional education process in your country?

3 Results

3.1 General Response to Our Proposal

We got answers of more than 14 countries concerning more than 20 education programs. Only 8 of them could finally attend Florence and will join the panel. Only 9 questionnaires were fully completed, 4 of them arriving from Chile, Canada, Switzerland and USA. The rest of the 20 education programs were not considered in the results because they either were not concerned by a Program that fulfilled IEA core competencies criteria, they couldn't attend the conference or didn't send any answers to our questions.

All the programs were described, we got all the links to their website, and the questions were answered. We will copy some of them in the next paragraphs.

3.2 Some of the Answers to the Questions

We will describe here the ones that won't be presented by the panelist. You will find panelists papers in the proceedings of the congress.

Difficulties Creating and Developping the Program and Contributing Factors
Concerning the difficulties they have encountered creating the program and developping it, and the contributing factors of these difficulties, here are some of the detailed mentioned ones:

Canada

1. There is always pressure within the university to justify the need for an ergonomics program. We have been able to do this by adjusting the program scope over the years. We have focused on Occupational Ergonomics as many of our students are interested in physical and psychosocial issues in general when they come into the Kinesiology program.
2. A challenge is introducing the option of Ergonomics early enough to generate interest in students.
3. Our enrollment in courses is adequate in lower level courses (about 40) as many students who are enrolled in the certificate also take these courses, but we could do with more students graduating with the certificate.
4. It would also be useful to have more faculty with expertise in Ergonomics.

It is stressful at times to be the main faculty member with ergonomics expertise and to champion the program.

Chile

1. Although it has not been a problem for our programs, the appearance of "overnight" courses that give high sounding titles in 40 or less hours is worrisome. It seems like a joke but it is a reality. In my opinion it is a very important topic to analyse.
2. One observes that many people and organizations are concerned about the certification of ergonomists, but the same care is not observed in the certification of study programs that are the most important to ensure the healthy development of our discipline.

Singapore

1. Lack of awareness of what human factors is as a scientific discipline: most people are not even aware that there is such a discipline; hence marketing the programme has been an uphill struggle.
2. Lack of industry support for the discipline and hence lack of HF related jobs in Singapore: As mentioned, many are not aware what human factors is. There aren't many companies who are hiring human factors professionals.
3. Lack of experienced HF professionals in Singapore who could be engaged to teach in the programme.

Switzerland

1. There was a fairly extensive Masters in Ergonomics up until about 5 years ago. It was cut as a result of funding issues. At the moment there are moves by some university teacher to try to put together a patchwork of training at various institutions but this is not really in line with the wishes of the individual universities who

would prefer people to train exclusively at their institution (even if they don't offer what is wanted).
2. Competition from other professions (medical, hygiene, psychology, safety) is rather a problem, as each feels that they can already offer all the ergonomics that is needed.

USA

The primary challenges associated with developing and maintaining safety and ergonomics programs in the U.S. are:

(1) recruiting and retaining high quality program faculty; and
(2) recruiting domestic students to graduate programs.

Contributing factors include a limited number of current PhD students pursuing careers in academia. The majority of our students pursue research careers in tech industries. Other contributing factors include very high salaries for undergraduates and high placement rates in industry. The majority of domestic students do not have interest in pursuing graduate studies. Approaches to motivating PhD student interest in academic careers could be another topic for discussion as part of the Roundtable. In addition, discussion of methods for recruiting graduates students might also be useful for participants.

Positive Aspects of the Development of the Program and Its Contributing Factors

Canada

The practicum component of our program has been a great success. Both students and employers have had very positive experiences. Students have said they feel very prepared to work in the field.

Chile

1. the fact that we have graduated about 170 Magister and more than 600 specialists with Diplomado, from all Latin America, shows that the program is very successful for our standards.
2. The most positive of our program is the response of the students who appreciate the fact that the Unit of Ergonomics of the University of Concepcion was created 46 years ago.
3. And the fact that we have been teaching undergraduates and carrying out research, without interruption in topics that are very relevant, not only for Chilean or Latin American workers, but also for workers from other parts of the world. I say this in a humble way, because nowadays many people talk about "models" of teaching ergonomics. We do not pretend, to have the "Chilean model", but we are certain that we have learn to build from the basis and the recognition of the students is very rewarding because they understand that the workers are not a software and they need actions to solve their problems. In other words, we do our best to train them for "action" to improve working conditions with a broad and socially oriented goal.

Singapore

4. The Ministry of Manpower has, over the years, increasingly emphasize the importance of human factors in the effort of improving workplace safety and health. Hence, this has raised the awareness and needs of human factors training. By working closer with the ministry, we have established an important niche of being the university of choice to provide human factors training especially in Workplace Safety and Health.
5. Through marketing the programme, either through workshops, talks, seminars and symposiums, we have managed to raise the level of awareness of what HF is (in Singapore) and that such a programme exists in our university. This has sometimes lead to requests for consultancy work or short duration workshops.
6. We have offered the first Full Human Factors undergraduate degree programme in Singapore and are currently training a whole new generation of human factors and safety professionals. That is a very commendable and satisfying achievement.
7. If you have experienced health issues related to the this activity (stress, etc.),
8. Ensuring that we have a healthy enrollment number is always stressful.

Switzerland

At the moment the changes in the recognition of health and safety experts by the authorities is helping to put some impetus into the development of the "patchwork" training scheme, as the authorities want trained ergonomists in the system.

USA

The positive aspects of the program include:

(1) making contributions to the human factors and ergonomics knowledge base;
(2) address industry and regional needs in terms of highly trained professionals in safety and ergonomics; and
(3) contributing to the reputation and ranking of the academic department housing the program.

Contributing factors include a steady stream of high quality international students into our graduate program as well as qualified faculty members for delivering high quality instruction. Other contributing factors include internal support from the coordinating center and department as well as external grants for sponsorship of student research. One topic for Roundtable discussion could be sources of external support for ergonomics academic program development and maintenance.

Concerning having experienced health issues related to this activity (stress, etc.): "I have not experienced any diagnosed health conditions that have been occupationally related. However, there are program management requirements that are demanding and create levels of mental and physical stress. These include student recruiting and advising, tracking of program and student progress, facilitating faculty and student collaboration in research, program reporting, and competing program funding renewals. Another topic for discussion at the Roundtable might be faculty strategies for coping with cognitive stress associated with academic program management.

3.3 Coded Answers

The former results were coded in the following categories and showed that:

- Many ergonomics/human factors education programs are taught in different universities around the world.
- Some of these programs are quite new (less than 5 years of existence), others emerged from older programs due to an adjustment process.
- Causes of adjusting scope of education programs are: (a) needs of industry in terms of highly trained ergonomists, (b) demands of students, (c) pressures of the university to adjust to constitutional constraints.
- Challenges facing education programs are: (a) Lack of awareness of ergonomics among stakeholders (b) Lack of industry support for the discipline (c) recruiting students to graduate programs (d) Lack of experienced HF professionals who could be engaged to teach in the programs.
- Permanent marketing of Ergonomic/HF programs are essential for their success.
- To be in charge of an ergonomics education programs is stressful in situations like: (a) being the main faculty member with ergonomics expertise (b) lack of awareness of ergonomics among stakeholders.

4 Discussion

The present study is not exhaustive, for many reasons:

1. It is only one of the first steps in a large perspective, which aims at enlarging the professional education processes in E/HF all over the world, as this world, due the globalization of products and services, is becoming closer. Ergonomics/Human Factors, therefore has to lead this global process, rather than to be led; education and training is the corner stone in a worldwide ergonomics strategy.
2. Particularly because of the timing pressure we could not reach all ergonomics education programs, especially those well-established in Europe, the present project aims at bringing more ergonomists in a collective effort to think together about Ergonomics Professional Education.
3. Although, the project tends to be ambitious, the scope of the present study is realistic, in that it gathered some of the training and education programs around the table in the same place, thanks to IEA 2018 for providing this opportunity.

Considering our study is not exhaustive we therefore believe it is only a first step in our necessary collective effort to think together Ergonomists Professional Education. Nevertheless, from our findings we can learn the following lessons and think about some ideas:

As a process, promoting ergonomics worldwide should start by the training and education aspects. The analysis of the sample of 8 ergonomics education programs who will present their case in our Symposium, pointed out the high success rate among these programs.

To fix some perspective milestones, the present study has pointed out three main difficulties that face most of the ergonomics training programs: (1) lack of ergonomics knowledge among companies, (2) lack of ergonomics knowledge among students, and (3) the lack of professionals to teach ergonomics (4) proliferation of regulations that open opportunities for other professions that compete with ergonomics in the market for the prevention of occupational risks, that weaken the efforts made by the few existing professional ergonomists in the design and real practice of professional ergonomists education programs.

5 Conclusions

To overcome these and other hurdles, we could plead for an "offensive" ergonomics education strategy, or an Ergonomics Education development based on the idea of offering and not just answering to the market's demand. This goes through the promotion of the discipline and the profession of ergonomics with all stakeholders. The company/university conventions are one way. Hence, the pedagogy could be revisited by emphasizing the field modules of ergonomic analysis of work and design.

As for professional speakers, video conferences between universities should be considered as one solution among many others. It is clear, at this level, only professional ergonomists from different backgrounds can bring lasting transformations. IEA could help in this sense, through facilitating expert's mobility and through the strengthening of the international certification system of ergonomists and education programs.

The present study shows the advantage of the mutual exchange between training programs worldwide, can further boost their mutual experience. We hope this research will add our "grain of sand" in the necessary collective work of IEA's "Professional Standards and Education Standing Committee".

References

Aslanides M, Cuenca G, Del Rosso R (2011) Modelo de certificación nacional del profesional ergónomo. Documento interno de la Comisión de certificación de ADEA

Aslanides M, Cuenca G, Del Rosso R (2010a) Documento de Reconocimiento del ergónomo profesional en la Argentina. Comisión de certificación ADEA

Aslanides M, Cuenca G, Del Rosso R (2010b) La Ergonomía en Argentina. Revista de la Cámara Argentina de Seguridad, pp 3–4

Arévalo N, Aslanides M, Ghram R, Mebarki B (2017) L'avenir de la Formation d'Ergonomes dans des contextes «arides» et «avides» d'ergonomie: enjeux d'une certification professionnelle internationale fiable. Actes du congres de la SELF 2017, Toulouse

Bridger R (2007) Critique of IEA basic document, guidelines and standards for accreditation of ergonomics education programmes at tertiary (university) level – Version 2, January 2003. Letter to David Caple (International Ergonomics Association president), 1 February 2007

Bridger RS (2009) Introduction to ergonomics, 3rd edn. CRC Press, Boca Raton

Dul I et al (2012) A strategy for human factors/ergonomics: developing the discipline and profession. Ergonomics 55(4):377–395

International Ergonomics Association Professional Standards and Education Committee (2001a) Criteria for IEA endorsement of certifying bodies, Version 4, October 2001. http://www.iea.cc/browse.php?contID=edu_criteria&phpMyAdmin=XPyBrlJQjtrNYKM50fpmCYvGm%2C8&phpMyAdmin=jLDUJrGUIxQ-3p3v5atPhaf1Xo8. Accessed 20 Aug 2012

International Ergonomics Association Professional Standards and Education Committee (2001b) Minimum criteria for the process of certification of an ergonomist – Version 4, October 2001. http://www.iea.cc/upload/IEAPSE_MinCriteriaErgonomistCertificationProcess_v4_1001.pdf. Accessed 20 Aug 2012

Japan Ergonomics Society (2007) Application for the IEA professional certification endorsement by Japan Ergonomics Society, Tokyo, Japan

Poy M (2006) Aspectos funcionales de los riesgos y desvíos de las normas de seguridad en el trabajo: un aporte a la comprensión de las relaciones entre actividad humana y seguridad. Tesis doctoral en psicología. Universidad de Palermo, Argentina

Poy M, Gomes JO, Soares M (2006) L'analyse de l'activité: expériences sud-américaines. Les cas de l'Argentine et du Brésil. In: L'analyse du travail en perspectives: Influences et évolutions. Valléry, G. et Amalberti R., éditeurs. Octarès Éditions, Toulouse, pp 79–96

Rookmaaker DP, Hurts CMM, Corlett EN, Queinnec Y, Schwier W (1992) Towards a European registration model for ergonomics. Final report of the working group Harmonising European Training Programs for the Ergonomics Profession (HETPEP), Leiden, Netherlands

Soares M (2009) Ergonomics in developing regions. Needs and applications. In: Scott PA (ed). CRC Press, Boca Raton. Print ISBN 978-1-4200-7911-1, eBook ISBN 978-1-4200-7912-8. https://doi.org/10.1201/9781420079128.ch23

Soares M (2006) Ergonomics in Latin America: background, trends and challenges. Appl Ergon 37(4):555–561

Smith TJ (2012) Certification of professional ergonomists a global perspective. Ergon Des Q Hum Factors Appl 20(4):22–28. https://doi.org/10.1177/1064804612455639

UQAM - DESS en intervention ergonomique en santé et sécurité au travail. https://etudier.uqam.ca/programme?code=1851

Creation of a Database for the Management of Working Gesture Rehabilitation of the Injured Worker

A. Resti[1(✉)], M. T. Covelli[1], C. Melai[2], M. Paoli[1], L. Pieroni[2], S. Tavolucci[2], C. Tonelli[2], and S. Verdesca[2]

[1] Sovrintendenza Medica Regionale, Via degli Orti Oricellari, 11, Florence, Tuscany, Italy
a.resti@inail.it
[2] Centro Polidiagnostico Regionale, Florence, Italy

Abstract. The International Classification of Functioning of the World Health Organization (ICF) is an innovative instrument of classification accepted by 191 countries as the international standard to measure and classify health and disability. Its use has an important impact on medical practice, research, population statistics and on social and health policies.

The substantial innovation of ICF is in the methodological approach to health that takes into account subject wholeness and his environment.

The rehabilitation treatment of the worker assumes a rehabilitative diagnosis by means of a multi-dimensional and social assessment, and it uses ICF bio-psycho-social model as a reference.

This model requires a methodological analysis in order to shape an operating profile after an accident at work or an occupation disease.

The aim of this study was to assess ICF classification use during the injured worker management, and to contribute to the definition of a specific Core Set for the evaluation of the working gesture, by providing a systematic coding scheme for Inail health information systems.

The primary goal was to create a database of INAIL working gesture, in order to obtain objective data for future research, fulfilling needs for knowledge and information. Another aim of the study was to evaluate the effectiveness of working gesture-oriented rehabilitation approach, that is for the wealth function of Inail an important challenge.

Between January 2015 and December 2017, we assessed 37 subjects referring to the outpatient clinic of a physical medicine and rehabilitation department of Inail. Of these 37 subjects, 30 underwent a specific working gesture-oriented rehabilitation, in addition to functional rehabilitation, totalizing 359 treatments on them.

Keywords: Rehabilitation · Working gesture · ICF core-set · Database

1 Introduction

Each rehabilitation intervention must integrate persons suffering an accident at work in a global manner, using all the tools and resources that facilitate their integration, including work.

S. Bagnara et al. (Eds.): IEA 2018, AISC 821, pp. 727–735, 2019.
https://doi.org/10.1007/978-3-319-96080-7_88

Workers who return to work after an injury often need to change the characteristics of their task or even work environment. Therefore, the set consisting of working gesture rehabilitation is an important tool for reintegration at work.

Since 2015 in our center we perform many rehabilitation programs aimed at working gesture recovery.

The aim of this study was to assess ICF classification use during the injured worker management, and to contribute to the definition of a specific "Core Set" for the evaluation of the working gesture, by providing a systematic coding scheme for Inail health information systems.

The primary goal was to create a database of INAIL working gesture, in order to obtain objective data for future research, fulfilling needs for knowledge and information (see Fig. 1).

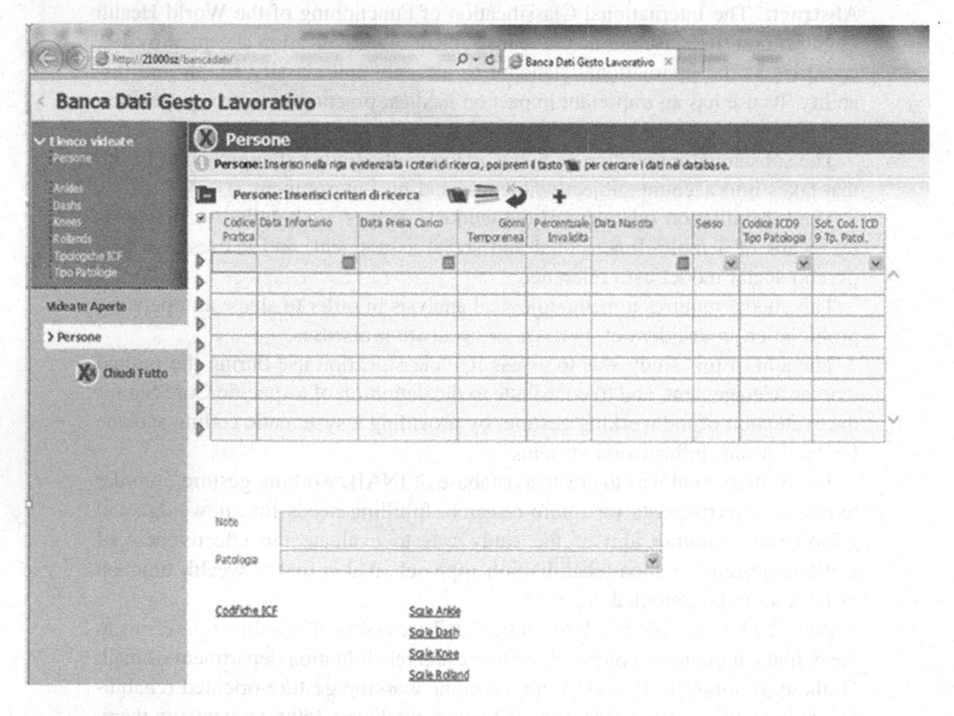

Fig. 1. Working gesture database.

2 Methods

We included 37 injured workers, based on the nature and the severity of the disability in the upper limb or the lower limb, which underwent an ICF evaluation to find how much their disability affected their work activity [1]. For this purpose we selected a specific ICF core-set (11 items) evaluating the inherent ability to perform a task or action during the work activity [2] (see Fig. 2). Each item is scored from 0 to 4 (0 = no problem; 4 = complete problem). We summed in the ICF final score the items that resulted as occasional or characterizing.

ICF	WORKING GESTURE	CHARACTERISTIC	OCCASIONAL	ICF óCAPACITY .	AID**	EVALUATION WITH TESTS***
d410	*Changing basic body position*					
d4103	Sitting					
d4104	Standing					
d4102	Kneeling					
d4101	Squatting					
d4100	Lying down					
d4105	Bending					
d4106	Shifting the bodyís centre of gravity					
d415	*Maintaining a body position*					
d4153	Maintaining a sitting position					
d4154	Maintaining a standing position					
d4152	Maintaining a kneeling position					
d4151	Maintaining a squatting position					
d4150	Maintaining a lying position					
d450	*Walking*					
d4500	Walking short distances					
d4501	Walking long distances					
d4502	Walking on different surfaces					
d4503	Walking around obstacles					
d455	*Moving around*					
d4551	Climbing					
d4550	crawling					
	Limbs					
d4401	grasping					
d4402	manipulating					
	Complex activities					
d4453	Turning or twisting the hands or arms					
d4450	Pulling					
d4451	Pushing					
d430	*Lifting and carrying objects*					
d4301	Carrying in the hands					
	Work situations					
d2202	Undertaking multiple tasks independently					
d4751	Driving motorized vehicles					
d4702	Using public motorized transportation					
b455	*Exercise tolerance functions*					

Fig. 2. ICF core-set.

Study population was divided according to professional category [3] (see Fig. 3).

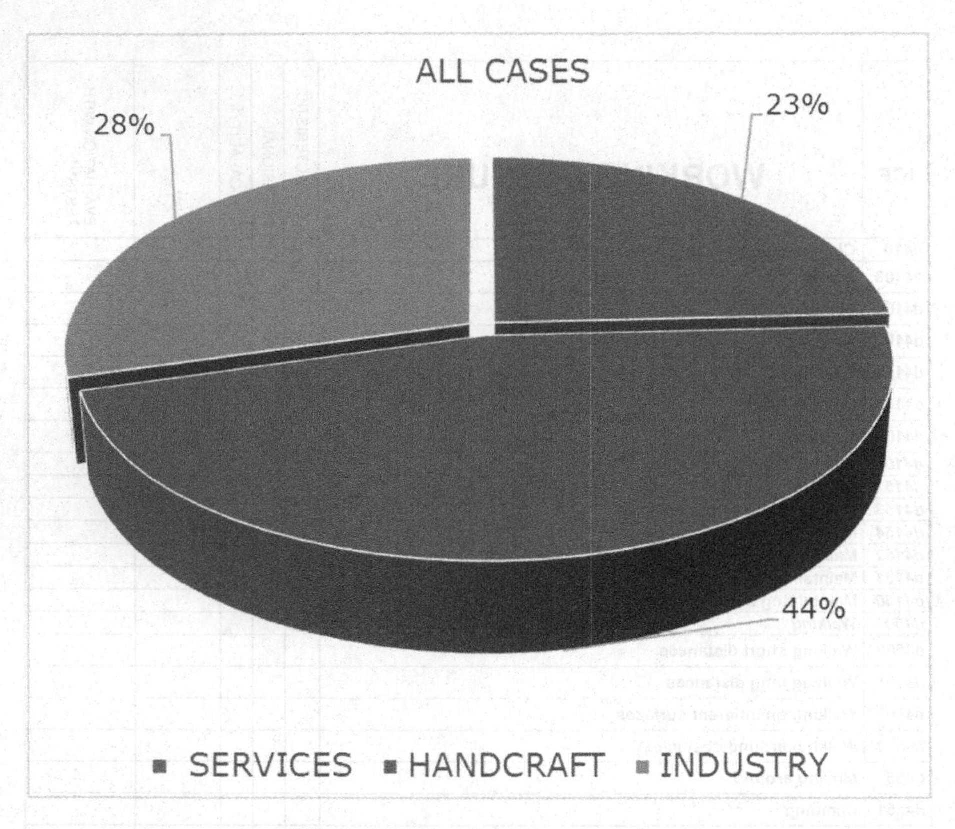

Fig. 3. Graphic of working categories.

3 Results

In the "services" and "industry" categories all the analyzed items were almost equally impaired, except for the items regarding the ability to reach the working site (see Figs. 4 and 5).

In the "handcraft" category the most impaired items were "work rate" and "to lift or lower loads" (see Fig. 6).

In all the three categories we found that the most impaired ICF items during work activities were "to lift and/or to lower loads" and "to tolerate work rate".

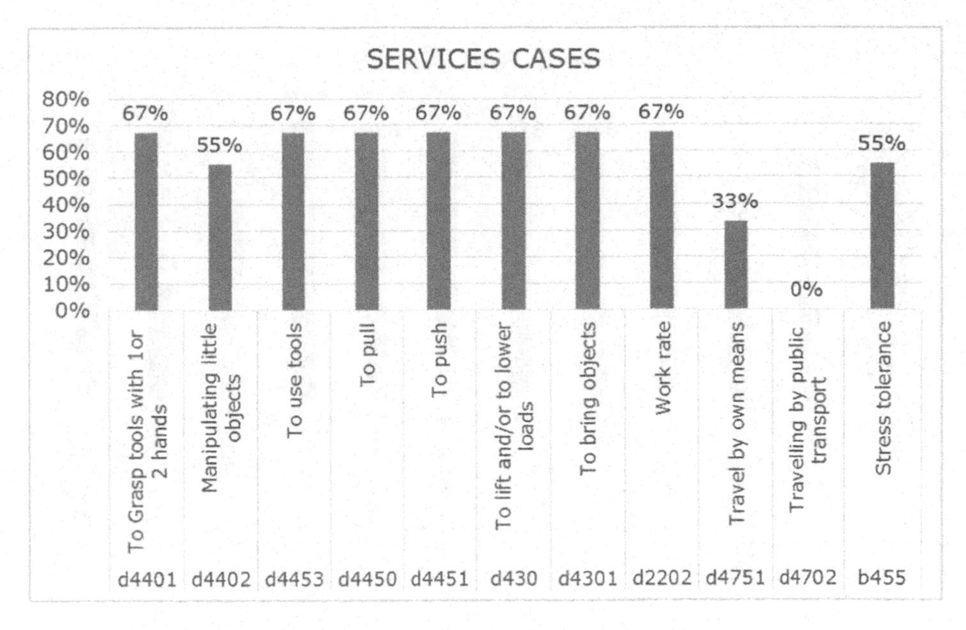

Fig. 4. Graphic of "Services cases" analysed items.

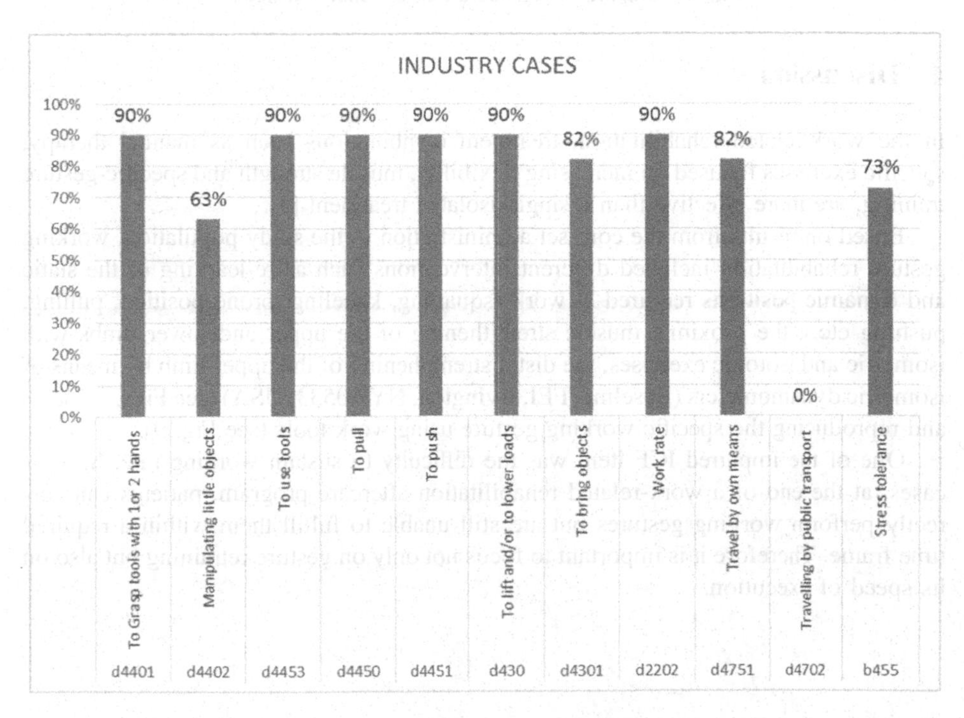

Fig. 5. Graphic of "Industry cases" analysed items.

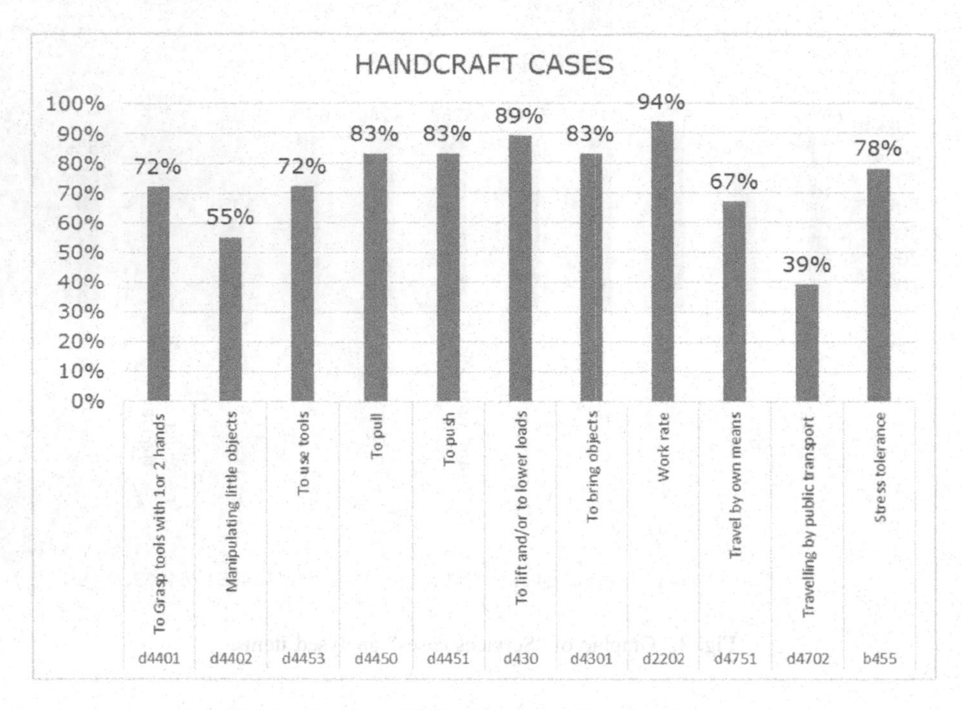

Fig. 6. Graphic of "Handcraft cases" analysed items.

4 Discussion

In the work-related rehabilitation, treatment combinations such as manual therapy, specific exercises focused on increasing flexibility, muscle strength and specific-gesture training, are more effective than a single isolated treatment [4].

Based on results from the core-set administration to the study population, working gesture rehabilitation included different interventions such as re-learning of the static and dynamic positions required at work (squatting, kneeling, prone position, pulling, pushing etc.), the proximal muscle strengthening of the upper and lower limbs with isometric and isotonic exercises, the distal strengthening of the upper limb by means of isometric dynamometers (Baseline, FEI, Irvington, NY 10533, USA) (see Figs. 7 and 8) and reproducing the specific working gesture using work tools (see Fig. 9).

One of the impaired ICF item was the difficulty to sustain working rate. In most cases, at the end of a work-related rehabilitation aftercare program, patients can correctly perform working gestures but are still unable to fulfill them within a required time frame. Therefore it is important to focus not only on gesture retraining but also on its speed of execution.

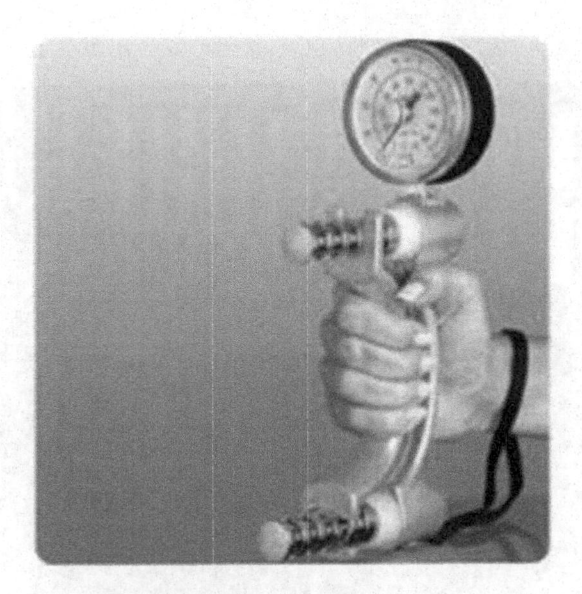

Fig. 7. Hydraulic handgrip gauge.

Fig. 8. Hydraulic pinch gauge.

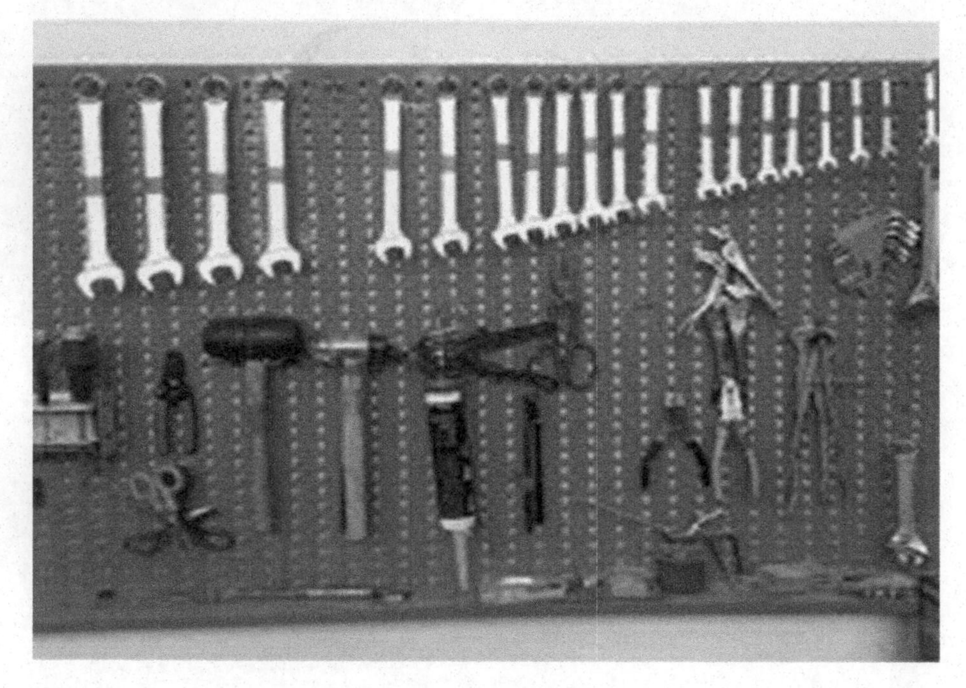

Fig. 9. Work station.

5 Conclusion

The ICF "core set" appears to be a suitable tool for exploring different rehabilitation needs among injured workers and for improving efficiency of work rehabilitation process.

It may be helpful to make more training choices, more worker input into the retraining goal, and to provide a better fit of the retraining goal with the workers' experience and abilities [5].

The work-focused rehabilitation provided by INAIL is therefore a comprehensive way of taking care of the injured worker. This approach requires the contribution of various professionals who work synergistically. The final aim is the recovery of the highest possible level of operation and participation of the worker.

Future studies would be required to evaluate the effectiveness of work-oriented rehabilitation approach on reducing chronic disability among injured workers and on improving efficiency of work rehabilitation process, that is an imperative challenge for the wealth function of INAIL.

Conflict of Interest
The authors declare that they have no conflict of interest.

References

1. World Health Organization (WHO) (2001) International Classification of Functioning, Disability and Health (ICF). Geneva. http://www.who.int
2. Aringer M, Stamm TA, Pisetsky DS, Yarboro CH, Cieza A, Smolen JS, Stucki G (2006) ICF core sets: how to specify impairment and function in systemic lupus erythematosus. Lupus 15 (4):248–253
3. American Physical Therapy Association-APTA (1992) Guidelines for programs in industrial rehabilitation. Mag Phys Therapy 1:69–72
4. Bertolini C et al (2010) Il ruolo della terapia occupazionale nella riabilitazione dell'infortunato sul lavoro. G Ital Med Lav Erg 32(Suppl. 172):4
5. Imbriani M, Bazzini G, Franchignoni F (2006) Argomenti di terapia occupazionale Aracne Srl, Rome

Forensic

Science and the Law

Clashing World Views in Search of a Common Understanding

P. A. Hancock(✉)

University of Central Florida, Orlando, FL 32826, USA
peter.hancock@ucf.edu

Abstract. The present work examines some of the clashes and contentions that exist between science (the way in which rational exposition seeks to understand the world around us) and the law (the encoded incarnation of sufficient levels of human agreement within specific societies). In theory, these ways of knowing should act in accord with each other. In practice, often they do not. The typical and traditional way to finesse any particular inconsistency between these realms of human activity is to assert that science and law occupy different 'magesteria' of thought. That is, they differ so radically that they may even necessarily be incommensurate one with the other. This assertion is tenable because science and the law are each framed by, react to, and serve sufficiently different constituencies of the human enterprise so as to reasonably be identified and separated in this way. However, I reject these so-called 'magesterial' divisions. To support such a rejection, I here compare science and law and how they each approach their mutual experience of a common reality. I look, albeit very briefly, to explore the historic antecedents of the current schism between the two and observe that not all cultures have rended apart these dual faces of humanity. Yet, unfortunately, this intellectual schizophrenia has burgeoned as a result of the pervasiveness of technology and the influence of the west in framing the global worldview. These concerns are expressed around questions as to ways in which scientific expertise is treated in the legal field. But, even as we view these cited interactions we can at the same time, also ask whether it id possible to actually enact a 'science' of law. The latter effort might, for example, specify when and where scientific understanding necessarily has to take priority over legal precedence. In conclusion, philosophical ways are sought through which the fundamental division between law and science can be healed to the betterment of our common future.

Keywords: Science · Law · Culture clash · Dissonance

1 Introduction

The following paper represents only a brief precis of a much more extensive work which is in the process of completion. The larger essay concerns and considers a full perspective on the disparate nature of science and the law [1]. The central purpose of that larger effort is to work toward some degree of acceptable and formalized reconciliation

S. Bagnara et al. (Eds.): IEA 2018, AISC 821, pp. 739–744, 2019.
https://doi.org/10.1007/978-3-319-96080-7_89

between the two, as opposed to the rather uneasy philosophical truce that they presently seem to bear one another. The principal examples that are used in the larger exposition come from the area of Forensic Human Factors and Ergonomics. However, the generality of the argument to many other domains of discourse where science and law meet and interact will, hopefully, be evident. That the present discussions are largely constrained to the perspective of English-speaking nations is a matter only of practical convenience and not a fault of either willful ignorance or neglect of other cultures and their own different approaches to this pervasive human question.

2 The Fundamental Schism

The first issue which founds all of the present discussion concerns the disparate philosophical foundations from which (western) science and the law (e.g., English Common Law) emanate. These basic differences are those from which many of the subsequent dissonances and contentions emerge. Some have argued that science and law occupy differing 'magesteria' [2]. That is, they are fundamentally different ways of knowing that produce their own disparate bodies of knowledge and understanding. As a vast generalization, it may be proposed that the law derives from a Manichean worldview in which opposite, but essentially equal and balanced sides oppose to each other. Hence aphorisms which we use today such as '*there are two sides to every argument.*' Parenthetically, this same ethos appears to underlie our media who are constantly looking to present '*both sides of the argument*' even if one side is total nonsense. And, with respect to the latter of course, in a celebrity-driven world there is *always* someone who can and will present the contrarian view. In contrast to this notion of equal and balanced forces, science is much more attuned to an Augustinian understanding of reality. In part, this latter notion acknowledges that nature is complex and our understanding of it is necessarily limited, but nature does not actively oppose our understanding of it in the same way that another opposing human agency might. Nature does not hack into computers at night to delete scientific data, at least not intentionally, in order to prevent discoveries being made. Also, in large part, nature does not 'change the rules of the game' partway through any exploration to hide solutions to difficult problems.

Lest it be thought that such philosophical contentions between such worldviews are simply arcane, academic disputes, it must be carefully understood that they are not. Such fundamental concerns still permeate the power structures of our world today and will almost certainly dominate it for the immediate foreseeable future. Take for example the following quotation: "*they knew who their enemies were and shouldn't fall into the trap of trying to make them their friends, because they wouldn't be; and so, from day one they should consider themselves on a war footing.*" This observation occurs on page forty-four of Wolff's recent book entitled '*Fire and Fury*' which recounts the first few months of the Trump Presidency [3]. It is redolent with the oppositional nature of such a worldview. It is imbued with the idea of fundamental adversity. This self-same principle underwrites other previous Presidential announcements about '*axes of evil.*' It is difficult to over-emphasize the impact of this vision of the world as epitomized by the self and the other, or more formally in psychological theory, the 'in-group' and the 'out-group' [4]. As politics and the law are essentially symbiotic partners, so the standpoint that you are "*either with us or against us*," also derives from this applicable and associated, us vs. them, mind-set [5]. This form of opponential mindset can easily dominate the politico-legal theater of human experience and discourse. It need not necessarily lead to implacable division but the perceptual processes which underlie mindset development tend toward this form of absolute division [6]. Immovable opinions, that respond to no appeal to fact or rational argument must eventually be to the detriment of all. As Fischer reported, Gandhi is reputed to have opined that the conception of an 'eye for an eye' eventually leaves the whole world blind.

To trace the philosophical division between what became science and law to its origins, we can reference the Manichean heresy. This was the proposition that the universe is divided between forces of light and forces of darkness that stand in direct and, largely, equal juxtaposition to each other [7]. The notion is neither unique to, nor original with Mani, but the term Manichean is a useful shorthand one to denote the critical dimension of division and opposition. Parenthetically, the Manichean heresy only became a heresy when it was outlawed by the Roman central government [8]. Thus, in psychological terms, if one is a Manichean, and one is then persecuted for that belief, the persecution itself serves to reinforce the fundamental truth of the forces of light vs. darkness contrast. Tragically then, this 'us versus them' paranoia can be very self-reinforcing if there are no rational constraints imposed upon it. The 'forces of darkness vs. forces of light' ethos underwrites many of the story themes, for example, of our modern-day entertainment [9]. Indeed, this is a common narrative which persists throughout all of human history and crosses virtually all cultures [10]. In such mythical worlds, there are heroes who are winners and there are villains who are losers. Often the role of hero or villain is only identified after the winner has emerged and history has been rewritten to accord with the new social reality [11]. Law formally introduces some rules into such domains, but often untamed savagery eschews even these conventions, as we see in transgressions involved in putative 'war crimes' [12]. It could be argued that the Manichean philosophical position is founded upon a purely pristine version of the Darwinian notion of survival of the fittest and, as such, is intrinsic to all predator-prey interaction. In this sense, the contrarian hypothesis offered in the Manichean heresy might be said to represent the foundational past of human social organization and may ever be one of the wellsprings of human ascendency.

In contrast, science does not represent a zero-sum situation in which inherently limited resources that are secured by the winners are necessary lost by the losers. In fact, rather the opposite. If one person does well in terms of making discoveries, another is not diminished by this success. In fact, someone else's discoveries can be the precise bases for others' advances. It is, of course, often true that the social hierarchy of science is not blind to the ambition entrenched within the Manichean heresy. We do witness instances in science of those who seek pre-eminence through *ad hominem* attack rather than the contribution of their own fundamental insights. Yet despite such occasions, the essential purpose of science is one of collective and mutual advance. The arbitrator of such advance is nature and should never be any powerful or privileged human agency. So much, pro tem, for foundational divisions. What do these differences mean in practice?

3 Winners and Losers

In legal disputes, which derive when humans argue with each other in some fashion, there are winners and there are losers. While it is true that it is possible to have processes such as arbitration and mediation. and so the winning and losing is not necessarily an absolute proposition, the fundamental object for each 'side' is to win. The law imposes some rules on this battle which are 'interpreted' by each side in their own most favorable fashion. Science cannot work like this. One cannot bargain with reality, nor submit appeals for it to change its mind. The great problem in recent years has been that individuals, imbued in the law, can come to believe that they 'make' their own reality and that natural constraints are not necessarily imposed upon them. However, as the physicist Richard Feynman opined; "*For a successful technology, reality must take precedence over public relations, for Nature cannot be fooled*" [13]. In law, as in politics, one can seek to silence one's opposition. There are legal procedures through which science, in the form of qualified expert witnesses, can be dismissed from proceedings and their opinion rendered null and void. In its essence, this strategy is a form of what might be more colloquially known as '*playing the man not the ball.*' It is this form of mindset that allows some in politics to believe that inconvenient or unwelcomed realities can be denied and/or nullified. It is why in politics some individuals find great frustration in the incompatibility with desired perceptions and actual realities [14]. In testing the veracity and vitality of any applicable scientific information, the law rightly seeks to impose a high standard of quality, such that not all individuals qualify as 'experts.' These are often referred to as the Daubert or Frye criteria in the United States [15]. Yet, most unfortunately, many if not most scientific experts are employed by one 'side' or the other. This partisanship almost inevitably leads to argumentation and dispute, and not necessarily clear inquiry and exposition. Such remarks do not apply ubiquitously since different aspects of the law, e.g., civil vs. criminal law, operate in somewhat differing ways.

The myth that truth emerges from such human disputes must surely be challenged as a ubiquitously true assertion. For even as those qualified and versed in the legal domain have asserted: "*But let no one pretend that our system of justice is a search for truth. It is nothing of the kind. It is a contest between two sides played according to certain rules, and if the truth happens to emerge as the result of the contest, then that is pure windfall. But it is unlikely to. It is not something with which the contestants are concerned. They are concerned only that the game should be played according to the rules. There are many rules and one of them is that some questions which might provide a shortcut to the truth are not allowed to be asked, and those that are asked are not allowed to be answered. The result is that verdicts are often reached haphazardly, for the wrong reasons, in spite of the evidence, and may or may not coincide with the literal truth. The tragedy of our courts is that means have come to count more than ends, form more than content, appearance more than reality*" [15]. Such are the frequency of statements concerning the above effect that I have provided only one example when so many others can readily be accessed.

4 What Has Gone Before

Science and the law differ also in the ways that they employ previous states of understanding. The law is highly influenced by precedent. That is, previous applicable rulings that have been recorded. Of course, precedent is necessarily of somewhat limited use. For, if there were a direct and clear precedent that could be applied then the dispute would be unequivocally resolved, as long as the specific precedent was adhered to. This might make us think that eventually law may become algorithmic and deterministic if a sufficiently broad and comprehensive body of precedent were collected together. However, lawyers are individuals who argue on both principle and interest, hence the legal world remains awash with disputed cases. Science accesses its own literature. Such reviews are equally an essential foundation for what work then follows. Yet, the collective assemblage of scientific knowledge at any one time is never fully determinative. Science practices an on-going mantra of doubt. There are evident overlaps between science and law here, since they are each accessing repositories of understanding. It is the subsequent utilization of that understanding that differentiates between them. However impressive, and whatever the standing of the prior scientist who utters the recorded words, the state of knowledge is and should be, constantly challenged in science. The same imperative does not necessarily apply to the law. In both domains, the body of understanding is cumulative, but the radical 'paradigm shifts' that characterize science do not appear to have had an equivalence in law; as yet.

5 Rapprochement and Reconciliation

In the foregoing I have presented an extremely restricted selection of issues that differ between science and law. I am at pains to formally emphasize that this is only a very limited set of concerns in what is a very much wider discussion. For example, legal procedures vary widely across nations and cultures and what I have observed here is certainly not globally applicable. In juxtaposing different facets of science and the law, I am in danger of adopting the Manichean heresy myself. And, as a scientist it is tempting to cast science as the force of light and thus law as its opponent. I eschew this temptation however, since I believe that science can serve the purposes of the law and also, that there can be a formal 'science' of law itself. Further, legal understanding has much to inform science in terms of its policies and procedures. In short, the two need not necessarily be oppositional in any way. One first step that I would advocate is for the scientific expert to be appointed by the court. This is not a new suggestion, nor is it one that has not been practiced in various contexts. However, its wider application may well be helpful. It would require the creation of a cadre of formally appointed scientific experts whose qualifications should perhaps be certified by the major applicable professional bodies. It would create a new professional categorization, but the acrimonious dimension of unending dispute might thus be mitigated, at least to some degree. Some people, especially those who are consistent 'winners' in the present system, might well reject such a proposal but to generate an essential accord science must exercise a rational and impartial role. It is true that scientific understanding is incomplete. Further, scientists as humans are also vulnerable to mistakes. Reviews of their testimony should be treated in the same skeptical manner as for all of the body of scientific understanding. But making scientists partisans in the black and white world of Mani's heresy fails to start us down the road to a much more sympathetic, informative, and just world to which, arguably, we should all aspire.

Acknowledgments. I am very grateful for the comments and recommendations provided by Professor Michael Wogalter in creating the final version of the present work. While the present opinions are solely my own, I remain grateful for his insights and directions.

References

1. Hancock PA (2018) Science in court. Manuscript in progress
2. Gould SJ (2002) Rocks of ages: science and religion in the fullness of life. Ballantine Books, New York
3. Wolff M (2018) Fire and fury: inside the Trump White House. Henry Holt, New York
4. Quattrone GA, Jones EE (1980) The perception of variability within in-groups and out-groups: implications for the law of small numbers. J Pers Soc Psychol 38(1):141–152
5. Dweck CS (2006) Mindset. Random House, New York
6. Lamme VA (1995) The neurophysiology of figure-ground segregation in primary visual cortex. J Neurosci 15(2):1605–1615
7. Wilson RM (1967) Mani and Manichaeism. In: Edwards P (ed) The encyclopedia of philosophy. Macmillan, New York
8. Gardner I, Lieu SN (eds) (2004) Manichaean texts from the Roman Empire. Cambridge University Press, Cambridge
9. Hancock PA, Hancock GM (2008) Is there a super-hero in all of us? In: Rosenberg RS, Canzoneri J (eds) The psychology of super-heroes. Benbella Books, Dallas, pp 105–117
10. Campbell J (2008) The hero with a thousand faces. New World Library, New York
11. Hancock PA (2009) Richard III and the murder in the tower. History Press, Stroud
12. Greenspan M (1959) The modern law of land warfare. University of California Press, Berkeley
13. https://www.edwardtufte.com/bboard/q-and-a-fetch-msg?msg_id=0002gi
14. Gore A (2006) An inconvenient truth: the planetary emergency of global warming and what we can do about it. Rodale, Emmaus
15. Groscup JL, Penrod SD, Studebaker CA, Huss MT, O'Neil KM (2002) The effects of Daubert on the admissibility of expert testimony in state and federal criminal cases. Psychol Pub Policy Law 8(4):339–372
16. Kennedy L (1965) The trial of Stephen Ward. Simon & Schuster, New York

Context Fidelity as a Factor in Forensic Site Inspection

Ian Y. Noy[(✉)]

Independent Consultant, 8116 Bibiana Way, Fort Myers, FL 33912, USA
iea2021chair@gmail.com

Abstract. Forensic HFE attempts to understand and explain factors that may have contributed to specific human-system failures or injury events. It involves, by definition, after-the-fact review of available evidence as well as activity analysis of all involved parties. If at all feasible, a comprehensive site inspection and evaluation can identify important contributing factors. Two case studies are described that illustrate the value of site inspection under conditions that correspond as much as reasonably possible to the actual conditions that existed at the time of the event of interest.

Keywords: Forensic Human Factors · Motor vehicle crash · Optical factors Driving · Vision

1 Introduction

Forensic Human Factors and Ergonomics (HFE) is the application of knowledge about human capabilities and limitations in an attempt to understand and explain factors that may have contributed to specific human-system failures or injury events (Noy and Karwowsky 2004). The science of HFE is broad in that it covers all aspects of human activity that can be influenced by cognitive, physical, physiological, behavioral and organizational factors. Forensic HF (Krauss and Olson 2015) involves, by definition, after-the-fact review of available evidence as well as activity analysis of all involved parties. An important aspect of gathering evidence is a site visit to gain a first-hand appreciation of the relevant activities and the conditions that were prevalent at the time of the injury event as well as other factors that may have contributed to the event.

During the data gathering stage, the importance of a comprehensive site inspection and evaluation cannot be overstated. Moreover, it is critical to perform the inspection under conditions that represent as far as possible the actual conditions that existed at the time of the event of interest. The reason is simply that some relevant factors may only become salient under conditions that existed at the time. To illustrate the value of context fidelity during site inspection, two case studies are described in which analysis of the conditions that existed at the time of the injury event led to an understanding of likely contributing factors. Both cases involved fatal vehicle crashes.

In both cases, the predominant factor was the visual environment at night and its effect on driver perception and response. The measure most commonly used to characterize driver performance is driver perception-reaction time (PRT). PRT is the total

S. Bagnara et al. (Eds.): IEA 2018, AISC 821, pp. 745–750, 2019.
https://doi.org/10.1007/978-3-319-96080-7_90

time taken by a driver to initiate a response to the appearance of a sudden hazard. There are four principal steps involved in this process: (1) Detection: the step during which the driver becomes aware that a hazard (object or condition) is present, (2) Identification: the step during which the driver acquires enough information about the hazard to be able to recognize and assess the risks associated with the hazard, (3) Decision: the step during which the driver decides what action, if any, is appropriate, and (4) Response: the step during which the brain issues the commands to the appropriate muscle groups to initiate the response.

The visual environment at night can have important effects on conspicuity, visual recognition, and response selection, which in turn can have adverse effects on PRT, as will be illustrated by the case studies.

2 Case 1 (Visual Illusion at a Crest in the Road)

In this case, the defendant was driving southbound on a paved two-lane country road. He changed to the northbound lane to pass two vehicles when he crashed head-on with a northbound vehicle. The collision occurred at the top of a crest in the road. According to the police report, it was clear and dark at the time of the collision. The principal question was whether the defendant should have seen or been aware of the approaching vehicle and whether he was careless in overtaking.

Site observations were made at night on the same date as the crash, during both daylight and nighttime, at time of the collision. Several drives were made in both directions, and while following other vehicles. Photographs were taken. The site inspection revealed that the terrain leading to the accident location was relatively flat, with the exception of the area of the collision, where south of the point of impact the road elevation drops markedly with a downgrade of approximately 8%. At night, the local topography is such that looking southbound from a vantage point at the point of impact (POI), the crest and the dip in the road beyond the crest are not discernable. In effect, a valley is created that is completely hidden where vehicles would not be seen. Headlights from approaching vehicles in the distance beyond the valley line up with the road alignment, creating the illusion that the road is flat and clear for several kilometers. The first cue of an approaching vehicle at night is the direct view of the headlights as the vehicle emerges from the valley.

Photos 1 and 2 show the view of the road looking southbound from a location about 1.5 s north of the point of impact. As can be seen in Photo 2, at night, the crest in the road ahead is not evident and there is no indication that the road drops in such a way as to completely obstruct the view of approaching vehicles.

Given the road profile and closing speeds of the two vehicles, it was calculated that the defendant would have been about 42 m north of the POI when he would have had the first direct view of the headlights of an approaching vehicle as it emerged from the valley. Figure 1 is the road profile, showing the relative location of the involved vehicles relative to the POI when the defendant had his first view of the headlights.

Observations were made of 10 approaching vehicles from a vantage point corresponding to the location of the defendant 42 m north of the POI. Time intervals during which the approaching vehicles disappeared into the valley and reappeared were

Photo 1. View looking south, 200 m from POI

Photo 2. Same view, at night

measured. The average length of time that a northbound vehicle was invisible was 43 s (range 38 s–50 s). During this period, there was no indication from a vantage point north of the crest that a vehicle was approaching. No light was visible until the vehicle was very close to the crest and then there was only faint light reflected from overhead power lines along the east shoulder.

Analyses based on the details available from the police report and data collected at the site, the two vehicles approached each other at a closing speed of 100–110 km/h and they were 72 m apart when the headlights first became conspicuous (refer to Fig. 1). From these data, the defendant saw the approaching vehicle 1.4–1.5 s prior to impact. The engineering reconstruction indicated that the defendant attempted to swerve to avoid impact. Thus, he reacted to the appearance of the approaching vehicle well within the normal perception-reaction-time of 1.5 s.

The opinion rendered was that the defendant drove within the bounds of reasonably alert and careful drivers (Krauss and Olson 2015). In particular, an unfamiliar driver

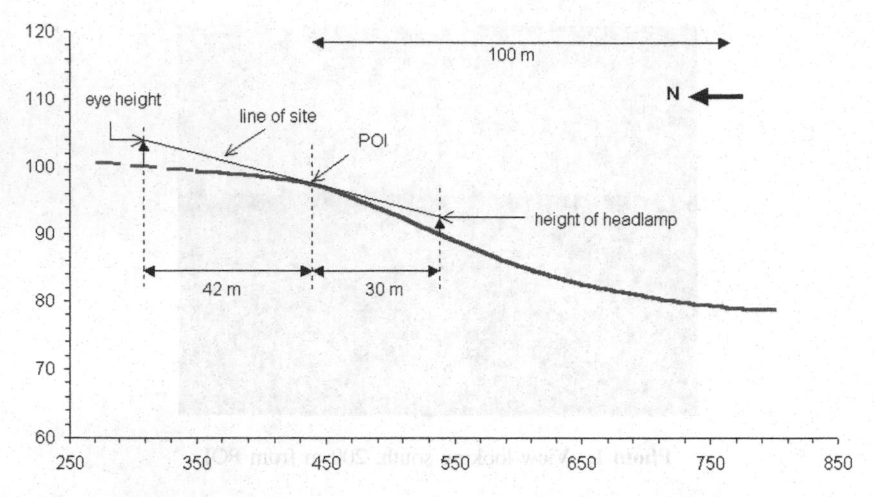

Fig. 1. Road profile and location of vehicles prior to collision

would not be able to discern the presence of the hill at night and would have a false sense that they can see far into the distance. Approaching vehicle headlamps were not visible until it was too late to react within normal perception-reaction time. Had there been a warning sign alerting drivers of the hidden roadway ahead, perhaps this collision might have been avoided.

3 Case 2 (Visual Confusion Due to Divided Road Curvature at Night)

This case concerned a station wagon that was involved in a head-on collision with a pick-up truck on a four-lane divided highway. The defendant (deceased) was driving the station wagon with 7 passengers eastbound in the passing lane while the pick-up truck was traveling westbound also in the eastbound passing lane. According to the Police Investigation Report, the driver of the truck had a blood alcohol concentration of 0.14%.

There is no question that the principal cause of the collision was the fact that the truck had been driving the wrong way on a divided highway. The issue addressed in the analysis was whether the defendant could possibly have detected and recognized the on-coming vehicle early enough to take effective avoidance action. According to the 'one percent rule' that was in effect in this particular jurisdiction, a defendant who is at least 1% liable for the collision may have to pay the plaintiff's entire claim if there are no other defendants able to satisfy judgment. In this case, the driver of the pickup truck had no insurance or other financial means. Accordingly, the forensic analysis assessed the visual factors associated with the collision and examined whether the defendant could reasonably have reacted in time to avoid the collision.

Site observations were made between the hours of 7:30 PM and 9:00 PM and again 11:00 PM to 00:30 the following morning. Then median was a ditch about 15 m wide in total with a guardrail running nominally along its centerline. The highway at night had light traffic, which was typical for this highway at midnight, corresponding to the time of collision. The site inspection revealed that looking eastbound from POI, the road curved to the left making a total angular displacement of 45°. From the defendant's viewpoint, both eastbound and westbound lanes appeared to converge at approximately 400 m ahead as they rounded the curve. It was estimated, however, that beyond 200 m, it was difficult to reliably judge that an on-coming vehicle was in the westbound passing lanes as opposed to the eastbound passing lanes.

Consistent with current knowledge about human perception and response (Krauss and Olson 2015) (Alexander and Lunenfeld 1975), the opinion rendered was that:

- It would have been very difficult for the defendant to ascertain that an on-coming vehicle was traveling in the wrong lane from more than 200 m due to the curvature in the road and lack of definition of the road lanes at night.
- The defendant's PRT was in the order of 3 s. This response was within the bounds of a normal, alert driver who is faced with factors such as poor conspicuity at night due to the optical merging of lanes at night, violation of expectation, and possible distraction from other vehicles (there was another vehicle adjacent to the defendant at the time traveling at about the same speed).
- The defendant also had to choose among response alternatives. He attempted to steer to the left, which was the most prudent avoidance reaction available to him under the circumstances. He may well have attempted to brake but had insufficient time (steering is faster than braking since the hands are already on the steering wheel) (Photograph 1).

Photograph 1. View eastbound from the driver's position in the passing lane

Lack of positive guidance, in this case in the form of roadway delineation, has been recognized as a critical factor in highway safety for decades (Alexander and Lunenfeld 1975). In this present case, the combination of road alignment ahead and visual factors at night creates ambiguity about which lane oncoming vehicles are in. Most of the time this is not an issue. This particular highway has gentle curves such as this all along its length of hundreds of miles. It is difficult to conceive of a cost-effective countermeasure that might have prevented this crash. As this collision is such a rare occurrence and it could potentially happen anywhere, the most viable countermeasure would be some form of electronic surveillance (Photograph 2).

Photograph 2. View at night of on-coming headlights in the westbound lanes

4 Conclusions

The two cases presented illustrate the value of context fidelity. Site observations at night were able to identify critical factors that likely contributed to the collisions. Of course, it may not always be possible to observe the site under conditions that replicate the event conditions for a number of reasons, including changes to the site, unavailability of critical elements or resources, trial timing, etc. Nonetheless, if feasible, viewing the site under fidelic conditions could reveal aspects that might otherwise not be apparent. In addition to helping identify contributing causes of injury events, forensic analyses can often inform potential countermeasures to improve safety.

References

Alexander GL, Lunenfeld H (1975) Positive guidance in traffic control. USDOT FHA, Washington, D.C.

Krauss D, Olson P (2015) Forensic aspects of driver perception and response. Lawyers and Judges Publishing Company, Tucson

Noy IA, Karwowski W (2004) Human factors in litigation. CRC Press, Boca Raton

A Forensic HF/E Analysis of a Trip and Fall Injury Event Involving a Wheel Stop in a Parking Lot

M. S. Wogalter(✉)

Psychology Department, North Carolina State University,
Raleigh, NC 27695-7650, USA
Wogalter@NCSU.edu

Abstract. Built (constructed) environments should have level surfaces to promote mobility and avoid injuries from trips and falls. Wheel stops (usually long concrete slabs with bolts to hold them in place) are sometimes placed in parking lots to limit and control how drivers position their vehicle in designated spaces. However, they present potential problems for pedestrians traversing through the area due to obstructed views and visual attention directed elsewhere. A forensic human factors and ergonomics (HFE) analysis of a specific scenario derived from a legal case is described. Issues and potential solutions are discussed.

Keywords: Wheel stop · Parking lot · Trip and fall · Forensic

1 Scenario

Mrs. Mary Lumens[1], age 68, was sitting comfortably at a booth at Snooty's Restaurant which is located near an Interstate 95 exit in South Carolina in the U.S. Her husband Elliott, age 71, had not yet arrived. She was also expecting to see her younger sister, Sally, and her husband, Bill, whom she had not seen for three years. They were driving through the area on their trip to Florida from Maryland. Her sister texted that they were about 15 miles (24 km) away and that was about 15 min ago. At that moment, she received a call from her husband indicating he had just pulled into the parking lot.

She picked Snooty's as the place to meet because it was located near their home, and it was the easiest way for her sister and husband to stop for an hour or two during their road trip since it was adjacent to the road that they would be traveling.

Snooty's was part of a huge gas station/truck stop combination, anchored and owned by Advanced Fuel, a division of Advanced Vehicle Fuel, Inc. They had 10 rows of pumps each able to handle 3 vehicles on each side, plus 8 additional lanes for tractor-trailers. The place was much larger than an average fueling station. The facility also included Snooty's restaurant and a small grocery/food store and touristy gift shop. The parking lot was huge, with spaces in the front, sides and back. Snooty's

[1] Names of persons, places and events are fictitious; resemblance to a specific event is coincidental.

S. Bagnara et al. (Eds.): IEA 2018, AISC 821, pp. 751–760, 2019.
https://doi.org/10.1007/978-3-319-96080-7_91

restaurant was on the right side of building complex and it included a drive-through lane for food orders from vehicles.

Still on the cell phone and unable to find a space, Elliot says to Mary “When I pulled into the parking lot, I thought I saw your sister’s vehicle. I think they are already here. Have you seen them?” She replies, “They haven’t come into the restaurant yet. I am holding a booth. I’ll wait a little longer before texting them again. I’ll walk out to see if I see them.” She leaves her coat piled up on the seat of the booth with part of it coming up and touching the table so it would be visible that booth was already taken. Before she reaches the door, she tells the reception girl that she will be right back. Having never hung up, Elliot says into the phone, “I found a space—pulling into it now.” Mary steps outside the restaurant’s side door but sees only a mass of cars and then walks across to the other side of the drive-thru lane. She says into her phone “I do not see you.” He says, “I’m just getting out. Hold on.” She is scanning the entire side of the parking lot and she does not see him or her sister and brother in law. She says into the phone, “Where are you?” “I said I’m just getting out.” He stands up and closes the car door. Elliot sees her standing in the area near the drive through and begins waving and then yelling her name. Finally, she sees Elliot and waves back to him. She begins to go around a parked vehicle. Just as she makes it around the front passenger side of the vehicle, she sees below that she will need to step over a yellow parking stop. She does so with her left foot but as she follows with her right foot, something catches her shoe and suddenly she is falling. She reported later that she was almost able to grab the car’s side view mirror but misses and slides by it. Her husband reported that after getting out of his car, he saw her at the edge of the parking lot and he tried to get her attention by waving and yelling in her direction. She finally saw him, waved back and then she disappeared between the parked vehicles. Since there were many vehicles parked between them so he was not exactly sure where she was. After looking down each row of cars in the vicinity, he sees her down one row on the ground between the fenders of two cars bleeding and sobbing. Soon thereafter, Mary was carried into the restaurant by two gas station employees and was seated along the length in the first booth.

That day, the 18th of November 2015, changed Mary Lumens’ life as well as her husband’s. She broke her hip and left arm and wrist. It took her 6 months before she could use a walker because of the substantial injury to both places on her body. She has had extensive physical therapy for which she has had to travel 45 miles each way to an approved orthopedic center. Two years after the fall, she is able to walk on her own, but reports daily pain in her wrist.

Shortly after the injury event, the restaurant manager on duty produced an incident report. The description was not very detailed but because most of the form’s questions were not applicable to the particular injury event. Two South Carolina State police officers who were in the area also arrived at the scene shortly after the incident. The troopers took some notes but did not produce a final report. A formal accident report was not filed but their notes were transcribed. Mary’s sister and husband arrived only minutes after Mary’s fall. They spent several hours at the emergency room with her and Elliot before continuing their trip. They stopped again on their trip back eight days later.

2 Additional Background Information

A few months later Mary was deposed in a legal case that she and Elliot filed against the owners of the facility. "I saw the parking bumper or wheel stop, or whatever they are called." She also said, "I can tell you that it was not as brightly colored yellow as it is today. We went out there last weekend to see if anything changed and found that some of the wheel stops had been repainted."

Recorded both in the establishment's and police reports, Mary stated that she tried to step over the wheel stop and believed that she had more than enough clearance but that her right foot got caught onto something in the process. The police report quoted her as saying, "After I fell I could see that there was a bolt sticking up out the parking stop and that the parking stop was not flat against the parking lot. I didn't notice those aspects *until* after I was already down on the ground."

Photos after Mary's fall confirmed the attributes that Mary mentioned. The wheel stop (also known as a bumper stop, parking stop, tire bumper and other names) was warped and the bolt (sometimes called a lag bolt or rebar spike) was partly sticking out above the parking stop.

Parts of the huge parking lot were in disrepair. There were bolts sticking out of the asphalt without any wheel stop nearby and numerous broken wheel stops were scattered at various places. Many of the wheel stops were broken at or near the bolt holes—an apparent weak point. These tripping hazards can be missed ("fail to be noticed") by pedestrians and have the potential to initiate a trip and fall.

The act of walking involves a motor-control system that is predominately automatic and unconscious. Of course parts of it are voluntarily and purposely initiated. For humans to react properly to obstacles in their way as they transverse the environment, visual and kinesthetic information needs to be processed as input. Unexpected or unusual aspects that are not prominent need to be enhanced and made salient so as not to be missed or else a misstep can be produced.

According to an engineering report produce by an expert retained in the resulting legal case, the subject wheel stop was composed of recycled plastic and opined that this material reacts to a greater extent to temperature changes than wheel stops traditionally composed of concrete. Wheel stops with recycled materials can result in substantial temperature-based distortion. With heat during the summer months it can expand and warp. These wheel stops may never completely return to the original horizontal shape when they contract in colder temperatures. Without this technical knowledge, the use of wheel stops made of recycled products would seem to be environmentally beneficial, i.e., doing the "green" thing by reducing landfill use. It was never determined in this case's discovery phase whether recycled-plastic wheel stops can be made (differently formulated) to limit warping and distortion. There is an additional relevant effect of the expansion and contraction cycles. When the wheel stop expands it causes the ends to rise up from the pavement due to being held and limited by the bolts. This has the effect of applying pressure on the bolts' heads and moving the bolts upwards. When the wheel stop contracts in cold temperatures, the bolts at each ends of the wheel stop may remain in a raised position. As a result, it sticks out above a flatter, contracted wheel stop. People navigating across the parking lot would have to notice that the bolts are raised above the wheels stop so as to clear them when stepping directly over them or adjusting their direction to avoid them. The problem is that only a portion of the wheel stops might be visible because they are partially obscured by the front or rear end of parked vehicles. When used as intended, drivers move into a parking space up to the wheel stop to the point where (in most cases depending on clearance) the vehicle rests on top of the wheel stops, sometimes touching or nearly touching the tread of the tires. When a vehicle is parked on top of a wheel stop, only the sides of wheel stops may be exposed to pedestrians while the middle is covered. The wheel stops including the one involved in Mary's fall was in an area that pedestrians would foreseeably be navigating to go to and from the restaurant to their parked vehicle. There were no dedicated pedestrian walkways or sidewalks within this particular parking lot.

3 Human Factors and Ergonomics (HF/E) Analysis

Mary Lumens (the plaintiff in the lawsuit) likely saw one or more wheel stops when she parked her car. Upon exiting the restaurant, there would be wheel stops in her visual field as she scanned the area but she probably did not give much (or any) attention to them. The bowing deformation and the end bolts sticking out are visible but they are not prominent features that would be quick to draw much or any attention (to an ordinary pedestrian). Much of Mary's attention was likely focused on looking for her husband and sister (and brother-in-law) and their cars. When Elliott got out of his vehicle, he saw Mary in the distance. He waved his arm back and forth and screamed her name to try to get her attention. During this time Mary's visual attention was focused on scanning the enormous lot, gazing outwards above the surfaces of hoods, roofs, racks, and trunks of different vehicles for family members. Thus, her gaze was mainly directed and focused on objects at varied distances in the parking lot, not down at her feet. Shortly after she seeing Elliott waving at her, she fell.

Thus there were several factors contributing to the trip and fall event. A lot of it concerns the relative use of visual attention and the partly obscured wheel stops. She was mainly looking elsewhere—focusing on areas above the vehicles. And many of the wheel stops in her immediate area could not be seen because parked vehicles were obstructing their view. Some of the wheel stops could not be seen until just coming upon them, just before a navigation response is needed. Only at these "near points" are they relevant to pedestrians to consider. Mary admitted seeing the subject wheel stop just before she was stepped over it to clear it. She also said, "I stepped plenty high to clear it," but also remarked that at that time, she did not see the bolt sticking out above it. It was the bolt head that caught her left shoe and initiated the trip and fall. She did not recall whether the bolt was black or painted partly or fully yellow. Photos taken a few months after the event showed that some bolt heads and necks in the parking lot were freshly painted yellow and others were only partly painted. See Figs. 1, 2, and 3.

People trip and fall over wheel stops even without the problems of warping and the bolts sticking out (Bell, Collins, Dalsey, Sublet et al. 2010; Washington State Department of Labor and Industries 2010). This is because the presence of wheel stops can be missed in part due to (a) vehicles obstructing views of them, (b) some are not distinguishable from the surrounding environment (e.g., not in a distinct color), and

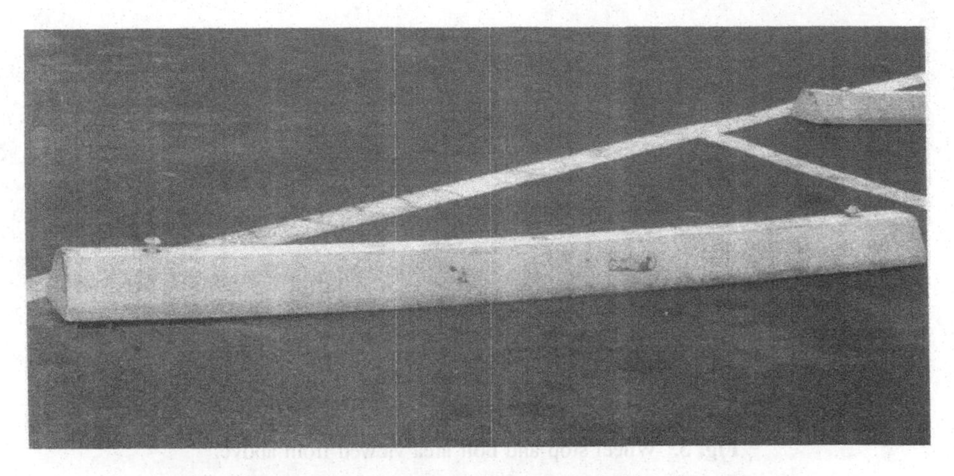

Fig. 1. Warped wheel stop with bolt raised above its surface.

Fig. 2. Close-up of a bolt sticking out.

their appearance in parking areas is so common and familiar that people pay little or no attention to them. Other potential issues include insufficient lighting, people looking elsewhere, people carrying or pushing large objects obscuring their view, attention distracted by cell phone use or by other tasks, among others.

3.1 Benefits of Wheel Stops

Wheel stops have intended benefits. One is that they can be used to control directional movement of vehicles in a parking lot, i.e., preventing vehicles from moving through certain areas. The main benefit is that they can control placement of parked vehicles.

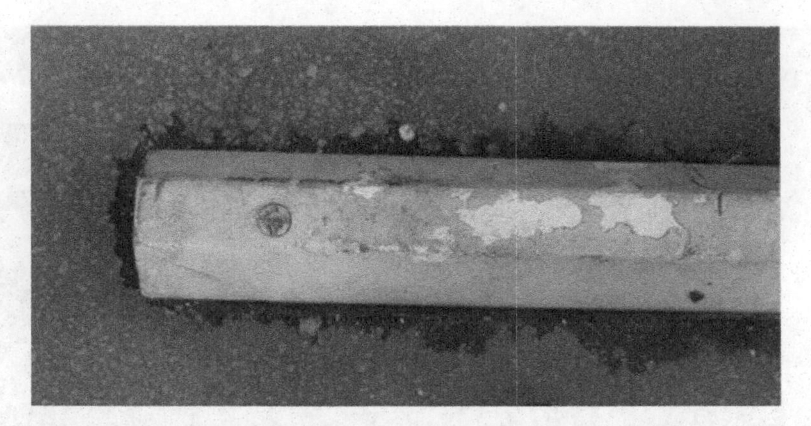

Fig. 3. Wheel stop and bolt area viewed from above.

They are a barrier or guard to prevent vehicles from encroaching beyond a certain point. They provide guidance and enable alignment of vehicles to maximize the number of vehicles in a lot and to prevent directional rollaways.

Parking lots are not made just for vehicles; they are made for people, too—drivers, occupants and pedestrians (who are often the same people in different roles at different times). People expect parking lots and decks to be reasonably safe (e.g., Mendat and Wogalter 2003). Because pedestrians and vehicles often traverse the same space, parking lots ought to be designed to avoid collisions, particularly between pedestrians and vehicles. Without dedicated walkways, pedestrians need to traverse and change direction around and between multiple vehicles or in vehicular lanes. In the scenario that started this article, Mary Lumens, the injured party, was looking for particular people and cars in the distance at or about the time she tripped and fell. But it should be noted that many of the same factors are involved in other common scenarios such when one traverses towards the entrance of a shopping center or big-box store or in looking for their vehicle after exiting. There are numerous instances where people will look in an approximately horizontal direction or somewhat lower, but wheels stops under foot are almost 90° below that. With vehicles blocking views, wheel stops might be only be seen just before a relevant response is needed to step over or aside to avoid them. Given this, it is foreseeable that people will sometimes trip on them. Not all trips lead to a fall (sometimes people can catch themselves and readjust in time) and not all falls lead to a serious injury. Trips and falls that do not produce serious injury are seldom reported, and if a non-serious injury is reported (a critical incident that should cue the potential for future serious injury), they are usually not acted upon to correct the "alleged" problem. The property owners, Advance Vehicle Fuels, reported they had no records of any falls at this property. They have a document retention policy of 7 years after which paperwork is destroyed. The manager's accident report was a new initiative that started earlier that same year. The corporate legal department receives all injury files.

3.2 Bowed Wheel Stop with Bolts Sticking Out

With a bowed wheel stop, a higher rise in the step is needed to go over it. An alternative is to step aside to avoid it. Warpage of the two ends of the wheel stop is not the full story about potential for height misapprehension. Several inches away from the ends of the subject wheel stop are large metal bolts intended to attach and retain the wheel stop on the parking lot surface. In the subject parking lot, many of the wheel stops, including the subject one, had bolt heads raised above the top surface. As described earlier, apparently the bolts were pulled up by the warping ends of the wheel stops and reached their apex in the hottest temperatures with the greatest expansion of the wheel stop's size. With cooler temperatures the wheel stop contracts in size and tends to straighten back out and down (but often not returning to their original state). However, with the contraction in cooler temperatures, the bolts do not return to their original position. The bolts do not come back down (or at least not all of the way) often staying in their apex position, protruding higher than the wheel stops themselves. Clearly, the bolt heads should never be higher than the wheel stop. It is a hazard when positioned over and above the wheel stop. People might not see them when stepping over the wheel stop and the bolt could "catch" the foot, shoe, or pant leg causing a misstep, a trip event, and potentially a serious fall.

Fractions of inches can be the difference between successful navigation or the catching of the foot in motion and causing a trip event. The Americans with Disabilities Act (ADA) of 1990 defines a "trip hazard" as any vertical change over 1/4 in. or more at any joint or crack. Protrusions of one-eighth inch can cause a misstep (e.g., Cohen and Pauls 2006). It can catch the foot, shoes, etc. and cause a tripping event. In most localities, there are no specific regulations or laws controlling the design and placement of wheel stops in parking lots. However, human factors/ergonomics (HF/E) principles can be applied to this domain. HF/E research and principles can be applied to predict and explain safety problems in built external environments such as parking lots. Perturbations on surfaces in built environments where the public traverse is a necessary concern for property owners and management. So in order to avoid a trip and fall incident and the serious injury that might result, small, less-prominent perturbations on the surfaces in which pedestrians traverse should be a concern and should be dealt with. Bolts sticking above the surface of wheel stops are a hazard that should be controlled. It should be removed, guarded by barrier, or at least, warned about.

At some point after the Mary's trip and fall, some of the wheel stops had been repainted the same bright yellow color. Some of the bolts heads and neck were coated with yellow paint (although not all of them). It is unclear whether the subject bolt was painted at the time of Mrs. Lumen's fall. Neither of the Lumens' remembered, and neither the accident report nor the police's notes mention the bolt's coloration. No photographs were taken immediately after the incident of the parking lot or the particular wheel stop. The lack of anyone noting the bolts having yellow paint on them at the time of the event suggests that the wheel stop bolt may not have been painted or incompletely covered in paint at the time of the fall. The bolt being painted *or not* could contribute to the fall. If it was painted, then under certain viewing angles it might not be seen because it was the same color of the wheel stop, i.e., there is a lack of figure-ground color contrast with the wheel stop. If the subject bolt was unpainted or mostly unpainted, and thus mostly a dark gray color, then it could blend in with the surrounding dark gray asphalt. Viewing angles could make it very difficult to see the bolts. The raised bolt in Fig. 3 is not visually prominent.

Clearly the owner/management should have better maintained the parking lot. After Mary's fall, the parking stops were likely repainted but other forms of maintenance and repair were not performed, such as removal of the broken pieces of wheel stops scattered in the lot and the removal of mostly-bent bolts sticking out of the pavement (without any wheel stop attached). Bolts were sticking up beyond the top surfaces of all unbroken wheel stops.

3.3 Hazard-Control Hierarchy

The concept of the hazard control is relevant to this incident. The basic hazard-control hierarchy is a generalized set of prioritized strategies of designing out, guarding against and warning about potential dangers.

A highly effective method of hazard control to prevent tripping from wheel stops is to not use them, i.e., eliminate them. If wheel stops are used, they should be made so they do no warp or change shape with expected temperature variations. There might be a way to make wheel stops with recycled material that does not warp, but if it cannot be done, then different material that does not warp such as conventional concrete should be used. Another set of design issues concerns the bolts and the associated holes. The wheels stop's bolt holes should have had a larger cylinder space for the bolt neck and a large inset space to bury the bolt head deeper into the wheel stop so that it stays below the surface. With heat and expansion, the bolt would not be pulled up, because with a wider hole, so that any expansion of the wheel stop would not move the bolt out of position, and any change in the wheel stop by expansion and contraction would slide by the bolt. A large inset space would allow some movement of the bolt without it having an apex height above the surface of the wheel stop. A deeper inset space would allow a bolt head to lie fully (and remain) below the surface of the wheel stop. At the original installation the rounded bolt head was too large to fit in a small inset space that was present, leaving part of the bolt head above wheel stop's top surface. This was an incorrect combination of wheel stop and bolt head type. The bolt head should always be below (or at least flush with) the top surface of the wheel stop.

Guarding is the second prioritized method of hazard control. Wheel stops themselves are a form of guarding for vehicle placement. There are other forms of guarding that are better and safer than wheel stops such as the use of taller guards/barriers that can be seen above vehicles. In fact, around the periphery of the subject gas station/restaurant building were tall, vertically mounted poles (pipes) to prevent vehicle encroachment and to enable people to see them above the vehicles. Also large bollards or planters could demark areas for parking. These barriers can control vehicular placement, and are more visible to pedestrians.

Warning about the hazard is the third method. Painting the parking stops yellow was a method of making the parking stops more conspicuous. However, the bolt heads —whether painted yellow or not—will still be a hazard because of inadequate contrast with the contextual environment. The relatively small size of the raised bolts makes its detection difficult. Furthermore, pedestrians only get a brief exposure to the wheel stop just before a relevant response is needed, and people are foreseeably looking in a different direction, i.e., above the vehicles. Thus, the hazard-control strategies of design changes to reduce the hazard and guarding against the hazard are likely to provide greater safety than markings for the wheel stop in this case. Even if the bolt were to be made a more distinguishable color, the hazard is not reduced by much.

4 Discussion and Conclusions

The injury from falling can be a life-changing event. Falling, as a general category leads to more emergency room visits than other accident causes, particularly in the younger and older age groups (e.g., NEISS 2017). This is why owners, operators, and managers of properties that invite the general public should be aware that they could be liable in a lawsuit for a serious injury if it is determined that their property's characteristics were causative for a fall. Thus, property owners, operators and managers need to scrutinize their space to determine if there are hazards to pedestrians, and if so to consider the prioritized strategies of redesign, guarding and warning to control the potential of a fall hazard. Clearly, they need to ensure public areas are maintained and repaired.

Parking stops are known to be involved in trip and fall accidents. Fortunately many people do not trip because they see the hazard and avoid it. Trips do not always result in a fall. Sometimes people can catch themselves and not fall. And those that fall may not get severely hurt. Undoubtedly, a lot of missteps involving wheel stops go unreported. The problem is that some people who trip and fall can get severely hurt, particularly older adults who tend to be less adept physically and perceptually than younger adults. Older adults are more fragile, tend to receive more severe injuries in these events, and they may never fully recover from the resultant injury. The hazard in this case is made worse because the wheel stops' ends were raised due to warping arising from anticipatable environmental exposure. The warping problem is made worse due to the bolts being forced out of the ground. Because the bolts that attach the wheel stops to the pavement are not a prominent feature (small size, low contrast), their position above the surface of the wheel stop may not be noticed.

People assume that the ground on a built/developed environment will be reasonably flat, and if not, then it should be guarded or well marked (e.g., Cohen and LaRue 2019). This is the reason that built environments need to be examined and analyzed in order to know what hazards need to be controlled. Property owners and managers need to inspect their property frequently to ensure there are no hazards due to bad configurations, poor maintenance, etc. The basic ideas of the hazard-control hierarchy of designing out (eliminating or reducing hazards), guarding against hazards (better barriers), and warning (marking) are useful. There are many ways to control the hazards associated with wheel stops. Wheel stops can be eliminated, or if used, be made of nonwarping material and use a better-designed connecting mechanism. Tall poles, bollards, and large planters could provide guarding. Using only warnings does not likely solve the bolt problem.

The case resolved in a confidential settlement before trial.

References

ADA (1990) American Disabilities Act of 1990. 42 U.S.C. Ch. 126§12101. U.S. Congress, Washington, DC

Bell J, Collins JW, Dalsey E, Sublet V (2010) Slip, trip and fall prevention for healthcare workers. Centers for Disease Control and Prevention, NIOSH, Department of Health and Human Services

Cohen HH, Pauls J (2006) Warnings and markings for stairways and pedestrian terrain. In: Wogalter MS (ed) Handbook of warnings, Chap 57, pp 711–722. Erlbaum, Mahwah, NJ and CRC Press, Boca Raton

Cohen HH, LaRue C (2019) A step in the right direction. In: Wogalter MS (ed) Forensic human factors and ergonomics: case studies and analyses. Taylor and Francis/CRC Press, Boca Raton, FL

Mendat CC, Wogalter MS (2003) Perceptions of parking facilities: factors to consider in design and maintenance. Proc Hum Factors Ergon Soc 47:918–921

NEISS (2017) National electronic injury surveillance system. U.S. Consumer Product Safety Commission, Washington, DC

Washington State Department of Labor & Industries (2010) Slips, trips, and falls, module one: powerpoint slide set, May 2010

Communication-Human Information Processing (C-HIP) Model in Forensic Warning Analysis

M. S. Wogalter[(✉)]

Psychology Department, North Carolina State University, Raleigh, NC 27695-7650, USA
Wogalter@NCSU.edu

Abstract. A model that combines human information processing with communication theory is described: the Communication-Human Information Processing (C-HIP) model. Emphasized are the factors that can influence processing at various stages. Bottlenecks in the process can reduce warning effectiveness. The C-HIP model can be used (e.g., by manufacturers) to assess warning utility. Human factors and ergonomics (HF/E) experts can use it as a method to systematically structure their warning analyses.

Keywords: Forensic · C-HIP model · Warning
Human information processing

1 Introduction

Warnings are used to convey hazard information to consumers for the purpose of reducing or avoiding injury or property damage (e.g., Laughery and Wogalter 2006; Wogalter et al. 2012. Typically, this information is visually displayed with text and/or graphics (e.g., symbols) on labels adhered to a product or a container, but it can also be in the form of inserts, product manuals, or on signs for environmental and facility hazards (Conzola and Wogalter 2001). Warnings need to be noticeable, legible, understandable, memorable, believable and motivating to facilitate goals of comprehension and compliance behavior. This article focuses its description of the C-HIP model as it relates to the processing of consumer product warnings. The basic principles are also applicable to signs and other kinds of warnings for environmental and facility hazards.

2 Communication-Human Information Processing (C-HIP) Model

When design and guarding do not control all of the hazards of a product (or environment or situation), warnings are needed. Warnings are intended to influence people and serve as an important means of hazard control. Because of this, it is important to describe the processes involved. A model is presented that combines the basic stages of

S. Bagnara et al. (Eds.): IEA 2018, AISC 821, pp. 761–769, 2019.
https://doi.org/10.1007/978-3-319-96080-7_92

a communication model (source, channel, and receiver) with human information processing approach (Wogalter 2006).

The basic C-HIP model is usually described as a linear, sequential process in which warning information should successfully flow from the beginning to the end, from the source to behavior and through the stages in between. This process is represented by the straight arrows going from the top to the bottom stages in Fig. 1. There can be "bottlenecks" in the process of moving down the stages; such processing difficulties would reduce or prevent the warning's effectiveness. The more complete C-HIP model is more complicated in that it includes arrows in the reverse direction to represent feedback loops in which the "later" stages can influence processing at "earlier" stages. The current version of C-HIP has two stages: Switch and Maintenance (e.g., cf. Wogalter 2006).

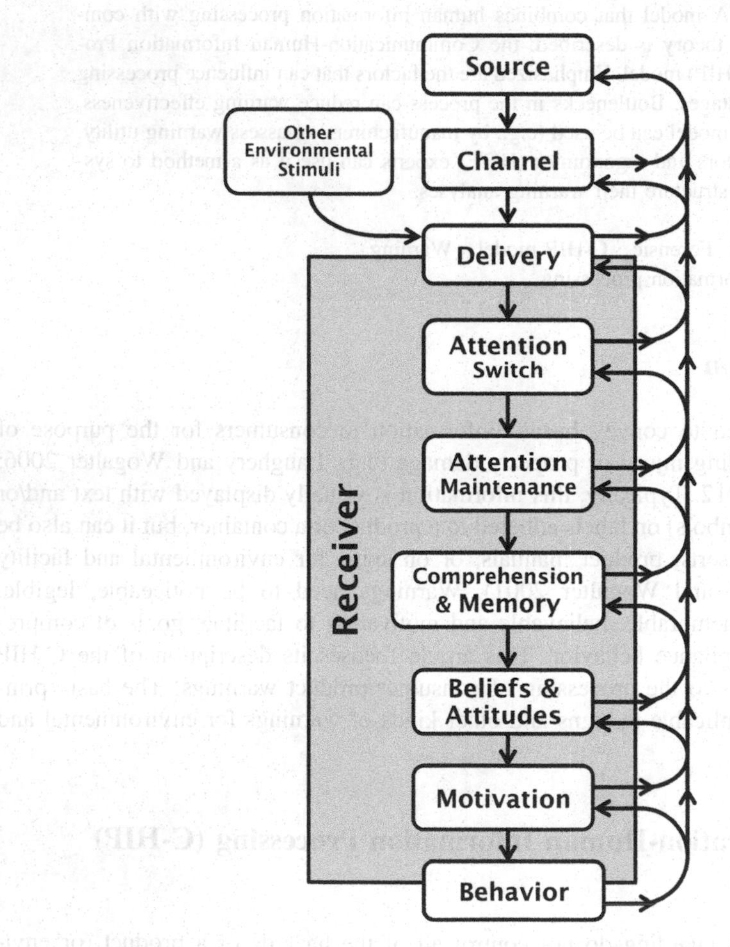

Fig. 1. Communication-human information processing (C-HIP) model.

Starting from the top of Fig. 1, the stages of the C-HIP model are described in separate sections below.

2.1 Source

The source is an entity that determines the necessity of warnings. That determination may be derived from the use of hazard analysis, data, industry standards, consumer reporting, physics/chemistry principles, or legal requirements) and if a warning is needed, the source transmits the warning. The source could be a government agency, a manufacturer, importer, trade group, or a particular person. The source is considered to have superior knowledge, at least in comparison to consumers. If hazards are incompletely controlled by design and guarding, then effective warnings should be provided so that people are informed about the hazards and what to do to avoid them.

2.2 Channel

Warnings can be given in many different ways including on-product labels, inserts, manuals, tags, web pages, public service announcement, etc. They can be given visually, auditorily or through other sensory modalities. The multiple modes and methods of dissemination for a product are together called the warning system (Laughery and Hammond 1999). Generally, providing information in more than one form (format and/or modality) is better because it can reach more people in more situations with greater impact (Cohen et al. 2006; Mazis and Morris 1999).

2.3 Delivery

Effective warnings need to reach the target audience who may be affected by the hazard. Warnings can be sent out but *never* succeed at arriving to relevant at-risk persons (Wogalter 2006b). For example, a company could print thousands of brochures with warnings but if those brochures are never distributed then their effectiveness is nil. Another example is public service announcements (PSAs) that are only broadcast in the early morning hours when most people are asleep; these warnings will have little impact on those not tuned in at the time. Delivery likelihood is greater when there is more than one presentation method (Cohen et al. 2006).

2.4 Environmental Stimuli

Other environmental stimuli (the context) can affect warning processing. Other stimuli compete with warnings for attention. A warning with a high level of salience (prominence or conspicuousness) makes it more likely that a warning will be attended to as opposed to attending to other things. Warnings can be salient in some environments and not in others. (Salience is discussed in more detail in the Attention Switch stage).

2.5 Receiver

Generally it is desirable to reach as many persons at risk as possible. Some of the persons at risk may require warnings with enhanced characteristics. For example, older adults with perceptual, cognitive and physical declines may not be able to read warnings in very small print or under low illumination, yet these characteristics may not present a problem for younger adults (Wogalter and Vigilante 2003; Mayhorn 2005). Warnings intended for trained, sophisticated healthcare professionals may be different than those given to ordinary consumers. The wide range of skills and abilities in the general population usually means that warnings for the ordinary consumer are capable of reaching the lowest denominator of capabilities (or those with the greatest limitations), inasmuch as feasible, so as to maximize its reach.

2.6 Attention Switch

Attention switch is the process where a person changes his/her attention *to* something else such as to a warning *from* something else (Wogalter and Vigilante 2006). It is related with the concepts of salience, conspicuousness, prominence, noticeability and attention-gettingness. Characteristics that benefit attention switch are increased size, high contrast, relative distinctiveness (e.g., different color from surroundings), apparent movement and other distinguishing characteristics. Graphics such as symbols can help promote attention switch.

In general, attention switch is directed to the most salient information at a given time. Warning processing competes with other ongoing task processing, including current and immediately-upcoming processing. If the warning is highly salient, it will be more likely to cause a switch to itself than if it were not salient. A warning in an environment with many "eye-catching" stimuli can reduce the likelihood that attention will switch to a warning, yet the same warning appearing in a plain, bland context is more likely to be seen.

2.7 Attention Maintenance

After switching to a warning stimulus, attention must be held for some length of time so that adequate information is acquired from it. Some of the main factors that enable maintenance attention include (a) having adequate print size (not extremely small or large), (b) high contrast (print to background), and (c) distinguishable important/relevant details so as to enable the person to read or see the warning. Other factors include brevity, white space, and having relatively low density/detail. Environmental exposure can cause degradation of the warning, reducing legibility.

The warning needs to be "attractive" and interesting enough so that people will stick with it long enough to extract adequate information instead of switching attention prematurely to other information. The reason for this is that the warning competes with other stimuli and processes or tasks, which could pull (switch) attention away from the warning through the attention switch mechanism discussed earlier. Good design makes it more likely that information is acquired quickly during the time attention is maintained on the warning. This relates directly with the next stage of processing.

2.8 Memory and Comprehension

While attention is being maintained on a warning, other processes may occur concurrently, including memory formation and comprehension. During this time, the material may be encoded which can produce new memory. If the material is highly technical or there is not much pre-existing knowledge of the subject matter, then the resulting memory formation may be minimal (at least in the short term), because people are likely to switch attention to something else. Another example of limited warning processing is when individuals are not skilled with the language used in the warning. If the individual already knows much of the information in the warning then processing it will be easier. In this case, the warning cues existing memory/knowledge (although not much new information will be acquired from the warning). Easy to process information is readily assimilable with the person's existing knowledge. If the information cannot be accommodated easily (without needing considerable time and effort), it will be less likely to hold attention. Generally, people will tend to maintain their attention longer when the warning has some moderate level of new and useful information. Warning designers should try to make the information easy to encode. This can be accomplished by ensuring some association with what the individual already knows.

With warning comprehension, the goal is to understand, in an adequate way, information about the nature of the hazard, what to do to avoid the hazard, and the consequences if the hazard is not avoided. Comprehension provides informed consent about risks.

The "gold standard" method for assessing comprehension is open-ended testing of the content. Probes or cues can be used to elicit other knowledge in memory (Brantley and Wogalter 1999).

If a warning does not produce adequate understanding, then there are methods to improve its performance. Usability type testing involving iterative prototype design-test development cycles can be used to improve comprehension (Wogalter et al. 2006).

Not every hazard needs to be warned about. One classic example is a knife. Virtually all adults know that knives can cut and cause harm, and so a warning is probably unnecessary [except that caretakers would need to warn young children].

Explicitness is another comprehension-related concept. In general, it is better to give specific information (e.g., can cause liver failure) than general information (e.g., may cause health problems) (Laughery and Smith 2006). Ambiguity and lack of clarity of text and graphics can slow processing and in some circumstances produce incorrect interpretations.

Habituation is a memory-related concept where repeated exposures over time to a stimulus produce memory (Thorley et al. 2001; Kim and Wogalter 2009), but as it does, attention is reduced. A negative effect of habituation (such as seeing a standardized warning format repetitively over time) is that attention may not be allocated to a similar-appearing warning for a different hazard. Accordingly, warnings that look similar to the habituated warning can evoke inadequate attention. Warning design standards, such as the ANSI (2007) Z535.4, promote uniformity, which could lead to a similar looking warning not eliciting adequate attention. Habituation is an example where a "later stage," in this case memory, influences an "earlier" stage, attention, as illustrated by the feedback loops in Fig. 1.

2.9 Beliefs and Attitudes

Beliefs are comprised of knowledge based on vast experiences gained throughout life (DeJoy 1999; Riley 2006). Attitudes are similar except they also have an affective component. Beliefs and attitudes are overall assumptions about how things work and are assumed to be true.

It is easier to process information that is consistent with existing beliefs. Inconsistent information is more difficult to accommodate and may result in incomplete processing.

People are more likely to read warnings for products believed as hazardous and the converse is true as well—they are less likely to read warnings concerning products that they believe are relatively safe (Wogalter et al. 1991; Wogalter et al. 1993). Another relevant factor is familiarity (Wogalter et al. 1991). Familiarity beliefs will tend to reduce the likelihood to look for or read warnings (Wogalter et al. 1991). This is an example of how a "later" stage in the C-HIP model affects an "earlier" stage (as in feedback loops in Fig. 1). To overcome existing, incongruous beliefs, the warning needs to be prominent and persuasive.

2.10 Motivation

Users might progress through all of the previous stages, yet compliance might not occur due to inadequate motivation. Several factors influence motivation. Cost of compliance is one factor. Warning-directed behavior might not be performed because it is too effortful, takes too much time or costs too much money (Wogalter et al. 1989; Wogalter et al. 1987).

Social influence is another factor. If other people comply with a warning, then individuals are more likely to comply as well. The converse is also true (Wogalter et al. 1989). Motivation is can also influenced by time stress (Wogalter et al. 1998), and mental workload (Wogalter and Usher 1999). Being in a rush or involved with other tasks tends to reduce compliance.

2.11 Behavior

Compliance behavior is sometimes considered the ultimate measure of warning effectiveness (e.g., Wogalter et al. 1987). Safe behavior can be increased in likelihood in the presence of well-designed warnings. Because measuring objective levels of compliance behavior can be difficult (e.g., Wogalter et al. 1987), subjective evaluations are sometimes used as indicators of compliance. Virtual and augmented reality can potential provide a realistic experience while not exposure to actual harm (Duarte et al. 2014; Vilar et al. 2014).

3 Relevance to Manufacturers and Forensic HF/E Experts

One of the basic goals of warnings is to convey safety information. In the C-HIP model, warning information must be processed through several stages without impediments or bottlenecks that would block its progression. The flow of information must be successfully completed but may be prevented at particular stages from doing so. C-HIP is useful in several ways. It serves to organize the considerable body of research that has accumulated in the last three decades in the human factors and ergonomics (HF/E) literature. It can be used as a tool to evaluate warnings. Influential factors for each stage of the C-HIP model can be used as an assessment checklist of warning effectiveness (see Lenorovitz et al. 2012; Wogalter 2019).

The C-HIP model can be helpful in determining why a warning is not working and can suggest improvements. Knowing what is causing a problem with a warning's processing saves money, effort, and time, which is useful because efforts can be focused on producing a better warning. A good way to assess the effectiveness is to ask several people to use (or assemble or install) a product (with a warning attached or given in an accompanying ancillary materials). If during this testing it turns out that participants only briefly gazed at the warning but soon thereafter look away to something else, then this finding would suggests a problem at the attention maintenance stage. The C-HIP model gives some insight on how the warning should be improved (at least in part). Here you know the warning was delivered and its presentation or availability led to a brief glance. This means that the warning had at least some effect at the switching attention stage (particularly if this pattern happens consistently with other participants). The problem appears to be that the warning did not hold attention after the switch event and thus efforts can be focused at improving the process at the maintenance stage.

To take this example further, participants might be asked later, "Why did you not pay attention to (read) the warning?" The participants' responses can offer some insight as to why they did not maintain attention to the warning. Suppose for this example, several participants say that they did not read it all because the print was too small and dense. If so, then how to fix the warning is straightforward. There may be other alternative responses and other potential fixes. Even after the attention maintenance problem is fixed, there may be other bottlenecks such as in the comprehension or motivation stages, but the C-HIP model offers guidance on fixing those problems as well. Detective work like this can ascertain and target specific problems, and thus potentially reducing costs associated with warning development.

The model also offers guidance in forensic HF/E analysis of injury events with regard to the warning system involved. It offers a systematic, structured way to analyze the warnings and their characteristics. It is a systematic method that can be applied to numerous situations, involving a multitude of product warnings and signage. It is therefore a useful tool for HF/E experts in forensic settings.

Although the C-HIP model was developed for warning processing, it is also a general model that can be applicable to other domains of risk disclosures such as informed consent forms, credit card terms, and software licenses (Wogalter et al. 1999b) and to warnings presented via other modalities, such as audition (Cohen et al. 2006).

It also could be used for explaining or structuring the processing of information in other HF/E domains, human-technology interaction.

References

ANSI (2007) American National Standard for Product Safety Signs and Labels. Z535.4. Rosslyn, National Electrical Manufacturers Association

Brantley KA, Wogalter MS (1999) Oral and written symbol comprehension testing: the benefit of cognitive interview probing. In: Proceedings of the human factors and ergonomics society, vol 43, pp 1060–1064

Cohen HH, Cohen J, Mendat CC, Wogalter MS (2006) Warning channel: modality and media. In: Wogalter MS (ed) Handbook of warnings. Lawrence Erlbaum Associates/CRC Press, Mahwah/Boca Raton, pp 123–134

Conzola VC, Wogalter MS (2001) A communication-human information processing (C–HIP) approach to warning effectiveness in the workplace. J Risk Res 4:309–322

DeJoy DM (1999) Beliefs and attitudes. In: Wogalter MS, DeJoy DM, Laughery KR (eds) Warnings and risk communication. Taylor & Francis, London, pp 183–219

Duarte E, Rebelo F, Teles J, Wogalter MS (2014) Behavioral compliance for dynamic versus static signs in an immersive virtual environment. Appl Ergon 45:1367–1375

Kim S, Wogalter MS (2009) Habituation, dishabituation, and recovery effects in visual warnings. In: Proceedings of the human factors and ergonomics society, vol 53, pp 1612–1616

Laughery KR, Hammond A (1999) Overview. In: Wogalter MS, DeJoy DM, Laughery KR (eds) Warnings and risk communication. Taylor & Francis, Philadelphia, pp 9–11

Laughery KR, Smith DP (2006) Explicit information in warnings. In: Wogalter MS (ed) Handbook of warnings. Lawrence Erlbaum Associates/CRC Press, Mahwah/Boca Raton, pp 419–428

Laughery KR, Wogalter MS (2006) Designing effective warnings. In: Williges R (ed) Reviews of human factors and ergonomics. Human Factors and Ergonomics Society, Santa Monica

Lenorovitz DR, Leonard SD, Karnes EW (2012) Ratings checklist for warnings: a prototype tool to aid experts in the adequacy evaluation of proposed or existing warnings. Work 41. https://doi.org/10.3233/wor-2012-0114-3616

Mayhorn CB (2005) Cognitive aging and the processing of hazard information and disaster warnings. Nat Hazards Rev 6:165–170

Mazis MB, Morris LA (1999) Channel. In: Wogalter MS, DeJoy DM, Laughery KR (eds) Warnings and risk communication. Taylor & Francis, London, pp 110–111

Riley DM (2006) Beliefs, attitudes, and motivation. In: Wogalter MS (ed) Handbook of warnings. Erlbaum, Mahwah, pp 289–300

Thorley P, Hellier E, Edworthy J (2001) Habituation effects in visual warnings. In: Hanson MA (ed) Contemporary ergonomics 2001. Taylor & Francis, London, pp 223–228

Vilar E, Rebelo F, Noriega P, Duarte E, Mayhorn CB (2014) Effects of competing environmental variables and signage on route choices in simulated everyday and emergency wayfinding situations. Ergonomics 57:511–524

Wogalter MS (2006) Communication-human information processing (C-HIP) model. In: Wogalter MS (ed) Handbook of warnings. Lawrence Erlbaum Associates/CRC PressMahwah/Boca Raton, Chap 5, pp 51–61

Wogalter MS (2019) Forensic human factors and ergonomics: case studies and analyses. Taylor & Francis/CRC Press, Boca Raton

Wogalter MS, Allison ST, McKenna N (1989) Effects of cost and social influence on warning compliance. Hum Factors 31:133–140

Wogalter MS, Brelsford JW, Desaulniers DR, Laughery KR (1991) Consumer product warnings: the role of hazard perception. J Saf Res 22:71–82

Wogalter MS, Brems DJ, Martin EG (1993) Risk perception of common consumer products: judgments of accident frequency and precautionary intent. J Saf Res 24:97–106

Wogalter MS, Conzola VC Vigilante, Jr WJ (2006) Applying usability engineering principles to the design and testing of warning text. In: Wogalter MS (ed) Handbook of warnings. Lawrence Erlbaum Associates/CRC Press, Mahwah/Boca Raton, pp 487–498

Wogalter MS, DeJoy DM, Laughery KR (eds) (1999a) Warnings and risk communication. Taylor & Francis, London

Wogalter MS, Godfrey SS, Fontenelle GA, Desaulniers DR, Rothstein PR, Laughery KR (1987) Effectiveness of warnings. Hum Factors 29:599–612

Wogalter MS, Howe JE, Sifuentes AH, Luginbuhl J (1999b) On the adequacy of legal documents: factors that influence informed consent. Ergonomics 42:593–613

Wogalter MS, Laughery KR, Mayhorn CB (2012) Warnings and hazard communications. In: Salvendy G (ed) Handbook of human factors and ergonomics, 4th edn. Wiley Interscience, New York, pp 868–894

Wogalter MS, Magurno AB, Rashid R, Klein KW (1998) The influence of time stress and location on behavioral compliance. Saf Sci 29:143–158

Wogalter MS, Usher M (1999) Effects of concurrent cognitive task loading on warning compliance behavior. In: Proceedings of the human factors ergonomics society, vol 43, pp 106–110

Wogalter MS, Vigilante WJ Jr (2003) Effects of label format on knowledge acquisition and perceived readability by younger and older adults. Ergonomics 46:327–344

Wogalter MS, Vigilante WJ Jr (2006) Attention switch and maintenance. In: Wogalter MS (ed) Handbook of warnings. Erlbaum/CRC Press, Mahwah/Boca Raton, pp 245–266

Relationship of Age on Responses to Automobile Warnings

S. David Leonard(✉) and Erica Medrano

University of Georgia, Athens, GA 30680, USA
dleonard11@juno.com

Abstract. A previous examination of knowledge of warnings using the present materials with college-age individuals did show some lack of knowledge about several specific warnings. Other studies have indicated differences in the age of the individuals participating in evaluation of warnings. Therefore, in the present study we examined the same materials that had been given to the undergraduate students with a sample of older individuals ranging in age from thirty to sixty-five and found that most of the responses of older individuals were similar to those of the younger individuals and their lack of knowledge of automobile warnings.

Keywords: Warnings · Age · Sex

1 Introduction

The primary concern of this paper is to determine information about what individuals know about the warnings and the appropriate responses to warnings used with vehicles. This study is a partial replication of a study that was presented to the ISOES in 2016, by Leonard et al. The subjects in that study were primarily college sophomores whose ages ranged from 18 to 23 years. Thus, the responses were based upon the knowledge level of individuals in that age range. There were several items in which the respondents may have made errors or may have been unable to provide an appropriate response. For example, the age at which tires should be replaced was not well known. In order to evaluate the knowledge of individuals who might have more experience in driving and particularly, more experience in examining the sorts of warnings that applied to children of various ages. The participants were obtained from personal relationships of friends and relatives of the author. Their ages ranged from XX to XX and their data were divided in terms of age ranges from a lower range of X to X and a higher range from Y to Y, as shown in Tables 1 and 2. The data presented within the relative age range is XX. Subsequent paragraphs, however, are indented.

S. Bagnara et al. (Eds.): IEA 2018, AISC 821, pp. 770–773, 2019.
https://doi.org/10.1007/978-3-319-96080-7_93

Table 1. Participant types.

Categories	Male	Female
1. New/used	New = 1; Used = 7	New = 5; Used = 9
2. Received manual	Yes = 6; No = 2	Yes = 13; No = 1
3. % Manual read	<5 = 4; 6–10 = 0; >10 = 4	<5 = 7; 6–10 = 3; >10 = 4
4. Driver training or parent	Driving training = 2 Parent = 6	Driving training = 2 Parent = 12
5. Drive time-wkly	<5 = 1; 5–10 = 3; >11 = 6	<5 = 1; 5–10 = 3; >11 = 9
6. Passenger recline	Never = 2; Long trips = 2; Often = 4	Never = 3; LT = 10; O = 1
7. Confusion w/door in unfamiliar car	Yes = 3; No = 4	Yes = 6; No = 8

Table 2. Responses.

Categories	Male	Female
1. Child/front seat	1–12 = 1; >13 = 1	1–12 = 7; >13 = 1
2. <40 lbs child in booster seat	Yes = 5; No = 2	Yes = 12; No = 1
3. Forward-facing child seat	Yes = 8; No = 0	Yes = 14; No = 0
4. Recline pass. Seat	Yes = 5; No = 2	Yes = 2; No = 12
5. Tire age replace	<6 = 4;6 = 0; >6 = 1	<6 = 3;6 = 0; >6 = 0
6. Read warning if hang tag on mirror	Very High = 2; Mod High = 2; Interm. = 3; Mod. Low = 0; Very Low = 1	Very High = 2; Mod. High = 7; Interm. = 3
7. Improper seat belt	Yes = 7; No = 1	Yes = 13; No = 1

2 Method

2.1 Participants

A group of respondents was obtained from among the friends and neighbors of the authors and their relatives. Participants were drawn from three different states. In order to avoid collaboration none of the males and females cane from the sane family. They were selected to have groups of males and females in the age groups of thirty to fifty and from fifty plus. After examining the data, the age groups were found to be sufficiently similar to be combined. Thus, the responses were examined for males and females, separately.

2.2 Materials

Two types of questions were asked and are listed separately. The first set related to the experience with types of vehicles they had driven, type of training, frequency of driving and experiences as driver and as passenger. In part, these may also have involved knowledge similar to that of the knowledge of what to do in attending to the vehicle and its passengers. The latter items were more concerned with specific activities and knowledge of items that could be found on the vehicle or in the manuals.

Two features of the experiences of passengers were briefly examined to evaluate how the general safety behavior (reclining the front passenger seat) and consistency of controls (knowledge of operation of passenger door locks).

The item regarding confusion about opening the passenger door was inserted to highlight the importance of consistency across controls across vehicles. Inability to open a door when the vehicle catches fire or gets in a flooded area might prove to be disastrous for the passenger.

3 Results

The responses for the questions regarding the individual's personal experiences are found in Table 1.

4 Discussion

In the survey by Leonard, et al. with undergraduates as participants, several warnings were found not to be known by the college students whose ages were primarily in the late teens. The ages of the participants in the present survey were from 30 up to the 70s. A preliminary analysis divided them into two groups, divided at age 50. Observation of the data found insignificant differences for the age groups, and the difference was the only difference examined. The reason for examining this factor is that traditionally, males are more likely to attend to maintenance and also are likely to be the major drivers on family excursions. As seen in Table 2, there are few differences between the sexes on most warnings.

Significantly, failures to recognize some problems, such as age of replacement for tires, were not different from the survey of the younger age group. Similarly, the individuals in this survey were also misled by the way in which the age limit at which children might be allowed as passengers in the front seat was presented. In Table 1, the question about a passenger reclining the front passenger seat resulted in the primary response of "only on long trips." This indicated a lack of knowledge about the danger involved. We might note that in Leonard, Karnes and Hill, approximately eighty percent of the respondents indicated that they did not know that reclining the passenger seat in motion was dangerous. In reference to the paper by Laughery and Wogalter's book (in press), a description of the seriousness of an accident involving a passenger lying down was described in morbid detail.

The problem exposed by the responses to difficulty in opening a passenger door in an unfamiliar vehicle substantiates the need for more standardization of automobile controls.

Similarity between the younger and older individuals is probably related to the fact that most families will trade in their cars somewhere between 5 and 10 years after they have purchased them. They often purchase another vehicle from the same manufacturer;

Thus, their attention to the manual will probably be more limited for later models of the same vehicle, or even a different vehicle, where they will anticipate the servicing and other information about the vehicle. This would include the information about where and how children should be located. Thus, the same responses to warnings for older people who were participants in this study would not necessarily have changed their approach to various warnings. For example, the warning about the age at which a child is allowed to sit in the front seat explains why the poor warnings are misunderstood, even though the driver may have seen them before. They are still badly phrased. It appears that the knowledge base is not greatly affected by the experience that comes with age.

The problem with lack of consistency of controls was shown by the fact that 50% of respondents had experienced difficulty in opening a door. Obviously, there are many other problems associated with the failure to have similarity in location and/or operation of controls.

In conclusion, it is worthwhile to examine particular areas of understanding warnings and directions from time to time to see if they produce the same errors. Future research would be useful to evaluate how some of these warnings and information about the problems with the vehicle should be modified.

References

Laughery KR, Wogalter MS (2016) Don't recline that seat. Forensic Hum Factors Ergon 2 (5):99–110

Leonard SD, Karnes EW, Hill GW (2005) Risk perception: has it changed? In: Proceedings of XIX annual international occupational ergonomics and safety conference, pp 21–25, 27-29 June 2005

Leonard SD, Lautenschlager G (2016) Knowledge of automobile warnings. In: Proceedings of XXVIII annual occupational ergonomics and safety conference, 9–10 June 2016

Botwinick J (1966) Cautiousness in advanced age. J Gerontol 21:347–353

PWC Off-Throttle Steering Hazards: Attempted Remedy; Failure; and then Success

Edward W. Karnes[1], S. David Leonard[2], and David R. Lenorovitz[3](✉)

[1] Metropolitan State University of Denver, Denver, CO 80201, USA
[2] University of Georgia, Athens, GA 30602, USA
[3] LENPRO Services, Inc., Littleton, CO 80123, USA
dave@lenproservices.com

Abstract. A major safety concern of personal watercraft (PWC) has been the hazard off-throttle-steering (OTS)—the lack of steering capability that results when the throttle is released in collision-avoidance situations. For three decades, PWC manufacturers have unsuccessfully used warnings in an attempt to remedy the OTS hazard. Subsequent engineering design changes (throttle reapplications and deployable rudders) successfully remedied the hazard. This article addresses the history of OTS hazard concerns and the OTS engineering control methods that were developed.

Keywords: Personal watercraft · Off-throttle steering
Collision avoidance maneuvers · Hazard mitigation efforts · Warnings
Engineering design remedies

1 Introduction/Background

Nearly all human factors or ergonomic (HFE) problems involve the interaction of three component parts: one or more human beings; some tool or piece of equipment; and some type of activity or task the human is supposed to perform using that tool or equipment item. This article will attempt to provide a historical perspective concerning a series of interconnected forensic human factors cases involving scores of people who were killed or seriously injured over the past three or four decades. The common device linking these occurrences was a recreational aquatic vehicle known as a Personal Watercraft (PWC) – sometimes generically referred to as a jet-ski.

The humans in these cases were the operators or pilots of the PWC in question, and the task of concern was attempting to pilot his/her PWC in some evasive maneuver intended to avoid a crash with an up-coming object. In each of these instances, the pilot, traveling at a relatively high rate of speed, suddenly realized that he/she was about to allide with some stationary object, or collide with some other moving object. The chief concern of the pilot was to slow down, steer away, or otherwise maneuver in some way so as to avoid crashing. Quite often, these last minute maneuvers were not successful, and the result was a horrendous crash resulting in serious injuries or death to one or more of the participants involved.

Sadly, the type of collision involved in each of these instances resulted from a type of operator control error that was largely caused by or attributable to design defects

S. Bagnara et al. (Eds.): IEA 2018, AISC 821, pp. 774–782, 2019.
https://doi.org/10.1007/978-3-319-96080-7_94

and/or operational characteristics of the vehicle itself. To a certain extent, these collisions were foreseeable and predictable hazardous events that the vehicle designers/manufactures could have anticipated, and should have taken steps to mitigate the problem. Such mitigations could have reduced or nullified the property damages and injuries that occurred.

1.1 Personal Watercraft (PWC) Operating Characteristics and Features

All Personal Watercraft (PWC) have an inboard engine driving a pump jet that creates thrust that is used for both the vehicle's propulsion and steering. PWCs, like most watercraft, have no brakes, per se, but instead rely on friction/water resistance to slow down and stop the vehicle's motion whenever the propulsive force is in turn slowed down or stopped. PWCs were first introduced by a single manufacturer in the late 1960s as single-rider models, and, in the 1970s, additional manufacturers entered the market with both operator-only and operator-plus-passenger models (see Fig. 1). Soon after their introduction, major safety concerns arose due to the inability of operators to steer the PWCs when the throttle was released during riding maneuvers. In collision-avoidance situations, when operators reflexively released the throttle and attempted to steer away from a perceived collision, the vehicles could not be successfully steered, contributing to an alarming number of collision incidents involving serious injuries and deaths.

Fig. 1. Generic representation of a multi-person Personal Watercraft (PWC).

The seriousness of the safety problem is analogous to designing an automobile so that when the driver encounters an imminent hazard and takes his or her foot off of the gas pedal (or releases brake pedal pressure), the driver loses the ability to steer.

The inability to steer a PWC without application of the throttle has been referred to as the off-throttle-steering (OTS) hazard. All manufacturers of PWCs clearly recognized the hazard and their initial attempts to remedy the problem consisted simply of providing the operators with warnings and operating instructions advising them "not to engage in that type of behavior". These written warnings/instructions were intended to inform operators of the hazard, and to instruct them as to how to proceed when encountering such a situation. These "warnings-only" approaches to trying to remedy the hazard continued – without appreciable success – for several decades.

1.2 PWC Manufacturer's Knowledge/Awareness of OTS Hazard and Consequences

Prior to the provision of OTS engineering design remedies (described below), the loss of directional control created by release of the throttle for water-jet-propelled PWCs resulted in numerous collision incidents. PWC manufacturers and safety researchers recognized this as a serious safety problem when the companies first began to design, manufacture, and market personal watercraft in the late 1960s. In response to that recognition, the manufacturers provided warnings and instructions on-product, in owners' manuals, in marketing materials, and within instructional videos. These warnings and instructions were intended to communicate the problem and to advise operators to not release the throttle when attempting steering maneuvers. In spite of those warnings and instructions, OTS-related collisions continued to occur.

The seriousness of the safety problem and the failure of warnings and instructions to remedy the problem were recognized by not only by PWC manufacturers but also numerous safety and government organizations. Publications addressing the OTS hazard and the need for remedies were documented in manufacturers' patent applications as well as in hazard analysis and safety research documents. A partial listing of these sources includes the following: Yamaha's 1990s patents for improvements in the steering capabilities of personal watercraft (Numbers 5,429,533, 4,744,325, 4,986,208, 5,113,777, and 4,949,622); instructions and warnings provided in PWC owner's manuals; manufacturers' marketing materials and TV tapes; S.R. Simmer's 1992 patent (5,094,182); Bombardier's 1999 Throttle Reapplication Patent Number 6,124,809; Arctic Cat's 1999 Throttle Reapplication Patent Number 6,124,410; the 1998 National Transportation Safety Board's Safety Study; the 1997 California Boating Report from the California Department of Boating and Waterways; the 1997 National Association of State Boating Law Administrators; and the 2000 U.S. Navy Patent No. 6,415,729.

Based largely on three decades of PWC manufacturers providing only OTS warnings and instructions, an abundance of lawsuits were filed involving PWC OTS-related collisions, injuries, and deaths. There were also repeated calls from various safety organizations to provide an engineering fix for these problems.

2 PWC Manufacturers' Re-design/Engineering Mitigation Approaches

Then, in the early 2000s, nearly all PWC manufacturers introduced engineering design changes to provide some form of steering even when the throttle was released (throttle reapplication systems and deployable rudders).

In 2001, Kawasaki developed and marketed a throttle reapplication system, the Kawasaki Smart Steer System (KSS). Within a year, Bombardier offered a retractable rudder system, O.P.A.S. (Off Power Assisted Steering system - 2002). The next year (2003) Yamaha offered a throttle reapplication system, Y.E.M.S. (Yamaha Engine Management System) followed by a Polaris throttle reapplication system, ESP (Enhanced Steering Performance system). PWC throttle reapplication systems use sensors to recognize when a rider abruptly releases the throttle in conjunction with a full-lock turn in either direction (the typical sign of a panicked rider trying to avoid a collision). Throttle reapplication systems then increase thrust slightly, just enough to provide some movement in the direction the driver has attempted to steer. Bombardier's O.P.A.S. uses a deployable dual rudder design that allows the operator of the PWC to steer when the throttle is released or the power is turned off and the handlebars are turned to one side. As of 2006, all sit-down personal watercraft on the market provided some device intended to improve off-throttle steering performance.

For many operators of PWCs, especially beginner and novice operators, releasing the throttle as a response to a perceived collision situation is a stimulus-driven, involuntary, reactive, learned protective motor response. All people learn very early in life that to avoid injury they should decrease the impact energy in any potential collision situation. While true human reflexes may range neurologically from simple monosynaptic reflexes (such as the "knee jerk" reflex) to polysynaptic reflexes (such as postural and righting reflexes), many learned protective motor responses are stimulus-based, involuntary, and reactive. Examples include:

- Extending the upper arms during the initiation of a fall event (the so-called parachute reflex).
- Quickly stepping on the brake to avoid an auto collision.
- Quickly moving out of the way when something falls.
- Reaching out to catch an object unexpectedly thrown in our direction (or turning away or ducking to avoid the thrown object).
- The "phantom-braking reflex" (the response of a passenger in an automobile who perceives an imminent collision and pushes his/her right leg to apply pressure to a nonexistent brake pedal).

A defining characteristic of all these types of stimulus-driven reactive responses (including acquired reflexes and programmed motor responses) is that the person who executes such a response doesn't spend much time deliberating or thinking about it—the person just very quickly does it. If a release of throttle in response to a perceived collision event occurs involuntarily, without conscious thought or deliberation, any warning instructing them "not to do it" is essentially useless.

The scientific and technical literature is replete with examples and discussions of the fact that collision-evasive responses of motorized vehicle operators involve release of throttle coupled with attempts to turn, slow down, and brake. (See Barrett, Kobayski and Fox 1968; Koppa and Hayes 1976; Olson and Sivak 1986; Evans 1991; Wierwillie 1984, 1991; Schmidt 1993; Mazzae, Baldwin and McGekee 1999, and NHTSA 2000). Release of throttle in response to a potential collision is a stimulus-driven human reactive/reflexive-type response. Reflexive, stimulus-driven, motor responses have been referred to or addressed as motor programs, conditioned reflexes, conditioned reactions, conditioned responses, conditioned avoidance responses, learned reactions, and acquired reflexes. Literature addressing reflexive responses has a long history in the behavioral science literature (cf. Ritche 1910; Breese 1917; Schniermann 1930; Evarts 1973; Kelly 2000 and Sherwood 2006). Recent research addressing the differences between stimulus-based and intention-based behavioral actions has been the topic of a considerable body of recent behavioral and neurophysiological research (cf. Waszak et al. 2005; Keller et al. 2006; Herwig, Prinz, and Waszak 2007; and

Welchman et al. 2010). Those research data clearly support a human factors conclusion that when a stimulus-driven behavioral response such as a collision-avoidance release of throttle occurs reactively, without conscious initiation, any warning instructing a person not to perform such a movement is of no benefit.

Based on the authors' PWC on-water riding research and involvement in scores of legal cases involving OTS collisions injuries and deaths, the failure of warnings to remedy the OTS hazard has been repeatedly addressed. In 1991, based on PWC on-water research, the off-throttle steering hazard was described as being comparable to trying to steer an automobile on ice (Burleson, Karnes, and Burton 1991). Five years later, Leonard and Karnes (2005) concluded that:

- Formal hands-on training and operating experience may tend to abate but will not totally eliminate the OTS hazard – even for highly experienced PWC operators.
- The development of collision-avoidance skills for the operator of a PWC, without OTS engineering design changes, requires constant, unwavering awareness of the fact that attempts to stop by a release of throttle would totally defeat the operator's ability to turn and avoid a collision and that even extensive PWC experience would be unlikely to overcome expectations and control responses to emergency, life-threatening collision avoidance situations that have been learned during life's experiences with all other vehicles.

PWC manufacturers should have realized that the provision of OTS warnings and instructions not to release the throttle while still wanting to maintain steering control – e.g., when suddenly needing to veer left or right in an effort to avoid a collision – would be an ineffective remedy for the OTS hazard. This was foreseeable given the facts that: PWCs had no brakes to slow the vehicle, and that the reducing speed could only be accomplished by releasing throttle; however, the act of releasing throttle would defeat the operators' ability to turn the vehicle in an effort to prevent a collision.

3 HFE and Human Behavioral Research Regarding Reactive or Reflexive Acts

PWC manufacturers certainly should have known that operators of their watercraft would not be likely to "just throttle straight ahead into a collision hazard". The fact that information in the form of PWC warnings and instructions do not provide an appropriate and effective remedy for the risk of serious injury or death created by loss of steering in collision-avoidance maneuvers has been provided by many sources: To name just a few:

The 1998 report by the National Transportation Safety Board, Safety Personal Watercraft Safety. NTSB/S5-98 01 identified the OTS hazard and recommended design action by personal watercraft manufacturers to correct that design deficiency within two years. That recommendation was based on the knowledge that the water-jet-powered PWC involves a totally unique vehicle control situation.

> "First, the vehicle has no brakes; the only way to stop is to release the throttle. While that particular control situation is unique for practically all surface vehicles, it is not unique for

boats. The PWC is, however, different from all vehicles, including boats, by the fact that releasing the throttle to stop causes a loss of steering control. In quick-response, emergency collision-avoidance situations where the operator's first instinct is to slow and stop (coupled with attempts to steer), the PWC creates a uniquely dangerous control problem."

U.S. Navy, Patent No. 6,415,729, filed December 14, 2000, states: "In many [PWCs], the only steering ability is that provided by steering the thruster jet nozzle. When an inexperienced rider wants to stop suddenly to avoid an unexpected obstacle, their first panic reaction is to let go of the throttle. When the throttle is off, the vehicle has no steerage and thus proceeds straight into the obstacle."

Bombardier Throttle Reapplication Patent, NO. 6,124,809, filed September 7, 1999, states: "In other words, in order to steer the personal watercraft, the operator must simultaneously operate the throttle control and the handlebar. To the inexperienced operator, this can sometimes seem to be paradoxical situation since a novice operator will typically back off or release the throttle. It would therefore be desirable to have a safety system which obviates the above conditions."

Arctic Cat Throttle Reapplication Patent, No. 6,231,410, Filed November 23, 1999, states: "This quick decrease in steering capability is particularly problematic in situations in which an inexperienced rider attempts to avoid an obstacle directly in front of the watercraft. To properly avoid the obstacle, the rider should apply a constant pressure on the throttle. However an inexperienced rider may release the throttle lever to slow the watercraft quickly while simultaneously turning the steering handle in an attempt to maneuver around the obstacle. In such a situation, the rider may not be able to maneuver around the obstacle since steering capability has been decreased."

The development of a new response (e.g., release throttle) to approximately the same stimulus situation (e.g., perception of an imminent collision) has been studied extensively in the behavioral sciences. That research has demonstrated that learning is impeded when one has to learn alternative responses to approximately the same situation. The interference with learning has been experimentally studied under the headings of habit interference, negative transfer, and proactive and retroactive interference. Early behavioral science research (for example, Viteles 1932and McGeoch 1942) showed that response interference is the greatest when a second response (e.g., maintaining throttle to avoid a collision) is in competition with a well-established habit (e.g., releasing throttle to avoid a collision) elicited by a similar stimulus situation (perception of an imminent collision).

The habit interference research clearly showed that the old instead of the new response is likely to occur. Behavioral science research findings regarding the precedence of old versus new responses to similar stimulus situations have been verified for operators of motor vehicles. For example, based on studies of driver's braking behavior in emergencies, Prynne and Martin (1995) reported: "In moments of extreme stress humans tend to revert to the response they have used most often to a particular stimulus so if a new response has been learned recently the older response will be used instead. This means that training cannot be expected to have much, if any, effect on behavior in emergencies."

Similar findings were reported in the investigations of driver reactions to unexpected tire tread separation incidents (NHTSA 2003). In those studies, the average driver responded to catastrophic tread separations by first steering followed by braking.

Those ingrained steering-first responses lead to significant losses of vehicle control. When drivers were instructed to provide appropriate responses that lessened loss of control by: (1) keep going straight; (2) gradually brake to slow down; and (3) pull off of the road, there was no reliable difference in the proportion of drivers that steered as a first response to tread separation, and there was no effect of instructions on the probability of loss of control.

Based on the abundance of human factors research cited above, it should have been apparent to manufacturers of PWCs that when a behavioral response such as a release of throttle occurs reactively, without conscious initiation, any warning instructing a person not to perform such a movement would be of little or no benefit especially for novice operators of PWCs reacting to a perceived collision situation.

4 Discussion/Conclusions

As discussed earlier, after several decades of using only warnings and instructions as a means to remedy the OTS hazard, PWC manufacturers developed and successfully used engineering design remedies to successfully address that hazard. PWCs have become common on American and other international waterways, and are likely to remain there for many years to come. Since 2006, all PWCs (except stand-up models) now come equipped with some form of off-throttle steering capability. But older models, without OTS engineering fixes, are still in use. PWC braking capabilities may also become helpful in an operator's reacting to some potential collisions, but they are currently available on only select models of one manufacture's watercraft.

As a result of the increasingly widespread proportion of OTS re-designed ("fixed") PWCs, there has been a noticeable decrease in the occurrence of OTS-type collisions. To date, we are unaware of any active legal complaint involving an OTS design defect claim on any PWC incorporating the OTS hazard design change.

Years ago, Christensen (1987a) offered two observations regarding hazard and safety systems analyses and the corresponding process of warning systems development. Those observations are still instructive today. One of these was that:

> "If the engineer/designer cannot eliminate the hazard or protect the user from exposure to it, he has a solemn duty to warn… [but, at the same time, he also cautioned:] A warning is an admission by the designer that hazards remain in the product that [he/she] has not [yet] been able to eliminate." (p. 6)

Lenorovitz, Karnes and Leonard (2012) extended that logic, by stating that:

> "…a warning may be viewed as an "IOU" from the product's designers and developers being extended to the product's users, effectively stating that "some better (safer) design version may be coming in the future, but for now, this potentially useful safety information is [all that is] being provided" (p. 3).

Christensen (1987b) also noted that:

> "Nothing gets the attention of an industry regarding the safe design of its products as effectively as the payment of a large amount of money due to legal action taken against them for personal injury or death" (p. 6).

In the case of the PWC OTS hazard, we can see that it took quite some time (over 40 years) for those initial designers'/engineers' "IOUs" to be settled, and that it took a large number of litigation cases to get the PWC industry to fully attend to the "an OTS engineering solution is what is needed" message.

References

Barrett G, Kobayski M, Fox B (1968) Feasibly of studying driver reaction to sudden pedestrian emergencies in an automobile simulator. Hum Factors 10:19–26

Breese BB (1917) Psychology. Scribner, New York

Burleson M, Karnes EW, Burton A (1991) Personal watercraft safety. SAE Technical Paper Series 911946. Society of Automotive Engineers

Christensen JM (1987a) Comments on product safety. In: 31st annual meeting, proceedings of the human factors society. Human Factors Society, Santa Monica, pp 1–12

Christensen JM (1987b) The human factors profession. In: Salvendy G (ed) Handbook of human factors. Wiley, New York, p 6 Chapter 1.1

Evarts EV (1973) Motor reflexes associated with learned movement. Science 179:501–503

Evans L (1991) Traffic safety and the driver. Van Nostrand Reinhold, New York

Herwig A, Prinz W, Waszak F (2007) Two modes of sensorimotor integration in intention-based and stimulus-based action. Q J Exp Psychol A: Hum Exp Psychol 60(11):1540–1554

Kelly L (2000) Essentials of human physiology for pharmacy. Oxford University Press, London

Keller PE, Wascher E, Prinz W, Waszak F, Koch I, Rosenbaum DA (2006) Differences between intention-based and stimulus-based actions. J Psychophysiol 20(1):1175–1195

Koppa R, Hayes G (1976) Driver inputs during emergency or extreme vehicle maneuvers. Hum Factors 18(4):361–370

Lenorovitz DR, Karnes EW, Leonard SD (2012) Mitigating product hazards via user warnings alone: when/why "warnings-only" approaches are likely to fail. Hum Factors Ergon Manuf Serv Indus 24(3):1–23

Leonard SD, Karnes EW (2005) Why some warnings don't work. In: Proceedings of the XIX annual international occupational ergonomics and safety conference, pp 15–19

Mazzae E, Baldwin GH, McGekee DV (1999) Driver crash avoidance behavior with ABS in an intersection incursion scenario on the Iowa driving simulator, National Highway Traffic Safety Administration, 1999-01-1290

McGeoch JA (1942) The psychology of human learning. An introduction. Longmans, Green, New York

NHTSA (2000) Light Vehicle Antilock Brake Systems Research Program Task 5, Part 1: Examination of Drivers' Collision Avoidance Behavior Using Conventional and Antilock Brake Systems on the Iowa Driving Simulator, U.S. Department of Transportation, August 2000

NHTSA (2003) Investigation of Driver Reactions to Tread Separation Scenarios in the National Advanced Driving Simulator (NADS), U.S. Department of Transportation, DOT H5 809 523, January 2003

Olson P, Sivak M (1986) Perception-response time to unexpected roadway hazards. Hum Factors 28(1):91–96

Prynne K, Martin P (1995) Braking behavior in emergencies, Lucas automotive, society of automotive engineers, SAE Technical Paper Series 950969

Ritche JW (1910) Sensation and physiology. World Book Co., London

Schmidt RA (1993) Unintended acceleration: human performance considerations. In: Automotive ergonomics. Taylor and Francis, New York Chapter 20

Schniermann AL (1930) Russian psychologies. Part VI in Psychologies of 1930. Clark University Press, Wordrester

Sherwood L (2006) Fundamentals of physiology. Thompson Brooks/Cole, Belmont

Viteles MS (1932) Industrial psychology. Norton, New York

Waszak F, Wascher E, Keller P, Koch I, Aschersleben G, Rosenbaum DA, Prinz W (2005) Intention-selection. Exp Brain Res 162:346–356

Wierwille WW (1984) Driver Steering performance. In: Automotive engineering and litigations, vol 1. Wiley, New York

Wierwille WW (1991) Driver error in using automotive pedals. In: Automotive engineering and litigation, vol 4. Wiley, New York (1991)

Welchman AE, Stanley J, Schomers MR, Miall RC, Bulthoff HH (2010) The quick and the dead: when reaction beats intention. In: Proceedings of the royal society of biological sciences, 02 March 2010

Allocation of Blame for Property Damage Originating in a Cigarette Receptacle Constructed from Flammable HDPE

Michael J. Kalsher(✉), Hayley McCullough, and William G. Obenauer

Rensselaer Polytechnic Institute, Troy, NY 12140, USA
{kalshm,mcculh,obenaw}@rpi.edu

Abstract. Human factors and Ergonomics (HFE) researchers have systematically investigated how people allocate responsibility in product liability cases in which people have been seriously injured or killed. Previous research has shown how blame is assigned in a broad range of user injury contexts and to an array of potentially blameworthy entities. We extend this research to investigate allocation of blame for an incident involving substantial property damage, but no user injuries or deaths. Participants read a realistic scenario based on an actual product liability case in which a warehouse sustained more than one-million dollars in damage from a fire that originated in a receptacle designed for disposal of lit smoking materials (e.g., cigarettes) positioned near the building. The scenario systematically varied quality of the product warning (original manufacturer's warning vs. a more comprehensive redesigned warning) and whether the manufacturer and the product's distributor lied or did not lie to purchasers and end-users (smokers) about design limitations of the receptacle. Overall, the pattern of results from the property-damage scenario closely mirrored previous research involving serious injury or death. Consistent with previous research in this area, the product manufacturer received significantly less blame when it provided a redesigned warning about the product's hazard. Interestingly, given the more comprehensive warning, blame was shifted disproportionately away from the manufacturer and toward the purchaser, not the other entities. Candor on the part of the manufacturer and distributor—when these entities provided accurate information—resulted in less blame assigned to them compared to when they either withheld this critical information or provided information that was misleading. Consequently, these findings add to the growing body of evidence indicating multiple benefits of safety.

Keywords: Warnings · Safety · Allocation of responsibility · Product liability Misleading

1 Introduction

A significant body of evidence has accrued concerning the process by which people explain the causes of behaviors and events in their lives—including harmful events [1]. Models developed to explain these processes have been consolidated under the

S. Bagnara et al. (Eds.): IEA 2018, AISC 821, pp. 783–789, 2019.
https://doi.org/10.1007/978-3-319-96080-7_95

umbrella term *attribution theory*. An important aim of attribution theory is to describe how people use information to arrive at causal explanations for these events. Findings generally indicate that contextual information surrounding life events plays an important role in shaping people's perceptions and attributions of blame for their occurrence [2]. These basic concepts have been extended to improve our understanding of how people *allocate* responsibility for events that involve multiple potentially blameworthy entities. Research on allocation of responsibility has spanned a broad array of social problems and contexts, including poverty and public assistance [3, 4], healthcare [5], and behavior problems at school [6].

HFE researchers have built on previous research in this area by systematically investigating how people allocate blame in product liability cases in which people have been seriously injured or killed. Findings to date confirm that decision makers (e.g., lay jurors) tend to allocate blame for these events on the basis of the amount and type of information made available to them [7–10]. The power of contextual variables in shaping culpability judgments is perhaps best illustrated by jury awards for consumer product injuries and/or deaths that contrast with the public's lay theories concerning assignment of blame and appropriate compensation. Consider, for example, the infamous McDonald's hot coffee case in which an elderly woman successfully sued McDonald's and received a large, but undisclosed award. Media reports of the public reaction to the outcome of this case revealed that most people were astounded by the decision. However, the verdict becomes more understandable if we consider the possibility that contextual information about the event made available to jurors during the trial, likely directed their attributions away from the injured person and toward McDonalds [11].

We extend the prior research in this area to investigate allocation of blame for an incident involving substantial property damage, but no user injuries or deaths. In this study, participants read a realistic scenario based on an actual product liability case (in the U.S.) in which a warehouse sustained more than one-million dollars in damage from a fire that originated in a receptacle designed for disposal of lit smoking materials (e.g., cigarettes) positioned near the building. The scenario systematically varied the quality of the product's warning (original manufacturer's warning vs. a redesigned warning) and whether the manufacturer and product distributor provided accurate vs. misleading information to purchasers and end-users (smokers) concerning the design limitations of the receptacle. The receptacles were constructed from high density polyethylene (HDPE), a type of plastic that is flammable. In some of the conditions, it was falsely claimed that the plastic material contained a flame retardant additive.

2 Method

2.1 Participants

A total of 147 undergraduate students (103 males and 44 females) at a private university in the Northeast U.S., who ranged from eighteen to twenty-seven years of age (M = 19.5, $S.D.$ = 1.40), served as study participants. The students received course credit for their participation. Eighteen of the participants identified as smokers.

2.2 Study Design and Procedure

After reading and signing a consent form, each participant read one of eight variants of a scenario in which a warehouse sustained substantial damage from a fire that originated in a receptacle designed for disposal of lit smoking materials (e.g., cigarettes) positioned near the building. The cigarette receptacle, termed the *SafetyCig*, is designed to accept discarded smoking materials (e.g., cigarettes) deposited via a small opening (.75-in/1.91-cm diameter) at the top of a vertical tube. Once inside, the materials are supposed to drop into a removable galvanized steel bucket positioned inside the base of the receptacle. The top and bottom halves of the receptacle can be secured using thumb screws. When tightened sufficiently, this configuration restricts the flow of oxygen into the unit and prevents the lit materials from igniting other flammable materials inside the unit (e.g., other discarded smoking materials or unauthorized paper waste). The steel bucket is intended to be manually removed and emptied regularly. The SafetyCig is marketed to the general public, and therefore, many of the people who will purchase and/or use this product are unlikely to know that the plastic material used for its construction is flammable and can be ignited by discarded smoking materials, or that the integrity of the unit's "oxygen starvation" feature can be defeated if the receptacle is breached (e.g., as a result of melting or puncture), if refuse is allowed to accumulate beyond the confines of the steel bucket or into the vertical tube, or if the thumb screws used to secure the bottom and top halves of the unit together are not replaced and tightened thereby allowing air to flow through the unit.

We systematically varied information provided to study participants as follows: (1) the quality of the product warning (original manufacturer warning vs. a redesigned warning); (2) whether the manufacturer provided accurate information about the design limitations of the receptacle (lying vs. not lying about the addition of a fire retardant to the flammable plastic material) to the distributor; and (3) whether the product distributor accurately passed along information provided by the manufacturer concerning the addition of a flame retardant to the flammable plastic materials to purchasers and end-users (smokers). In all variants, the fact that the cigarette receptacle was made of a flammable material was disclosed. Table 1 shows how the levels of the three independent variables were combined to form the eight groups and describes what information (or misinformation) was provided by the manufacturer and the distributor.

After reading the scenario, participants were asked to allocate blame to each of four entities (the manufacturer, distributor, purchaser and the smoker who discarded the lit smoking material into the receptacle that started the fire) such that the total blame assigned to the four entities summed to 100%. Participants were also asked to rate how dangerous they perceived the cigarette receptacle to be and the effectiveness of the product ads on seven-point verbally anchored rating scales.

3 Results

3.1 Overall, the manufacturer received significantly more blame (M = 44.79, SE = 2.02) than the distributor (M = 22.57, SE = 1.55), the purchaser (M = 24.96, SE = 1.74) or the smoker (M = 7.04, SD = 1.15), respectively. However, the pattern of blame

Table 1. Summary of the study design.

Group	Independent variables Warning	Manufacturer's actions	Distributor's actions
1	Original	Lied by telling the distributor there was a fire retardant in the flammable plastic material	Did not lie to the purchaser and end-users about fire retardant
2			Lied to purchaser and end-users about fire retardant
3		Did not lie to the distributor	Lied to purchaser and end-users about fire retardant
4			Did not lie to the purchaser and end-users about fire retardant
5	Redesigned	Lied by telling the distributor there was a fire retardant in the flammable plastic material	Did not lie to the purchaser and end-users about fire retardant
6			Lied to purchaser and end-users about fire retardant
7		Did not lie to the distributor	Lied to purchaser and end-users about fire retardant
8			Did not lie to the purchaser and end-users about fire retardant

allocated to these entities depended on the quality of the warning. Specifically, participants assigned more than half (57.29%) of the blame to the manufacturer when it provided the original warning, but substantially less blame (32.28%) when it provided the redesigned warning. Given the redesigned warning, blame was shifted disproportionately to the purchaser. The purchaser received a relatively small percentage (10.25%) of the blame given the original manufacturer's warning, but more than three times this amount (39.67%) when the manufacturer provided the redesigned warning. Table 2 presents the percentage blame assigned to the manufacturer, distributor, purchaser and smoker overall, and as a function of the type of warning (original vs. the redesigned warning).

3.2 Separate 2 (Warning: Original vs. Redesigned warning) × 2 (Manufacturer's actions: lying vs. not lying about the addition of a fire retardant to the flammable plastic material) × 2 (Distributor's actions: passed along the information concerning the flame retardant vs. did not pass this information along) three-way between-subjects ANOVAs were conducted for each of the four blameworthy entities (manufacturer, distributor, purchaser, smoker). Participant gender did not play a significant role in determining blame ratings (all p's > .05).

Table 2. Effect of type of warning on allocation of blame.

Type of warning		Manufacturer	Distributor	Purchaser	Smoker
Original warning	*Mean*	57.29	24.62	10.25	6.96
	S.E.	2.83	2.17	2.44	1.62
Redesigned warning	*Mean*	32.28	20.53	39.67	7.11
	S.E.	2.88	2.22	2.49	1.65
Overall	*Mean*	44.79	22.57	24.96	7.04
	S.E.	2.02	1.55	1.74	1.15

3.2.1 *Manufacturer Blame*. There were no significant interactions, all p's > .05. There was a significant main effect of warning, $F(1,139) = 38.39$, $p < .01$, *partial eta square* = .22. Consistent with previous research, the manufacturer received greater blame ($M = 57.29$, $SE = 2.83$) when it provided a poor warning than when it provided the redesigned warning ($M = 32.28$, $SE = 2.88$). There was a significant main effect associated with the manufacturer's actions, $F(1,139) = 5.51$, $p < .05$, *partial eta square* = .04. The manufacturer received greater blame when it lied about the addition of a flame retardant ($M = 49.53$, $SE = 2.84$) than when it did not ($M = 40.05$, $SE = 2.87$). The main effect associated with the distributor's actions was non-significant, $p > .05$. Whether the distributor lied about the addition of a flame retardant, or not, did not significantly affect the amount of blame ascribed to the manufacturer.

3.2.2 *Distributor Blame*. There was a significant three-way interaction, $F(1,139) = 4.13$, $p < .05$, *partial eta-square* = .03. As shown in Table 3, the distributor received the greatest percentage of blame given the redesigned warning and when the manufacturer did not lie about the addition of a flame retardant, but it (the distributor) did.

Table 3. Allocation of blame to the distributor as a function of warning quality, the manufacturer's actions, and the distributor's actions.

Warning quality	Did the manufacturer lie?	Did the distributor lie?	Mean	Std. error
Original warning	No	No	27.10	4.20
		Yes	28.79	4.43
	Yes	No	22.89	4.43
		Yes	19.68	4.31
Redesigned warning	No	No	11.28	4.43
		Yes	35.59	4.55
	Yes	No	20.53	4.31
		Yes	14.72	4.43

3.2.3 *Purchaser Blame*. There was a significant main effect of Warning, $F(1,139) = 71.22$, $p < .01$, *partial eta-square* = .34, a large effect size. Quality of warning appeared to play a substantial role in determining how participants allocated blame to purchasers. Purchasers received significantly more blame when the manufacturer provided the redesigned warning ($M = 39.67$, $SE = 2.49$) as compared to the original warning ($M = 10.25$, $SE = 2.49$). No other effects were significant, all p's > .05.

3.2.4 *Smoker Blame*. There were no significant effects, all p's > .05. In general, the smoker received very little blame, regardless of condition.

3.3 *Perceptions of Danger*. Participants were asked to rate how dangerous they believed the cigarette receptacle to be on a 7-pt. Likert-type scale where 0 = Not at all dangerous and 6 = Extremely dangerous. In general, participants perceived the receptacle to be moderately dangerous ($M = 3.10$, $SE = 0.11$). An independent-samples t-test conducted to determine if there were any differences in perceptions of danger as a function of warning condition was non-significant, $t(145) = 1.01$, $p > .05$.

3.4 *Perceptions of Effectiveness*. Participants were asked to rate the effectiveness of the product warnings for their intended purpose of informing end-users of the hazards of this product on a 7-pt. Likert-type scale where 0 = Not at all effective and 6 = Extremely effective. An independent-samples t-test conducted to determine if there were any differences in perceptions of effectiveness of the warnings was significant, $t(145) = -2.82$, $p < .05$. Participants who received the redesigned warning gave higher effectiveness ratings ($M = 2.65$, $SE = 1.62$) than participants who received the original warning ($M = 1.92$, $SE = 1.53$).

4 Summary

Overall, the pattern of results of this research, which involved only property damage, closely mirrored previous research involving serious injury or death. Consistent with previous findings, the product manufacturer received significantly less blame when it provided a more comprehensive redesigned warning about the product's hazards. Interestingly, given the redesigned warning, blame was shifted disproportionately away from the manufacturer and toward the purchaser, rather than the other entities. The results also showed that candor on the part of the manufacturer and distributor matters. Specifically, when these entities provided accurate information, they received less blame as compared to when they provided information that was misleading. Consequently, these findings add to the growing body of evidence indicating multiple benefits of safety. Future research should continue to investigate additional factors that affect how people assign blame for injuries and/or property damage sustained through the use of or exposure to products, equipment and product-use environments.

References

1. Shaver KG (1985) The attribution of blame: causality, responsibility, and blameworthiness. Springer, New York
2. Karlovac M, Darley JM Attribution of responsibility for accidents: a negligence law analogy. Soc Cogn 6(4):287–318
3. Iyengar S (1990) Framing responsibility for political issues: the case of poverty. Polit Behav 12(1):19–40
4. Skitka LJ, Tetlock PE (1993) Of ants and grasshoppers: the political psychology of allocating public assistance. In: Mellers BA, Baron J (eds) Psychological perspectives on justice. Cambridge University Press, Cambridge, pp 205–233
5. Lane RE (2000) Moral blame and causal explanation. J Appl Psychol 17(1):45–58
6. Pottick KJ, Davis DM (2001) Attributions of responsibility: parents and professionals at odds. Am J Orthopsychiatry 71(3):66–84
7. Kalsher MJ, Viale AJ, Williams KJ (2003) Separating the effects of warning and information distribution practices: a case of cascading responsibility. In: Proceedings of the human factors and ergonomics society, vol 47, pp 1721–1725
8. Kalsher MJ, Williams KJ, Murphy S (2001) Allocating blame for airbag deployment injuries: separating manufacturers' blame from personal responsibility. In: Proceedings of the human factors and ergonomics society, vol 45, pp 1458–1462
9. Laughery KR, Lovvoll DR, McQuilkin ML (1996) Allocation of responsibility for child safety. In: Proceedings of the human factors and ergonomics society, vol 40, pp 810–813
10. Williams KJ, Kalsher MJ, Maru M, Wogalter MS (2000) Emphasizing non-obvious hazards using multi-frame pictorials and color on allocation of blame. In: Proceedings of the international ergonomics association and the human factors and ergonomics society congress, vol 4, pp 124–127
11. Kalsher MJ, Phoenix GM, Wogalter MS, Braun CC (1998) How do people attribute blame for burns sustained from hot coffee? The role of causal attributions. In: Proceedings of the human factors and ergonomics society, vol 42, pp 651–655

Author Index

S. Bagnara et al. (Eds.): IEA 2018, AISC 821, pp. 791–794, 2019.
https://doi.org/10.1007/978-3-319-96080-7

Printed in the United States
By Bookmasters